# The Kinetochore

Peter De Wulf • William C. Earnshaw
Editors

# The Kinetochore

## From Molecular Discoveries to Cancer Therapy

Springer

*Editors*
Peter De Wulf
European Institute of Oncology
Milan, Italy

William C. Earnshaw
University of Edinburgh
Edinburgh, UK

ISBN: 978-1-4419-2398-1 e-ISBN: 978-0-387-69076-6

DOI 10.1007/978-0-387-69076-6



*Cover illustration*: The main image of the mitotic cell was taken by Ana Carvalho (University of Edinburgh). The image of the chromosomes stained for kinetochore proteins was taken by Peter Warburton (University of Edinburgh).

Printed on acid-free paper

springer.com

# Preface

The accurate segregation of replicated chromosomes (sister chromatids) during cell division guarantees a correct number of chromosomes in subsequent generations of cells. Importantly, errors made during this process lead to aneuploid progeny with abnormal chromosome numbers, which can either cause genetic diseases, or, in the case of somatic cells, cause diseases such as cancer. For example, virtually all solid tumors known to date are aneuploid, suggesting that chromosome missegregation underlies or contributes to the initiation and/or progression of cancer.

Kinetochores are highly conserved multi-protein structures that form on the centromeric regions of sister chromatid pairs. Kinetochores orchestrate sister chromatid segregation and ensure that cellular ploidy is maintained. Following the identification of the first three kinetochore proteins in 1985 by one of us, 80–100 proteins (depending on the species) have now been localized to centromeres. These proteins act either as structural kinetochore or centromere components, or as regulators of centromere establishment, and kinetochore formation or activity. Arguably, the kinetochore is one of the most dynamic and complex protein structures known to date.

Recent years have witnessed an outpouring of studies on kinetochore components and centromeres. During the last five years alone, a yearly average of 200 and 500 papers cite the kinetochore and centromere, respectively (Pubmed). This research avalanche has resulted in an almost unmanageable amount of data. Unfortunately, since the publication of his outstanding book by Andy Choo over a decade ago (K. Choo, *The Centromere*, Oxford University Press, New York, 1997), so much has been discovered and written that the non-expert is once again overloaded and bewildered. We therefore decided to create an up-to-date reference that provides a firm basis for understanding the past, current, and also future research on kinetochores and centromeres. To do this, we decided to bring together leading researchers in kinetochore and (neo)centromere biology to share their past experiences during development of the field, to summarize the current state of the art, and to offer hypotheses and predictions that will set the framework for future research.

Chapter 1 gives an historical account of how kinetochore proteins and centromeric regions were discovered. Chapter 2 details the chromosome segregation process and the players involved in it. Chapters 3–5 discuss the chromosomal regions onto which kinetochores assemble (Chapter 3: centromeres, Chapter 4: neocentromeres, Chapter 5: artificial centromeres). Chapter 6 summarizes the composition, formation, and organization of kinetochores, while Chapter 7 reconstructs how kinetochores and centromeres developed during evolution. Chapter 8 discusses the mitotic spindle with which kinetochores interact and within which they segregate into the daughter cells. Chapter 9 describes how kinetochores establish firm contact with and bi-orient on the spindle. Chapter 10 details essential enzyme activities that regulate kinetochore assembly and function. Chapter 11 describes how the proof-reading mitotic checkpoint ensures that incorrect attachments of kinetochores to spindle microtubules are detected and corrected prior to sister chromatid segregation at anaphase. Chapter 12 explains how certain kinetochore complexes (most notably the chromosomal passenger complex) relocalize to the spindle midzone at anaphase onset, thereby regulating sister chromatid segregation and triggering cytokinesis. Chapter 13 concentrates on the roles of kinetochores and centromere-bound cohesin in meiosis. Chapter 14 describes the ongoing efforts of mapping mutations in genes encoding kinetochore proteins and measuring kinetochore protein expression levels in tumor tissues. Last, but surely not least, Chapter 15 outlines how kinetochore proteins and their regulators can be turned into targets of anti-mitotic anti-cancer drugs.

We hope that with this book we have created a useful reference that will benefit experienced researchers in the field and provide an inspiration for those younger aspiring scientists and students who may wish to understand how kinetochores and centromeres orchestrate the fascinating processes of chromosome segregation that form a crucial underpinning for the continuation of life.

We would like to express our gratitude to the panel of international experts who donated their valuable time to help us review the contents of this book:

**Dr. William R. Brinkley** Department Molecular and Cellular Biology, Baylor College of Medicine, Houston, U.S.A.

**Dr. Iain M. Cheeseman** Whitehead Institute for Biomedical Research, Cambridge, U.S.A.

**Dr. Brenda R. Grimes** Department of Medical and Molecular Genetics, Indiana University-Purdue University, Indianapolis, U.S.A.

**Dr. Silke Hauf** Friedrich Miescher Laboratory of the Max Planck Society, Tübingen, Germany.

**Dr. Jeffrey R. Jackson** Oncology Centre for Excellence in Drug Discovery, GlaxoSmithKline, Collegeville, U.S.A.

**Dr. Marko Kallio** Turku Centre for Biotechnology, Turku, Finland.

**Dr. Katsumi Kitagawa** St. Jude Children's Research Hospital, Memphis, U.S.A.

**Dr. Song-Tao Liu** Department of Biological Sciences, University of Toledo, Toledo, U.S.A.

**Dr. Helder Maiato** Institute for Molecular and Cell Biology, University of Porto, Porto, Portugal.

**Dr. Alison L. Pidoux** Wellcome Trust Centre for Cell Biology, University of Edinburgh, Edinburgh, U.K.

**Dr. Ingo Schubert** Leibniz Institute of Plant Genetics and Crop Plant Research, Gatersleben, Germany.

**Dr. Stephen S. Taylor** University of Manchester, Cancer Research U.K., Manchester, U.K.

**Dr. Frank Uhlmann** London Research Institute, Cancer Research U.K., London, U.K.

**Dr. Peter E. Warburton** Mount Sinai School of Medicine, New York, U.S.A.

We also thank those of our consultants who preferred to remain anonymous.

Milan, Italy — Dr. Peter De Wulf
Edinburgh, U.K. — Professor William C. Earnshaw

# Contents

# Contributors

**Joshua J. Bayes** Molecular Cellular Biology Program, University of Washington, Fred Hutchinson Cancer Research Center, 1100 Fairview Avenue N, Seattle, WA 98109-1024, U.S.A.

**Sue Biggins** Division of Basic Sciences, Fred Hutchinson Cancer Research Center, 1100 Fairview Ave N, Seattle, WA 98109

**Kerry S. Bloom** Department of Biology, University of North Carolina at Chapel Hill, Chapel Hill, North Carolina 27599-3280, U.S.A.

**Daniel J. Burke** Department of Biochemistry and Molecular Genetics, University of Virginia Medical Center, 1300 Jefferson Park Avenue, Charlottesville VA 22908-0733, U.S.A.

**K.H. Andy Choo** Laboratory of Chromosome and Chromatin Research, Murdoch Children's Research Institute, Royal Children's Hospital, Flemington Road, Parkville Victoria 3052, Australia

**Lesley Clayton** College of Life Sciences, University of Dundee, Wellcome Trust Biocentre, Dow Street, Dundee DD1 5EH, U.K.

**Peter De Wulf** Department of Experimental Oncology, European Institute of Oncology, Via Adamello 16, 20139 Milan, Italy

**Tatsuo Fukagawa** Department of Molecular Genetics, National Institute of Genetics and The Graduate University for Advanced Studies, Mishima, Shizuoka 411-8540, Japan

**Rebecca Heald** Molecular and Cell Biology Department, University of California, Berkeley, Berkeley, CA 94720-3200, U.S.A.

**Steven Henikoff** Howard Hughes Medical Institute and Fred Hutchinson Cancer Research Center, 1100 Fairview Avenue N, Seattle, WA 98109-1024, U.S.A.

**Kenneth B. Kaplan** Department of Molecular and Cellular Biology, University of California, Davis, One Shields Ave., Briggs Hall 149, Davis, CA 95616, U.S.A.

**Chitra V. Kotwaliwale** Lawrence Berkeley National Labs, Building 84 (MS84R0171), 1 Cyclotron Road, Berkeley, CA 94720, U.S.A.

**Song-Tao Liu** Department of Biological Sciences, University of Toledo, Toledo, OH43606, U.S.A.

**Owen J. Marshall** Laboratory of Chromosome and Chromatin Research, Murdoch Children's Research Institute, Royal Children's Hospital, Flemington Road, Parkville Victoria, VA 3052, Australia

**Adele L. Marston** The Wellcome Trust Centre for Cell Biology, University of Edinburgh, School of Biological Sciences, Michael Swann Building, Mayfield Road, Edinburgh, EH9 3JR, U.K.

**Hiroshi Masumoto** Lab of Call Function and Regulation, Department of Human Genome Research, Kazusa DNA Research Institute, 2-6-7 Kazusa-Kamatari, Kisarazu, Chiba 292-0818, Japan

**Teruaki Okada** Division of Biological Science, Graduate School of Science, Nagoya University, Chikusa-ku, Nagoya 464-8602, Japan

**Yasuhide Okamoto** Division of Biological Science, Graduate School of Science, Nagoya University, Chikusa-ku, Nagoya 464-8602, Japan

**P. Todd Stukenberg** Department of Biochemistry and Molecular Genetics, University of Virginia Medical Center, 1300 Jefferson Park Avenue, Charlottesville VA 22908-0733, U.S.A.

**Beth A. Sullivan** Institute for Genome Sciences & Policy and Department of Molecular Genetics and Microbiology, Duke University, Durham, NC, U.S.A.

**Paul B. Talbert** Howard Hughes Medical Institute and Fred Hutchinson Cancer Research Center, 1100 Fairview Avenue N, Seattle, Washington 98109-1024, U.S.A.

**Tomoyuki U. Tanaka** College of Life Sciences, University of Dundee, Wellcome Trust Biocentre, Dow Street, Dundee DD1 5EH, U.K.

**Scott Thomas** Department of Molecular and Cellular Biology, University of California, Davis, One Shields Ave., Briggs Hall 149, Davis, CA 95616, U.S.A.

**Takeshi Tomonaga** Department of Molecular Diagnosis, Graduate School of Medicine, Chiba University, 1-8-1 Inohana, Chuo-ku, Chiba 260-8670, Japan

**Claire E. Walczak** Medical Sciences Program, Indiana University, Myers Hall 262, 915 East 3rd Street, Bloomington, IN 47405, U.S.A.

**Mitsuhiro Yanagida** CREST Research Program, Japan Science and Technology Corporation (JST), Graduate School of Biostudies, Kyoto University, Sakyo-ku, Kyoto 606-8501, Japan and Initial Research Program (IRP), Okinawa Institute of Science and Technology (OIST) Promotion Corporation, Uruma 904-2234, Okinawa, Japan

**Tim J. Yen** Fox Chase Cancer Center, Philadelphia, PA19111, U.S.A.

# Chapter 1
# Centromeres and Kinetochores: An Historical Perspective

**Kerry S. Bloom**

## 1.1 Identification of Yeast Centromere DNA

As a preface of tribute to Centromere and Kinetochore function, it is interesting to reflect upon the discovery of chromosomes in the late 1880s when chromosomes were named (HWG von Waldeyer, 1888) and their function in heredity proposed by Boveri and Sutton's "Chromosome Theory of Inheritance" to almost 100 years later when the elements of chromosome propagation, namely centromere, telomere, and origins of replication were clearly identified. One can only imagine that the excitement in the field in the early 1880s was matched by the bold proposal that chromosomes were the unit of inheritance by Thomas Hunt Morgan in 1915 and contained the hereditary material. The DNA was discovered by Freidrich Meischer in 1869. It is noteworthy that it took almost 30 years after the determination of the double helical DNA structure, in 1953, to identify the sequence elements of chromosome structure. Identifying genes was child's play in comparison. The centromere does not encode protein, and therefore could not simply be cloned by complementing auxotrophic mutations. Mutations in centromeres should result in the loss of an entire chromosome; there is nothing conditional about that, and even if one managed to introduce a centromere into another site on a chromosome, Barbara McClintock showed us that this would trigger a breakage fusion bridge cycle that is catastrophic to the cell (McClintock, 1939). From the genetic perspective, the centromere is readily identified; it is the genetic locus that exhibits first division segregation in organisms with ordered or linear tetrads. The centromere is the primary constriction of condensed mitotic chromosome and provided a reference point for construction of genetic maps.

From a cytological perspective, the centromere is readily identified as the site of kinetochore assembly. The first description of the specialized disc-shaped kinetochore, a proteinaceous structure found at the periphery of the centromere

K.S. Bloom (✉)
Department of Biology, University of North Carolina at Chapel Hill, Chapel Hill, North Carolina 27599-3280
e-mail: kerry_bloom@unc.edu

P. De Wulf, W.C. Earnshaw (eds.), *The Kinetochore*,
DOI 10.1007/978-0-387-69076-6_1, © Springer Science+Business Media, LLC 2009

came from electron micrographs of fixed specimens (Brinkley and Stubblefield, 1966; Jokelainen, 1967). These studies revealed alternating electron-dense, -lucent, and -dense layers. A recent review of these papers along with a reproduction of the early drawings can be found in Rieder (2005).

To appreciate the cloning of the first centromere in 1980 by Louise Clarke and John Carbon (UCSB; Clarke and Carbon, 1980), we have to step back in time to realize the difficulty and daring required to undertake this project. While the tools for cloning were available (e.g., restriction enzymes, ligase, plasmid vectors for growth in bacteria, etc.), there were no shuttle vectors for plasmid amplification in eukaryotes, no PCR, no genome sequences and no software for DNA data analysis. The only personal computers available at the time were available from Radio Shack or the Apple IIes, introduced in 1980, prior to the introduction of IBM PCs in 1981. Needless to say, there were none in the beautific campus nestled between mountains and the beach on the Pacific coast.

To embark on the isolation of the centromere one had to assume that the chromosome was a single linear DNA duplex. What was the evidence for this bold premise? Chromosomes were visible in stained preparations, but the definitive separation of linear chromosomal sized DNA molecules by gel electrophoresis was not performed until 1984 (Schwartz and Cantor, 1984). The assumption was based upon kinetics of deoxyribonuclease cleavage (Gall, 1963) and nucleic acid reassociation kinetics (Britten and Kohne, 1968; Wetmur and Davidson, 1968). The rate of nucleic acid hybridization allowed these investigators to determine genome size of various organisms and therefore the amount of DNA/chromosome. Bacterial genomes were in the range of several million base pairs and in the case of *Escherichia coli*, were contained in a single circular molecule. If eukaryotic chromosomes were also linear, then one should be able to isolate genes on either side of the centromere (its position defined by the patterns of first or second-division in meiosis) and walk through the centromere.

Chromosome walking was based on a technique known as overlap hybridization. Clarke and Carbon were the first to construct a library of *E. coli* chromosomal DNA fragments by shearing the genome into small fragments and cloning these into the Col E1 plasmid vector (Clarke and Carbon, 1976). One of the insights in this paper was the number of colonies that had to be isolated to ensure that the cloned fragments covered the entirety of the genome. They needed the data from nucleic acid reassociation kinetics to know the size of the genome and estimate how many clones would ensure a greater than 99% probability that the entire genome would be represented in the collection. The cloning of DNA libraries was a cottage industry in the early 1980s with the *Saccharomyces cerevisiae* one of the first eukaryotic libraries to be constructed (Chinault and Carbon, 1979; Nasmyth and Reed, 1980). The tools were now in hand, namely DNA reassociation and clone libraries.

To start this bold adventure, Louise Clarke and John Carbon isolated genes on either side of the centromere on yeast chromosome III (*LEU2* and *PGK1*). LEU2 was isolated by complementation of an auxotrophic mutant in *E. coli* (*leuB6*; Chinault and Carbon, 1979) and PGK1 by an innovative immunological detection system for protein expression from bacterial colonies (Hitzeman et al., 1980). The walk entailed radiolabeling the DNA encoding *LEU2*, and using it as a hybridization probe to find *E. coli* colonies that contain some or all of the *LEU2* fragment. The strategy depends upon the fact that the library was made by A/T tailing with randomly sheared fragments of the genomic DNA, and thus on average, different colonies will contain differing pieces of the same gene. As one can see, this strategy is blind to direction and thus the walk in one direction is half the rate of walking (in this case, toward and away from the centromere). Clones containing overlapping fragments are then identified, picked, and DNA amplified by shuttling back to *E. coli*. Yeast DNA inserts were radiolabeled and used in a second round of overlap hybridization. Yeoman's work indeed and the bulk of the work were performed by A. Craig Chinault, a talented postdoctoral fellow at UCSB (Chinault and Carbon, 1979). One of the first landmarks discovered in this walk was the retrotransposon (yeast TY2; Kingsman et al., 1981). Transposable elements had just been discovered in yeast (Cameron et al., 1979). There are about 30 Ty1 elements dispersed in the genome, and over 100 of the "delta" sequences repeated at the termini of TY elements. Repeated DNA is the bane of the overlap hybridization strategists. Once a repeated region is encountered, many colonies "light up," and there is little hope that one can "walk across" the repeat with the tools in hand.

As often the case with science, serendipity interceded. A new faculty member, Dr. Steven Reed arrived at UCSB, hailing from Lee Hartwell's laboratory. Dr. Hartwell was busy determining the logic circuitry of the cell cycle, and several ambitious students were busy cloning a number of these cell division cycle mutants (aka *cdc* mutants; Hartwell et al., 1970). One of these, *CDC10*, was very closely linked to the centromere on chromosome III. Louise Clarke isolated a clone complementing the temperature-sensitive *cdc10* mutation. This clone contained an 8 kb fragment that overlapped with clones in the laboratory from the *LEU2* region. The hunt was on. *CDC10* is so close to the centromere that it was hard from genetic crossing-over data to distinguish whether it was on the side of *LEU2* or the other side. It was possible that this clone contained the elusive centromere.

Now we have to bear in mind that there was no complementation, or other routine assay for centromere function. It was not a simple step from *cdc10* to the centromere. It is necessary to take another step back to understand and appreciate the isolation of the first centromere. Transformation into yeast was just being developed (Hinnen et al., 1978; timeline Table 1.1), and opened the door for gene identification. One could readily use hybrid plasmids (yeast and *E. coli*) to complement yeast auxotrophic mutations. The major laboratories each had their favorite gene (Botstein- *URA3*; Fink- *HIS3*, *LEU2*; Davis – *TRP1*, Carbon – *ARG4*, *PGK1*), and rapidly shuttled them into the respective auxotrophic yeast mutants. The problems for Clarke and Carbon were that transformation was inefficient and depended upon genomic integration. This is incompatible with centromere function, as integration of a second centromere will

**Table 1.1** Timeline of historic achievements in chromosome structural elements

| | |
|---|---|
| 1966 | Trilaminar structure of the Kinetochore (Brinkely and Stubblefield, 1966; Jokelainen, 1967) |
| 1976 | *E. coli* library (Clarke and Carbon, 1976) |
| 1978 | Yeast transformation (Hinnen et al., 1978) |
| 1979 | Yeast origin of replication (ARS; Stinchcomb et al., 1979) |
| 1980 | Isolation of centromere DNA (Clarke and Carbon, 1980) |
| 1980 | Identification of autoantibody to centromere proteins (Moroi et al., 1980) |
| 1981 | Fragment-mediated transformation (Orr-Weaver et al., 1981) |
| 1981 | Direct selection for centromeres (Hsiao and Carbon, 1981) |
| 1982 | Sequence of centromere DNA (Fitzgerald-Hayes et al., 1982) |
| 1982 | Isolation of yeast telomeres (Szostak and Blackburn, 1982) |
| 1982 | Chromatin structure of a yeast centromere (Bloom and Carbon, 1982) |
| 1983 | Genetic substitutions (Clarke and Carbon, 1983) |
| 1983 | Pedigree analysis of chromosome segregation, Construction of artificial chromosomes (Murray and Szostak, 1983b and 1985) |
| 1985 | First identification of centromere proteins (Earnshaw and Rothfield, 1985) |
| 1986 | *S. pombe* centromere (Clarke et al., 1986) |
| 1987 | First cloning of a centromere protein, CENP-B (Earnshaw et al., 1987) |
| 1991 | Identification of yeast centromere DNA binding proteins (Lechner and Carbon, 1991) |
| 1992 | Complete sequence of chromosome III (Oliver et al., 1992) |
| 1995 | DIC microscopy to visualize yeast chromosome movements (Yeh et al., 1995) |
| 1995 | GFP fusions of cytoskeletal components in yeast (Kahana et al., 1995; Doyle and Botstein, 1996; Fleig et al., 1996; Carminati and Stearns, 1997; Shaw et al., 1997a, b). |

create dicentric chromosomes, which were known to be unstable (McClintock, 1939, 1941, 1942). The next hurdle in the field was to solve the efficiency problem. It turned out that high frequency transformation depended on providing an origin of replication to the transforming plasmid (ARS, autonomously replicating sequence). This was first accomplished in Ron Davis's laboratory (Stinchcomb et al., 1979; Struhl et al., 1979). The *TRP1* gene was very closely linked to an ARS element (*TRP1-ARS1* on a 1.4 kb *EcoRI* fragment) and when introduced into yeast gave high frequency transformation. Shortly thereafter, there followed a breakthrough from the world of recombination, where R. Rothstein and colleagues realized that linearizing transforming DNA fragments enhanced the frequency of homologous recombination in mitosis >1000-fold (Orr-Weaver et al., 1981). These were very heady times for the field, indeed. The world of "gene therapy" was opening before our very eyes.

Meanwhile, Clarke and Carbon were busy introducing various fragments from the plasmid that complemented a temperature-sensitive *cdc10* mutant into replicating vectors. The purpose was to efficiently transform yeast. Here is where persistence and the adage, "chance favors a prepared mind" (L. Pasteur) are relevant. The game at the time was identifying genes. Prior to the discovery of ARS elements, clones that complemented metabolic defects all involved plasmids that had integrated into the genomes, and all transformed cells had the same phenotype i.e., they were protrophic for the mutation in question. However Stinchcomb et al. (1979) noted that unlike the "low frequency" transformants, all the transformants resulting from these autonomously replicating plasmids were unstable. When the cells were grown in the absence of genetic selection for the complementing plasmid, the plasmids were lost from the population. In retrospect, we understand this because the plasmids lacked a centromere, and were not actively partitioned to the daughter cells. However in 1979, this was perplexing to students and postdoctoral fellows trying to learn yeast molecular biology.

Unlike the students and fellows in the laboratory, Clarke and Carbon were undaunted, and realized that the centromere should provide an active partitioning function. When cells containing the plasmid that complemented a *cdc10* mutant were grown in the absence of genetic selection, the plasmids were not lost from the population. Furthermore, the plasmids exhibited classical Mendelian segregation in meiosis (Clarke and Carbon, 1980). The criterion of first division segregation in meiosis was met, and Clarke and Carbon had unequivocally identified the first centromere. The game was afoot to isolate the remaining 15 centromeres and identify the DNA sequence that conferred centromere function.

Only a few laboratories had the expertise for DNA sequencing in the early 1980s. Two methods for sequencing DNA appeared on the scene, chain termination (F. Sanger) and chemical sequencing (base-specific chemical cleavage, A. Maxam and W. Gilbert). J. Carbon realized that sequencing would be key to understanding centromere function and sought a postdoctoral fellow, Molly Fitzgerald-Hayes with expertise in the methodology. Between DNA sequencing and continuing to transform yeast with plasmids containing successively smaller and smaller pieces of the centromere, Molly, Louise, and John discovered that the entire centromere was encoded on a piece of DNA approximately 120 bp in length. The characteristics of this fragment were several conserved sequence elements, denoted as centromere DNA element I, II, III, and VI. CDEIII is 25 bp and partially palindromic. A single base change of CDEIII could completely compromise centromere segregation function (McGrew et al., 1986). CDEII is 76 bp and >90%AT and CDEI is 8 bp.

On the other front of identifying the remaining centromeres, the pace quickened. Both the Carbon and Davis laboratories quickly realized that the genetic selection of loss of plasmids under "no selection" was a powerful strategy. Hsiao and Carbon (Hsiao and Carbon, 1981) introduced a yeast DNA library into a strain. Once transformants were isolated, they grew the cells in the absence of selection for the complementing gene on the plasmid. This would have been heretical just 2 years back. However, after many rounds of non-selective growth, they plated cells on selective media for the complementing plasmid, and quickly found cells that contained autonomously replicating plasmids. Thus in one transformation, Hsiao and Carbon directly isolated several additional centromeres. A few years later, Phil Hieter and colleagues from the

Davis laboratory developed a colony color assay that allowed them to isolate 11 centromeres in one genetic screen (Hieter et al., 1985b).

The centromeres from all 16 chromosomes in budding yeast have similar CDEI, II, and III sequence motifs. Moreover, Clarke and Carbon demonstrated that centromere DNA sequences from different chromosomes are interchangeable (Clarke and Carbon, 1983). There is no chromosome specificity for centromere DNA sequence, nor is there positional specificity within the chromosome. This result had important implications for chromosome pairing in meiosis and pointed to sites outside the centromere as important for this function. The centromere confers genetic stability in a variety of topologies (i.e., linear chromosomes, plasmids) or sequence contexts (Lambie and Roeder, 1986). When the author was a postdoctoral fellow in the Carbon laboratory one of the models for centromere function invoked a tRNA like adaptor molecule built at each centromere to ensure accurate segregation. I am sure this reflected John's thinking from the days of the genetic code, not many years prior to 1980. This model has remained in the bowels of UCSB, and was quickly dispelled by the genomic substitutions of centromeres in different chromosomes.

The isolation of centromeres and origins of replication only shortly preceded the cloning of the first yeast telomere. In a very elegant cloning strategy, Elizabeth Blackburn and Jack Szostack took a linear DNA fragment with the telomere from Tetrahymena on one end, and yeast fragments that could function in yeast as stable linear chromosomes were selected. In this way, they cloned and characterized the first yeast telomere (Szostak and Blackburn, 1982). This laid the foundation for creating the first artificial chromosome (Clarke and Carbon, 1980; Murray and Szostak, 1983a) and serves as a paradigm to the present day for linear artificial chromosomes that function with high fidelity in mammalian cells. The four elements of chromosome structure are the gene, centromere, telomere, and origin of replication. Remarkably, we can construct an entire chromosome no bigger that a few kilobase pairs in budding yeast.

One of the last mysteries of chromosome segregation (in terms of genetic segregation and not the mechanism of motility) was tackled by a young graduate student, Andrew Murray in Jack Szostak's laboratory. Andrew was perplexed by the asymmetry of partitioning of acentric plasmids. To follow the segregation of these acentric plasmids, Andrew performed a pedigree analysis of cells and their plasmids in yeast (Murray and Szostak, 1983b), and discovered that ARS plasmids remained predominantly in the mother cell (Fig. 1.1). The centromere provided an active partitioning function, and overrides the default asymmetric pattern of segregation. The simplest model that the acentric plasmids were biased toward the mother due to catenation of DNA strands during replication was disproven by Koshland and Hartwell (Koshland and Hartwell, 1987). We still do not know why the default pathway leads to accumulation of acentric plasmids in mother cells.

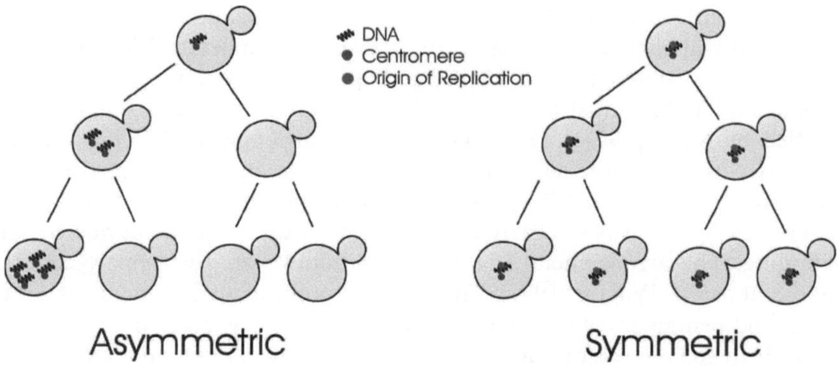


**Fig. 1.1** Pedigree analysis of chromosome segregation in budding yeast. The DNA, original or replication (*red*), and centromere (*blue*) are indicated. (*See* Color Insert)

This rapid advance from cloning the first centromere to identification, sequence analysis and genomic substitution opened the door for the next generation of questions.

## 1.2 Point Versus Regional Centromeres

Louise Clarke next initiated studies on centromeres of the fission yeast, *S. pombe*. Much to everyone's surprise, Louise, together with the Yanagida laboratory in a simultaneous quest across the ocean discovered a considerably more complex centromere DNA sequence (Chikashige et al., 1989; Clarke et al., 1986). The complexity of the centromere sequence as well as the different nomenclatures adopted by each laboratory gave students of centromere severe headaches at the time. While several of the same strategies were applicable in *S. pombe*, and facilitated progress toward centromere identification, the initial studies revealed a highly complex array of repeated sequences and a considerably larger centromere. The centromeres in *S. pombe* were on the order of 50–100 kb (Baum et al., 1994), as opposed to 125 bp in *S. cerevisiae*. The major difference in sequence organization of centromeres from budding to fission yeast led Pluta and Earnshaw (Pluta et al., 1995) to distinguish point centromeres (*S. cerevisiae* type) from regional centromeres (*S. pombe* type).

Since that time, centromeres have been identified in a variety of organisms including yeast *Candida albicans* (Sanyal et al., 2004), bread mold *Neurospora crassa* (Centola and Carbon, 1994), plants, *Arabidopsis thaliana* (Copenhaver et al., 1999), flies *Drosophila melanogaster* (Sun et al., 2003), and humans *Homo sapiens* (Schueler et al., 2001).

The difficulty in cloning and identifying centromeres from organisms with regional centromeres is 2-fold. One is the sheer size of the centromere, in humans the centromere region can be greater than 5 Mb, and two is the lack of a reliable artificial chromosome segregation assay. The most striking characteristic of human centromeres is the abundance of tandem repeats of simple sequence DNA. The most abundant repeat, alphoid satellite ($\alpha$-satellite) was identified by Maio (1971). The monomeric repeat length of 171 bp was very provocative in light of the fact that nucleosomes were just being described. The relationship between the 171 bp $\alpha$-satellite and the nucleosome core – linker (146 bp + 25 bp) – was the early evidence for a sequence code in nucleosome positioning. The arrangement of these repeats however is extremely complex (Waye et al., 1987; Willard, 1991). There are hierarchical arrangements of dimer, trimers, and pentamers of ($\alpha$-satellite) that in turn are organized in higher-order arrays. With the advent of complete genome sequencing, we now have a very good understanding of the human centromere (Schueler et al., 2001), and as such there has been considerable progress on the construction and use of human artificial minichromosomes (Basu and Willard, 2006; Ren et al., 2006; Suzuki et al., 2006; Tsuduki et al., 2006; see the chapter by Masumoto in this book).

## 1.3 Conditional Centromeres, Conditional ARS

The major developments in yeast that led to its prominence as a genetic model system included the ease of isolating a variety of auxotrophic mutations and temperature-sensitive mutations. Conditional mutants provide the opportunity to maintain cell populations with defects in essential genes. Conditionally mutant gene products are typically altered in one or more amino acid, and render the protein defective under sub-optimal growth conditions. It is equally advantageous to have a conditional centromere, and thereby conditionally regulate individual chromosome segregation. Conditional mutants in kinetochore protein result in loss of the entire chromosome set, and not useful for studying one chromosome. Hill and Bloom (1987) discovered that strong transcriptional promoter adjacent to the centromere can inactivate centromere function, producing a conditional centromere. Similarly, a transcriptional promoter is conditionally disruptive for origin of replication function (Snyder et al., 1988). Very recently, it has been possible to produce a conditional centromere in a human artificial chromosome (HAC), by targeting histone modification activities into the kinetochore DNA array (Nakano et al., 2008). Interestingly, both induction of open chromatin or closed heterochromatin can inactivate the kinetochore.

## 1.4 Epigenetic Specification of Centromere Function

An epigenetic phenomenon is one in which the heritable phenotype is conferred by something in addition to the DNA genotype. This phenomenon was originally identified in Drosophila as variegated eye color by Hermann Muller in 1938. The genetic control of variegated phenotypes was dissected in yeast by examining gene expression of telomere-linked genes (Gottschling et al., 1990). Through the use of the colony color sectoring assay, similar to that described above (see Hieter et al., 1985a) Gottschling and coworkers demonstrated variegation in gene expression. This variegation was found in centromere-linked genes adjacent to the *S. pombe* centromere (Allshire et al., 1994). Later that same year, an epigenetic feature regulating centromere function was discovered by Steiner and Clarke in their attempts to clone the minimal segregation unit from *S. pombe* centromeres (Steiner and Clarke, 1994). Much to their surprise, they found that nonfunctional centromeres on small circular minichromosomes could be converted to functional centromeres on the same chromosome. Interestingly, Earnshaw and Migeon first noted the differential segregation capacity of two centromeres on a dicentric chromosome by the presence or absence of staining with autoimmune sera that recognizes centromere proteins (CENPs; Earnshaw and Migeon, 1985) and proposed that there must be some "alteration of the chromatin conformation at the second centromere, preventing binding of the CENP species or sequestering them in an internal region of the chromosome where they are inaccessible for binding to antibodies in vitro or microtubules in vivo." There is now evidence for epigenetic phenomena in centromere function in *S. cerevisiae* (Mythreye and Bloom, 2003), *C. albicans* (Mishra et al., 2007), *S. pombe* (Folco et al., 2008), and humans (Morris and Moazed, 2007).

## 1.5 Centromere Proteins

Well before the yeast geneticists or biochemists identified centromere DNA binding and/or kinetochore proteins, the cytogeneticists were well on their way in this endeavor. The discovery of centromere-specific proteins came from clinical studies on patients with progressive systemic sclerosis. In particular, patients with symptoms of **c**alcinosis, **R**aynaud's phenomena, **e**sophageal dysmotility, **s**celodactyly, **t**elangiectasia (known as CREST), contain anticentromere antibodies (Earnshaw and Rothfield, 1985; Moroi et al., 1980). The use of CREST antisera led to the first identification of CENPs in any species (Earnshaw and Rothfield, 1985), now including CENP-A through –U (reviewed in (Maiato et al., 2004)). CENP-A is a 17 kd protein that was subsequently shown to a histone H3 variant (Sullivan et al., 1994), conserved from fungi to mammals. Interestingly, the early studies revealed that CENP-A was associated with the inner domain of the kinetochore in mammalian cells (Warburton et al., 1997), and worms (Moore et al., 1999), indicative of DNA in the outermost region of the kinetochore. The distribution of this histone variant in the inner plate led to early attempts to map the path of kinetochore DNA. These studies utilized electron spectroscopy for the distribution of phosphorous (Rattner and Bazett-Jones, 1988, 1989). Unfortunately, electron spectroscopy does not distinguish phosphorylation of protein versus phosphate in the DNA backbone and evidence for DNA fibers in the kinetochore outer plate could not be confirmed (Cooke et al., 1993).

Once the centromere DNA sequence had been discovered in fungal system, the next step for the yeast geneticists was to identify centromere DNA binding proteins. Again, it is important to consider the state of scientific progress at the time. The knowledge of DNA-binding proteins was influenced by studies on operons in lambda phage and *E. coli* that taught us about high-affinity sequence-specific binding proteins and in particular motifs like helix-loop-helix. The nucleosome structure of chromatin was only recently published (Kornberg and Thomas, 1974; Olins and Olins, 1974), much less accepted in the literature (for recent historical account, see Olins and Olins, 2003). With this in mind, the early studies in the Carbon laboratory involved isolating biochemical quantities of protein and reconstitution with CEN DNA. Many a postdoctoral fellow worked diligently at this project. The breakthrough in biochemical isolation of the DNA binding complex came when J. Lechner and Carbon realized that a chaperone function (in the first experiments provided artifactually by casein) was required to facilitate sequence-specific binding of the complex to centromere DNA. They published the isolation of a 240 kd complex, denoted CBF3 for the three proteins in the complex (Lechner and Carbon, 1991). Two years later, the gene for the large subunit (110 kd) of the complex was identified simultaneously by Jiang et al. (Jiang et al., 1993) and Goh and Kilmartin (Goh and Kilmartin, 1993). Jiang et al. (1993) subjected the protein complex to tryptic digests, sequenced the peptides and synthesized degenerate oligonucleotides to screen DNA libraries. Goh and Kilmartin had been studying nuclear division cell mutants (*ndc10*) in the spirit of Hartwell's high successful cell division cycle mutant screen. Goh and Kilmartin (1993) isolated mutants that failed to segregate entire chromosome sets based on cytological screening of nuclear division. Thus almost 13 years after the identification of centromere DNA did we have the first gene for a bona fide centromere binding protein in yeast, and only 6 years after the first centromere protein, CENP-B was cloned (Earnshaw et al., 1987). It is safe to say that we do not yet have the structure of the Centromere binding factor (CBF3), and do not fully understand how this protein complex recognizes the centromere DNA.

Once the gene was in hand, the identification of genes encoding the other members of the complex was forthcoming. The breakthrough in this effort came from a genetic screen performed several years earlier that took advantage of the colony sectoring assay developed by Hieter and Koshland (Hieter et al., 1985a; Koshland et al., 1985). Spencer et al. (1990) isolated many of the kinetochore components in their screen for defects in chromosome transmission fidelity (*ctf*). The functional characterization of genes from this inspired screen continues to this day.

## 1.6 Organization of Centromere in Chromatin

The DNA sequence of the centromere has revealed remarkably little about its function, beyond providing a binding site in the case of budding yeast for a core centromere DNA-binding factor 3 (CBF3). To gain insights into centromere function, determining the organization of these sequences in chromatin was the next logical step. The first analysis of centromere chromatin structure was performed in the Carbon laboratory (Bloom et al., 1984). Bloom and Carbon found that the centromere was organized into a unique structure, protected from nuclease, and slightly larger than a canonical nucleosome (centromere 220–250 bp vs. nucleosome 160 bp). Bloom went on to demonstrate that the centromere was indeed built upon a core of histone proteins containing histone H2B and H4 (Saunders et al., 1990). This led to the idea that the nonhistone binding proteins likely bind a centromere DNA-histone substrate. Later it was found that one of the centromere specific proteins (Cse4) was indeed a histone H3 variant as well as a component of the yeast kinetochore (Meluh et al., 1998). Cse4 is conserved throughout phylogeny (aka CENP-A) and present in centromeres from yeast to human. There is renewed attention to the question of the centromeric histone as a novel protein; Scm3 (Camahort et al., 2007; Mizuguchi et al., 2007; Stoler et al., 2007) has recently been discovered to reside at the budding yeast centromere.

## 1.7 Centromeres in Living Cells

The power of yeast as a pioneer model system included its ease of genetics, molecular manipulation of the genome, construction of yeast artificial chromosomes (YACs) and an early entry in the queue for genome sequence (chromosome III sequence, Oliver et al., 1992). However, one limitation of genetics as a method to map protein circuitry and function is the lack of mechanistic insight into a given function. The earliest reflection on the demand for mechanism can be traced to 1861, where Brucke wrote in the minutes of the meeting of the mathematical-scientific Classe of the imperial Academy of Sciences, Die Elementarorganismen, that cell histologists address mechanism (living cells, apart from which molekularstructur of the organic compounds they contain, still another and in other wise complicated structural attribute, the name organization designates) see Thompson, 1917. The definition of mechanism from the late nineteenth to early twentieth century is instructive "From a physical point of view, we understand by a 'mechanism' whatsoever checks or controls, and guides into determinate paths, the workings of energy: in other word, whatsoever leads in the degradation of energy to its manifestation in some form of work, at a stage short of that ultimate degradation which lapses in uniformly diffused heat" (Thompson, 1917).

An important step toward understanding centromere function was to visualize its form as it exists in living cells, in mitosis in particular. The first live cell analysis of yeast was performed by Koning (Koning et al., 1993) through the use of fluorescent lipophilic dyes to stain internal membranes. The challenge with yeast for differential interference contrast (DIC) microscopy was the high refractive index of the cell wall. Tim Stearns (Stanford U) had been experimenting with using gelatin and other agents to match the refractive index of the cell wall and enable live cell analysis. Yeh et al. (1995) took advantage of this to determine the morphological changes and kinetics of nuclear migration and spindle elongation as cells progress through anaphase. This work showed us that cytoplasmic dynein was not a kinetochore protein, but rather contributed to spindle orientation toward the bud neck in mitosis, and secondly that there are specific kinetic phases of anaphase, which later were shown to be under the control of specific plus end-directed motor proteins (Straight et al., 1998). Shortly thereafter, several groups fused their favorite cytoskeletal proteins (actin, tubulin, spindle components) to GFP to visualize microtubule and spindle dynamics in live cells (Carminati and Stearns, 1997; Doyle and Botstein, 1996; Fleig et al., 1996; Kahana et al., 1995; Shaw et al., 1997a b). As often with breakthrough experiments, even more interesting features can be seen in retrospect. One of the most amazing aspects of budding yeast centromere function was staring us in the face in Kahana et al. (1995), namely that sister kinetochores are separated in metaphase. Why didn't we see this in 1995? There were several reasons, one was that the Nuf2 protein fused to GFP was thought to be a spindle pole component (Osborne et al., 1994). This was by virtue of a two-hybrid interaction with the nucleoporin Nup1, and secondly by its "co-localization" with spindle poles (Figs. 1.2 and 1.3). The idea that kinetochores were separated was far from anyone's mind. Second was that initial attempts to label centromeres used markers inserted at *LEU2*, 23 kb from the centromere. Straight et al. (1996) inserted 256 copies of the lac operator from *E. coli* into yeast and visualized the chromosome containing these sequences with lac repressor fused to GFP. The *LEU2* was the infamous centromere-linked gene used in chromosome walking strategy to identify the centromere (7.6 cm from CEN on chromosome III). This visualization strategy for yeast centromeres was breakthrough work, and revealed for the first time a bonafide anaphase A chromosome to pole movement, as discovered almost 20 years earlier through the use of DIC microscopy (Inoue and Ritter, 1978). From a physical perspective however 23 kb is 7.6 μm of B-form DNA, quite away from the centromere, and quite distant in an organism requiring only 125 bp for chromosome segregation.

These early papers in live cell microscopy opened the door for quantitative analysis of dynamic processes, and brought an important technical advance toward our quest for mechanism. As additional laboratories brought their questions and expertise to these problems, it became apparent that sister centromeres were indeed separated in mitosis, prior to anaphase (Goshima and Yanagida, 2000; He et al., 2000; Pearson et al., 2001; Tanaka et al., 2000). Using closely linked lac operators to the centromere (within a kilobase pair or so) it was shown that tension across the centromere results in separation of sister centromeres and proximal chromatin prior to the onset of anaphase. There is a bit of controversy still associated with this view. Goshima and Yanagida addressed the issue directly in a very nice quantitative study (Goshima and Yanagida, 2001). Several additional lines of evidence support

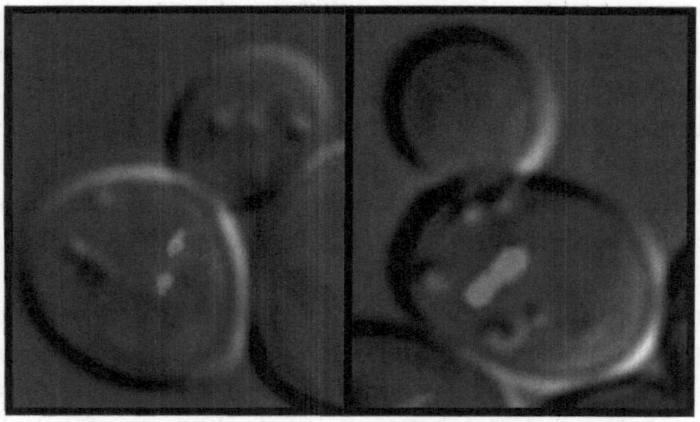

**Fig. 1.2** Centromere DNA, spindle poles and microtubules in mitosis. (*left*) Centromere (CEN) DNA visualized with lac-repressor-GFP bound to lac operator integrated 1 kb from CEN3 on chromosome III (*green*) and spindle poles (*red*). Sister centromeres are separated in mitosis (*right*) Microtubules in the mitotic spindle visualized with tubulin-GFP. The spindle is thicker at the ends versus the middle. Sixteen kinetochore microtubules originate from each spindle pole and extend approximately a quarter the length of the spindle. (*See* Color Insert)

this view. One is the distribution of kinetochore markers, such as Nuf2 (Kahana et al., 1995). Nuf2 and other kinetochore proteins are visualized as two separated spots in mitosis. Furthermore, the variance of fluorescence in the two kinetochore spots (Cse4, Joglekar et al., 2006) is very low, indicative of the fact that the number of kinetochores in each of the separated spots is the same. Finally, upon anaphase, the fluorescence in separated kinetochore clusters segregate to opposite poles, as determined by fluorescence recovery after photobleaching (Molk and Bloom, unpublished).

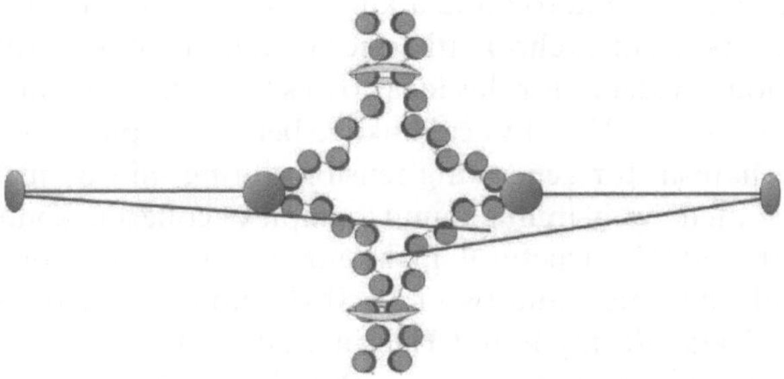

**Fig. 1.3** Organization of a mitotic chromosome. Chromosome arms are closely apposed and held together via cohesin (*yellow rings*). Sister kinetochores (in *blue*) are attached to kinetochore microtubules (*red*) and the pericentric chromatin is stretched toward the spindle poles. There are 16 chromosomes in yeast, and 16 kinetochore microtubules in each spindle half. Cohesin between sister chromatids provides a mechanism to resist microtubule pulling forces and generate tension at centromeres. The function of cohesin in pericentric chromatin is not well understood. (*See* Color Insert)

This observation of clustered centromeres is consistent with electron microscopy of the yeast spindle (Peterson and Ris, 1976). Peterson and Ris found kinetochore microtubules to be discontinuous in the spindle with approximately 16 microtubules emanating from each spindle pole. However, the kinetochore microtubules are fairly short and only span roughly 25% of the spindle length. There is a gap between the two ends of the microtubules, remarkably similar to the distance between separated centromeres as seen by fluorescence microscopy 25 years later. Thus the EM and fluorescence inform us that kinetochore microtubules extend about ¼ the length of the bipolar spindle, and that they are organized in parallel arrays relative to the central interpolar microtubules centromeres, at kinetochore microtubule plus ends, and are under tension and span the distance between kinetochore microtubules emanating from each spindle pole. Based on the position of the lac operator markers, centromere proximal DNA is elongated by an average distance of 0.16 μm per kb prior to anaphase onset. Naked B-form DNA is 0.34 μm per kb and a 7-fold nucleosome compaction predicts that 1 kb of mitotic chromatin covers a distance of 0.05 μm. These results indicate that the level of DNA compaction at the centromere and surrounding chromatin is 3-fold less than that of nucleosomal DNA. Upon anaphase onset, this stretching of DNA is further amplified due to the spindle forces generated during spindle elongation.

## 1.8 What is the Minimal Chromosome Segregation Unit?

The question raised by the small size of CEN DNA in budding yeast is whether the knowledge gleaned from *S. cerevisiae* is instructive for understanding larger eukaryotic centromeres. The minimal centromere in *S. pombe* is 40–60 kb, and of the order of megabase pairs in mammalian cells. While 125 bp of CEN DNA is the minimal size required to build a kinetochore, there may be considerably more DNA (i.e., pericentric chromatin) recruited to the spindle during chromosome segregation. Evidence for this idea comes from the distribution of cohesin along the chromosome. The physical linkage between replicated sister chromatids is the mechanism for generating tension during mitotic metaphase. This linkage is mediated by a multisubunit complex, cohesin, composed of two members of the SMC (structural maintenance of chromosomes) family of ATPases, Smc1 and Smc3, and two non-SMC subunits, Mcd1/Scc1 and Scc3 (Huang et al., 2005; Nasmyth and Haering, 2005). It has been assumed that cohesin promotes association between sister chromatids (intermolecular linkage), and that is the basis for tension when sister chromatids are oriented to opposite spindle pole bodies. The Scc1 subunit disappears from chromosomes when sisters separate at the metaphase/anaphase transition. Scc1 is cleaved by separase upon anaphase onset. The key experiment demonstrating that loss of cohesin is sufficient for sister chromatid separation was artificial cleavage of a modified form of Scc1 by a foreign protease (TEV, Tobacco Etch Virus;

Uhlmann et al., 2000). Activation of TEV protease promotes sister chromatid separation in budding yeast cells when arrested in metaphase. The discovery of cohesin dispelled the view that sister chromatids might be held via intercatenation of sister DNAs that was resolved at anaphase due to microtubule pulling forces (Murray and Szostak, 1985).

Genome-wide chromatin immunoprecipitation (ChIP) in budding yeast has revealed the predominant sites of cohesin binding (Blat and Kleckner, 1999; Weber et al., 2004). Most notable is the finding that cohesin is enriched ~3-fold in a 20–50 kb domain flanking the centromere, relative to the concentration of cohesin on chromosome arms. Although the location of cohesin along the length of the yeast chromosome has been established, little is known about how the concentration of cohesin within pericentric chromatin contributes to the fidelity of chromosome segregation, or whether the cohesin in the pericentric region is indicative for a role of a larger chromosomal domain in kinetochore function. It has recently been demonstrated that cohesin is organized into a cylindrical array encompassing the mitotic spindle in budding yeast (Yeh et al., 2008).

## 1.9 Future Questions

We do not know the nature of chromatin platform on which the centromere is built. Unlike the microtubule, where it appears that the yeast Dam1 kinetochore complex encircles the plus end of the microtubule in a way that is permissive for tubulin addition and loss, there is very little understanding about the molecular structure of the chromatin platform. We have recently proposed that the core centromere (120 bp DNA wrapped around a Cse4 containing nucleosome) and flanking chromatin adopts a cruciform configuration in metaphase (Bloom et al., 2006; Yeh et al., 2008). Whether this or another structure represents the chromatin platform remains to be seen. However it is extremely likely that interesting features of the organization of pericentric chromatin will be forthcoming in future endeavors.

## References

Allshire, R.C., J.P. Javerzat, N.J. Redhead, and G. Cranston. 1994. Position effect variegation at fission yeast centromeres. *Cell*. 76:157–69.

Basu, J., and H.F. Willard. 2006. Human artificial chromosomes: potential applications and clinical considerations. *Pediatr Clin North Am*. 53:843–53, viii.

Baum, M., V.K. Ngan, and L. Clarke. 1994. The centromeric K-type repeat and the central core are together sufficient to establish a functional Schizosaccharomyces pombe centromere. *Mol Biol Cell*. 5:747–61.

Blat, Y., and N. Kleckner. 1999. Cohesins bind to preferential sites along yeast chromosome III, with differential regulation along arms versus the centric region. *Cell*. 98:249–59.

Bloom, K., E. Amaya, and E. Yeh. 1984. Centromeric DNA structure in yeast. *Mol Biol Cytoskeleton*. 175–84.

Bloom, K., S. Sharma, and N.V. Dokholyan. 2006. The path of DNA in the kinetochore. *Curr Biol*. 16:R276–8.

Bloom, K.S., and J. Carbon. 1982. Yeast centromere DNA is in a unique and highly ordered structure in chromosomes and small circular minichromosomes. *Cell*. 29:305–17.

Brinkley, B.R., and E. Stubblefield. 1966. The fine structure of the kinetochore of a mammalian cell in vitro. *Chromosoma*. 19:28–43.

Britten, R.J., and D.E. Kohne. 1968. Repeated sequences in DNA. Hundreds of thousands of copies of DNA sequences have been incorporated into the genomes of higher organisms. *Science*. 161:529–40.

Camahort, R., B. Li, L. Florens, S.K. Swanson, M.P. Washburn, and J.L. Gerton. 2007. Scm3 is essential to recruit the histone h3 variant cse4 to centromeres and to maintain a functional kinetochore. *Mol Cell*. 26:853–65.

Cameron, J.R., E.Y. Loh, and R.W. Davis. 1979. Evidence for transposition of dispersed repetitive DNA families in yeast. *Cell*. 16:739–51.

Carminati, J.L., and T. Stearns. 1997. Microtubules orient the mitotic spindle in yeast through dynein-dependent interactions with the cell cortex. *J Cell Biol*. 138:629–41.

Centola, M., and J. Carbon. 1994. Cloning and characterization of centromeric DNA from Neurospora crassa. *Mol Cell Biol*. 14:1510–9.

Chikashige, Y., N. Kinoshita, Y. Nakaseko, T. Matsumoto, S. Murakami, O. Niwa, and M. Yanagida. 1989. Composite motifs and repeat symmetry in S. pombe centromeres: direct analysis by integration of NotI restriction sites. *Cell*. 57:739–51.

Chinault, A.C., and J. Carbon. 1979. Overlap hybridization screening: isolation and characterization of overlapping DNA fragments surrounding the leu2 gene on yeast chromosome III. *Gene*. 5:111–26.

Clarke, L., H. Amstutz, B. Fishel, and J. Carbon. 1986. Analysis of centromeric DNA in the fission yeast Schizosaccharomyces pombe. *Proc Natl Acad Sci USA*. 83:8253–7.

Clarke, L., and J. Carbon. 1976. A colony bank containing synthetic Col El hybrid plasmids representative of the entire E. coli genome. *Cell*. 9:91–9.

Clarke, L., and J. Carbon. 1980. Isolation of a yeast centromere and construction of functional small circular chromosomes. *Nature*. 287:504–9.

Clarke, L., and J. Carbon. 1983. Genomic substitutions of centromeres in Saccharomyces cerevisiae. *Nature*. 305:23–8.

Cooke, C.A., D.P. Bazett-Jones, W.C. Earnshaw, and J.B. Rattner. 1993. Mapping DNA within the mammalian kinetochore. *J Cell Biol*. 120:1083–91.

Copenhaver, G.P., K. Nickel, T. Kuromori, M.I. Benito, S. Kaul, X. Lin, M. Bevan, G. Murphy, B. Harris, L.D. Parnell, W.R. McCombie, R.A. Martienssen, M. Marra, and D. Preuss. 1999. Genetic definition and sequence analysis of Arabidopsis centromeres. *Science*. 286:2468–74.

Doyle, T., and D. Botstein. 1996. Movement of yeast cortical actin cytoskeleton visualized in vivo. *Proc Natl Acad Sci USA*. 93:3886–91.

Earnshaw, W.C., and B.R. Migeon. 1985. Three related centromere proteins are absent from the inactive centromere of a stable isodicentric chromosome. *Chromosoma*. 92:290–6.

Earnshaw, W.C., and N. Rothfield. 1985. Identification of a family of human centromere proteins using autoimmune sera from patients with scleroderma. *Chromosoma*. 91:313–21.

Earnshaw, W.C., K.F. Sullivan, P.S. Machlin, C.A. Cooke, D.A. Kaiser, T.D. Pollard, N.F. Rothfield, and D.W. Cleveland. 1987. Molecular cloning of cDNA for CENP-B, the major human centromere autoantigen. *J Cell Biol*. 104:817–29.

Fitzgerald-Hayes, M., L. Clarke, and J. Carbon. 1982. Nucleotide sequence comparisons and functional analysis of yeast centromere DNAs. *Cell*. 29:235–44.

Fleig, U., M. Sen-Gupta, and J.H. Hegemann. 1996. Fission yeast mal2+ is required for chromosome segregation. *Mol Cell Biol*. 16:6169–77.

Folco, H.D., A.L. Pidoux, T. Urano, and R.C. Allshire. 2008. Heterochromatin and RNAi are required to establish CENP-A chromatin at centromeres. *Science*. 319:94–7.
Gall, J.G. 1963. Kinetics of deoxyribonuclease action on chromosomes. *Nature*. 198:36–8.
Goh, P.Y., and J.V. Kilmartin. 1993. NDC10: a gene involved in chromosome segregation in Saccharomyces cerevisiae. *J Cell Biol*. 121:503–12.
Goshima, G., and M. Yanagida. 2000. Establishing biorientation occurs with precocious separation of the sister kinetochores, but not the arms, in the early spindle of budding yeast. *Cell*. 100:619–33.
Goshima, G., and M. Yanagida. 2001. Time course analysis of precocious separation of sister centromeres in budding yeast: continuously separated or frequently reassociated? *Genes Cells*. 6:765–73.
Gottschling, D.E., O.M. Aparicio, B.L. Billington, and V.A. Zakian. 1990. Position effect at S. cerevisiae telomeres: reversible repression of Pol II transcription. *Cell*. 63:751–62.
Hartwell, L.H., J. Culotti, and B. Reid. 1970. Genetic control of the cell-division cycle in yeast. I. Detection of mutants. *Proc Natl Acad Sci USA*. 66:352–9.
He, X., S. Asthana, and P.K. Sorger. 2000. Transient sister chromatid separation and elastic deformation of chromosomes during mitosis in budding yeast. *Cell*. 101:763–75.
Hieter, P., C. Mann, M. Snyder, and R.W. Davis. 1985a. Mitotic stability of yeast chromosomes: a colony color assay that measures nondisjunction and chromosome loss. *Cell*. 40:381–92.
Hieter, P., D. Pridmore, J.H. Hegemann, M. Thomas, R.W. Davis, and P. Philippsen. 1985b. Functional selection and analysis of yeast centromeric DNA. *Cell*. 42:913–21.
Hill, A., and K. Bloom. 1987. Genetic manipulation of centromere function. *Mol Cell Biol*. 7:2397–405.
Hinnen, A., J.B. Hicks, and G.R. Fink. 1978. Transformation of yeast. *Proc Natl Acad Sci USA*. 75:1929–33.
Hitzeman, R.A., L. Clarke, and J. Carbon. 1980. Isolation and characterization of the yeast 3-phosphoglycerokinase gene (PGK) by an immunological screening technique. *J Biol Chem*. 255:12073–80.
Hsiao, C.L., and J. Carbon. 1981. Direct selection procedure for the isolation of functional centromeric DNA. *Proc Natl Acad Sci USA*. 78:3760–4.
Huang, C.E., M. Milutinovich, and D. Koshland. 2005. Rings, bracelet or snaps: fashionable alternatives for Smc complexes. *Philos Trans R Soc Lond B Biol Sci*. 360:537–42.
Inoue, S., and H. Ritter, Jr. 1978. Mitosis in Barbulanympha. II. Dynamics of a two-stage anaphase, nuclear morphogenesis, and cytokinesis. *J Cell Biol*. 77:655–84.
Jiang, W., J. Lechner, and J. Carbon. 1993. Isolation and characterization of a gene (CBF2) specifying a protein component of the budding yeast kinetochore. *J Cell Biol*. 121:513–9.
Joglekar, A.P., D.C. Bouck, J.N. Molk, K.S. Bloom, and E.D. Salmon. 2006. Molecular architecture of a kinetochore-microtubule attachment site. *Nat Cell Biol*. 8:581–5.
Jokelainen, P.T. 1967. The ultrastructure and spatial organization of the metaphase kinetochore in mitotic rat cells. *J Ultrastruct Res*. 19:19–44.
Kahana, J.A., B.J. Schnapp, and P.A. Silver. 1995. Kinetics of spindle pole body separation in budding yeast. *Proc Natl Acad Sci USA*. 92:9707–11.
Kingsman, A.J., R.L. Gimlich, L. Clarke, A.C. Chinault, and J. Carbon. 1981. Sequence variation in dispersed repetitive sequences in Saccharomyces cerevisiae. *J Mol Biol*. 145:619–32.
Koning, A.J., P.Y. Lum, J.M. Williams, and R. Wright. 1993. DiOC6 staining reveals organelle structure and dynamics in living yeast cells. *Cell Motil Cytoskeleton*. 25:111–28.
Kornberg, R.D., and J.O. Thomas. 1974. Chromatin structure; oligomers of the histones. *Science*. 184:865–8.
Koshland, D., and L.H. Hartwell. 1987. The structure of sister minichromosome DNA before anaphase in Saccharomyces cerevisiae. *Science*. 238:1713–6.
Koshland, D., J.C. Kent, and L.H. Hartwell. 1985. Genetic analysis of the mitotic transmission of minichromosomes. *Cell*. 40:393–403.

Lambie, E.J., and G.S. Roeder. 1986. Repression of meiotic crossing over by a centromere (CEN3) in Saccharomyces cerevisiae. *Genetics*. 114:769–89.
Lechner, J., and J. Carbon. 1991. A 240 kd multisubunit protein complex, CBF3, is a major component of the budding yeast centromere. *Cell*. 64:717–25.
Maiato, H., J. DeLuca, E.D. Salmon, and W.C. Earnshaw. 2004. The dynamic kinetochore-microtubule interface. *J Cell Sci*. 117:5461–77.
Maio, J.J. 1971. DNA strand reassociation and polyribonucleotide binding in the African green monkey, Cercopithecus aethiops. *J Mol Biol*. 56:579–95.
McClintock, B. 1939. The Behavior in Successive Nuclear Divisions of a Chromosome Broken at Meiosis. *Proc Natl Acad Sci USA*. 25:405–16.
McClintock, B. 1941. The Stability of Broken Ends of Chromosomes in Zea Mays. *Genetics*. 26:234–82.
McClintock, B. 1942. The Fusion of Broken Ends of Chromosomes Following Nuclear Fusion. *Proc Natl Acad Sci U S A*. 28:458–63.
McGrew, J., B. Diehl, and M. Fitzgerald-Hayes. 1986. Single base-pair mutations in centromere element III cause aberrant chromosome segregation in Saccharomyces cerevisiae. *Mol Cell Biol*. 6:530–8.
Meluh, P.B., P. Yang, L. Glowczewski, D. Koshland, and M.M. Smith. 1998. Cse4p is a component of the core centromere of Saccharomyces cerevisiae. *Cell*. 94:607–13.
Mishra, P.K., M. Baum, and J. Carbon. 2007. Centromere size and position in Candida albicans are evolutionarily conserved independent of DNA sequence heterogeneity. *Mol Genet Genomics*.
Mizuguchi, G., H. Xiao, J. Wisniewski, M.M. Smith, and C. Wu. 2007. Nonhistone Scm3 and histones CenH3-H4 assemble the core of centromere-specific nucleosomes. *Cell*. 129:1153–64.
Moore, L.L., M. Morrison, and M.B. Roth. 1999. HCP-1, a protein involved in chromosome segregation, is localized to the centromere of mitotic chromosomes in Caenorhabditis elegans. *J Cell Biol*. 147:471–80.
Moroi, Y., C. Peebles, M.J. Fritzler, J. Steigerwald, and E.M. Tan. 1980. Autoantibody to centromere (kinetochore) in scleroderma sera. *Proc Natl Acad Sci USA*. 77:1627–31.
Morris, C.A., and D. Moazed. 2007. Centromere assembly and propagation. *Cell*. 128:647–50.
Murray, A.W., and J.W. Szostak. 1983a. Construction of artificial chromosomes in yeast. *Nature*. 305:189–93.
Murray, A.W., and J.W. Szostak. 1983b. Pedigree analysis of plasmid segregation in yeast. *Cell*. 34:961–70.
Murray, A.W., and J.W. Szostak. 1985. Chromosome segregation in mitosis and meiosis. *Annu Rev Cell Biol*. 1:289–315.
Mythreye, K., and K.S. Bloom. 2003. Differential kinetochore protein requirements for establishment versus propagation of centromere activity in Saccharomyces cerevisiae. *J Cell Biol*. 160:833–43.
Nakano, M., S. Cardinale, V.N. Noskov, R. Gassmann, P. Vagnarelli, S. Kandels-Lewis, V. Larionov, W.C. Earnshaw, and H. Masumoto. 2008. Inactivation of a human kinetochore by specific targeting of chromatin modifiers. *Developmental Cell*. 14:507–522.
Nasmyth, K., and C.H. Haering. 2005. The structure and function of SMC and kleisin complexes. *Annu Rev Biochem*. 74:595–648.
Nasmyth, K.A., and S.I. Reed. 1980. Isolation of genes by complementation in yeast: molecular cloning of a cell-cycle gene. *Proc Natl Acad Sci USA*. 77:2119–23.
Olins, A.L., and D.E. Olins. 1974. Spheroid chromatin units (v bodies). *Science*. 183:330–2.
Olins, D.E., and A.L. Olins. 2003. Chromatin history: our view from the bridge. *Nat Rev Mol Cell Biol*. 4:809–14.
Oliver, S.G., Q.J. van der Aart, M.L. Agostoni-Carbone, M. Aigle, L. Alberghina, D. Alexandraki, G. Antoine, R. Anwar, J.P. Ballesta, P. Benit, and et al. 1992. The complete DNA sequence of yeast chromosome III. *Nature*. 357:38–46.

Orr-Weaver, T.L., J.W. Szostak, and R.J. Rothstein. 1981. Yeast transformation: a model system for the study of recombination. *Proc Natl Acad Sci USA*. 78:6354–8.

Osborne, M.A., G. Schlenstedt, T. Jinks, and P.A. Silver. 1994. Nuf2, a spindle pole body-associated protein required for nuclear division in yeast. *J Cell Biol*. 125:853–66.

Pearson, C.G., P.S. Maddox, E.D. Salmon, and K. Bloom. 2001. Budding yeast chromosome structure and dynamics during mitosis. *J Cell Biol*. 152:1255–66.

Peterson, J.B., and H. Ris. 1976. Electron-microscopic study of the spindle and chromosome movement in the yeast Saccharomyces cerevisiae. *J Cell Sci*. 22:219–42.

Pluta, A.F., A.M. Mackay, A.M. Ainsztein, I.G. Goldberg, and W.C. Earnshaw. 1995. The centromere: hub of chromosomal activities. *Science*. 270:1591–4.

Rattner, J.B., and D.P. Bazett-Jones. 1988. Electron spectroscopic imaging of the centrosome in cells of the Indian muntjac. *J Cell Sci*. 91 (Pt 1):5–11.

Rattner, J.B., and D.P. Bazett-Jones. 1989. Kinetochore structure: electron spectroscopic imaging of the kinetochore. *J Cell Biol*. 108:1209–19.

Ren, X., C.G. Tahimic, M. Katoh, A. Kurimasa, T. Inoue, and M. Oshimura. 2006. Human artificial chromosome vectors meet stem cells: new prospects for gene delivery. *Stem Cell Rev*. 2:43–50.

Rieder, C.L. 2005. Kinetochore fiber formation in animal somatic cells: dueling mechanisms come to a draw. *Chromosoma*. 114:310–8.

Sanyal, K., M. Baum, and J. Carbon. 2004. Centromeric DNA sequences in the pathogenic yeast Candida albicans are all different and unique. *Proc Natl Acad Sci USA*. 101:11374–9.

Saunders, M.J., E. Yeh, M. Grunstein, and K. Bloom. 1990. Nucleosome depletion alters the chromatin structure of Saccharomyces cerevisiae centromeres. *Mol Cell Biol*. 10:5721–7.

Schueler, M.G., A.W. Higgins, M.K. Rudd, K. Gustashaw, and H.F. Willard. 2001. Genomic and genetic definition of a functional human centromere. *Science*. 294:109–15.

Schwartz, D.C., and C.R. Cantor. 1984. Separation of yeast chromosome-sized DNAs by pulsed field gradient gel electrophoresis. *Cell*. 37:67–75.

Shaw, S.L., E. Yeh, K. Bloom, and E.D. Salmon. 1997a. Imaging green fluorescent protein fusion proteins in Saccharomyces cerevisiae. *Curr Biol*. 7:701–4.

Shaw, S.L., E. Yeh, P. Maddox, E.D. Salmon, and K. Bloom. 1997b. Astral microtubule dynamics in yeast: a microtubule-based searching mechanism for spindle orientation and nuclear migration into the bud. *J Cell Biol*. 139:985–94.

Snyder, M., R.J. Sapolsky, and R.W. Davis. 1988. Transcription interferes with elements important for chromosome maintenance in Saccharomyces cerevisiae. *Mol Cell Biol*. 8:2184–94.

Spencer, F., S.L. Gerring, C. Connelly, and P. Hieter. 1990. Mitotic chromosome transmission fidelity mutants in Saccharomyces cerevisiae. *Genetics*. 124:237–49.

Steiner, N.C., and L. Clarke. 1994. A novel epigenetic effect can alter centromere function in fission yeast. *Cell*. 79:865–74.

Stinchcomb, D.T., K. Struhl, and R.W. Davis. 1979. Isolation and characterisation of a yeast chromosomal replicator. *Nature*. 282:39–43.

Stoler, S., K. Rogers, S. Weitze, L. Morey, M. Fitzgerald-Hayes, and R.E. Baker. 2007. Scm3, an essential Saccharomyces cerevisiae centromere protein required for G2/M progression and Cse4 localization. *Proc Natl Acad Sci U S A*. 104:10571–6.

Straight, A.F., A.S. Belmont, C.C. Robinett, and A.W. Murray. 1996. GFP tagging of budding yeast chromosomes reveals that protein-protein interactions can mediate sister chromatid cohesion. *Curr Biol*. 6:1599–608.

Straight, A.F., J.W. Sedat, and A.W. Murray. 1998. Time-lapse microscopy reveals unique roles for kinesins during anaphase in budding yeast. *J Cell Biol*. 143:687–94.

Struhl, K., D.T. Stinchcomb, S. Scherer, and R.W. Davis. 1979. High-frequency transformation of yeast: autonomous replication of hybrid DNA molecules. *Proc Natl Acad Sci USA*. 76:1035–9.

Sullivan, K.F., M. Hechenberger, and K. Masri. 1994. Human CENP-A contains a histone H3 related histone fold domain that is required for targeting to the centromere. *J Cell Biol*. 127:581–92.

Sun, X., H.D. Le, J.M. Wahlstrom, and G.H. Karpen. 2003. Sequence analysis of a functional Drosophila centromere. *Genome Res.* 13:182–94.

Suzuki, N., K. Nishii, T. Okazaki, and M. Ikeno. 2006. Human artificial chromosomes constructed using the bottom-up strategy are stably maintained in mitosis and efficiently transmissible to progeny mice. *J Biol Chem.* 281:26615–23.

Szostak, J.W., and E.H. Blackburn. 1982. Cloning yeast telomeres on linear plasmid vectors. *Cell.* 29:245–55.

Tanaka, T., J. Fuchs, J. Loidl, and K. Nasmyth. 2000. Cohesin ensures bipolar attachment of microtubules to sister centromeres and resists their precocious separation. *Nat Cell Biol.* 2:492–9.

Thompson, D.A.W. 1917. On Growth and Form. Cambridge University Press.

Tsuduki, T., M. Nakano, N. Yasuoka, S. Yamazaki, T. Okada, Y. Okamoto, and H. Masumoto. 2006. An artificially constructed *de novo* human chromosome behaves almost identically to its natural counterpart during metaphase and anaphase in living cells. *Mol Cell Biol.* 26:7682–95.

Uhlmann, F., D. Wernic, M.A. Poupart, E.V. Koonin, and K. Nasmyth. 2000. Cleavage of cohesin by the CD clan protease separin triggers anaphase in yeast. *Cell.* 103:375–86.

Warburton, P.E., C.A. Cooke, S. Bourassa, O. Vafa, B.A. Sullivan, G. Stetten, G. Gimelli, D. Warburton, C. Tyler-Smith, K.F. Sullivan, G.G. Poirier, and W.C. Earnshaw. 1997. Immunolocalization of CENP-A suggests a distinct nucleosome structure at the inner kinetochore plate of active centromeres. *Curr Biol.* 7:901–4.

Waye, J.S., S.J. Durfy, D. Pinkel, S. Kenwrick, M. Patterson, K.E. Davies, and H.F. Willard. 1987. Chromosome-specific alpha satellite DNA from human chromosome 1: hierarchical structure and genomic organization of a polymorphic domain spanning several hundred kilobase pairs of centromeric DNA. *Genomics.* 1:43–51.

Weber, S.A., J.L. Gerton, J.E. Polancic, J.L. DeRisi, D. Koshland, and P.C. Megee. 2004. The kinetochore is an enhancer of pericentric cohesin binding. *PLoS Biol.* 2:E260.

Wetmur, J.G., and N. Davidson. 1968. Kinetics of renaturation of DNA. *J Mol Biol.* 31:349–70.

Willard, H.F. 1991. Evolution of alpha satellite. *Curr Opin Genet Dev.* 1:509–14.

Yeh, E., J. Haase, L.V. Paliulis, A. Joglekar, L. Bond, D. Bouck, E.D. Salmon, and K.S. Bloom. 2008. Pericentric chromatin is organized into an intramolecular loop in mitosis. *Curr Biol.* 18:81–90.

Yeh, E., R.V. Skibbens, J.W. Cheng, E.D. Salmon, and K. Bloom. 1995. Spindle dynamics and cell cycle regulation of dynein in the budding yeast, Saccharomyces cerevisiae. *J Cell Biol.* 130:687–700.

# Chapter 2
# The Basics of Chromosome Segregation

**Mitsuhiro Yanagida**

## 2.1 Scope of this Chapter

During cell division, chromosomes carrying thousands of genes are correctly transmitted to daughter cells via a motile apparatus named the mitotic spindle (a schematic outline of the cell (division) cycle is shown in Fig. 2.1). In post-replicative (post S phase) cells, chromosomes comprise duplicated sister chromatids. In the cell cycle stage called mitotic metaphase, all sister chromatid pairs are aligned and bi-oriented to the spindle apparatus. In anaphase, all sister chromatids separate in concert and segregate oppositely along the anaphase spindle (towards the spindle poles/centrosomes) into the two daughter cells (Fig. 2.2). The once-in-a-cell-cycle occurrence of the chromosome-segregation process suggests that this event should be studied with respect to cell cycle control (reviewed in Morgan 2006).

Our current understanding of chromosome segregation was greatly advanced by the discovery of cyclin-dependent protein kinases (CDKs; Doree and Hunt 2002). CDKs promote cell-cycle transitions and are the main engines of the cell cycle (Sánchez and Dynlacht 2005). Mitotic CDKs are inactivated when bound cyclin is degraded by the 26S proteasome through the ubiquitin pathway (Hershko 2005) thereby promoting the transition from metaphase to anaphase. Simultaneously, securin, a key inhibitor of separase, the enzyme whose action triggers chromosome segregation, is degraded by the same mechanism (Yanagida 2000, 2005; Fig. 2.1). Cell cycle control and chromosome segregation are temporally coordinated using the same destruction motif. In meiosis, the sexual reproduction cycle, there are two types of metaphase. In metaphase I, homologous chromosomes are associated, and in anaphase I they segregate without sister chromatid separation (Clarke and Orr-Weaver 2006).

M. Yanagida (✉)
CREST Research Program, Japan Science and Technology Corporation (JST), Graduate School of Biostudies, Kyoto University, Japan and Initial Research Program (IRP), Okinawa Institute of Science and Technology (OIST) Promotion Corporation, Uruma 904-2234, Okinawa, Japan
e-mail: yanagida@kozo.lif.kyoto-u.ac.jp

P. De Wulf, W.C. Earnshaw (eds.), *The Kinetochore*,
DOI 10.1007/978-0-387-69076-6_2, © Springer Science+Business Media, LLC 2009

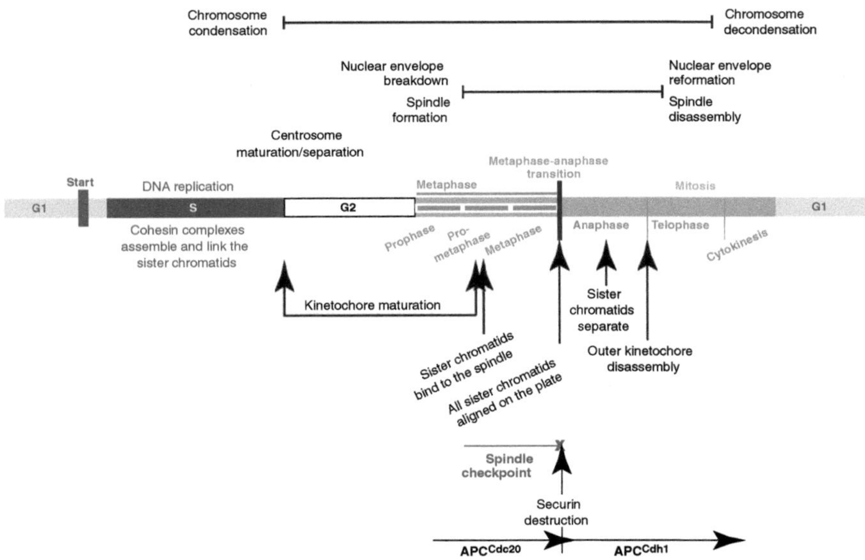


**Fig. 2.1** The mitotic cell division, kinetochore and chromosome cycles. Schematic outline of the cell cycle stages (G1: *blue*, S-phase: *red*, G2: white, mitosis: *green*), and of the state/activity of chromosomes, the spindle, kinetochores and sister chromatids (shown in *black*). The activities of the mitotic spindle checkpoint and the anaphase-promoting complex (APC) are shown in pink and brown, respectively. In metazoan, kinetochores undergo a cycle of maturation and partial disassembly. In the yeasts, kinetochores are tethered to microtubules throughout the cell cycle and shortly detach from the microtubule tips during replication of the centromeric regions in S phase. (*See* Color Insert)

In metaphase II and anaphase II, chromosomes are segregated, as they are in somatic cells (Morgan 2006).

The chromosome segregation process is elaborate, with checkpoints and error-correction mechanisms, as the transmission of chromosomes requires high fidelity. Errors in chromosome segregation cause aneuploidy, cancer, and various diseases (Epstein 2007, Stallings 2007, Weaver and Cleveland 2007). Of the ~5,000 genes in simple eukaryotes, ~500 genes are presumed to be required for proper chromosome segregation. Two eukaryotic microbes, the budding yeast *Saccharomyces cerevisiae* and the fission yeast *Schizosaccharomyces pombe*, have proven to be excellent model species for studying chromosome segregation. These organisms are evolutionarily distant, and thus mechanisms that are conserved between them are generally also conserved in vertebrates. Studies on the chromosome behavior of worm, fly, and vertebrates are largely consistent with the notions developed in yeast studies (Oegema et al., 2001, Herzig et al., 2002). In sum, firm evidence suggests that the basic mechanisms underlying eukaryotic chromosome segregation are the same in all eukaryotes as many of the genes involved in the process have been evolutionary conserved (Yanagida 2005).

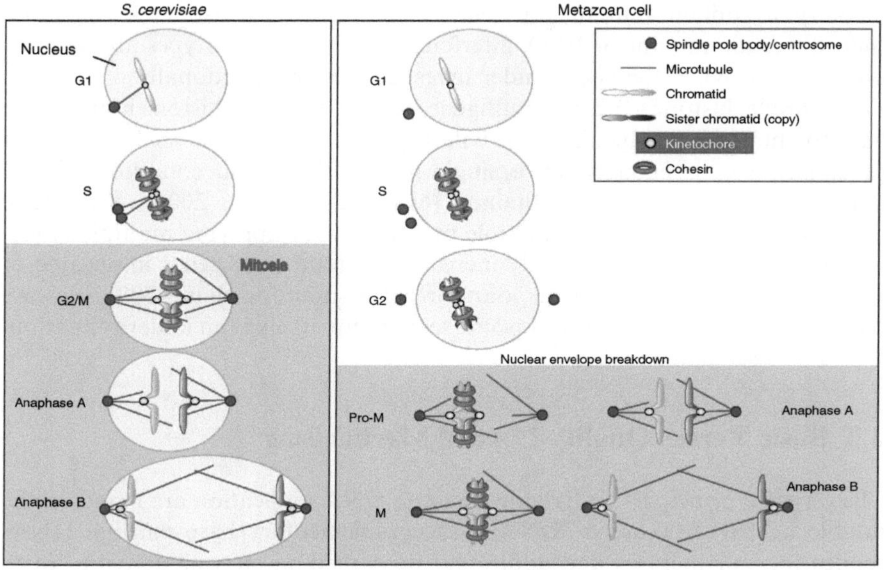


**Fig. 2.2** The eukaryotic chromosome cycle. Schematic representation of chromosome replication and segregation in the budding yeast *Saccharomyces cerevisiae* and in a metazoan cell. The processes involved are highly conserved. A notable difference is that the yeast cell cycle is "closed" (occurs inside the nucleus), whereas in higher eukaryotes the cell cycle is "open." More specifically, chromosomes in higher cells are duplicated inside the nucleus but can only establish contact with the spindle apparatus following the breakdown of the nuclear membrane. Chromosome segregation then takes place in the cytoplasm.

This chapter focuses on the basic mechanisms that underlie the transmission of DNA from one generation to the next, and ensure that this process occurs with the highest fidelity. The main factors involved in chromosome segregation will be discussed and special attention will given to players whose roles in mitosis are not discussed elsewhere in this book (e.g., condensin, cohesin).

## 2.2 Gene Identification in Chromosome Segregation is Incomplete

Classic genetic analyses that led to the phenotypic identification of many genes (and processes) involved in chromosome segregation have been recently extended using comprehensive, high-throughput methods. In higher eukaryotes, defects in chromosome segregation are examined using small inhibitory dsRNA oligonucleotides that knockdown individual gene products (versus mutations of particular genes obtained by mutagenesis). The identification of whole suites of genes required for chromosome segregation, however, is far from complete. The redundancy of gene function is the principal reason for our inability to obtain mutants in genes mediating chromosome segregation. Indeed, certain mutant or RNA interference (RNAi) phenotypes are weak or not apparent when the genes under investigation are functionally redundant. For example, histone genes are multigene systems such that histone mutants are hard to obtain by mutagenizing the wild-type strain. The *S. pombe* histone H2B gene, however, is encoded by a single gene (*htb1*$^{+}$) and conditional-lethal mutants of *htb1* are easily obtained (Maruyama et al., 2006). Systematic approaches for constructing multiple mutants or for applying multiple RNAi will be necessary to substantially increase the number of genes implicated in chromosome segregation. Thus, our current assumption that ~500 genes are required for high-fidelity chromosome segregation likely is an underestimation.

## 2.3 Basic Versus Quality Control Mechanisms

The physical principals underlying accurate DNA replication are based on the double helical structure of DNA with complementary base pairing. DNA-replicating polymerases and additional proofreading enzyme activities (e.g., DNA-damage repairing proteins) ensure correct replication of DNA. Although there are currently no known central physical principles underlying chromosome segregation, distinctions exist between the basic and quality-control mechanisms of chromosome segregation. The basic mechanisms may comprise a relatively small number of proteins and complexes. However, if these were to function on their own, then the fidelity of segregation would be relatively low. The high-fidelity segregation observed in cells may require the combination of several quality control mechanisms.

Quality-control genes are involved in numerous cell activities, including the cell cycle, signal transduction, macromolecule and metabolite trafficking, organelle formation and segregation, metabolic turnover, ubiquitylation, phosphorylation, sumoylation, and proteolysis. In addition, the spindle with the spindle poles (mitotic centrosomes) and the kinetochore contain a large number of proteins, many of whose functions are not well understood at the molecular level. Some of these proteins may be required for quality control rather than for basic segregation activities, particularly in higher eukaryotes. We predict that many gene functions dispensable for cell division and viability may serve to improve chromosome segregation. Hence, chromosomal abnormalities and human diseases may be due more to abnormalities of quality control genes than to the loss of the basic genes, as the latter would likely lead to cell and embryonic death.

Essentially nothing is currently known about the metabolic regulation of chromosome segregation. Mechanisms that produce cellular energy may be part of the chromosome segregation process and may be crucial for mitotic progression (Pederson 2003). In *S. pombe*, reducing the concentration of glucose in the medium immediately inhibits mitosis. Cell growth in *S. pombe* (e.g., increase in cell length) is abruptly blocked during the G2-mitosis transition suggesting the presence of a nutrient switch from growth to mitosis. While the defect in energy metabolism may block the entire program of cell cycle progression, including chromosome segregation, mitosis itself requires significant amounts of energy. Entry into and exit from mitosis require phosphorylation and ubiquitin-mediated protein degradation, respectively, processes that consume ATP. Chromosome segregation requires the assembly and dynamic movements of the spindle apparatus, which requires a number of proteins with ATPase and GTPase activities (Morgan 2006).

## 2.4 Gene Nomenclature for Chromosome Segregation

The nomenclature used for genes involved in chromosome segregation is a serious problem in communicating results obtained in different organisms. Many genes are initially identified through the use of mutants, antibodies, or amino acid sequences of purified proteins and their molecular functions are not known. Thus, many of the gene names do not give functional clues and are difficult to remember. Although similar proteins exist in other organisms, researchers tend to use their own organism's nomenclature, as it is often unclear whether these genes are functionally equivalent to similar genes in other organisms. Indeed, genes with analogous sequences but distinct functions are not uncommon. It is therefore very difficult for researchers in other fields and for newcomers to the field to understand the functions of a particular gene by reading the literature.

A number of protein complexes essential for chromosome segregation, however, have been given common names across organisms. The presence of multiple subunits that all share sequence similarity in different organisms is convincing evidence of the functional similarity of these complexes, such as condensin, cohesin, anaphase-promoting complex (APC/C), and mitotic checkpoint complex (MCC). The use of a common nomenclature for these complexes promotes integrated studies. For example, condensin is a hetero-pentameric complex required for mitotic chromosome architecture. It consists of two subunits belonging to the structural maintenance of chromosome (SMC) ATPase protein family, and three non-SMC components (reviewed in Nasmyth and Haering 2005, Belmont 2006, Hirano 2006). Frog condensin contains XCAP-C (SMC4) and XCAP-E (SMC2), two heterodimeric coiled-coil SMCs and three non-SMC proteins: XCAP-H, -G, and -D2. In *S. cerevisiae*, the dimeric Smc2 and Smc4 associate with three non-SMC subunits, Ycg1, Ycs4, and Brn1. Similarly, two SMC proteins of *S. pombe*, Cut3 and Cut14, form a heterodimer and bind to three non-SMC subunits, Cnd1, Cnd2, and Cnd3 (Nasmyth and Haering 2005, Belmont 2006, Hirano 2006). The sequences of each of these sets of five subunits are similar from fungi to human, indicating that they are functionally conserved. Although different names remain for individual subunits, they are less important than those of complexes.

Complexes required for chromosome segregation are often multifunctional. Condensin (see above) is also required for interphase activities, such as

DNA-damage repair (Heale et al., 2006). Cohesin, the multiprotein complex that holds sister chromatids together following DNA replication, is also required for DNA-damage repair (Ström et al., 2007, Unal et al., 2007, Ball and Yokomori 2008) and developmental transcriptional regulation (Dorsett et al., 2005, Dorsett 2007, Gullerova and Proudfoot 2008, Wendt et al., 2008). The name, usually based on the initially discovered function, might only partially represent the functions mediated by the complex and could be misleading. Therefore, biologists and geneticists should use caution when naming a complex according to its originally discovered function.

The anaphase-promoting complex/cyclosome (APC/C) has an instructive history with regard to the naming. The APC/C was discovered as a complex and called a cyclosome (Sudakin et al., 1995), as it is essential for the degradation of mitotic cyclin in vitro. This same complex was also called the APC, as it was defined as an anaphase-promoting complex (King et al., 1995). The APC/C, which contains ~15 subunits (Passmore et al., 2005), is the E3 ubiquitin ligase that poly-ubiquitylates mitotic cyclin and securin for degradation in a destruction-box (DB)-dependent manner (reviewed in Sullivan and Morgan 2007). APC/C activation is inhibited by the spindle assembly checkpoint (also called the spindle checkpoint or mitotic checkpoint; see Chapter 11). Poly-ubiquitylated cyclin and securin are rapidly degraded by the 26S proteasome, leading to the activation of separase, the cleavage of cohesin, the separation of the sister chromatids, and the onset of anaphase (Morgan 2006, Fig. 2.1). Because the abbreviation APC also refers to the frequently cited tumor suppressor protein *a*denomatous *p*olyposis *c*oli, it is currently recommended that the abbreviation APC/C be used to avoid confusion. This distinction has become particularly necessary as the tumor suppressor APC interacts with the plus ends of the microtubules and is implicated in the spindle checkpoint (Draviam et al., 2006). While the APC/C regulates the exit of mitosis in dividing cells (Sullivan and Morgan 2007), it is also abundant in non-dividing cells such as neurons and muscles (van Roessel et al., 2004, Zarnescu and Moses 2004). The APC/C seems to have a postmitotic role at *Drosophila* neuromuscular synapses: in neurons, the APC/C controls synaptic size, and in muscles, it regulates synaptic transmission (van Roessel et al., 2004). The roles of the APC/C in non-dividing differentiated cells are elusive, but clearly different from its role in mitotic progression and exit. Thus, a new name, particularly one based on a single function, could cause misconceptions concerning the roles of these complexes.

## 2.5 Basic Mutant Phenotypes

Defects in chromosome segregation are variable but classifiable. Mutations in the subunits of the same complexes often produce similar phenotypes. For example, more than 10 APC/C subunits are essential for mitotic progression and their mutants inhibit degradation of mitotic cyclin and securin, leading to similar mitotically arrested phenotypes. A typical mutant phenotype is the arrest or delay that occurs with condensed chromosomes and continued activation of mitotic CDK. Poly-ubiquitylation mediated proteolysis by the APC/C and the 26S proteasome is necessary to inactivate mitotic CDK (Sullivan and Morgan 2007). The spindle checkpoint inhibits mitotic progression through APC/C inactivation (Hwang et al., 1998, Kim et al., 1998; see below). A failure to form the proper spindle apparatus or kinetochore structure activates the spindle checkpoint (Rieder et al., 1994, see Chapter 11). Therefore, many genes implicated in the structure and function of the mitotic spindle, the kinetochore, APC/C, and the 26S proteasome, produce similar mitotically arrested or delayed phenotypes. Only detailed phenotypic analyses can reveal molecular differences underlying these similar phenotypes (e.g., aberrantly formed kinetochore, kinetochores misattaching the chromatid pair to the spindle, non-dynamic microtubules, etc).

Another principal mitotic phenotype is the *cut* (*c*ell *u*ntimely *t*orn) phenotype in which chromosome segregation is physically impaired but cytoplasmic events, such as cytokinesis, are not. Mutants exhibiting such apparently uncoupled phenotypes occur in various genes (reviewed in Yanagida 1998). In *S. pombe*, for example, *cut* mutants are defective in Top2/DNA topoisomerase II, *cut1*/separase, *cut2*/securin, *cut3* and *cut14*/condensin subunits, etc. Events following anaphase appear to take place in which only a portion of the chromosomes, such as the centromere/kinetochore, is separated and segregated and moves to the spindle poles, while the bulk of the chromosomes remains associated and stuck near the spindle equator. The mitotic checkpoint is not activated so that mitotic cell cycle progression is not inhibited in these mutant cells (Yanagida 1998). The reason for this failure is unclear.

Another characteristic mitotic phenotype represents unequal chromosome segregation, which can be visualized by the sizes of the nuclear chromatin or the actual number of chromosomes in the daughter nuclei. In *S. pombe,* almost all mutants defective in essential centromere-binding proteins that are bound to the central core domain of centromeres (see below) show the phenotype of large and small daughter nuclei (Takahashi et al., 1994). Whether these mutants (many of them are called *mis* mutants) are defective in spindle checkpoint control remains to be determined. There are other missegregation phenotypes, such as aneuploidy or ploidy changes. Changes in chromosome number are important diagnostic features of cancer and other diseases, and are due to a wide variety of causes.

## 2.6 Simple Analogies of the Chromosome Segregation Process

In order to better understand the complex chemo-mechanical processes that underlie the chromosome segregation process, simple but useful analogies can be proposed.

### 2.6.1 Cooking Analogy

The period of chromosome segregation in anaphase is short (in the order of minutes and comparable to the "meal time"). However, many steps must occur prior to chromosome segregation. The term "chromosome cooking" refers to the long and careful preparatory steps (in human cells in the order of hours) that culminate in anaphase. The chromosome-cooking phase is under the control of the cell cycle. As in cooking, significant changes occur in the chromosome structures during the preparatory period. Cellular structures implicated in chromosome segregation either form (e.g., maturation of the kinetochore) or are greatly altered (e.g., spindle dynamics) or even disappear (e.g., cleavage/removal of cohesin complexes).

### 2.6.2 Festival Analogy

Chromosome segregation and, in general, mitosis resemble a "festival." Indeed, festivals typically occur on a seasonal basis and do not last very long. Similarly, chromosome segregation and mitosis are the shortest of the four cell cycle stages: G1, G2, S, and M (G, S, and M stand for gap, DNA synthesis, and mitosis, respectively; Fig. 2.1). The spindle apparatus and the kinetochores appear only during mitosis for the movement of the chromosomes (Fig. 2.1). A great deal of energy is used to push mitosis toward the festival of segregation and completion of cell division. Of note, like human festival participants and supporters, many of the mitotic molecules do not live just for mitotic events, but have additional functions during the non-festive interphase. For example, DNA topoisomerase II (Top2) is required during interphase (G1, G2, and S) for replication and transcription as a housekeeping enzyme (reviewed in Larsen et al., 2003). It becomes essential during mitosis for the final condensation and chromosome segregation, perhaps because of its ATP-driven enzymatic ability of catenation–decatenation (reviewed in Bates and Maxwell 2007).

The cohesin subunit Rad21/Scc1/Mcd1 was initially determined to be involved in DNA damage repair (Birkenbihl and Subramani 1992) but is also cleaved to allow sister chromatid separation in anaphase (Tomonaga et al., 2000, Sonoda et al., 2001).

Condensin and separase–securin are required for DNA damage repair (Nagao et al., 2004, Heale et al., 2006) as well as for proper chromosome segregation. These mitotic complexes also have roles in the ordinary interphase stages of the cell cycle.

### 2.6.3 Freight Train Analogy

Chromosomes are like freight trains carrying thousands of genes: the centromere-associated kinetochore acts as the "locomotive," a powered vehicle. The shortest freight train may carry only a single gene but nevertheless needs a locomotive. Indeed, *S. cerevisiae* minichromosomes may contain only a short piece of centromeric DNA, the replication origin, and a single marker gene (Clarke and Carbon 1980). *S. pombe* minichromosomes, even those containing a single marker gene, require a much longer centromeric DNA that contains replication origins for proper segregation (Niwa et al., 1987). In the freight train analogy, the locomotive vehicle is far more important than the cargo vehicles so that enigmas in chromosome segregation may be more efficiently understood through investigating the segregation mechanisms of sister kinetochores. This analogy is oversimplified and may not address interesting problems such as the segregation of telomeres, rDNAs, and the non-coding DNA sequences in the arm regions of chromosomes. It is a useful analogy, however, for understanding the problems of chromosome segregation. Centromeres/kinetochores are very complicated molecular machines consisting of hundreds of proteins. The process of sister kinetochore segregation itself raises many difficult questions, nearly equivalent to the number of questions raised regarding the control of chromosome segregation.

### *2.6.4 Glue-Cohesion Analogy*

Postreplicative chromosome DNAs in eukaryotic cells are held together by "glue" until metaphase. The presence of glue on the chromosomes is a marker of postreplicative chromosomes and should not be present on the chromosomes in anaphase. In postreplicative (G2 stage) haploid cells, there is one set of glued (paired) sister chromatids, whereas in prereplicative (G1 stage) diploid cells, there are two sets of non-glued single duplex chromosomes. The latter cells, which contain the same amount of chromosomal DNA, can be distinguished by the presence of the glue. The currently accepted terminology is cohesion rather than glue, and the cohesin complex, which is hypothesized to form a large ring that embraces the sister chromatids, is thought to be the sole link present between sister chromatids (reviewed in Nasmyth and Haering 2005). Other glues, however, may be present, and may include proteins (chromatid binding through protein–protein interaction) or RNA that exists between duplicated DNAs or any topologic bond of DNA, or may even consist of non-covalent macromolecular bonds.

### *2.6.5 Cleansing Analogy*

In interphase cells, numerous proteins and RNAs are associated with the chromosomes and participate in various DNA metabolic activities, such as replication, transcription, recombination, damage repair, etc. In mitosis, prior to anaphase, such components bound to interphase chromosomes are thought to be removed as they may inhibit the segregation mechanism or reduce the fidelity of segregation. This removal process summons an analogy to cleansing. If sister chromatids are catenated, anaphase cannot proceed properly until the obstacles are removed. The interphase components, which remain present in metaphase, must be cleaned up or degraded in anaphase. Certain proteins essential for segregation may actually function in cleansing processes rather than in chromatid separation and segregation. Little is known about the molecular cleansing machinery. Top2 is a potential candidate as it may abolish catenation and topological entanglement prior to anaphase. Polo kinase or separase may also have a role in removing cohesin from the chromosomes. Many components might be involved in such presumed cleansing functions or be the targets of the cleansing molecular machines, as so many components are bound to interphase chromosomes.

### *2.6.6 Chromosome–Corpse Analogy*

In this analogy, the spindle apparatus is dynamic and lively, but mitotic chromosomes are condensed and gene expression is thought to be minimal in mitosis. Investigators first interested in mitosis half-century ago studied how the spindle moved and worked, and thought the chromosomes looked like corpses carried by the spindle. This analogy of mitotic chromosomes as corpses was proposed long before the discovery of many mitotic components that dynamically regulate the structure and behavior of chromosomes in mitosis (see p. 212, Mazia 1961).

In reality, *de novo* RNA transcription during mitosis has been scarcely investigated that so little is known about how the transcriptional machinery acts on the chromosomes from early to late mitosis.

## 2.7 Centromere and Kinetochore

The centromere and kinetochore that assembles on it orchestrate chromosome segregation in eukaryotes. The use of the artificial minichromosome first constructed by Clarke and Carbon (1980) was immensely helpful for analyzing the segregation mechanism and led to identification of the functional centromeric DNA sequence (Chapters 1 and 3) and kinetochore proteins (Chapter 6). The size of the minimal functional centromere, determined using the *S. cerevisiae* circular minichromosome, is rather small (~125 bp). A large number of centromere/kinetochore proteins interact with this small centromere region to ensure correct segregation (Chapter 6). The linear minichromosomes of *S. cerevisiae* (Murray and Szostak 1983), which are faithfully transmitted during cell division, attach to the telomeric sequences at their ends and are much longer than the circular minichromosomes.

The centromere DNAs of *S. pombe* are 300–1,000 times longer than those of *S. cerevisiae* (Takahashi et al., 1992). Fly and higher eukaryotic centromeres are longer (up to 7 Mb) and more complex (Chapter 3). The centromeric regions of *S. pombe*, identified based on the mitotic stability of artificial linear and circular minichromosomes (Clarke et al., 1986, Nakaseko et al., 1986) have two functionally distinct domains. The central domains (*ctr* and *imr*) are roughly constant in length (~15 kb), while the outer heterochromatic repeat region (otr) that surrounds the central domain varies in length (20~100 kb). The central domains and a small portion of the outer region are necessary for correct mitotic segregation of minichromosomes (Chapter 3). Known kinetochore proteins such as CENP-A, Mis6, Mis12, Mis18, and Mad2 (Hayashi et al., 2004) are bound to the central regions of the *S. pombe* centromere, while the otr regions contain heterochromatin and small inhibitory RNA transcribed in the otr outer region (Volpe et al., 2002). Cohesin and aurora kinase are also enriched in the otr region. The otr-like heterochromatin is also present in the mating-type and telomeric regions (Grewal and Klar 1997), and is thus not specific to the centromere (Chapter 3).

## 2.8 Basics of Centromere-Kinetochore Proteins

The highly diverged centromeric DNA sequences of different organisms first seemed to eliminate any chance of obtaining a unified concept for the centromere (Chapters 1 and 7). When the first proteins binding to the centromere were identified (e.g., CENP-A/Cse4/CID/Cnp1, CENP-C/Mif2), their evolutionary conservation from fungi to humans was a pleasant surprise (Chapter 7). Centromeric DNA together with a number of centromere-binding proteins constitute the centromeric chromatin, which underlies the kinetochore, the proteinaceous structure that attaches each sister centromere to the spindle microtubules. Kinetochore formation is initiated by CENP-A/Cse4/CID/Cmnp1 and occurs in a complex, epigenetically regulated (Chapter 10) and hierarchical manner from more than 100 proteins (Chapter 6). Kinetochore components are classified based either on their location within the kinetochore (e.g., inner centromere-bound proteins versus outer microtubule-binding components), their activity at the kinetochore (structural versus enzymatic components; e.g., epigenetic factors) or on their stability (non-dynamic or dynamic, e.g., spindle checkpoint proteins). Human kinetochores start to assemble (mature) on centromeric chromatin at the initiation of prometaphase, following nuclear envelope breakdown (Fig. 6.1, Chapter 6). The formation of the kinetochore and its subsequent binding to the spindle is monitored by the spindle checkpoint (Chapter 11). Following binding to the spindle and mediation of sister chromatid segregation (via outer kinetochore motor proteins and spindle depolymerization) the outer kinetochore components leave the centromere. In many cases, these components become degraded following epigenetic marking (e.g., ubiquitylation, Chapter 10).

## 2.9 Generation of Force Required to Segregate Separated Chromatids Towards the Poles

How the spindle generates the force required to bring the separated chromatids toward the opposing spindle poles in anaphase and whether that force is regulated are important questions (also see Chapter 8). Although there are, as yet, no firm answers to these questions, some experimental results allow for speculation. Legendary experiments using grasshopper spermatocytes (Nicklas 1983) indicated that the force needed for pulling chromosomes is in the range of 50 pN per kinetochore microtubule, though the precise value remains to be determined (Rieder et al., 1986). Factors that depolymerize microtubules may generate the spindle force. Is chromosome a heavy cargo for a microtubule to carry? The average weight of a single *S. pombe* chromosome is roughly 30 femtograms equivalent to ~50 $T_{\mathbf{even}}$ bacteriophages or 10,000 ribosome particles. The weight of the average human chromosome is 1 picogram, ~30 times greater than that of a fission yeast chromosome. As 2–3 kinetochore microtubules are bound to each kinetochore in *S. pombe* and 20–30 are bound to each kinetochore in mammalian cells, the chromosome mass pulled by a single kinetochore microtubule in fission yeast and human cells may be around 15–50 femtograms. Nicklas (1983) stated that the spindle apparatus and kinetochore microtubules are powerful enough to generate $10^4$ times the force required for free movement of the chromosomes, and showed that a single microtubule can move a newt chromosome, which is much bigger than a human chromosome (Nicklas and Kubai 1985). The force required for the anaphase spindle to separate the chromosomes is thus rather small relative to the maximal force that the spindle can produce. The maximal spindle force may never be used in normal chromosome segregation. Alternatively, such force may be required very briefly to overcome the built-in sister chromatid-linking in metaphase and even anaphase chromosomes (e.g., Baumann et al., 2007). There may be force-sensitive association and dissociation reactions between sister chromatids or sister kinetochores.

The pulling and/or pushing forces (the latter is also called the polar ejection force (Rieder et al., 1986, Salmon 1989) or -more poetically- the polar wind (Carpenter 1991)) of the spindle are created in prometaphase to metaphase spindles through the dynamic properties of microtubules and/or motors. There is a common belief among investigators that the forces are generated primarily by microtubule dynamics that occur in a spontaneous as well as in a regulated manner, and that these forces are spatially and temporally controlled directly by microtubule interacting proteins (Chapter 8). Little is known, however, about how these forces are exerted on kinetochores and microtubules, which produce their sequential changes during mitotic progression under the surveillance of the spindle checkpoint.

Our main focus here lies on the regulated behavior of spindle and kinetochore microtubules and whether these properties abruptly change in anaphase.

Again, there is little data relating to this question. The rate of chromosome movement during anaphase A in *S. pombe* and HeLa cells is ~1–3 μm/min (16–50 nm/s). In *S. pombe,* the path of anaphase chromosome movement is ~1 μm; thus, anaphase A is completed within 1 min. In HeLa cells, anaphase A chromosomes move ~5 μm within 2 min (2,5 μm/min, ~40 nm/s). Thus, the duration of anaphase A is rather brief. Anaphase A movement of chromosomes in living yeast and HeLa cells is smooth and steady, although when examined at high resolution in PtK1 cells the movements remain oscillatory, even during anaphase (Salmon, E.D., personal communication). Note, however, that kinetochore microtubule shortening is 10 times slower than the maximum shortening speed of cytoplasmic or isolated microtubules, which is 30 μm/min (Mitchison and Kirschner 1984).

Microtubule- and kinetochore-interacting proteins that critically regulate the dynamic properties to shorten kinetochore and spindle microtubules during metaphase–anaphase progression have not yet been established. Certain microtubule-binding proteins such as Dis1/Stu2/XMAP215/ch-TOG are microtubule polymerizers (Brouhard et al., 2008), while others including kinesin-13 family members MCAK/XKCM1, depolymerize microtubules (Walczak et al., 1996, Helenius et al., 2006). The activity of the latter is suppressed by tau and, to a lesser extent, by XMAP215 (Noetzel et al., 2005). In anaphase A, only kinetochore microtubules are shortened, whereas the pole-to-pole microtubules that are interdigitated at the middle of the spindle begin to elongate in anaphase B. In addition to these two classes of microtubules, the aster microtubules radiating from the spindle poles and their plus ends associating with the cellular cortex determine the positions of the spindle apparatus within the cell (for details, see Chapter 8).

## 2.10 Key Players in Chromosome Segregation

A few of the key players in chromosome segregation, including CENP-A, cohesin, condensin, components that are required for mitotic checkpoint and anaphase, are described below.

### *2.10.1 CENP-A and Its Recruitment Factors*

The presence of centromere-specific histone CENP-A/CID/Cse4/Cnp1, which replaces canonical histone H3 at centromric nucleosomes (Earnshaw and Rothfield 1985, Valdivia and Brinkley 1985, Palmer et al., 1991), suggests that the centromere chromatin is specific, but its nature is not well understood. In *S. pombe*, the centromere region containing Cnp1 (hereafter referred to as CENP-A) has a smeared micrococcal digestion pattern, while the surrounding heterochromatic regions show regular digestion patterns, but this difference is not due solely to the presence of CENP-A (Takahashi et al., 2000). As CENP-A and its family members are present in all functional centromeres, CENP-A is an appropriate marker for centromere identity in eukaryotic organisms (Chapter 7). Understanding how CENP-A is recruited to the centromere and how CENP-A containing nucleosomes form are important questions (for details, see Chapter 6).

A surprising finding that was recently reported is that a centromere-specific nucleosome core in *S. cerevisiae* lacks the canonical histone H2A-H2B dimer, which is replaced by Scm3, a non-histone protein that interacts with *S. cerevisiae* CENP-A/Cse4 (Mizuguchi et al., 2007). Scm3, conserved across fungi, stoichiometrically binds to H4-CENP-A in vitro, and forms the centromere specific "nucleosome" containing CENP-A/Cse4. These results suggest that centromeric histones of *S. cerevisiae* do not form octamer nucleosomes, but instead non-histone Scm3 may serve to assemble and maintain CENP-A-H4 at the centromeres in *S. cerevisiae*. Whether the same type of centromeric chromatin forms in other organisms remains to be determined. Scm3 may also be a recruitment factor rather than the replacement of histone H2A and H2B (for details, see Chapter 6).

The recruitment of CENP-A to the centromere requires multiple protein complexes (for a discussion in detail, see Chapter 6). In *S. pombe*, Mis16 and Mis18 are the most upstream factors for centromeric loading of CENP-A, whereas Mis6 (and its partner proteins) is located downstream (Hayashi et al., 2004). In their absence, CENP-A is not recruited to the centromere. In human cells, homologues of the above CENP-A-loading proteins exist. RbAp46/48, similar to Mis16, may be a chaperone for histone H4 and is implicated in chromatin assembly through histone H4 acetylation (Fujita et al., 2007). In the greater part of the cell cycle, centromeric acetylation of H4 is minimal, while acetylation may occur during mitosis. A striking property of Mis18 is its cell cycle-dependent localization at the centromere: Mis18 is absent from mitotic kinetochores, but is located transiently at the centromere from telophase to early G1. This localization is essential for subsequent recruitment of CENP-A (Hayashi et al., 2004). The Mis18–Mis16/RbAp complex seems to "prime" the pre-existing centromeric chromatin in the telophase-G1 phase for the loading of newly made CENP-A (Fujita et al., 2007). Priming is related to protein acetylation, as trichostatin A, an inhibitor of histone deacetylases, suppresses the loss of recruitment factors. Once primed, newly synthesized CENP-A is recruited onto centromeres with flexible timing prior to mitosis. The histone acetylation–deacetylation cycle seems to strongly affect CENP-A loading (Fujita et al., 2007, see also Chapter 10).

### *2.10.2 Cohesin for Cohesion, DNA-Break Repair, and Transcriptional Regulation*

Cohesin, which consists of two SMC (Smc1 and Smc3) ATPases and two non-SMC subunits (reviewed in Nasmyth 2005a, Nasmyth and Haering 2005), is required for the formation of sister chromatid cohesion, mitotic and meiotic chromosome segregation, and the repair of double-strand breaks (Guacci et al., 1997, Ciosk et al., 1998, Uhlmann et al., 1999). In addition, cohesin is involved in developmental transcriptional regulation (Dorsett et al., 2005, Dorsett 2007, Gullerova and Proudfoot 2008, Wendt et al., 2008). In the normal cell division cell cycle, cohesin is associated with the chromosome during replication, supposedly to hold sister chromatids together, and removed from the chromosome during mitosis. The cohesin complex is highly dynamic, as it is associated with postreplicative chromosomes by a single DNA break to create the cohesion necessary for DNA repair (Ström et al., 2007, Unal et al., 2007, Ball and Yokomori 2008).

Cohesin is removed from the chromosome in a two-step manner. The first step is phosphorylation by polo-like kinase and the second step is proteolysis by separase. Polo is the principal kinase that regulates cohesin, but other kinases are also involved under different physiologic conditions. In the absence of cohesin, sister chromatids separate prematurely, the mitotic checkpoint is activated, and chromosome segregation is greatly delayed and abnormal (reviewed in Nasmyth and Haering 2005, Belmont 2006, Hirano 2006). Loading of cohesin to the chromosome is intimately linked to DNA replication and requires evolutionarily conserved recruitment factors, including deposition factors (adherins) Scc2 and Scc4, helicases, DNA polymerases and the Ctf7–RFC–PCNA complex, which pairs cohesin complexes together, thereby linking the sister chromatids (reviewed in Skibbens 2005). A defect in the human gene responsible for loading cohesin (the Scc2 homologue named NIPBL) causes the Cornelia de Lange syndrome (Krantz et al., 2004, Tonkin et al., 2004, reviewed in Dorsett 2004), a congenital disorder associated with delays in physical development and mental retardation, among other abnormalities.

### 2.10.3 Condensin for Condensation, Segregation, and DNA-Damage Repair

Condensin consists of two SMC ATPases (Smc2, Smc4) and three non-SMC subunits (reviewed in Nasmyth and Haering 2005, Belmont 2006, Hirano 2006). None of the subunits is shared with cohesin. Condensin can induce positive supercoiling of DNA in the presence of ATP (Kimura and Hirano 1997). The heterodimeric SMC complex has the ability to re-anneal single strand DNAs in the absence of ATP (Sutani and Yanagida 1997). Although the requirement for condensin in both mitotic condensation and chromosome segregation is well established (Hirano and Mitchison 1994, Hudson et al., 2003), the actual molecular role of condensin is not understood (Gassmann et al., 2004, Hirano 2005). The interphase chromosome is already 1,000 to 2,000-fold compacted, and further mitotic condensation of the chromosome is only ~5-fold. As condensin does not seem to have any role in interphase chromosome compaction, the actual role of condensin in chromosome condensation *per se* should not be overestimated. Instead, condensin may function during preseparation of the sister chromatids in the metaphase chromosome. Condensin appears to increase the integrity of individual sister chromatids in metaphase chromosomes to maintain the "state of being whole" against the pulling force exerted on the kinetochores (Oliveira et al., 2005, Gerlich et al., 2006). This integrity increase may result from being "condensed" in mitosis. If condensin is absent, the integrity is lost so that only the kinetochore portions may be pulled out and moved towards the opposite poles, leaving the bulk of the remaining sister chromatids in the middle of the nuclei. In vertebrates, there are two kinds of condensin complexes (I and II), and condensin II is enriched at kinetochores (Ono et al., 2003). These two condensins have distinct tasks for increasing the integrity of the kinetochores and the arms of chromosomes in vertebrate cells (reviewed in Hirano 2005, Belmont 2006).

In *S. pombe*, the amount of condensin associated with the interphase chromosome is low. Condensin is needed for DNA damage repair at the intra S-phase checkpoint (Aono et al., 2002). Condensin is also required for DNA repair in vertebrates (Heale et al., 2006). Upon entry into mitosis, the accumulation of condensin at mitotic chromosomes occurs in multiple steps, and this accumulation at kinetochores and rDNAs appears to be necessary for normal chromosome segregation (Nakazawa et al., 2008). First, condensin is mobilized into the nucleus through Cdc2 phosphorylation of an SMC subunit with the aid of importin $\alpha$. In addition, Ark1/Aurora B kinase and Survivin/Bir1/Cut17, a subunit of the Aurora B kinase chromosomal passenger complex, are required for condensin to locate on the mitotic chromosome docking sites. Further, certain centromere- and rDNA-binding proteins are necessary to dock condensin at both the kinetochores and rDNAs. Several proteins that are necessary to form the chromatin architecture of the kinetochores and rDNAs are required for condensin to accumulate specifically at these sites. The mechanism of condensin accumulation at the kinetochores may be conserved, as human condensin II fails to accumulate at kinetochores in cells treated with RNAi for the same CENP (Nakazawa et al., 2008).

### 2.10.4 Components Required for the Mitotic Checkpoint

Proper attachment/alignment of the mitotic chromosomes and a correct timing of sister chromatid separation are assured by the mitotic checkpoint (spindle checkpoint; reviewed in Musacchio and Salmon 2007, Chapter 11). Checkpoint proteins called mitotic arrest deficient proteins (MADs) and budding uninhibited by benzimidazole proteins (BUBs) are necessary for establishing the mitotic checkpoint. They are conserved from fungi to human, and are temporarily recruited to the kinetochores. The failure of checkpoint proteins to locate at kinetochores leads to chromosome misalignment and/or premature sister chromatid separation, thus raising the question of how checkpoint proteins control the progression of mitosis. Checkpoint proteins are linked to the APC/C complex by Cdc20/Slp1/Fizzy, which directly interacts with APC/C. Mad2 and Mad3/BubR1 are physically bound to Cdc20 and are thought to inactivate APC/C through this interaction, suggesting that the activation of APC/C requires the inactivation of Mad2 (Musacchio and Salmon 2007, Chapter 11).

Currently there are two mechanistic models of the spindle checkpoint. First, the two-state model is based on biochemical identification of the in vitro inhibitor of mitotic APC/C, which was purified from HeLa cells, and is called the mitotic checkpoint complex (MCC). The MCC contains BubR1 (similar to Mad3 in fungi), Bub3, Cdc20, and Mad2 in near equal stoichiometry (Sudakin et al., 2001, Sudakin and Yen 2004, Chapter 11). The MCC inhibition of APC/C is 3,000-fold greater than that of recombinant Mad2, which also inhibits APC/C in vitro. MCC is present and unexpectedly active in interphase cells. Only APC/C isolated from mitotic cells, however, is sensitive to inhibition by MCC. The majority of APC/C in mitotic lysates is associated with the MCC, which likely contributes to the lag in ubiquitin ligase activity. The preformed interphase pool of MCC may allow for the rapid inhibition of APC/C when cells enter mitosis. This inhibition by MCC is independent of the kinetochores, and the proposed role of unaligned kinetochores is to sensitize APC/C to MCC-mediated inhibition. In the original model, the MCC is stable across the cell cycle, but may dissociate upon exit from the checkpoint, suggesting that MCC is dynamically regulated (Braunstein et al., 2007).

Second, an alternative model proposes that Mad1-Mad2 at kinetochores acts as a template to change the conformation of another molecule of Mad2 (discussed in Nasmyth 2005b). This templated change in Mad2 conformation is proposed as a mechanism for the amplification of the "wait anaphase" signal (DeAntoni et al., 2005). In this model, Mad2 acts to sequester Cdc20 to halt anaphase. Mad2 is recruited to the kinetochores in prometaphase, and is activated with the help of Mad1 and binds to Cdc20. Mad2 has two conformers, a closed form that is bound to its kinetochore receptor Mad1 or its target in the checkpoint Cdc20 and an open form that is not bound to these ligands. A closed conformer of Mad2 constitutively bound to Mad1 is the kinetochore receptor for cytosolic open Mad2. In this model, the interaction between the open and closed forms of Mad2 is essential to sustain the spindle assembly checkpoint. The closed Mad2 bound to Mad1 is proposed to represent a template for the conversion from the open Mad2 to closed Mad2 bound to Cdc20 (discussed in Mapelli and Musacchio 2007). This predicts a mechanism for cytosolic propagation of the spindle checkpoint signal away from the kinetochores. The cause of the loss of this ability of Mad2 is p31Comet (Cmt2; Habu et al., 2002), which is required for exit from the spindle checkpoint. p31Comet associates with Mad2 and blocks Mad2 activation through structural mimicry (Mapelli et al., 2007, Yang et al., 2007).

What is the role of kinetochores in the mitotic checkpoint? The HEC1/Ndc80 and Blinkin/Spc105/Spc7/KNL-1 kinetochore complexes are involved in spindle checkpoint activity (Martin-Lluesma et al., 2002, Gillett et al., 2004, Kiyomitsu et al., 2007). The HEC1 complex has not yet been shown to directly interact with components of the spindle checkpoint. Recently, Blinkin was revealed to interact via two-hybrid analysis with BubR1 and Bub1 (Kiyomitsu et al., 2007). The depletion of both HEC1 and Blinkin abolishes the kinetochore localization of the spindle checkpoint, causing premature entry into anaphase and chromosome misalignment (discussed in Chapter 11). How both kinetochore complexes dovetail with the two-state and/or the template model of the checkpoint remains to be determined.

### *2.10.5 Components Required for Anaphase*

The current prevailing concept for the mechanism triggering anaphase onset is the silencing of the mitotic checkpoint, which allows for the activation of APC/C, which destroys mitotic cyclin and securin to promote anaphase (reviewed in Sullivan and Morgan 2007). The loss of securin or the inactivation of CDK activates separase, which removes cohesin from chromosome. In human cells, CDK1/cyclin B phosphorylation inhibits separase activation. CDK1/cyclin B1 and securin interact with separase in a mutually exclusive manner (Gorr et al., 2005). In addition, type 2A phosphatase (PP2A) containing the B56 subunit is bound to separase and regulates the timing of separase activation (Holland et al., 2007). Securin is a chaperone and inhibitor of separase, and a part of its structure may mimic the Rad21 cleavage site recognized by separase. The chaperonic role of securin/Cut2 may be regulated by phosphorylation under different stress conditions. High stress induces more phosphorylation and stabilizes securin. In *S. cerevisiae*, Pds1/securin is the target of the Rad53/Chk2 damage checkpoint protein to arrest the cell cycle for repair (Cohen-Fix and Koshland 1997).

In addition to its role as a site-specific protease that cleaves Rad21/Mcd1, separase seems to have various other roles. In *S. pombe*, separase/Cut1 is required for normal duplication and separation of the spindle pole bodies (SPBs, equivalent to the centrosome). It binds to interphase microtubules, metaphase and extending anaphase mitotic spindles, and to the centrosome-equivalent of SPBs (Nakamura et al., 2002). The protease-dead carboxy fragment can strongly affect SPB positioning. Separase is also required for duplication of the centrosome (Tsou and Stearns 2006). A single centrosome that contains a pair of centrioles duplicates once before mitosis. During duplication, new centrioles form orthogonally to existing ones and remain engaged with those centrioles until late mitosis or early G1 (Nigg 2007). Centriole disengagement appears to require separase, and the disengagement is hypothesized to control centriole duplication in the next cell cycle. The involvement of separase in both centriole disengagement and sister chromatid separation would prevent premature centriole disengagement before anaphase onset, and thereby inhibit the formation of multipolar spindles as well as genomic instability.

## 2.11 Future Prospects

Studies on chromosome segregation began to have a real cellular basis after the discovery of CDKs in 1988, and a significant impact after finding the connection between cell cycle control and chromosome segregation through the shared destruction mechanisms of poly-ubiquitylated cyclin and securin. Importantly, the research area is still expanding. Indeed, a PubMed search for chromosome segregation hit 4,135 original papers; half of which were published since 2002. We hope that the fundamental issues discussed here will inspire young researchers interested in chromosome and cellular biology. Their contributions during the next 10 years promises the discovery of novel genes and gene functions that underlie chromosome segregation and ensure the fidelity of the process in various organisms.

## References

Aono, N., Sutani, T., Tomonaga, T., Mochida, S., and Yanagida, M. 2002. Cnd2 has dual roles in mitotic condensation and interphase. Nature 417: 197–202.

Ball, A.R. Jr., and Yokomori, K. 2008. Damage-induced reactivation of cohesin in postreplicative DNA repair. Bioassays 30: 5–9.

Bates, A.D., and Maxwell, A. 2007. Energy coupling in type II topoisomerases: why do they hydrolyze ATP? Biochemistry 46: 7929–7941.

Baumann, C., Körner, R., Hofmann, K., and Nigg, E.A. 2007. PICH, a centromere-associated SNF2 family ATPase, is regulated by Plk1 and required for the spindle checkpoint. Cell 128: 101–114.

Belmont, A.S. 2006. Mitotic chromosome structure and condensation. Curr. Opin. Cell Biol. 18: 632–638.

Birkenbihl, R.P., and Subramani, S. 1992. Cloning and characterization of rad21 an essential gene of *Schizosaccharomyces pombe* involved in DNA double-strand-break repair. Nucleic Acids Res. 20: 6605–6611.

Braunstein, I., Miniowitz, S., Moshe, Y., and Hershko, A. 2007. Inhibitory factors associated with anaphase-promoting complex/cylosome in mitotic checkpoint. Proc. Natl. Acad. Sci. U.S.A. 104: 4870–4875.

Brouhard, G.J., Stear, J.H., Noetzel, T.L., Al-Bassam, J., Kinoshita, K., Harrison, S.C., Howard, J., and Hyman, A.A. 2008. XMAP215 is a processive microtubule polymerase. Cell 132: 79–88.

Carpenter, A.T. 1991. Distributive segregation: motors in the polar wind? Cell 64: 885–890.

Ciosk, R., Zachariae, W., Michaelis, C., Shevchenko, A., Mann, M., and Nasmyth, K. 1998. An ESP1/PDS1 complex regulates loss of sister chromatid cohesion at the metaphase to anaphase transition in yeast. Cell 93: 1067–1076.

Clarke, A., and Orr-Weaver, T.L. 2006. Sister chromatid cohesion at the centromere: confrontation between kinases and phosphatases? Dev. Cell 10: 544–547.

Clarke, L., Amstutz, H., Fishel, B., and Carbon, J. 1986. Analysis of centromeric DNA in the fission yeast *Schizosaccharomyces pombe*. Proc. Natl. Acad. Sci. U.S.A. 83: 8253–8257.

Clarke, L., and Carbon, J. 1980. Isolation of a yeast centromere and construction of functional small circular chromosomes. Nature 287: 504–509.

Cohen-Fix, O., and Koshland, D. 1997. The anaphase inhibitor of *Saccharomyces cerevisiae* Pds1p is a target of the DNA damage checkpoint pathway. Proc. Natl. Acad. Sci. U.S.A. 94: 14361–14366.

DeAntoni, A., Sala, V., and Musacchio, A. 2005. Explaining the oligomerization properties of the spindle assembly checkpoint protein Mad2. Philos. Trans. R. Soc. Lond. B Biol. Sci. 360: 637–647, discussion 447–638.
Doree, M., and Hunt, T. 2002. From Cdc2 to Cdk1: when did the cell cycle kinase join its cyclin partner? J. Cell Sci. 115: 2461–2464.
Dorsett, D. 2004. Adherin: key to the cohesin ring and Cornelia de Lange syndrome. Curr Biol. 14: R834-R836.
Dorsett, D. 2007. Roles of the sister chromatid cohesion apparatus in gene expression, development, and human syndromes. Chromosoma 116: 1–13.
Dorsett, D., Eissenberg, J.C., Misulovin, Z., Martens, A., Redding, B., and McKim, K. 2005. Effects of sister chromatid cohesion proteins on cut gene expression during wing development in *Drosophila*. Development 132: 4743–4753.
Draviam, V.M., Shapiro, I., Aldridge, B., and Sorger, P.K. 2006. Misorientation and reduced stretching of aligned sister kinetochores promote chromosome missegregation in EB1- or APC-depleted cells. EMBO J. 25: 2814–2827.
Earnshaw, W.C., and Rothfield, N. 1985. Identification of a family of human centromere proteins using autoimmune sera from patients with scleroderma. Chromosoma 91: 313–321.
Epstein, C.J. 2007. The consequences of chromosome imbalance. Principles, mechanisms and models. Cambridge University Press. pp. 508.
Fujita, Y., Hayashi, T., Kiyomitsu, T., Toyoda, Y., Kokubu, A., Obuse, C., and Yanagida, M. 2007. Priming of centromere for CENP-A recruitment by human hMis18alpha, hMis18beta, and M18BP1. Dev. Cell 12: 17–30.
Gassmann, R., Vagnarelli, P., Hudson, D., and Earnshaw, W.C. 2004. Mitotic chromosome formation and the condensin paradox. Exp. Cell Res. 296: 35–42.
Gerlich, D., Hirota, T., Koch, B., Peters, J. M., and Ellenberg, J. 2006. Condensin I stabilizes chromosomes mechanically through a dynamic interaction in live cells. Curr. Biol. 16: 333–344.
Gillett, E.S., Espelin, C.W., and Sorger, P.K. 2004. Spindle checkpoint proteins and chromosome-microtubule attachment in budding yeast. J. Cell Biol. 164: 535–546.
Gorr, I. H., Boos, D., and Stemmann, O. 2005. Mutual inhibition of separase and Cdk1 by two-step complex formation. Mol. Cell 19: 135–141.
Grewal, S.I., and Klar, A.J. 1997. A recombinationally repressed region between mat2 and mat3 loci shares homology to centromeric repeats and regulates directionality of mating-type switching in fission yeast. Genetics 146: 1221–1238.
Guacci, V., Koshland, D., and Strunnikov, A. 1997. A direct link between sister chromatid cohesion and chromosome condensation revealed through the analysis of MCD1 in *S. cerevisiae*. Cell 91: 47–57.
Gullerova, M., and Proudfoot, N.J. 2008. Cohesin complex promotes transcriptional termination between convergent genes in *S. pombe*. Cell 132: 983–995.
Habu, T., Kim, S.H., Weinstein, J., and Matsumoto, T. 2002. Identification of a MAD2-binding protein, CMT2, and its role in mitosis. EMBO J. 21: 6419–6428.
Hayashi, T., Fujita, Y., Iwasaki, O., Adachi, Y., Takahashi, K., and Yanagida, M. 2004. Mis16 and Mis18 are required for CENP-A loading and histone deacetylation at centromeres. Cell 118: 715–729.
Heale, J.T., Ball, A.R., Jr., Schmiesing, J.A., Kim, J.S., Kong, X., Zhou, S., Hudson, D.F., Earnshaw, W.C., and Yokomori, K. 2006. Condensin I interacts with the PARP-1-XRCC1 complex and functions in DNA single-strand break repair. Mol. Cell 21: 837–848.
Helenius, J., Brouhard, G., Kalaidzidis, Y., Diez, S. and Howard, J. 2006. The depolymerizing kinesin MCAK uses lattice diffusion to rapidly target microtubule ends. Nature 441: 115–119.
Hershko, A. 2005. The ubiquitin system for protein degradation and some of its roles in the control of the cell division cycle. Cell Death Differ. 12: 1191–1197.
Herzig, A., Lehner, C.F., and Heidmann, S. 2002. Proteolytic cleavage of the THR subunit during anaphase limits *Drosophila* separase function. Genes Dev. 16: 2443–2454.

Hirano, T. 2005. Condensins: organizing and segregating the genome. Curr. Biol. 15: R265-R275.

Hirano, T. 2006. At the heart of the chromosome: SMC proteins in action. Nat. Rev. Mol. Cell. Biol. 7: 311–322.

Hirano, T., and Mitchison, T.J. 1994. A heterodimeric coiled-coil protein required for mitotic chromosome condensation in vitro. Cell 79: 449–458.

Holland, A. J., Bottger, F., Stemmann, O., and Taylor, S.S. 2007. Protein phosphatase 2A and separase form a complex regulated by separase autocleavage. J. Biol. Chem. 282: 24623–24632.

Hudson, D.F., Vagnarelli, P., Gassmann, R., and Earnshaw, W.C. 2003. Condensin is required for nonhistone protein assembly and structural integrity of vertebrate mitotic chromosomes. Dev. Cell 5: 323–336.

Hwang, L.H., Lau, L.F., Smith, D.L., Mistrot, C.A., Hardwick, K.G., Hwang, E.S., Amon, A., and Murray, A.W. 1998. Budding yeast Cdc20: a target of the spindle checkpoint. Science 279: 1041–1044.

Kim, S.H., Lin, D.P., Matsumoto, S., Kitazono, A., and Matsumoto, T. 1998. Fission yeast Slp1: an effector of the Mad2-dependent spindle checkpoint. Science 279: 1045–1047.

Kimura, K., and Hirano, T. 1997. ATP-dependent positive supercoiling of DNA by 13S condensin: a biochemical implication for chromosome condensation. Cell 90: 625–634.

King, R.W., Peters, J.M., Tugendreich, S., Rolfe, M., Hieter, P., and Kirschner, M.W. 1995. A 20S complex containing CDC27 and CDC16 catalyzes the mitosis-specific conjugation of ubiquitin to cyclin B. Cell 81: 279–288.

Kiyomitsu, T., Obuse, C., and Yanagida, M. 2007. Human Blinkin/AF15q14 is required for chromosome alignment and the mitotic checkpoint through direct interaction with Bub1 and BubR1. Dev. Cell 13: 663–676.

Krantz, I.D., McCallum, J., DeScipio, C., Kaur, M., Gillis, L.A., Yaeger, D., Jukovsky, L., Wassarman, N., Bottani, A., Morris, C.A., Nowaczyk, M.J., Toriello, H., Bamshad, M.J., Carey, J.C., Rappaport, E., Kawauchi, S., Lander, A.D., Calof, A.L., Li, H.H., Devoto, M., and Jackson, L.G. 2004. Cornelia de Lange syndrome is caused by mutations in NIPBL, the human homolog of the *Drosophila* Nipped-B gene. Nat. Genet. 36: 631–635.

Larsen, A.K., Escargueil, A.E., and Skladanowski, A. 2003. From DNA damage to G2 arrest: the many roles of topoisomerase II. Prog. Cell Cycle Res. 5: 295–300.

Mapelli, M., Massimiliano, L., Santaguida, S., and Musacchio, A. 2007. The Mad2 conformational dimer: structure and implications for the spindle assembly checkpoint. Cell 131: 730–743.

Mapelli M, and Musacchio, A. 2007. MAD contortions: conformational dimerization boosts spindle checkpoint signaling. Curr. Opin. Struct. Biol. 17: 716–725.

Martin-Lluesma, S., Stucke V.M., and Nigg, E.A. 2002. Role of Hec1 in spindle checkpoint signaling and kinetochore recruitment of Mad1/Mad2. Science 297: 2267–2270.

Maruyama, T., Nakamura, T., Hayashi, T., and Yanagida, M. 2006. Histone H2B mutations in inner region affect ubiquitination, centromere function, silencing and chromosome segregation. EMBO J. 25: 2420–2431.

Mazia, D. 1961. Mitosis and the physiology of cell division. In: The Cell. Biochemistry, Physiology, Morphology. Brachet, J., and Mirsky, A.E., eds. Academic press, New York, pp. 77–412.

Mitchison, T., and Kirschner, M. 1984. Dynamic instability of microtubule growth. Nature 312: 237–242.

Mizuguchi, G., Xiao, H., Wisniewski, J., Smith, M.M., and Wu, C. 2007. Nonhistone Scm3 and histones CenH3-H4 assemble the core of centromere-specific nucleosomes. Cell 129: 1153–1164.

Morgan, D.O. 2006. The Cell Cycle: principles of control. New Science Press, Ltd. London, U.K. 327 pp.

Murray, A.W., and Szostak, J.W. 1983. Construction of artificial chromosomes in yeast. Nature 305: 189–193.

Musacchio, A., and Salmon, E.D. 2007. The spindle-assembly checkpoint in space and time. Nature Rev. Mol. Cell. Biol. 8: 379–393.

Nagao, K., Adachi, Y., and Yanagida, M. 2004. Separase-mediated cleavage of cohesin at interphase is required for DNA repair. Nature 430: 1044–1048.

Nakamura, T., Nagao, K., Nakaseko, Y., and Yanagida, M. 2002. Cut1/separase C-terminus affects spindle pole body positioning in interphase of fission yeast: pointed nuclear formation. Genes Cells 7: 1113–1124.

Nakaseko, Y., Adachi, Y., Funahashi, S., Niwa, O., and Yanagida, M. 1986. Chromosome walking shows a highly homologous repetitive sequence present in all the centromere regions of fission yeast. EMBO J. 5: 1011–1021.

Nakazawa, N., Nakamura, T., Kokubu, A., Ebe, M., Nagao, K., and Yanagida, M. 2008. Dissection of the essential steps for condensin accumulation at kinetochores and rDNAs during fission yeast mitosis. J. Cell Biol. 180: 1115–1131.

Nasmyth, K. 2005a. How might cohesin hold sister chromatids together? Philos. Trans. R. Soc. Lond. B Biol. Sci. 360: 483–496.

Nasmyth, K. 2005b. How do so few control so many? Cell 120: 739–746.

Nasmyth, K., and Haering, C.H. 2005. The structure and function of SMC and kleisin complexes. Annu. Rev. Biochem. 74: 595–648.

Nicklas, R.B. 1983. Measurements of the force produced by the mitotic spindle in anaphase. J. Cell Biol. 97: 542–548.

Nicklas, R.B., and Kubai, D.F. 1985. Microtubules, chromosome movement, and reorientation after chromosomes are detached from the spindle by micromanipulation. Chromosoma 92: 313–324.

Nigg, E.A. 2007. Centrosome duplication: of rules and licenses. Trends Cell. Biol. 17: 215–221.

Niwa, O., Matsumoto, T., Chikashige, Y., and Yanagida, M. 1987. Characterization of *Schizosaccharomyces pombe* minichromosome deletion derivatives and a functional allocation of their centromere. EMBO J. 8: 3045–3052.

Noetzel, T.L., Drechsel, D.N., Hyman, A.A., and Kinoshita, K. 2005. A comparison of the ability of XMAP215 and tau to inhibit the microtubule destabilizing activity of XKCM1. Philos. Trans. R. Soc. Lond. B Biol. Sci. 360:591–594.

Oegema, K., Desai, A., Rybina, S., Kirkham, M., and Hyman, A.A. 2001. Functional analysis of kinetochore assembly in *Caenorhabditis elegans*. J. Cell Biol. 153: 1209–1226.

Oliveira, R.A., Coelho, P. A., and Sunkel, C.E. 2005. The condensin I subunit Barren/CAP-H is essential for the structural integrity of centromeric heterochromatin during mitosis. Mol. Cell. Biol. 25: 8971–8984.

Ono, T., Losada, A., Hirano, M., Myers, M.P., Neuwald, A.F., and Hirano, T. 2003. Differential contributions of condensin I and condensin II to mitotic chromosome architecture in vertebrate cells. Cell 115: 109–121.

Palmer, D.K., O'Day K., Trong, H.L., Charbonneau, H., and Margolis, R.L. 1991. Purification of the centromere-specific protein CENP-A and demonstration that it is a distinctive histone. Proc. Natl. Acad. Sci. U.S.A. 88: 3734–3738.

Passmore, L.A., Booth, C.R., Vénien-Bryan, C., Ludtke, S.J., Fioretto, C., Johnson, L.N., Chiu, W., and Barford, D. 2005. Structural analysis of the anaphase-promoting complex reveals multiple active sites and insights into polyubiquitylation. Mol. Cell 20: 855–866.

Pederson, T. 2003. Historical review: An energy reservoir for mitosis, and its productive wake. Trends Biochem. Sci. 28: 125–129.

Rieder, C.L., Davison, E.A., Jensen, L.C., Cassimeris, L., and Salmon, E.D. 1986. Oscillatory movements of monooriented chromosomes and their position relative to the spindle pole result from the ejection properties of the aster and half-spindle. J. Cell Biol. 103: 581–591.

Rieder, C.L., Schultz, A., Cole, R., and Sluder, G. 1994. Anaphase onset in vertebrate somatic cells is controlled by a checkpoint that monitors sister kinetochore attachment to the spindle. J. Cell Biol. 127: 1301–1310.

Salmon, E.D. 1989. Microtubule dynamics and chromosome movement. In: Mitosis. Hyams, J.S., and Brinkley, B.R., eds. Academic Press, New York, pp. 119–181.
Sánchez, I., and Dynlacht, B.D. 2005. New insights into cyclins, CDKs, and cell cycle control. Semin. Cell. Dev. Biol. 16: 311–21.
Skibbens, R.W. 2005. Unzipped and loaded: the role of DNA helicases and RFC clamp-loading complexes in sister chromatid cohesion. J. Cell Biol. 169: 841–846.
Sonoda, E., Matsusaka, T., Morrison, C., Vagnarelli, P., Hoshi, O., Ushiki, T., Nojima, K., Fukagawa, T., Waizenegger, I.C., Peters, J.M., Earnshaw, W.C., and Takeda, S. 2001. Scc1/Rad21/Mcd1 is required for sister chromatid cohesion and kinetochore function in vertebrate cells. Dev. Cell 6: 759–770.
Stallings, R.L. 2007. Are chromosomal imbalances important in cancer? Trends Genet. 23: 278–283.
Ström, L., Karlsson, C., Lindroos, H.B., Wedahl, S., Katou, Y., Shirahige, K., and Sjogren, C. 2007. Postreplicative formation of cohesion is required for repair and induced by a single DNA break. Science 317: 242–245.
Sudakin, V., Chan, G.K., and Yen, T.J. 2001. Checkpoint inhibition of the APC/C in HeLa cells is mediated by a complex of BUBR1, BUB3, CDC20, and MAD2. J. Cell Biol. 154: 925–936.
Sudakin, V., Ganoth, D., Dahan, A., Heller, H., Hershko, J., Luca, F.C., Ruderman, J.V., and Hershko, A. 1995. The cyclosome, a large complex containing cyclin-selective ubiquitin ligase activity, targets cyclins for destruction at the end of mitosis. Mol. Biol. Cell 6: 185–197.
Sudakin, V., and Yen, T.J. 2004. Purification of the mitotic checkpoint complex, an inhibitor of the APC/C from HeLa cells. Methods Mol. Biol. 281: 199–212.
Sullivan, M., and Morgan, D.O. 2007. Finishing mitosis one step at a time. Nat. Rev. Mol. Cell. Biol. 8: 894–903.
Sutani, T., and Yanagida, M. 1997. DNA renaturation activity of the SMC complex implicated in chromosome condensation. Nature 388: 798–801.
Takahashi, K., Chen, E.S., and Yanagida, M. 2000. Requirement of Mis6 centromere connector for localizing a CENP-A-like protein in fission yeast. Science 288: 2215–2219.
Takahashi, K., Murakami, S., Chikashige, Y., Funabiki, H., Niwa, O., and Yanagida, M. 1992. A low copy number central sequence with strict symmetry and unusual chromatin structure in fission yeast centromere. Mol. Biol. Cell 3: 819–835.
Takahashi, K., Yamada, H., and Yanagida, M. 1994. Fission yeast minichromosome loss mutants mis cause lethal aneuploidy and replication abnormality. Mol. Biol. Cell 5: 1145–1158.
Tomonaga, T., Nagao, K., Kawasaki, Y., Furuya, K., Murakami, A., Morishita, J., Yuasa, T., Sutani, T., Kearsey, S.E., Uhlmann, F., Nasmyth, K., and Yanagida, M. 2000. Characterization of fission yeast cohesin: essential anaphase proteolysis of Rad21 phosphorylated in the S phase. Genes Dev. 14: 2757–2770.
Tonkin, E.T., Wang, T.J., Lisgo, S., Bamshad, M.J., and Strachan, T. 2004. NIPBL, encoding a homolog of fungal Scc2-type sister chromatid cohesion proteins and fly Nipped-B, is mutated in Cornelia de Lange syndrome. Nat. Genet. 36: 636–641.
Tsou, M.F., and Stearns, T. 2006. Mechanism limiting centrosome duplication to once per cell cycle. Nature 442: 947–951.
Uhlmann, F., Lottspeich, F., and Nasmyth, K. 1999. Sister-chromatid separation at anaphase onset is promoted by cleavage of the cohesin subunit Scc1. Nature 400: 37–42.
Unal, E., Heidinger-Pauli, J.M., and Koshland, D. 2007. DNA double-strand breaks trigger genome-wide sister-chromatid cohesion through Eco1 (Ctf7). Science 317: 245–248.
Valdivia, M.M., and Brinkley, B.R. 1985. Fractionation and initial characterization of the kinetochore from mammalian metaphase chromosomes. J. Cell Biol. 101: 1124–1134.
van Roessel, P., Elliott, D.A., Robinson, I.M., Prokop, A., and Brand, A.H. 2004. Independent regulation of synaptic size and activity by the anaphase-promoting complex. Cell 119: 707–718.

Volpe, T.A., Kidner, C., Hall, I.M., Teng, G., Grewal, S.I., and Martienssen, R.A. 2002. Regulation of heterochromatic silencing and histone H3 lysine-9 methylation by RNAi. Science 297: 1833–1837.

Walczak, C.E., Mitchison, T.J., and Desai, A. 1996. XKCM1: a *Xenopus* kinesin-related protein that regulates microtubule dynamics during mitotic spindle assembly. Cell 84: 37–47.

Weaver, B.A., and Cleveland, D.W. 2007. Aneuploidy: instigator and inhibitor of tumorigenesis. Cancer Res. 67: 10103–10105.

Wendt, K.S., Yoshida, K., Itoh, T., Bando, M., Koch, B., Schirghuber, E., Tsutsumi, S., Nagae, G., Ishihara, K., Mishiro, T., Yahata, K., Imamoto, F., Aburatani, H., Nakao, M., Imamoto, N., Maeshima, K., Shirahige, K., and Peters, J.M. 2008. Cohesin mediates transcriptional insulation by CCCTC-binding factor. Nature 451: 796–801.

Yanagida, M. 1998. Fission yeast cut mutations revisited: control of anaphase. Trends Cell Biol. 8: 144–149.

Yanagida, M. 2000. Cell cycle mechanisms of sister chromatid separation; roles of Cut1/separin and Cut2/securin. Genes Cells 5: 1–8.

Yanagida, M. 2005. Basic mechanism of eukaryotic chromosome segregation. Trans. R. Soc. Lond. B Biol. Sci. 360: 609–621.

Yang, M., Li, B., Tomchick, D.R., Machius, M., Rizo, J., Yu, H., and Luo, X. 2007. p31comet blocks Mad2 activation through structural mimicry. Cell 131: 744–755.

Zarnescu, D.C., and Moses, K. 2004. Born again at the synapse: a new function for the anaphase promoting complex/cyclosome. Dev. Cell 7: 777–778.

# Chapter 3
# The Centromere

**Beth A. Sullivan**

**Abstract** Centromeres are chromosomal loci that assemble the proteinaceous kinetochore, maintain sister chromatid cohesion, regulate chromosome attachment to the spindle, and direct chromosome movement during cell division. Although the function of centromeres and proteins that contribute to their complex structure are conserved in eukaryotes, centromeric DNAs are strikingly divergent. In this chapter, I review centromere organization in a range of organisms, including unicellular eukaryotes, fruit flies, plants, and mammals. Sequence features and epigenetic mechanisms of centromere identity and regulation, including DNA–protein interactions, post-translational modifications, RNA, and protein dosage that influence centromere-specific chromatin architecture are discussed. Understanding the assembly and organization of centromeres and the contributions of sequence and epigenetic features in centromere identity and diversity remain important areas of study in chromosome biology.

## 3.1 Introduction

The centromere is an essential chromosomal locus that is important for normal genome inheritance. It is typically an area of reduced recombination and is found in regions of the genome with few genes. The centromere is the foundation for kinetochore assembly, sister chromatid cohesion, and chromosome movements during cell division. Cytologically, it is visible on metaphase chromosomes as the primary constriction, the pinched, "waist-like" area that holds replicated sister chromatids together (Fig. 3.1). Defective centromere architecture or function results in cell cycle arrest, chromosome structural abnormalities, and aneuploidy. Centromeres play such an important and conserved role in chromosome architecture and genome stability, yet their underlying

B.A. Sullivan (✉)
Institute for Genome Sciences & Policy and Department of Molecular Genetics and Microbiology, Duke University, Durham, NC, U.S.A.
e-mail: beth.sullivan@duke.edu

P. De Wulf, W.C. Earnshaw (eds.), *The Kinetochore*,
DOI 10.1007/978-0-387-69076-6_3, © Springer Science+Business Media, LLC 2009

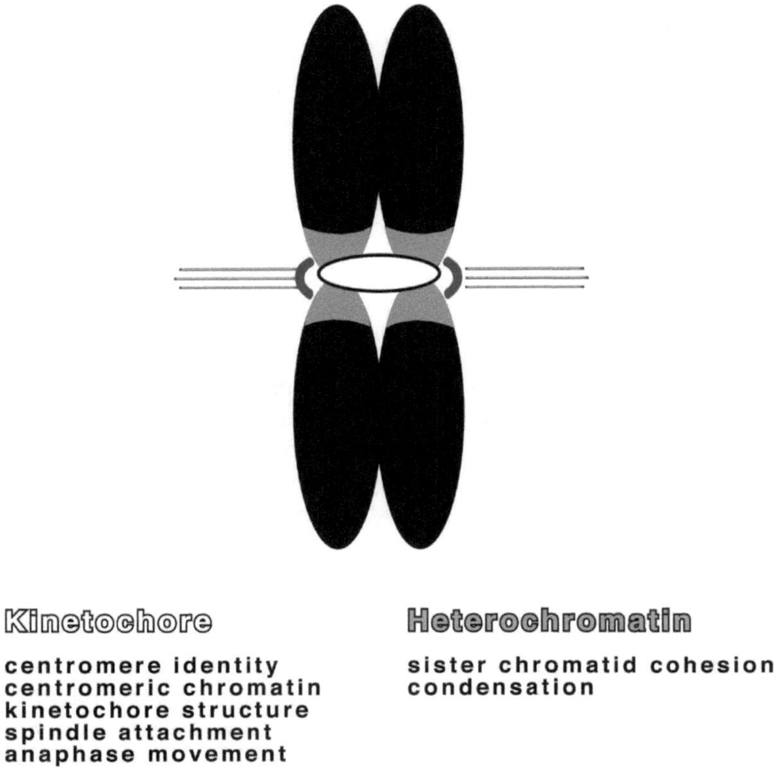


**Fig. 3.1** Role of the centromere. The centromere is the primary constriction on metaphase chromosomes. It is composed of two functional domains that contain DNA and proteins that are arranged in a unique higher order organization. The centromere is the site of kinetochore formation, the proteinaceous structure that makes contact with spindle microtubules and facilitates anaphase movements. The centromere region is also the site of heterochromatin formation that is important for sister centromere cohesion and chromosomal condensation

sequences are diverse within and between species. Many structural and functional protein components are highly conserved, suggesting that epigenetic, or sequence independent components are important for centromere architecture and function. This chapter will cover recent advances that highlight the many similarities (and a few differences) in structural and organizational features of eukaryotic centromeres, ranging from the small "point" centromere of budding yeast to the immense, "regional" centromeres of plants and mammals. The roles that DNA, RNA, and protein (histone and nonhistone) composition have in establishing the foundation for the centromere will be discussed. The reader is directed to other chapters in this volume for detailed explanations of the kinetochore complex and its role in mitosis, checkpoint signaling, and interactions with spindle microtubules.

## 3.2 Centromeric DNA: Essential Points and Regional Differences

Eukaryotic centromeres of monocentric chromosomes are typically classified as either point centromeres such as those of budding yeast or regional centromeres found in fission yeast, Drosophila, plants and mammals. Point centromeres that span over a hundred basepairs are packaged into a single nucleosome and interact with a single microtubule. Regional centromeres span several kilobases to several hundred kilobases and interact with many microtubules (Table 3.1). Both point and regional centromeres display unique nucleosomal and chromatin structures compared to bulk chromatin that may distinguish them for their specialized organization and function.

### *3.2.1 Saccharomyces Cerevisiae*

The point centromeres of budding yeasts are only a few hundred basepairs in size while the large, regional centromeres of fission yeast, flies, plants, and humans can span 40–4,000 kb (Sullivan et al., 2001). The fact that the centromere forms consistently on the same region of a chromosome in almost every organism implies that underlying DNA sequence confers centromere assembly or function. In the budding yeast, *Saccharomyces cerevisiae*, this is indeed the case. The 16 simple "point" centromeres consist of a conserved, single copy minimal 125 bp CEN sequence that is necessary and sufficient for chromosome segregation. The CEN contains three functional centromere DNA elements, CDEI, CDEII, and CDEIII (Fig. 3.2). The central element CDEII (80 bp) is AT-rich and flanked by two highly conserved palindromic sequences, CDEI (8 bp) and CDEIII (26 bp; McAinsh et al., 2003). The CDEIII is required for kinetochore assembly, and recruits proteins that interact with each element

**Table 3.1** Eukaryotic centromeric DNAs

| Organism | Type of CEN sequence | Size of CEN DNA (haploid chromosome #) |
|---|---|---|
| *S. cerevisiae* | AT-rich | 125 bp (16) |
| *S. pombe* | unique core flanked by inverted repeats | 40–100 kb (3) |
| *C. albicans* | unique sequence | 3–4.5 kb (8) |
| *D. melanogaster* | dodeca satellite? AATAT, AAGAG | 200–450 kb (4) |
| *A. thaliana* | 180 bp satellite | 0.4–1.4 Mb (5) |
| *O. sativa* | CentO 155 bp satellite; CRR retrotransposon | 2–3 Mb (12) |
| *Z. mays* | CentC 156 bp satellite; CRM retrotransposon | 0.3–3 Mb (10) |
| *M. musculus* | minor satellite 120 bp repeat | 0.1–1 Mb (20) |
| *H. sapiens* | alpha satellite 171 bp repeat | 0.1–5 Mb (23) |

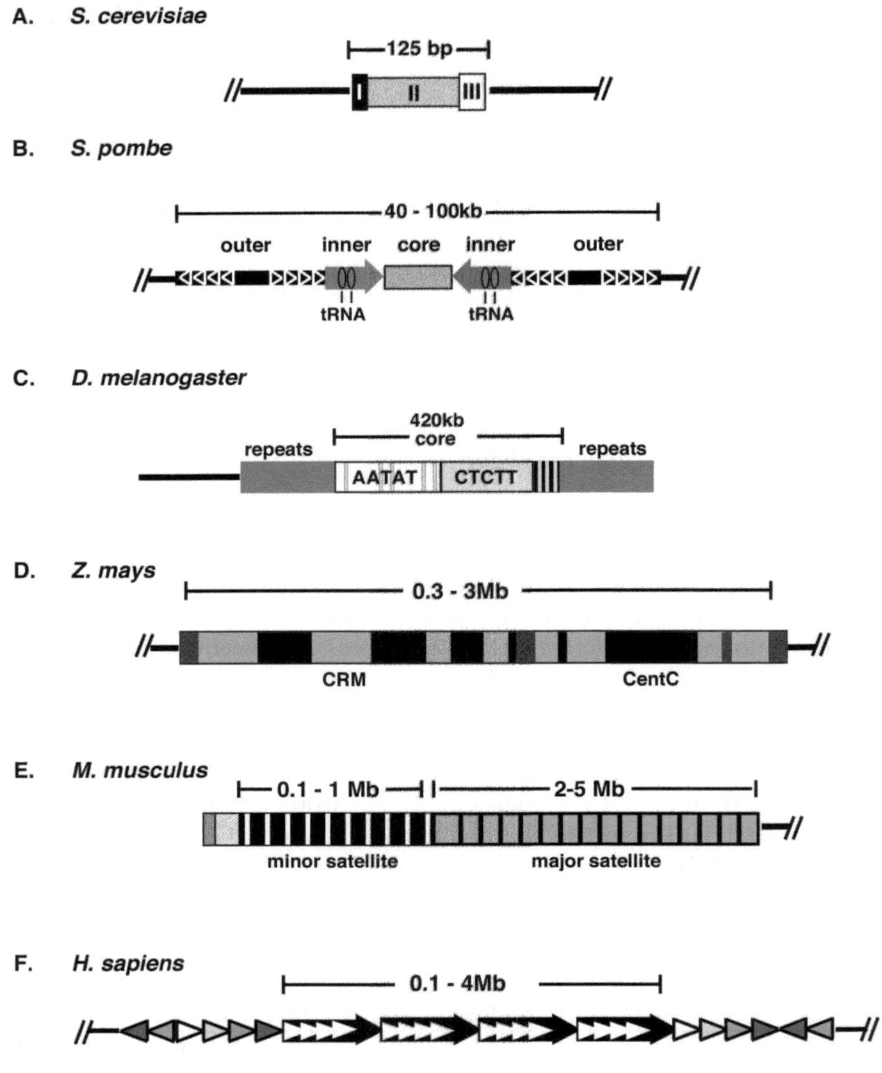


**Fig. 3.2** Eukaryotic centromeric DNAs and genomic centromere structure. The DNA sequence of centromeres is different among species. (A) In *Saccharomyces cerevisiae*, the centromere is defined by a 125 bp region that contains three conserved elements (I, II, and III), which are incorporated into a single centromeric nucleosome. (B) *Schizosaccharomyces pombe* centromeres range in size from 40 to100 kb and contain a unique central core sequence (*light gray*) flanked by inverted inner and outer repeats (*black boxes, white arrowheads*). The central core region is separated from the flanking heterochromatic repeats by sequence elements, two tRNA genes. (C) The only characterized centromere in Drosophila is on the X-derived minichromosome, Dp1187, which consists of a core of 5 bp satellites AATAT (*white*) and TTCTC (*light gray*), and transposable elements (*gray and black vertical lines*) flanked by other satellite repeats (*dark gray*). (D) Plant centromeres are also composed of large satellite DNA arrays. In *Z. mays*, the centromeres are several megabases long and

to create multicomponent complexes that organize the region into intricate but cytologically invisible kinetochores (De Wulf et al., 2003; McAinsh et al., 2003).

### *3.2.2 Schizosaccharomyces Pombe*

This dependence on primary sequence is notably absent in other eukaryotes, although some functional and structural characteristics of centromeric domains are conserved. Generally, centromeres are located near heterochromatin, in regions containing reiterated (repetitive) DNA and typically few genes. Both point and regional centromeres are located in regions that are AT-rich, however, no common sequence motif has been identified. In the fission yeast *Schizosaccharomyces pombe*, the centromeres span 35–100 kb and are composed of a central core region of nonrepetitive, nonhomologous 4–5 kb sequences. The central core on each chromosome is flanked by regions of inverted repeats: innermost (imr) repeats that are perfect inverted repeats, and the outer repeats (otr) that vary in size and orientation (Pidoux and Allshire 2005; Fig. 3.2). Two tRNA genes located on either side of the central core functionally separate the core and flanking repeats (Scott et al., 2006; see below). As little as 25 kb containing the nonrepetitive central core, inner repeats and a portion of the outer repeats, is required for centromere function and for stable chromosome transmission (Baum et al., 1994; Hahnenberger et al., 1991). The outer repeats are also important for other centromeric functions, such as heterochromatin formation, establishing sister chromatid cohesion and ensuring proper chromosome segregation (Bernard et al., 2001; Nonaka et al., 2002; Partridge et al., 2002). Dissection of the fission yeast centromere into functional domains has been achieved by inserting marker genes (*ura4* or *ade6*) across pombe centromere 1 and assaying for gene expression within various structural and functional contexts (Allshire et al., 1994, 1995; Ekwall et al., 1997). These studies have provided the framework for our understanding of how chromatin domains are functionally partitioned and regulated within complex centromere regions and will be discussed in detail in the following sections.

**Fig. 3.2** (continued) contain the CentC satellite and the interspersed retroelement CRM. Each maize centromere has different amounts of CentC (*black*) and CRM (*gray*). (E) Mouse centromere regions contain two types of satellite DNAs, minor (*black*) and major satellite (*gray*) that are interspersed with shorter repeats MS3 (*white vertical bars*) and MS4 (*black vertical bars*). (F) Human centromeres are composed of alpha satellite DNA arrays. Monomeric 171 bp repeat units (*white arrowheads*) are tandemly arranged into higher order repeats (*black arrowed boxes*). Higher order arrays are flanked by unordered monomeric repeats (*white, light gray, medium gray, and dark gray arrowheads*) that span the region between the higher order array core and the euchromatic chromosome arms

### 3.2.3 *Candida Albicans*

In other related yeasts, centromeric sequences are quite different both in sequence and in size from those of *S. cerevisiae*. *Candida albicans* is a common human fungal pathogen that is diploid and exists either in unicellular budding form or filamentous forms. Although this organism has been completely sequenced, the location and biology of its centromeres has only been recently explored, uncovering exciting mechanisms of epigenetic centromere identity that have similarities to larger eukaryotic centromeres (Baum et al., 2006). The eight centromeres in *C. albicans* are located in gene-poor regions of the chromosome and are flanked by euchromatin (Sanyal et al., 2004). Each centromere is encoded by a 3–4.5 kb sequence that is different and unique on each chromosome (Mishra et al., 2007). Although these CEN sequences have a high AT content, they lack conserved DNA motifs as well as repetitive elements, including transposons. Pericentric inverted repeats or long terminal repeats, elements that are present at other eukaryotic centromeres, are present in the pericentromeric regions of some but not all *C. albicans* centromeres. These remarkable findings suggest that epigenetic factors, such as DNA conformation, DNA–protein interactions and/or chromatin modifications, are responsible for conferring centromere identity in *C. albicans*. The fact that *C. albicans* centromeres, while different in sequence, are fixed and similarly sized indicates that epigenetic factors closely regulate the CEN regions and their borders in this organism.

### 3.2.4 *Drosophila Melanogaster*

*Drosophila* centromeres are also located within repetitive DNA. The only molecularly and functionally defined *Drosophila* centromere is on a 1.3 Mb X-derived minichromosome, *Dp1187*, progressive deletions of which have defined a 420 kb region that genetically confers normal chromosome inheritance (Murphy and Karpen 1995; Sun et al., 2002; Sun et al., 1997). The 1.3 Mb chromosome contains two large blocks of 5 bp repeats, other satellites, and both intact and fragmented transposable elements. The 420 kb CEN region contains AATAT and TTCTC repeat satellite sequences. (Sun et al., 2002; Fig. 3.2). The block of AATAT is interspersed with transposons, but the TTCTC region is largely uninterrupted. The blocks of AATAT and TTCTC are flanked by other repetitive DNAs, including 1.688 satellite. Although DNA sequences at the individual, endogenous Drosophila centromeres have yet to be identified, there is evidence to suggest that *Drosophila melanogaster* centromeres contain significant blocks of repetitive DNA (Abad et al., 2000; Andreyeva et al., 2007). A recent study using a suppressor of underreplication mutant (SuUR) reported that the Drosophila centromere-specific histone

(CENP)-A$^{CID}$ co-localizes on polytene chromosomes with a BAC clone containing a block of dodeca satellite DNA derived from the 3rd chromosome. This finding supports the (potential) location of endogenous insect centromeres to regions of repetitive DNA, as observed in larger eukaryotes, and emphasizes a common genomic feature of metazoan centromeres (see below).

### *3.2.5 Plant Centromeres*

Plant centromeres have largely been identified as regions of low recombination. They are typically composed of large arrays of centromeric satellite repeats and centromeric retrotransposons. Depending on the organism, and even within the same organism, the abundance and distribution of these centromere elements vary. *Arabidopsis thaliana* centromeres can span 0.4–1.4 Mb and contain several repetitive elements but the most abundant is a 180 bp satellite repeat that comprises the functional centromeric core (Copenhaver et al., 1999; Heslop-Harrison et al., 2003; Nagaki et al., 2003). Maize (*Zea mays*) centromeres have long arrays of the tandem repeat CentC and are enriched for centromeric retroelements (CRM; Jiang et al., 2003; Fig. 3.2). Like centromeres in mice and humans, only a portion of the CentC arrays and CRMs in maize and the 180 bp satellite arrays in *A. thaliana* actually participate in kinetochore formation (Jin et al., 2004), suggesting that the flanking regions of plant centromeres are derived from degenerate or discarded repeats. Rice (*Oryza sativa*) centromeres are defined by stretches of the CentO 155 bp satellite that are interrupted by centromere-specific CCR retrotransposons. CentO arrays span 60 kb–2 Mb. Interestingly, rice centromeres also contain genes that are actively transcribed (Nagaki et al., 2004).

### *3.2.6 Mouse Centromeres*

Mouse chromosomes, with the exception of the Y chromosome, are telocentric, meaning that the centromeres are located in close proximity to the telomere at one end of chromosome. Depending on the chromosome, the telomere can be located 0.5–2.4 kb from the centromere. In *Mus musculus*, the telomere is separated from the centromere by tL1 sequences and/or TLC tandem repeats (Kalitsis et al., 2006). The functional mouse centromere, as defined by CENP binding, is located at minor satellite DNA (Kalitsis et al., 2006; Wong and Rattner 1988; Zeng et al., 2004). Minor satellite DNA arrays are composed of ~120 bp monomers and range up to 1 megabase (Mb) in size, much smaller than chromosome-specific blocks of human higher order alpha satellite (see below; Aker and Huang 1996; Kipling et al., 1991; Fig. 3.2). The minor satellite array contains the CENP-B binding motif, CENP-B box, and the monomers are highly homogenous, even at the edges of the array (satellite junctions) Major satellite DNA, based on a 234 bp monomeric repeat, is located in the pericentromere, immediately adjacent to minor satellite and spans over 2 Mb (Vissel and Choo 1989). It does not contain CENP-B boxes and does not bind CENPs. Both minor and major satellite sequences are AT-rich, a common feature of eukaryotic centromeric DNAs. While minor and major satellite sequences predominate at mouse centromeres, other repeats are interspersed within these arrays. Mouse satellite 3 (MS3) and mouse satellite 4 (MS4) are GC-rich repeats that are distributed within minor and major satellite, respectively (Kuznetsova et al., 2005, 2006). MS3 is based on a 100 bp monomer while a MS4 monomer is 300 bp long. The roles of these subsets of satellites within the minor and major satellite arrays on centromere assembly and function are currently unknown. Attempts to generate mouse artificial chromosomes in mouse cells from cloned arrays of minor satellite DNA have been largely unsuccessful, so it is possible that MS3 and MS4 are important for centromere organization or chromosome stability.

The abundance and location of minor and major satellite varies among mouse species. In *M. spretus* and *M. caroli*, major satellite DNA is found at other genomic sites, while in *M. caroli*, minor satellite is completely absent from centromeres (Wong et al., 1990). In fact, in *M. caroli*, a completely unique sequence is present at the centromere (Kipling et al., 1995). Although the *M. caroli* centromeric sequence has no homology to minor or major satellite, it does contain the CENP-B binding motif, suggesting a common structural and functional link among centromeres in different mouse species that appear to be otherwise epigenetically defined.

### *3.2.7 Human Centromeres*

The sequence present at all primate chromosomes is alpha satellite DNA. It is based upon divergent 171 bp repeat units arranged in a tandem, head-to-tail fashion (Willard 1990). Alpha satellite that exists in an unordered fashion is termed monomeric and is present within the pericentromeric regions of most human chromosomes (Alexandrov et al., 1993). Each human chromosome is also characterized by a chromosome-specific higher order array of alpha satellite (Fig. 3.2; Willard 1985). A higher order repeat (HOR) array is a repeat unit consisting of a defined number of alpha satellite monomers; each chromosome exhibits a chromosome-specific number of repeating units (Willard and Waye 1987). The HOR arrays are tandemly repeated and are highly homogenous. The same HOR array is reiterated hundreds or thousands of times within a given centromere locus and can span 0.1–5 Mb (Lee et al., 1997; Willard 1990).

Sequence contigs at the edges of chromosome specific arrays exist for only a few chromosomes (8, 17, and the X; Nusbaum et al., 2006; Ross et al., 2005;

Rudd and Willard 2004). These contigs reveal stretches of monomeric alpha satellite on either side of the higher order array as well as other satellites that join arm sequences to the higher order array. These monomeric alpha satellite blocks are interrupted by interspersed elements (LINE, SINE, LTR) and lack overall directionality such that distinctive satellite blocks of monomers oriented in one direction are distinguishable from adjacent blocks that are arranged in opposite orientation. Distinct transitions separate monomeric alpha satellite from higher order alpha satellite. Genomic sequence contigs for only a few human centromeres have been placed onto genomic assemblies extending past the arm-centromere junctions. Thus, it is unclear if, like in fission yeast, specific sequence elements function as boundaries between the centromere and pericentromere or the pericentromere and chromosome arms (Noma et al., 2006; Scott et al., 2006), or if monomeric satellite forms a regional heterochromatic boundary between higher order alpha satellite and euchromatin.

A given HOR alpha satellite array spans many megabases, however, the entire array is not involved in kinetochore assembly. Immunocytochemistry on extended chromatin fibers using antibodies to kinetochore proteins, such as the centromere-specific histone CENP-A (see Table 3.1 and below), localize to only a portion of alpha satellite DNA arrays (Lam et al., 2006; Spence et al., 2002). Furthermore, endogenous chromosomes that are naturally or artificially deleted for large amounts of alpha satellite still assemble a kinetochore and segregate normally (Brown et al., 1994; Farr et al., 1995; Wevrick et al., 1990). *De novo* chromosome assembly of human artificial chromosomes is achieved only when higher order alpha satellite DNA is introduced into human cell lines, implying that this repetitive DNA is the most efficient template for seeding new centromere formation in cultured cells (Grimes et al., 2002; Harrington et al., 1997). It is possible that higher order alpha satellite contains sequence cues, such as specific protein binding motifs or nucleotide composition that enhances nucleosome arrangements or phasing that is most proficient for nucleating and establishing centromere identity and function (Ohzeki et al., 2002). However, not all alpha satellite arrays efficiently form centromeres on artificial chromosomes, suggesting that other sequences or epigenetic factors are required to assemble and maintain functional human centromeres.

### *3.2.8 Caenorhabditis Elegans*

The chromosomal location of centromeres in the nematode *C. elegans* is strikingly different from the point and regional centromeres in other eukaryotes. In *C. elegans*, the chromosomes are holocentric, meaning that a primary constriction is not observed and spindle microtubules attach along the entire length of the chromosome. Despite this difference in location of the kinetochore, most of the centromere and kinetochore proteins found in other eukaryotes are conserved in the nematode. CENP-A (CeCENP-A, originally named Holocentric Centromere Protein 3, HCP-3), CENP-C (CeCENP-C, originally named HCP-4), and CENP-F (HCP-1, HCP-2) localize along the length of mitotic chromosomes, appearing as two ribbons, or paired lines on opposing faces of the chromosomes. Over 30 kinetochore proteins have been identified in *C. elegans*, and half of these were identified by homology to known kinetochore proteins in other organisms. Genetic and RNAi screens and biochemical assays have also been important for identifying the remaining portion of kinetochore proteins. The genetic and knockdown studies have used immunofluorescence as well as chromosome segregation defects or a "kinetochore null" phenotype (KNL) as a functional measure of recruitment of kinetochore proteins and their effects on kinetochore structure and function. These elegant studies have revealed complex protein pathways that regulate kinetochore assembly and mitotic chromosome structure, such as CENP-$A^{HCP\text{-}3}$ in establishment of the inner kinetochore, KNL-1 in the assembly of the outer kinetochore, and CENP-$C^{HCP\text{-}4}$ that bridges the two domains (Maddox et al., 2004). These types of studies have also been important for uncovering unexpected dynamics of CENP-$A^{HCP\text{-}3}$ between meiosis and mitosis in *C. elegans* (Monen et al., 2005; see below).

## 3.3 Domain Organization of the Centromere

Eukaryotic centromeres are commonly organized into multiple functional domains (Fig. 3.1). In the case of fission yeast centromeres, these domains are correlated with underlying DNA sequence, but even mammalian centromeres that have large arrays of repetitive DNA are functionally partitioned into distinctive structural and functional domains. Eukaryotic centromere regions are broadly classified into domains encoding: (1) centromeric (CEN) chromatin and (2) heterochromatin (Choo 2001; Pidoux and Allshire 2004; Sullivan 2002; Fig. 3.1). CEN chromatin, a unique type of chromatin containing a centromere-specific histone variant, is the core of the centromere, while blocks of heterochromatin flank one or both sides of CEN chromatin (Fig. 3.3). CEN chromatin is the foundation for the kinetochore and is the site to which spindle interacting and checkpoint proteins are recruited. Heterochromatin flanking the CEN chromatin core contributes to sister chromatid cohesion and condensation (Bernard et al., 2001; Giet and Glover 2001; Hagstrom et al., 2002; Hendzel et al., 1997; Jager et al., 2005; Maddox et al., 2007). CEN chromatin and heterochromatin are assembled independently, yet each is important for chromosome segregation and genome stability (Bernard et al., 2001; Blower and Karpen 2001; Blower et al., 2002). How these domains are defined and regulated is an area of intense study.

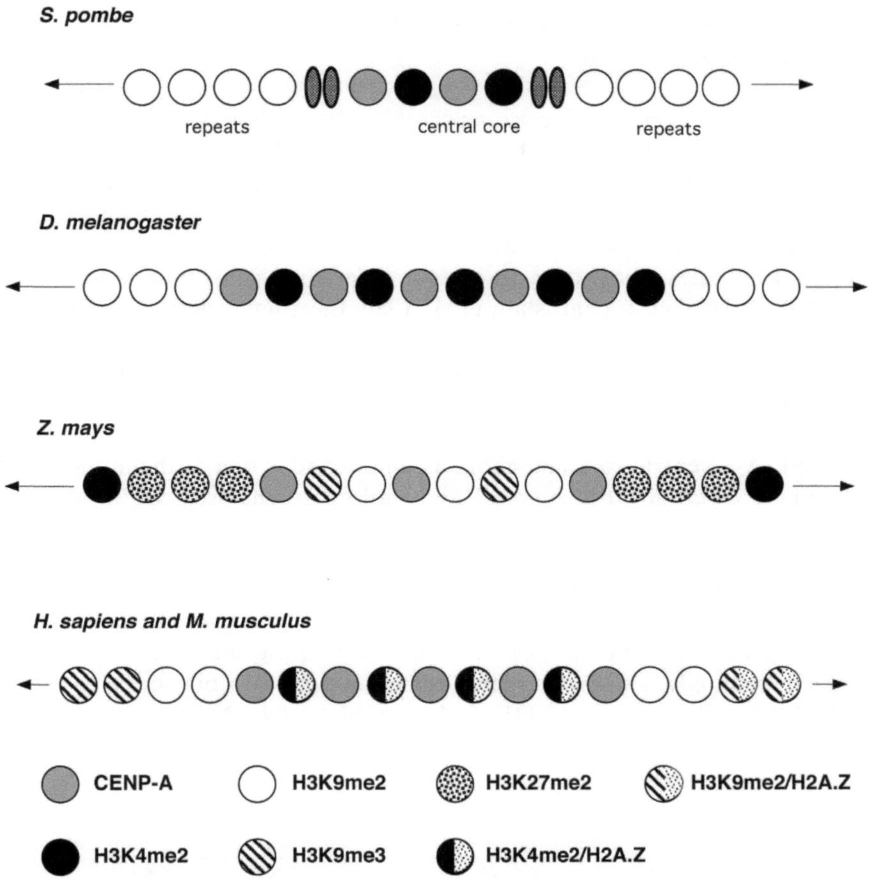


**Fig. 3.3** Chromatin organization within eukaryotic centromere regions. Eukaryotic centromeres contain a spectrum of histone modifications that functionally partition the region. *S. pombe* centromeres have a central core of centromeric (CEN) chromatin that gives rise to the kinetochore and contains the histone H3 variant CENP-$A^{Cnp1}$ and nucleosomes that contain canonical H3 dimethylated at lysine 4 (H3K4me2). This region is flanked by heterochromatin that contains H3K9me2. Drosophila centromeres contain CEN chromatin characterized by interspersion of CENP-$A^{CID}$ nucleosomes and H3K4me2 nucleosomes. The CEN chromatin domain is flanked by heterochromatic nucleosomes containing H3K9me2. Maize (*Z. mays*) centromeres, are enriched for CENP-$A^{ZmcenH3}$ nucleosomes, as well as nucleosomes containing H3K9me2 and H3K9me3. Heterochromatin enriched for H3K27me2 is found near the centromeres, while H3K4me2 is located outside of the centromere. Human and mouse centromeres are defined by cores of CEN chromatin in which interspersed CENP-A and H3K4me2 nucleosomes are largely restricted. CEN chromatin is flanked by H3K9me2 and H3K9me3 that are located within the pericentromere. Mammalian centromere regions also contain H2A.Z, a variant histone that is found within CEN chromatin associated with nucleosomes containing H3K4me2 and within the flanking heterochromatin in nucleosomes containing H3K9me3

## 3.4 CENP-A, a Variant Histone, is the Foundation of the Kinetochore

Despite divergent centromeric sequences, CENPs are highly conserved across species (Talbert et al., 2004). The first human CENPs were isolated over two decades ago from patients with autoimmune disorders, such as calcinosis, Raynaud's phenomena, esophageal dysmotility, scelodactyly, telangiectasia (CREST) syndrome (Earnshaw and Rothfield 1985). Sera from these patients contain antibodies recognizing three centromeric antigens, named CENP-A, CENP-B, and CENP-C (Earnshaw et al., 1986; Earnshaw and Rothfield 1985). These three proteins are associated with centromeres throughout the cell cycle, and form the foundation of the vertebrate kinetochore (or the pre-kinetochore) in concert with several recently identified proteins (Foltz et al., 2006; Izuta et al., 2006).

CENP-A is central to kinetochore assembly and is one of the initiators of centromere assembly. Biochemical and molecular analyses confirmed that CENP-A is a centromere-specific histone protein related to histone H3 through a common histone fold domain (Palmer et al., 1987, 1989, 1991). CENP-A, also referred to as cenH3s, is present in mammals as well as yeasts (*S. cerevisiae* and *S. pombe*), *C. elegans* and *Drosophila*, *A. thaliana*, *Oryzis* (rice) and *Z. mays* (maize (Table 3.2), representing an evolutionary link between seemingly widely divergent centromeric DNA sequences (Cleveland et al., 2003; Sullivan et al., 2001; Talbert et al., 2004). Nucleosomes can be

**Table 3.2** Eukaryotic proteins within centromeric chromatin and the prekinetochore

| Organism | CENP-A/ CenH3 proteins | CENP-A loading proteins | CENP-B proteins | CENP-C proteins |
|---|---|---|---|---|
| *S. cerevisiae* | Cse4 | Scm3 | | Mif2 |
| *S. pombe* | Cnp1 | Mis6, Mis16, Mis18 | Abp1, Cbh1, Cbh2 | Cnp3 |
| *C. albicans* | CaCse4 | ? | ? | CaMif2 |
| *D. melanogaster* | CID | RpAp48? | ? | DmCenp-C |
| *A. thaliana* | HTR12 | AtMSL1(RbAp46/ 48)? | CENP-B | AtCENP-C |
| *O. sativa* | rice CENH3 | ? | ? | CENP-C1 |
| *Z. mays* | ZmCENH3 | ? | ? | ZmCENPcA, ZmCENPcB, ZmCENPcC |
| *C. elegans* | CeCENP-A/ HCP-3 | KNL-2 (MIS18BP1) | ? | CeCENP-C/ HCP-4 |
| *M. musculus* | CENP-A | mis16, mis18? | CENP-B | CENP-C |
| *H. sapiens* | CENP-A | Mis16 (RbAp48, RbAp46); hKNL2 | CENP-B | CENP-C |

? = unknown

assembled in vitro from purified CENP-A and histones H2A, H2B, and H4 (Yoda et al., 2000), indicating that CENP-A can replace both copies of H3 in centromeric nucleosomes. In vivo, CENP-A replaces one or both copies of H3 (Foltz et al., 2006; Shelby et al., 1997).

CENP-A shares homology to H3 in the C-terminal histone fold domain, but is distinct from H3 due to a unique N-terminal region and a centromere targeting domain (CATD; Black et al., 2004, 2007; Shelby et al., 1997). Structural features of CENP-A itself are thought to fundamentally contribute to centromere identity. Nucleosomes containing CENP-A and H4 are biochemically and structurally compact and have a more rigid conformation than H3–H4 nucleosomes (Black et al., 2004). Two domains within the CENP-A amino acid sequence are responsible for selective centromeric targeting, and provoke a conformation change that also involves H4, restricting CENP-A from other genomic regions and creating a platform for assembling downstream kinetochore components.

CENP-A is present at the active centromere of dicentric chromosomes and on neocentromeres that form on regions of noncentromeric DNA (reviewed in Sullivan et al., 2001). Inactivation of CENP-A by knockout, deletion, or RNAi results in mitotic defects, including cell cycle arrest and chromosome segregation abnormalities. CENP-A is thought to be one of the first epigenetic marks placed upon chromatin to specify the genomic site of centromere identity and function. As such, CENP-A is required to recruit many other centromere/kinetochore proteins. Once incorporated into nucleosomes, CENP-A initiates assembly of a complex of other proteins (CENP-C, CENP-H, CENP-U, CENP-M, CENP-N, and CENP-T) that contribute to the inner kinetochore region of the centromere, closest to the DNA and centromeric (CEN) chromatin (Foltz et al., 2006). Many other kinetochore components that are located distal to the CENP-A region rely on assembly of the CENP-A nucleosome-associated complex for their localization and for structural and functional maturation of the kinetochore (Foltz et al., 2006; Okada et al., 2006). CENP-A and its role in establishing the foundation of the kinetochore are discussed in the following section and in other chapters of this book.

Most studies have focused on the role of CENP-A in mitotic kinetochore assembly. However, a recent study in *C. elegans* has revealed unexpected and disparate roles for CENP-A in chromosome segregation in meiosis versus mitosis. Although CeCENP-A is required to recruit CeCENP-C, KNL-1 and other proteins to form mitotic kinetochores, it is not required for meiotic kinetochore formation (Monen et al., 2005). In fact, in oocytes depleted for CeCENP-A, KNL-1 was recruited to the chromosome prior to meiosis I independent of CeCENP-A. Other outer kinetochore proteins were normally assembled onto the chromosomes in CeCENP-A-depleted embryos. However, the depleted embryos failed to complete embryonic mitotic segregation after the second meiotic division. An interpretation of these results is that kinetochore assembly and chromosome segregation in meiosis are coordinately controlled by recombination and cohesion. A single crossover occurs on homologous chromosomes in *C. elegans*, the position of which determines sites of cohesion. Loss of cohesion after each meiotic division may directly recruit outer components to meiotic chromosomes in the place of inner kinetochore chromatin and proteins. In *C. elegans*, the nucleus undergoes dramatic reorganization at the onset of prophase, with chromosomes clustering at one side of the nucleus. Perhaps this large-scale meiotic reorganization of the nucleus and/or the formation of monopolar kinetochores require the absence of inner kinetochore proteins. Notably, meiotic nuclear re-organization also occurs in larger eukaryotes, but how kinetochore proteins behave during meiosis in these organisms is unknown.

Although CENP-A replaces one or both copies of H3 in centromeric nucleosomes, not every centromeric nucleosome contains CENP-A. Over 15 years ago, Zinkowski and Brinkley showed that antibodies to CENPs on mechanically stretched chromosomes were visualized as re-iterated blocks separated by intervening spaces that lacked CENP antibody staining (Zinkowski et al., 1991). It has since been demonstrated that blocks of CENP-A containing nucleosomes are interspersed with blocks of nucleosomes that contain canonical H3 (Blower et al., 2002). This type of chromatin is referred to hereafter as centromeric chromatin. Although this unique pattern of histones was originally described at human and *Drosophila* centromeres, it has since been demonstrated that neocentromeres, worm, plant, and fission yeast centromeres are similarly organized as interspersed domains of CENP-A and H3 nucleosomes (Alonso et al., 2007; Cam et al., 2005; Chueh et al., 2005; Nagaki et al., 2004; Yan et al., 2005). The only exception to date is the budding yeast centromere, which contains a single nucleosome containing CENP-A (Furuyama and Biggins 2007). Nevertheless, this nucleosome at the budding yeast centromere is unique and lacks canonical H2A and H2B. Instead, Scm3, a nonhistone protein is present. Scm3 loads CENP-A$^{Cse4p}$, creating Cse4p-H4 dimers and replacing H2A and H2B in centromeric nucleosomes (Camahort et al., 2007; Mizuguchi et al., 2007; Stoler et al., 2007).

## 3.5 CENP-B, a DNA Binding Protein that Positions Centromeric Nucleosomes and Participates in Heterochromatin Assembly

CENP-B is an 80 kd DNA binding protein that recognizes a 17 bp motif, called the CENP-B box, found in many centromeric satellite monomers in primates and mice. The role of CENP-B in centromere assembly and function has been widely debated. Two lines of evidence argue against an essential role for CENP-B in centromere function. First, CENP-B knockout mice are viable and exhibit no centromeric defects or chromosome segregation abnormalities, implying that CENP-B is not required for centromere function and chromosome stability (Hudson et al., 1998). Second, CENP-B is absent from the normal human Y centromere, even though other CENPs bind here. The

Y centromere lacks CENP-B boxes in all monomers of its higher order array. Furthermore, CENP-B does not bind to neocentromeres that are formed on noncentromeric DNA (Depinet et al., 1997; du Sart et al., 1997). Lastly, on dicentric chromosomes in which one centromere is inactivated, CENP-B is usually present at both the active and inactive centromeres (Sullivan and Schwartz 1995).

However, there are several arguments for a key role for CENP-B in centromere organization and identity. In human artificial chromosome assembly assays, alpha satellite arrays that lack CENP-B boxes are unable to form *de novo* centromeres (Harrington et al., 1997; Ikeno et al., 1998). When CENP-B boxes are mutated within a higher order alpha satellite array, the alpha satellite is incapable of supporting formation of an autonomous chromosome/centromere (Ohzeki et al., 2002). These results suggest that CENP-B may be necessary for de novo centromere formation, but not epigenetic maintenance. However, a new study has uncovered an important role for CENP-B in centromere assembly and function. Alpha satellite arrays containing CENP-B boxes are able to form autonomous chromosomes in mouse cells, and the human arrays are recognized by mouse CENP-A and CENP-B. Chromatin on these mammalian artificial chromosomes is partitioned into regions containing CENP-A and regions containing H3K9me3 (Okada et al., 2007). However, in mouse embryonic fibroblasts (MEFs) that lack CENP-B, artificial chromosome assembly using alpha satellite sequences, even those containing CENP-B boxes, is extremely inefficient and results predominantly in integration of the input alpha satellite sequences into endogenous mouse chromosomes (Okada et al., 2007). The sites of integration are particularly enriched for heterochromatin, presumably to inhibit recruitment of CENP-A and formation of a second functional centromere. From these studies, the authors inferred that CENP-B controls the assembly of CENP-A chromatin, depending on the overall chromatin context of the surrounding environment. Interestingly, in some dicentric chromosomes, CENP-B is present only at the active, but not inactivated centromeres (Earnshaw and Migeon 1985). Centromere switching, in which some cells containing a dicentric chromosome have one centromere active while in other cells the opposite centromere is active, is well documented (Fisher et al., 1997; Haaf and Schmid 1990; Higgins et al., 2005; Ing and Smith 1983). Perhaps the absence of CENP-B at an inactive centromere that contains CENP-B binding sites, in addition to associated chromatin changes, creates a more complete and irreversible inactivation of centromere that prevents the assembly of CENP-A and other kinetochore proteins.

At human centromeres, the CENP-B DNA binding motif (CENP-B box) occurs every other 171 bp monomer (Ikeno et al., 1994). Two CENP-B molecules dimerize so that two CENP-B boxes are brought into proximity (Kitagawa et al., 1995). This results in nonrandom positioning, or phasing, of nucleosomes between CENP-B boxes (Tanaka et al., 2005). This conformation would allow CENP-A and core histones to be assembled into nucleosomes between the CENP-B binding regions. The combination of the two proteins at different regions of the satellite DNA is thought to establish centromere-specific higher order structure. Unlike CENP-A and CENP-C that bind to a small portion of alpha satellite, CENP-B is located along the length of the satellite arrays, binding at least several hundred kilobases and up to several megabases (Ando et al., 2002; Haaf and Ward 1994). The functional significance of this broad binding domain is unclear. CENP-B homologues are present in primates, mammals, and plants (Barbosa-Cisneros et al., 1997). Although CENP-B has not been identified in Drosophila or nematodes, three proteins that are homologous to CENP-B have been identified in the fission yeast *S. pombe*. These functionally redundant proteins, Abp1p, Cbh1p, and Cbh2p, bind centromeric DNA at fission yeast centromeres and are involved in chromosome segregation and replication (Baum and Clarke 2000; Locovei et al., 2006). Genetic studies have shown that the *S. pombe* CENP-B also nucleate heterochromatin formation via post-translational modifications of histone tails and recruitment of heterochromatin proteins (Nakagawa et al., 2002). It is tempting to speculate that the binding of CENP-B throughout an entire satellite array may influence centromeric structure through nucleosomal spacing or replication of centromere DNA, and create a functional bridge between domains of CEN chromatin and heterochromatin.

## 3.6 CENP-C, -H, and -I: a Trilogy of Proteins Within the Pre-Kinetochore

Assembly of pre-kinetochore chromatin containing CENP-A and CENP-B is completed by recruitment of additional CENPs. CENP-H and CENP-I are required to recruit newly synthesized CENP-A to the centromere (Okada et al., 2006). This complex of proteins then recruits CENP-C, a constitutive CENP that is part of the pre-kinetochore chromatin and is a structural component of the 3D inner kinetochore at metaphase (Fukagawa et al., 2001a; Nishihashi et al., 2002; Saitoh et al., 1992; Sugata et al., 2000). CENP-C has been identified in yeasts, flies, plants, and mammals and are required for chromosome segregation, proper kinetochore size, sister centromere resolution, and mitotic progression (Dawe et al., 1999; Fukagawa et al., 2001b; Moore and Roth 2001; Ogura et al., 2004; Schuh et al., 2007; Tomkiel et al., 1994). CENP-C has several domains that target it to the centromere, promote protein–protein interactions and dimerization, and confer DNA binding activity (Politi et al., 2002; Song et al., 2002; Trazzi et al., 2002). The DNA binding activity of CENP-C is relatively nonspecific, but is thought to anchor CENP-C to the centromere and incorporate it into chromatin once it is recruited by other components (Ando et al., 2002). CENP-C monomers or dimers are distributed across repetitive units of CENP-A/CENP-B nucleosome complexes. The assembly of this chromatin is thought to confer a conformational change that is the platform for recruitment of outer kinetochore proteins and maturation of the kinetochore.

## 3.7 Centromeric Chromatin Contains Histone Variants, Core Histone Modifications, and NonHistone Proteins

Higher order packaging achieved through incorporation of CENP-A, CENP-B and/or nonhistone accessory proteins like Scm3 is thought to distinguish the centromere from the rest of the genome. The *S. cerevisiae* 120 bp CEN region containing a single CENP-$A^{Cse4p}$ nucleosome that is nuclease-resistant is surrounded by nucleosomes that are regularly spaced (Saunders et al., 1988). The recruitment of Scm3 followed by the removal of core H2A and H2B, resulting in a unique centromeric nucleosome, may be sufficient to mark the region as centromeric. In fission yeast, the chromatin containing CENP-$A^{Cnp1}$ (cnt) and in the innermost repeats (imr) also has distinctive nuclease sensitivity, while the outer repeat (otr) region maintain the regular nucleosome spacing as observed in bulk chromatin (Clarke et al., 1993; Takahashi et al., 1992). Mammalian centromeres too demonstrate distinctive nuclease digestion and sedimentation properties, consistent with higher order folding of chromatin fibers compared to bulk chromatin (Gilbert and Allan 2001; Guenatri et al., 2004). While the incorporation of CENP-A into centromeric nucleosomes might create more compact structures or unique chromatin, other cellular signals may promote higher order, 3D folding of centromeric chromatin.

Interspersion of subdomains of CENP-A and H3 nucleosomes in two-dimensions is thought to create a unique 3D structure in which CENP-A nucleosomes and H3 nucleosomes oppositely coalesce. In this model, stacks of CENP-A nucleosomes are positioned toward the outward face of the chromosome to recruit additional kinetochore proteins and biorient sister kinetochores toward opposite spindle poles (Blower et al., 2002). Conversely, H3 subdomains are positioned toward the interior of the chromosome to establish a platform for heterochromatin assembly and centromeric cohesion protein recruitment. Most centromeres are located in or near heterochromatin, suggesting that repressive or distinctive surrounding chromatin environment, or centromere-specific remodeling complexes may characterize the centromere and promote its structure and function.

Modifications (methylation, acetylation, and phosphorylation) of amino acids in core histone N-terminal tails integrates a complex system of epigenetic modifications and chromosomal proteins that establish and maintain distinctive types of chromatin. Distinctive combinations of histone modifications, known as the "histone code," partition the genome into functional domains, such as transcriptionally silent heterochromatin and transcriptionally active euchromatin. In humans, mice, flies, and fission yeast, H3-containing nucleosomes that are interspersed with CENP-A nucleosomes within CEN chromatin are methylated at K4 (H3K4me2), while regions outside the CEN chromatin core are enriched for H3K9me2 and H3K9me3 (Guenatri et al., 2004; Sullivan and Karpen 2004; Fig. 3.3). In mice, the major satellite DNA within the pericentromere, and to a lesser extent, the minor satellite within the centromere are enriched for the heterochromatic modifications H3K9me3, H3K27me1, and H4K20me3 (Martens et al., 2005; Peters et al., 2003). Although both minor and major satellites are enriched for similar heterochromatic modifications, they form distinctive types of chromatin, defined by different micrococcal nuclease sensitivities (Guenatri et al., 2004). Minor satellite chromatin is also enriched for H3K4me2, although to a lesser extent than what is observed for heterochromatic histone modifications (Greaves et al., 2007). These findings are in agreement with the interpersion of CENP-A and H3K4me2 within fission yeast, fly, and human centromeres (Cam et al., 2005; Sullivan and Karpen 2004). Yeast, human, mouse, and Drosophila centromeres are also typically depleted for acetylated histones H3 and H4, histone modifications that correlate with transcriptionally active chromatin. These studies suggest that centromeric chromatin is functionally distinguished from both classical euchromatin and heterochromatin (Sullivan and Karpen 2004).

Other histone variants in addition to CENP-A also appear to play a role in centromere specificity and function. In mouse and human cells, the histone H2 variant H2A.Z is found in centromeric nucleosomes containing H3K4me2, while CENP-A nucleosomes contain only H2A (Greaves et al., 2007). In addition, a small fraction of H3K9me3 nucleosomes near CEN chromatin also contain H2A.Z. Thus, both histone modifications and histone variants may be important for centromere assembly and function in animal cells by creating higher order organization during mitosis (Fig. 3.4).

It is important to mention that while clusters of interspersed H3 and CENP-A nucleosomes have also been observed at plant centromeres, there are notable differences in the post-translational modifications of H3 that are present within plant CEN chromatin. In rice and Arabidopsis, centromeric nucleosomes that are interspersed with CENP-A contain H3K4me2 or H3K9me2 (May et al., 2005; Nagaki et al., 2004; Yan et al., 2005). Furthermore, they are enriched for H4 acetylation, and this modification in addition to H3K4me2 coincides with areas of active gene expression within the rice centromeres (Yan et al., 2005). Flanking heterochromatin is marked by H3K9me1, H3K9me2, H3K27me1, H3K27me2, and H4K20me1 (May et al., 2005). Several of these modifications (H3K9me2 and H3K27me1) are also enriched within pericentromeric heterochromatin in mammals (Martens et al., 2005; Peters et al., 2003).

In maize, it has been surprising to discover that the reading of euchromatin and heterochromatin based on histone modifications is vastly different than that for animal cells. H3K9me2 and H3K9me3, which are heterochromatic modifications in fission yeast and vertebrates, are associated with euchromatic regions of the maize genome. These modifications are also found within maize

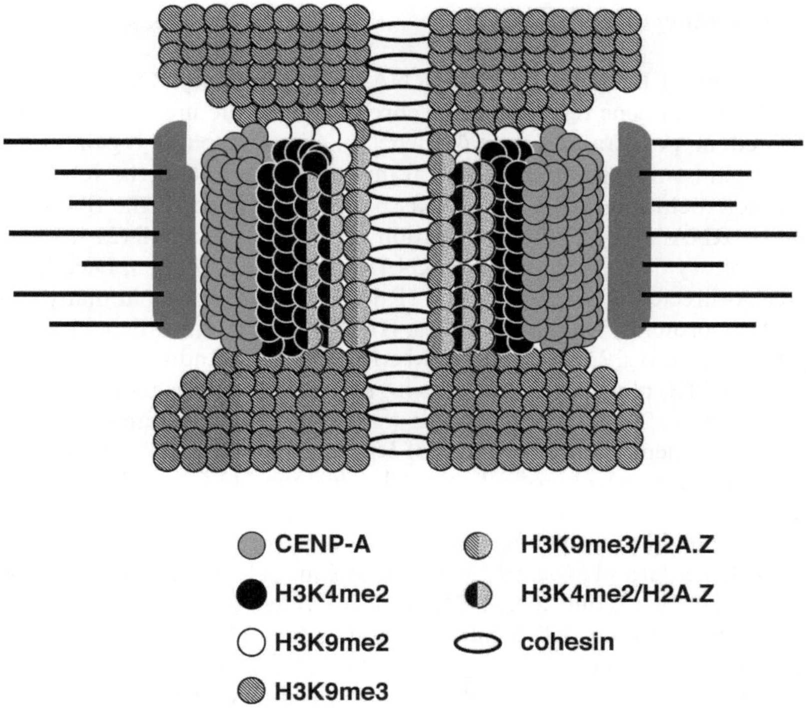


**Fig. 3.4** A model for 3D organization of centromeric (CEN) chromatin into the kinetochore at metaphase in Drosophila and mammals. Histone modification patterns on linear chromatin fibers suggests that interspersion of CENP-A and H3 nucleosomes during chromosome compaction blocks of CENP-A nucleosomes are sequestered to the poleward face of the chromosome, while H3K4me2 containing nucleosomes are stacked toward the paired sister kinetochores. Nucleosomes containing both H3K4me2 and H2A.Z may provide a functional distinction between closely located nucleosomes that contain H3K9me3 and/or H3K9me2/H2A.Z. Chromatin containing H3K9me2 and H3K9me3 is located between sister chromatids and extends out into a broader region. H3K9 methylation is crucial for recruitment of heterochromatin and cohesion proteins. H2A.Z may be important for spatially defining the distinctive functional domains (CEN chromatin, heterochromatin) within the centromere region. Model adapted from Sullivan and Karpen (2004) and Greaves et al., (2007)

centromeres (May et al., 2005). H3K27me1 and H3K27me2 are associated with classical heterochromatin and are enriched in pericentromeric regions (Shi and Dawe 2006; Fig. 3.3). H4K20me2 and H4K20me3, marks of poised and classical heterochromatin in vertebrates, are virtually absent from the maize genome. Despite these differences in correlative histone modification-chromatin states, it is clear that chromatin at maize centromeres, like that at other eukaryotic centromeres, is functionally distinct from flanking pericentromeric heterochromatin.

## 3.8 Centromeres and RNA

The discovery of RNA interference (RNAi) and regulatory RNAs has revolutionized views of gene regulation and chromosome structure. It was originally implicated in post-transcriptional silencing, developmental regulation, and defense against transposition and viral invasion. The role of the RNAi pathway in heterochromatin formation has been well-defined in the fission yeast *S. pombe*. RNAi involves the degradation of mRNA into small (21–23 nucleotide) inhibitory dsRNA duplexes (siRNAs). Argonaute (Ago1), Dicer (Dcr1), and RNA-directed RNA polymerase (RdRP) mediate RNAi in many organisms. Dicer proteins, which are RNase III class enzymes, mediate cleavage of long dsRNAs into siRNAs. The RNAi complex, RNA-Induced Transcription Silencing (RITS) physically interacts with the siRNAs (Dykxhoorn et al., 2003; Motamedi et al., 2004). RITS contains Ago1, Tas3 and Chp1 and siRNA. The siRNA component directs targeting of RITS to homologous, nascent centromeric transcripts, which recruits the Clr4 histone H3K9 methyltransferase. H3K9 methylation creates a binding site for the chromodomain proteins Swi6/HP1 and Chp1, which tether RITS to centromeric nucleosomes (Verdel et al., 2004; Verdel and Moazed 2005). The actions of histone deacetylases occur at this time, since it is only after H3K9 is deacetylated that Clr4/Su(var)3–9 can methylate this residue. RITS interacts with other RNAi complexes such as RDRC (*R*NA *D*irected *R*NA polymerase *C*omplex) enforce the association of RISC with centromeric RNAs. RDRC also produces additional centromeric dsRNA that are subsequently processed into siRNAs, creating a self-enforcing loop of siRNA production and heterochromatin assembly/maintenance at the centromere.

Based upon the fact that there are known RNAi pathway in other eukaryotes, it is likely that RNA molecules and/or the RNAi machinery be associated with the complex centromeric repeats in these larger eukaryotes. In fact, transcripts originating from centromeric satellites in *Arabidopsis*, maize, mice, and humans have been described (Bouzinba-Segard et al., 2006; Fukagawa et al., 2004; Lee et al., 2006; May et al., 2005; Rudert et al., 1995; Topp et al., 2004). In *Arabidopsis*, transcripts originating from one strand of the 180 bp satellite are processed by the RNAi pathway and regulated by histone modifications, suggesting a role in heterochromatin formation and/or epigenetic inheritance of centromeres (May et al., 2005). In maize, siRNAs are not observed, but long transcripts from the 156 bp satellites and the CRM retroelement are associated with CENP-A/maize CENH3 (Topp et al., 2004). In mammals, transcripts arising from both major and minor satellite have been identified (Bouzinba-Segard et al., 2006; Rudert et al., 1995). Forced accumulation of the minor satellite transcripts results in chromosome segregation defects and aberrant localization of CENPs. Human alpha satellite transcripts are produced from both higher order and monomeric arrays, are processed by Dicer, and are involved in heterochromatin formation (Fukagawa et al., 2004). There is also evidence to support a role for alpha satellite RNA in kinetochore assembly and nuclear dynamics, although the precise mechanisms of these pathways are currently unclear (Wong et al., 2007).

## 3.9 Dynamics of CEN Chromatin and Mechanisms for Regulating the Centromere Region

Since there is no obvious sequence-specificity for CENP-A binding and CENP-A is present on different DNA sequences in various organisms, it is reasonable to assume that the deposition of CENP-A, and even CEN chromatin itself, is malleable. CENP-A is present on euchromatic sequences of neocentromeres, and presumably, the *C. elegans* genome is assembled onto noncentromeric DNA inserted into centromeres of fission yeast, and it spreads into euchromatic or noncentromeric DNA that flanks centromeres on fly minichromosomes and human artificial chromosomes (Blower and Karpen 2001; Castillo et al., 2007; Lam et al., 2006; Nakashima et al., 2005; Okamoto et al., 2007). Furthermore, when CENP-A is overexpressed, it spreads throughout the centromere region and/or is mistargeted to euchromatic regions, triggering chromosome instability and producing multiple ectopic kinetochores to which other CENPs are recruited (Castillo et al., 2007; Collins et al., 2007; Heun et al., 2006; Lam et al., 2006; Van Hooser et al., 2001).

Spreading of CENP-A over noncentromeric sequences raises the question of whether CEN chromatin affects gene expression. Does CEN chromatin behave like heterochromatin and silence genes, or is it permissive for gene expression since CEN chromatin in many organisms exhibits histone modifications associated with euchromatin? In humans, genes that are present in neocentromeres are expressed when assembled into chromatin that contains CENP-A (Saffery et al., 2003; Wong et al., 2006). On human artificial chromosomes, CENP-A spreads over selectable marker DNA that is interspersed with alpha satellite sequences and does not prevent expression of the selectable marker even when cells are grown off of drug selection (Lam et al., 2006; Nakashima et al., 2005). Similarly at rice centromeres, CENP-A binding does not prevent expression of genes interspersed within the satellite arrays (Nagaki et al., 2004). Therefore, in some organisms, CENP-A chromatin creates a chromatin environment that is permissive for gene expression. Interestingly, CENP-A$^{Cnp1}$ chromatin in *S. pombe* has an opposite effect on gene expression. When the ura4 gene is placed within the central core of *S. pombe* chromosome 1, CENP-A$^{Cnp1}$ binds to this noncentromeric DNA and represses the expression of ura4 (Castillo et al., 2007). The repression is even stronger when CENP-A$^{Cnp1}$ is overexpressed, but ura4 is robustly expressed when CENP-A$^{Cnp1}$ is deleted or if levels of H3 are increased (see below). One explanation for the differential effects of CENP-A chromatin in *S. pombe* versus other eukaryotes may be the centromere size. The smaller central core region in *S. pombe* may have a limited nucleosomal or spatial capability for achieving the higher order structure necessary for kinetochore function during mitosis and binding of transcription factors may impede the packaging of the kinetochore.

Intense investigation is focused on how CENP-A is inherited at the same genomic site throughout subsequent cell divisions. CENP-A in mammals and *Arabidopsis* is synthesized in G2 and can be loaded onto centromeres in G2/M by a replication-independent mechanism (Lermontova et al., 2006; Shelby et al., 2000). However, several studies argue against this timing for new CENP-A deposition. In fission yeast, newly synthesized Cnp1$^{\text{CENP-A}}$ in fission yeast is loaded onto centromeres by Mis6p during G1 (Saitoh et al., 1997). This finding is supported by a recent study in mammalian cells showing that new CENP-A is loaded during a limited window of time at the end of mitosis and during G1 in humans and *Drosophila* (Jansen et al., 2007; Schuh et al., 2007). Furthermore, a complete mitosis is necessary for newly synthesized CENP-A to be incorporated into centromeres at G1. This waiting period may be required for the production or availability of CENP-A loading factors, such as Mis6 in yeast and hMis18$\alpha$, HMis18$\beta$, and hKNL2 proteins in mammals (Fujita et al., 2007; Hayashi et al., 2004). It is unclear what targets the CENP-A loading factors to the centromere, although it has been hypothesized that existing CENP-A or centromeric histone modifications may direct this process.

Unlike H3 which is incorporated into chromatin in early S phase during replication, existing CENP-A nucleosomes are not removed and replaced, but are inherited semiconservatively, such that "old" CENP-A is divided between daughter strands (Sullivan et al., 2001). However, new CENP-A must be targeted correctly and monitored for removal if it is incorporated into the wrong part of the genome. How does new CENP-A know where to go? Since CENP-A recruits additional CENPs and is a fundamental component of centromeric chromatin, CENP-A retention and recruitment to the centromere is thought to be self-directed. New evidence to support a self-replenishing model has arisen from studies of Drosophila centromeric nucleosomes (Dalal et al., 2007a). Because CENP-A closely resembles H3 and is capable of replacing H3 in nucleosomes reconstituted in vivo, it has been assumed that the CENP-A nucleosome is an octamer. However, in *Drosophila* interphase cells, CENP-A$^{\text{CID}}$ nucleosomes exist as tetramers containing one copy of CENP-ACID, H2A, H2B, and H4 (Dalal et al., 2007b). The tetramers are smaller than bulk nucleosomes and more stable than CENP-A octamers reconstituted in vitro (Conde e Silva et al., 2007). These half-nucleosomes may be stabilized at centromeres by surrounding pericentromeric heterochromatin but may be highly unstable within euchromatin, allowing CENP-A to be readily removed from inappropriate genomic locations via proteolysis (Collins et al., 2004; Moreno-Moreno et al., 2006). It is unclear if CENP-A nucleosomes exist as octamers during mitosis. Since CENP-A containing nucleosomes are spatially distributed into stacks, it is possible that the tetramers coalesce into octamers during G2/M.

## 3.10 Identification of CENP-A Loading/Interacting Factors

Proteins that assemble nucleosomes containing CENP-A and their general roles in chromatin assembly may shed light on the timing, properties, and dynamics of CENP-A chromatin. In budding yeast, Scm3 recruitment to the centromere, interaction with Ndc10, and the subsequent eviction of H2A and H2B allows a CENP-$A^{Cse4p}$:H4:Scm3 nucleosome to be assembled (Mizuguchi et al., 2007). In fission yeast, CENP-$A^{Cnp1}$ is loaded by a complex of factors, including Mis15, Mis16, Mis17, and Mis18 (Hayashi et al., 2004). Mis16 and Mis18 in particular are part of an assembly cascade in which deacetylated nucleosomes within the centromere are maintained before CENP-$A^{Cnp1}$ is loaded. In other eukaryotes, many proteins have been shown to affect loading of CENP-A, such as MIS16, MIS18, KNL-2 (MIS18BP1), CENP-H and CENP-I (Foltz et al., 2006; Fujita et al., 2007; Maddox et al., 2007; Oegema et al., 2001; Okada et al., 2006). Further research will determine which are involved in direct versus indirect and/or conserved interactions with CENP-A in different organisms. Many of these proteins are highly conserved with equally conserved functions in chromatin remodeling or nucleosome interactions. For example, the human homologues of Mis16 and Mis18, RbAp48 and RbAp46, are also involved in assembly of CENP-A at centromeres. RbAp48 and RbAp46 are part of other chromatin remodeling/chaperone complexes including nucleosome remodeler (NURF), Polycomb group silencing complex, and complexes that specifically load all three isoforms of H3 (Dalal et al., 2007a). Although homologues of budding yeast Scm3 have not been described in larger eukaryotes, it is possible that an Scm3-like protein may load CENP-A at multiple sites along the expansive lengths of larger eukaryotic centromeres and create epigenetic memory for future CENP-A loading. How this putative protein, and the other proteins mentioned previously, participates in the temporal loading of CENP-A during or after mitosis remains to be explored.

## 3.11 Centromeric Boundaries: Sequence Elements or Regional Boundaries?

The question still remains, though, as to how the extent of CEN chromatin at the centromere remains relatively constant. It is clear that tRNA boundary elements in yeast are able to limit the spread of CENP-$A^{Cnp1}$ to the central core region and to prevent heterochromatin from spreading from flanking repeats into the core (Castillo et al., 2007; Scott et al., 2006). However, at large regional centromeres, sequence elements that restrict the spread of CENP-A chromatin or heterochromatin remain unidentified. In lieu of, or in combination with, defined boundary elements, epigenetic factors are thought to regulate centromeric domains. Genetic and chromatin studies in *S. pombe*, *Drosophila,* and mammalian cultured cells have shown that CEN chromatin exhibits remarkable flexibility and that the density of CENP-A, core histones, and the enzymes that modify certain amino acid residues on histone tails are important for assembly and maintenance of chromatin domains. When CENP-A is depleted in various organisms, the concentration of H3-containing nucleosomes within the centromere region increases and chromosomes fail to segregate properly (Blower and Karpen 2001; Blower et al., 2002). Conversely, when the ratio of H3:H4 is altered either by increasing gene number or by overexpressing H3, more H3 and less CENP-A is deposited into centromere regions (Castillo et al., 2007). These latter experiments, performed in *S. pombe*, suggest that increased amounts of H3 out-compete CENP-A for H4, leading to more H3–H4 containing nucleosomes within the centromere and chromosome segregation failure.

In noncentromeric regions of the genome in various organisms, a dynamic equilibrium exists between euchromatin and heterochromatin (Di Stefano et al., 2007; Ebert et al., 2004; Kimura and Horikoshi 2004). Since protein dosage, and more specifically histone H3 and H4 dosage, is a factor in maintaining centromere structure and function (Castillo et al., 2007), it is possible that dosage of enzymes that perform the chemical modifications may also affect formation and function of eukaryotic centromeres. Basically, it may be necessary to maintain a balance between heterochromatin and CEN chromatin. This model may be particularly relevant for insects, plants and mammals. The yeast genome is small, so strict distinctions between expressed and silenced regions may be necessary to ensure genome stability and organism viability. But at more complex eukaryotic centromeres that are orders of magnitude larger than those in yeast, the boundaries between euchromatin and heterochromatin may be more flexible. It has been suggested from Drosophila studies that heterochromatin serves as a physical boundary, and when it is removed by structural rearrangement, CENP-A spreads into euchromatin (Maggert and Karpen 2001). Furthermore, overexpression of CENP-A in human and fly cells promotes expansion of CEN chromatin into euchromatic sites (Heun et al., 2006; Van Hooser et al., 2001). Thus, a balance between CEN chromatin and heterochromatin exists in human cells, even without physically removing or rearranging subdomains. Future studies will be important in determining which heterochromatin or euchromatic regulators, or perhaps unique chromatin modifiers, regulate the borders of chromatin domains within large regional centromeres.

## 3.12 Concluding Remarks

Over the past two and half decades, major advances in our understanding of centromere architecture and function have revealed the complexity of this essential chromosomal locus. Eukaryotic centromeres are defined by unique chromatin typified by histone variants and unusual nucleosomal properties, rather than common genomic sequences. The organization and regulation of chromatin domain within the centromere, the roles of RNA, and the propagation of CENP-A and the histone modifications that are associated with CEN chromatin and flanking heterochromatin are areas that await more extensive study. Furthermore, the contribution of centromere to cancer or aneuploidy is still unclear. Studies that address the regulation of centromeric dynamics throughout the cell cycle and its dependence on processes such as DNA replication, repair, and transcription are likely to uncover potential links between centromere architecture and behavior and relevant human diseases.

## References

Abad JP et al. (2000) Pericentromeric regions containing 1.688 satellite DNA sequences show anti-kinetochore antibody staining in prometaphase chromosomes of Drosophila melanogaster. Mol Gen Genet 264: 371–377.

Aker M, Huang HV (1996) Extreme heterogeneity of minor satellite repeat arrays in inbred strains of mice. Mamm Genome 7: 62–64.

Alexandrov IA et al. (1993) Definition of a new alpha satellite suprachromosomal family characterized by monomeric organization. Nucleic Acids Res 21: 2209–2215.

Allshire RC et al. (1994) Position effect variegation at fission yeast centromeres. Cell 76: 157–169.

Allshire RC et al. (1995) Mutations derepressing silent centromeric domains in fission yeast disrupt chromosome segregation. Genes Dev 9: 218–233.

Alonso A et al. (2007) Co-localization of CENP-C and CENP-H to discontinuous domains of CENP-A chromatin at human neocentromeres. Genome Biol 8: R148.

Ando S et al. (2002) CENP-A, -B, and -C chromatin complex that contains the I-type alpha-satellite array constitutes the prekinetochore in HeLa cells. Mol Cell Biol 22: 2229–2241.

Andreyeva EN et al. (2007) High-resolution analysis of Drosophila heterochromatin organization using SuUR Su(var)3-9 double mutants. Proc Natl Acad Sci USA 104: 12819–12824.

Barbosa-Cisneros O et al. (1997) CENP-B autoantigen is a conserved protein from humans to higher plants: identification of the aminoterminal domain in Phaseolus vulgaris. Rev Rhum Engl Ed 64: 368–374.

Baum M, Clarke L (2000) Fission yeast homologs of human CENP-B have redundant functions affecting cell growth and chromosome segregation. Mol Cell Biol 20: 2852–2864.

Baum M et al. (1994) The centromeric K-type repeat and the central core are together sufficient to establish a functional Schizosaccharomyces pombe centromere. Mol Biol Cell 5: 747–761.

Baum M et al. (2006) Formation of functional centromeric chromatin is specified epigenetically in Candida albicans. Proc Natl Acad Sci USA 103: 14877–14882.

Bernard P et al. (2001) Requirement of Heterochromatin for Cohesion at Centromeres. Science 294: 2539–2542.

Black BE et al. (2004) Structural determinants for generating centromeric chromatin. Nature 430: 578–582.

Black BE et al. (2007) Centromere identity maintained by nucleosomes assembled with histone H3 containing the CENP-A targeting domain. Mol Cell 25: 309–322.

Blower MD, Karpen GH (2001) The role of Drosophila CID in kinetochore formation, cell-cycle progression and heterochromatin interactions. Nat Cell Biol 3: 730–739.

Blower MD et al. (2002) Conserved organization of centromeric chromatin in flies and humans. Dev Cell 2: 319–330.

Bouzinba-Segard H et al. (2006) Accumulation of small murine minor satellite transcripts leads to impaired centromeric architecture and function. Proc Natl Acad Sci USA 103: 8709–8714.

Brown KE et al. (1994) Dissecting the centromere of the human Y chromosome with cloned telomeric DNA. Hum Mol Genet 3: 1227–1237.

Cam HP et al. (2005) Comprehensive analysis of heterochromatin- and RNAi-mediated epigenetic control of the fission yeast genome. Nat Genet 37: 809–819.

Camahort R et al. (2007) Scm3 is essential to recruit the histone h3 variant cse4 to centromeres and to maintain a functional kinetochore. Mol Cell 26: 853–865.

Castillo AG et al. (2007) Plasticity of fission yeast CENP-A chromatin driven by relative levels of histone H3 and H4. PLoS Genet 3: e121.

Choo KH (2001) Domain organization at the centromere and neocentromere. Dev Cell 1: 165–177.

Chueh AC et al. (2005) Variable and hierarchical size distribution of L1-retroelement-enriched CENP-A clusters within a functional human neocentromere. Hum Mol Genet 14: 85–93.

Clarke L et al. (1993) Structure and function of Schizosaccharomyces pombe centromeres. Cold Spring Harb Symp Quant Biol 58: 687–695.

Cleveland DW et al. (2003) Centromeres and kinetochores: from epigenetics to mitotic checkpoint signaling. Cell 112: 407–421.

Collins KA et al. (2007) The overexpression of a Saccharomyces cerevisiae centromeric histone H3 variant mutant protein leads to a defect in kinetochore biorientation. Genetics 175: 513–525.

Collins KA et al. (2004) Proteolysis contributes to the exclusive centromere localization of the yeast Cse4/CENP-A histone H3 variant. Curr Biol 14: 1968–1972.

Conde e Silva N et al. (2007) CENP-A-containing nucleosomes: easier disassembly versus exclusive centromeric localization. J Mol Biol 370: 555–573.

Copenhaver GP et al. (1999) Genetic definition and sequence analysis of Arabidopsis centromeres. Science 286: 2468–2474.

Dalal Y et al. (2007a) Structure, dynamics, and evolution of centromeric nucleosomes. Proc Natl Acad Sci U S A. 104:15974–15981.

Dalal Y et al. (2007b) Tetrameric structure of centromeric nucleosomes in interphase Drosophila cells. PLoS Biol 5: e218.

Dawe RK et al. (1999) A maize homolog of mammalian CENPC is a constitutive component of the inner kinetochore. Plant Cell 11: 1227–1238.

De Wulf P et al. (2003) Hierarchical assembly of the budding yeast kinetochore from multiple subcomplexes. Genes Dev 17: 2902–2921.

Depinet TW et al. (1997) Characterization of neo-centromeres in marker chromosomes lacking detectable alpha-satellite DNA. Hum Mol Genet 6: 1195–1204.

Di Stefano L et al. (2007) Mutation of Drosophila Lsd1 disrupts H3-K4 methylation, resulting in tissue-specific defects during development. Curr Biol 17: 808–812.

du Sart D et al. (1997) A functional neo-centromere formed through activation of a latent human centromere and consisting of non-alpha-satellite DNA. Nat Genet 16: 144–153.

Dykxhoorn DM et al. (2003) Killing the messenger: short RNAs that silence gene expression. Nat Rev Mol Cell Biol 4: 457–467.

Earnshaw W et al. (1986) Three human chromosomal autoantigens are recognized by sera from patients with anti-centromere antibodies. J Clin Invest 77: 426–430.

Earnshaw WC, Migeon BR (1985) Three related centromere proteins are absent from the inactive centromere of a stable isodicentric chromosome. Chromosoma 92: 290–296.

Earnshaw WC, Rothfield N (1985) Identification of a family of human centromere proteins using autoimmune sera from patients with scleroderma. Chromosoma 91: 313–321.

Ebert A et al. (2004) Su(var) genes regulate the balance between euchromatin and heterochromatin in Drosophila. Genes Dev 18: 2973–2983.

Ekwall K et al. (1997) Transient inhibition of histone deacetylation alters the structural and functional imprint at fission yeast centromeres. Cell 91: 1021–1032.
Farr CJ et al. (1995) Generation of a human X-derived minichromosome using telomere-associated chromosome fragmentation. EMBO J 14: 5444–5454.
Fisher AM et al. (1997) Centromeric inactivation in a dicentric human Y;21 translocation chromosome. Chromosoma 106: 199–206.
Foltz DR et al. (2006) The human CENP-A centromeric nucleosome-associated complex. Nat Cell Biol 8: 458–469.
Fujita Y et al. (2007) Priming of centromere for CENP-A recruitment by human hMis18alpha, hMis18beta, and M18BP1. Dev Cell 12: 17–30.
Fukagawa T et al. (2001a) CENP-H, a constitutive centromere component, is required for centromere targeting of CENP-C in vertebrate cells. Embo J 20: 4603–4617.
Fukagawa T et al. (2004) Dicer is essential for formation of the heterochromatin structure in vertebrate cells. Nat Cell Biol 6: 784–791.
Fukagawa T et al. (2001b) Creation and characterization of temperature-sensitive CENP-C mutants in vertebrate cells. Nucleic Acids Res 29: 3796–3803.
Furuyama S, Biggins S (2007) Centromere identity is specified by a single centromeric nucleosome in budding yeast. Proc Natl Acad Sci USA 104: 14706–14711.
Giet R, Glover DM (2001) Drosophila aurora B kinase is required for histone H3 phosphorylation and condensin recruitment during chromosome condensation and to organize the central spindle during cytokinesis. J Cell Biol 152: 669–682.
Gilbert N, Allan J (2001) Distinctive higher-order chromatin structure at mammalian centromeres. Proc Natl Acad Sci USA 98: 11949–11954.
Greaves IK et al. (2007) H2A.Z contributes to the unique 3D structure of the centromere. Proc Natl Acad Sci USA 104: 525–530.
Grimes BR et al. (2002) alpha-Satellite DNA and Vector Composition Influence Rates of Human Artificial Chromosome Formation. Mol Ther 5: 798–805.
Guenatri M et al. (2004) Mouse centric and pericentric satellite repeats form distinct functional heterochromatin. J Cell Biol 166: 493–505.
Haaf T, Schmid M (1990) Y isochromosome associated with a mosaic karyotype and inactivation of the centromere. Hum Genet 85: 486–490.
Haaf T, Ward DC (1994) Structural analysis of alpha-satellite DNA and centromere proteins using extended chromatin and chromosomes. Hum Mol Genet 3: 697–709.
Hagstrom KA et al. (2002) C. elegans condensin promotes mitotic chromosome architecture, centromere organization, and sister chromatid segregation during mitosis and meiosis. Genes Dev 16: 729–742.
Hahnenberger KM et al. (1991) Identification of DNA regions required for mitotic and meiotic functions within the centromere of Schizosaccharomyces pombe chromosome I. Mol Cell Biol 11: 2206–2215.
Harrington JJ et al. (1997) Formation of *de novo* centromeres and construction of first-generation human artificial microchromosomes. Nat Genet 15: 345–355.
Hayashi T et al. (2004) Mis16 and Mis18 are required for CENP-A loading and histone deacetylation at centromeres. Cell 118: 715–729.
Hendzel MJ et al. (1997) Mitosis-specific phosphorylation of histone H3 initiates primarily within pericentromeric heterochromatin during G2 and spreads in an ordered fashion coincident with mitotic chromosome condensation. Chromosoma 106: 348–360.
Heslop-Harrison JS et al. (2003) Tandemly repeated DNA sequences and centromeric chromosomal regions of Arabidopsis species. Chromosome Res 11: 241–253.
Heun P et al. (2006) Mislocalization of the Drosophila centromere-specific histone CID promotes formation of functional ectopic kinetochores. Dev Cell 10: 303–315.
Higgins AW et al. (2005) Engineered human dicentric chromosomes show centromere plasticity. Chromosome Res 13: 745–762.

Hudson DF et al. (1998) Centromere protein B null mice are mitotically and meiotically normal but have lower body and testis weights. J Cell Biol 141: 309–319.
Ikeno M et al. (1998) Construction of YAC-based mammalian artificial chromosomes. Nat Biotechnol 16: 431–439.
Ikeno M et al. (1994) Distribution of CENP-B boxes reflected in CREST centromere antigenic sites on long-range alpha-satellite DNA arrays of human chromosome 21. Hum Mol Genet 3: 1245–1257.
Ing PS, Smith SD (1983) Cytogenetic studies of a patient with mosaicism of isochromosome 13q and a dicentric (Y;13) translocation showing differential centromeric activity. Clin Genet 24: 194–199.
Izuta H et al. (2006) Comprehensive analysis of the ICEN (Interphase Centromere Complex) components enriched in the CENP-A chromatin of human cells. Genes Cells 11: 673–684.
Jager H et al. (2005) The Drosophila melanogaster condensin subunit Cap-G interacts with the centromere-specific histone H3 variant CID. Chromosoma 113: 350–361.
Jansen LE et al. (2007) Propagation of centromeric chromatin requires exit from mitosis. J Cell Biol 176: 795–805.
Jiang J et al. (2003) A molecular view of plant centromeres. Trends Plant Sci 8: 570–575.
Jin W et al. (2004) Maize centromeres: organization and functional adaptation in the genetic background of oat. Plant Cell 16: 571–581.
Kalitsis P et al. (2006) Mouse telocentric sequences reveal a high rate of homogenization and possible role in Robertsonian translocation. Proc Natl Acad Sci USA 103: 8786–8791.
Kimura A, Horikoshi M (2004) Partition of distinct chromosomal regions: negotiable border and fixed border. Genes Cells 9: 499–508.
Kipling D et al. (1991) Mouse minor satellite DNA genetically maps to the centromere and is physically linked to the proximal telomere. Genomics 11: 235–241.
Kipling D et al. (1995) CENP-B binds a novel centromeric sequence in the Asian mouse Mus caroli. Mol Cell Biol 15: 4009–4020.
Kitagawa K et al. (1995) Analysis of protein-DNA and protein-protein interactions of centromere protein B (CENP-B) and properties of the DNA-CENP-B complex in the cell cycle. Mol Cell Biol 15: 1602–1612.
Kuznetsova I et al. (2006) High-resolution organization of mouse centromeric and pericentromeric DNA. Cytogenet Genome Res 112: 248–255.
Kuznetsova IS et al. (2005) New types of mouse centromeric satellite DNAs. Chromosome Res 13: 9–25.
Lam AL et al. (2006) Human centromeric chromatin is a dynamic chromosomal domain that can spread over noncentromeric DNA. Proc Natl Acad Sci USA 103: 4186–4191.
Lee C et al. (1997) Human centromeric DNAs. Hum Genet 100: 291–304.
Lee HR et al. (2006) Transcription and evolutionary dynamics of the centromeric satellite repeat CentO in rice. Mol Biol Evol 23: 2505–2520.
Lermontova I et al. (2006) Loading of Arabidopsis centromeric histone CENH3 occurs mainly during G2 and requires the presence of the histone fold domain. Plant Cell 18: 2443–2451.
Locovei AM et al. (2006) The CENP-B homolog, Abp1, interacts with the initiation protein Cdc23 (MCM10) and is required for efficient DNA replication in fission yeast. Cell Div 1: 27.
Maddox PS et al. (2007) Functional genomics identifies a Myb domain-containing protein family required for assembly of CENP-A chromatin. J Cell Biol 176: 757–763.
Maddox PS et al. (2004) "Holo"er than thou: chromosome segregation and kinetochore function in C. elegans. Chromosome Res 12: 641–653.
Maggert KA, Karpen GH (2001) The activation of a neocentromere in Drosophila requires proximity to an endogenous centromere. Genetics 158: 1615–1628.

Martens JH et al. (2005) The profile of repeat-associated histone lysine methylation states in the mouse epigenome. Embo J 24: 800–812.
May BP et al. (2005) Differential regulation of strand-specific transcripts from Arabidopsis centromeric satellite repeats. PLoS Genet 1: e79.
McAinsh AD et al. (2003) Structure, function, and regulation of budding yeast kinetochores. Annu Rev Cell Dev Biol 19: 519–539.
Mishra PK et al. (2007) Centromere size and position in Candida albicans are evolutionarily conserved independent of DNA sequence heterogeneity. Mol Genet Genomics 278:455–465.
Mizuguchi G et al. (2007) Nonhistone Scm3 and histones CenH3-H4 assemble the core of centromere-specific nucleosomes. Cell 129: 1153–1164.
Monen J et al. (2005) Differential role of CENP-A in the segregation of holocentric C. elegans chromosomes during meiosis and mitosis. Nat Cell Biol 7: 1248–1255.
Moore LL, Roth MB (2001) HCP-4, a CENP-C-like protein in Caenorhabditis elegans, is required for resolution of sister centromeres. J Cell Biol 153: 1199–1208.
Moreno-Moreno O et al. (2006) Proteolysis restricts localization of CID, the centromere-specific histone H3 variant of Drosophila, to centromeres. Nucleic Acids Res 34: 6247–6255.
Motamedi MR et al. (2004) Two RNAi complexes, RITS and RDRC, physically interact and localize to noncoding centromeric RNAs. Cell 119: 789–802.
Murphy TD, Karpen GH (1995) Interactions between the nod+ kinesin-like gene and extracentromeric sequences are required for transmission of a Drosophila minichromosome. Cell 81: 139–148.
Nagaki K et al. (2004) Sequencing of a rice centromere uncovers active genes. Nat Genet 36: 138–145.
Nagaki K et al. (2003) Chromatin immunoprecipitation reveals that the 180-bp satellite repeat is the key functional DNA element of Arabidopsis thaliana centromeres. Genetics 163: 1221–1225.
Nakagawa H et al. (2002) Fission yeast CENP-B homologs nucleate centromeric heterochromatin by promoting heterochromatin-specific histone tail modifications. Genes Dev 16: 1766–1778.
Nakashima H et al. (2005) Assembly of additional heterochromatin distinct from centromere-kinetochore chromatin is required for *de novo* formation of human artificial chromosome. J Cell Sci 118: 5885–5898.
Nishihashi A et al. (2002) CENP-I is essential for centromere function in vertebrate cells. Dev Cell 2: 463–476.
Noma K et al. (2006) A role for TFIIIC transcription factor complex in genome organization. Cell 125: 859–872.
Nonaka N et al. (2002) Recruitment of cohesin to heterochromatic regions by Swi6/HP1 in fission yeast. Nat Cell Biol 4: 89–93.
Nusbaum C et al. (2006) DNA sequence and analysis of human chromosome 8. Nature 439: 331–335.
Oegema K et al. (2001) Functional analysis of kinetochore assembly in Caenorhabditis elegans. J Cell Biol 153: 1209–1226.
Ogura Y et al. (2004) Characterization of a CENP-C homolog in Arabidopsis thaliana. Genes Genet Syst 79: 139–144.
Ohzeki J et al. (2002) CENP-B box is required for *de novo* centromere chromatin assembly on human alphoid DNA. J Cell Biol 159: 765–775.
Okada M et al. (2006) The CENP-H-I complex is required for the efficient incorporation of newly synthesized CENP-A into centromeres. Nat Cell Biol 8: 446–457.
Okada T et al. (2007) CENP-B controls centromere formation depending on the chromatin context. Cell 131: 1287–1300.

Okamoto Y et al. (2007) A minimal CENP-A core is required for nucleation and maintenance of a functional human centromere. Embo J 26: 1279–1291.

Palmer DK et al. (1989) Biochemical analysis of CENP-A, a centromeric protein with histone-like properties. Prog Clin Biol Res 318: 61–72.

Palmer DK et al. (1991) Purification of the centromere-specific protein CENP-A and demonstration that it is a distinctive histone. Proc Natl Acad Sci USA 88: 3734–3738.

Palmer DK et al. (1987) A 17-kD centromere protein (CENP-A) copurifies with nucleosome core particles and with histones. J Cell Biol 104: 805–815.

Partridge JF et al. (2002) cis-acting DNA from fission yeast centromeres mediates histone H3 methylation and recruitment of silencing factors and cohesin to an ectopic site. Curr Biol 12: 1652–1660.

Peters AH et al. (2003) Partitioning and plasticity of repressive histone methylation states in mammalian chromatin. Mol Cell 12: 1577–1589.

Pidoux AL, Allshire RC (2004) Kinetochore and heterochromatin domains of the fission yeast centromere. Chromosome Res 12: 521–534.

Pidoux AL, Allshire RC (2005) The role of heterochromatin in centromere function. Philos Trans R Soc Lond B Biol Sci 360: 569–579.

Politi V et al. (2002) CENP-C binds the alpha-satellite DNA in vivo at specific centromere domains. J Cell Sci 115: 2317–2327.

Ross MT et al. (2005) The DNA sequence of the human X chromosome. Nature 434: 325–337.

Rudd MK, Willard HF (2004) Analysis of the centromeric regions of the human genome assembly. Trends Genet 20: 529–533.

Rudert F et al. (1995) Transcripts from opposite strands of gamma satellite DNA are differentially expressed during mouse development. Mamm Genome 6: 76–83.

Saffery R et al. (2003) Transcription within a functional human centromere. Mol Cell 12: 509–516.

Saitoh H et al. (1992) CENP-C, an autoantigen in scleroderma, is a component of the human inner kinetochore plate. Cell 70: 115–125.

Saitoh S et al. (1997) Mis6, a fission yeast inner centromere protein, acts during G1/S and forms specialized chromatin required for equal segregation. Cell 90: 131–143.

Sanyal K et al. (2004) Centromeric DNA sequences in the pathogenic yeast Candida albicans are all different and unique. Proc Natl Acad Sci USA 101: 11374–11379.

Saunders M et al. (1988) Chromatin structure of altered yeast centromeres. Proc Natl Acad Sci USA 85: 175–179.

Schuh M et al. (2007) Incorporation of Drosophila CID/CENP-A and CENP-C into centromeres during early embryonic anaphase. Curr Biol 17: 237–243.

Scott KC et al. (2006) A heterochromatin barrier partitions the fission yeast centromere into discrete chromatin domains. Curr Biol 16: 119–129.

Shelby RD et al. (2000) Chromatin assembly at kinetochores is uncoupled from DNA replication. J Cell Biol 151: 1113–1118.

Shelby RD et al. (1997) Assembly of CENP-A into centromeric chromatin requires a cooperative array of nucleosomal DNA contact sites. J Cell Biol 136: 501–513.

Shi J, Dawe RK (2006) Partitioning of the maize epigenome by the number of methyl groups on histone H3 lysines 9 and 27. Genetics 173: 1571–1583.

Song K et al. (2002) Mutational analysis of the central centromere targeting domain of human centromere protein C, (CENP-C). Exp Cell Res 275: 81–91.

Spence JM et al. (2002) Co-localization of centromere activity, proteins and topoisomerase II within a subdomain of the major human X alpha-satellite array. EMBO J 21: 5269–5280.

Stoler S et al. (2007) Scm3, an essential Saccharomyces cerevisiae centromere protein required for G2/M progression and Cse4 localization. Proc Natl Acad Sci USA 104: 10571–10576.

Sugata N et al. (2000) Human CENP-H multimers colocalize with CENP-A and CENP-C at active centromere–kinetochore complexes. Hum Mol Genet 9: 2919–2926.

Sullivan BA (2002) Centromere round-up at the heterochromatin corral. Trends Biotechnol 20: 89–92.
Sullivan BA et al. (2001) Determining centromere identity: cyclical stories and forking paths. Nat Rev Genet 2: 584–596.
Sullivan BA, Karpen GH (2004) Centromeric chromatin exhibits a histone modification pattern that is distinct from both euchromatin and heterochromatin. Nat Struct Mol Biol 11: 1076–1083.
Sullivan BA, Schwartz S (1995) Identification of centromeric antigens in dicentric Robertsonian translocations: CENP-C and CENP-E are necessary components of functional centromeres. Hum Mol Genet 4: 2189–2197.
Sun X et al. (2002) Sequence analysis of a functional *Drosophila* centromere. Genome Res 13: 182–188.
Sun X et al. (1997) Molecular structure of a functional Drosophila centromere. Cell 91: 1007–1019.
Takahashi K et al. (1992) A low copy number central sequence with strict symmetry and unusual chromatin structure in fission yeast centromere. Mol Biol Cell 3: 819–835.
Talbert PB et al. (2004) Adaptive evolution of centromere proteins in plants and animals. J Biol 3: 18.
Tanaka Y et al. (2005) Human centromere protein B induces translational positioning of nucleosomes on alpha-satellite sequences. J Biol Chem 280: 41609–41618.
Tomkiel J et al. (1994) CENP-C is required for maintaining proper kinetochore size and for a timely transition to anaphase. J Cell Biol 125: 531–545.
Topp CN et al. (2004) Centromere-encoded RNAs are integral components of the maize kinetochore. Proc Natl Acad Sci U S A 101: 15986–15991.
Trazzi S et al. (2002) In vivo functional dissection of human inner kinetochore protein CENP-C. J Struct Biol 140: 39–48.
Van Hooser AA et al. (2001) Specification of kinetochore-forming chromatin by the histone H3 variant CENP-A. J Cell Sci 114: 3529–3542.
Verdel A et al. (2004) RNAi-mediated targeting of heterochromatin by the RITS complex. Science 303: 672–676.
Verdel A, Moazed D (2005) RNAi-directed assembly of heterochromatin in fission yeast. FEBS Lett 579: 5872–5878.
Vissel B, Choo KH (1989) Mouse major (gamma) satellite DNA is highly conserved and organized into extremely long tandem arrays: implications for recombination between nonhomologous chromosomes. Genomics 5: 407–414.
Wevrick R et al. (1990) Partial deletion of alpha satellite DNA associated with reduced amounts of the centromere protein CENP-B in a mitotically stable human chromosome rearrangement. Mol Cell Biol 10: 6374–6380.
Willard HF (1985) Chromosome-specific organization of human alpha satellite DNA. Am J Hum Genet 37: 524–532.
Willard HF (1990) Centromeres of mammalian chromosomes. Trends Genet 6: 410–416.
Willard HF, Waye JS (1987) Hierarchical order in chromosome-specific human alpha satellite DNA. Trends Genet 3: 192–198.
Wong AK et al. (1990) The chromosomal distribution of the major and minor satellite is not conserved in the genus Mus. Chromosoma 99: 190–195.
Wong AK, Rattner JB (1988) Sequence organization and cytological localization of the minor satellite of mouse. Nucleic Acids Res 16: 11645–11661.
Wong LH et al. (2007) Centromere RNA is a key component for the assembly of nucleoproteins at the nucleolus and centromere. Genome Res 17: 1146–1160.
Wong NC et al. (2006) Permissive transcriptional activity at the centromere through pockets of DNA hypomethylation. PLoS Genet 2: e17.
Yan H et al. (2005) Transcription and histone modifications in the recombination-free region spanning a rice centromere. Plant Cell 17: 3227–3238.

Yoda K et al. (2000) Human centromere protein A (CENP-A) can replace histone H3 in nucleosome reconstitution in vitro. Proc Natl Acad Sci U S A 97: 7266–7271.
Zeng K et al. (2004) Localisation of centromeric proteins to a fraction of mouse minor satellite DNA on a mini-chromosome in human, mouse and chicken cells. Chromosoma 113: 84–91.
Zinkowski RP et al. (1991) The centromere-kinetochore complex: a repeat subunit model. J Cell Biol 113: 1091–1110.

# Chapter 4
# Neocentromeres

**Owen J. Marshall and K.H. Andy Choo**

## 4.1 Introduction

Almost all eukaryotic centromeres are marked by their accumulation of repetitive DNA. Such repeats are specific to the organism, and indicate some form of sequence sharing between centromeres. Generally, the pattern at the core of the centromere is of a tandem repeat unit, organised into higher order repeats. In humans, the repeat has been termed alpha-satellite (or alphoid) DNA and a consensus sequence exists between centromeres (Choo et al. 1991). As such, the concept of a relationship between DNA sequence and centromere formation was compelling.

All this changed, however, with the discovery in 1993 of an ectopic centromere, or 'neocentromere', formed on a marker chromosome without any alpha-satellite DNA (Voullaire et al. 1993). The centromere on this marker chromosome, designated mardel (10), was shown to have formed at 10q25—a euchromatic region of a chromosome arm that had not undergone any rearrangement or sequence change (du Sart et al. 1997; Lo et al. 2001a). This was the discovery of a striking epigenetic phenomenon—the ability of a structure as complex as a centromere to spontaneously form at a seemingly random genomic location was unprecedented.

Such neocentromeres are very different from the 'classical' plant neocentromeres first described by Rhoades and Vilkomerson (1942), which lack fundamental centromere proteins and interact with microtubules in a very different manner to normal centromeres (for review, see Dawe and Hiatt (2004)). In contrast, human neocentromeres have been shown to bind all known centromere proteins and behave identically in mitosis and meiosis to their satellite-DNA-based cousins.

K.H. Andy Choo (✉)
Laboratory of Chromosome and Chromatin Research, Murdoch Children's Research Institute, Royal Children's Hospital, Flemington Road, Parkville Victoria 3052, Australia
e-mail: andy.choo@mcri.edu.au

P. De Wulf, W.C. Earnshaw (eds.), *The Kinetochore*,
DOI 10.1007/978-0-387-69076-6_4, © Springer Science+Business Media, LLC 2009

Since the initial discovery in 1993, almost one hundred cases of neocentromere formation in humans have been described in the literature (for a compiled table to clinical neocentromere cases, see Marshall et al. 2008). These cases, together with research from other organisms, have led not only to a greater understanding of the processes of neocentromere formation itself but also to a better appreciation of the structure and function of all centromeres.

## 4.2 Human Neocentromeres

Almost all initial information on neocentromeres stems from human clinical data gathered by cytogenetic screening. Generally, neocentric marker chromosomes form when an acentric chromosomal fragment is rescued via the formation of a neocentromere, and these marker chromosomes result from two main classes of rearrangement. These are either an inverted duplication of the distal part of a chromosome arm resulting in an unbalanced karyotype (Class I), or a balanced chromosomal rearrangement into linear and circular marker chromosomes (Class II; Fig. 4.1).

Of the two classes of chromosomal rearrangements, Class I marker chromosomes are by far the most commonly reported, representing over two-thirds of all cases. Inverted duplications of terminal chromosomal segments may arise through several means, either during meiosis or mitosis (Voullaire et al. 2001).

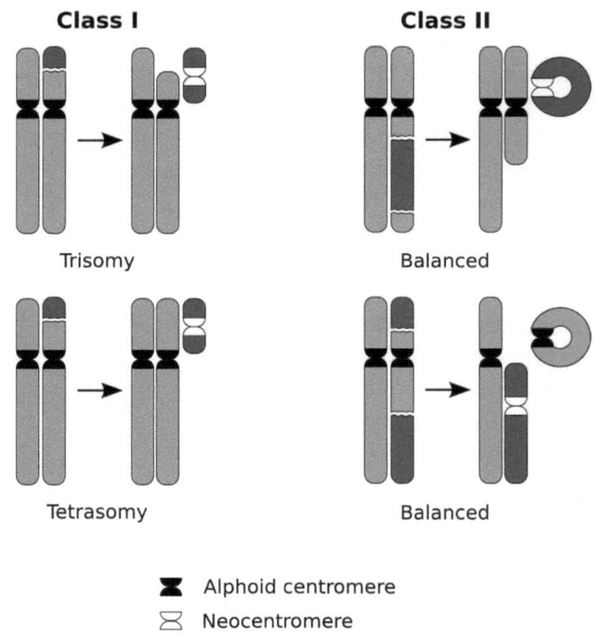


**Fig. 4.1** Chromosome rearrangements following chromatid breaks that are the common cause of neocentromere formation in humans. Single, unreplicated chromatids of a homologous chromosome pair are depicted; regions of the chromosome involved in formation of the derivative neocentric marker chromosome are shaded *dark gray*. The resulting effect on the karyotype is listed underneath each alternative rearrangement

In the mitotic formation of Class I neocentric marker chromosomes, either trisomy or tetrasomy can result following a break on the chromatid arms, depending on the segregation of the acentric fragment (Fig. 4.1). Subsequent replication of the fragment and joining of the replicated strands results in the inverted duplication, with neocentromerisation occurring either at this point or after the next round of cell division. Alternatively, tetrasomy may also result from a distal U-type exchange at meiosis I, with subsequent neocentromere formation within the inverted duplication allowing the rescue of the marker chromosome. Regardless of how the inverted duplication arises, the result is an unbalanced karyotype with either trisomy or tetrasomy for the region of duplication.

Such inverted duplicated marker chromosomes are almost always present in the individual in mosaic form. This mosaicism may be due to the mechanisms of marker chromosome formation or some intrinsic mitotic instability of the marker chromosome, but the selective disadvantage of trisomy/tetrasomy must be a contributing factor in some tissues.

The other common form of neocentric marker chromosomes are balanced rearrangements, whereby a chromosome has been rearranged to form a ring chromosome and a linear marker (Fig. 4.1). The neocentromere may form on either the rod or ring derivative, whichever is left acentric from the initial rearrangement. Precisely how and when this rearrangement occurs is unclear—the general assumption is that this process occurs via the chromosome breaking twice and the ends rejoining (Kosztolányi 1987; Warburton et al. 2003), although an alternative explanation would be looping and homologous recombination within a sister chromatid at meiosis I.

In contrast to the Class I marker chromosomes above, these balanced chromosomal rearrangements are generally marked by the stability of the linear chromosome derivative, and some degree of mosaicism with the ring derivative (as is common with ring chromosomes), regardless of which fragment contains the neocentromere. Such stability of the neocentromere-containing fragment, compared to the inherent mosaicism of the Class I neocentric chromosomes, presumably relates to the balanced nature of the chromosomal rearrangement, where loss of the fragment would create partial monosomy and be deleterious to cell survival.

The phenotype associated with these rearrangements is thus limited to the disrupted region of the chromosome at the breakpoints and the slight aneuploidy of the ring derivative caused through ring behaviour. Since such genotypic changes can be relatively minor, it is possible that many such rearrangements have not been detected through clinical screening. Indeed, there are at least two examples of Class-II neocentric marker chromosome only being detected serendipitously in an individual considered to be phenotypically normal: once where the rearrangement was ascertained in the offspring of an individual (Wandall et al. 1998) and once where an individual was only discovered to possess a rearrangement due to a high proportion of miscarriages (Knegt et al. 2003).

### *4.2.1 Frequency of Neocentromere Formation in Humans*

A key question in understanding neocentromere formation is how frequent an event the phenomenon is. It is difficult to make a clear estimate of frequency, but an investigation of the statistics of reported small Supernumerary Marker Chromosomes (sSMCs) in the literature may give an indication of the relative frequency of deleterious neocentromere formation. Such sSMC chromosomes are rearranged markers that are smaller than chromosome 21, often featuring rearrangements of the short arms of the acrocentric chromosomes; however, the inverted duplications of chromosomes that feature neocentromeres are also grouped within this category. An ongoing compilation of published sSMC cases made available online (markerchromosomes.ag.vu/) has catalogued 2480 sSMC cases in the literature, of which 81 features confirmed or putative neocentromeres—suggesting neocentromeres represent around 3% of published sSMC cases. Naturally, considering the novelty of neocentromere cases and thus the greater likelihood of publication, this number may be an overestimate. However, a study of 241 unpublished sSMC cases in 2005 suggested a similar frequency—the study found only three putative occurrences of neocentromere formation (Dalprà et al. 2005), suggesting that neocentromeres represent around 1% of total sSMC cases. Considering that sSMCs are found in 0.043% of live births (Liehr et al. 2004), these numbers give us an estimate of neocentromere formation on inverted duplicated chromosomes occurring in approximately 0.0005–0.0014%, or once in every 70,000–200,000 live births.

However such an estimate does not give a complete picture of neocentromere formation. These studies do not include the incidence of balanced rearrangements (Class-II neocentromere markers), which owing to the less severe phenotype associated with such cases may be under reported in the literature. Furthermore, we can gain no clear idea as to the frequency of 'centromere repositioning' events (see Section 4.3, below) from these statistics, as such rearrangements have only been detected serendipitously—individuals with such rearrangements show no detrimental phenotype at all.

The overall picture, then, is that neocentromerisation is a rare, but by no means infrequent event.

## 4.3 Centromere Repositioning and Speciation

In the clinical literature, neocentromere formation provides little evolutionary advantage. Although the formation of a neocentromere rescues the carrier from embryonic lethality, the chromosomal rearrangements associated with neocentromere formation are generally deleterious, resulting in either partial trisomy/tetrasomy, or in a ring chromosome that is subject to aneuploidy. Why, then, does the process of neocentromere formation occur at all, and with reasonable frequency?

The answer to this question may lie with a rare, third type of neocentromere formation reported in the clinical data. These neocentromeres are formed on an intact chromosome with the pre-existing, repetitive-DNA-based centromere still present, but inactivated. In essence, the active centromere on these pseudo-dicentric chromosomes has been repositioned. Such neocentromeres are uncommon, with only five cases reported in the literature (Rivera et al. 1996; Bukvic et al. 1996; Tyler-Smith et al. 1999; Amor et al. 2004; Ventura et al. 2004). However, considering that no obvious defect is associated with these repositioned centromeres, such neocentromeres may be more common than the statistics suggest.

Three examples of this phenomenon were neocentromeres that formed in the heterochromatic long arm of the Y chromosome (Rivera et al. 1996; Bukvic et al. 1996; Tyler-Smith et al. 1999). Whilst one example was present with a high degree of mosaicism and instability (Bukvic et al. 1996), the other two examples were stably transmitted through at least three generations of males, with the alpha-satellite DNA of the pre-existing centromere still present on the chromosome arms, but failing to form a constriction. Quantitative FISH analysis of the alpha-satellite remaining at one inactive centromere suggested that there may have been a partial deletion of the alphoid DNA, although the amount was only slightly outside the range of normal variation seen in the population (Tyler-Smith et al. 1999).

Whilst the heterochromatic long arm of the Y chromosome may be particularly predisposed to neocentromere formation, a further two cases of centromere repositioning have been described on the autosomes (Amor et al. 2004; Ventura et al. 2004). In one example on chromosome 3, the neocentromere was only recorded to have been transmitted through one generation, having formed *de novo* in the father (Ventura et al. 2004). However, with the other example on chromosome 4 the original progenitor of the chromosome could not be determined, with the chromosome stably inherited through at least two generations without any alpha-satellite being present at the constriction (Amor et al. 2004). The levels of alpha-satellite DNA remaining at the old centromere in this last case were quantified via FISH, but without the original progenitor chromosome 4 available for study it was unclear whether the amount of satellite DNA had been reduced. Although the amount of satellite DNA was low for chromosome 4, the quantity (~1.3 Mb) was within the range of variation seen within the population (Amor et al. 2004).

The fact that there may have been a reduction in the amount of satellite DNA at the old centromere in two cases raises an interesting possibility: could neocentromere formation in these examples have been induced by the weakening or deactivation of the old centromere by partial deletion? Even if all cases had involved deletions of alpha-satellite at the old centromeres, though, it would not necessarily mean that such deletions were the cause of neocentromere formation. In each instance where the amount of alpha-satellite remaining on the centromere was quantified the original progenitor of the neocentromere could not be traced, meaning that the neocentromere could potentially be many generations old (Tyler-Smith et al. 1999; Amor et al. 2004). In such cases, a gradual deletion and mutation of the alpha-satellite DNA at the inactive centromere would be expected.

Such cases are especially interesting considering the well-documented process of centromere-repositioning seen in vertebrates. Comparative studies of chromosomes in primates, other mammals and birds have demonstrated that the positioning of centromeres changes over the course of evolution, in a means that is unrelated to the surrounding pericentric DNA markers (Montefalcone et al. 1999; Ventura et al. 2001; Eder et al. 2003; Ventura et al. 2004; Kasai et al. 2003; Cardone et al. 2006; Ventura et al. 2007). These observations were first demonstrated for the evolution of chromosome IX in primates (Montefalcone et al. 1999), but have since been reproduced through the observations of other chromosomes in primates (Ventura et al. 2007), birds (Kasai et al. 2003) and other mammals (Cardone et al. 2006; Szpirer et al. 2005). In all such cases, the order of the DNA markers surrounding the new centromeric location had remained unchanged, and the most parsimonious series of chromosomal rearrangements suggested that one centromere had been de-activated, and a new centromere formed *de novo* at a new location.

Could such repositioning come about through a neocentric intermediate? Although the possibility cannot be excluded that repositioned centromeres are formed through the ectopic incorporation of centromeric satellite DNA, the examples of neocentromere formation presented above strongly suggest that neocentromeres are indeed the means of centromere repositioning. In each of these cases, the neocentromere in question was able to be stably transmitted through multiple generations, with the old, alphoid centromere remaining inactivated. Furthermore, there seems to be a relationship between regions known to form neocentromeres on human chromosomes, and the sites of evolutionary centromere repositioning events in other organisms. On chromosomes 15, 3 and 13, centromere repositioning events have occurred several times in regions known to favour neocentromere formation (Ventura et al. 2003, 2004; Cardone et al. 2006).

Of course, such repositioned centromeres do not remain devoid of repetitive satellite DNA, but must acquire it during subsequent evolution. Most intriguing in this respect are the recent studies of two rice centromeres (Nagaki et al. 2004; Yan et al. 2006). Whilst most rice chromosomes contain typical centromeres demarcated by long stretches of satellite repeats embedded within heterochromatin, the centromere on chromosome 8, Cen8, is unusual, containing an extremely low quantity (40 kb) of satellite repeats (Nagaki et al. 2004). Mapping of the binding domain of CENH3 (the rice paralogue of CENP-A, a fundamental centromere protein—see Section 4.4, below) by chromatin immunoprecipitation (ChIP) demonstrated that while this block of satellite DNA lay within the centromeric chromatin, most of the CENH3 domain was occupied by actively transcribed genes (Nagaki et al. 2004) and was thus very similar to mapped human neocentromeres (see Section 4.5, below). Cen8 may thus represent the next step in the centromere repositioning process: an example of a neocentromere that has become fixed within the species, and is beginning to slowly incorporate satellite repeat sequences. A second mapped rice centromere, Cen3, possibly represents a step further in this evolutionary process. On this centromere, the satellite DNA occupied a much larger block of 450 kb, but the remainder of the 1.8 Mb region of CENH3 binding was again occupied by actively transcribed genes (Yan et al. 2006).

Interestingly, it appears that the satellite repeat sequences do not gradually integrate throughout the active genes of a neocentromere, but rather expand outwards from a single location. In the case of a centromere repositioning event within chromosome 6 in Old World Monkeys, large amounts of satellite and repetitive sequences associated with the new centromere appear to have been introduced at the new centromere site without any change to the surrounding sequences at a Bacterial Artificial Chromosome (BAC) resolution level (Eder et al. 2003). The incorporation thus appears to be more in the nature of an initial insertion followed by expansion (through mechanisms such as unequal crossing over), rather than a gradual accumulation of satellite DNA over multiple regions within the neocentromere. If this is indeed the case then the centromere signal must be capable of a gradual shift, from the euchromatic DNA of the initial neocentromere to the inserted satellite sequences.

There is some evidence, though, that the two gene-containing rice centromeres described above may be unusual cases in the process of centromere repositioning. Recent studies of the phenomenon on chromosome 13 and on the Macaque genome have suggested that centromere repositioning events generally occur in gene 'deserts'—regions of the genome that are completely devoid of genes (Cardone et al. 2006; Ventura et al. 2007). It has been suggested that lack of genes within the site of neocentromere formation on pseudo-dicentric chromosomes may be an important factor in determining whether the neocentromere becomes subsequently fixed in the population and incorporates satellite sequences (Cardone et al. 2006; Ventura et al. 2007).

Precisely why the subsequent incorporation of satellite sequences at repositioned centromere sites occurs, however, remains a mystery. Clearly there must be an evolutionary advantage in having repetitive DNA at centromeres, since neocentromeres are not known to have become fixed in any organism studied to date. One possibility is that repetitive satellite DNA may help to increase the loading of constitutive centromere proteins such as CENP-A at centromeres. This theory has been supported by studies of the levels of CENP-A present at centromeres, which has shown significantly less amounts of the protein to be present at neocentromeres (Irvine et al. 2004, and see Section 4.4, below) If this was the case, the presence of satellite DNA could be rapidly selected for via the phenomenon of meiotic drive (Henikoff et al. 2001).

An alternate possibility for the incorporation of satellite DNA at neocentromeres is that the repetitive DNA, devoid of active genes, helps to promote a heterochromatic environment more favourable for sister chromatid cohesion. Centromeric heterochromatin is known to be necessary for sister chromatid cohesion in yeast (Bernard et al. 2001; Nonaka et al. 2002), and an increase in the amount of heterochromatin could thereby allow the centric region to better withstand microtubule tension and aid the stability of the chromosome during mitosis and meiosis. Considering the generally high mitotic stability of balanced neocentromere marker chromosomes this possibility may be regarded as less likely. However, significantly greater sister chromatid separation has been documented for at least one neocentromere (the pseudo-dicentric chromosome 4 mentioned above) indicating a possible reduction in chromatid cohesion (Amor et al. 2004), and this may be a contributing factor to the eventual fixation of satellite sequences.

Naturally, repositioning the centromere on a chromosome provides an effective mechanism of reproductive isolation and thus evolutionary speciation. If neocentromere formation were indeed the cause of this process—something which from the above discussion appears highly likely—this would provide a simple explanation as to why mechanisms exist to drive neocentromere formation.

## 4.4 Protein Studies at Neocentromeres

It is now well documented that neocentromeres bind all known centromere proteins, with the exception of the sequence-specific and apparently redundant protein CENP-B (Saffery et al. 2000; Craig et al. 2003a). Of great interest, however, is the precise location of these proteins within the centromere. One of the great advantages that neocentromeres provide is their lack of repetitive DNA, and this has allowed detailed mapping of the binding domains of centromere proteins—something which has been impossible to achieve with normal centromeres.

### *4.4.1 CENP-A*

Most extensively mapped is the fundamental centromere protein CENP-A. This protein, a histone H3 paralogue found only at the nucleosomes of active centromeres, has been mapped to eight neocentromeres at BAC or better resolution via ChIP (Lo et al. 2001a, b; Alonso et al. 2003; Cardone et al. 2006; Alonso et al. 2007; Capozzi et al. 2008). From these mapping experiments, it is clear that a large degree of variability in the size of the CENP-A domain exists between neocentromeres (Fig. 4.2). Although resolution using BAC arrays is relatively poor, the extent covered by CENP-A ranges from 54 to 464 kb.

On three neocentromeres—the mardel (10) neocentromere at 10q25 (Chueh et al. 2005), the BBB neocentromere at 13q33 (Alonso et al. 2007) and a neocentromere at 9q33.1 (Capozzi et al. 2008)—CENP-A has been mapped at higher resolution using an array of PCR or oligo fragments. The first study to achieve

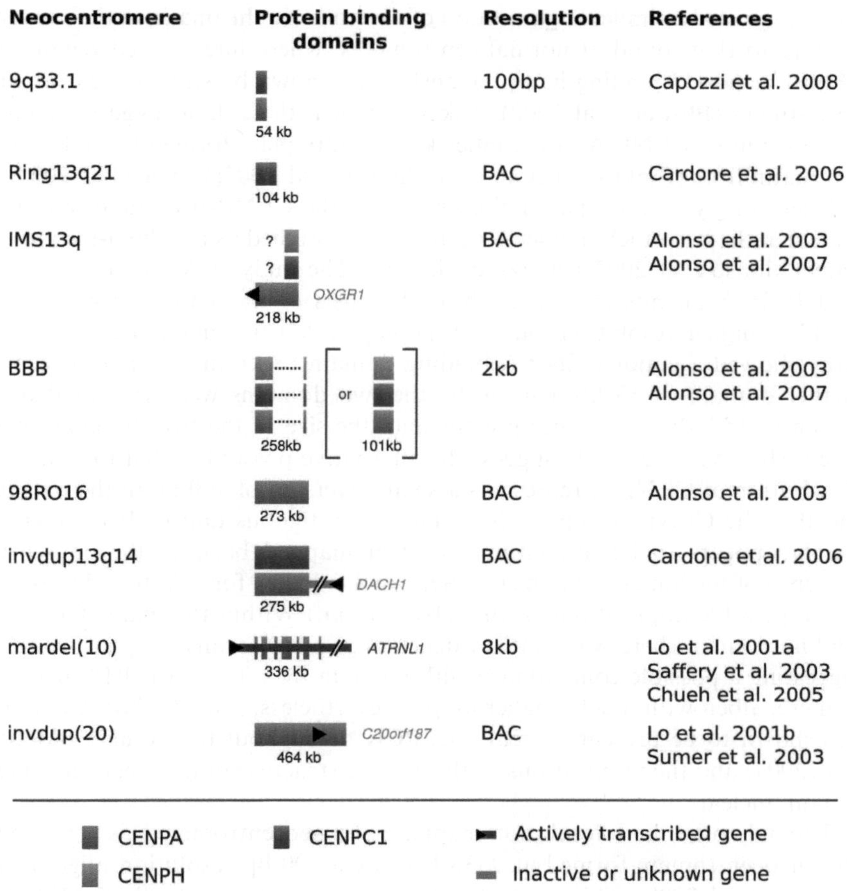


**Fig. 4.2** The size of mapped protein domains at neocentromeres (derived from BAC data from Lo et al. (2001a, b); Alonso et al. (2003); Cardone et al. (2006) updated against Ensembl release 44 from the Ensembl project (Hubbard et al. 2002)). Size of the protein-binding domain for each neocentromere is listed. Known protein-coding genes present within these domains are shown (derived from Ensembl release 44). The discontinuous nature of the mardel(10) (Chueh et al. 2005) and BBB (Alonso et al. 2007) neocentromeres are illustrated. Two possibilities of the layout of the BBB neocentromere are provided—see text for details. The IMS13q CENP-C/CENP-H domains are marked "?"to represent an uncertainty as to the size of these domains—the authors of this study suggested that the lower efficiency of ChIP with anti-CENP-C and CENP-H antibodies might prevent the full extent of these domains from being detected (Alonso et al. 2007). (Adapted from Marshall et al. 2008)

this used PCR fragments with an average size of 8 kb to map the mardel (10) neocentromere (Chueh et al. 2005). At this resolution, the CENP-A domain was shown to be composed of seven separate regions interspersed with histone H3 (Fig. 4.2). The regions of CENP-A binding were regularly spaced, with an average of 54 kb peak-to-peak distance (standard deviation: 10 kb). Such a result suggested a similar organisation of centromeric chromatin at the neocentromere to that found at normal centromeres, where interspersed regions of CENP-A and H3 binding had previously been shown by stretched chromatin fibre studies (Blower et al. 2002). Taken together, these data suggested a contiguous unit of CENP-A at the inner kinetochore plate formed by coiling the chromatin fibre (Zinkowski et al. 1991; Blower et al. 2002; Chueh et al. 2005).

Interestingly, evenly spaced clusters within the CENP-A domain were less evident at the two much smaller neocentromeres studied using a high-resolution array (Alonso et al. 2007; Capozzi et al. 2008). The study by Alonso et al. (2007) used PCR fragments around 2 kb as the basis of the array—providing an eightfold higher resolution, but with perhaps a lower signal:noise ratio. The array showed a major CENP-A binding domain 88 kb in size, and a second smaller domain of 13 kb. Curiously, the two domains were separated by a stretch of 157 kb—a distance greater than the size of the two domains combined. However, we would suggest the alternative possibility that this curious distribution of CENP-A represents a small inversion of 160 kb in the patient, and that the CENP-A domain is, in fact, a contiguous unit of 101 kb which merely appears to be discontinuous when mapped back to the consensus sequence of the human genome (see Section 4.6, below for a further discussion of the possible implications of this observation.) Within the major CENP-A binding domain there was some evidence of regular, localised signal troughs, suggesting a possible concordance with the data from the mardel(10) neocentromere, albeit with much smaller loops. Nevertheless, some CENP-A binding was shown to be present by real-time PCR throughout the domain (Alonso et al. 2007) and the implications of this for the structure of this neocentromere remain unclear.

The only other high resolution mapping of a neocentromere was performed on a neocentromere formed at 9q33.1, using a 100 bp resolution oligo array (Capozzi et al. 2008). This neocentromere was formed on a small (12 Mb) ring chromosome resulting from a balanced chromosomal rearrangement, and possesses the smallest CENP-A domain yet reported at only 54 kb (Fig. 4.2). Once again, no evidence of a discontinuous CENP-A domain was apparent, although the domain was not analysed at a subnucleosome level resolution as with Alonso et al. (2007).

Most recently, the distribution of CENP-A has been studied at normal centromeres and at a neocentromere using electron microscopy (EM). Using flow cytometry to isolate individual populations of chromosomes, the distribution of CENP-A at the mardel(10) neocentromere could be directly compared to that found at normal alphoid centromeres (O. Marshall, A. Marshall and K. H. A. Choo, unpublished data). The results suggested a surprising similarity in the CENP-A binding domain size between the two types of centromeres: in both cases there was no significant difference between the physical size of the CENP-A domain or in the proportion of chromatin occupied by CENP-A relative to the constriction (Fig. 4.3).

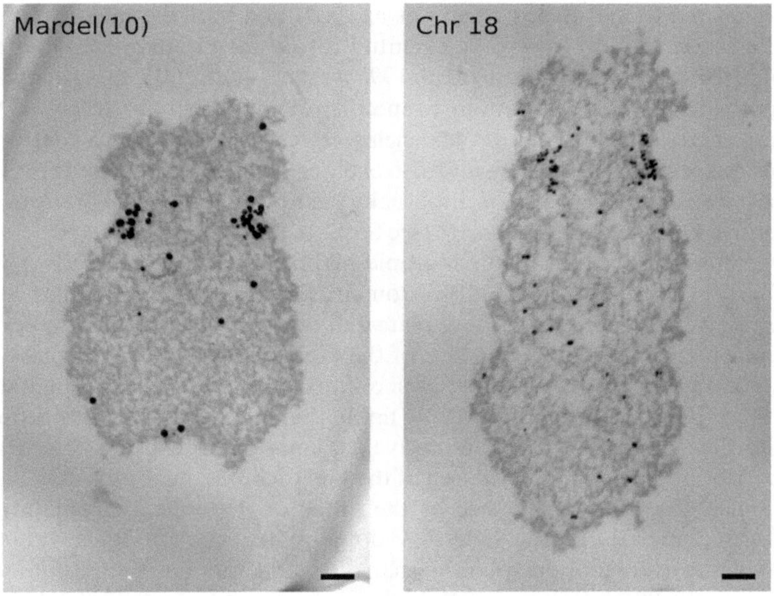


**Fig. 4.3** Fine structural localisation of CENP-A at a neocentromere and an alphoid centromere. A 45 nm-thick section through each chromosome is shown. Chromosomes were sorted by flow cytometry, fixed in acetone and labelled with a mouse monoclonal anti-human CENP-A primary antibody (MBL) and a Ultra-small gold anti-mouse secondary antibody (Aurion), before post-fixation, embedding and sectioning. Scale bar, 200 nm. (Adapted from Marshall et al. 2008)

The amount of CENP-A present on neocentromeres, relative to that found on normal centromeres, has also been investigated. Using a cell line expressing a GFP-CENP-A fusion protein and measuring the relative levels of fluorescence at the centromere, two separate neocentromeres (the mardel(10) neocentromere and an invdup(20) neocentromere) were shown to bind one-third the amount of CENP-A as most other human centromeres (a result which was verified using immunofluorescence; Irvine et al. 2004). Considering the EM data above, this may suggest that the loading of CENP-A at the inner plate is less frequent at neocentromeres. Furthermore, recent ultra high resolution ChIP at a subnucleosome level has demonstrated that at one neocentromere CENP-A is not present in contiguous, uniform blocks, but rather that individual CENP-A nucleosomes are interspersed with H3 nucleosomes (Alonso et al. 2007)—also supporting a theory of less frequent incorporation.

Precisely why the incorporation of CENP-A at neocentromeres should be less frequent is unclear. It is interesting to note, however, that both the neocentromeres and the Y chromosome centromere—which also exhibits significantly reduced CENP-A binding (Irvine et al. 2004)—fail to bind CENP-B. Considering that CENP-A has been shown to be strongly associated with alpha-satellite sequences containing the CENP-B box (Ando et al. 2002) and that the presence of such satellite sequences appears to be essential for *de novo* human artificial chromosome formation (Masumoto et al. 1998; Ohzeki et al. 2002; Basu et al. 2005; Okamoto et al. 2007), it is tempting to speculate a non-essential role for CENP-B at centromeres in which the protein helps to recruit CENP-A to centromeric regions. Such an observation would also explain why neocentromeres are merely a transient structure in evolution, eventually incorporating satellite DNA to form a repetitive DNA based centromere (see Section 4.3, above).

One interesting aspect of the multiple mapping results for CENP-A is the variability in the size of the CENP-A domain. Does this reflect a similar change in the size of the inner-kinetochore plate and the primary constriction between neocentromeres? If the physical size of the kinetochore is variable, this would again point to the transient nature of neocentromeres and perhaps suggest that such structures only reach maturity upon the subsequent incorporation of satellite DNA. However, an alternative explanation is also possible. Rather than reflecting a variation in the size of the kinetochore, the results may instead represent a variation in the size of the loops of the coiled chromatin fibre that form the basis of the CENP-A domain (see above). By a variety of methods, the average loop size in human chromosomes has been estimated to be between 30–90 kb (Filipski et al. 1990; Paulson and Laemmli 1977; Laemmli et al. 1978; Jackson et al. 1990). If the seven peaks of CENP-A binding at the mardel(10) neocentromere represent an end-to-end distance of six loops (five full loops and two half-loops), it could be expected that a CENP-A binding domain made of seven binding peaks could range between 180 and 540 kb. Such a prediction fits well with five of the eight mapped neocentromeres, but fails to adequately explain the size of the three smallest neocentromeres (including the BBB neocentromere if it is assumed to include an inversion (Fig. 4.2)).

### 4.4.2 CENP-C and CENP-H

Two other vital inner kinetochore proteins, CENP-C and CENP-H, have also been mapped by ChIP on multiple neocentromeres. Both proteins are known to interact with CENP-A (Foltz et al. 2006), and CENP-A and CENP-H are known to cooperate in recruiting CENP-C (Fukagawa et al. 2001). The two proteins would therefore be expected to occupy overlapping locations with CENP-A on neocentromeres, and indeed this appears to be largely the case. Co-localisation between CENP-C and CENP-A on ChIP arrays has been now shown for four neocentromeres (Cardone et al. 2006; Alonso et al. 2007; Allshire 2002; Capozzi et al. 2008), and between CENP-H and CENP-A on two neocentromeres (Alonso et al. 2007), although on one the overlap was only partial, with CENP-C and CENP-H occupying a subset of the CENP-A domain (Alonso et al. 2007; Fig. 4.2). Furthermore, precise co-localisation of CENP-A and CENP-C was shown using high-resolution ChIP at two neocentromeres (Alonso et al. 2007; Capozzi et al. 2008), and also with CENP-H at the BBB neocentromere (Alonso et al. 2007).

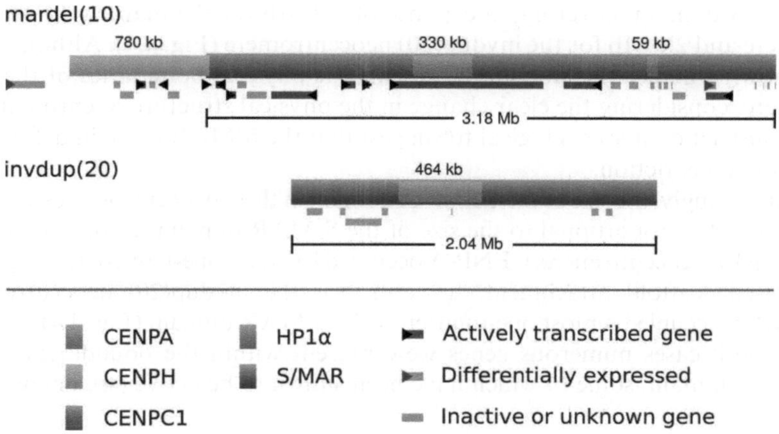


**Fig. 4.4** Scaffold domains, protein binding domains and genes present within two mapped neocentromeres. (derived from BAC positional data from Lo et al. (2001a,b); Sumer et al. (2003); Saffery et al. (2003) updated against Ensembl release 44 from the Ensembl project (Hubbard et al. 2002)). Known protein-coding genes present within these domains are shown (derived from Ensembl release 44), with expression data from cell lines derived from Saffery et al. (2003) and Wong et al. (2006). Differentially expressed genes denote two protein-coding genes found to be activated following neocentromere formation (Wong et al. 2006). (Adapted from Marshall et al. 2008)

Curiously, no co-localisation was seen between CENP-A and CENP-H on the mardel(10) neocentromere – CENP-H was found to be present in a large, 900 kb domain over 1 Mb distant from the CENP-A domain (Saffery et al. 2003; Fig. 4.4). The implications of this last result are unclear, but it may perhaps imply a higher order chromatin folding at this particular neocentromere, bringing the two separate regions into closer proximity to enable protein–protein interactions.

### *4.4.3 The Chromosome Scaffold*

The presence of scaffold or matrix proteins at the cores of condensed, mitotic chromosomes has long been demonstrated (Adolph et al. 1977; Tavormina et al. 2002; Christensen et al. 2002; Maeshima and Laemmli 2003; Coelho et al. 2003; Kireeva et al. 2004). At alphoid centromeres, the frequency of sites of scaffold attachment increases dramatically (Strissel et al. 1996), suggesting a tighter compaction of centromeric chromatin. The same observation has also been shown for two neocentromeres (Sumer et al. 2003; Saffery et al. 2003), and the non-repetitive nature of neocentromeres has allowed the extent of the enhanced scaffold/matrix attachment region (S/MAR) to be defined. In both cases, the S/MAR domain was found to be much larger than the associated

CENP-A domain, covering an expanse of 3.2 Mb for the mardel(10) neocentromere and 2.0 Mb for the invdup(20) neocentromere (Fig. 4.4). Although it is unknown whether this domain represents the physical boundaries of the centromere, considering the clear change in the physical structure of chromatin at the constriction it seems logical to suggest that the S/MAR domain defines the primary constriction.

Interestingly, the size of the CENP-A domain in these neocentromeres does not appear to be proportional to the size of the S/MAR domain. In the case of the mardel(10) neocentromere, CENP-A occupied less than one-tenth of the region of increased scaffold attachment; however, for the invdup(20) neocentromere CENP-A occupied almost one-quarter of the S/MAR domain (Fig. 4.4).

In both cases numerous genes were present within the boundaries of the scaffold domain, some of which have been shown to be active (see below).

### *4.4.4 HP1α*

The positioning of the heterochromatin protein marker HP1$\alpha$ has also been studied on a single neocentromere, the mardel(10). Curiously, considering the strong requirement of heterochromatin for sister chromatid cohesion at the centromeres of fission yeast (Bernard et al. 2001; Nonaka et al. 2002), the protein showed only slight enrichment at the neocentromere compared to the progenitor 10 chromosome, at a single BAC position 800 kb distant from the CENP-A domain (Saffery et al. 2003; Fig. 4.4). Although this region represents an extremely small domain of heterochromatin, it is important to note that the ChIP study was a comparative one, only measuring the levels of enrichment of the protein. Thus it is possible that an extant domain of heterochromatin already existed at the normal 10q25.3 region, which has merely been augmented following neocentromere formation.

## 4.5 Gene Expression Within Neocentromeres

One of the more fascinating aspects of neocentromeres is their location within euchromatic regions of the genome. This is particularly evident in the case of the mardel(10) neocentromere, which has a long gene transcript spanning the entire length of the CENP-A domain (Figs. 4.2 and 4.4). However, three other neocentromeres also have known protein-coding genes within their mapped CENP-A domains (Fig. 4.2)—meaning that of the seven mapped inner-kinetochore plates, the majority contain protein-coding genes.

Whether such euchromatic genes could be expressed within the boundaries of centromeric chromatin has always been an area of interest. The CENP-A N-terminal tail lacks a lysine amino acid at residue 4, preventing the methylation marks that denote euchromatin and active genes, and consequently it might be thought that centromeric chromatin was silent by default. It has also been shown that CENP-A forms a tighter nucleosome structure (Black et al. 2004, 2007) than the H3-containing alternative, which might also form a barrier to transcription.

By comparing the expression levels of known protein-coding genes within the mardel(10) neocentromere domain to the same genes on the progenitor chromosome 10 from which the neocentric marker chromosome had formed, it has been possible to investigate this problem. Surprisingly, ATRNL1, the gene that runs through the CENP-A domain on the mardel(10) neocentromere, was found to be actively expressed (Saffery et al. 2003). Furthermore, it appears that the formation of the mardel(10) neocentromere at 10q25 did not significantly change the expression levels of this gene (Saffery et al. 2003). Most recently, this phenomenon of active transcription through a region of CENP-A containing chromatin has also been demonstrated in alpha-satellite containing human artificial chromosomes, where the CENP-A domain was shown to not be restricted to the alpha-satellite repeats, but to have spread over the active selective marker gene (Lam et al. 2006) adjacent to these repeats. Transcription of centromeric chromatin has also been shown within the euchromatic CENH3 binding domain of two rice centromeres (Nagaki et al. 2004; Yan et al. 2006; see Section 4.3, above). From these results, it appears that CENP-A-containing chromatin represents no barrier to gene transcription.

Within the S/MAR domain of the mardel(10) neocentromere there are eight other actively expressed genes which were similarly unaffected by the formation of a neocentromere (Saffery et al. 2003; Fig. 4.4). Indeed, the only differences in gene expression detected following neocentromere formation was the activation of two protein-coding genes on either end of the S/MAR domain (Wong et al. 2006), where in both instances the genes were only expressed following neocentromere formation (Fig. 4.4). These genes corresponded with regions of hypomethylation at the mardel(10) neocentromere, and their activation may have been induced as a by-product of the epigenetic remodelling that accompanies neocentromerisation (see Section 4.7, below).

Thus despite the increased scaffold attachment sites and a corresponding tighter packing of chromatin, gene transcription can continue regardless within the primary constriction and is occasionally even promoted. All the evidence therefore points to the centromeric structure being largely irrelevant to gene transcription, and again raises the question as to why satellite repeats are necessary at all at eukaryotic centromeres.

## 4.6 Neocentromere Formation

### *4.6.1 Neocentromerisation*

Although much is now known about the structural characteristics of neocentromeres, comparatively little is known about how they actually form. Currently, there are no reports of neocentromeres forming experimentally in human cell lines, and the little information that we have stems from studies undertaken in flies and plants.

The first report of neocentromerisation occurring experimentally was from work undertaken in Drosophila, investigating the germ-line transmission of fragments of noncentromeric DNA following radiation damage (Williams et al. 1998; Maggert and Karpen 2001). In an initial study, a small (<300 kb) subtelomeric fragment of a rearranged Drosophila X minichromosome ($\gamma$238) was shown to spontaneously form a neocentromere after release by radiation (Williams et al. 1998). Significantly, the rearrangements within the $\gamma$238 minichromosome had placed this subtelomeric fragment very close to the normal, repetitive centromere of the chromosome. When the same subtelomeric fragment was released from a normal X chromosome, where it is separated from the active centromere by 40 Mb of DNA, no neocentromere formation was detected (Williams et al. 1998). A follow-up study released the same subtelomeric fragment from various sites within the Drosophila genome, and demonstrated that it was this close association to a normal centromere in the $\gamma$238 minichromosome that was responsible for conferring neocentromeric potential on the fragment—a neocentromere was only formed when the subtelomeric DNA was located next to an active centromere, with breakpoints falling within 20 kb of the active centromere domain (Maggert and Karpen 2001). Such results suggest that neocentromere formation—in flies at least—is reliant upon spreading of a centromeric 'signal' from an active centromere. It should be noted, however, that these studies were based upon the neocentromere-forming capacity of a single, short length of subtelomeric DNA. It is possible that other regions of the Drosophila genome may be more predisposed to neocentromere formation without requiring proximity to an active centromere.

A second example of neocentromeres forming close to the site of active centromeres has been observed on plant chromosomes (Nasuda et al. 2005). This study used a gametocidal system in wheat to induce structural changes in added barley chromosomes. Two fragments of barley chromosome 7 were recovered that lacked the original primary constriction and any form of centromeric repeats. Both chromosome derivatives were shown to form telocentric neocentromeres at the pericentric breakpoint from the original chromosome, suggesting that spreading of a centromeric signal from the original chromosome was the cause of neocentromere formation (Nasuda et al. 2005).

The potential of centromeric spreading to generate a neocentromere cannot account for the neocentromere cases reported in humans, where there is no record of neocentromeres forming near active centromeres. Considering the chromosomal rearrangements known to result in neocentromere formation, though, this is not entirely surprising. Neocentric inverted duplications generally occur far from the centromere on the distal ends of chromosome arms; balanced chromosomal rearrangements contain one fragment with the centromere and the surrounding pericentric DNA. Thus whilst it may be possible for the centromeric signal to spread in humans just as in flies, this will rarely, if ever, be seen in vivo. Nevertheless, it is worth noting the balanced neocentromere rearrangement on chromosome 3 described by Wandall et al. (1998), where the break points for the rearrangement were very close to the original alphoid centromere (Wandall et al. 1998; Ventura et al. 2004). Even in this scenario, no spreading of the centromeric signal was observed: the neocentromere was formed at 3q26.1, more than 70 Mb distant from the break point, rather than in a region close to the original alphoid centromere.

If the centromere signals do not possess the capability to spread in humans, how then are neocentromeres formed? One possibility is that aberrant incorporation of CENP-A at ectopic sites within the genome during DNA replication may lead to the formation of a new centromere at mitosis. It is known that overexpression of CENP-A in cell lines leads to mis-incorporation of the protein within the chromosome arms (Van Hooser et al. 2001; Heun et al. 2006). There has also been the suggestion that CENP-A mis-incorporation is common, and that subsequent protease digestion is required to remove the protein from the chromosome arms during the normal cell cycle (Collins et al. 2004; Moreno-Moreno et al. 2006). Could neocentromerisation, therefore, simply occur via small pockets of CENP-A retained on acentric fragments? The only evidence that this may be possible stems from a study in Drosophila, where overexpression of the Drosophila CENP-A orthologue CID was shown to occasionally give rise to regions with ectopic kinetochore activity, which were able to successfully bind microtubules at mitosis (Heun et al. 2006). However, a similar study in humans reported no such kinetochore formation (Van Hooser et al. 2001) despite observing a high degree of CENP-A incorporation within the chromosome arms. It seems likely, therefore, that overexpression of the protein needs to be inordinately high to drive the formation of ectopic kinetochores in humans at least, and that this is therefore unlikely to be a main method of neocentromerisation (although it may well play a role in cancer formation – see Section 4.8, below).

Related to this theory is the idea that CENP-A is incorporated in potential neocentromeric sites at high efficiency, but is rapidly removed on chromosomes that already possess a centromere by factors acting in cis (Amor and Choo 2002). In this hypothesis, each acentric fragment has a strong potential to form a neocentromere at multiple locations, and each active centromere plays a role in silencing alternative centromeres. Removing the controlling centromere—by a chromosome rearrangement, break or epigenetic silencing—would thus allow one of these alternative regions to form a neocentromere. There is some evidence in support of the idea of centromere silencing in cis: whilst dicentric chromosomes with two pairs of active kinetochores can form when the distance between the two centromeres is small, once this distance becomes greater than 12 Mb two active centromeres are never observed (Sullivan and Willard 1998). However, this phenomenon may simply result from the instability and loss of chromosomes with a large degree of separation between the two active centromeres, rather than from the involvement of cis-acting silencing factors. Furthermore, studies in Drosophila have demonstrated that a satellite-rich region in a chromosome can bind centromere proteins and occasionally form an active centromere, even in the presence of the wild-type active centromere (Platero et al. 1999)—an event which suggests a lack of centromeric silencing in cis, or at least that certain regions can escape silencing.

The above hypotheses suggest that a proto-kinetochore is already present at the site of neocentromere formation before breakage of the chromosome (or silencing of the initial centromere) occurs. An alternative possibility is that the neocentromere signals are acquired after chromosome breakage or rearrangement, via aberrant association of the acentric fragment with centromeric DNA during CENP-A loading (Maggert and Karpen 2001). Considering that loading of CENP-A occurs immediately following cell division and before DNA replication (Jansen et al. 2007; Schuh et al. 2007), this would imply that neocentromerisation would occur only after the cell has progressed through one round of division with an acentric fragment.

Perhaps most intriguing of all is the suggestion that chromosome rearrangements may be responsible for inducing neocentromere formation, possibly through a change in the epigenetic state of the chromatin following DNA repair (Ventura et al. 2003, 2007). The evidence for this theory is based upon the high-resolution FISH mapping of two 15q25 neocentromeres on inverted-duplicated marker chromosomes. The mapping showed a close association (to within 500 kb) between the site of neocentromere formation and the duplication and rearrangement boundaries of the marker chromosomes (Ventura et al. 2003). Indeed, a significant proportion of all known invdup marker chromosomes have a centromere at the approximate position of the site of recombination, suggesting a link between the two events. In this respect, the possibility of a small inversion at the mapped BBB neocentromere (see Section 4.4, above) is particularly interesting, and it is tempting to suggest that the inversion and formation of the neocentromere in this case may be related.

### *4.6.2 Neocentromere Hotspots*

A surprising but clearly recognisable trend from the human clinical data is the clustering of sites of neocentromere formation at 'hotspots' (Amor and Choo 2002; Warburton 2004). Certain regions of chromosomes—e.g. 15q, 8p, 13q—seem to be particularly prone to forming neocentromeres. In part, this trend must be ameliorated by the recognition that these sites cluster in chromosomes that are inverted duplications, and as such it is logical to see a predisposition of the ends of chromosomes to form neocentromeres. There is also the issue of the tolerance of trisomy or tetrasomy for the region, which may prevent inverted duplications from forming on several chromosomes. However, even considering the above points it is clear that neocentromere formation in certain regions is favoured (Fig. 4.5), and these hotspots are of considerable interest in studying the process of neocentromere formation.

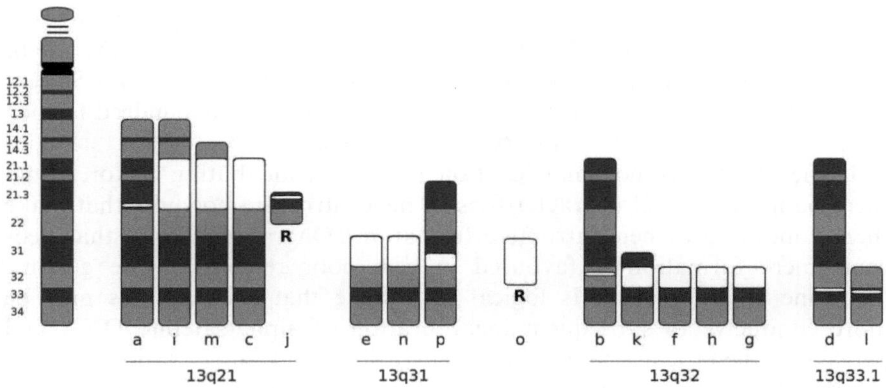


**Fig. 4.5** Neocentromere 'hotspots' on 13q. Sites of neocentromere formation are shown in white within the length of the marker chromosomes, with markers grouped by neocentromere formation within cytogenetic bands. All reported neocentromere cases from chromosome 13 are illustrated: (a–h) are as described by Warburton et al. (2000), (i) Morrissette et al. (2001), (j) Knegt et al. (2003), (k–m) are cases BC, MG and JM respectively, as described by Li et al. (2002), (n) Barwell et al. (2004), (o) Amor et al. (2005) and (p) Tonnies et al. (2006). Additional mapping data of (a and j) from Cardone et al. (2006) and (b, d and g) from Alonso et al. (2003). All marker chromosomes are inverted duplications (for the sake of simplicity, the inversion is not illustrated for these chromosomes), with the exception of two ring chromosomes designated 'R'. (Adapted from Marshall et al. 2008)

Whether neocentromeres form at the same location within each hotspot has been an important issue in understanding neocentromerisation—if this were indeed the case, it would suggest a fundamental relationship between DNA sequence (or a highly localised chromatin environment) and centromere formation. This question has been the subject of studies on two hotspots on chromosome 13, 13q32 (Alonso et al. 2003) and 13q21 (Cardone et al. 2006). Both studies mapped the CENP-A (or CENP-C) binding domain of several neocentromeres known to be located within the same cytogenetic band, and found that each binding domain was localised to a different region of the chromosome (Alonso et al. 2003; Cardone et al. 2006).

These data by themselves are not necessarily proof against a common sequence basis for the primary constriction within these hotspots. The combined neocentromere mapping studies (described in Section 4.4, above) suggest that the exact location of the CENP-A domain within the primary constriction (as indicated by the S/MAR domain) may be variable, raising the possibility that the same region of the chromosome forms a primary constriction each time, within which the position of the CENP-A domain may change. However, the distance between the 13q neocentromeres investigated in these studies suggests that this scenario is unlikely. In the case of three 13q32 neocentromeres, the CENP-A domains of each centromere were found to be located within a stretch of more than 6 Mb of DNA (Alonso et al. 2003), with two domains separated by over 5 Mb; similarly, the two 13q21 neocentromeres were separated by over 3 Mb of DNA. These distances are greater than the size of either previously mapped neocentric constriction (Sumer et al. 2003; Saffery et al. 2003), and suggest that the neocentromeres studied have indeed formed at different locations within the cytogenetic bands.

If the exact chromosomal location is not a contributing factor, could there be more general characteristics of neocentromere hotspots that make them conducive to neocentromere formation? One possibility is that neocentromere formation is favoured in gene-poor regions of the genome (Cardone et al. 2006). It is logical to assume that such regions may be more conducive to subsequent incorporation of alpha-satellite DNA and pericentric heterochromatin, and there are indeed some regions of the genome that have a low gene density and appear to be favoured for neocentromere formation (e.g. 9p23). However, there are also apparent hotspots in relatively gene-rich cytogenetic bands (e.g. 13q32), implying that there is no clear trend as to where neocentromeres are formed. Similarly, the observation that the CENP-A domains of the 13q32 neocentromeres observed by Alonso et al. (2003) were found between active genes appears not to be a general trend, judging by the presence of active genes within the CENP-A domains of three other neocentromeres (see Section 4.4, above). Finally, both mapped S/MAR domains (on the mardel(10) and invdup(20) neocentromeres) occupy regions filled with known protein-coding genes, at least some of which are known to be actively expressed (Saffery et al. 2003; Fig. 4.4). From the above evidence, therefore, it seems unlikely that gene density is a deciding factor in the positioning of neocentromeres.

An alternative explanation for the existence of neocentromere hotspots relates to the possible link between genome rearrangement and neocentromere formation discussed above (Ventura et al. 2003, 2007). This theory suggests that regions of the genome with a high content of duplications are predisposed to rearrangements, which then lead to neocentromere formation through an epigenetic change in the chromatin following DNA repair. Such a hypothesis would also explain the link between several neocentromere hotspots and the locations of ancestral centromeres (which have a high number of duplicons; Ventura et al. 2003). There has not yet been a comprehensive analysis of duplicons within the genome, so it is currently impossible to assess whether this theory can explain all known hotspots. However, there is a good correlation seen at neocentromere hotspots from 15q (Ventura et al. 2003), 13q21 (Cardone et al. 2006) and 3q26 (Ventura et al. 2004), suggesting that such a relationship between the two may be a distinct possibility.

One final point of interest regarding neocentromere hotspots is the accuracy of conventional cytogenetic mapping of the primary constriction. Two of the supposed 13q32 neocentromeres mentioned above were shown through ChIP to have CENP-A domains within the next cytogenetic band, 13q33 (Alonso et al. 2003), suggesting that some previously reported band locations of neocentric constrictions may be inaccurate. In other words, the clustering of neocentromeres at particular cytogenetic bands may not be as extensive as the literature would suggest.

### *4.6.3 DNA Sequence Similarities*

Considering the homology of repetitive satellite sequences at normal centromeres, it is perhaps not surprising that numerous studies have sought to find a correlation in the DNA sequences of neocentromeres, in particular in the sequence of known CENP-A binding domains (Lo et al. 2001a, b; Alonso et al. 2003; Chueh et al. 2005; Cardone et al. 2006). Currently, most analyses of these domains have failed to find significant deviations from the genome average, in terms of various centromere motifs or repetitive elements. However, one observable trend is the AT base-pair content of the CENP-A binding regions: this has been shown to be consistently higher than normal in all seven neocentromeres observed, ranging from 59.9% content to 66.1%, with the genome average being 59% (Lo et al. 2001a, b; Alonso et al. 2003; Cardone et al. 2006). Whether this is a major factor for determining the site of neocentromere formation is less clear, though. The three neocentromeres investigated by Alonso et al. (2003) were located within 6 Mb of each other, yet the AT content of their CENP-A binding domains varied between 59.9 and 65.0%. If AT content was really a deciding factor in determining the site of neocentromerisation, it is surprising that all three neocentromeres did not form within the region of highest AT content. Similarly, a comprehensive analysis of the AT content of the seven mapped neocentromeres and the surrounding DNA suggested that there was little correlation between regions of extremely high AT content and neocentromere formation (Alonso et al. 2007). Nevertheless, a threshold AT content requirement of at least the genome average is a possibility that cannot be discounted.

The only other clear significant difference in DNA content within a neocentromere was reported in the high-resolution CENP-A mapping study of Chueh et al. (2005). Although no difference was noted over the entire CENP-A binding domain of the mardel(10) neocentromere, when analysing the separate peaks of CENP-A binding within this domain a significant increase was observed in the number of L1 LINE elements compared to the surrounding H3-containing regions. Analysis of the BBB neocentromere, the other neocentromere mapped at PCR-fragment resolution, also showed a higher frequency of L1 elements (and MaLR LTR elements) within the peaks of CENP-A binding compared to the intervening DNA (Alonso et al. 2007), and a 70 bp motif associated with young L1 elements was found to be the only common feature shared between the DNA sequence of the seven mapped neocentromere domains (Alonso et al. 2007). An association between L1 content and neocentromere formation is therefore a distinct possibility, although the exact basis of this association is at this stage unclear.

## 4.7 Epigenetic Maintenance of Neocentromeres

Once a new centromere has been formed there remains the question of how the position of the centromere is subsequently maintained. Without evidence of DNA sequence specificity at neocentromeres, it appears that epigenetic marks are the means through which this is achieved. CpG island methylation has been an obvious candidate for this process, and a comprehensive study of CpG methylation at the mardel(10) neocentromere was recently carried out by Wong et al. (2006). This study looked not only at CpG islands (stretches of DNA longer than 200 bp enriched for CpG dinucleotides) but also at smaller regions ('islets') enriched for CpG residues. The findings of this study were striking: although islets and islands within the extended S/MAR region were in general hypermethylated at the neocentromere when compared to the progenitor chromosome 10, at the boundaries of the CENP-H, HP1 and S/MAR domains there was clear hypomethylation of islands or islets.

Such results could, of course, be coincidental—only a similar correlation seen at the boundary elements of other neocentromeres would confirm this uncategorically. Nevertheless, the results are intriguing, especially when seen in the light of a trend of hypermethylation of CpG residues located elsewhere within the neocentromere domain. Furthermore, the study showed that the hypomethylated state of these islands and islets was preserved even after chromosomal transfer from a Chinese Hamster Ovary background to two different mouse cell lines, suggesting that the maintenance of such epigenetic marks is strong (Wong et al. 2006). However, it is currently unknown whether such marks actively demarcate the boundaries of neocentromeres, or are simply created in response to the changes in chromatin caused by the formation of a neocentromere.

An alternative means of epigenetic regulation of neocentromeres lies in the modification of histones through methylation or acetylation. The response of both the CENP-A binding domain and the S/MAR domain (see Section 4.4, above) to the hyperacetylation of histones caused by trichostatin A (TSA) has been investigated at the mardel(10) neocentromere (Craig et al. 2003b; Sumer et al. 2004). These studies have demonstrated that hyperacetylation causes a striking change in both domains: the CENP-A domain was found to shift in position in response to histone hyperacetylation (Craig et al. 2003b), yet at the same time the S/MAR domain was observed to shrink to less than half of its original size (Sumer et al. 2004). Both studies thus suggested a link between histone acetylation and the position of the neocentromere.

Notably, however, the shift in position of the CENP-A domain observed in response to TSA treatment was found to be reversible following removal of the drug (Craig et al. 2003b), suggesting that other epigenetic marks such as CpG methylation discussed above may play a role in the maintenance of neocentromeres. Further studies are needed to investigate whether TSA treatment affects the hypomethylated regions found at this neocentromere, and whether induced changes to the DNA methylation environment can perturb the formation of the S/MAR domain.

## 4.8 Neocentromeres and Cancer

Although neocentromere formation is in general a rare occurrence, there are several types of cancer that are associated with the formation of a complex, rearranged chromosome containing a neocentromere. These instances are especially interesting considering the clear somatic nature of neocentromere formation in these tumours, and the suggestion that neocentromeres have formed in order to stabilise complex rearranged acentric chromosomes.

### *4.8.1 Lipomatous Tumours*

The best characterised link between neocentromeres and a specific form of cancer is found in the atypical lipomas and well-differentiated liposarcomas (ALP-WDLPS) class of lipomatous tumours. These cancers are marked by a supernumerary ring chromosome or large marker chromosome devoid of alpha-satellite DNA, primarily comprised of amplified 12q14-15 sequences that contain oncogenes (Pedeutour et al. 1999; Gisselsson et al. 1999; Sirvent et al. 2000; Forus et al. 2001) and possessing a neocentromere at the primary constriction (Sirvent et al. 2000; Forus et al. 2001). This is in direct contrast to other liposarcomas, which contain the same amplified 12q14-15 sequences but on chromosomes with alphoid centromeres (Sirvent et al. 2000). Thus the presence of a marker chromosome with a neocentromere appears to be a defining characteristic of this tumour class.

With the strong link between an amplified sequence and the formation of a neocentromere, it might be expected that the neocentromere was formed from the amplified DNA. Curiously, however, this appears not to be the case. Although the neocentric marker chromosomes consist largely of the 12q14-15 region, immuno-FISH showed the kinetochore proteins to be bound to DNA from other regions present in these chromosomes (Sirvent et al. 2000).

The formation of these marker chromosomes may thus be a multi-step process, involving amplification of 12q14-15 and capture/amplification of a different sequence involved in neocentromere formation. Precisely why there should be a selective advantage conferred by the formation of a new marker chromosome complete with a neocentromere, as opposed to the amplification of these sequences within an existing chromosome, is unclear. Interestingly, though, similar lipomatous tumours where the region from 12q is amplified on marker chromosomes with alphoid centromeres are aggressive, metastatic tumours, and thus have very different behaviour to the ALP-WDLPS tumours – suggesting a difference between alphoid centromeres and neocentromeres in this instance (Sirvent et al. 2000). One intriguing possibility would be that these more aggressive lipomatous tumours are derived from ALP-WDLPS tumours where the neocentromere has rapidly evolved into a repetitive, alphoid centromere, in a process similar to centromere repositioning described above. However, whether there is a link between the aggressive nature of a tumour and the presence of alpha-satellite at a marker centromere is at this stage unclear.

### *4.8.2 Other Cancers*

Neocentromeres have also been reported on amplified marker chromosomes in one lung carcinoma patient (Italiano et al. 2006). These marker chromosomes contain an amplification of the 9p23-24 region and a neocentromere formed at 9p23. The fact that neocentromeres have been twice previously reported at 9p23 suggests that amplification of this region may perhaps have facilitated neocentromere formation and stabilised the marker chromosome.

Neocentromeres have been described in two AML tumours, although these two cases involved different chromosome rearrangements. In one instance the neocentromere was present on a standard inverted duplication of 10q (Abeliovich et al. 1996), and in the other on a complex, rearranged ring chromosome (Gisselsson et al. 1999).

However, it is possible that the frequency of neocentromere formation in cancers is much higher than the literature suggests, as most cancers are not subjected to karyotype analysis that would detect neocentromere formation. There are also the intriguing reports that CENP-A and another centromere protein, CENP-H, are overexpressed in all colorectal cancer cell lines (Tomonaga et al. 2003, 2005). Considering the possible ectopic kinetochore formation associated with CENP-A overexpression (Heun et al. 2006; see Section 4.6, above), it is possible that upregulation of centromere proteins that promote neocentromerisation is a common step in promoting the genome rearrangements and abnormalities that lead to cancer.

## 4.9 Conclusion

Neocentromeres are thus an astonishing example of epigenetic change within the genome. Although most commonly observed on re-arranged fragments of chromosomes and occasionally in cancer cells, they appear to possess a broader role in providing a means of directing centromere repositioning and thereby speciation. By nature of their formation within non-repetitive DNA sequences, neocentromeres have provided many insights into the processes of centromere formation and maintenance, and into the structure of the centromere itself.

**Acknowledgments** We thank E. Northrop and M. Pertile for stimulating discussions and critical comments on the manuscript. This work was supported by the National Health and Medical Research Council of Australia and the National Institutes of Health.

## References

Abeliovich, D., Yehuda, O., Ben-Neriah, S., Kapelushnik, Y., and Ben-Yehuda, D., 1996. dup(10q) lacking alpha-satellite DNA in bone marrow cells of a patient with acute myeloid leukemia. Cancer Genet Cytogenet 89:1–6.

Adolph, K. W., Cheng, S. M., Paulson, J. R., and Laemmli, U. K., 1977. Isolation of a protein scaffold from mitotic HeLa cell chromosomes. Proc Natl Acad Sci U S A 74:4937–4941.

Allshire, R., 2002. Molecular biology. RNAi and heterochromatin–a hushed-up affair. Science 297:1818–9.

Alonso, A., Fritz, B., Hasson, D., Abrusan, G., Cheung, F., Yoda, K., Radlwimmer, B., Ladurner, A., and Warburton, P., 2007. Co-localization of CENP-C and CENP-H to discontinuous domains of CENP-A chromatin at human neocentromeres. Genome Biol 8:R148.

Alonso, A., Mahmood, R., Li, S., Cheung, F., Yoda, K., and Warburton, P. E., 2003. Genomic microarray analysis reveals distinct locations for the CENP-A binding domains in three human chromosome 13q32 neocentromeres. Hum Mol Genet 12:2711–2721.

Amor, D. J., Bentley, K., Ryan, J., Perry, J., Wong, L., Slater, H., and Choo, K. H. A., 2004. Human centromere repositioning "in progress". Proc Natl Acad Sci USA 101:6542–6547.

Amor, D. J. and Choo, K. H. A., 2002. Neocentromeres: role in human disease, evolution, and centromere study. Am J Hum Genet 71:695–714.

Amor, D. J., Voullaire, L., Bentley, K., Savarirayan, R., and Choo, K. H. A., 2005. Mosaic monosomy of a neocentric ring chromosome maps brachyphalangy and growth hormone deficiency to 13q31.1-13q32.3. Am J Med Genet A 133:151–157.

Ando, S., Yang, H., Nozaki, N., Okazaki, T., and Yoda, K., 2002. CENP-A, -B, and -C chromatin complex that contains the I-type alpha-satellite array constitutes the prekinetochore in HeLa cells. Mol Cell Biol 22:2229–41.

Barwell, J., Mazzaschi, R., Bint, S., Ogilvie, C. M., and Elmslie, F., 2004. A new neocentromere locus on chromosome 13 resulting in mosaic tetrasomy for distal 13q and an asymmetric phenotype. Am J Med Genet A 130:295–298.

Basu, J., Stromberg, G., Compitello, G., Willard, H. F., and Bokkelen, G. V., 2005. Rapid creation of BAC-based human artificial chromosome vectors by transposition with synthetic alpha-satellite arrays. Nucleic Acids Res 33:587–596.

Bernard, P., Maure, J. F., Partridge, J. F., Genier, S., Javerzat, J. P., and Allshire, R. C., 2001. Requirement of heterochromatin for cohesion at centromeres. Science 294:2539–2542.

Black, B. E., Foltz, D. R., Chakravarthy, S., Luger, K., Woods, V. L., and Cleveland, D. W., 2004. Structural determinants for generating centromeric chromatin. Nature 430:578–582.

Black, B. E., Jansen, L. E. T., Maddox, P. S., Foltz, D. R., Desai, A. B., Shah, J. V., and Cleveland, D. W., 2007. Centromere Identity Maintained by Nucleosomes Assembled with Histone H3 Containing the CENP-A Targeting Domain. Mol Cell 25:309–322.

Blower, M. D., Sullivan, B. A., and Karpen, G. H., 2002. Conserved organization of centromeric chromatin in flies and humans. Dev Cell 2:319–30.

Bukvic, N., Susca, F., Gentile, M., Tangari, E., Ianniruberto, A., and Guanti, G., 1996. An unusual dicentric Y chromosome with a functional centromere with no detectable alpha-satellite. Hum Genet 97:453–456.

Capozzi, O., Purgato, S., di Cantogno, L. V., Grosso, E., Ciccone, R., Zuffardi, O., Valle, G. D., and Rocchi, M., 2008. Evolutionary and clinical neocentromeres: two faces of the same coin? Chromosoma 117:339–44.

Cardone, M. F., Alonso, A., Pazienza, M., Ventura, M., Montemurro, G., Carbone, L., de Jong, P. J., Stanyon, R., D'Addabbo, P., Archidiacono, N., She, X., Eichler, E. E., Warburton, P. E., and Rocchi, M., 2006. Independent centromere formation in a capricious, gene-free domain of chromosome 13q21 in Old World monkeys and pigs. Genome Biol 7:R91.

Choo, K. H., Vissel, B., Nagy, A., Earle, E., and Kalitsis, P., 1991. A survey of the genomic distribution of alpha satellite DNA on all the human chromosomes, and derivation of a new consensus sequence. Nucleic Acids Res 19:1179–1182.

Christensen, M. O., Larsen, M. K., Barthelmes, H. U., Hock, R., Andersen, C. L., Kjeldsen, E., Knudsen, B. R., Westergaard, O., Boege, F., and Mielke, C., 2002. Dynamics of human DNA topoisomerases IIalpha and IIbeta in living cells. J Cell Biol 157:31–44.

Chueh, A. C., Wong, L. H., Wong, N., and Choo, K. H. A., 2005. Variable and hierarchical size distribution of L1-retroelement-enriched CENP-A clusters within a functional human neocentromere. Hum Mol Genet 14:85–93.

Coelho, P. A., Queiroz-Machado, J., and Sunkel, C. E., 2003. Condensin-dependent localisation of topoisomerase II to an axial chromosomal structure is required for sister chromatid resolution during mitosis. J Cell Sci 116:4763–4776.

Collins, K. A., Furuyama, S., and Biggins, S., 2004. Proteolysis contributes to the exclusive centromere localization of the yeast Cse4/CENP-A histone H3 variant. Curr Biol 14:1968–1972.

Craig, J. M., Earle, E., Canham, P., Wong, L. H., Anderson, M., and Choo, K. H. A., 2003a. Analysis of mammalian proteins involved in chromatin modification reveals new metaphase centromeric proteins and distinct chromosomal distribution patterns. Hum Mol Genet 12:3109–3121.

Craig, J. M., Wong, L. H., Lo, A. W. I., Earle, E., and Choo, K. H. A., 2003b. Centromeric chromatin pliability and memory at a human neocentromere. EMBO J 22:2495–2504.

Dalprà, L., Giardino, D., Finelli, P., Corti, C., Valtorta, C., Guerneri, S., Ilardi, P., Fortuna, R., Coviello, D., Nocera, G., Amico, F. P., Martinoli, E., Sala, E., Villa, N., Crosti, F., Chiodo, F., di Cantogno, L. V., Savin, E., Croci, G., Franchi, F., Venti, G., Donti, E., Migliori, V., Pettinari, A., Bonifacio, S., Centrone, C., Torricelli, F., Rossi, S., Simi, P., Granata, P., Casalone, R., Lenzini, E., Artifoni, L., Pecile, V., Barlati, S., Bellotti, D., Caufin, D., Police, A., Cavani, S., Piombo, G., Pierluigi, M., and Larizza, L., 2005. Cytogenetic and molecular evaluation of 241 small supernumerary marker chromosomes: cooperative study of 19 Italian laboratories. Genet Med 7:620–625.

Dawe, R. K. and Hiatt, E. N., 2004. Plant neocentromeres: fast, focused, and driven. Chromosome Res 12:655–669.

du Sart, D., Cancilla, M. R., Earle, E., Mao, J. I., Saffery, R., Tainton, K. M., Kalitsis, P., Martyn, J., Barry, A. E., and Choo, K. H., 1997. A functional neo-centromere formed through activation of a latent human centromere and consisting of non-alpha-satellite DNA. Nat Genet 16:144–153.

Eder, V., Ventura, M., Ianigro, M., Teti, M., Rocchi, M., and Archidiacono, N., 2003. Chromosome 6 phylogeny in primates and centromere repositioning. Mol Biol Evol 20:1506–1512.

Filipski, J., Leblanc, J., Youdale, T., Sikorska, M., and Walker, P. R., 1990. Periodicity of DNA folding in higher order chromatin structures. EMBO J 9:1319–1327.

Foltz, D. R., Jansen, L. E. T., Black, B. E., Bailey, A. O., Yates, J. R., and Cleveland, D. W., 2006. The human CENP-A centromeric nucleosome-associated complex. Nat Cell Biol 8:458–469.

Forus, A., Bjerkehagen, B., Sirvent, N., Meza-Zepeda, L. A., Coindre, J. M., Berner, J. M., Myklebost, O., and Pedeutour, F., 2001. A well-differentiated liposarcoma with a new type of chromosome 12-derived markers. Cancer Genet Cytogenet 131:13–18.

Fukagawa, T., Mikami, Y., Nishihashi, A., Regnier, V., Haraguchi, T., Hiraoka, Y., Sugata, N., Todokoro, K., Brown, W., and Ikemura, T., 2001. CENP-H, a constitutive centromere component, is required for centromere targeting of CENP-C in vertebrate cells. EMBO J 20:4603–4617.

Gisselsson, D., Höglund, M., Mertens, F., Johansson, B., Cin, P. D., den Berghe, H. V., Earnshaw, W. C., Mitelman, F., and Mandahl, N., 1999. The structure and dynamics of ring chromosomes in human neoplastic and non-neoplastic cells. Hum Genet 104:315–325.

Henikoff, S., Ahmad, K., and Malik, H. S., 2001. The centromere paradox: stable inheritance with rapidly evolving DNA. Science 293:1098–102.

Heun, P., Erhardt, S., Blower, M. D., Weiss, S., Skora, A. D., and Karpen, G. H., 2006. Mislocalization of the Drosophila centromere-specific histone CID promotes formation of functional ectopic kinetochores. Dev Cell 10:303–315.

Hubbard, T., Barker, D., Birney, E., Cameron, G., Chen, Y., Clark, L., Cox, T., Cuff, J., Curwen, V., Down, T., Durbin, R., Eyras, E., Gilbert, J., Hammond, M., Huminiecki, L., Kasprzyk, A., Lehvaslaiho, H., Lijnzaad, P., Melsopp, C., Mongin, E., Pettett, R., Pocock, M., Potter, S., Rust, A., Schmidt, E., Searle, S., Slater, G., Smith, J., Spooner, W., Stabenau, A., Stalker, J., Stupka, E., Ureta-Vidal, A., Vastrik, I., and Clamp, M., 2002. The Ensembl genome database project. Nucleic Acids Res 30:38–41.

Irvine, D. V., Amor, D. J., Perry, J., Sirvent, N., Pedeutour, F., Choo, K. H. A., and Saffery, R., 2004. Chromosome size and origin as determinants of the level of CENP-A incorporation into human centromeres. Chromosome Res 12:805–815.

Italiano, A., Attias, R., Aurias, A., Pérot, G., Burel-Vandenbos, F., Otto, J., Venissac, N., and Pedeutour, F., 2006. Molecular cytogenetic characterization of a metastatic lung sarcomatoid carcinoma: 9p23 neocentromere and 9p23-p24 amplification including JAK2 and JMJD2C. Cancer Genet Cytogenet 167:122–130.

Jackson, D. A., Dickinson, P., and Cook, P. R., 1990. The size of chromatin loops in HeLa cells. EMBO J 9:567–571.

Jansen, L. E. T., Black, B. E., Foltz, D. R., and Cleveland, D. W., 2007. Propagation of centromeric chromatin requires exit from mitosis. J Cell Biol 176:795–805.

Kasai, F., Garcia, C., Arruga, M. V., and Ferguson-Smith, M. A., 2003. Chromosome homology between chicken (Gallus gallus domesticus) and the red-legged partridge (Alectoris rufa); evidence of the occurrence of a neocentromere during evolution. Cytogenet Genome Res 102:326–330.

Kireeva, N., Lakonishok, M., Kireev, I., Hirano, T., and Belmont, A. S., 2004. Visualization of early chromosome condensation: a hierarchical folding, axial glue model of chromosome structure. J Cell Biol 166:775–785.

Knegt, A. C., Li, S., Engelen, J. J. M., Bijlsma, E. K., and Warburton, P. E., 2003. Prenatal diagnosis of a karyotypically normal pregnancy in a mother with a supernumerary neocentric 13q21 –>13q22 chromosome and balancing reciprocal deletion. Prenat Diagn 23:215–220.

Kosztolányi, G., 1987. Does "ring syndrome" exist? An analysis of 207 case reports on patients with a ring autosome. Hum Genet 75:174–179.

Laemmli, U. K., Cheng, S. M., Adolph, K. W., Paulson, J. R., Brown, J. A., and Baumbach, W. R., 1978. Metaphase chromosome structure: the role of nonhistone proteins. Cold Spring Harb Symp Quant Biol 42 Pt 1:351–360.

Lam, A. L., Boivin, C. D., Bonney, C. F., Rudd, M. K., and Sullivan, B. A., 2006. Human centromeric chromatin is a dynamic chromosomal domain that can spread over noncentromeric DNA. Proc Natl Acad Sci USA 103:4186–4191.

Li, S., Malafiej, P., Levy, B., Mahmood, R., Field, M., Hughes, T., Lockhart, L. H., Wu, Z., Huang, M., Hirschhorn, K., Velagaleti, G. V. N., Daniel, A., and Warburton, P. E., 2002. Chromosome 13q neocentromeres: molecular cytogenetic characterization of three additional cases and clinical spectrum. Am J Med Genet 110:258–267.

Liehr, T., Claussen, U., and Starke, H., 2004. Small supernumerary marker chromosomes (sSMC) in humans. Cytogenet Genome Res 107:55–67.

Lo, A. W., Craig, J. M., Saffery, R., Kalitsis, P., Irvine, D. V., Earle, E., Magliano, D. J., and Choo, K. H., 2001a. A 330 kb CENP-A binding domain and altered replication timing at a human neocentromere. EMBO J 20:2087–96.

Lo, A. W., Magliano, D. J., Sibson, M. C., Kalitsis, P., Craig, J. M., and Choo, K. H., 2001b. A novel chromatin immunoprecipitation and array (CIA) analysis identifies a 460-kb CENP-A-binding neocentromere DNA. Genome Res 11:448–57.

Maeshima, K. and Laemmli, U. K., 2003. A two-step scaffolding model for mitotic chromosome assembly. Dev Cell 4:467–480.

Maggert, K. A. and Karpen, G. H., 2001. The activation of a neocentromere in Drosophila requires proximity to an endogenous centromere. Genetics 158:1615–1628.

Marshall, O. J., Chueh, A. C., Wong, L. H., and Choo, K. H. A., 2008. Neocentromeres: new insights into centromere structure, disease development, and karyotype evolution. Am J Hum Genet 82:261–282.

Masumoto, H., Ikeno, M., Nakano, M., Okazaki, T., Grimes, B., Cooke, H., and Suzuki, N., 1998. Assay of centromere function using a human artificial chromosome. Chromosoma 107:406–416.

Montefalcone, G., Tempesta, S., Rocchi, M., and Archidiacono, N., 1999. Centromere repositioning. Genome Res 9:1184–1188.

Moreno-Moreno, O., Torras-Llort, M., and Azorín, F., 2006. Proteolysis restricts localization of CID, the centromere-specific histone H3 variant of Drosophila, to centromeres. Nucleic Acids Res 34:6247–6255.

Morrissette, J. D., Celle, L., Owens, N. L., Shields, C. L., Zackai, E. H., and Spinner, N. B., 2001. Boy with bilateral retinoblastoma due to an unusual ring chromosome 13 with activation of a latent centromere. Am J Med Genet 99:21–28.

Nagaki, K., Cheng, Z., Ouyang, S., Talbert, P. B., Kim, M., Jones, K. M., Henikoff, S., Buell, C. R., and Jiang, J., 2004. Sequencing of a rice centromere uncovers active genes. Nat Genet 36:138–145.

Nasuda, S., Hudakova, S., Schubert, I., Houben, A., and Endo, T. R., 2005. Stable barley chromosomes without centromeric repeats. Proc Natl Acad Sci USA 102:9842–9847.

Nonaka, N., Kitajima, T., Yokobayashi, S., Xiao, G., Yamamoto, M., Grewal, S. I. S., and Watanabe, Y., 2002. Recruitment of cohesin to heterochromatic regions by Swi6/HP1 in fission yeast. Nat Cell Biol 4:89–93.

Ohzeki, J., Nakano, M., Okada, T., and Masumoto, H., 2002. CENP-B box is required for *de novo* centromere chromatin assembly on human alphoid DNA. J Cell Biol 159:765–775.

Okamoto, Y., Nakano, M., Ohzeki, J.-I., Larionov, V., and Masumoto, H., 2007. A minimal CENP-A core is required for nucleation and maintenance of a functional human centromere. EMBO J 26:1279–1291.

Paulson, J. R. and Laemmli, U. K., 1977. The structure of histone-depleted metaphase chromosomes. Cell 12:817–828.

Pedeutour, F., Forus, A., Coindre, J. M., Berner, J. M., Nicolo, G., Michiels, J. F., Terrier, P., Ranchere-Vince, D., Collin, F., Myklebost, O., and Turc-Carel, C., 1999. Structure of the supernumerary ring and giant rod chromosomes in adipose tissue tumors. Genes Chromosomes Cancer 24:30–41.

Platero, J. S., Ahmad, K., and Henikoff, S., 1999. A distal heterochromatic block displays centromeric activity when detached from a natural centromere. Mol Cell 4:995–1004.

Rhoades, M. M. and Vilkomerson, H., 1942. On the Anaphase Movement of Chromosomes. Proc Natl Acad Sci USA 28:433–436.

Rivera, H., Vassquez, A. I., Ayala-Madrigal, M. L., Ramirez-Dueñas, M. L., and Davalos, I. P., 1996. Alphoidless centromere of a familial unstable inverted Y chromosome. Ann Genet 39:236–239.

Saffery, R., Irvine, D. V., Grifiths, B., Kalitsis, P., Wordeman, L., and Choo, K. H., 2000. Human centromeres and neocentromeres show identical distribution patterns of >20 functionally important kinetochore-associated proteins. Hum Mol Genet 9:175–185.

Saffery, R., Sumer, H., Hassan, S., Wong, L. H., Craig, J. M., Todokoro, K., Anderson, M., Stafford, A., and Choo, K. H. A., 2003. Transcription within a functional human centromere. Mol Cell 12:509–516.

Schuh, M., Lehner, C. F., and Heidmann, S., 2007. Incorporation of Drosophila CID/CENP-A and CENP-C into centromeres during early embryonic anaphase. Curr Biol 17:237–243.

Sirvent, N., Forus, A., Lescaut, W., Burel, F., Benzaken, S., Chazal, M., Bourgeon, A., Vermeesch, J. R., Myklebost, O., Turc-Carel, C., Ayraud, N., Coindre, J. M., and Pedeutour, F., 2000. Characterization of centromere alterations in liposarcomas. Genes Chromosomes Cancer 29:117–129.

Strissel, P. L., Espinosa, R., Rowley, J. D., and Swift, H., 1996. Scaffold attachment regions in centromere-associated DNA. Chromosoma 105:122–133.

Sullivan, B. A. and Willard, H. F., 1998. Stable dicentric X chromosomes with two functional centromeres. Nat Genet 20:227–228.

Sumer, H., Craig, J. M., Sibson, M., and Choo, K. H. A., 2003. A rapid method of genomic array analysis of scaffold/matrix attachment regions (S/MARs) identifies a 2.5-Mb region of enhanced scaffold/matrix attachment at a human neocentromere. Genome Res 13:1737–1743.

Sumer, H., Saffery, R., Wong, N., Craig, J. M., and Choo, K. H. A., 2004. Effects of scaffold/matrix alteration on centromeric function and gene expression. J Biol Chem 279:37631–37639.

Szpirer, C., Rivière, M., VanVooren, P., Moisan, M.-P., Haller, O., and Szpirer, J., 2005. Chromosome evolution of MMU16 and RNO11: conserved synteny associated with gene order rearrangements explicable by intrachromosomal recombinations and neocentromere emergence. Cytogenet Genome Res 108:322–327.

Tavormina, P. A., Côme, M.-G., Hudson, J. R., Mo, Y.-Y., Beck, W. T., and Gorbsky, G. J., 2002. Rapid exchange of mammalian topoisomerase II alpha at kinetochores and chromosome arms in mitosis. J Cell Biol 158:23–29.

Tomonaga, T., Matsushita, K., Ishibashi, M., Nezu, M., Shimada, H., Ochiai, T., Yoda, K., and Nomura, F., 2005. Centromere protein H is up-regulated in primary human colorectal cancer and its overexpression induces aneuploidy. Cancer Res 65:4683–4689.

Tomonaga, T., Matsushita, K., Yamaguchi, S., Oohashi, T., Shimada, H., Ochiai, T., Yoda, K., and Nomura, F., 2003. Overexpression and mistargeting of centromere protein-A in human primary colorectal cancer. Cancer Res 63:3511–3516.

Tonnies, H., Gerlach, A., Heineking, B., Starke, H., Neitzel, H., and Neumann, L. M., 2006. Molecular cytogenetic identification and characterization of a *de novo* supernumerary neocentromeric derivative chromosome 13. Cytogenet Genome Res 114:325–329.

Tyler-Smith, C., Gimelli, G., Giglio, S., Floridia, G., Pandya, A., Terzoli, G., Warburton, P. E., Earnshaw, W. C., and Zuffardi, O., 1999. Transmission of a fully functional human neocentromere through three generations. Am J Hum Genet 64:1440–1444.

Van Hooser, A. A., Ouspenski, I. I., Gregson, H. C., Starr, D. A., Yen, T. J., Goldberg, M. L., Yokomori, K., Earnshaw, W. C., Sullivan, K. F., and Brinkley, B. R., 2001. Specification of kinetochore-forming chromatin by the histone H3 variant CENP-A. J Cell Sci 114:3529–42.

Ventura, M., Antonacci, F., Cardone, M. F., Stanyon, R., D'Addabbo, P., Cellamare, A., Sprague, L. J., Eichler, E. E., Archidiacono, N., and Rocchi, M., 2007. Evolutionary formation of new centromeres in macaque. Science 316:243–246.

Ventura, M., Archidiacono, N., and Rocchi, M., 2001. Centromere emergence in evolution. Genome Res 11:595–599.

Ventura, M., Mudge, J. M., Palumbo, V., Burn, S., Blennow, E., Pierluigi, M., Giorda, R., Zuffardi, O., Archidiacono, N., Jackson, M. S., and Rocchi, M., 2003. Neocentromeres in

15q24-26 map to duplicons which flanked an ancestral centromere in 15q25. Genome Res 13:2059–2068.

Ventura, M., Weigl, S., Carbone, L., Cardone, M. F., Misceo, D., Teti, M., D'Addabbo, P., Wandall, A., Björck, E., de Jong, P. J., She, X., Eichler, E. E., Archidiacono, N., and Rocchi, M., 2004. Recurrent sites for new centromere seeding. Genome Res 14:1696–1703.

Voullaire, L., Saffery, R., Earle, E., Irvine, D. V., Slater, H., Dale, S., du Sart, D., Fleming, T., and Choo, K. H., 2001. Mosaic inv dup(8p) marker chromosome with stable neocentromere suggests neocentromerization is a post-zygotic event. Am J Med Genet 102:86–94.

Voullaire, L. E., Slater, H. R., Petrovic, V., and Choo, K. H., 1993. A functional marker centromere with no detectable alpha-satellite, satellite III, or CENP-B protein: activation of a latent centromere? Am J Hum Genet 52:1153–1163.

Wandall, A., Tranebjaerg, L., and Tommerup, N., 1998. A neocentromere on human chromosome 3 without detectable alpha-satellite DNA forms morphologically normal kinetochores. Chromosoma 107:359–65.

Warburton, P. C., Barwell, J., Splitt, M., Maxwell, D., Bint, S., and Ogilvie, C. M., 2003. Class II neocentromeres: a putative common neocentromere site in band 4q21.2. Eur J Hum Genet 11:749–753.

Warburton, P. E., 2004. Chromosomal dynamics of human neocentromere formation. Chromosome Res 12:617–626.

Warburton, P. E., Dolled, M., Mahmood, R., Alonso, A., Li, S., Naritomi, K., Tohma, T., Nagai, T., Hasegawa, T., Ohashi, H., Govaerts, L. C., Eussen, B. H., Van Hemel, J. O., Lozzio, C., Schwartz, S., Dowhanick-Morrissette, J. J., Spinner, N. B., Rivera, H., Crolla, J. A., Yu, C., and Warburton, D., 2000. Molecular cytogenetic analysis of eight inversion duplications of human chromosome 13q that each contain a neocentromere. Am J Hum Genet 66:1794–806.

Williams, B. C., Murphy, T. D., Goldberg, M. L., and Karpen, G. H., 1998. Neocentromere activity of structurally acentric mini-chromosomes in Drosophila. Nat Genet 18:30–7.

Wong, N. C., Wong, L. H., Quach, J. M., Canham, P., Craig, J. M., Song, J. Z., Clark, S. J., and Choo, K. H. A., 2006. Permissive transcriptional activity at the centromere through pockets of DNA hypomethylation. PLoS Genet 2:e17.

Yan, H., Ito, H., Nobuta, K., Ouyang, S., Jin, W., Tian, S., Lu, C., Venu, R. C., Wang, G.-L., Green, P. J., Wing, R. A., Buell, C. R., Meyers, B. C., and Jiang, J., 2006. Genomic and geneticcharacterization of rice Cen3 reveals extensive transcription and evolutionary implications of acomplex centromere. Plant Cell 18:2123–2133.

Zinkowski, R. P., Meyne, J., and Brinkley, B. R., 1991. The centromere-kinetochore complex: a repeat subunit model. J Cell Biol 113:1091–1110.

# Chapter 5
# Human Artificial Centromeres: *De novo* Assembly of Functional Centromeres on Human Artificial Chromosomes

**Hiroshi Masumoto, Teruaki Okada, and Yasuhide Okamoto**

**Abstract** The centromere is a chromosomal domain that is required for correct segregation of eukaryotic chromosomes during mitotic and meiotic cell division. CENP-A is a centromere-specific histone H3 variant highly conserved among eukaryotes, and thus has been used as a biochemical marker of centromeric chromatin. Maintenance of centromere structure and function is thought to involve epigenetic chromatin assembly mechanisms. However, the mechanism by which CENP-A is targeted to a specific region of the chromosome remains unclear. $\alpha$-satellite (alphoid) DNA is a characteristic feature of the human centromere. *De novo* assembly of functional centromeres on human alphoid DNA has been demonstrated on Human/Mammalian Artificial Chromosomes (HAC/MACs) in human HT1080 cells and mouse embryonic cells. Thus, specific genomic sequences, as well as epigenetic mechanisms, appear to be able to support centromere assembly in human and mouse cells. This study focuses on *de novo* centromere assembly on HACs as a model system for understanding assembly of the human centromere.

## 5.1 Introduction

The centromere plays several important roles during cell division and chromosome segregation (McIntosh et al. 2002; Cleveland et al. 2003; Amor et al. 2004; Maiato et al. 2004), which includes the following events and steps. (i) Centromere/kinetochore structural components including CENP-A, CENP-B, CENP-C, CENP-H, hMis6 (CENP-I), hMis12, CENP-F and CENP-K$\sim$U, as well as microtubule motor proteins (CENP-E and dynein–dynactin) and mitotic check point proteins (e.g. Mad2 and BubR1) assemble and form the kinetochore structure at the lateral side of the centromere

H. Masumoto (✉)
Lab of Cell Function and Regulation, Department of Human Genome Research, Kazusa DNA Research Institute, 2-6-7 Kazusa-kamatari, Kisarazu, Chiba 292-0818, Japan
e-mail: masumoto@kazusa.or.jp

P. De Wulf, W.C. Earnshaw (eds.), *The Kinetochore*, 
DOI 10.1007/978-0-387-69076-6_5, 

(Foltz et al. 2006; Izuta et al. 2006; Okada et al. 2006). (ii) Spindle microtubules are captured at the outer surfaces of the kinetochore, and this interaction facilitates alignment and retention of chromosomes at the metaphase plate. (iii) Sister chromatids, which are linked by the cohesin complex, resolve and separate at the metaphase/anaphase transition. (iv) The resolved chromatids move toward their respective spindle poles. (v) Each separated set of chromatids adheres and connects to the spindle pole via spindle microtubules until the nuclear envelope regenerates during telophase.

Centromere structure has been studied extensively in the yeast *Saccharomyces cerevisiae*. The yeast centromere is likely to be simpler than the centromere in higher eukaryotic cells; nevertheless, more than 65 proteins involved in centromere structure and function have been identified in *S. cerevisiae*, and many of these proteins are conserved in human cells (Kitagawa and Hieter 2001; Chan et al. 2005). In addition, nuclear pore components (Nup107-160) associate with the kinetochore throughout mitosis (Loiodice et al. 2004; Zuccolo et al. 2007). Eukaryotic centromeres include the centromere-specific histone H3 variant, CENP-A (Palmer et al.1991; Sullivan et al. 1994; Blower et al. 2002). CENP-A is co-purified with nucleosomes and in vitro assembly analyses indicate that histone H3 can be replaced with CENP-A in these nucleosome structures (Palmer et al. 1991; Yoda et al. 2000; Black et al. 2004). Loss of CENP-A blocks assembly of other many centromere components, indicating that CENP-A chromatin plays an essential role in assembling a functional kinetochore at the centromere (Howman et al. 2000; Goshima et al. 2003; Amor et al. 2004). The role of CENP-A appears to be highly conserved in multiple species (Sullivan et al. 1994; Howman et al. 2000; Sullivan 2001; Oegema et al. 2001; Van Hooser et al. 2001; Goshima et al. 2003). However, it is not yet known clearly how CENP-A chromatin itself is loaded onto the centromere. It was shown that CENP-A loading is uncoupled from DNA replication (Shelby et al. 2000). Recent analyses in yeast, fly, nematode and human cells revealed that loading of CENP-A onto the centromere is restricted to early G1 phase (Schuh et al. 2007; Jansen et al. 2007). Furthermore, loading of CENP-A requires CENP-H and -I (Okada et al. 2006), hMis16, RbAp48-like histone chaperones (Hayashi et al. 2004; Furuyama et al. 2006), Mis18$\alpha$, $\beta$, and Mis18BP (KLN-2; Fujita et al. 2007; Maddox et al. 2007), and depends on the acetylation state of histones (Fujita et al. 2007). *S. cerevisiae* Cse4, a homologue of CENP-A, associates with the Scm3 protein in the nucleosome instead of histone H2A and H2B (Stoler et al. 2007; Camahort et al. 2007; Mizuguchi et al. 2007). Overexpression of CENP-A resulted in mistargeting of CENP-A and misassembly of CENP-C at ectopic loci without accompanying centromere function in mammalian cells (Van Hooser et al. 2001). Interestingly, CENP-A nucleosomes alone do not assemble preferentially on human centromeric alphoid DNA in vitro (Black et al. 2007; Conde e Silva et al. 2007). These results are consistent with the hypothesis that assembly of CENP-A chromatin on DNA is not on its own sufficient to drive normal centromere assembly in vitro or in vivo.

Specific centromeric DNA sequences have the capacity to direct centromere assembly more efficiently than DNA from any other part of the genome. This was first shown by the demonstration that naked yeast centromeric DNA drives *de novo* assembly of a functional centromere in *S. cerevisiae* or *Schizosaccharomyces pombe* (Fitzgerald-Hayes et al. 1982; Hahnenberger et al. 1989). In human cells, centromeric $\alpha$-satellite (alphoid) DNA also appears to play an important role in maintaining truncated minichromosomes (Heller et al. 1996; Mills et al. 1999; Spence et al. 2002) and establishing functional human artificial chromosomes (HACs; Harrington et al. 1997; Ikeno et al. 1998; Ebersole et al. 2000 2005; Mejia et al. 2001; Grimes et al. 2001). The exact mechanism by which $\alpha$-satellite DNA sequences promote centromere assembly is not known. It is important to note that centromere assembly is modulated in a species-specific manner, which depends on specific centromeric DNA sequences.

Many questions concerning centromere assembly in eukaryotic cells still need to be answered. For example: What factors direct centromere-specific assembly of CENP-A chromatin? What factors regulate the timing and location of centromere/kinetochore assembly? What role does primary DNA sequence play in centromere assembly? In addition, centromeres in many species associate with pericentromeric heterochromatin, which is maintained by epigenetic mechanisms (Grewal and Moazed 2003; Lachner et al. 2003). What role does the pericentromeric heterochromatin play in centromere formation? It has not been easy to address these questions using native mammalian chromosomes, because of their large size, enrichment for repetitive DNA and involvement of epigenetic phenomena. Recent studies have exploited artificial chromosomes with defined DNA sequences to study the human centromere structure and function. This review reports on recent progress in dissecting the functional human centromere using HACs. We suggest that the centromere activity of centromeric satellite DNA is determined by a dynamic balance between the nucleation, spreading and maintenance of CENP-A chromatin, and the opposing processes of nucleation, spreading and maintenance of heterochromatin.

## 5.2 Role of Repetitive Centromeric DNA in Kinetochore Assembly

Human centromeres contain tandem repeats of an AT-rich 171 bp DNA sequence known as $\alpha$-satellite DNA or alphoid DNA. Human alphoid DNA arrays range in size from 200 kb to 5 Mb, some of which are organized into chromosome-specific higher order repeats (Willard and Waye 1987). The repetitive structure of alphoid DNA can be classified into 2 types of repeats (Ikeno et al. 1994, Alexandrov et al. 2001). Type-I alphoid DNA includes several monomer repeats, which are organized into higher order chromosome-specific

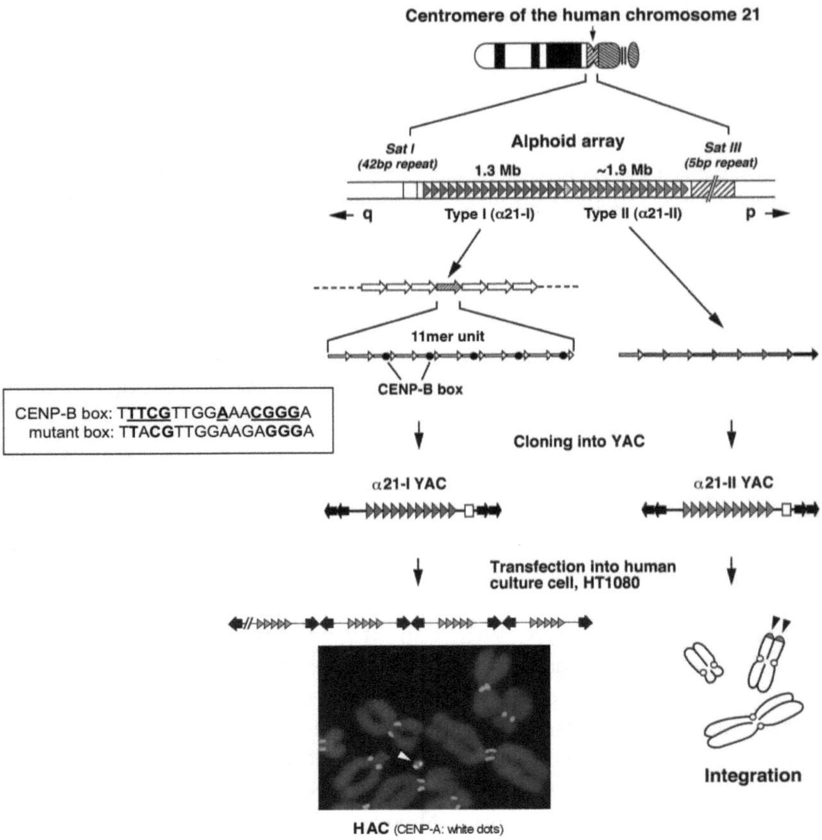


**Fig. 5.1** Schematic representation of the centromere of human chromosome 21. The centromere of human chromosome 21 includes a 1.3 Mb type I alphoid array ($\alpha$21-I) and a 1.9 Mb type II alphoid DNA array ($\alpha$21-II) The type I array is composed of 11mer repeats with five (centromere protein) CENP-B boxes and the type II array, located toward the short arm of chromosome 21, is composed of divergent alphoid repeats, lacks CENP-B boxes and does not have a higher order repeat structure (Trowell et al. 1993; Ikeno et al. 1994). Type-I alphoid repeats ($\alpha$21-I) are associated with both the existence of functional centromere components on the endogenous human chromosome 21 and with the capacity for *de novo* centromere assembly in cultured cells by concatemer formation with the input linear DNA (Ikeno et al. 1998; Okamoto et al. 2007)

repeating units (e.g. $\alpha$21-I alphoid DNA on chromosome 21; Fig. 5.1). In contrast, type II alphoid DNA consists of diverged alphoid monomers with no higher order repeat structure (e.g. $\alpha$21-II on chromosome 21; Fig. 5.1, Trowell et al. 1993; Ikeno et al. 1994; Alexandrov et al. 2001). CENP-B binds to the CENP-B box, which is found in type-I alphoid DNA on human autosomes and the X chromosomes but not in type-II alphoid DNA or the Y chromosome alphoid DNA (Earnshaw et al. 1987; Masumoto et al. 1989, Alexandrov et al. 2001). Thus, CENP-B exists on all the centromeres of these chromosomes excepting the Y chromosome (Earnshaw et al. 1989). Immunocytological studies on human chromosomes 7 and 21 show that type-I alphoid DNA co-localizes with antigenic sites for anti-centromere antiserum (ACA), which reacts with centromere/kinetochore components including CENP-A, -B and -C (Earnshaw and Rothfield 1985; Haaf and Ward 1994; Ikeno et al. 1994). These components are co-immunoprecipitated with each other and with type-I alphoid DNA when exposed to specific antibodies against CENP-A or CENP-C (Ando et al. 2002; Politi et al. 2002; Suzuki et al. 2004). However, the presence of type I alphoid DNA is not sufficient to generate a functional centromere (see below).

## 5.3 Epigenetic Mechanisms in Forming a Functional Centromere

Several lines of evidence suggest that epigenetic mechanisms play an important role in establishing and maintaining centromeric structure. For example, stable dicentric human chromosomes, which are generated by chromosome rearrangement, have one active and one inactive centromere, despite the presence of two regions containing centromere-specific alphoid DNA (Earnshaw and Migeon 1985; Earnshaw et al. 1989; Sullivan and Schwartz 1995, Sullivan et al. 2001). In addition, in rare cases, a functional centromere assembles on a chromosome fragment that completely lacks centromeric alphoid DNA (Amor and Choo 2002; Warburton 2004) and is referred to as a neocentromere. CENP-A is present at active centromeres and neocentromeres, but not at the inactive centromere on a dicentric chromosome (Warburton et al. 1997). Fine DNA sequence mapping of neocentromeres at chromosome 10q25 and13q32 and 13q21 using chromatin immunoprecipitation (ChIP) on CHIP analysis revealed no DNA sequence similarity between the neocentromere sites (Lo et al. 2001; Alonso et al. 2003, 2007; Chueh et al. 2005). This suggests that a functional centromere can be assembled and maintained without centromere-specific DNA sequences, and that this may involve an epigenetic mechanism. However, neocentromeres and inactive centromeres on dicentric chromosomes form on rearranged or fragmented chromosomes, which may lead to relocation of CENP-A chromatin (Sullivan 2001; Sullivan et al. 2001). Assembly of neocentromeres at ectopic sites on rearranged chromosomal fragments presumably causes chromosome instability. Rearranged chromosomal fragments carrying neocentromeres at ectopic sites can be highly unstable, generating additional acentric and dicentric chromosomes during meiotic recombination. Furthermore, assembly of CENP-A chromatin at ectopic sites also may affect the transcriptional control of genes which are normally regulated by histone modifications. Therefore, formation of neocentromeres may be restricted to specific areas of the genome, and stable inheritance of chromosomes with neocentromeres presumably requires suppression of meiotic recombination or mating between congeners. Stable inheritance of a neocentromere on the Y chromosome has been reported (Tyler-Smith et al. 1999). This may reflect the fact that meiotic recombination between X and Y chromosomes occurs in pseudoautosomal regions near the end of the X and Y, but is lacking in the remainder of the Y chromosome, which is composed primarily of satellite DNAs. Recently, a second example of an inherited neocentromere was reported (Amor et al. 2004).

The DNA sequence of the centromere is more divergent and less well conserved than the protein components of the centromere. Therefore, it is unclear why centromeric DNA sequences are conserved in a species-specific manner (e.g. human, mouse, rice, maize, and budding and fission yeast chromosomes have species-specific centromeric DNA sequences or structures; Henikoff et al. 2001; Sullivan et al. 2001; Kitagawa and Hieter 2001; Cleveland et al. 2003; Mellone and Allshire 2003; Nagaki et al. 2004). One possible explanation for this phenomenon is that epigenetic mechanisms regulate assembly of CENP-A chromatin, and these mechanisms may be tightly associated with species-specific centromeric DNA (Masumoto et al. 2004; Ventura et al. 2006). In other words, the epigenetic mechanisms that regulate centromere function may be implicitly linked to a specific chromosomal position and/or a specific DNA sequence in a species-specific manner. Furthermore, it is possible that the development of this mechanism may be correlated with the process of speciation.

## 5.4 Human Artificial Chromosomes

### *5.4.1 Role of Alphoid DNA in* De Novo *Assembly of Centromeres on Human Artificial Chromosomes*

The development and use of HACs to study human chromosome structure has led to important insights into the structure and function of centromeres in human chromosomes. Harrington et al. (1997) first isolated stable HACs in human HT1080 cells which had been co-transfected with type I alphoid DNA, cloned telomere repeats, human genomic DNA fragments and a selectable marker. Ikeno et al. (1998) also demonstrated that DNA molecules containing only alphoid DNA and YAC vector sequences can form centromeres with high efficiency. For example, a linear YAC clone containing 70 kb $\alpha$21-I alphoid DNA and terminal human telomere sequences formed a stable single copy HAC in cultured cells without acquisition of detectable host sequences, bound CENP-A, -B, -C and -E, and segregated with an efficiency of 98.4–99.5% per cell division in the absence of selection, (Fig. 5.1; Ikeno et al. 1998; Masumoto et al. 1998). Physical analyses including pulse-field gel electrophoresis, Southern blot and real-time PCR suggest that these HACs include 30–50 concatenated copies of YAC DNA and are ~3 Mb in size (Ikeno et al. 1998; Masumoto et al. 1998; Nakano et al. 2003;

Fig. 5.1). HACs also form efficiently from circular constructs containing type I alphoid DNA (Ebersole et al. 2000) indicating that telomeric DNA sequences are not required for the stable transmission of circular HACs. Circular constructs frequently recombine during chromosome formation, leading to variable size and variable amounts of alphoid DNA on the HACs generated.

Assembly and stability of these HACs required the presence of centromere-specific segments of type I alphoid DNA (e.g. DNA from the centromere of chromosomes 5, 17, 21, or X, or a synthetic repeat derived from type I alphoid DNA) and the efficiency of HAC formation varied from 9 to 100%, depending on the specific alphoid DNA and vector constructs (Laner et al. 2004; Kaname et al. 2005; Ebersole et al. 2005; Moralli et al. 2006). In contrast, type II alphoid DNA, Y-chromosome alphoid DNA, (Harrington et al. 1997; Ikeno et al. 1998; Grimes et al. 2002; Mejia et al. 2002; Kouprina et al. 2003) and DNA from a neocentromere at 10q25 (Saffery et al. 2001) did not support *de novo* centromere formation under any conditions tested to date. This may reflect the fact that CENP-B binding sites (CENP-B boxes) are present in type I alphoid DNA, but are lacking in type II alphoid DNA, Y alphoid DNA and in the 10q25 neocentromere sequence. Other substrates that lack the ability to form HACs include synthetic type-I alphoid DNA with mutant CENP-B boxes and GC-rich synthetic repetitive DNA containing CENP-B boxes (Ohzeki et al. 2002). In summary, these data indicate that DNA fragments containing type I alphoid repeats with CENP-B boxes support *de novo* assembly of functional centromeres on HACs in human cells. Additional factors or requirements that promote *de novo* centromere assembly remain to be determined.

### 5.4.2 Functional Centromeres in Stable HACs

HACs are mitotically stable and have a functional centromere that includes CENP-A, -B, -C and -E and Mad2 proteins (Ikeno et al. 1998; Ohzeki et al. 2002; Tsuduki et al. 2006). HACs also bind heterochromatin proteins HP1$\alpha$ and Aurora B and have modifications of histone H3 such as heterochromatin-associated trimethylation of lysine 9 (H3K9me3; Grimes et al. 2004; Nakashima et al. 2005; Lam et al. 2006). Recently, the GFP-Lac repressor fusion protein (GFP-LacR) / Lac operator system (Robinett et al. 1996) has been used to follow HAC behaviour during mitosis. These studies demonstrate that HACs composed of multimerized alphoid DNA and YAC or BAC vector arms segregate properly during mitosis (Tsuduki et al. 2006). Real-time imaging showed that GFP-tagged HACs align accurately at the spindle midzone in HT1080 cells and undergo frequent oscillations around the spindle midzone. One of the major functions of the kinetochore is to generate and control the dynamic tension between spindle microtubules and the kinetochores at each lateral side of the centromere (Nicklas 1997; McIntosh et al. 2002; Cleveland et al. 2003; Maiato et al. 2004; Salmon et al. 2005; Ruchaud et al. 2007). Interestingly, HACs appear to undergo more frequent and faster oscillations than native human chromosomes during prometaphase to metaphase. The frequent oscillations of the HAC around the spindle midzone are consistent with an increased dynamics of polymerization and depolymerization at microtubule plus ends of the HAC kinetochores (Tsuduki et al. 2006). These results suggest that the kinetochore and inner centromere on these HACs function normally in HT1080 cells. HACs may demonstrate more lateral (pole-to-pole) and longitudinal movement during prometaphase and metaphase, because they are smaller than native human chromosomes.

HACs progress through mitosis with the same timing as native chromosomes, indicating normal control of sister chromatid cohesion and resolution. In addition, the observation that all HAC sister chromatids (whether derived from the YAC-based linear or BAC-based circular constructs) hold together in mitotic cells arrested by treatment with an inhibitor of microtubule assembly, also strongly supports the conclusion that sister chromatid cohesion is controlled by mitotic checkpoints and is coupled to assembly of a functional HAC centromere (our unpublished observations). The motor activity of centromeres on HACs and on native human chromosomes is similar: average velocity $\approx 0.55$ μm/min and maximum velocity 0.71–1.79 μm/min for HACs vs. average velocity $\approx 0.56$ μm/min and maximum velocity 0.50–1.32 μm/min for native human chromosomes during anaphase. The obvious short length of HAC arms does not appear to influence their behaviour during mitosis, including telophase. The role played by heterochromatin (trimethyl-H3K9 and HP1$\alpha$) in regulating cohesion of HAC sister chromatids is discussed below.

These observations indicate that HACs form and maintain normal functional centromeres, which allow HACs to be functionally similar to native human chromosomes during all stages of the mitotic cell cycle, even in the absence of selective conditions. Nevertheless, a small fraction of HACs are or become unstable (loss rate of 0.1–1.6%). HAC instability, including nondisjunction and lagging, increased up to 10% when cells were treated with nocodazole and dihydrocytochalasin B, inhibitors of microtubule polymerization and cytokinesis, respectively (Rudd et al. 2003). However, elevated segregation error rates were detected similarly with naturally occurring small marker chromosomes, suggesting that smaller chromosomes may be particularly susceptible if the cellular and chromatin environment is drastically changed by such treatments. Indeed, even native chromosomes displayed a 15-fold increase in rate of merotelic kinetochores (17%) after treatment with an inhibitor of microtubule polymerization (Cimini et al. 2003), and mechanisms exist to correct improper kinetochore–microtubule attachments under normal conditions (Salmon et al. 2005; DeLuca et al. 2006).

### *5.4.3 The Role of CENP-A Chromatin in Establishing and Maintaining a Functional Human Centromere*

As mentioned above, the availability of stable HACs has made it possible to systematically investigate the structure and function of the human centromere

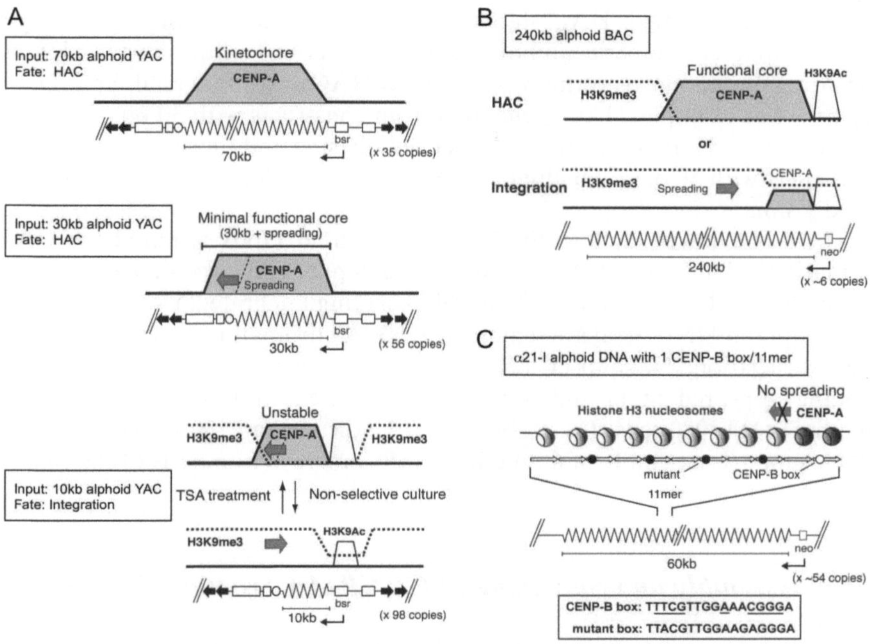


**Fig. 5.2** Hypothetical models for assembly and spreading of CENP-A chromatin. (**A**) Chromatin assembly on a human artificial chromosome (HAC) generated from 70 kb or 30 kb a21-I alphoid arrays and at an ectopic site following transfection of a 10 kb a21-I alphoid array (Okamoto et al. 2007). CENP-A chromatin (*solid thick line and gray area*) preferentially assembled on the 70 kb alphoid DNA array. However, CENP-A chromatin spread (*gray arrow*) was maintained on the YAC vector arm on HACs formed from the 30 kb a21-I construct. The multimer of the integrated 10 kb a21-I alphoid YAC was enriched for H3K9me3 heterochromatin (*dotted line*). Treatment with TSA allowed formation of unstable CENP-A chromatin, which exists in dynamic balance with spreading H3K9me3 heterochromatin in the absence of selection. (**B**) Chromatin assembly on a HAC or at an integration site following transfection of a 240 kb a21-I alphoid DNA array. The 240 kb alphoid BAC can exist as an episomal HAC with a functional centromere, but undergoes heterochromatization as at the ectopic integration site. (**C**) CENP-A nucleosomes can assemble on alphoid DNA integration site containing only one CENP-B box per 11mer a21-I higher order repeat unit (60 kb alphoid insert). However such a CENP-A nucleosomes cannot spread and do not generate a functional centromere core

in cultured cells. For example, the influence of the length of alphoid DNA on HAC formation and stability has been studied in HT1080 cells (Fig. 5.2). These experiments demonstrated that a minimum of 30 kb type I alphoid DNA is required to form a stable HAC, resulting in $\approx 9\%$ efficiency of forming a stable HAC (Okamoto et al. 2007). This efficiency improves to $\approx 35\%$ for constructs with 50–70 kb type I alphoid DNA, but then falls to $\approx 5\%$ when the array size is increased to 240 kb. Grimes et al (2002) reported similar results using circular constructs with 35 or 80 kb isogenic alphoid DNA from chromosome 17. In contrast, once formed, HAC stability is not strictly dependent on the length of the input centromeric alphoid DNA, and HACs with shorter alphoid DNA regions can be as or more stable than HACs with longer alphoid DNA regions. For example, HACs generated from constructs with 30 kb of alphoid DNA exhibited a loss rate of 0.01% per cell division and those generated from constructs with 50 or 70 kb alphoid DNA inserts had loss rates of 0.49–0.86% per division.

There is a lower limit to the length of alphoid DNA required for HAC formation. HACs did not form when the input YAC vector carried only 10 kb alphoid DNA. Ten to 30% of cells carrying integrated YACs with 30 to 70 kb alphoid DNA arrays assembled CENP-A chromatin to a varying degree (i.e., variegation, Nakano et al 2003; Okamoto et al. 2007). This CENP-A binding appeared to be independent of the integration site or YAC copy number in HT1080 cells. These observations indicate that $\geq$30 kb type I alphoid DNA supports assembly of a functional centromere in human HT1080 cells.

### *5.4.4 Assembly and Spreading of CENP-A Chromatin in HACs*

A number of studies indicate that CENP-A chromatin nucleates on type-I alphoid DNA and spreads into the vector DNA to cover a total of approximately 30 kb, and that this size may represent a threshold for establishing a stable CENP-A chromatin domain and a functional centromere core unit in a HAC. CENP-A chromatin can spread from alphoid DNA into flanking YAC/BAC vector DNA on HACs (Lam et al. 2006; Okamoto et al. 2007). However, YAC/BAC vectors with either no insert, type II alphoid DNA or type I alphoid DNA with mutated CENP-B boxes, do not form HACs or assemble CENP-A chromatin at integration sites in human cells (Ikeno et al. 1998; Ohzeki et al. 2002; Nakano et al. 2003). In our experience, spreading of CENP-A chromatin onto adjacent vector sequences is only observed on HACs generated from vectors with $\leq$ 30 kb of alphoid DNA (Fig. 5.2A). This spreading of CENP-A extends only partially (several kb) into the 26 kb vector arm. This result also clearly demonstrates that CENP-A chromatin can spread onto flanking vector DNA, if the length of alphoid DNA is below 30 kb and suggests that under these conditions, CENP-A spreading onto vector DNA is required to support centromere function. Such extended CENP-A chromatin domains are presumably maintained by epigenetic mechanisms, which are likely to be mediated by the CENP-H –CENP-I complex (Okada et al. 2006) and Mis18$\alpha$, $\beta$ and Mis18BP1 (Fujita et al. 2007). It is worth noting that the reported size of CENP-A chromatin domains on neocentromeres in a hybrid hamster background (10.5–51.8 kb; Chueh et al. 2005), in native mouse centromeres (10–20 kb; Greaves et al. 2007) and in *Drosophila* centromeres (15–40 kb; Blower et al. 2002) are similar to the $\approx$ 30 kb minimal CENP-A core seen on HACs.

Centromere spreading has also been observed in *Drosophila* minichromosomes formed by gamma irradiation (Maggert and Karpen 2007). This study suggested that centromeric chromatin spreads into flanking chromosomal sequences after removal of heterochromatin by an inversion within a centromere domain. It has also been proposed that centromere expansion/evolution may involve insertion of multiple transposons which disrupt a functional centromere in or near centromeric domains followed by centromere shifting (Csink and Henikoff 1998; Laurent et al.1999; Henikoff et al. 2001). The results from HAC analysis support these hypothetical centromere spreading phenomena especially in HACs with insufficient alphoid DNA to form a functional contiguous centromere core unit. However, it should be noted that excess CENP-A binds non-specifically to entire chromosome arms but does not form functional centromeric chromatin in cells engineered to overexpress this protein (Sullivan et al. 1994; Van Hooser et al. 2001). Thus, spreading of CENP-A centromere chromatin is a specialized mechanism that supports functional centromere assembly.

### *5.4.5 H3K9me3 Chromatin Formation Inhibits CENP-A Deposition on Transfected 10 kb Alphoid Arrays*

The formation of CENP-A chromatin appears to be balanced between the opposing processes of heterochromatin formation and histone acetylation. CENP-A chromatin does not assemble at ectopic integration sites of YAC/BACs with 10 kb arrays of type I alphoid DNA. However, a high level of trimethyl histone H3-K9 (H3K9me3) has been observed at such sites (Nakano et al. 2003; Okamoto et al. 2007). This suggests that enrichment for H3K9me3, which is characteristic of heterochromatin, may inhibit and/or preclude formation of CENP-A chromatin in introduced constructs that have 10 kb alphoid DNA (Fig. 5.2A). This idea is consistent with the observation that treatment of cells with trichostatin A (TSA), which inhibits histone deacetylation, breaks down pericentromeric heterochromatin in mouse and *S. pombe* cells and leads to centromere inactivation (Ekwall et al. 1997; Taddei et al. 2001) or reactivation in human isodicentric chromosomes (Higgins et al. 2005). Indeed, TSA-treatment can generate a functional centromere with ectopic CENP-A chromatin reassembly at sites of integrated 30–70 kb alphoid DNA arrays and thus increases the formation of fragmented chromosomes via chromosome breakage events (‘reformed’ minichromosomes; Nakano et al. 2003; Okamoto et al. 2007). TSA also destabilizes or decreases formation of H3K9me3 heterochromatin and stimulates binding of CENP-A and CENP-C to ectopic 10 kb alphoid DNA arrays. However, functional centromeres do not assemble at the ectopic sites with the 10 kb alphoid DNA arrays, and CENP-A chromatin is unstable in the absence of selection for *bsr* (Fig. 5.2A), a selectable marker on the vector arm that encodes resistance to Blasticidin S (BS). These results suggest that chromatin remodelling associated with transcriptional activation of *bsr* may have a synergistic effect on assembly of CENP-A chromatin in alphoid DNA, while H3K9me3 heterochromatin may preferentially form on shorter alphoid DNA arrays and be antagonistic to assembly or maintenance of CENP-A chromatin, as mentioned above (Fig. 5.2A). Interestingly, recent studies indicate that centromeric regions of native rice chromosome 8 and a human neocentromere are transcriptionally active and therefore must contain a transcriptionally permissive chromatin structure (Nagaki et al. 2004; Saffery et al. 2003).

### 5.4.6 CENP-B Box Density and Alphoid Length Influence Formation of CENP-A Chromatin

Large alphoid DNA arrays (>30 kb) with above a threshold density of CENP-B boxes are required for *de novo* formation of a functional centromere on a HAC (Ohzeki et al. 2002; Okamoto et al. 2007). Recent studies indicate that HACs form more efficiently on synthetic alphoid DNA containing CENP-B boxes in every alphoid monomer (Basu et al. 2005) and that the presence of CENP-B boxes correlates with promoting CENP-A binding in native human centromeres (Irvine et al. 2004). CENP-A chromatin assembled inefficiently on transfected DNA containing one CENP-B box per 11mer (32 CENP-B boxes per 60 kb of alphoid DNA), and $\alpha$-satellite DNA with this density of CENP-B boxes fails to form a functional centromere and a HAC (Fig. 5.2C, Okamoto et al. 2007). Although CENP-A chromatin may nucleate at a low level, it fails to spread across alphoid DNA lacking a sufficient density of CENP-B boxes or carrying mutant CENP-B boxes (Fig. 5.2C). These results indicate that a minimum density of CENP-B boxes is required to establish and maintain a CENP-A chromatin core in alphoid DNA. Nevertheless, short alphoid DNA arrays including a higher density of 25–27 CENP-B boxes in 10 kb also did not maintain CENP-A chromatin (Okamoto et al. 2007). These results suggest that both the overall length of alphoid DNA and the density of CENP-B boxes influence the efficiency of assembly and disassembly of a functional centromere core. A high density of CENP-B boxes is also conserved in mouse centromeric minor satellite DNA (Masumoto et al. 1989; Kipling et al. 1995).

### 5.4.7 A Dynamic Balance Between CENP-A Chromatin and Heterochromatin in Alphoid DNA

Recent studies show that constructs containing 120 or 240 kb arrays of alphoid DNA do not exhibit an increased efficiency for the formation of functional centromeres on HACs (Okamoto et al. 2007). For example, CENP-A chromatin is not stably maintained when constructs with 240 kb of alphoid DNA integrate at ectopic sites in human chromosomes. However, these long alphoid

DNA arrays assemble H3K9me3 heterochromatic modifications with high efficiency (Fig. 5.2B). This observation and other studies with shorter alphoid arrays suggest that there may be a dynamic balance between the nucleation, spreading and maintenance of CENP-A chromatin and nucleation, spreading and maintenance of heterochromatin. Histone acetylation and chromatin unfolding or selection for active transcription can alter this dynamic balance, disfavouring heterochromatization and promoting nucleation and spreading of CENP-A chromatin (Fig. 5.2).

Nevertheless, it is not clear why long alphoid DNA arrays (120 and 240 kb) support HAC formation and CENP-A assembly less efficiently than 50–70 kb alphoid DNA arrays. An interesting answer to this question arises from the observation that CENP-B may play two antagonistic roles when human alphoid DNA is introduced into mouse embryonic cells (Okada et al. 2007). CENP-B/CENP-B box interactions are required for *de novo* assembly of CENP-A chromatin and for formation of stable artificial chromosomes following transfection of human alphoid DNA into mouse cells. However, they also promote heterochromatin formation by enhancing H3K9me3 hypermodification and CpG hypermethylation in ectopic integrated human alphoid DNA in mouse chromosomes (Okada et al. 2007). These conflicting roles of CENP-B may act as a binary switch, ensuring either de novo centromere or heterochromatin assembly on contiguous centromeric satellite DNA. The latter process may prevent formation of excess centromeres, allowing only one centromere to be active on dicentric chromosomes (Earnshaw et al 1989, Sullivan and Schwartz 1995; Warburton et al. 1997) and on long regions of repetitive satellite DNA (Okamoto et al. 2007).

### 5.4.8 The Role of Vector Sequences in Heterochromatization and HAC Formation

An additional transcriptional unit on the vector arm may change the dynamic chromatin balance between HAC formation and/or hetero-chromatization. Vector sequences can also nucleate heterochromatin containing H3K9me3 and/or HP1$\alpha$ on episomal HACs and at HAC integration sites. Assembly of HP1$\alpha$ heterochromatin appears to be proportional to the size of HACs derived from circular constructs, with less heterochromatin forming on 1–3 Mb HACs than on 3–10 Mb HACs (Grimes et al. 2004). HACs derived from linear YAC vectors comprise regular concatemers of the input DNA; ChIP analysis of HACs derived from the linear 70 kb-SV HAC vector demonstrated that H3K9me3 and HP1$\alpha$ were enriched on the left arm of the vector, which includes a promoterless $\beta$*geo* gene (Fig. 5.3A and B, Nakashima et al. 2005). In contrast, the 70 kb-SV/CMV vector does not form episomal HACs. This vector is identical to 70 kb-SV, except that additional transcription of $\beta$*geo* is driven by the CMV promoter on the left vector arm (Fig. 5.3A and C). However, transcription of $\beta$*geo* per se does not inhibit assembly of a functional centromere as indicated below.

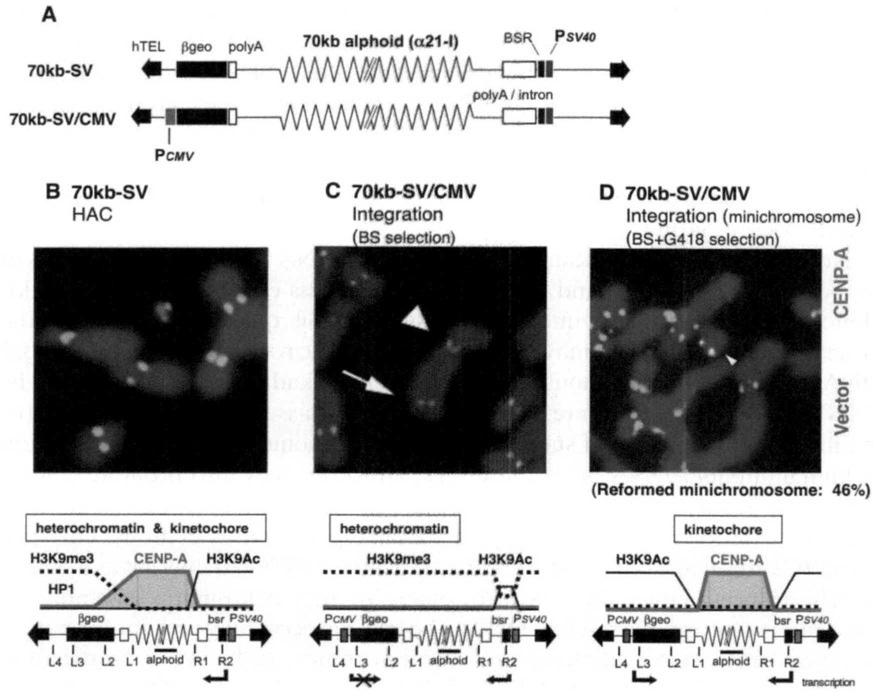


**Fig. 5.3** Heterochromatin assembly during *de novo* Human Artificial Chromosome (HAC) formation. (**A**) Schematic diagrams of 70 kb-SV and 70 kb-SV/CMV are shown (Nakashima et al. 2005). (**B**, **C**, and **D**) HT1080 cells were transfected with the indicated constructs and grown in the absence or presence of BS and/or G418, as indicated. Metaphase chromosomes were prepared and analyzed by FISH or immunocytology using probes for vector DNA (*red*) or CENP-A (*green*) to demonstrate the fate of the input DNA. In (D), the initial fate is integration (as shown in C). Subsequent growth on BS and G418 double selection results in a chromosome breakage event and formation of a minichromosome in 46% of cells. An interpretation of the chromatin structure of the transfected DNA based on ChIP analysis is shown below the chromosome images (see text for discussion)

A cell line carrying chromosomally integrated copies of the 70 kb-SV/CMV vector shown in Fig. 5.3C was cultured under conditions selecting for expression of the *bsr* gene (BS selection). The same line was subsequently cultured in the presence of BS and G418 (Fig. 5.3D), which selects for transcriptional activity in both the right and left vector arms (*bsr* and *βgeo* genes). When cells were selected only for expression of *bsr*, CENP-A and -C staining was low (at 6–26% of ectopic integration sites, Fig. 5.3C). In contrast, when cells were selected for expression of *bsr* and *βgeo*, CENP-A, -B and -C staining increased (62.2–98%, Fig. 5.3D) and episomally stable minichromosomes stably associated with CENP-A formed via chromosome fragmentation events (Fig. 5.3D). These results indicate that a functional centromere/kinetochore can assemble on the

70 kb-SV/CMV construct under conditions selecting for transcription of genes flanking the alphoid insert following the initial integration event.

On the other hand, heterochromatin structures are incompatible with the transcriptional activation in the left vector arm in 70 kb-SV/CMV by an additional promoter (Fig. 5.3C). H3K9me3 was predominantly enriched at the transcriptionally silent left arm region of the HAC derived from 70 kb-SV (Fig. 5.3B) and at the integration site in cells with 70 kb-SV/CMV cultured in the presence of BS but not G418 (Fig. 5.3C). Consistent with this, more CENP-A chromatin formed on the alphoid DNA and less H3K9me3-rich heterochromatin formed on the left vector arm when cells were cultured in the presence of BS and G418 (Fig. 5.3D). These data suggest that H3K9me3 assembled on the promoterless left vector arm (70 kb-SV) of the stable HAC and that formation of H3K9me3 heterochromatin is antagonistic to transcriptional activity and to assembly of CENP-A chromatin at integrated sites (70 kb-SV/CMV).

### *5.4.9 The Role of Heterochromatin in* De Novo *HAC Formation*

Pericentromeric heterochromatin is required for *de novo* formation and maintenance of HACs. HP1$\alpha$ and H3K9me3 are enriched at the pericentromeric regions of mitotic human and mouse chromosomes (Fig. 5.3; Hayakawa et al. 2003; Peters et al. 2003; Guenatri et al. 2004). Swi6 is an HP1 homologue in fission yeast that is enriched in pericentromeric heterochromatin and that forms a complex with the cohesin protein Rad21. Because this complex is essential for sister chromatid cohesion and chromosome segregation, Swi6 may help recruit Rad21 to pericentromeric heterochromatin (Bernard et al. 2001; Nonaka et al. 2002). In higher eukaryotes, pericentromeric heterochromatin domains and cohesion domains overlap on mitotic chromosomes and agents that disturb heterochromatin structure also interfere with chromatid cohesion (Peters et al. 2001; Valdeolmillos et al. 2004; Guenatri et al. 2004). Moreover, the cohesin complex also plays a role in recruiting the chromosomal passenger complex (CPC) to the inner centromeric region (Sonoda et al. 2001; Vagnarelli and Earnshaw 2001; Vass et al. 2003). For example, Aurora B kinase forms the CPC complex with INCENP, survivin and Borealin (Gassmann et al. 2004, Ruchaud et al. 2007). INCENP binds to pericentromeric heterochromatin in a cohesin–dependent manner early in mitosis (Ainsztein et al. 1998; Sonoda et al. 2001).

Assembly of HP1$\alpha$ and Aurora B on HACs and reformed minichromosomes correlated with the presence of a functional centromere; in contrast, these factors were not enriched on integrated alphoid DNA on metaphase chromosome arms, and these structures lacked centromere function (Fig. 5.3C). In contrast, when the same cells were selected for expression of *bsr* and *βgeo*, episomally stable minichromosomes formed via chromosome fragmentation events with CENP-A assembly on the integrated alphoid DNA (Fig. 5.3D). It is possible that acquisition of a host chromosomal fragment might compensate for a deficiency in heterochromatin/cohesion on the left vector arm of 70 kb-SV/CMV on the reformed minichromosome (Nakashima et al. 2005). Similarly, HP1$\alpha$ and HP1$\gamma$ co-purify with hMis12 (Obuse et al. 2004), a human centromere protein (Goshima et al. 2003; Kline et al. 2006), suggesting that pericentromeric heterochromatin contributes to centromere function. These results suggest that as well as the assembly of a kinetochore from naked DNA *de novo*, at least a transient acquisition of heterochromatin is necessary during the early stages of stable HAC formation; thus, introduced naked DNA needs to supply an appropriate platform both for kinetochore formation and for assembly of a heterochromatin domain (Figs. 5.2 and 5.3B). These data also suggest that multiple structural domains, including transcriptionally active chromatin, transcriptionally permissive centromeric chromatin, and transcriptionally silent heterochromatin, may compete with one another during *de novo* HAC formation, and may co-exist in a dynamic balance on stable HACs (Fig. 5.4). Heterochromatin assembly and heterochromatin spreading on vector sequences or those on alphoid DNA

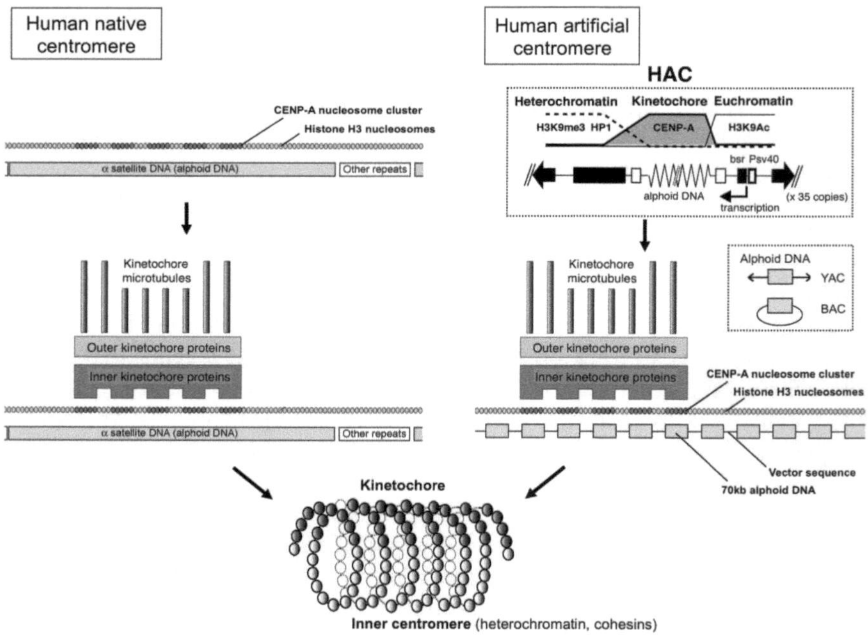


**Fig. 5.4** Models of centromeres on Human Artificial Chromosomes (HACs) and native human chromosomes. Clusters of CENP-A chromatin, acetylated histone H3 (H3K9Ac), and H3K9me3 are interspersed in functional centromeres in HACs correlating with alphoid DNA and vector DNA. CENP-A chromatin cores assemble CENP-C, -H, -I and other centromere components and may form a higher order helical structure (Zinkowski et al. 1991; Sullivan and Karpen 2004; Amor et al. 2004; Greaves et al. 2007). A higher order helical centromeric structure is consistent with the complexity and flexibility of centromeres on HACs and native chromosomes. See text for additional discussion of this model

containing CENP-B binding sites may involve different mechanisms. It is not clear at present, and thus needs to be addressed in future studies, how heterochromatin assembles and spreads on HACs derived from circular BACs with one selectable marker gene on a short vector arm.

## 5.5 Models for Centromere Structure on HACs and Native Chromosomes

Zinkowski et al. (1991) proposed that centromeres are composed of modular repeated subunits. Recent evidence lends further support to this model as the kinetochore domains of human, mouse and *Drosophila* chromosomes are characterized by interspersed clusters of H3 nucleosomes that are modified by dimethylation of lysine 4 (H3K4me2), a neutral chromatin modification, and CENP-A chromatin (Blower et al. 2002; Sullivan and Karpen 2004; Greaves et al. 2007). The idea that centromeres may contain transcriptionally permissive chromatin is further supported by the demonstration that genes can be expressed in centromere regions (Saffery et al. 2003; Nagaki et al. 2004). These observations suggest a centromere structure that is sufficiently flexible to permit formation of a kinetochore domain that assembles transcriptionally active chromatin and an inner centromere region that assembles heterochromatin. A modified version of Zinkowski's model proposes that interspersed clusters of H3K4me2 and CENP-A chromatin form a 3D helical higher order chromatin in native centromeres (Amor et al. 2004; Sullivan and Karpen 2004; Greaves et al. 2007, Fig. 5.4).

Chromatin analyses of HACs demonstrate that CENP-A, acetylated histone H3 (H3K9Ac on the right vector arm in Figs. 5.3 and 5.4) and H3K9me3 are enriched on alphoid DNA and vector DNA. Furthermore, assembly of H3K4me2 has been demonstrated on HACs (Grimes et al. 2004). Thus, the functional structure of a HAC centromere is composed of interspersed clusters of these chromatin domains(Figs. 5.2, 5.3, 5.4). For example, CENP-A clusters are detected as closely spaced double dots at the surface of the HAC centromere in conventional metaphase chromosome spreads (Figs. 5.1, 5.3), but on contiguous extended HAC chromatin fibres, 9–33% (4–12 CENP-A clusters) of alphoid YAC units exist as CENP-A chromatin dots (Supplementary Figure S4 in Okamoto et al. 2007). At the outer surface of the centromere (kinetochore), CENP-A chromatin cores containing CENP-C, -H, -I and other centromere components may be bundled into a higher order helical structure, similar to the model proposed for native centromeres (Fig. 5.4). A higher order helical centromeric structure is consistent with the complexity and flexibility of functional centromeres on HACs and native chromosomes, and would accommodate proposed features of centromeres, including the observed limit on size of CENP-A chromatin domains, the influence of length of alphoid DNA on HAC stability, the dynamic balance between CENP-A and H3K9me3 chromatin during HAC formation, and the roles of CENP-B in promoting assembly of CENP-A and/or H3K9me3 chromatin (Fig. 5.4).

## 5.6 Summary

*De novo* centromere assembly occurs on human alphoid DNA in human HT1080 and mouse embryonic cells transfected with naked YAC/BAC constructs. For stable HACs to form with high efficiency, at least 30 kb of alphoid DNA and a high density of CENP-B boxes are required (Ohzeki et al. 2002; Okamoto et al. 2007), cells must be grown under selective conditions before establishing transformants and YAC/BAC input constructs form large concatemers during HAC formation. When the density of CENP-B boxes is low or $\leq$10 kb alphoid DNA is present on the input construct, CENP-A chromatin assembles at a low level. Even if heterochromatin assembly is inhibited, a functional centromere does not form on these constructs. In addition, functional centromeres do not form on ectopically integrated alphoid DNA, regardless of the presence of CENP-B and CENP-B binding sites; instead, assembly of H3K9me3 chromatin, DNA CpG hypermethylation, and heterochromatization is favoured under these conditions, especially in the presence of CENP-B and CENP-B binding sites (Okada et al. 2007).

CENP-B is not required for chromosome segregation in CENP-B knockout mice, or on human neocentromeres, or the Y chromosome centromere (Earnshaw et al. 1989; Hudson et al. 1998; Kapoor et al. 1998; Perez-Castro et al. 1998). This is surprising, because CENP-B appears to play an essential role in promoting formation of artificial chromosomes in human and mouse cells and *de novo* CENP-A chromatin assembly on naked DNA. Nucleation and spreading of CENP-A chromatin is in dynamic balance with nucleation, spreading, assembly and disassembly of H3K9me3 heterochromatin during formation and maintenance of HACs. Assembly and maintenance of H3K9me3 heterochromatin is partly dependent on HP1, histone metyltransferase Suv39, and RNA interference (RNAi; Lachner et al. 2003; Grewal and Moazed 2003) and it may also occur via a CENP-B dependent mechanism.

The current body of evidence suggests that highly specific conditions must be met in order for stable HACs to form in cultured cells. This may explain why *de novo* centromere assembly occurs in specific cell lines and/or only with specific satellite DNA sequences. Other factors, including cellular environment, chromatin context (euchromatin, centromeric chromatin, and heterochromatin) and DNA primary sequence may also play a role in the process. Chromatin structures are maintained epigenetically and also changed dynamically. Therefore, it is experimentally challenging to analyze the properties and functions of native chromosomes and centromeres. However, the availability of HACs has begun to allow systematic analyses of human centromere structure and function. Future studies will address the flexible and dynamic nature of human chromosomes, chromatin and centromere structure and explore the mechanisms that regulate and/or modulate centromere assembly and function. Experiments in which chromatin structure is actively manipulated may yield strong insights into the flexible and dynamic mechanisms that regulate centromere function (Nakano et al. 2008). Ultimately, it is hoped that these studies will facilitate development of 'next-generation' HACs as well as vectors for gene therapy, thus contributing to development of novel methods to treat and/or prevent human disease.

**Acknowledgments** We would like to express our gratitude to Drs. Peter De Wulf and William C. Earnshaw for giving us the opportunity to write this review article. We also thank Dr. Brenda R. Grimes for critical reading. Due to space constraints, it was not possible to cite all of the excellent work in this area, and we regret omitting mention of many valuable studies in the field of human artificial chromosome technology/biology. This work was supported by Grants-In-Aid to HM for Scientific Research on Priority Areas and Special Coordination Funds for Promoting Science and Technology from the Ministry of Education, Science, Sports and Culture of Japan.

## References

Ainsztein, A.M., Kandels-Lewis, S.E., Mackay, A.M. and Earnshaw, W.C. (1998) INCENP centromere and spindle targeting: identification of essential conserved motifs and involvement of heterochromatin protein HP1. J. Cell Biol. 143, 1763–1774.

Alexandrov, I., Kazakov, A., Tumeneva, I., Shepelev, V. and Yurov, Y. (2001) Alpha-satellite DNA of primates: old and new families. Chromosoma. 110, 253–266.

Alonso, A., Mahmood, R., Li, S., Cheung, F., Yoda, K. and Warburton, P.E. (2003) Genomic microarray analysis reveals distinct locations for the CENP-A binding domains in three human chromosome 13q32 neocentromeres. Hum. Mol. Genet. 12, 2711–2721.

Alonso, A., Fritz, B., Hasson, D., Abrusan, G., Cheung, F., Yoda, K., Radlwimmer, B., Ladurner, A.G. and Warburton, P.E. (2007) Co-localization of CENP-C and CENP-H to discontinuous domains of CENP-A chromatin at human neocentromeres. Genome Biol. 8, R148

Amor, D.J. and Choo, K.H. (2002) Neocentromeres: role in human disease, evolution, and centromere study. Am. J. Hum. Genet. 71, 695–714.

Amor, D.J., Bentley, K., Ryan, J. Perry, J., Wong, L., Stater, H. and Choo, K.H. (2004) Human centromere repositioning "in progress". Proc. Natl. Acad. Sci. USA. 101, 6542–6547.

Amor, D.J., Kalitsis, P., Sumer, H. and Choo, K.H. (2004) Building the centromere: from foundation proteins to 3D organization. Trends Cell Biol. 14, 359–368.

Ando, S., Yang, H., Nozaki, N., Okazaki, T. and Yoda, K. (2002) CENP-A, -B, and -C chromatin complex that contains the I-type alpha-satellite array constitutes the prekinetochore in HeLa cells. Mol. Cell Biol. 22, 2229–2241.

Basu, J., Stromberg, G., Compitello, G., Willard, H.F. and Van Bokkelen, G. (2005) Rapid creation of BAC-based human artificial chromosome vectors by transposition with synthetic alpha-satellite arrays. Nucleic Acids Res. 33, 587–596.

Bernard, P., Maure, J.F., Partridge, J.F., Genier, S., Javerzat, J.P. and Allshire, R.C. (2001) Requirement of heterochromatin for cohesion at centromeres. Science. 294, 2539–2542.

Black, B.E., Foltz, D.R., Chakravarthy, S., Luger, K., Woods, V.L. and Cleveland, D.W. (2004) Structural determinants for generating centromeric chromatin. Nature, 430, 578–582.

Black, B.E., Brock, M.A., Bédard, S., Woods, V.L. Jr and Cleveland, D.W. (2007) An epigenetic mark generated by the incorporation of CENP-A into centromeric nucleosomes. Proc. Natl. Acad. Sci. USA. 104, 5008–5013.
Blower, M.D., Sullivan, B.A. and Karpen, G.H. (2002) Conserved organization of centromeric chromatin in flies and humans. Dev. Cell. 2, 319–330.
Camahort, R., Li, B., Florens, L., Swanson, S.K., Washburn, M.P. and Gerton, J.L. (2007) Scm3 is essential to recruit the histone h3 variant cse4 to centromeres and to maintain a functional kinetochore. Mol. Cell. 26, 853–865.
Chan, G.K., Liu, S.T. and Yen, T.J. (2005) Kinetochore structure and function. Trends Cell Biol. 15, 589–598.
Chueh, A.C., Wong, L.H., Wong, N. and Choo, K.H. (2005) Variable and hierarchical size distribution of L1-retroelement-enriched CENP-A clusters within a functional human neocentromere. Hum. Mol. Genet. 14, 85–93.
Cimini, D., Moree, B., Canman, J.C. and Salmon, E.D. (2003) Merotelic kinetochore orientation occurs frequently during early mitosis in mammalian tissue cells and error correction is achieved by two different mechanisms. J. Cell Sci. 116, 4213–4225.
Cleveland, D.W., Mao, Y. and Sullivan, K.F. (2003) Centromeres and kinetochores: from epigenetics to mitotic checkpoint signaling. Cell. 112, 407–421.
Conde e Silva, N., Black, B.E., Sivolob, A., Filipski, J., Cleveland, D.W. and Prunell, A. (2007) CENP-A-conaining nucleosomes: easier disassembly versus exclusive centromeric localization. J. Mol. Biol. 370, 555–573.
Csink, A.K. and Henikoff, S. (1998) Something from nothing: the evolution and utility of satellite repeats. Trends Genet. 14, 200–204.
DeLuca, J.G., Gall, W.E., Ciferri, C., Cimini, D., Musacchio, A. and Salmon, E.D. (2006) Kinetochore microtubule dynamics and attachment stability are regulated by Hec1. Cell. 127, 969–982.
Earnshaw, W.C. & Migeon, B. (1985). Three related centromere proteins are absent from the latent centromere of a stable isodicentric chromosome. Chromosoma. 92, 290–296.
Earnshaw, W.C. and Rothfield, N. (1985). Identification of a family of human centromere pro¬teins using autoimmune sera from patients with scleroderma. Chromosoma. 91, 313–321.
Earnshaw, W.C., Sullivan, K.F., Machlin, P.S., Cooke, C.A., Kaiser, D.A., Pollard, T.D., Rothfield, N.F. and Cleveland, D.W. (1987) Molecular cloning of cDNA for CENP-B, the major human centromere autoantigen. J. Cell Biol. 104, 817–829.
Earnshaw, W.C., Ratrie, H. 3rd and Stetten, G. (1989) Visualization of centromere proteins CENP-B and CENP-C on a stable dicentric chromosome in cytological spreads. Chromosoma. 98, 1–12.
Ebersole, T.A., Ross, A., Clark, E., McGill, N., Schindelhauer, D., Cooke, H. and Grimes, B. (2000) Mammalian artificial chromosome formation from circular alphoid input DNA does not require telomere repeats. Hum. Mol. Genet. 9, 1623–1631.
Ebersole, T., Okamoto, Y., Noskov, V.N., Kouprina, N., Kim, J.H., Leem, S.H., Barrett, J.C., Masumoto, H. and Larionov, V. (2005) Rapid generation of long synthetic tandem repeats and its application for analysis in human artificial chromosome formation. Nucleic Acids Res. 33, e130
Ekwall, K., Olsson, T., Turner, B.M., Cranston, G. and Allshire, R.C. (1997) Transient inhibition of histone deacetylation alters the structural and functional imprint at fission yeast centromeres. Cell. 91, 1021–1032.
Fitzgerald-Hayes, M., Clarke, L. and Carbon, J. (1982) Nucleotide sequence comparisons and functional analysis of yeast centromere DNAs. Cell. 29, 235–244.
Foltz, D.R., Jansen, L.E., Back, B.E., Bailey, A.O., Yates, J.R., 3rd and Cleveland, D.W. (2006) The human CENP-A centromeric nucleosome-associated complex. Nat. Cell Biol. 8, 458–469.
Fujita, Y., Hayashi, T., Kiyomitsu, T., Toyoda, Y., Kokubu, A., Obuse, C. and Yanagida, M. (2007) Priming of centromere for CENP-A recruitment by human hMis18alpha, hMis18-beta, and M18BP1. Dev. Cell. 12, 17–30.

Furuyama, T., Dalal, Y. and Henikoff, S. (2006) Chaperone-mediated assembly of centromeric nucleosome assembly *in vitro*. Proc. Natl. Acad. Sci. USA. 103, 6172–6177.
Gassmann, R., Carvalho, A., Henzing, A.J., Ruchaud, S., Hudson, D.F., Honda, R., Nigg, E. A., Gerloff, D.L. and Earnshaw, W.C. (2004) Borealin: a novel chromosomal passenger required for stability of the bipolar mitotic spindle. J. Cell Biol. 166, 179–191.
Goshima, G., Kiyomitsu, T., Yoda, K. and Yanagida, M. (2003) Human centromere chromatin protein hMis12, essential for equal segregation, is independent of CENP-A loading pathway. J. Cell Biol. 160, 25–39.
Greaves, I.K,, Rangasamy, D., Ridgway, P. and Tremethick, D.J. (2007) H2A.Z contributes to the unique 3D structure of the centromere. Proc. Natl. Acad. Sci. USA. 104, 525–530.
Grewal, S.I. and Moazed, D. (2003) Heterochromatin and epigenetic control of gene expression. Science. 301, 798–802.
Grimes, B.R., Babcock, J., Rudd, M.K., Chadwick, B. and Willard, H.F. (2004) Assembly and characterization of heterochromatin and euchromatin on human artificial chromosomes. Genome Biol. 5, R89
Grimes, B.R., Rhoades, A.A. and Willard, H.F. (2002) Alpha-satellite DNA and vector composition influence rates of human artificial chromosome formation. Mol. Ther. 5, 798–805.
Grimes, B.R., Schindelhauer, D., McGill, N. I., Ross, A., Ebersole, T.A. and Cooke, H.J. (2001) Stable gene expression from a mammalian artificial chromosome. EMBO Rep. 2, 910–914.
Guenatri, M., Bailly, D., Maison, C. and Almouzni, G. (2004) Mouse centric and pericentric satellite repeats form distinct functional heterochromatin. J. Cell Biol. 166, 493–505.
Haaf, T. and Ward, D.C. (1994) Structural analysis of alpha-satellite DNA and centromere proteins using extended chromatin and chromosomes. Hum. Mol. Genet. 3, 697–709.
Hahnenberger, K.M., Baum, M.P., Polizzi, C.M., Carbon, J. and Clarke, L. (1989) Construction of functional artificial minichromosomes in the fission yeast *Schizosaccharomyces pombe*. Proc. Natl. Acad. Sci. USA. 86, 577–581.
Harrington, J.J., Van Bokkelen, G., Mays, R.W., Gustashaw, K. and Willard, H.F. (1997) Formation of *de novo* centromeres and construction of first-generation human artificial microchromosomes. Nat. Genet. 15, 345–355.
Hayakawa, T., Haraguchi, T., Masumoto, H. and Hiraoka, Y. (2003) Cell cycle behavior of human HP1 subtypes: distinct molecular domains of HP1 are required for their centromeric localization during interphase and metaphase. J. Cell Sci. 116, 3327–3338.
Hayashi, T., Fujita, Y., Iwasaki, O., Adachi, Y., Takahashi, K. and Yanagida, M. (2004) Mis16 and Mis18 are required for CENP-A loading and histone deacetylation at centromeres. Cell. 118, 715–729.
Heller, R., Brown, K.E., Burgtorf, C. and Brown, W.R. (1996) Mini-chromosomes derived from the human Y chromosome by telomere directed chromosome breakage. Proc. Natl. Acad. Sci. USA. 93, 7125–7130.
Henikoff, S., Ahmad, K. and Malik, H.S. (2001) The centromere paradox: stable inheritance with rapidly evolving DNA. Science. 293, 1098–1102.
Higgins, A.W., Gustashaw, K.M. and Willard, H.F. (2005) Engineered human dicentric chromosomes show centromere plasticity. Chromosome Res. 13, 745–762.
Howman, E.V., Fowler, K.J., Newson, A.J., Redward, S., MacDonald, A.C., Kalitsis, P. and Choo, K.H. (2000) Early disruption of centromeric chromatin organization in centromere protein A (Cenpa) null mice. Proc. Natl. Acad. Sci. USA. 97, 1148–1153.
Hudson, D.F., Fowler, K.J., Earle, E., Saffery, R., Kalitsis, P., Trowell, H., Hill, J., Wreford, N.G., de Kretser, D.M., Cancilla, M.R., Howman, E., Hii, L., Cutts, S.M., Irvine, D.V. and Choo, K.H. (1998) Centromere protein B null mice are mitotically and meiotically normal but have lower body and testis weights. J. Cell Biol. 141, 309–319.
Ikeno, M., Grimes, B., Okazaki, T., Nakano, M., Saitoh, K., Hoshino, H., McGill, N.I., Cooke, H. and Masumoto, H. (1998) Construction of YAC-based mammalian artificial chromosomes. Nat. Biotechnol. 16, 431–439.

Ikeno, M., Masumoto, H. and Okazaki, T. (1994) Distribution of CENP-B boxes reflected in CREST centromere antigenic sites on long-range alpha-satellite DNA arrays of human chromosome 21. Hum. Mol. Genet. 3, 1245–1257.

Irvine, D.V., Amor, D.J., Perry, J., Sirvent, N., Pedeutour, F., Choo, K.H. and Saffery, R. (2004) Chromosome size and origin as determinants of the level of CENP-A incorporation into human centromeres. Chromosome Res. 12, 805–815.

Izuta, H., Ikeno, M., Suzuki, N., Tomonaga, T., Nozaki, N., Obuse, C., Kisu, Y., Goshima, N., Nomura, F., Nomura, N. and Yoda, K. (2006) Comprehensive analysis of the ICEN (Interphase Centromere Complex) components enriched in the CENP-A chromatin of human cells. Genes Cells. 11, 673–684.

Jansen, L.E., Black, B.E., Foltz, D.R. and Cleveland, D.W. (2007) Propagation of centromeric chromatin requires exit from mitosis. J. Cell Biol. 176, 795–805.

Kaname, T., McGuigan, A., Georghiou, A., Yurov, Y., Osoegawa, K., De Jong, P.J., Ioannou, P. and Huxley, C. (2005) Alphoid DNA from different chromosomes forms *de novo* minichromosomes with high efficiency. Chromosome Res. 13, 411–422.

Kapoor, M., Montes, de Oca Luna, R., Liu, G., Lozano, G., Cummings, C., Mancini, M., Ouspenski, I., Brinkly, B.R. and May, G.S. (1998) The cenpB gene is not essential in mice. Chromosoma. 107, 570–576.

Kipling, D., Mitchell, A.R., Masumoto, H., Wilson, H.E., Nicol, L. and Cooke, H.J. (1995) CENP-B binds a novel centromeric sequence in the Asian mouse Mus caroli. Mol. Cell. Biol. 15, 4009–4020.

Kitagawa, K. and Hieter, P. (2001) Evolutionary conservation between budding yeast and human kinetochores. Nat. Rev. Mol. Cell Biol. 2, 678–687.

Kline, S.L., Cheeseman, I.M., Hori, T., Fukagawa, T. and Desai, A. (2006) The human Mis12 complex is required for kinetochore assembly and proper chromosome segregation. J. Cell Biol. 173, 9–17.

Kouprina, N, Ebersole, T., Koriabine, M., Pak, E., Rogozin, I.B., Katoh, M., Oshimura, M., Ogi, K., Peredelchuk, M., Solomon, G., Brown, W., Barrett, J.C. and Larionov, V. (2003) Cloning of human centromeres by transformation-associated recombination in yeast and generation of functional human artificial chromosomes. Nucleic Acids Res. 31, 922–934.

Lachner, M., O'Sullivan, R.J. and Jenuwein, T. (2003) An epigenetic road map for histone lysine methylation. J. Cell Sci. 116, 2117–2124.

Lam, A.L., Boivin, C.D., Bonney, C.F., Rudd, M.K. and Sullivan, B.A. (2006) Human centromeric chromatin is a dynamic chromosomal domain that can spread over noncentromeric DNA. Proc. Natl. Acad. Sci. USA. 103, 4186–4191.

Laner, A., Schwarz, T., Christan, S. and Schindelhauer, D. (2004) Suitability of a CMV/EGFP cassette to monitor stable expression from artificial chromosomes but not transient transfer in the cells forming viable clones. Cytogenet. Genome Res. 107, 9–13.

Laurent, A.M., Puechberty, J. and Roizès, G. (1999) Hypothesis: for the worst and for the best L1Hs retrotransposons actively participate in the evolution of the human centromeric alphoid sequences. Chromosome Res. 7, 305–317.

Lo, A.W., Magliano, D.J., Sibson, M.C., Kalitsis, P., Craig, J.M. and Choo, K. H. (2001) A novel chromatin immunoprecipitation and array (CIA) analysis identifies a 460-kb CENP-A binding neocentromere DNA. Genome Res. 11, 448–457.

Loiodice, I., Alves, A., Rabut, G., Van Overbeek, M., Ellenberg, J., Sibarita, J.B. and Doye, V. (2004) The entire Nup107-160 complex, including three new members, is targeted as one entity to kinetochores in mitosis. Mol. Biol. Cell. 15, 3333–3344.

Maddox, P.S., Hyndman, F., Monen, J., Oegema, K. and Desai, A. (2007) Functional genomics identifies a Myb domain-containing protein family required for assembly of CENP-A chromatin. J. Cell Biol. 176, 757–763.

Maggert, K.A. and Karpen G.H. (2007) The activation of a neocentromere in *Drosophila* requires proximity to an endogenous centromere. Genetics. 158, 1615–1628.

Maiato, H., DeLuca, J., Salmon, E.D. and Earnshaw, W.C. (2004) The dynamic kinetochore-microtubule interface. J. Cell Sci. 117, 5461–5477.
Masumoto, H., Ikeno, M., Nakano, M., Okazaki, T., Grimes, B., Cooke, H. and Suzuki, N. (1998) Assay of centromere function using a human artificial chromosome. Chromosoma. 107, 406–416.
Masumoto, H., Masukata, H., Muro, Y., Nozaki, N. and Okazaki, T. (1989) A human centromere antigen (CENP-B) interacts with a short specific sequence in alphoid DNA, a human centromeric satellite. J. Cell Biol. 109, 1963–1973.
Masumoto, H., Nakano, M. and Ohzeki, J. (2004) The role of CENP-B and alpha-satellite DNA: *de novo* assembly and epigenetic maintenance of human centromeres. Chromosome Res. 12, 543–556.
McIntosh, J.R., Grishchuk, E.L. and West, R.R. (2002) Chromosome-microtubule interactions during mitosis. Annu. Rev. Cell Dev. Biol. 18, 193–219.
Mejia, J.E., Willmott, A., Levy, E., Earnshaw, W.C. and Larin, Z. (2001) Functional complementation of a genetic deficiency with human artificial chromosomes. Am. J. Hum. Genet. 69, 315–326.
Mejia, J.E., Alazami, A., Willmott, A., Marschall, P., Levy, E., Earnshaw, W.C. and Larin, Z. (2002) Efficiency of *de novo* centromere formation in human artificial chromosomes. Genomics. 79, 297–304.
Mellone, B.G. and Allshire, R.C. (2003) Stretching it: putting the CEN(P-A) in centromere. Curr. Opin. Genet. Dev. 13, 191–198.
Mills, W., Critcher, R., Lee, C. and Farr, C.J. (1999) Generation of an approximately 2.4 Mb human X centromere-based minichromosome by targeted telomere-associated chromosome fragmentation in DT40. Hum. Mol. Genet. 8, 751–761.
Mizuguchi, G., Xiao, H., Wisniewski, J., Smith, M.M. and Wu, C. (2007) Nonhistone Scm3 and histones CenH3-H4 assemble the core of centromere-specific nucleosomes. Cell. 129, 1153–1164.
Moralli, D., Simpson, K.M., Wade-Martins, R. and Monaco, Z.L. (2006) A novel human artificial chromosome gene expression system using herpes simplex virus type 1 vectors. EMBO Rep. 7, 911–918.
Nagaki, K., Cheng, Z., Ouyang, S., Talbert, P.B., Kim, M., Jones, K.M., Henikoff, S., Buell, C.R. and Jiang, J. (2004) Sequencing of a rice centromere uncovers active genes. Nat. Genet. 36, 138–145.
Nakano, M., Okamoto, Y., Ohzeki, J. and Masumoto, H. (2003) Epigenetic assembly of centromeric chromatin at ectopic alpha-satellite sites on human chromosomes. J. Cell Sci. 116, 4021–4034.
Nakano, M., Cardinale, S., Noskov, V.N., Gassmann, R., Vagnarelli, P., Kandels-Lewis, S., Larionov, V., Earnshaw W.C. and Masumoto. H. (2008). Inactivation of a human kinetochore by specific targeting of chromatin modifiers. Dev. Cell. 14, 507–522.
Nakashima, H., Nakano, M., Ohnishi, R., Hiraoka, Y., Kaneda, Y., Sugino, A. and Masumoto, H. (2005) Assembly of additional heterochromatin distinct from centromere-kinetochore chromatin is required for *de novo* formation of human artificial chromosome. J. Cell Sci. 118, 5885–5898.
Nicklas, R.B. (1997) How cells get the right chromosomes. Science. 275, 632–637.
Nonaka, N., Kitajima, T., Yokobayashi, S., Xiao, G., Yamamoto, M., Grewal, S.I. and Watanabe, Y. (2002) Recruitment of cohesin to heterochromatic regions by Swi6/HP1 in fission yeast. Nat. Cell Biol. 4, 89–93.
Obuse, C., Iwasaki, O., Kiyomitsu, T., Goshima, G., Toyoda, Y. and Yanagida, M. (2004) A conserved Mis12 centromere complex is linked to heterochromatic HP1 and outer kinetochore protein Zwint-1. Nat. Cell Biol. 6, 1135–1141.
Okada, M., Cheeseman, I.M., Hori, T., Okawa, K., McLeod, I.X., Yates, J.R. 3rd, Desai, A. and Fukagawa, T. (2006) The CENP-H-I complex is required for the efficient incorporation of newly synthesized CENP-A into centromeres. Nat. Cell Biol. 8, 446–457.

Okada, T., Ohzeki, J., Nakano, M., Yoda, K., Brinkly, B.R., Larionov, V. and Masumoto, H. (2007) CENP-B controls centromere formation depending on the chromatin context. Cell. 131, 1287–1300.

Okamoto, Y., Nakano, M., Ohzeki, J., Larionov, V. and Masumoto, H. (2007) A minimal CENP-A core is required for nucleation and maintenance of a functional human centromere. EMBO J. 26, 1279–1291.

Oegema, K., Desai, A., Rybina, S., Kirkham, M. and Hyman, A.A. (2001) Functional analysis of kinetochore assembly in Caenorhabditis elegans. J. Cell Biol. 153, 1209–1226.

Ohzeki, J., Nakano, M., Okada, T. and Masumoto, H. (2002) CENP-B box is required for *de novo* centromere chromatin assembly on human alphoid DNA. J. Cell Biol. 159, 765–775.

Palmer, D. K., O'Day, K., Trong, H. L., Charbonneau, H. & Margolis, R. L. (1991) Purification of the centromere-specific protein CENP-A and demonstration that it is a distinctive histone. Proc. Natl Acad. Sci. USA. 88, 3734–3738.

Perez-Castro, A.V., Shamanski, F.L., Meneses, J.J., Lovato, T.L., Vogel, K.G., Moyzis, R.K. and Pedersen, R. (1998) Centromeric protein B null mice are viable with no apparent abnormalities. Dev. Biol. 201, 135–143.

Peters, A.H., Kubicek, S., Mechtler, K., O'Sullivan, R.J., Derijck, A.A., Perez Burgos, L., Kohlmaier, A., Opravil, S., Tachibana, M., Shinkai, Y., Martens, J.H. and Jenuwein, T. (2003) Partitioning and plasticity of repressive histone methylation states in mammalian chromatin. Mol. Cell. 12, 1577–1589.

Peters, A.H., O'Carroll, D., Scherthan, H., Mechtler, K., Sauer, S., Schofer, C., Weipoltshammer, K., Pagani, M., Lachner, M., Kohlmaier, A., Opravil, S., Doyle, M., Sibilia, M. and Jenuwein, T. (2001) Loss of the Suv39h histone methyltransferases impairs mammalian heterochromatin and genome stability. Cell. 107, 323–337.

Politi, V., Perini, G., Trazzi, S., Pliss, A., Raska, I., Earnshaw, W.C. and Della Valle, G. (2002) CENP-C binds the alpha-satellite DNA *in vivo* at specific centromere domains. J. Cell Sci. 115, 2317–2327.

Robinett, C.C., Straight, A., Li, G., Willhelm, C., Sudlow, G., Murray, A. and Belmont, A.S. (1996) *In vivo* localization of DNA sequences and visualization of large-scale chromatin organization using lac operator/repressor recognition. J. Cell Biol. 135, 1685–1700.

Ruchaud, S., Carmena, M. and Earnshaw, W.C. (2007) Chromosomal passenger: conducting cell division. Nat. Rev. Mol. Biol. 8, 798–812.

Rudd, M. K., Mays, R.W., Schwartz, S. and Willard, H.F. (2003) Human artificial chromosomes with alpha satellite-based *de novo* centromeres show increased frequency of nondisjunction and anaphase lag. Mol. Cell. Biol. 23, 7689–7697.

Saffery, R., Sumer, H., Hassan, S., Wong, L.H., Craig, J.M., Todokoro, K., Anderson, M., Stafford, A. and Choo, K.H. (2003) Transcription within a functional human centromere. Mol. Cell. 12, 509–516.

Saffery. R., Wong, L.H., Irvine, D.V., Bateman, M.A., Griffiths, B., Cutts, S.M., Cancilla, M.R., Cendron, A.C., Stafford, A.J. and Choo, K.H. (2001) Construction of neocentromere-based human minichromosomes by telomere-associated chromosomal truncation. Proc. Natl. Acad. Sci. USA. 98, 5705–5710.

Salmon, E.D., Cimini, D., Cameron, L.A. and DeLuca, J.G. (2005) Merotelic kinetochores in mammalian tissue cells. Philos. Trans. R. Soc. Lond. B Biol. Sci. 360, 553–568.

Schuh, M., Lehner, C.F. and Heidmann, S. (2007) Incorporation of *Drosophila* CID/CENP-A and CENP-C into centromere during early embryonic anaphase. Curr. Biol. 17, 237–243.

Shelby, R.D., Monier, K. and Sullivan, K.F. (2000) Chromatin assembly at kinetochore is uncoupled from DNA replication. J. Cell Biol. 151, 1113–1118.

Sonoda, E., Matsusaka, T., Morrison, C., Vagnarelli, P., Hoshi, O., Ushiki, T., Nojima, K., Fukagawa, T., Waizenegger, I.C., Peters, J.M., Earnshaw, W.C. and Takeda, S. (2001) Sccl/Rad21/Mcdl is required for sister chromatid cohesion and kinetochore function in vertebrate cells. Dev. Cell. 1, 759–770.

Spence, J.M., Critcher, R., Ebersole, T.A., Valdivia, M.M., Earnshaw, W.C., Fukagawa, T. and Farr, C.J. (2002) Co-localization of centromere activity, proteins and topoisomerase II within a subdomain of the major human X alpha-satellite array. EMBO J. 21, 5269–5280.

Stoler, S., Rogers, K., Weitze, S., Morey, L., Fitzgerald-Hayes, M. and Baker, R.E. (2007) Scm3, an essential *Saccharomyces cerevisiae* centromere protein required for G2/M progression and Cse4 localization. Proc. Natl. Acad. Sci. USA. 104, 10571–10576.

Sullivan, B.A., Blower, M.D. and Karpen, G.H. (2001) Determining centromere identity: cyclical stories and forking paths. Nat. Rev. Genet. 2, 584–596.

Sullivan, B.A. and Karpen, G.H. (2004) Centromeric chromatin exhibits a histone modification pattern that is distinct from both euchromatin and heterochromatin. Nat. Struct. Mol. Biol. 11, 1076–1083.

Sullivan, B.A. and Schwartz, S. (1995) Identification of centromeric antigens in dicentric Robertsonian translocations: CENP-C and CENP-E are necessary components of functional centromeres. Hum. Mol. Genet. 4, 2189–2197.

Sullivan, K.F., Hechenberger, M. and Masri, K. (1994) Human CENP-A contains a histone H3 related histone fold domain that is required for targeting to the centromere. J. Cell Biol. 127, 581–592.

Sullivan K.F., (2001) A solid foundation: functional specialization of centromeric chromatin. Curr. Opin. Genet. Dev. 11, 182–188.

Suzuki, N., Nakano, M., Nozaki, N., Egashira, S., Okazaki, T. and Masumoto, H. (2004) CENP-B interacts with CENP-C domains containing Mif2 regions responsible for centromere localization. J. Biol. Chem. 279, 5934–5946.

Taddei, A., Maison, C., Roche, D. and Almouzni, G. (2001) Reversible disruption of pericentric heterochromatin and centromere function by inhibiting deacetylases. Nat. Cell Biol. 3, 114–120.

Trowell, H.E., Nagy, A., Vissel, B. and Choo, K.H. (1993) Long-range analyses of the centromeric regions of human chromosomes 13, 14 and 21: identification of a narrow domain containing two key centromeric DNA elements. Hum. Mol. Genet. 2, 1639–1649.

Tsuduki, T., Nakano, M., Yasuoka, N., Yamazaki, S., Okada, T., Okamoto, Y. and Masumoto, H. (2006) An artificially constructed *de novo* human chromosome behaves almost identically to its natural counterpart during metaphase and anaphase in living cells. Mol. Cell. Biol. 26, 7682–7695.

Tyler-Smith, C., Gimelli, G., Giglio, S., Floridia, G., Pandya, A., Terzoli, G., Warburton, P. E., Earnshaw, W.C. and Zuffardi, O. (1999) Transmission of a fully functional human neocentromere through three generations. Am. J. Hum. Genet. 64, 1440–1444.

Vagnarelli, P.B. and Earnshaw, W.C. (2001) INCENP loss from an inactive centromere correlates with the loss of sister chromatid cohesion. Chromosoma. 110, 393–401.

Valdeolmillos, A., Rufas, J.S., Suja, J.A., Vass, S., Heck, M.M., Martinez, A.C. and Barbero, J.L. (2004) *Drosophila* cohesins DSA1 and Drad21 persist and colocalize along the centromeric heterochromatin during mitosis. Biol. Cell. 96, 457–462.

Van Hooser, A.A., Ouspenski, II, Gregson, H.C., Starr, D.A., Yen, T.J., Goldberg, M.L., Yokomori, K., Earnshaw, W.C., Sullivan, K.F. and Brinkley, B.R. (2001) Specification of kinetochore-forming chromatin by the histone H3 variant CENP-A. J. Cell Sci. 114, 3529–3542.

Vass, S., Cotterill, S., Valdeolmillos, A.M., Barbero, J.L., Lin, E., Warren, W.D. and Heck, M.M. (2003) Depletion of Drad21/Scc1 in *Drosophila* cells leads to instability of the cohesin complex and disruption of mitotic progression. Curr. Biol. 13, 208–218.

Ventura, M., Antonacci, F., Cardone, M.F., Stanyon, R., D'Addabbo, P., Cellamare, A., Sprague, L.J., Eichler, E.E., Archidiacono, N.. and Rocchi, M. (2006), Evolutionary formation of new centromeres in macaque. Science. 316, 243–246.

Warburton, P.E. (2004) Chromosomal dynamics of human neocentromere formation. Chromosome Res. 12, 617–626.

Warburton, P. E., Cooke, C.A., Bourassa, S., Vafa, O., Sullivan, B.A., Stetten, G., Gimelli, G., Warburton, D., Tyler-Smith, C., Sullivan, K.F., Poirier, G.G. and Earnshaw, W.C. (1997) Immunolocalization of CENP-A suggests a distinct nucleosome structure at the inner kinetochore plate of active centromeres. Curr. Biol. 7, 901–904.

Willard, H.F. and Waye, J.S. (1987) Hierarchical order in chromosome-specific human alpha satellite DNA. Trends Genet. 3, 192–198.

Yoda, K., Ando, S., Morishita, S., Houmura, K., Hashimoto, K., Takeyasu, K. and Okazaki, T. (2000) Human centromere protein A (CENP-A) can replace histone H3 in nucleosome reconstitution in vitro. Proc. Natl. Acad. Sci. USA. 97, 7266–7271

Zinkowski, R.P., Meyne, J. and Brinkley, B.R. (1991) The centromere-kinetochore complex: a repeat subunit model. J. Cell Biol. 113, 1091–1110.

Zuccolo, M., Alves, A., Galy, V., Bolhy, S., Formstecher, E., Racine, V., Sibarita, J. B., Fukagawa, T., Shiekhattar, R., Yen, T. and Doye, V. (2007) The human Nup107-160 nuclear pore subcomplex contributes to proper kinetochore functions. EMBO J. 26, 1853–1864.

# Chapter 6
# Kinetochore Composition, Formation, and Organization

**Tatsuo Fukagawa and Peter De Wulf**

**Abstract** Highly conserved multiprotein assemblies known as kinetochores orchestrate chromosome segregation during cell division and ensure that DNA is correctly transmitted from one generation to the next. Kinetochores act directly by assembling onto chromatin regions named centromeres. Activities governed by kinetochores include establishing centromeric heterochromatin, holding sister chromatids together via centromeric cohesin, nucleating kinetochore microtubules, attaching and orienting sister chromatids to spindle microtubules, sensing the tension or occupancy resulting from chromatid–microtubule attachment, recruiting spindle checkpoint components to induce a mitotic delay when tension/occupancy is absent, separating sister chromatids along the spindle, and regulating the initiation of cytokinesis.

Kinetochore defects generate cells with abnormal numbers of chromosomes resulting in genetic disease, cancer formation/progression or even death of the cell or organism. Likely because they manage so many roles, kinetochores are some of the most complex intracellular structures known today. Understanding how kinetochores coordinate chromosome segregation requires the identification of their components and the characterization of how they assemble into a competent mechanochemical structure. This chapter discusses our current understanding of how kinetochore components recognize and assemble onto centromeric regions. We will discuss similarities and differences between both processes in various species. However, as the budding yeast and vertebrate kinetochores have been studied most intensely, their kinetochores will be discussed in greatest detail.

P. De Wulf (✉)
Department of Experimental Oncology, European Institute of Oncology, 20139 Milan, Italy
e-mail: peter.dewulf@ifom-ieo-campus.it

P. De Wulf, W.C. Earnshaw (eds.), *The Kinetochore*,
DOI 10.1007/978-0-387-69076-6_6, 

## 6.1 Organization and Formation of Kinetochores

Electron microscopic analysis of vertebrate cells following fixation and staining (Brinkley and Stubblefield 1966, Jokelainen 1967, Comings and Okada 1971) or high-pressure freezing/freeze substitution (McEwen et al. 1998) showed that the vertebrate kinetochore exists as a button-like structure consisting of three layers of proteins (Fig. 6.1A). First, an electron-dense inner plate of proteins anchors the kinetochore on centromeric heterochromatin and recruits all other kinetochore components. Second, an electron-lucent interzone viewed using certain fixation protocols contains proteins that stain with 3F3/2 antibodies and act in tension sensing and spindle checkpoint signaling. Third, an electron-dense outer plate harbors components that make end-on contact with up to 20–30 microtubules. It consists of microtubule-associated proteins (MAPs), kinesins and structural units involved in the regulation of spindle dynamics and checkpoint signaling. The outer plate is surrounded by a prominent fibrous corona (width of 100–150 nm) that is only seen clearly when microtubules are absent. The outer plate and corona form the outer kinetochore. The corona contains resident and transient kinetochore components that are involved in recruiting spindle checkpoint proteins, establishing kinetochore–microtubule attachment, and regulating microtubule dynamics and chromosome segregation (Maiato et al. 2004).

Possibly because of their small(er) sizes and lack of an electron-dense organized structure, kinetochores from unicellular eukaryotes (e.g., the budding yeast *Saccharomyces cerevisiae*) cannot be visualized directly in the electron microscope. However, based on the structure of vertebrate kinetochores, kinetochores of yeasts (Fig. 6.1B) are generally described as composed of three distinct regions; an inner region of centromere-bound proteins, a middle region of components that bridge the distance between the centromere and microtubule tip, and an outer region of MAPs and kinesins that regulate microtubule stability and dynamics, and contribute to the segregation of separated sister chromatids along the anaphase spindle.

Besides discrepancies in size, kinetochores of metazoans and yeasts also differ with respect to how and when they assemble on centromeres. In budding and fission yeasts, kinetochores form on centromeres immediately after DNA replication and keep chromatids attached to the spindle microtubules throughout the cell cycle (Kilmartin 1994). In contrast, vertebrate kinetochores undergo a pathway of structural maturation following nuclear envelope breakdown (Brinkley and Stubblefield 1966, Roos 1973, Ris and Witt 1981, McEwen et al. 1993). Numerous vertebrate kinetochore components undergo cycles of assembly and disassembly during mitosis (Maiato et al. 2004, Liu et al. 2006).

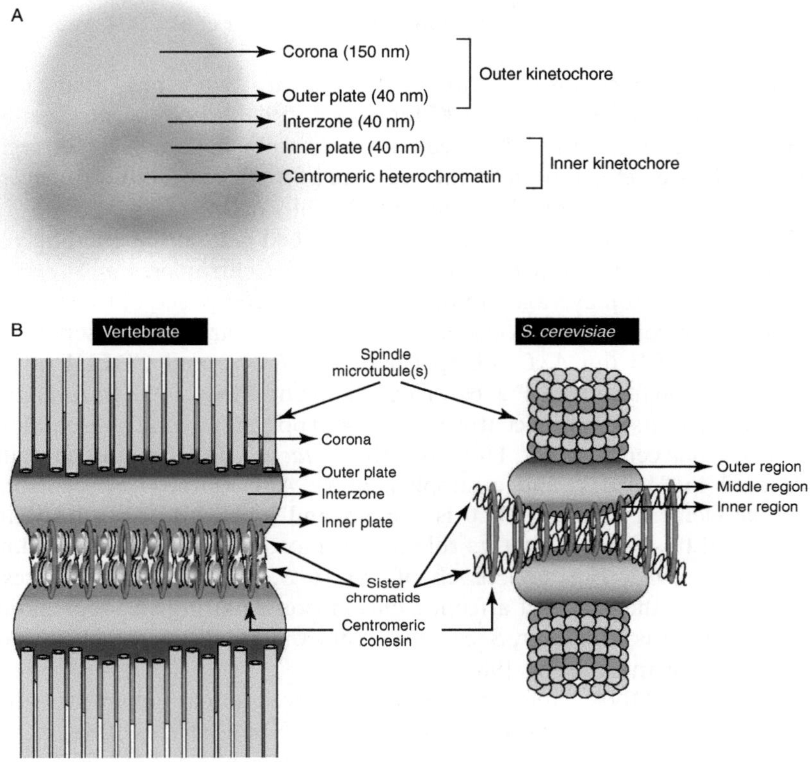


**Fig. 6.1** Kinetochores (**A**) Model of the vertebrate kinetochore (not bound to spindle microtubules) based on electron micrographs (Brinkley and Stubblefield 1966, Jokelainen 1967, Comings and Okada 1971, McEwen et al. 1998). (**B**) Models of metaphase vertebrate and budding yeast kinetochores bipolarly attached to spindle microtubules. Intra-kinetochore regions are shown in color. (*See* Color Insert)

## 6.2 The Centromeric Sequence does not Define Centromere Identity

The most commonly studied centromeres (budding and fission yeast, human centromeres) occupy a single primary constriction on the mitotic chromosome. Such chromosomes are said to be monocentric. In contrast, some plants (e.g., *Luzula spp.*), insects (Lepidoptera, Hemiptera) and nematodes (e.g., *Caenorhabditis elegans*) have chromosomes that are holocentric – the centromere determinants are distributed all along the chromosome (Pimpinelli and Goday 1989, Maddox et al. 2004). If holocentric chromosomes are fragmented, all fragments segregate normally in mitosis. Monocentric chromosomes interact with the spindle via attachments of a single kinetochore to one or more spindle microtubules (1 microtubule in budding yeast, 3–4 microtubules in fission yeast, 20–30 microtubules in human cells; Amor et al. 2004, Westermann et al. 2007).

Monocentric chromosomes can be divided into two types—point centromeres and regional centromeres (Pluta et al. 1990). Point centromeres are defined by highly specific DNA sequences, whereas the much larger regional centromeres are defined both by the presence of preferred DNA sequences and by epigenetic determinants that remain to be fully defined.

To date, point centromeres have been described only in the budding yeasts, including *S. cerevisiae* and *Kluyveromyces lactis*. Centromeres of *S. cerevisiae* are short regions (on average 115 bp long) that are composed of three sequence elements; centromere DNA element (CDE)I (a partially conserved 8 bp sequence), CDEII (an A/T rich spacer of 78–86 bp), and CDEIII (a highly conserved 25 bp fragment; Fig. 6.2, Hegemann and Fleig 1993). In *K. lactis*, the organization of its six point centromeres (~450 bp long) strongly resembles that of *S. cerevisiae* centromeres. However, the *K. lactis* CDEII region contains 80 additional nucleotides, and a supplementary A/T rich 100 bp centromeric sequence element named CDE0 is present and located 150 bp upstream of CDEI. CDE0 is essential but its role in chromosome segregation is unknown (Heus et al. 1993; Fig. 6.2). Each of the 16 *S. cerevisiae* centromeres can substitute the other without affecting chromosome pairing and segregation in mitosis or meiosis. However, *S. cerevisiae* centromeres do not function in other species (Clarke and Carbon 1985).

Centromeric DNA sequences are not conserved across eukaryotes and are some of the fastest evolving regions in the genome (Henikoff et al. 2001, Chapter 7). Consequently, centromere activity does not depend on a primary DNA sequence but rather on an organizational theme consisting of the presence of repetitive DNA (Choo 2001). The fission yeast *Schizosaccharomyces pombe* centromeres are 30–110 kb long (regional centromeres) and comprise a central core sequence that is flanked by various inverted repeats (Pidoux and Allshire 2004; Fig. 6.2). In multicellular eukaryotes, centromeres are made up of repeating DNA sequences that are arranged in long head-to-tail tandem arrays (Fig. 6.2). The monomeric sequences that make up these arrays are not conserved and range in size from 5–7 bp (*Drosophila*) to 340 bp (pigs). In humans, the repeating motif is a 171 bp monomer known as $\alpha$-satellite (alphoid) DNA. Arrays formed from $\alpha$-satellite repeats range in size from less than 200 kb to up to 7 Mb. Alphoid DNA has a complex hierarchical structure and shows chromosome-specific variations both in sequence and in the arrangement of its higher-order repeats (Choo 2001, Amor et al. 2004).

Although $\alpha$-satellite DNA is found at the centromeres of all human chromosomes, its presence does not guarantee centromeric activity. Indeed, in stable dicentric chromosomes (rearranged chromosomes containing two centromeres) only one of the two $\alpha$-satellite blocks forms a functional kinetochore (Earnshaw and Migeon 1985, Merry et al. 1985; see Chapters 3 and 4 for details). Indeed, analysis of one such centromere led to the first proposal that centromere

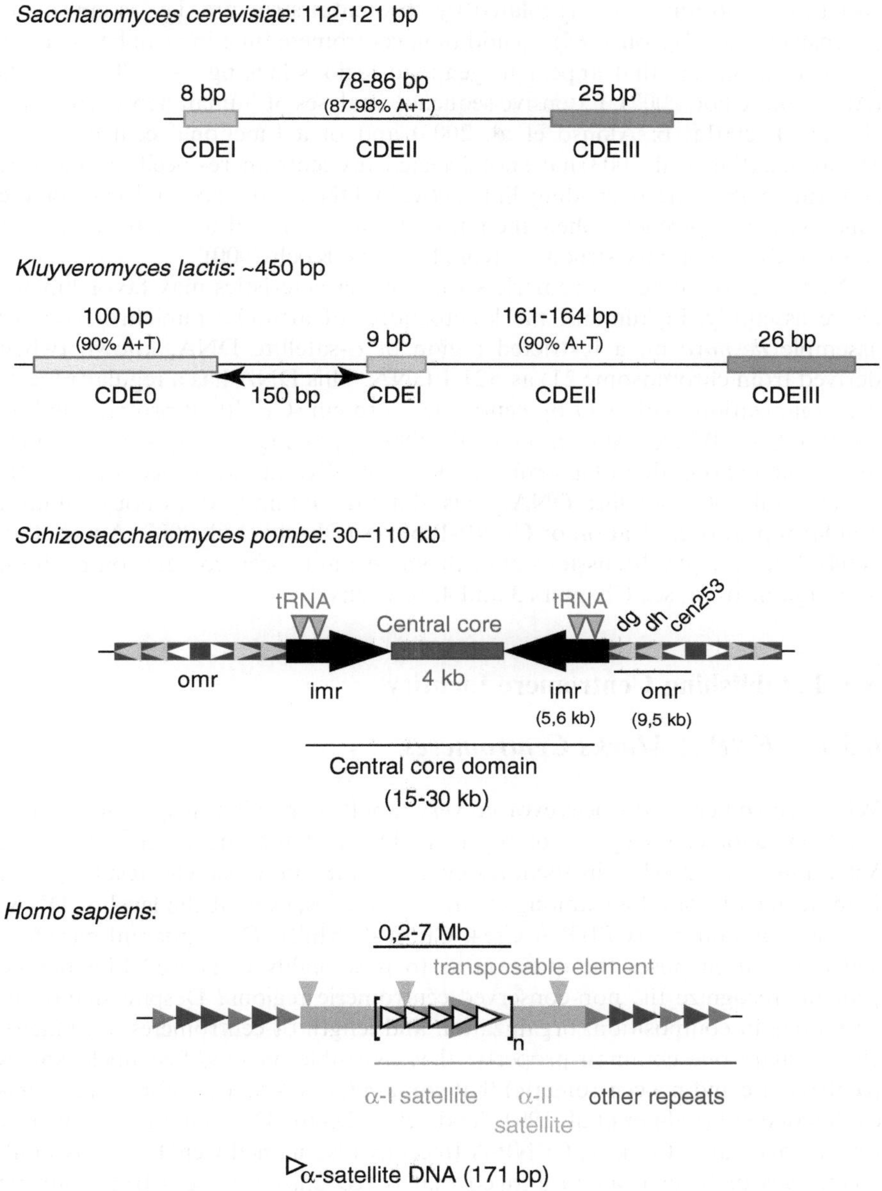


**Fig. 6.2** Centromeres. The centromeric sequence elements of *Saccharomyces cerevisiae, Kluyveromyces lactis, Schizosaccharomyces pombe,* and human chromosomes. Based on Malik and Henikoff (2002), and Pidoux and Allshire (2004). For explanation, see text

specification in humans is regulated by chromatin structure, i.e., is epigenetic (Earnshaw and Migeon 1985). In addition, centromere function can be assumed by neocentromeres that appear in genomic regions lacking $\alpha$-satellite repeats (Amor and Choo 2002). Extensive sequence analyses of human neocentromeres (Lo et al. 2001a, b; Alonso et al. 2003) and of a functional centromere in *Drosophila* (Sun et al. 2003) have not detected any centromere-specific sequences, confirming that the overriding link between DNA sequence and centromere function is an epigenetic phenomenon that may be linked to the formation of recognizable secondary structures (e.g., hairpins; Koch 2000).

Nonetheless, some centromeric sequence characteristics may favor kinetochore assembly: in humans, the kinetochores of artificial minichromosomes assemble *de novo* on a restricted region of $\alpha$-satellite DNA, known (when derived from chromosome 21) as $\alpha$21-I DNA. This DNA has a regular repeating organization, with a 17 bp centromere protein (CENP)-B protein-binding motif (CENP-B box – Masumoto et al. 1989) appearing in every second monomer. Kinetochores do not assemble on the more diverged sequences of the $\alpha$21-II subfamily of $\alpha$-satellite DNA. This alphoid subfamily does not contain a regular repeat organization or CENP-B boxes (Ohzeki et al. 2002, Amor et al. 2004). For in-depth discussions of centromere and neocentromere composition and organization, see Chapters 3 and 4, respectively.

## 6.3 Establishing Centromere Identity

### *6.3.1 CENP-A Marks Centromeres*

While centromeric sequences evolved very rapidly (e.g., via transposition) possibly from telomeres (regions of repetitive DNA at the end of chromosomes; Villasante et al. 2007), kinetochores evolved much more slowly, resulting in a high degree of homology among eukaryotic species, even at the level of DNA-binding components (CENP-A/Cse4, CENP-C/Mif2). This apparent paradoxical observation poses the question as to how highly conserved kinetochore proteins recognize the non-conserved centromeric regions? Despite significant variations in composition, organization and length of centromeres, all kinetochores share one common property: they assemble on modified nucleosomes (centromeric and neocentromeric) that contain CENP-A, an isoform of canonical histone H3 (Palmer et al. 1991, Yoda et al. 2000). The founding member of this histone-variant family, CENP-A (recently also named CenH3; Malik et al. 2002), was discovered as an antigen that is recognized by sera from patients suffering from scleroderma spectrum disease (Earnshaw and Rothfield 1985, Valdivia and Brinkley 1985). CENP-A homologs have been identified in mouse (Cenpa), chicken (CENP-A), *Drosophila melanogaster* (CID), *C. elegans* (HCP-3), *S. pombe* (Cnp1) and *S. cerevisiae* (Cse4; Table 6.1). If CENP-A is depleted or mutated, other kinetochore components are not recruited to the centromere. Loss of CENP-A homologs is lethal for every organism tested to date.

**Table 6.1** Conservation of proteins acting at the centromere/kinetochore

| Human | Mouse | Chicken | *D. melanogaster* | *C. elegans* | *S. pombe* | *S. cerevisiae* |
|---|---|---|---|---|---|---|
| **Chromatin/Inner kinetochore components** | | | | | | |
| CENP-A | Cenpa | CENP-A | CID | HCP-3 | Cnp1/Sim2 | Cse4 |
| CENP-B | Cenpb | CENP-B | – | – | Abp1, Cbh1, Cbh2 | – |
| CENP-C | Cenpc | CENP-C | CENP-C | HCP-4 | Cnp3 | Mif2 |
| CENP-D | – | – | – | – | – | – |
| CENP-G | – | – | – | – | – | – |
| CENP-H | Cenph | CENP-H | – | – | Fta3 | Mcm16 |
| CENP-I | Cenpi | CENP-I | – | – | Mis6 | Ctf3 |
| CENP-K | – | CENP-K | – | – | Sim4 | – |
| CENP-50/CENP-U | MLF1IP/PBIP1 | CENP-50/-U | – | – | Fta4 | – |
| CENP-O/Mcm21R | – | CENP-O | – | – | Mal2 | Mcm21 |
| CENP-P | – | CENP-P | – | – | Fta2 | Ctf19 |
| CENP-Q | – | CENP-Q | – | – | Fta7 | – |
| CENP-R | – | CENP-R | – | – | – | – |
| CENP-L | – | CENP-L | – | – | Fta1 | – |
| CENP-M | – | CENP-M | – | – | Mis17 | Iml3 |
| CENP-N/Chl4R | – | CENP-N | – | – | Mis15 | Chl4 |
| CENP-S | – | CENP-S | – | – | – | YOL86-A |
| CENP-T | – | CENP-T | – | – | SPBC800 | – |
| SGOL2/Tripin | Sgol2 | – | – | – | Sgo2 | – |
| CAP-D2 | – | – | – | HCP-6 | – | – |
| DDB-1 | – | – | DDB-1 | – | – | – |
| BMI-1 | Bmi-1 | BMI-1 | Psc1 | – | – | – |
| – | – | – | – | – | – | Cbf1 |
| – | – | – | – | – | – | Ndc10 |
| – | – | – | – | – | – | Cep3 |
| – | – | – | – | – | – | Ctf13 |

**Table 6.1** (continued)

| Human | Mouse | Chicken | *D. melanogaster* | *C. elegans* | *S. pombe* | *S. cerevisiae* |
|---|---|---|---|---|---|---|
| SKP1A | Skp1a | SKP1 | SKP-1 | Skp-1 | Skp1 | Skp1 |
| – | – | – | – | – | NP593516 | Scm3 |
| – | – | – | – | pqn-85 | Mis4 | Scc2 |
| CBX5 | Cbx5 | – | HP1a | – | – | – |
| CBX1 | Cbx1 | – | – | – | – | – |
| SGOL1/Shugoshin | Sgol1 | LOC460220 | MEI-S332 | Sgo-1 | Sgo1 | Sgo1 |
| HPC1 | – | – | – | pcs-1 | Pcs1 | Csm1 |
| – | – | – | – | – | Mde4 | Lrs4 |
| CSNK1D | Csnk1d | LOC417378 | – | Kin-19/ Kin20 | Hhp1 | Hrr25 |
| – | – | – | – | – | – | Mam1 |
| – | – | – | – | – | Moa1 | – |
| RAD54B | E130016E0 3RIK | RAD54B | – | Rad-54 | Rdh54 | Rdh54 |
| Cohesin | | | | | | |
| RAD21 | Rad21 | Rad21 | RAD21 | Scc-1 | Rad21 | Scc1/Mcd1 |
| REC8 | Rec8L1 | – | Rec8 | Rec-8 | Rec8 | Rec8 |
| STAG1 | Stag1 | STAG1 | SA/stromalin | Scc-3 | Psc3 | Scc3/Irr1 |
| SMC1A/SMC1B | Smc1a/Smc1b | SMC1L1 | SMC1 | Him-1 | Psm1/Smc1 | Smc1 |
| SMC3 | Smc3 | CSPG6 | SMC3 | Smc-3 | Smc3/Psm3 | Smc3 |
| PDS5/Aprin | Aprin | – | CG17509 | Evl-14 | Pds5 | Pds5 |
| Condensin | | | | | | |
| SMC2/CAP-E | Smc2 | SMC2L1 | Smc2 | Mix-1 | Cut14 | Smc2 |
| SMC4/CAP-C | Smc4 | SMC4L1 | Glu/Smc4 | Smc-4 | Cut3 | Smc4 |
| BRRN1/CAP-H | Brrn1 | LOC425304 | barr | – | Cnd2 | Brn1 |
| CAP-G | – | LOC422822 | Cap-G | – | Cnd3 | Ycg1 |
| CAP-D2 | Cap-D2 | LOC418275 | CAP-D2 | dpy-28 | Cnd1 | Ycs4 |

**Table 6.1** (continued)

| Human | Mouse | Chicken | *D. melanogaster* | *C. elegans* | *S. pombe* | *S. cerevisiae* |
|---|---|---|---|---|---|---|
| **Middle kinetochore components** | | | | | | |
| 3F3/2 mAb antigen | – | – | – | – | – | – |
| – | – | – | – | – | – | Okp1 |
| – | – | – | – | – | – | Ame1 |
| MIS12 | Mis12 | Mis12 | Mis12 | Mis-12 | Mis12 | Mtw1 |
| Nnf1R/PMF1 | PMF1 | Nnf1 | Nnf1R1<br>NNf1R2 | KBP-1 | Nnf1 | Nnf1 |
| Nsl1R/DC8/DC31 | Nsl1 | Nsl1 | Nsl1R | KBP-2 | Mis14 | Nsl1 |
| Dsn1R/C20orf172 | Dsn1 | Dsn1 | – | KNL-3 | Mis13 | Dsn1 |
| – | – | – | – | KBP-1 | – | – |
| – | – | – | – | KBP-2 | – | – |
| – | – | – | – | KBP-5 | – | – |
| KNL-3 | – | – | – | KNL-3 | – | – |
| – | – | – | – | – | – | Mcm22 |
| – | – | – | – | – | – | Cnn1 |
| – | – | – | – | – | – | Nkp1 |
| – | – | – | – | – | – | Nkp2 |
| SUPT4H1 | Supt4h1 | – | Spt4 | Spt-4 | Spt4 | Spt4 |
| – | – | – | – | – | Fta5 | – |
| – | – | – | – | – | Fta6 | – |
| – | – | – | – | – | Cnl2 | – |
| **Outer kinetochore components** | | | | | | |
| CENP-E | Cenpe | LOC422716 | CENP-meta<br>CENP-ana | – | – | – |
| CENP-F | Cenpf/Lek1 | – | – | Hcp-1<br>Hcp-2 | – | – |
| ZW10 | Zw10 | ZW10 | ZW10/mit(1)15 | Czw-1 | – | – |

**Table 6.1** (continued)

| Human | Mouse | Chicken | *D. melanogaster* | *C. elegans* | *S. pombe* | *S. cerevisiae* |
|---|---|---|---|---|---|---|
| ZWINT-1/ZWINT | Zwint-1/Zwint | ZWINT-1 | – | – | – | – |
| ZWILCH | Zwilch | ZWILCH | Zwilch | ZWL-1 | – | – |
| ROD | Kntc1 | LOC416865 | Rod | – | – | – |
| dynein | Dynein | DNCl2 | dynein/sw | Dyci-1 | dynein | Dyn1 Dyn2 |
| Dynactin complex | | | | | | |
| p150 glued/DCTN1 | Dctn1 | DCTN1 | Glued/GL | Dynactin/ Dnc-1 | – | Nip100/Pac13 |
| ARP1/Centracin | Arp1 | ACTR1A | Arp1 | Arp-1 | – | Arp1/Act3/ Act5 |
| ARP11/ACTR3B | Arp11/Actr3B | – | ACTR3B | Arx-1 | Arx1 | Arp3 |
| CapZ$\alpha$/$\beta$ | Capza1 Capza2 | CapZA1 CapZA2 | Cpb | Cap-1 Cap-2 | Cap1 Cap2 | Cap1 Cap2 |
| $\beta$ actin | $\beta$ actin | $\beta$ actin | $\beta$ actin | $\beta$ actin | $\beta$ actin | $\beta$ actin |
| p62/DCTN4 | Dctn4 | DCTN4 | CG12042 | C26B2.1 | – | – |
| p50/dynamitin/DCTN2 | Dctn2 | P50 | Dmn | – | – | – |
| Jnp24/p22/DCTN3 | Dctn3 | – | – | – | – | – |
| p25/DCTN5 | Dctn5 | DCTN5 | Dyn-p25 | | Y71F9AL.14 | – |
| – | | | | | | |
| p27/DCTN6 | Dctn6 | – | CG17347 | | Y54E10A.5 | – |
| – | | | | | | |
| PAFAH1B1/Lis1 | Pafah1b1 | – | Lis1 | Lis-1 | – | Pac1 |
| CLASP1 | Clasp1 | LOC429042 | Chb | C07H6.3 | – | – |
| CLASP2 | Clasp2 | LOC420723 | – | Cls-2 | – | – |
| APC | Apc | APC | Apc/Apc2 | Apr-1 | – | – |
| Blinkin/SPC105/A14q15 | – | Spc105 | Spc105 | KNL-1 | Spc7 | Spc105 |
| ZWINT-1/ZWINT | Zwint-1/Zwint | ZWINT-1 | – | – | – | Kre28 |
| HEC1/NDC80 | Ndc80 | Hec1 | Ndc80 | Ndc-80 | Ndc80 | Ndc80 |

**Table 6.1** (continued)

| Human | Mouse | Chicken | *D. melanogaster* | *C. elegans* | *S. pombe* | *S. cerevisiae* |
|---|---|---|---|---|---|---|
| NUF2/Nuf2R | Nuf2 | Nuf2 | Nuf2 | Nuf2/ Him-10 | Nuf2 | Nuf2 |
| SPC24/SPBC24 | Spbc24 | Spc24 | – | KBP-4 | Spc24 | Spc24 |
| SPC25/SPBC25 | Spbc25 | Spc25 | Spc25/Mitch | KBP-3 | Spc25 | Spc25 |
| EB1 | EB1 | EB1 | Eb1 | Ebp-2 | Mal3 | Bim1 |
| – | – | – | – | – | Alp7 | Slk19 |
| – | – | – | – | – | – | Spc19 |
| – | – | – | – | – | Spc34 | Spc34 |
| – | – | – | – | – | Dam1 | Dam1 |
| – | – | – | – | – | – | Duo1 |
| – | – | – | – | – | Dad1 | Dad1 |
| – | – | – | – | – | Hos2 | Dad2/Hsk1 |
| – | – | – | – | – | – | Dad3 |
| – | – | – | – | – | – | Dad4 |
| – | – | – | – | – | Ask1 | Ask1 |
| KIF11 | Kif11 | – | Klp61F | Bmk-1 | Cut7 | Kip1 |
| – | – | – | – | – | Klp5 | Kip3 |
| – | – | – | – | Klp-20 | Klp2 | Kar3 |
| CLIP-170 | CLIP-170 | – | CLIP-170 | – | Tip1 | Bik1 |
| hEg5 (?) | – | – | KLP61F | – | – | Cin8 |
| TOG1 | Ckap5 | CKAP5 | Msps | Zyg-9 | Dis1 | Stu2 |
| RSN | Rsn | RSN | CLIP-190 | – | – | – |
| – | – | – | – | – | Alp14/Mtc1 | – |
| – | – | LOC428578 | Klp98A | Klp-6 | Klp6 | – |
| MCAK/KIF2C | Kif2c | KIF2C | Klp59D | Klp-7 | – | – |
| Non-checkpoint components that translocate from nuclear pores | | | | | | |
| RanGAP1 | Rangap1 | RANGAP1 | – | – | – | – |

**Table 6.1** (continued)

| Human | Mouse | Chicken | *D. melanogaster* | *C. elegans* | *S. pombe* | *S. cerevisiae* |
|---|---|---|---|---|---|---|
| RanBP2/Nup358 | Ranbp2 | – | Nup358 | Npp-9 | – | – |
| ELYS/AHCTF1 | Elvs/Ahctf1 | AHCTF1 | – | – | – | – |
| Nup107-160 complex | | | | | | Nup84 complex |
| NUP160 | Nup160 | | RCJMB04_1d22 | CG4738 | – | Nup120 |
| Nup120 | | | | | | |
| NUP133 | Nup133 | LOC421535 | CG6958 | Npp-15 | Nup133a/ Nup133b | Nup133 |
| NUP107 | Nup107 | LOC417841 | Nup107 | Npp-5 | Nup107 | Nup84 |
| NUP96/NUP98 | NUP98 | Nup98 | Nup96/Nup98 | Npp-10 | Nup145C | Nup145C |
| NUP85 | Nup85 | NUP85 | CG5733 | Npp-2 | Nup85 | Nup85 |
| NUP43 | Nup43 | NUP43 | CG7671 | – | – | – |
| NUP37 | Nup37 | NUP37 | CG11875 | – | – | – |
| SEC13 | Sec13l | SEC13L1 | Sec13 | Npp-20 | Sec13 | Sec13 |
| SEH1 | Seh1l | SEH1L | Nup44A | Npp-18 | Seh1 | Seh1 |
| **Regulatory components** | | | | | | |
| HSP90 | Hsp90 | Hsp90A | Hsp90 | Hsp90/ Daf-21 | Hsp90/Swo1 | Hsp90 |
| SGT1A | Sgt1 | SGT1 | Sgt1/ecd | D1054.3 | Sgt1 | Sgt1 |
| SUPT4H1 | Supt4h1 | – | Spt4 | Spt-4 | Spt4 | Spt4 |
| – | – | – | – | – | Ams2 | Dal80 |
| Rbbp7/RbAp46 | Rggp7/RbpA46 | – | CAF1 | Lin-53 | Mis16 | Msi1/Cac3 |
| Rbbp4/RbAp48 | Rbbp4/RbpA48 | – | – | – | Mis16 | Msi1/Cac3 |
| Mis18$\alpha$ | – | – | – | – | Mis18 | – |
| Mis18$\beta$ | – | – | – | – | Mis18 | – |
| M18BP1/KNL-2 | – | – | – | Kbp-2 | – | – |
| CAF1p150/CHAF1A | Chaf1a | CHAF1A | CAF1-180 | T06D10.2 | Spcc330.04cp | Rlf2/Cac1 |

**Table 6.1** (continued)

| Human | Mouse | Chicken | *D. melanogaster* | *C. elegans* | *S. pombe* | *S. cerevisiae* |
|---|---|---|---|---|---|---|
| CAF1p60/CHAF1B | CAF1p60/ Chaf1b | CHAF1B | Caf1-105 | Y71G12B | SPAC26H5.03 | Cac2 |
| HIRA | Hira | HIRA | Hira | K10D2.1 | Hir1 | Hir1 |
| – | – | – | – | – | SPBC15D4.03 | Hir2 |
| – | – | – | – | – | SPBC31F10.14c | Hir3 |
| CBX4 | Elac2 | – | – | – | – | Hpc2 |
| FACTp80/SSRP1 | Ssrp1 | SSRP1 | Ssrp | Hmg-4 | – | – |
| FACTp140 | Supt16h | – | Dre4 | F55A3.3 | SPBP8B7.19 | Spt16 |
| SUPT3H | Supt3h | | RCJMB04_23d3 | Spt3 | – | Spt3 |
| Spt3 | | | | | | |
| TOP2A/B | Top2a/b | TOP2A/B | Top2 | K12D12.1 | Top2 | Top2 |
| PLK1 | Plk1 | – | Plk1 | Plk-1 | Cdc5 | Cdc5 |
| NUDC | Nudc | NUDC | NudC | Nud-1 | NudC | – |
| ERK | Erk | ERK | Rolled | Mek-2 | – | – |
| SUGT1/SGT1 | Sugt1 | – | ecd | Sgt-1 | Git7 | Sgt1 |
| PP1γ | Ppp1cc | – | Pp1-87 | Gsp-1 | Dis2 | Glc7 |
| PPP1R7 | Ppp1r7 | – | Sds2 | – | Sds22 | Sds22 |
| – | – | – | – | – | Rdp1 | Sin3 |
| DcR1 | Dicer1 | – | Dcr1 | Dcr-1 | Dcr1 | Dcr1/Gid8 |
| – | – | – | – | – | Clr3 | – |
| HDAC1 | HDAC1 | HD1 | Hdac1 | Had-1 | Clr6/Hda1 | Rpd3 |
| SIRT2 | Sirt2 | SIR2 | Sir2 | Sir-2.1 | Sir2 | Sir2 |
| CHORDC1 | Chordc1 | – | – | Chp-1 | Chp1 | – |
| – | | – | – | – | Cbp1 | Pdc2 |
| AGO1/EIF2C1 | Ago1/EIF2C1 | – | AGO1 | Alg-1/ Alg-2 | Ago1 | – |
| DSPP | DSPP | – | – | – | Tas3 | Ddr48 |
| SETD1A | Setd1a | – | CG17396 | Set-1 | Set1 | Set1 |

**Table 6.1** (continued)

| Human | Mouse | Chicken | *D. melanogaster* | *C. elegans* | *S. pombe* | *S. cerevisiae* |
|---|---|---|---|---|---|---|
| SUV39H1 | SUV39H1 | – | SU(VAR)3-9 | Set-2 | Clr4/Set2 | Set2 |
| – | – | – | – | – | Rik1 | – |
| – | – | – | – | – | Raf2 | – |
| GCN5L2 | Gcn5l2 | GCN5L2 | PCAF | Pcaf-1 | Gcn5 | Gcn5 |
| TIP60/HTATIP | Tip60/Htatip | TIP60 | Tip60 | Vc5.4 | Mst1 | Esa1 |
| HP1 | Hp1 | CBX1 | HP1/SU(VAR)2-5 | Hpl-1Hpl-2 | Swi6/Hp1 | Swi6 |
| CHEK1 | Chek1 | CHEK1 | Grp | Chk-1 | Chk1 | Chk1 |
| MOBKL1A | Mobkl1A | – | Mob1 | – | Mob1 | Mob1 |
| NDR/warts | – | – | Ndr | – | Sid2 | Dbf2 |
| CDC14 | Cdc14a/b | – | – | Cdc-14 | Cdc14 | Cdc14 |
| CDK2 | Cdk2 | CDK2 | Cdc2c | K03E5.3 | Cdc2 | Cdk1/Cdc28 |
| GSG2/Haspin | Gsg2 | GSG2 | Haspin | C01H6.9 | SPAC23C4.03 | Alk2 |
| Aurora A/AURKA | AurkA | AURORA A | aurora-A | Air-1 | – | – |
| MAP kinase/MAPK1 | Mapk1 | MAPK1 | rl/CG12559-PC.3 | Mpk-1 | Spk1 | Kss1 |
| GSK3 | Gsk3 | GSK3 | gskt | Gsk-3 | Gsk3 | Mck1 |
| NEK2/NEK2A | Nek2 | NEK2 | Nek2 | – | Fin1 | Kin3 |
| PPP2R5C | Ppp2r5c | LOC423460 | PP2A-B' | | C13G3.3A | SPCC188.02 |
| Rts1 | | | | | | |
| UFD1L | Ufd1l | UFD1L | Ufd1-like | Ufd-1 | Ufd1 | Ufd1 |
| UBE2I | Ube2i | UBE2I | lwr | Ubc-9 | Hus5 | Ubc9 |
| PIAS2 | Pias2 | PIAS2 | Su(var)2-10 | Gei-17 | SPAC1687.05 | Siz1 |
| PARP-1 | Parp-1 | PARP1 | Parp | Pme-1 | C2A9.07c | – |
| SUMO-1 | Sumo-1 | SUMO-1 | Smt3 | Smo-1 | Pmt3 | Smt3 |
| SUMO-2/3 | Sumo2/3 | SUMO-2/3 | – | – | – | – |
| PLCG1 | Plcg1 | LOC419175 | Sl | Plc-1 | Plc1 | Plc1 |
| Spindle checkpoint | | | | | | |
| BUB1 | Bub1 | BUB1 | Bub1 | Bub-1 | Bub1 | Bub1 |
| BUB3 | Bub3 | BUB3 | Bub3 | – | Bub3 | Bub3 |

**Table 6.1** (continued)

| Human | Mouse | Chicken | *D. melanogaster* | *C. elegans* | *S. pombe* | *S. cerevisiae* |
|---|---|---|---|---|---|---|
| MAD1 | Mad1 | MAD1 | Mad1 | Mdf-1/ Mad-1 | Mad1 | Mad1 |
| MAD2 | Mad2 | Mad2 | Mad2 | Mdf-2/ Mad-2 | Mad2 | Mad2 |
| Rae1 | Rae1 | – | Rae1 | – | Rae1 | Gle2 |
| COMET/p31/Cmt2 | p31/Cmt2B | – | p31 | – | – | – |
| BUBR1/BUB1B | Bub1b | – | – | San-1/ Mad3 | Mad3 | Mad3 |
| TAO1/TAOK1/ MARKK | Tao1/Taok1 | TAOK1 | Tao1 | Kin-18 | – | – |
| MPS1/TTK | Mps1/ttk | – | Mps1/ald | – | Mph1 | Mps1 |
| CDC20 | Cdc20 | CDC20 | Cdc20/Fizzy | Fzy-1/ Fizzy | Cdc20/Slp1 | Cdc20 |
| Spindly | – | – | Spindly | – | – | – |
| Chromosomal passenger complex (CPC) | | | | | | |
| Aurora B/AIRK2 | Aurkb | – | ial | Air-2 | Ark1 | Ipl1 |
| INCENP | Incenp | INCENP | Incenp | Icp-1 | Pic1 | Sli15 |
| Survivin | Birc5 | – | Deterin | Bir-1 | Cut17 | Bir1 |
| Borealin/Dasra B/ CDCA8 | Cdca8 | – | Borr | Csc-1 | – | – |
| – | Dasra A | Dasra A | Australin | – | – | – |
| TD60/RCC2 | Rcc2 | – | – | – | – | – |

(–): Unknown/not yet identified. The proteins are classified in the table based on their location and function. In some cases also the intra-kinetochore localization of conserved components differs slightly from species to species.

The homology between CENP-A and histone H3 is situated at the $\alpha$-helical C-terminal histone fold domain of the protein (60% identity; Luger et al. 1997). In contrast, the N-terminal tails of histone H3, CENP-A, and CENP-A homologs are highly diverged among species. This tail extends from the core of the nucleosome and is proposed to recruit and associate with other kinetochore proteins (Chen et al. 2000). Protein structure studies have revealed that CENP-A contains a short CENP-A targeting domain (CATD) within the histone fold region. Importantly, histone H3 provided with the CATD can target to the centromere and function as a centromere-specific histone (Black et al. 2004, 2007). CENP-A localizes to the inner plate of mammalian kinetochores (Warburton et al. 1997) where it binds to the 171 bp $\alpha$-satellite DNA. In budding yeast, *cse4* mutants genetically interact with CDEI and CDEII (Smith et al. 1996, Keith and Fitzgerald-Hayes 2000) while Cse4 chromatin immunoprecipitates (ChIPs) with the CDEIII region of the centromere (Meluh et al. 1998).

Live-cell studies in numerous organisms indicated that the deposition of CENP-A onto centromeric DNA is an early event in kinetochore assembly (Meluh et al. 1998, Howman et al. 2000, Blower and Karpen 2001, Oegema et al. 2001, Goshima et al. 2003, Regnier et al. 2005). However, the presence of CENP-A alone does not trigger kinetochore formation: overexpressing CENP-A leads to its ectopic incorporation along chromosome arms and to an aberrant localization of kinetochore-protein subsets, but not to functional kinetochores at those sites (Van Hooser et al. 2001), with the exception of certain sites in *Drosophila* (Heun et al. 2006). This observation suggests that next to the presence of CENP-A, additional kinetochore-foundation proteins (Amor et al. 2004) and epigenetic marks underlie kinetochore maturation.

The presence of CENP-A itself may provide a primary mark for the self-propagation of centromeric chromatin. Indeed, a biophysical comparison of $[H2A{:}H2B]_2$-$[CENP\text{-}A{:}H4]_2$ and $[H2A{:}H2B]_2$-$[H3{:}H4]_2$ nucleosomes revealed that the former are structurally more compact than the latter (despite CENP-A being four residues longer than histone H3) because CENP-A is more rigidly associated with histone H4 (Black et al. 2004). Furthermore, in budding yeast, $[H2A\text{-}H2B]_2$ are replaced by an Scm3 homodimer resulting in a hexameric $[Scm3\text{-}Cse4\text{-}H4]_2$ nucleosome at centromeres (Mizuguchi et al. 2007). Recently, Dalal et al. (2007) reported that in *D. melanogaster*, the centromeric nucleosome is organized as a heterotypic tetramer of [CENP-A, H4, H2A and H2B]. They refer to this nucleosome as a hemisome. It is presently unclear if tetrameric nucleosomes exist at centromeres in other organisms. Taken together, the special composition and compact configuration of CENP-A – containing nucleosomes may provide a physical platform endowed with extra strength to withstand the forces to which kinetochores are subjected during mitosis.

### 6.3.2 Incorporation of CENP-A in Centromeric Nucleosomes

In human cells, the assembly of CENP-A nucleosomes into nascent chromatin is uncoupled from DNA replication. CENP-A produced in G2 following centromere replication (Shelby et al. 2000), is deposited into chromatin during telophase/early G1. In fact, exit from mitosis is required for the loading of nascent CENP-A into centromeric chromatin (Jansen et al. 2007). When CENP-A expression is limited to S phase by cloning the coding region into an mRNA with classical histone 3' processing motifs, the protein is not targeted to centromeres (Shelby et al. 1997). Thus, a replication-independent assembly of centromeric nucleosomes seems essential to maintain the specificity of CENP-A incorporation (Heit et al. 2006). Similar conclusions were reached for the loading of CENP-A/CID in *D. melanogaster* (Schuh et al. 2007). By contrast, conventional H3 nucleosomes are assembled concomitantly with DNA replication.

During G1 and G2, histone H3 variant H3.3 is incorporated into chromatin in conjunction with active transcription, and serves as an example of replication-independent nucleosome assembly (Hendzel and Davie 1990, Ahmad and Henikoff 2001). This requires a chromatin assembly factor (CAF) called HIRA, which specifically recognizes histone H3.3 (Ray-Gallet et al. 2002, Tagami et al. 2004). Foltz et al. (2006) observed that chromatin containing nucleosomes harboring histone H3 variant H3.1 is associated with heterotrimeric chromatin assembly complex CAF-1 (p60/Mis16/RbAp46-48; Verreault et al. 1996). Both HIRA and CAF-1 activities depend on RbAp48. Its homolog in *S. pombe* (Mis16) has histone deacetylation activity at centromeres (Hayashi et al. 2004; see below). As shown by purification, however, CAF-1 and HIRA do not associate with CENP-A chromatin in human cells (Foltz et al. 2006). The absence of HIRA in CENP-A purifications was very surprising given that CENP-A nucleosomes also assemble independently of DNA replication (Shelby et al. 2000, Jansen et al. 2007). Indeed, CENP-A is not incorporated into centromeres under conditions of an RbAP48 knockdown (Hayashi et al. 2004). Furthermore, mutations in the fission yeast RbAp46/48 homolog Mis16 causes defects in CENP-A deposition. In addition, RbAp48 was found to associate with CENP-A purified from *Drosophila* cells (Furuyama and Henikoff 2006). On the other hand, CENP-A assembly in human cells could well be mediated by (an) alternative chromatin assembly factor(s).

Human MIS18 isoforms (MIS18$\alpha$ and MIS18$\beta$), which are homologs of the *S. pombe* Mis18 (Hayashi et al. 2004), form a complex with M18BP1 (MIS18 Binding Protein 1) and were shown to be required for CENP-A loading in human cells (Fujita et al. 2007). Thus, the epigenetic preparation of the centromeric nucleosome seems very important for the inclusion of CENP-A (Chapter 10). Depletion of the *C. elegans* homolog of M18BP1, known as Kinetochore-null (KNL)-2, results in the loss of CENP-A from centromeres, resulting in a phenotype that is indistinguishable from that of a CENP-A depletion (Oegema et al. 2001). Indeed, KNL-2 is required for CENP-A localization to centromeres (Maddox et al. 2007). Interestingly, KNL-2/M18BP1 and MIS18$\alpha/\beta$ localize transiently to kinetochores from telophase to early G1 (Fujita et al. 2007, Maddox et al. 2007), the period during which CENP-A deposition occurs (Jansen et al. 2007). Thus, KNL-2 functions at chromatin during CENP-A loading, possibly as a targeting element for a CENP-A containing histone–chaperone complex which remains to be identified (Carrol and Straight 2007).

Recently, the proteins FACTp140 and FACTp80, which are involved in chromatin remodeling and transcription, were isolated in CENP-A but not in H3.1 purifications (Obuse et al. 2004b, Foltz et al. 2006, Izuta et al. 2006). This finding suggests that transcription may pass through CENP-A chromatin or that chromatin remodeling has a role in CENP-A nucleosome formation and/or maintenance. It will be very interesting to see whether FACT and other additional factors (Hayashi et al. 2004) are required for centromere formation or function.

Recently, Okada et al. (2006) demonstrated that knockouts of chicken CENP-H, -I, -K, -L and -M proteins, which constitute the CENP-H/I complex that co-purifies with CENP-A (Foltz et al. 2006, Izuta et al. 2006), block the centromeric incorporation of CENP-A–GFP expressed from a constitutive promoter. In these experiments, however, the localization of endogenous CENP-A was unaffected. One possibility is that the CENP-H/I complex is required for incorporating newly synthesized rather than previously assembled CENP-A nucleosomes, particularly during interphase. Alternatively, CENP-A–GFP may be slightly defective. Mis6 (CENP-I) is required for the loading of CENP-A–GFP to centromeres in *S. pombe*, but not for the loading of untagged CENP-A. Indeed, the CENP-H/I complex is reciprocally dependent on CENP-A for its centromere localization. Because the localizations of CENP-H and CENP-I require CENP-A, this deposition mechanism makes the loading of newly synthesized CENP-A dependent on pre-existing CENP-A nucleosomes, with CENP-H/I acting as an intermediate. Because the members of the CENP-H/I complex do not share any homology with known nucleosome assembly factors, the complex likely acts as a recruitment rather than assembly factor. Because of the overall complexity, key molecular mechanisms and components responsible for specifying and propagating CENP-A localization and centromere identity remain elusive.

In *S. pombe*, CENP-A homolog Cnp1 is incorporated both in S phase and in G2 via two phase-dependent inclusion pathways. S-phase incorporation of Cnp1 requires transient GATA-transcription factor Ams2 (Takahashi et al. 2005). The G2 pathway, which is of particular relevance to the replication-independent assembly that operates in mammalian cells (see above) depends on the *S. pombe* Mis16–Mis18 complex and more downstream on the Mis15 (CENP-N)-Mis17(CENP-M)-Mis6(CENP-I)-Sim4(CENP-K) complex (Hayashi et al. 2004). As mentioned above, Mis6 was reported to be required for the centromeric localization of Cnp1-GFP in *S. pombe* (Takahashi et al. 2000), however its budding yeast homolog Ctf3 is recruited to centromeres by Cse4, and not vice versa (Measday et al. 2002).

At *S. cerevisiae* centromeres, the CBF3 complex initiates kinetochore formation through sequence-specific binding to centromeric element CDEIII. Upstream events of kinetochore assembly require the protein Scm3. Scm3, which was originally identified as a high-copy suppressor of mutations in the Cse4 histone fold (Chen et al. 2000), is required for recruitment of Cse4 and CBF3 subunit Ndc10 to centromeres (Mizuguchi et al. 2007, Camahort et al. 2007, Stoler et al. 2007). Scm3 is chromatin-associated and immunoprecipitates with Cse4 but not with histone H3 illustrating a selective localization to centromeric nucleosomes (Mizuguchi et al. 2007). Furthermore, centromeric chromatin containing Cse4 and Scm3 lacks histones H2A and H2B in vivo. This led to the suggestion that the homodimeric Scm3 replaces $[H2A\text{-}H2B]_2$ thereby creating an atypical hexameric nucleosome that marks the centromere for kinetochore assembly. Indeed, Scm3 forms a stoichiometric complex with Cse4:H4 tetramers in vitro and in the absence of DNA. Scm3 can also replace H2A and H2B when mixed with preformed octamers containing H2A, H2B, Cse4, and H4 (Mizuguchi et al. 2007). Thus, Scm3 may be involved in the delivery of newly synthesized Cse4 to centromeres, the incorporation of Cse4 into chromatin, and/or the maintenance of Cse4 centromeres. It is presently unclear whether Cse4:H4 tetramers are initially associated with H2A and H2B tetramers at some point in the cell cycle but are then replaced by Scm3. The requirement for Scm3 in early S phase for the centromeric localization of kinetochore initiation factor Ndc10 suggests that Scm3 is necessary for CBF3 to bind to the CDEIII element of budding yeast centromeres.

An Scm3 homolog has been identified in *S. pombe* (spScm3, NP593516) but has not yet been analyzed in detail. Correlating SpScm3 activity to that of GATA factor Ams2 (required for S phase incorporation of Cnp1) may reveal at a molecular level how Scm3 contributes to creating centromeric chromatin following DNA replication in *S. pombe*. Orthologs of Scm3 have not been identified in eukaryotes outside of fungi. Indeed, *Drosophila* and human centromeric nucleosomes contain H2A and H2B (Blower et al. 2002, Foltz et al. 2006). However, a divergent Scm3 ortholog could be present in a small subset of CENP-A:H4 nucleosomes in these organisms (Mizuguchi et al. 2007).

Just as important as the selective incorporation of Cse4 at centromeres is the prevention of its deposition at noncentromeric regions. The latter is taken care of by chromatin assembly complexes CAF-1 (Cac1-3) and Hir (Hir1-3, Hpc2) because mutants in these complexes still incorporate Cse4 at centromeres, also at new extra-centromeric locations (Sharp et al. 2002).

Using conditional dicentric chromosomes (next to the endogenous *CEN3*, a *GAL1*-regulated *CEN3* cassette was introduced into chromosome III), Mythreye and Bloom (2003) assessed kinetochore formation in *S. cerevisiae*. In the presence of galactose, the conditional *CEN3* was actively transcribed from the upstream *GAL1* thereby prohibiting the formation of a kinetochore on this centromere. In the presence of glucose, transcription of *GAL1-CEN3* was repressed and a kinetochore formed on this centromere resulting in a chromosome containing two kinetochores. In mitosis, the chromosome broke and rearranged into a monocentric derivate resulting in colonies that were heterogenous in size and shape. Using this assay, kinetochore component Chl4 was identified as a factor required for the *de novo* formation of kinetochores on newly activated centromeres (Mythreye and Bloom 2003). Importantly, Cse4 and Ndc10 did not assemble on the newly activated centromere in the absence of Chl4 (Mythreye and Bloom 2003). Chl4 is a non-essential protein in *S. cerevisiae*, implying that pre-existing centromeres are normally propagated in an epigenetic manner. In contrast, the propagation of kinetochores that require de novo assembly (e.g., those on centromeric plasmids) appears more stringent and is driven by Chl4. Of note, the budding yeast Chl4 is the homolog of CENP-N, a member of the human CENP-H/I complex (Meraldi et al. 2006, Okada et al. 2006) suggesting a similar role for the protein/complex in metazoan kinetochore formation.

Although CENP-A establishes centromere identity in all eukaryotes, the question remains how the diverged centromeric sequence is recognized as the site for CENP-A recognition. A model proposed by Mellone and Allshire (2003) suggested that tension applied across the centromere as a result of bipolar attachment to the spindle could condition the centromeric chromatin for CENP-A loading, thereby ensuring that only active centromeres are targeted for the deposition of new CENP-A. However, Jansen et al. (2007) disproved this model as human cells forced to exit from mitosis in the absence of kinetochore–microtubule attachments still loaded CENP-A normally.

## 6.4 Identification and Characterization of Kinetochore Proteins and Complexes

### *6.4.1 Identification of Vertebrate Kinetochore Proteins*

The identification of kinetochore proteins was first accomplished in vertebrates via the use of autoantibodies (Earnshaw and Rothfield 1985), a few years before the first yeast kinetochore proteins were identified (Cai and Davis 1989, Baker et al. 1989, Cai and Davis 1990, Lechner and Carbon 1991). However, the subsequent dissection of vertebrate kinetochores lagged behind the yeast work. This was because vertebrate centromeres are difficult to manipulate genetically and because vertebrate kinetochore proteins are tightly bound to centromeric chromatin (Earnshaw et al. 1984) making metazoan kinetochores hard to enrich in solution. In contrast, budding yeast kinetochores easily detach from centromeres and spindle microtubules during cell extraction, and spontaneously dissociate into their stable subunits (proteins and complexes; De Wulf et al. 2003).

Originally, vertebrate kinetochore components were identified by virtue of anti-centromere antibodies present in antisera of patients suffering from scleroderma spectrum disease. The first of these patients had the CREST (Calcinosis, Raynaud's phenomenon, Esophageal dysmotility, Sclerodactyly, and Telangiectasia) syndrome (Moroi et al. 1980), but extensive study showed that the only clinical feature common to almost all patients with anti-centromere antibodies was Raynaud's phenomenon (a vasospastic disorder causing discoloration of the fingers and toes; Weiner et al. 1991). Similarly, antiserum of a patient with the GAVE (Gastric Antral Vascular Ectasia) syndrome lead to the identification of CENP-G (He et al. 1998), a kinetochore component that is yet to be described at a molecular level. Analysis of cell extracts with antisera derived from patients suffering from the CREST and GAVE syndromes lead to the identification of CENP-A, -B, -C, -D, -F, and -G (Earnshaw and Rothfield 1985, Kingwell and Rattner 1987, Liao et al. 1995, Gimelli et al. 2000). Antibody microinjection experiments demonstrated that CENPs -A, -B and -C are essential for kinetochore assembly and stability (Bernat et al. 1990, 1991). The first of these proteins to be cloned, CENP-B (Earnshaw et al. 1987) was subsequently shown to be a component of the centromeric heterochromatin, rather than the kinetochore (Cooke et al. 1990). CENP-B was also shown to bind to a specific 17 bp sequence in $\alpha$-satellite DNA (Masumoto et al. 1989). Cloned CENP-C was found to be located in the inner kinetochore plate (Saitoh et al. 1992); antibody microinjection studies subsequently proved that it is essential for the assembly of the trilaminar kinetochore (Tomkiel et al. 1994).

Two-hybrid interaction analyses, affinity purifications, database sequence searches and the localization of proteins involved in cell division to the centromere further extended the number of vertebrate kinetochore proteins (Sugata et al. 1999, Nishihashi et al. 2002, Liu et al. 2003, Goshima et al. 2003, Cheeseman et al. 2004, Obuse et al. 2004 a,b, Mikami et al. 2005, Minoshima et al. 2005, Foltz et al. 2006, Izuta et al. 2006, Okada et al. 2006). Many of these components are conserved between eukaryotes (Table 6.1), albeit sometimes to a low degree.

In an attempt to identify kinetochore proteins that are associated specifically with centromeric chromatin, two groups identified complexes that co-purified with CENP-A (Obuse et al., 2004b, Foltz et al., 2006). In a parallel study, Okada et al. (2006) isolated a kinetochore complex (named the CENP-H/I complex) CENP-H and CENP-I. These studies revealed overlapping sets of components, numerous of which proved to be novel members of the vertebrate kinetochore. Four known centromeric proteins (CENP-B, -C, -H, and -U/-50) and three novel components (CENP-M, -N, and -T) were isolated by their association with CENP-A nucleosomes, and were designated members of the "CENP-A NAC" (nucleosome-associated complex; Foltz et al. 2006). Purification of NAC components CENP-M, -N, and -U then led to the identification of six additional proteins: CENP-I, -K, -L, -O, -P, -Q, -R, and -S. These proteins were not identified in the CENP-A–TAP purifications, suggesting that they form a complex distinct from centromeric chromatin and NAC proteins. Consequently, this complex was named "CENP-A CAD" (*C*ENP-*A d*istal).

Okada et al. (2006) performed gene knockouts with chicken DT40 cells (a pre-B lymphocyte cell line with a high degree of homologous recombination – Buerstedde and Takeda 2006) and RNA interference (RNAi) knockdown analyses in human HeLa cells for each component of the CENP-H/I complex. Based on phenotypes, the large CENP-H/I complex was divided into three complexes named the CENP-H class (consisting of CENP-H, -I, -K), CENP-M class (CENP-M), and CENP-O class (CENP-O, -P, -Q, -R, -U/-50). The two other subunits, CENP-L and CENP-N require further characterization. Homologs of CENP-K through -R have been identified in some vertebrate organisms (Table 6.1), but few orthologs have been detected via sequence comparison in more distant eukaryotes.

Chicken DT40 cells with genetic knockouts of CENP-H-class proteins show severe mitotic defects. In addtion, the kinetochore localization of all CENP-H/I complex members and many outer kinetochore proteins is abolished, suggesting that CENP-H class members act early in kinetochore assembly (Okada et al. 2006). The mitotic phenotypes of CENP-M-deficient cells are less severe than those exhibited by CENP-H-class deficient cells. Consistent with the knockout phenotype, unpublished biochemical data (T. Fukagawa) further suggest that the CENP-M class differs from the CENP-H and CENP-O classes. DT40 cells bearing knockouts of CENP-O class proteins are viable.

Bioinformatic data mining has recognized homologs of kinetochore proteins in *Drosophila melanogaster* (Henikoff et al. 2000, Przewloka et al. 2007), a species whose centromeres have been studied extensively (Murphy and Karpen 1995, Le et al. 1995, Sun et al. 1997) but whose kinetochore proteins have only recently begun to be identified. Identification of the kinetochore proteins and their associated components should not only lead to detailed kinetochore protein-interaction maps of the *Drosophila* kinetochore and but may also allow the identification of their homologs in other eukaryotes.

### 6.4.2 Identification of Yeast Kinetochore Proteins

Following the isolation and characterization of yeast centromeres (Clarke and Carbon 1985) was the identification of the proteins that are bound to these sequences. Although the first centromere proteins had been identified and characterized in humans (CENP-B (Earnshaw et al. 1987), CENP-C (Saitoh et al. 1992)) no homology was discovered with any yeast proteins, so strategies to identify yeast kinetochore proteins employed the specific binding of proteins to yeast centromere DNA. Based on the sequence organization and mitotic activities of *S. cerevisiae* centromeres, functional and non-functional centromere variants were used in DNA-affinity purifications leading to the discovery of centromere-associated proteins Cbf1 (Centromere binding factor 1, binds to CDEI as a homodimer (Cai and Davis 1989,1990, Baker et al. 1989) and the kinetochore-initiating complex CBF3, which binds to CDEIII and comprises the Ndc10, Cep3, Ctf13 and Skp1 dimers (Lechner and Carbon 1991, Lechner 1994). Carried out in parallel with these biochemical efforts were screens for *trans*-acting mutations that affect chromosome stability in *S. cerevisiae*. Several mutant collections using the phenotypic criterion of chromosome missegregation were isolated (e.g., via a color-based colony sectoring assay; Hieter et al. 1985, Koshland et al. 1985). These mutants were designated: *ctf* (*c*hromosome *t*ransmission *f*idelity; Spencer et al. 1990), *cin* (*c*hromosome *in*stability; Hoyt et al. 1990), *chl* (*ch*romosome *l*oss; Kouprina et al. 1993b), *mcm* (*m*ini*c*hromosome *m*aintenance), *mif* (*mi*totic *f*idelity; Meeks-Wagner et al. 1986), *cse* (*c*hromosome *se*gregation; McGrew et al. 1989), *cep* (*ce*ntromere *p*roteins; Strunnikov et al. 1995), and *ndc* (*n*uclear *d*ivision *c*ycle; Goh and Kilmartin 1993) mutants. Since these mutations lie in genes that are required for a variety of processes contributing to faithful chromosome transmission (e.g., DNA replication, spindle formation, kinetochore assembly, cohesion) the identification of centromere/kinetochore-specific mutations required secondary screens. These screens were based on an increased sensitivity of the mutants to transcription through a conditional centromere (Hill and Bloom 1989, Doheny et al. 1993), a decreased breakage of a dicentric chromosome in the mutants (Koshland et al. 1987, Brock and Bloom 1994; both of these methods detected a weakened kinetochore in the presence of the given mutant), or the direct analysis of spindle defects and chromosome transmission (Goh and Kilmartin 1993). Analogously, *cut* (*c*ell *u*ntimely *t*orn; Hagan and Yanagida 1990), *dis* (*d*efective *i*n *s*ister chromatid disjoining; Ohkura et al. 1988), *mal* (*m*inichromosome *a*ltered *l*oss; Fleig et al. 1996), *mis* (*mi*nichromosome *s*tability; Takahashi et al. 1994), *mlo* (*m*issegregation and lethal when *o*verexpressed; Javerzat et al. 1996), and *nda* (*n*uclear *d*ivision *a*rrest; Toda et al. 1983) mutant collections were generated in *S. pombe*. Components uncovered in both species subsequently lead to the identification of additional yeast kinetochore subunits (Table 6.1) via synthetic lethality/rescue screens (e.g., Wigge and Kilmartin 2001, Measday et al. 2005, Pinsky et al. 2006), dosage lethality/rescue experiments (e.g., Kroll et al. 1996, Kopski and Huffaker 1997, Hyland et al. 1999, Yoon and Carbon 1999, Measday et al. 2002), phenotypic enhancement/suppression analyses (e.g., Meluh and Koshland 1995, Winkler et al. 2000), 1- and 2-hybrid interaction studies (e.g., Ortiz et al. 1999, Janke et al. 2001, Wong et al. 2007), immunoprecipitation of epitope-tagged kinetochore proteins followed by mass spectrometry (De Wulf et al. 2003, Nekrasov et al. 2003, Westermann et al. 2003, Liu et al. 2005), and bioinformatic sequence mining (Meraldi et al. 2006).

The currently known yeast kinetochore components are listed in Table 6.1. Of note, the above discussed approaches have covered all components of the budding yeast kinetochore many times over, arguing that all subunits of the *S. cerevisiae* kinetochore have now been identified. The list comprises more than 120 potential components including structural and transiently residing regulatory units. Note that some are deduced to be associated with the kinetochore via sequence homology to kinetochore proteins of other eukaryotes and need to be confirmed.

### 6.4.3 Biophysical and Structural Characterization of Kinetochore Proteins and Complexes

The affinity-based isolation of kinetochore proteins followed by mass spectrometry is a powerful means to identify new kinetochore subunits (see above). However, the number and identity of proteins co-purifying with a tagged bait varies from experiment to experiment (De Wulf et al. 2003, Nekrasov et al. 2003, Westermann et al. 2003). Consequently, it is hard to determine which proteins form a stable complex with the tagged protein, which interactions are secondary, or which proteins act transiently at the kinetochore. For proteins that can be solubilized as components of native complexes, hydrodynamic analysis (gel filtration combined with glycerol/sucrose gradient velocity sedimentation ultracentrifugation) of the tagged protein and those purifying with it can yield a clear picture of how co-precipitating components relate to each other. Indeed, each protein has a unique hydrodynamic fingerprint that is comprised of its diffusion coefficient (gel filtration) and Svedberg constant (ultracentrifugation). Two or more proteins that form a stable complex will share the same hydrodynamic fingerprint as defined by the diffusion and Svedberg constants. Indeed, high-throughput hydrodynamic analysis of all known kinetochore proteins in budding yeast cell extracts suggested that yeast kinetochore proteins are typically organized in complexes (De Wulf et al. 2003). The same is true for kinetochores of other species (fission yeast; Liu et al. 2005, *Xenopus laevis*; Emanuele et al. 2005, chicken cells; Hori et al, 2008, human cells; Kline et al. 2006). The composition of potential complexes can be confirmed by eliminating one protein (via gene deletion, degron-tagging, temperature-sensitive mutation, immunodepletion, conditional knock out, or siRNA) followed by analyzing the hydrodynamic behavior of its potential partner protein(s), which will exhibit altered diffusion and Svedberg coefficients. Besides establishing the relationship between kinetochore proteins, hydrodynamics can also reveal how proteins assemble into a complex as subunits and subassemblies of complexes are found in extracts of cycling populations (De Wulf et al. 2003). By feeding the measured coefficients into well-established equations one can obtain an estimate of the mass, size and shape of the examined protein/complex (Schuyler and Pellman 2002). Hydrodynamic analysis of kinetochore proteins in cell extracts of budding yeast (De Wulf et al. 2003), fission yeast (Liu et al. 2005), *Xenopus laevis* (Emanuele et al. 2005) chicken cells (Hori et al., 2008) and human cells (Kline et al. 2006) revealed that most kinetochore components are organized in complexes with an average weight of 200–250 kDa (budding yeast) or 1 MDa or more (higher eukaryotes; e.g., the RZZ [ROD-ZW10-ZWILCH], nuclear pore and dynactin complexes). The assembly of kinetochores from complexes rather than individual proteins may ensure their stability and maximize the efficiency and rate with which kinetochores form following DNA replication.

Of note, most inner and middle kinetochore complexes are highly elongated with frictional coefficients around 2.0 and axial ratios of 20:1. If such complexes were oriented vertically, they might contribute to bridging the distance between the centromere and microtubule tip(s).

Because many yeast and metazoan kinetochore complexes are stable in solution (cell extract) these complexes can be produced recombinantly in *E. coli* or insect cells. This allows the recombinant complexes to be observed via electron microscopy (Cheeseman et al. 2006, Miranda et al. 2005, 2007, Westerman et al. 2005, Wei et al. 2005), atomic force microscopy (Pietrasanta et al. 1999, Ciferri et al. 2005), or high-resolution fluorescence imaging (Ciferri et al. 2008) leading to a better understanding of the complex's architecture, organization and even function. For example, in vitro microtubule-binding assays with recombinant complexes followed by electron or fluorescence microscopy revealed that the budding yeast DASH complex encloses and stabilizes microtubules by forming rings around micortubules (Miranda et al. 2005, Westermann et al. 2005). Recombinant human outer kinetochore Highly Expressed in Cancer (HEC)1 complex was shown to localize to the kinetochore–microtubule interface in an angular and cooperative manner possibly forming a wide HEC1 sleeve that attaches kinetochores to the microtubule tip (Ciferri et al. 2008). Significant progress is also being made in the crystallographic analysis of kinetochores components (e.g., CENP-A nucleosome (Black et al. 2004, 2007), the HEC1 complex (Ciferri et al. 2008), EB1 (Hayashi and Ikura 2003), and Cep3; a member of the budding yeast CBF3 complex (Bellizzi et al. 2007, Purvis and Singleton 2008). Structural analysis of recombinant proteins/complexes has already revealed important information as to how certain proteins interact with the centromere (e.g., Cep3; Bellizzi et al. 2007, Purvis and Singleton 2008) or spindle microtubules. For instance, calponin-homology domains identified in the structures of EB1, Ndc80 (HEC)1, and Nuf2 were shown via mutational dissection to mediate the binding of these proteins to microtubules (Hayashi and Ikura 2003, Ciferri et al. 2008).

## 6.5 Hierarchical Assembly of Kinetochores

Early electron microscopic studies revealed that the trilaminar structure of the mature vertebrate kinetochore is only visible from late prophase till the end of mitosis, suggesting that kinetochores undergo a cycle of assembly and disassembly (Roos 1973, Brenner et al. 1981, He and Brinkley 1996, Maiato et al. 2004, Chan et al. 2005, Liu et al. 2006; Fig. 6.3). Indeed, many inner plate components localize to the kinetochore in early G2, whereas other components are recruited in prophase and prometaphase. At the end of metaphase, many components that do not belong to the inner kinetochore leave the kinetochore in a well-orchestrated manner (Fig. 6.3). In budding and fission yeast, the

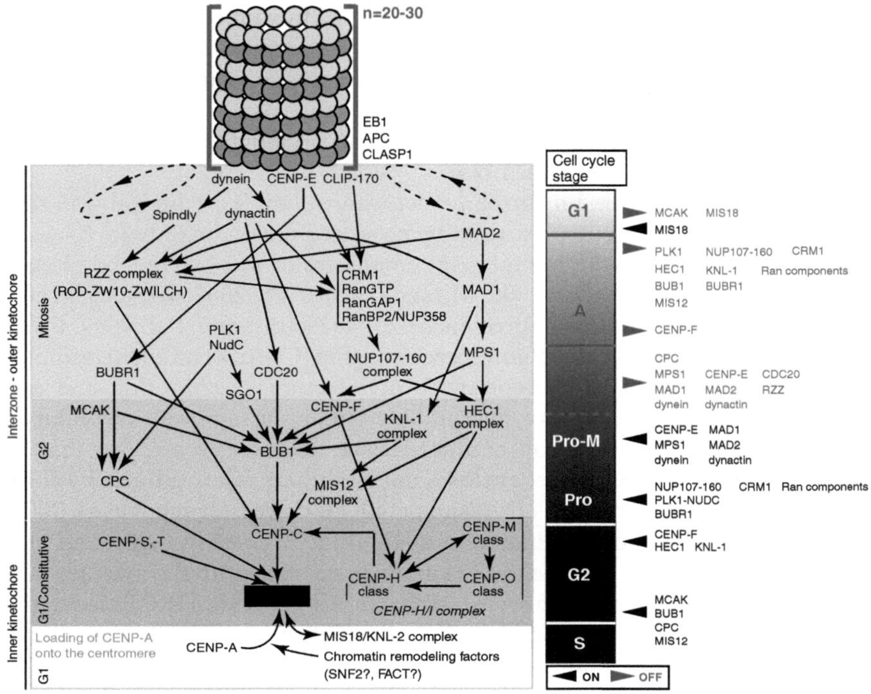


**Fig. 6.3** Assembly of vertebrate kinetochores in time and space. Only components whose recruitment to the centromere have been studied are shown

kinetochores keep sister chromatids tethered to spindle microtubules throughout the cell cycle. Only during DNA replication do yeast kinetochores dissociate from the centromere allowing the replication machinery to pass through (centromeres are early replicated regions). Following centromere replication, yeast kinetochores quickly reassemble to bind the sister chromatids to the microtubules.

In order to determine which DNA–protein, protein–protein and protein–tubulin interactions underlie the hierarchical formation of kinetochores, a target protein is first eliminated or depleted (via gene deletion, degron tagging, temperature-sensitive mutation, immunodepletion, conditional knock out, or siRNA) and the effects of its loss on the localization of other kinetochore proteins is examined epistatically via fluorescence imaging or ChIP. Kinetochore assembly has been studied most extensively in budding yeast. Our understanding of kinetochore formation in metazoans is still incomplete because many more proteins are involved (and many kinetochore proteins likely remain to be identified) and because numerous kinetochore subunits remain to be mapped onto the centromere. Nevertheless, a fair outline of the pathway of kinetochore formation can be presented for a number of organisms (Fig. 6.4).

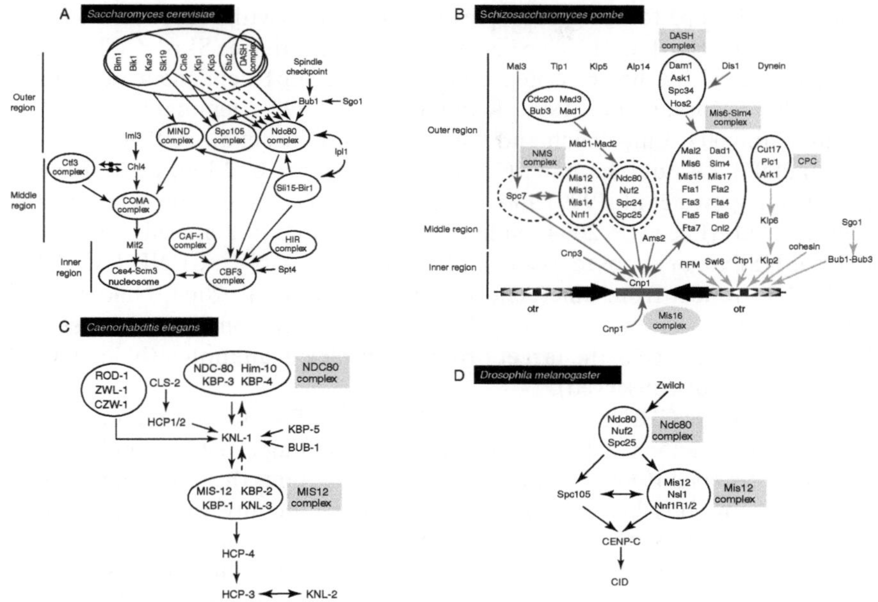


**Fig. 6.4** Assembly of nonvertebrate kinetochores. (**A**) Assembly of the *Saccharomyces cerevisiae* kinetochore. The centromere is shown unwrapped from its nucleosome. Middle layer proteins (e.g., Nkp1, Nkp2, Cnn1, Plc1, etc.) whose mode of recruitment is presently unclear are not shown. (**B**) Assembly of the *Schizosaccharomyces pombe* kinetochore. The red arrows indicate proteins that are recruited to the central core region, the green arrows those that are recruited to the outer repeats. Lines with double arrows symbolize protein interdependencies for localization to the centromere. RFM: proteins required for methylation (Rdp1, Dcr1, Ctr4, Chp1, Tas3, Ago1, Clr3, Clr6, Sir2, Abp1, Cbh1, Cbh2, Rik1). RFM proteins Rdp1, chp1, Tas3, and Abp1 were shown to physically associate with a certain domain via ChIP. The other RFM proteins are required for the proper function of the domain but for them no ChIP data have been published. Imr: inner most repeat, otr: outer repeat. (**C**) Assembly of the *Caenorhabditis elegans* kinetochore. The dashed lines represent partial recruitment dependencies. (**D**) Assembly of the *Drosophila melanogaster* kinetochore. Conserved Ndc80 complex member Spc24 has not yet been identified in *Drosophila*. Nnf1R1 and Nnf1R2 are two Nnf1 homologues. It is currently unclear which one of the two is part of the Mis12 complex. (*See* Color Insert)

## *6.5.1 Inner Components of the Budding Yeast Kinetochore*

In budding yeast, histone variants Cse4 and Scm3 establish centromere identity. Kinetochore formation, however, is initiated by the CBF3 complex (Ndc10, Ctf13, Cep3, Skp1) what may give the point centromere (more) independence from epigenetic effects on kinetochore assembly. Conserved Sgt1 and chaperone Hsp90 help CBF3 formation by activating Skp1 and Ctf13. More specifically, Skp1 binds to the F-box of Ctf13, an event that requires Hsp90 to correctly fold Ctf13. Ctf13 is then activated by Skp1-dependent phosphorylation and the transient interaction of Sgt1 with Skp1. Skp1 and phospho-Ctf13 associate with Cep3 forming a complex that associates with Ndc10 and binds to the centromere. Skp1-Ctf13-Cep3 is in equilibrium with free Skp1-Ctf13, which can be degraded by the 26 proteasome following Skp1-dependent ubiquitination by SCF (Kaplan et al. 1997, Kitagawa et al. 1999, Stemmann et al. 2002, Bansal et al. 2004, Lingelbach and Kaplan 2004, Gillis et al. 2005, Catlett and Kaplan 2006). Of note, human SGT1A can rescue the budding yeast *sgt1*-null mutant (Kitagawa et al. 1999) and associates with heat shock protein (HSP)90 in human cells (Lee et al. 2004, Niikura et al. 2006). In human cells depleted of SGT1 or HSP90, or exposed to an HSP90 inhibitor; CEN-C, CENP-I, CENP-F and HEC1 are lost from kinetochores (Steensgaard et al. 2004, Niikura et al. 2006) suggesting that the roles of Sgt1 and Hsp90 at centromeres are conserved, although in human cells the target protein/complex is unknown (Ndc10, Cep3 and Ctf13 are not conserved).

Together with CBF3, Cse4 recruits Mif2/CENP-C to yeast centromeres. CDEI-binding factor Cbf1, the first identified budding yeast kinetochore protein (Cai and Davis 1989, Baker et al. 1989), also requires the CBF3 complex for its recruitment, but has not been shown to recruit any other kinetochore component. Cbf1 is not conserved in higher eukaryotes. In budding yeast, Cbf1 acts as a transcription factor (Cai and Davis 1990) and the analogy with the binding of transcriptional regulator ARS binding Protein (ABF)1 (Buchman and Kornberg 1990) to the 5' region of the yeast origin of replication autonomously replicating sequence (ARS)1 is intriguing (Marahrens and Stillman 1992). However, its role at the kinetochore is still unclear. Domain mapping of Cbf1 has suggested that the proximity of negatively charged domains to the centromere may play a role in facilitating the binding of key non-histone proteins or other components to the highly compact centromere by countering competition from nucleosomes (Mellor et al. 1990).

### *6.5.2 Middle Components of the Budding Yeast Kinetochore*

Following DNA replication; Cse4, Scm3, CBF3, Mif2, and Cbf1 associate with sister centromeres and orchestrate kinetochore formation by hierarchically recruiting middle and outer components, a process that is completed within a few minutes (Tanaka et al. 2005). The budding yeast middle kinetochore components represent distinct and evolutionary conserved complexes, including the MIND (Mtw1/MIS12, Nnf1/PMF1, Nsl1/Nsl1R, Dsn1/Dsn1R), COMA (Ctf19/CENP-P, Okp1, Mcm21/CENP-O, Ame1), and Ctf3 (Ctf3/CENP-I, Mcm16/CENP-H, Mcm22) complexes, next to the Chl4 (CENP-N) and Iml3 (CENP-M) subunits.

Importantly, the vertebrate homologs of Mcm16 and Ctf3 (CENP-H and CENP-I, respectively), Mcm21 and Ctf19 (CENP-O and CENP-P, respectively), Iml3 and Chl4 (CENP-M and CENP-N, respectively) have been localized to the inner kinetochores of human and chicken cells (Foltz et al. 2006, Okada et al. 2006). In budding yeast, the position of these components relative to the centromere is still unclear. For example, Chl4 may well be an inner kinetochore component as it shares a low degree of homology with bacterial RecA proteins (Kouprina et al. 1993a) and contains an $\alpha$-helix-turn-helix motif indicative of a potential DNA binding property (Mythreye and Bloom, 2003).

All budding yeast middle kinetochore complexes assemble onto Cse4-Scm3-CBF3-Mif2 in a COMA-dependent manner (Fig. 6.4A). However, the MIND and Ctf3 complexes, as well as the Chl4/Iml3 proteins are recruited independently of each other to COMA. Because the Cft3 complex and Iml3/Chl4 are non-essential in mitotic yeast, COMA and MIND mutants phenotypically mimic both each other and *scm3*, *cse4* and *mif2* mutants. The mutants exhibit an unstable spindle and monopolar attachment phenotype, and are delayed in mitosis by the spindle checkpoint (Brown et al. 1993, Stoler et al. 1995, Ortiz et al. 1999, Scharfenberger et al. 2003, Stoler et al. 2007). The roles of the Ctf3 complex, Iml3 and Chl4 in mitosis are unclear, but Iml3 and Chl4 are essential for the maintenance of centromere cohesin in meiosis I (see Chapter 13).

It should be noted that the term middle component is something of a misnomer, implying as it does, a rather passive role of the component(s) in bridging the inner and outer kinetochore proteins. The conserved budding yeast MIND complex likely has no direct microtubule-binding capacity, as quantified via in vitro experiments with complexes isolated from *C. elegans* (Cheeseman et al. 2006). Indeed, in budding yeast, MIND mutants have kinetochores that are still attached to the spindle tips (De Wulf et al. 2003, Scharfenberger et al. 2003). Importantly, in budding yeast, the MIND complex is recruited to the kinetochore independently of the Ndc80 and Spc105 complexes. In contrast, the latter complexes depend on the MIND/MIS12 complex for their localization to the vertebrate kinetochore. The yeast MIND complex is not passive in that it is crucial for the recruitment of MAPs and kinesins to the kinetochore–microtubule interface (De Wulf P., unpublished data, Fig. 6.4A) and may maximize the affinity of the outer kinetochore for the microtubule tip.

### *6.5.3 Outer Components of the Budding Yeast Kinetochore*

The budding yeast outer kinetochore Ndc80 (Ndc80/HEC1, Nuf2/hNuf2, Spc24/hSpc24, Spc25/hSpc25), Spc105 (Spc105/SPC105/Blinkin/KNL, Kre28/Zwint-1), and chromosomal passenger complex ((CPC); Ipl1/Aurora B, Sli15/INCENP, Bir1/Survivin) are essential and conserved. Both in budding yeast and metazoans, the Ndc80/HEC1 complex attaches kinetochores to the microtubules of the mitotic spindle (Janke et al. 2001, Wigge and Kilmartin 2001, DeLuca et al. 2006, Wei et al. 2007). In budding yeast, loss of the Spc105 complex leads to monopolar attachment (De Wulf P., Marco E., and Sorger P.K.; unpublished data), whereas in human cells, loss of SPC105/Blinkin results in an inability of kinetochores to bind and align onto the spindle, despite the presence of the HEC1 complex, which is recruited independently of the SPC105 complex

(Kiyomitsu et al. 2007, Cheeseman et al. 2008). In budding yeast, the Bir1-Sli15 complex makes contact with the spindle via Sli15 and may (as is the case in human cells) phosphorylate the Ndc80 complex by recruiting Ipl1, thereby coupling the mechanical state of kinetochore–microtubule attachment to the local control of Ipl1 kinase activity (Sandall et al. 2006). Ipl1 would then detach wrongly attached sister kinetochores via phosphorylation of Ndc80 and allow the outer kinetochore to reattach correctly to the spindle, thereby satisfying the spindle checkpoint.

Of significance, the budding yeast Ndc80 and Spc105 complexes are recruited to the centromere directly by CBF3 and not via the Cse3-Cse4-Mif2-COMA-MIND pathway (Fig. 6.4A). In higher species, the HEC1 and SPC105 complexes are recruited by MIND/MIS12 and the CENP-H class of inner kinetochore proteins (Ctf3 complex; Fig. 6.3).

ChIP analysis of MAPs (non-motor MAPs and motors) in budding yeast mutants that disrupt middle and outer complexes have revealed that multiple middle and outer complexes are in many cases required for kinetochore binding of a single MAP or motor (De Wulf P., unpublished data; Fig. 6.4A). Among the middle complexes, MIND proved to be the most important as centromere binding by all known MAPs and motors was reduced to background levels in MIND mutants implying that MIND and the MAPs it recruits are not critical for kinetochore–microtubule binding per se. Inactivating the Ndc80 complex also disrupted much of the outer kinetochore, except that Cin8, Kip1, and Kip3 remained centromere-associated at about one-half wild-type levels. Finally, in *spc105* mutants, the Bim1, Bik1, and Slk19 MAPs, and the Cin8 and Kar3 motors were lost from kinetochores whereas the other outer kinetochore components remained associated. Taken together, these data seem inconsistent with simple structural models in which a single outer kinetochore complex forms high-affinity associations with specific MAP or motor. Rather, complexes act in aggregate to create a network onto which microtubule-binding proteins assemble. It seems likely that one key component in this network is the microtubule plus end itself: many yeast kinetochore subunits probably make their primary physical interactions with tubulin subunits rather than with outer kinetochore proteins.

### 6.5.4 Constitutively Associated Components in Vertebrate Kinetochores

Next to CENP-A, which specifies the formation of a functional centromere, a collection of proteins localizes constitutively to the centromere during the vertebrate cell cycle. CENP-A directs three assembly pathways that are specified by CENP-C, the chromosomal passenger complex (Aurora B complex), and the CENP-S/T proteins, respectively (Fig. 6.3). Some of these proteins are conserved, including CENP-C (ScMif2, SpCnp3, CeHCP-4), CENP-H

(ScMcm16, SpFta3), CENP-I (ScCtf3, SpMis6), CENP-K (SpSim4), CENP-L (SpFta1), CENP-N (ScChl4, SpMis15), CENP-O (ScMcm21, SpMal2), and CENP-P (ScCtf19, SpFta2; Amor et al. 2004, Chan et al. 2005, Meraldi et al. 2006, Okada et al. 2006; Table 6.1).

In contrast to the CPC, which undergoes a cell cycle-dependent localization at the centromere/kinetochore (Cooke et al. 1987, Liu et al. 2006; Fig. 6.3), CENP-C and CENP-S/T are constitutive kinetochore components (Foltz et al. 2006, Okada et al. 2006). CENP-M class proteins depend for their localization on CENP-O and CENP-H class proteins (Okada et al. 2006). The dependency of CENP-S and CENP-T with respect to CENP-C, and the CPC is currently unclear.

All vertebrate kinetochore components, except for CENP-B and the members of the CENP-O complex, are essential and their depletion results in increased mitotic indices, lagging and misaligned chromosomes, and, in some cases, hypercondensed chromosomes (Foltz et al. 2006, Okada et al. 2006, Mellone et al. 2006). CENP-B may not be required for the maintenance of functional centromeres (Kipling and Warburton 1997) because mice lacking CENP-B have normal centromeres and do not exhibit chromosome segregation problems (Hudson et al. 1998, Perez-Castro et al. 1998, Kapoor et al. 1998). Furthermore, kinetochores of the normal human Y chromosome do not contain CENP-B (Earnshaw et al. 1991), nor do kinetochores found at human neocentromeres (Craig et al. 1999). *S. pombe* has three CENP-B homologs (Abp1, Cbh1, and Cbh2; Baum and Clarke 2000, represented by RFM (proteins required for methylation; in Fig. 6.4B) that bind to the outer repeat DNA sequences (otr in Fig. 6.4B). The activities of the CENP-B homologs are important, though, as the proteins assist in stable chromosome segregation possibly via involvement in heterochromatin formation (Nakagawa et al. 2002). Furthermore, the importance of the non-essential CENP-B became evident when CENP-B binding was found to be required for *de novo* kinetochore formation when naked $\alpha$-satellite DNA was introduced into human cells (Ohzeki et al. 2002). This was recently shown to result from CENP-B modulating the balance of heterochromatin and euchromatin during CENP-A deposition on this DNA (Okada et al. 2007, Okamoto et al. 2007).

Chicken cells depleted of CENP-O class proteins are viable and mislocalize kinetochore proteins of the CEN-M class. CENP-50/-U deficient cells (CENP-50/-U belongs to the CENP-O complex) show premature sister chromatid separation in nocodazole and a slow recovery from spindle damage (Minoshima et al. 2005). The CENP-O complex consequently may have a redundant role in sister chromatid cohesion or in the regulation of spindle checkpoint activity. Recently, Kang et al. (2006) reported that human CENP-U/-50 (referred as PBIP1) associates with polo-like kinase 1 (PLK1). The activity of CENP-O class proteins may in fact be regulated by PLK1 (Hori et al., 2008).

In general, CENP-A NAC (*n*ucleosome-*a*ssociated *c*omplex) components recruit CENP-A CAD (*C*ENP-*A d*istal) proteins, which in turn recruit subsets of outer kinetochore proteins. However, NAC component CENP-H and CAD protein CENP-I (both belong to the CENP-H class, Fig. 6.3) depend on each other for proper localization to the centromere (Fukagawa et al. 2001, Nishihashi et al. 2002; Fig. 6.3). In addition, a recent study indicates that depleting CENP-H and CENP-I from DT40 chicken cells results in a failure to incorporate newly synthesized CENP-A (Okada et al. 2006). Thus, the interdependency between CENP-A and components of the NAC and CAD complexes is likely more complex than is currently understood.

Foltz et al. (2006) observed that nucleolar protein nucleophosmin-1 is associated with CENP-A chromatin. Centromeres are known to associate with nucleoli during interphase in fly and human cells, and centromere proteins have been identified in purified nucleoli (Ochs and Press 1992). In addition, nucleolar transcription factors UBF1 and UBF2 interact with CENP-C (Pluta and Earnshaw 1996). The nucleolar anchoring of centromeres may protect centromeres during centromeric chromatin assembly or may either promote or reduce its accessibility to other factors. At the onset of mitosis, the centromere could then be released via nucleolar disassembly, allowing the association of other proteins resulting in a mature kinetochore (Mellone et al. 2006). Alternatively, the association of centromeres with human nucleoli could have a more prosaic explanation and simply arise from the fact that on human acrocentric chromosomes (chromosomes with their centromere near the very end of the chromatids), the centromere and ribosomal repeats are physically adjacent (acrocentric chromosomes contain the genes encoding ribosomal RNAs).

### *6.5.5 G2-Associating Components in Vertebrate Kinetochores*

Although kinetochore structures are not visible in interphase, pre-kinetochore structures persist, as shown by light and immunoelectron microscopy with antibodies to kinetochore components (Moroi et al. 1981, Brinkley et al. 1984, Cooke et al. 1990). One particularly elegant study used through-lens densitometry to show that pre-kinetochores resolve into double structures only late during the G2 phase of the cell cycle (Brenner et al. 1981). More recent studies have shown that a series of inner and outer proteins hierarchically accumulate at the kinetochore in G2. Constitutive members underlie several G2 assembly pathways (Fig. 6.3).

CENP-H class proteins recruit the HEC1/Ndc80 complex and CENP-F independently of one another (Hori et al. 2003, Liu et al. 2006). These subunits promote end-on microtubule contact and engage spindle checkpoint proteins, which oversee kinetochore–microtubule binding. Importantly, when the HEC1 protein is depleted from kinetochores by 10-fold, CENP-I levels drop by 40% suggesting the existence of a negative feedback loop between HEC1/Ndc80 and CENP-H class components (Liu et al. 2006).

Originally, the targeting of CENP-A and the MIS12 complex was reported to be mutually exclusive (Goshima et al. 2003). However, recent data have now placed MIS12 downstream of CENP-A in human cells (Liu et al. 2006),

*C. elegans* (Cheeseman et al. 2004) and budding yeast (Scharfenberger et al. 2003). These differences may arise from the varying efficiencies of RNAi in different experiments. Indeed, depleting kinetochore proteins by ~10-fold may not be sufficient to yield a true null phenotype (Meraldi et al. 2004).

CENP-C recruits the conserved MIS12/MIND complex in human (Liu et al. 2006) and chicken cells (Kwon et al. 2007). The vertebrate MIS12 complex independently recruits the Spc105/KNL-1/Blinkin and HEC1/Ndc80 complexes to kinetochores (Liu et al. 2006). In budding yeast and vertebrates, Spc105/KNL-1/Blinkin localizes to kinetochores independently of Ndc80/HEC1 (Kiyomitsu et al. 2007, Cheeseman et al. 2008, De Wulf et al.; in preparation). However, in *C. elegans*, the centromeric recruitment of the HEC1 complex depends on KNL-1 (Fig. 6.4C). In all eukaryotes, the HEC1/Ndc80, Spc105/KNL-1/Blinkin, and MIS12/MIND complexes collaborate to establish robust kinetochore–microtubule contacts (Desai et al. 2003, Cheeseman et al. 2006, Kline et al. 2006, Kiyomitsu et al. 2007) in order to align and bi-orient chromosomes at the metaphase plate, and to create a functional kinetochore–microtubule interface by selectively recruiting MAPs and kinesins (P. De Wulf et al. in preparation; see also in the next sessions).

CENP-C recruits spindle checkpoint protein BUB1(budding uninhibited by benzimidazole proteins) via a second G2 assembly pathway. BUB1 then recruits Mitotic Centromere-Associated Kinesin (MCAK; also known as XKCM1), CENP-F and the Spc105/KNL-1/Blinkin complex in G2. In mitosis, CENP-C recruits SGO1, the NUP107–160 complex as well as spindle checkpoint components BUBR1, monopolar spindle (MPS)1 and cell division cycle (CDC)20 (Loïodice et al. 2004, Liu et al. 2006, Orjalo et al. 2006). Thus, BUB1 is thought to oversee kinetochore–microtubule attachment via the MIS12, KNL-1 and HEC1 complexes. CENP-C also recruits the minus-end kinesin motor CENP-E and the RZZ (ROD-ZW10-ZWILCH) complex, respectively, in mitosis. CPC (Chromosome Passenger Components: Aurora B, INCENP, survivin, Borealin/Dasra B) recruitment has been reported to require CENP-A, and then cooperates with BUB1 to recruit MCAK and BUBR1 (Maiato et al. 2004, Liu et al. 2006). In *C. elegans*, the CPC targets to centromeres independently of CENP-A (Oegema et al. 2001). It is worth noting that in prometaphase and metaphase, the great majority of the CPC is localized in the heterochromatin beneath the kinetochore, and not associated with the kinetochore plates themselves (Earnshaw and Cooke 1991).

### 6.5.6 Mitosis-Associated Components in Vertebrate Kinetochores

Although vertebrate kinetochore proteins exhibit a temporal, hierarchical fashion of centromeric recruitment, cumulative data show that they associate via intersecting routes that form an intricate network (Fig. 6.3). This allows an optimal control of the presence and activities of kinetochore components in time and space.

For a successful mitosis to occur, the microtubules of the mitotic spindle must gain access to the chromosomes and the kinetochores formed on them. Microtubules are cytoplasmic structures in metazoans, such that kinetochore–microtubule contact necessitates the dissolution of the nuclear envelope (NE) and dispersion of the major structural elements of the NE, including the nuclear lamina and nuclear pore complexes (NPCs). Immediately following NE breakdown, kinetochores form and their trilaminar structure becomes visible (Roos 1973).

During mitotic prophase, PLK-NUDC (Nishino et al. 2006) and BUBR1 bind to the immature kinetochores via SGO1, BUB1 and the CPC (Maiato et al. 2004, Chan et al. 2005, Pouwels et al. 2007). In parallel, the NUP107-160 nuclear pore complex (NUP160, NUP133, NUP107, NUP96, NUP85, NUP43, NUP37, SEC13, and SEH1) is recruited by the HEC1 complex, and less substantially, by CENP-F (Zuccolo et al. 2007).

During pro-metaphase, the minus-end motor dynein, the nine-protein dynactin complex, CENP-E, MPS1, MAD1, MAD2, RZZ, CDC20, chromosomal region maintenance (CRM)1, RanGTP, RanGAP1, RanBP2/NUP358 join the kinetochore (Fig. 6.3). More specifically, CENP-E association requires CENP-C. Spindle checkpoint kinase MPS1 joins the kinetochore together with mitotic arrest deficient proteins MAD1 and MAD2 in a process that requires the HEC1 complex. Consequently, the CENP-A → MAD2 assembly pathway links the inner and outer kinetochore plates. In a separate CENP-H based branch, CENP-F recruits the dynactin microtubule-binding complex, followed by the dynein motor, to the kinetochore–microtubule interface. This recruitment of dynein is further supported by the RZZ complex and Spindly, which associate with kinetochores via CENP-C. RZZ also binds to MAD2. Spindly and dynactin then cooperatively recruit dynein (Griffis et al. 2007). Both dynein and CENP-E have been implicated in the initial encounters of kinetochores with microtubules (Rieder et al. 1990, Cooke et al. 1997, Yao et al. 1997). These encounters result in the transport of the chromosome towards the centrosomes along the lateral surface of microtubules (Rieder et al. 1990, Tanaka et al. 2005). Spindle assembly checkpoint signaling at the kinetochore is silenced when end-on kinetochore– microtubule binding is established.

Outer kinetochore Ran pathway proteins RanGTP, CRM1 (RanGTP-binding nuclear export receptor), RanGAP1 (GTPase activating protein that stimulates the conversion of RanGTP to RanGDP), and Ran-binding nucleoporin RanBP2/NUP358 are recruited to the kinetochore by guanine-exchange factor called Regulator of Chromatin Condensation 1 (RCC1; is chromatin-bound throughout the cell cycle; Ohtsubo et al. 1989, Moore et al. 2002) and the NUP107-160 nuclear pore complex (Fig. 6.3). This recruitment occurs only after initial kinetochore–microtubule attachments have been made (Joseph et al. 2002; 2004, Salina et al. 2003). Hence, the Ran pathway components are not indispensable for all kinetochore–microtubule connections but their arrival on kinetochores may promote some restructuring of kinetochore–microtubule connections to form mature attachments. This could occur via the destabilization of incorrect attachments made during early prometaphase likely in concert with other mitotic regulators (e.g., Aurora B, MCAK). The Ran components may also stabilize plus end kinetochore–microtubule attachments, promote the conversion of lateral to end-on attachments, and promote kinetochore fiber assembly by activating the microtubule-nucleating properties of kinetochores (Arnaoutov and Dasso 2005).

The NUP107–160 complex relocates to kinetochores, together with nucleoporin RanBP2/NUP358, upon NE breakdown and may serve to coordinate NE breakdown with spindle assembly. In budding yeast, a nearly identical NUP107-160 complex exists (Siniossoglou et al. 1996, 2000; Teixeira et al. 1997, Lutzmann et al. 2002) but it lacks NUP43 and NUP37, which have no yeast counterparts (Table 6.1). Depleting NUP107-160 leads to impaired chromosome congression, defects in kinetochore–microtubule binding, and reduced kinetochore tension (with a consequent increase in spindle length) resulting in a mitotic arrest. Furthermore, loss of the complex causes a delay in or failure to complete cytokinesis (Rasala et al. 2006). Putative transcription factor ELYS was recently shown to co-purify with the NUP107–160 complex in vertebrates, and to localize to NPCs during interphase and to kinetochores throughout mitosis (Rasala et al. 2006). Its activity at the kinetochore, however, is unknown.

CRM1, RanGAP1 and RanBP2/NUP358 recruit dynein, dynactin, the RZZ complex and the plus end-tracking protein CLIP-170 to the outer kinetochore, components that behave dynamically at the kinetochore–microtubule interface (see next section).

Following the attachment of kinetochores to microtubules and the alignment of the sister chromatids at the metaphase plate, the CPC, spindle checkpoint components (MPS1, MAD1, MAD2) and dynein–dynactin leave the kinetochore outer plate. Subsequently, CENP-F and, following chromosome segregation, PLK1-NUDC, HEC1, BUB1, BUBR1, MIS12, CRM1 and the Ran pathway components dissociate from the kinetochore. Most components of these complexes subsequently are degraded. At telophase, MCAK dissociates from the kinetochore leaving only the constitutive components at the centromere. The MIS18 complex is recruited to centromeres for allowing histone deacetylation and newly synthesized CENP-A to become incorporated. Later in G1 the MIS18 complex leaves the centromeres (Hayashi et al. 2004; Fig. 6.3).

### *6.5.7 Stable Versus Dynamic Components in Vertebrate Kinetochores*

Next to classifying kinetochore proteins based on their position within the organelle or when they are recruited to the centromere (see above), kinetochore components can also be grouped based on whether their concentration remains constant or varies at kinetochores during mitosis, or whether they turn over slowly (stable) or rapidly (dynamic) at their kinetochore location (Maiato et al. 2004). Dynamics of protein turnover have been measured by FRAP (Fluorescence Recovery After Photobleaching) of fluorescently tagged kinetochore members. Proteins that remain nearly constant during mitosis (slow turnover over 10 minutes) include the constitutive components of the inner plate (e.g., CENP-A, CENP-H, CENP-I) and the stable kinetochore components including the HEC1 complex, MCAK, PLK1, CENP-F, BUB1, and BUBR1 (Hori et al. 2003, Andrews et al. 2004, Howell et al. 2004, Mikami et al. 2005). These proteins form the core of the inner-, inter- and outer-plate structures. Dynamic components that change in concentration at the kinetochore during mitosis (residence half-life of 30 s or less) include CENP-E, RZZ, Spindly, dynein, dynactin, the CPC and members of the spindle assembly checkpoint MPS1, MAD1, MAD2, and CDC20 (Howell et al. 2000, Kallio et al. 2002, Howell et al. 2004, Shah et al. 2004). The latter proteins assemble at high concentrations at kinetochores in the absence of microtubules. This generates the signal that inhibits the anaphase promoting complex APC/C (Chapter 2). Checkpoint protein levels at the kinetochore are quickly reduced following the binding of spindle microtubules resulting in the establishment of an active kinetochore–microtubule interface (Hoffman et al. 2001). By metaphase, CENP-E, BUB3 and BUB1 levels at the kinetochore drop 3–4-fold relative to their levels at unattached kinetochores, whereas dynein, dynactin, MAD1, MAD2, and BUBR1 levels fall by a factor of up to 100-fold (King et al. 2000, Hoffman et al. 2001, Howell et al. 2004, Shah et al. 2004, Maiato et al. 2004).

A constant streaming of dynamic components between kinetochores and centrosomes seems to occur along the spindle microtubules. When detectable at kinetochores, MAD1, MAD2, Spindly and RZZ are seen to cycle continuously between kinetochores and spindle poles in a dynein- and ATP-dependent manner (Howell et al. 2001, Wojcik et al. 2001, Basto et al. 2004). Whereas these dynamic outer kinetochore components are removed from the kinetochore when microtubules attach (Hoffman et al. 2001), other outer components including EB1, TOG, LIS1, Eg5, APC, CLIP-170 and Ran pathway proteins (CRM1, RanGap1, RanGap2/NUP358) associate with kinetochores only when microtubules are attached (Kaplan et al. 2001, Tirnauer et al. 2002, Fodde et al. 2001, Joseph et al. 2002, Salina et al. 2003). These components might regulate kinetochore dynamics only when kinetochores are firmly bound to the microtubule plus ends. Alternatively, they may associate primarily with the plus ends of bundled kinetochore microtubules rather than with other kinetochore components themselves.

Photobleaching studies of various kinetochore proteins in *S. cerevisiae* have shown that after kinetochore assembly and formation of their bi-oriented metaphase configuration, fluorescently labeled kinetochore proteins show virtually no fluorescence recovery after photobleaching. This suggests that a fully assembled and bi-oriented kinetochore is relatively stable (Joglekar et al. 2006).

### 6.5.8 Kinetochore Assembly in Other Eukaryotes

The assembly of metazoan and yeast kinetochores occur through complex non-linear processes (Fig. 6.4A–D). Because many of the kinetochore proteins examined to date are evolutionarily conserved (Table 6.1) some degree of similarity in how they are recruited to centromeres is expected. However, a few differences are noteworthy.

In budding and fission yeast, Spc105/Spc7/KNL-1 is recruited independently of Mif2/CENP-C (Fig. 6.4A,B), whereas in *C. elegans* and *D. melanogaster* CENP-C directs the recruitment of KNL-1, which co-localizes with CENP-C to the inner kinetochore at the start of prophase (Desai et al. 2003, Cheeseman et al. 2004; Fig. 6.4D). Furthermore, in *C. elegans* and *D. melanogaster*, KNL-1/Spc105 recruits the HEC1/Ndc80 complex, whereas in budding yeast the Ndc80 and Spc105 complexes are recruited to centromeres independently of each other (Fig. 6.4C,D). In *C. elegans*, the KNL-1 complex recruits the MIS12/MIND complex, but in budding yeast both complexes are recruited independently of one another to the centromere (Fig. 6.4A). In fission yeast and *Drosophila*, the Spc105/Spc7 and MIS12 complexes interdepend for their localization to the kinetochore (Kerres et al. 2007). In addition, the fission yeast *N*dc80, *Mis*12, and *S*pc7 proteins/complexes form a larger, so-called NMS complex (Liu et al. 2005; Fig. 6.4B).

In sum, kinetochore assembly pathways in higher cells have evolved slightly differently in response to the rapid evolution of their underlying centromeres. Because the longer regional centromeres require a larger number of kinetochore proteins to establish and regulate contact with spindle microtubules, metazoan kinetochores may have preferably assembled via interwoven rather than independent pathways to ensure their quick and error-free formation in mitosis. The paradox is that this strategy may not have actually worked all that well since error rates during chromosome segregation are about 100-fold higher in humans than they are in budding yeast.

### 6.5.9 Kinetochore Assembly in Meiosis

During the first nuclear division in meiosis (meiosis I) homologous chromosomes are separated from each other, but sister chromatids stay together at their centromeres. During the second nuclear division (meiosis II), the sister chromatids are separated and segregated to opposite poles, as occurs in mitosis. In addition to linkages between homologs (usually chiasmata; chromatin structures assembled at the sites of cross overs), two further major modifications underlie this specialized segregation pattern. First, the orientation of sister kinetochore attachments to microtubule changes during meiosis. During meiosis I, sister kinetochores bind to microtubules emanating from the same spindle pole (co-orientation). This is in contrast to mitosis and meiosis II where sister kinetochores attach to microtubules emanating from opposite poles (bi-orientation). This creates the profound difference that in mitosis, tension across the centromeric chromatin between sister kinetochores is important to regulate chromosome–microtubule interactions, whereas in meiosis this tension must instead be propagated along the axis of the chromosome arms between homologous kinetochore pairs. Second, and in contrast to mitosis where all cohesion is lost during the metaphase-to-anaphase transition, the loss of cohesin in meiosis is spread out over two nuclear divisions. Cohesion on chromosome arms is lost during meiosis I, thereby triggering the segregation of homologs to the opposite poles. Cohesion around centromeric regions, however, is maintained until meiosis II. Loss of centromere cohesion during meiosis II triggers the segregation of sister chromatids to the opposite poles (Nasmyth 2005, Nasmyth and Haering 2005). Events at the kinetochore direct both the mono-orientation of kinetochores and the retention of centromere cohesion (for details see chapter 13).

A conserved feature of meiosis is that mitotic cohesin subunit Scc1/Mcd1/RAD21 is replaced by Rec8 (Klein et al. 1999, Watanabe and Nurse 1999, Pasierbek et al. 2001, Xu et al. 2005). However, in mammalian cells and grasshoppers, REC8- and RAD21-containing cohesin complexes co-exist during meiosis. In some organisms, the composition of cohesin complexes at centromeres and along chromosome arms differs. For example, in fission yeast and mammals the mitotic Scc3-like subunits are found in cohesin complexes around centromeres but are replaced on chromosome arms with a meiosis-specific variant named SpRec11 and STAG3, respectively (Prieto et al. 2001, Kitajima et al. 2003). This specialization of arm cohesin seems to facilitate meiotic recombination but could also contribute to the differential timing with which arm and centromere cohesins are lost in meiosis (meiosis I and II, respectively). The cleavage of cohesin by separase triggers chromatid segregation. Kinetochore component SGO1, originally described as MEI-S332 (*Drosophila*; Kerrebrock et al. 1992), Shugoshin/Sgo1 (*S. pombe*; Kitajima et al. 2004, Rabitsch et al. 2004), Sgo1 (*S. cerevisiae*; Katis et al. 2004, Marston et al. 2004) or zmSgo1 (*Z. mais*; Hamant et al. 2005) protects centromere cohesin from premature cleavage by separase. In budding yeast, this protection occurs in part via the association of Sgo1 with protein phosphatase PP2A. This complex inhibits separase activity at centromere cohesin in meiosis I via dephosphorylation of at least one cohesin subunit, e.g., Rec8 and Scc3, which are targets of the Polo kinase Cdc5 (Brar et al. 2006, Kitajima et al. 2006, Riedel et al. 2006). In mitosis, Cdc5 phosphorylation is required to render Scc1 an efficient substrate for separase (Alexandru et al. 2001).

In mitosis, Shugoshin family members promote the bi-orientation of sister chromatids and are required to sense a lack of tension at kinetochores in budding yeast (Indjeian et al. 2005), fission yeast (Kawashima et al. 2007), and *Drosophila* (Vanoosthuyse et al. 2007). This can explain the activity of these proteins at meiosis II kinetochores. However, Sgo1 is not an essential component of the budding yeast kinetochore in mitosis.

How Sgo1 is recruited to the kinetochores is poorly understood. In budding yeast, Sgo1 partially depends on cohesin because Rec8 is required for Sgo1 association with the pericentromere but not the core centromere (Kiburz et al. 2005). Indeed, in the absence of kinetochore proteins Iml3 or Chl4, which maintain centromere cohesin during meiosis I (Marston et al. 2004), Sgo1 still associates with the centromere but is reduced at pericentromeric regions (Kiburz et al. 2005). Spindle checkpoint protein Bub1 and its kinase activity are instrumental for the localization of Sgo1 to the centromere in budding yeast, fission yeast, and human cells (Kitajima et al. 2004, Rabitsch et al. 2004, Vaur et al. 2005, Pouwels et al. 2007). In *Drosophila*, INCENP is required for MEI-S332 localization at centromeres both during mitosis and meiosis, and MEI-S332 is phosphorylated by the chromosomal passenger Aurora B kinase (Resnick et al. 2006). In addition, the chromosomal passenger kinase complex appears to be required for the full localization of Sgo1. Importantly, the release of MEI-S332/Sgo1 from centromeres is regulated via phosphorylation by Polo kinase in *Drosophila* (Clarke et al. 2005) and mammalian cells (Tang et al. 2006).

During the transition from mitosis to meiosis, the kinetochore undergoes significant reorganization, switching from a bipolar to a monopolar orientation. In budding yeast, the mono-orientation of sister kinetochores in meiosis I is achieved by the four-protein complex named monopolin (Mam1, casein kinase 1 Hrr25, Csm1, Lrs4), which may act to clamp together the microtubule binding sites of the sister chromatids (Rabitsch et al. 2003, Gregan et al. 2007). Csm1 and Lrs4 generally reside in the nucleolus but promote mono-orientation upon their release by polo kinase Cdc5. This ensures that monopolin forms and then accumulates at kinetochores in meiosis I only. New data further suggest that Mam1 and/or Rec8 promote mono-orientation of sister kinetochores following phosphorylation by Hrr25 (Petronczki et al. 2006). Interestingly, overexpression of *MAM1* together with *CDC5* causes mono-orientation of kinetochores in budding yeast during mitosis (Monje-Casas et al. 2007) suggesting that Mam1 may be the only critical subunit of monopolin that is meiosis-specific.

The monopolin complex is found only in fungi (Rabitsch et al. 2003). However, in fission yeast, the Csm1 and Lrs4 homologs, known as Pcs1 and Mde4, respectively, are not required for mono-orientation in meiosis I but rather contribute to chromosome segregation in both mitosis and meiosis II (Rabitsch et al. 2003, Gregan et al. 2007). In fission yeast, Rec8-containing cohesin complexes and meiosis-specific factor Moa1 localize to the core of the centromere and promote mono-orientation in meiosis I (Yokobayashi and Watanabe 2005).

The role of Rec8 in monopolar attachment appears to be conserved in maize and *Arabidopsis* (Chelysheva et al. 2005, Hamant et al. 2005) but the process by which Rec8 establishes monopolar binding of sister kinetochores is still unclear.

In order to determine how *S. pombe* kinetochores assemble in meiotic prophase, Hayashi et al. (2006) examined the localization of 22 mitotic and 2 meiotic (Sgo1, Moa1) kinetochore proteins during meiosis. They found that kinetochore proteins can be classified into three groups based on when they are recruited to the centromere during meiosis. The Mis6-like group (protein members are shown in Fig. 6.4B) was found at the kinetochore from karyogamy to meiosis II. Moa1 is then recruited to the Mis6-loaded centromere. Members of the NMS group (Ndc80 complex – Mis12/MIND complex – Spc7/Spc105 complex) together with Sgo1 were recruited at the end of prophase, followed by the DASH-Dam1 complex, allowing the mono-orientation of mature sister kinetochores to the spindle tips and protection of centromere cohesin (Hayashi et al. 2006). The observed recruitment hierarchy in meiosis reflects the pattern of kinetochore formation during mitosis (Fig. 6.4B), although no biochemical dissection of kinetochore assembly has been done on meiotic *S. pombe*.

Electron microscopic analyses of holocentric chromosomes from *Oncopeltus* (milkweed bug) and *C. elegans* first indicated that the trilaminar kinetochore structure is absent in meiosis as microtubules appeared to embed directly into chromatin (Comings and Okada 1972, Albertson and Thomson 1993). However, the ultrastructural examination of pig and mouse oocyte kinetochores via immunoelectron microscopy subsequently revealed the typical trilaminar organization of the organelle (Schatten et al. 1988, Lee et al. 2000), suggesting that kinetochore formation during mitosis and meiosis differs only in holocentromeric species (see below). Importantly, how kinetochores assemble during meiosis is largely unknown. Indeed, most studies of meiotic kinetochores have to date focused on the localization and recruitment dynamics of spindle checkpoint proteins. Kallio et al. (2000) showed that MAD2 is present continuously at kinetochores during mouse female meiosis. In contrast, dynein removes MAD2 from vertebrate mitotic kinetochores in metaphase when firm contact with microtubules is established. The same is true for CENP-E; the latter leaves kinetochores at the end of mitotic metaphase and becomes degraded, whereas in meiosis (pig oocytes) CENP-E remains at kinetochores during the meiosis I/ meiosis II transition (Lee et al. 2000). The differences in recruitment patterns may reflect differences in spindle formation between mitosis (the spindle is established before nuclear breakdown) and meiosis (tubulin polymerizes following germinal vesicle breakdown; Ma et al. 2003). Alternatively, this may reflect differences in "reading" spindle tension across centromere chromatin (mitosis) or along chromosome axes (meiosis).

During mitosis in *C. elegans*; CENP-A/HCP-3, CENP-C/HCP-4 and KNL-1 follow a linear assembly pathway that is crucial for forming the kinetochore–microtubule interface (Fig. 6.4C; Desai et al. 2003, Cheeseman et al. 2004). Remarkably, a close examination of HCP-3, HCP-4 and KNL-1 localization during oocyte meiosis showed that in *C. elegans* the targeting of outer kinetochore components was uncoupled from HCP-3 and HCP-4 containing chromatin (Monen et al. 2005). Indeed, KNL-1 was targeted normally to centromeres during meiosis I and II in HCP-3 or HCP-4 depleted embryos. In addition, chromosome alignment and segregation were not significantly affected during either meiotic division in HCP-3 minus embryos. Although the timing of the meiotic divisions and the interval between metaphase of meiosis II and the metaphase of the first mitotic division did not change in the HCP-3 depleted embryos, the embryos failed to segregate their chromosomes during mitosis (Monen et al. 2005). In *C. elegans*, the restriction of crossovers to only one per homologous pair, the crossover position-dependent remodeling of the bivalent chromosome pair in diakinesis, and the uncoupling of the targeting of outer-kinetochore proteins (e.g., KNL-1 complex) from HCP-3/-4 chromatin seems to act coordinately to ensure chromosome segregation during meiosis (Monen et al. 2005).

## 6.6 Structural Organization of Kinetochores

Epistasis analysis of kinetochore protein recruitment to centromeres has yielded valuable insight into how kinetochores form, both in time and space (Figs. 6.3, 6.4). However, it is important to note that protein-recruitment patterns do not per se define protein-protein interactions as numerous kinetochore components likely wait to be identified, especially in metazoans.

In higher eukaryotes, protein recruitment maps and electron tomography followed by a computational reconstitution of the images in 3D have revealed that the outer kinetochore of vertebrate cells is organized as a series of anastomosing fibers (Dong et al. 2007). This work extended previous tomography studies by McEwen et al. (1993) as well as studies localizing components to various domains within the kinetochore by immunoelectron microscopy (Cooke et al. 1990, Steuer et al. 1990, Saitoh et al. 1992, Cooke et al. 1997). For budding yeast, structural information on kinetochore architecture has yet to be obtained via electron microscopy (Peterson and Ris 1976, Winey et al. 1995). However, the kinetochore recruitment map combined with subunit quantification data derived from microtubule-bound kinetochores (Joglekar et al. 2006) allows us to propose a hypothetical model of the budding yeast kinetochore (Fig. 6.5). For *S. pombe*, electron microscopy does reveal specialized structures where the 2–4 microtubules bind to each chromosome (Ding et al. 1993), and subsequent analysis is consistent with a layered structure reminiscent of that seen in metazoan kinetochores (Kniola et al. 2001). Ultimately, the production, crystallization and/or *in vitro* reconstitution of recombinant kinetochore proteins and complexes should reveal what kinetochores look like and how their biochemical activities are organized in three-dimensional space.

### *6.6.1 Structural Organization of Budding Yeast Kinetochores*

Significant progress in our understanding of budding yeast kinetochore architecture was made when Joglekar and colleagues (2006) measured the copy numbers of *S. cerevisiae* proteins in microtubule-bound kinetochores. This

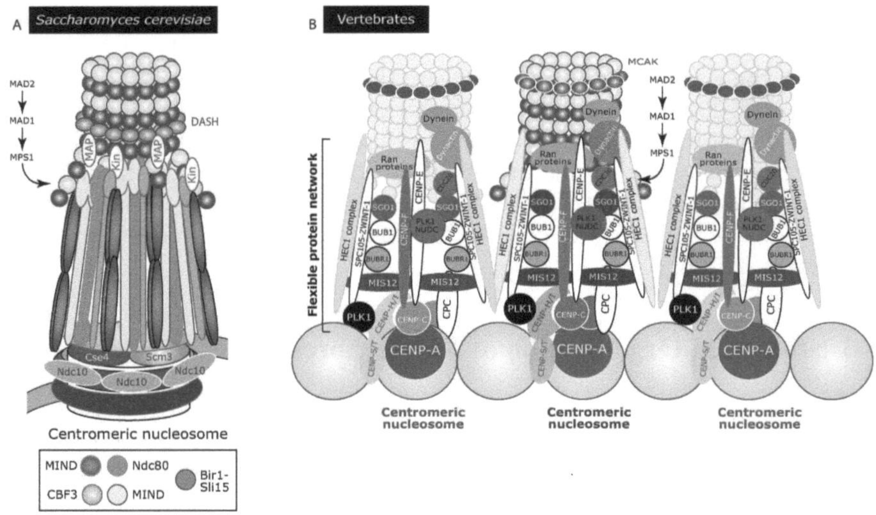


**Fig. 6.5** Kinetochore organization (A) Model of the *S. cerevisiae* kinetochore. Based on the assembly map of the budding yeast kinetochore (Fig. 6.4A), copy numbers of kinetochore proteins and complexes (Joglekar et al. 2006), the in vitro reconstitution of homologues of yeast kinetochore proteins and complexes (Cheeseman et al. 2006), and in vitro data showing that the Ndc80 complex binds to microtubules in an angle (Cheeseman et al. 2006, Ciferri et al. 2008) and likely forms a sleeve around the edge of a single microtubule due to the cooperative binding of Ndc80 complexes (Ciferri et al. 2008), we represent the kinetochore as a pyramid- or cone-like assembly that attaches a single centromeric nucleosome to a single microtubule. (B) Model of the vertebrate kinetochore. The model was constructed based on assembly maps of vertebrate kinetochores (Fig. 6.3), copy numbers of kinetochore proteins and complexes (Emanuele et al. 2005), and data showing that the Ndc80 complex binds to microtubules in an angle (Cheeseman et al. 2006, Ciferri et al. 2008) and likely forms a sleeve around each microtubule due to the cooperative binding of Ndc80 complexes (Ciferri et al. 2008). Vertebrate kinetochores contain 20–30 microtubules. Recent observations by Dong et al. (2007) suggest that kinetochore proteins and complexes form dense parallel bundles in the absence of microtubules. In the presence of microtubules, the complexes and proteins arrange into a packed network that firmly encapsulates the microtubules. See text for details

was done by quantitative fluorescence microscopy, and was based on the observation that there are two copies of Cse4/CENP-A per kinetochore. DNA-bound Mif2/CENP-C, for instance, is present in one or two copies, whereas the KNL-1/Spc105, MIS12/MIND, CPC/Sli15-Bir1, HEC1/Ndc80 complexes are present in five to eight copies. The presence of the microtubule-binding HEC1/Ndc80 complex in eight copies, which likely encircle the outer edge of a single microtubule (25 nm diameter; Ciferri et al. 2008), may produce a cylindrical or cone-like kinetochore structure that fits on top of a centromeric nucleosome (10 nm diameter; Fig. 6.5A). A symmetric spacing of the complexes within the kinetochore would evenly distribute the metaphase forces exercised by the microtubule tips onto the sister centromeres.

### *6.6.2 Structural Organization of Vertebrate Kinetochores*

The significant size difference between vertebrate and yeast kinetochores, the fact that kinetochore components are well conserved among eukaryotes (Table 6.1), and the observation that some metazoan kinetochore proteins are present in very high copy number (some in more than 1000 copies) lead to the "repeat-subunit kinetochore model", which states that vertebrate kinetochores assemble forming an array of identical units (docking sites) that are spread along the centromere (Zinkowski et al. 1991).

Dong et al. (2007) used electron tomography to construct a high-resolution structural map of the kinetochore outer plate. They found that before microtubules bind to a kinetochore, the kinetochore outer plate comprises cross-linked fibers that consist of, among other, the rod-shaped HEC1, Blinkin/KNL-1, MIS12 and CENP-F subunits (Fig. 6.5B). When microtubules bind, the outer-plate meshwork becomes less organized and thicker in appearance as fibers extend from the outer plate and bind to the sides of microtubules. Other fibers change their orientation within the plane of the outer plate to form a radial array of attachments to the microtubule tips. Based on these findings, a "flexible network model" of the kinetochore outer plate was proposed (Dong et al. 2007).

Individually, outer kinetochore members HEC1, KNL-1 and CENP-F exhibit low microtubule-binding affinities. However, there is strong synergy when more than one putative outer plate component is present (Cheeseman et al. 2006, Feng et al. 2006). It has been postulated that each microtubule is bound by an array of low-affinity components, rather than by a few high-affinity subunits (Cheeseman et al. 2006). It has been reported that vertebrate kinetochores have approximately 800–1,200 copies of the HEC1 and MIS12 complexes (Emanuele et al. 2005), which is a 3–10-fold excess over what is required to bind the 20–30 microtubules. This observation supports the idea that microtubule-binding activity in vertebrates is dispersed over a network, rather than packed into singular binding sites (Dong et al. 2007). The network model thus postulates an efficient system for microtubule capture because the microtubules do not need to orient precisely into the outer kinetochore. Microtubules can contact the sticky network at a variety of angles, as is frequently observed. This model thus solves the problem of how a kinetochore converts from lateral to end-on binding without falling off the spindle, because lateral attachments can simply remain intact while end-on attachments are added.

## 6.7 Final Comment

Chromosome segregation in mitosis and meiosis depends on the assembly of functional kinetochores on centromeres. Despite their universal roles, centromeres and kinetochore proteins have diverged rapidly among eukaryotes. The continuing identification of kinetochore proteins in numerous species and the analysis of how these proteins form mature assemblies have already revealed a common theme of formation via networking. Despite the significant advances that have been made since the first cloning of centromere DNA and proteins in the 1980s, we still lack a clear understanding of the basic mechanisms that underlie kinetochore formation and its regulation, including for example how CENP-A is incorporated at the centromere. This knowledge will be invaluable to develop the kinetochore into an almost inexhaustible target for new anti-cancer drugs (Chapter 15). However, because progress in kinetochore and centromere biology is being made at an astounding pace, new insights in kinetochore formation and activity can be expected in the very near future.

**Acknowledgments** We wish to thank the members of the Fukagawa and De Wulf laboratories for discussions and comments. Research in the Fukagawa laboratory is supported by Grants-in-aid for Scientific Research from the Ministry of Education, Science, Sports and Culture of Japan (MEXT). Research in the De Wulf laboratory is supported by the European Institute of Oncology, the Italian Association for Cancer Research, the European Molecular Biology Organization, the Association for International Cancer Research (U.K.), and the National Cancer Institute (U.S.A.).

## References

Ahmad, K., and Henikoff, S. 2001. Centromeres are specialized replication domains in heterochromatin. J. Cell Biol. 153: 101–110.

Albertson, D.G., and Thomson, J.N. 1993. Segregation of holocentric chromosomes at meiosis in the nematode, *Caenorhabditis elegans*. Chromosome Res. 1: 15–26.

Alexandru, G., Uhlmann, F., Mechtler, K., Poupart, M.A., and Nasmyth, K. 2001. Phosphorylation of the cohesin subunit Scc1 by polo/Cdc5 kinase regulates sister chromatid separation in yeast. Cell 105: 459–472.

Alonso, A., Mahmood, R., Li, S., Cheung, F., Yoda, K., and Warburton, P.E. 2003. Genomic microarray analysis reveals distinct locations for the CENP-A binding domains in three human chromosome 13q32 neocentromeres. Hum. Mol. Genet. 12: 2711–2721.

Amor, D.J., and Choo, K.H.A. 2002. Neocentromeres: role in human disease, evolution, and centromere study. Am. J. Hum. Genet. 71: 695–714.

Amor D.J., Kalitsis, P., Sumer, H., and Choo, K.H.A. 2004. Building the centromere: from foundation proteins to 3D organization. Trends Cell Biol. 14: 359–368.

Andrews, P.D., Ovechkina, Y., Morrice, N., Wagenbach, M., Duncan, K., Wordeman, L., and Swedlow, J.R. 2004. Aurora B regulates MCAK at the mitotic centromere. Dev. Cell 6: 253-268.

Arnaoutov, A., and Dasso, M. 2005. The Ran GTPase regulates kinetochore attachment in somatic cells. Cell Cycle 4: 1161–1165.

Baker, R.E.M., Fitzgerald-Hayes, M., and O'Brien, T.C. 1989. Purification of the yeast centromere binding protein CP1 and a mutational analysis of its binding site. J. Biol. Chem. 264: 10843–10850.

Bansal, P.K., Abdulle, R., and Kitagawa, K. 2004. Sgt1 associates with Hsp90: an initial step of assembly of the core kinetochore complex. Mol. Cell. Biol. 24: 8069–8079.

Basto, R., Scaerou, F., Mische, S., Wojcik, E., Lefebvre, C., Gomes, R., Hays, T., and Karess, R. 2004. In vivo dynamics of the rough deal checkpoint protein during *Drosophila* mitosis. Curr. Biol. 14: 56–61.

Baum, M. and Clarke, L. 2000. Fission yeast homologs of human CENP-B have redundant functions affecting cell growth and chromosome segregation. Mol. Cell. Biol. 20: 2852–2864.

Bellizzi, J.J. 3rd, Sorger, P.K., and Harrison, S.C. 2007. Crystal structure of the yeast inner kinetochore subunit Cep3p. Structure 15: 1422–1430.

Bernat, R.L., Borisy, G.G., Rothfield, N.F., and Earnshaw, W.C. 1990. Injection of anticentromere antibodies in interphase disrupts events required for chromosome movement at mitosis. J. Cell Biol. 111: 1519–1533.

Bernat, R.L., Delannoy, M.R., Rothfield, N.F., and Earnshaw, W.C. 1991. Disruption of centromere assembly during interphase inhibits kinetochore morphogenesis and function in mitosis. Cell 66: 1229–1238.

Black, B.E., Foltz, D.R., Chakravarthy, S., Luger, K., Woods, V.L. Jr., and Cleveland, D.W. 2004. Structural determinants for generating centromeric chromatin. Nature 430: 578–582.

Black, B.E., Jansen, L.E., Maddox, P.S., Foltz, D.R., Desai, A.B., Shah, J.V., and Cleveland, D.W. 2007. Centromere identity maintained by nucleosomes assembled with histone H3 containing the CENP-A targeting domain. Mol. Cell 25: 309–322.

Blower, M.D., and Karpen, G.H. 2001. The role of *Drosophila* CID in kinetochore formation, cell-cycle progression and heterochromatin interactions. Nat. Cell Biol. 3: 730–739.

Blower, M.D., Sullivan, B.A., and Karpen, G.H. 2002. Conserved organization of centromeric chromatin in flies and humans. Dev. Cell 2: 319–330.

Brar, G.A., Kiburz, B.M., Zhang, Y., Kim, J.E., White, F., and Amon, A. 2006. Rec8 phosphorylation and recombination promote the step-wise loss of cohesins in meiosis. Nature 441: 532–536.

Brenner, S., Pepper, D., Berns, M.W., Tan, E., and Brinkley, B.R. 1981. Kinetochore structure, duplication and distribution in mammalian cells: analysis by human autoantibodies from scleroderma patients. J. Cell Biol. 91: 95–102.

Brinkley, B.R., and Stubblefield, E. 1966. The fine structure of the kinetochore of a mammalian cell in vitro. Chromosoma 19: 28–43.

Brinkley, B.R., Valdivia, M.M., Tousson, A., and Brenner, S.L. 1984. Compound kinetochores of the Indian muntjac: Evolution by linear fusion of unit kinetochores. Chromosoma 91: 1–11.

Brock, J.K., and Bloom, K. 1994. A chromosome breakage assay to monitor mitotic forces in budding yeast. J. Cell Sci. 107: 891–902.

Brown, M.T., Goetsch, L., and Hartwell, L.H. 1993. MIF2 is required for mitotic spindle integrity during anaphase spindle elongation in *Saccharomyces cerevisiae*. J. Cell Biol. 123: 387–403.

Buchman, A.R., and Kornberg, R.D. 1990. A yeast ARS-binding protein activates transcription synergistically in combination with other weak activating factors. Mol. Cell. Biol. 10: 887–897.

Buerstedde, J.M., and Takeda, S. 2006. Reviews and Protocols in DT40 Research, Vol. 40, Springer Publ. New York.

Cai, M., and Davis, R.W. 1989. Purification of a yeast centromere-binding protein that is able to distinguish single base-pair mutations in its recognition site. Mol. Cell. Biol. 9: 2544–2550.

Cai, M., and Davis, R.W. 1990. Yeast centromere binding protein CBF1, of the helix-loop-helix protein family, is required for chromosome stability and methionine prototrophy. Cell 61: 437–446.

Camahort, R., Li, B., Florens, L., Swanson, S.K., Washburn, M.P., and Gerton, J.L. 2007. Scm3 is essential to recruit the histone H3 variant Cse4 to centromeres and to maintain a functional kinetochore. Mol. Cell 26: 853–865.

Carrol, C.W., and Straight, A.F. 2007. Centromeric chromatin gets loaded. J. Cell Biol. 176: 735–736.

Catlett, M.G., and Kaplan, K.B. 2006. Sgt1p is a unique co-chaperone that acts as a client adaptor to link Hsp90 to Skp1p. J. Biol. Chem. 281: 33739–33748.

Chan, G.K., Liu, S.T., and Yen, T.J. 2005. Kinetochore structure and function. Trends Cell Biol. 15: 589–598.

Cheeseman, I.M., Niessen, S., Anderson, S., Hyndman, F., Yates, J.R. III, Oegema, K., and Desai, A. 2004. A conserved protein network controls assembly of the outer kinetochore and its ability to sustain tension. Genes Dev. 18: 2255–2268.

Cheeseman, I.M., Chappie, J.S., Wilson-Kubalek, E.M., and Desai, A. 2006. The conserved KMN network constitutes the core microtubule-binding site of the kinetochore. Cell 127: 983–997.

Cheeseman, I.M, Hori, T., Fukagawa, T., and Desai, A. 2008. KNL1 and the CENP-H/I/K complex coordinates direct kinetochore assembly in vertebrates. Mol. Biol. Cell. 19: 587–594.

Chelysheva, L., Diallo, S., Vezon, D., Gendrot, G., Vrielynck, N., Belcram, K., Rocques, N., Marquez-Lema, A., Bhatt, A.M., Horlow, C., Mercier, R., Mezard, C., and Grelon, M. 2005. AtREC8 and AtSCC3 are essential to the monopolar orientation of the kinetochores during meiosis. J. Cell Sci. 118: 4621–4632.

Chen, Y., Baker, R.E., Keith, K.C., Harris, K., Stoler, S., and Fitzgerald-Hayes, M. 2000. The N terminus of the centromere H3-like protein Cse4p performs an essential function distinct from that of the histone fold domain. Mol. Cell. Biol. 20: 7037–7048.

Ciferri, C., De Luca, J., Monzani, S., Ferrari, K.J., Ristic, D., Wyman, C., Stark, H., Kilmartin, J., Salmon, E.D., and Musacchio, A. 2005. Architecture of the human Ndc80-Hec1 complex, a critical constituent of the outer kinetochore. J. Biol. Chem. 280: 29088–29095.

Ciferri, C., Pasqualato, S., Screpanti, E., Varettti, G., Santaguida, S., Maiolica, A., Polka, J., De Luca, J.G., Dos Reis, G., De Wulf, P., Salek, M., Rappsilber, J., Moores, C.A., Salmon, E.D., and Musacchio, A. 2008. Implications for kinetochore-microtubule attachment from the structure of an engineered Ndc80 complex. Cell 133: 427–439.

Choo, K.H. 2001. Domain organization at centromere and neocentromere. Dev. Cell 1: 165–177.

Clarke, L., and Carbon, J. 1985. The structure and function of yeast kinetochores. Ann. Rev. Genet. 19: 29–56.

Clarke, A.S., Tang, T.T., Ooi, D.L., and Orr-Weaver, T.L. 2005. POLO kinase regulates the *Drosophila* centromere cohesion protein MEI-S332. Dev. Cell 8: 53–64.

Comings, D.E., and Okada, T.A. 1971. Fine structure of kinetochore in Indian muntjac. 1971. Exp. Cell Res. 67: 97–110.

Comings, D.E., and Okada, T.A. 1972. Holocentric chromosomes in *Oncopeltus*: kinetochore plates are present in mitosis but absent in meiosis. Chromosoma 37: 177–192.

Cooke, C.A., Bernat, R.L., and Earnshaw, W.C. 1990. CENP-B: A major human centromere protein located beneath the kinetochore. J. Cell Biol. 110: 1475–1488.

Cooke, C.A., Heck, M.M.S., and Earnshaw, W.C. 1987. The INCENP antigens: movement from inner centromere to midbody during mitosis. J. Cell Biol. 105: 2053–2067.

Cooke, C.A., Schaar, B., Yen, T.J., and Earnshaw, W.C. 1997. Localization of CENP-E in the fibrous corona and outer plate of mammalian kinetochores from prometaphase through anaphase. Chromosoma 106: 446–455.

Craig, J.M., Earnshaw, W.C., and Vagnarelli, P. 1999. Mammalian centromeres: DNA sequence, protein composition, and role in cell cycle progression. Exp. Cell Res. 246: 249–262.

Dalal, Y., Furuyama, T., Vermaak, D., and Henikoff, S. 2007. Inaugural Article: Structure, dynamics, and evolution of centromeric nucleosomes. Proc. Natl. Acad. Sci. U.S.A. 104: 15974–15981.

DeLuca, J.G., Gall, W.E., Ciferri C., Cimini, D., Musacchio, A., and Salmon, E.D. 2006. Kinetochore microtubule dynamics and attachment stability are regulated by Hec1. Cell 127: 969–982.

Desai, A., Rybina, S., Muller-Reichert, T., Shevchenko, A., Shevchenko, A., Hyman, A., and Oegema, K. 2003. KNL-1 directs assembly of the microtubule-binding interface of the kinetochore in *C. elegans*. Genes Dev. 17: 2421–2435.

De Wulf, P., McAinsh, A.D., and Sorger, P.K. 2003. Hierarchical assembly of the budding yeast kinetochore from multiple subcomplexes. Genes Dev. 17: 2902–2921.

Ding, R., McDonald, K. L., and McIntosh, J. R. 1993. Three-dimensional reconstruction and analysis of mitotic spindles from the yeast *Schizosaccharomyces pombe*. J. Cell Biol. 120: 141–151.

Doheny, K.F., Sorger, P.K., Hyman, A.A., Tugendreich, S., Spencer, F., and Hieter, P. 1993. Identification of essential components of the *S. cerevisiae* kinetochore. Cell 73: 761–774.

Dong, Y., Vanden Beldt, K.J., Meng X., Khodjakov, A., and McEwen, B.F. 2007. The outer plate in vertebrate kinetochores is a flexible network with multiple microtubule interactions. Nat. Cell Biol. 9: 516–522.

Earnshaw, W.C., Bernat, R.L., Cooke, C.A., and Rothfield, N.F. 1991. The role of the centromere/ kinetochore in cell cycle control. Cold Spring Harbor Symp. Quant. Biol. 56: 675–685.

Earnshaw, W.C., and Cooke, C.A. 1991. Analysis of the distribution of the INCENPs throughout mitosis reveals the existence of a pathway of structural changes in the chromosomes during metaphase and early events in cleavage furrow formation. J. Cell Sci. 98: 443–461.

Earnshaw, W.C., Halligan, N., Cooke, C., and Rothfield, N. 1984. The kinetochore is part of the chromosome scaffold. J. Cell Biol. 98: 352–357.

Earnshaw, W.C., and Migeon, B. 1985. A family of centromere proteins is absent from the latent centromere of a stable isodicentric chromosome. Chromosoma 92: 290–296.

Earnshaw, W.C., and Rothfield, N. 1985. Identification of a family of human centromere proteins using autoimmune sera from patients with scleroderma. Chromosoma 91: 313–321.

Earnshaw, W.C., Sullivan, K.F., Machlin, P.S., Cooke, C.A., Kaiser, D.A., Pollard, T.D., Rothfield, N.F., and Cleveland, D.W. 1987. Molecular cloning of cDNA for CENP-B, the major human centromere autoantigen. J. Cell Biol. 104: 817–829.

Emanuele, M.J., McCleland, M.L., Satinover, D.L., and Stukenberg, P.T. 2005. Measuring the stoichiometry and physical interactions between components elucidates the architecture of the vertebrate kinetochore. Mol. Biol. Cell 16: 4882–4892.

Feng, J., Huang, H., and Yen, T.J. 2006. CENP-F is a novel microtubule-binding protein that is essential for kinetochore attachments and affects the duration of the mitotic checkpoint delay. Chromosoma 115: 320–329.

Fleig, U., Sen-Gupta, M., and Hegemann, J.H. 1996. Fission yeast mal2+ is required for chromosome segregation. Mol. Cell Biol. 16: 6169–6177.

Fodde, R., Kuipers, J., Rosenberg, C., Smits, R., Kielman, M., Gaspar, C., van Es, J.H., Breukel, C., Wiegant, J., Giles, R.H., and Clevers, H. 2001. Mutations in the APC tumour suppressor gene cause chromosomal instability. Nat. Cell Biol. 3: 433–438.

Foltz, D., Jansen, L.E., Black, B.E., Bailey, A.O., Yates, J.R. III, and Cleveland, D.W. 2006. The human CENP-A centromeric nucleosome-associated complex. Nature Cell Biol. 8: 458–469.

Fukagawa, T., Mikami, Y., Nishihashi, A., Regnier, V., Haraguchi, T., Hiraoka, Y., Sugata, N., Todokoro, K., Brown, W., and Ikemura, T. 2001. CENP-H, a constitutive centromere component, is required for centromere targeting of CENP-C in vertebrate cells. EMBO J. 20: 4603–4617.

Fujita, Y., Hayashi, T., Kiyomitsu, T., Toyoda, Y., Kokubu, A., Obuse, C., and Yanagida, M. 2007. Priming of centromere for CENP-A recruitment by human hMis18alpha, hMis18beta, and M18BP1. Dev. Cell 12: 17–30.

Furuyama, T., and Henikoff, S. 2006. Biotin-tag affinity purification of a centromeric nucleosome assembly complex. Cell Cycle 5: 1269–1274.

Gillis, A.N., Thomas, S., Hansen, S.D., and Kaplan, K.B. 2005. A novel role for the CBF3 kinetochore-scaffold complex in regulating septin dynamics and cytokinesis. J. Cell. Biol. 171: 773–784.

Gimelli, G., Zuffardi, O., Giglio, S., Zeng, C., and He, D. 2000. CENP-G in neocentromeres and inactive centromeres. Chromosoma 109: 328–333.

Goh, P.Y., and Kilmartin, J.V. 1993. NDC10: a gene involved in chromosome segregation in *Saccharomyces cerevisiae*. J. Cell Biol. 121: 503–512.

Goshima, G., Kiyomitsu, T., Yoda, K., and Yanagida, M. 2003. Human centromere chromatin protein hMis12, essential for equal segregation, is independent of CENP-A loading pathway. J. Cell Biol. 160: 25–39.

Gregan, J., Riedel, C.G., Pidoux, A.L., Katou, Y., Rumpf, C., Schleiffer, A., Kearsey, S.E., Shirahige, K., Allshire, R.C., and Nasmyth, K. 2007. The kinetochore proteins pcs1 and mde4 and heterochromatin are required to prevent merotelic orientation. Curr. Biol. 17: 1190–1200.

Griffis, E.R., Stuurman, N., and Vale, R.D. 2007. Spindly, a novel protein essential for silencing the spindle assembly checkpoint, recruits dynein to the kinetochore. J. Cell Biol. 177: 1005–1015.

Hagan, I., and Yanagida, M. 1990. Novel potential mitotic motor protein encoded by the fission yeast cut7+ gene. Nature 347: 563–566.

Hamant, O., Golubovskaya, I., Meeley, R., Fiume, E., Timofejeva, L., Schleiffer, A., Nasmyth, K., and Cande, W.Z. 2005. A REC8-dependent plant Shugoshin is required for maintenance of centromeric cohesion during meiosis and has no mitotic functions. Curr. Biol. 15: 948-954.

Hayashi, A., Asakawa, H., Haraguchi, T., and Hiraoka, Y. 2006. Reconstruction of the kinetochore during meiosis in fission yeast *Schizosaccharomyces pombe*. Mol. Biol. Cell 17: 5173–5184.

Hayashi, T., Fujita, Y., Iwasaki, O., Adachi, Y., Takahashi, K., and Yanagida, M. 2004. Mis16 and Mis18 are required for CENP-A loading and histone deacetylation at centromeres. Cell 118: 715–729.

Hayashi, I., and Ikura M. 2003. Crystal structure of the amino-terminal microtubule-binding domain of end-binding protein 1 (EB1). J. Biol. Chem. 278: 36430–36434.

He, D., and Brinkley, B.R. 1996. Structure and dynamic organization of centromeres/prekinetochores in the nucleus of mammalian cells. J. Cell Sci. 109: 2693–2704.

He, D., Zeng, C., Woods, K., Zhong, L., Turner, D., Busch, R.K., Brinkley, B.R., and Busch, H. 1998. CENP-G: a new centromeric protein that is associated with the alpha-1 satellite DNA subfamily. Chromosoma 107: 189–197.

Hegemann, J.N., and Fleig, U.N. 1993. The centromere of budding yeast. BioEssays 15: 451–460.

Heit R., Underhill, D.A., Chan, G., and Hendzel, M.J. 2006. Epigenetic regulation of centromere formation and kinetochore function. Biochem. Cell Biol. 84: 605–618.

Hendzel, M.J., and Davie, J.R. 1990. Nucleosomal histones of transcriptionally active/competent chromatin preferentially exchange with newly synthesized histones in quiescent chicken erythrocytes. Biochem. J. 271: 67–73.

Henikoff, S., Ahmad, K., and Malik, H.S. 2001. The centromere paradox: stable inheritance with rapidly evolving DNA. Science 293: 1098–1102.

Henikoff, S., Ahmad, K., Platero, J.S., and van Steensel, B. 2000. Heterochromatic deposition of centromeric histone H3-like proteins. Proc. Natl. Acad. Sci. U.S.A. 97: 716–721.

Heun, P., Erhardt, S., Blower, M.D., Weiss, S., Skora, A.D., and Karpen, G.H. 2006. Mislocalization of the *Drosophila* centromere-specific histone CID promotes formation of functional ectopic kinetochores. Dev. Cell 10: 303–315.

Heus, J.J., Zonneveld, B.J., de Steensma, H.J., and van den Berg, J.A. 1993. The consensus sequence of *Kluyveromyces lactis* centromeres shows homology to functional centromeric DNA from *Saccharomyces cerevisiae*. Mol. Gen. Genet. 236: 355–362.

Hieter, P., Mann, C., Snyder, M., and Davis, R.W. 1985. Mitotic stability of yeast chromosomes: a colony color assay that measures nondisjunction and chromosome loss. Cell 40: 381–392.

Hill, A., and Bloom, K. 1989. Acquisition and processing of a conditional dicentric chromosome in *Saccharomyces cerevisiae*. Mol. Cell. Biol. 9: 1368–1370.

Hoffman, D.B., Pearson, C.G., Yen, T.J., Howell, B.J., and Salmon, E.D. 2001. Microtubule-dependent changes in assembly of microtubule motor proteins and mitotic spindle checkpoint proteins at ptk1 kinetochores. Mol. Biol. Cell 12: 1995–2009.

Hori, T., Haraguchi, T., Hiraoka, Y., Kimura, H., and Fukagawa, T. 2003. Dynamic behavior of Nuf2-Hec1 complex that localizes to the centrosome and centromere and is essential for mitotic progression in vertebrate cells. J. Cell Sci. 116: 3347–3362.

Hori, T., Okada, M., Maenaka, K., and Fukagawa, T. 2008. CENP-O-class proteins form a stable complex and are required for proper kinetochore function. Mol. Biol. Cell 19: 843–854.

Howell, B.J., Hoffman, D.B., Fang, G., Murray, A.W., and Salmon, E.D. 2000. Visualization of Mad2 dynamics at kinetochores, along spindle fibers, and at spindle poles in living cells. J. Cell Biol. 150: 1233–1250.

Howell, B.J., McEwen, B.F., Canman, J.C., Hoffman, D.B., Farrar, E.M., Rieder, C.L., and Salmon, E.D. 2001. Cytoplasmic dynein/dynactin drives kinetochore protein transport to the spindle poles and has a role in mitotic spindle checkpoint inactivation. J. Cell Biol. 155: 1159–1172.

Howell, B.J., Moree, B., Farrar, E.M., Stewart, S., Fang, G., and Salmon, E.D. 2004. Spindle checkpoint protein dynamics at kinetochores in living cells. Curr. Biol. 14: 953–964.

Howman, E.V., Fowler, K.J., Newson, A.J., Redward, S., MacDonald, A.C., Kalitsis, P., and Choo, K.H. 2000. Early disruption of centromeric chromatin organization in centromere protein A (Cenpa) null mice. Proc. Natl. Acad. Sci. U.S.A. 97: 1148–1153.

Hoyt, M.A., Stearns, T., and Botstein, D. 1990. Chromosome instability mutants of *Saccharomyces cerevisiae* that are defective in microtubule mediated processes. Mol. Cell Biol. 10: 223–234.

Hudson, D.F., Fowler, K.J., Earle, E., Saffery, R., Kalitsis, P., Trowell, H., Hill, J., Wreford, N.G., de Kretser, D.M., Cancilla, M.R., Howman, E., Hii, L., Cutts, S.M., Irvine, D.V., and Choo, K.H. 1998. Centromere protein B null mice are mitotically and meiotically normal but have lower body and testis weights. J. Cell Biol. 141: 309–319.

Hyland, K.M., Kingsbury, J., Koshland, D., and Hieter, P. 1999. Ctf19p: A novel kinetochore protein in *Saccharomyces cerevisiae* and a potential link between the kinetochore and mitotic spindle. J. Cell Biol. 145: 15–28.

Indjeian, V.B., Stern, B.M., and Murray, A.W. 2005. The centromeric protein Sgo1 is required to sense lack of tension on mitotic chromosomes. Science 307: 130–133.

Izuta, H., Ikeno, M., Suzuki, N., Tomonaga, T., Nozaki, N., Obuse, C., Kisu, Y., Goshima, N., Nomura, F., Nomura, N., and Yoda, K. 2006. Comprehensive analysis of the ICEN (Interphase Centromere Complex) components enriched in the CENP-A chromatin of human cells. Genes Cells 11: 673–684.

Janke, C., Ortiz, J., Lechner, J., Shevchenko, A., Shevchenko, A., Magiera, M.M., Schramm, C., and Schiebel, E. 2001. The budding yeast proteins Spc24p and Spc25p interact with Ndc80p and Nuf2p at the kinetochore and are important for kinetochore clustering and checkpoint control. EMBO J. 20: 777–791.

Jansen, L.E., Black, B.E., Foltz, D.R., and Cleveland, D.W. 2007. Propagation of centromeric chromatin requires exit from mitosis. J. Cell Biol. 176: 795–805.

Javerzat, J.P., Cranston, G., and Allshire, R.C. 1996. Fission yeast genes which disrupt mitotic chromosome segregation when overexpressed. Nucleic Acids Res. 24: 4676–4683.

Joglekar, A.P., Bouck, D.C., Molk, J.N., Bloom, K.S., and Salmon, E.D. 2006. Molecular architecture of a kinetochore-microtubule attachment site. Nat. Cell Biol. 8: 581–585.

Jokelainen, P.T. 1967. The ultrastructure and spatial organization of the metaphase kinetochore in mitotic rat cells. J. Ultras. Res. 19: 19–44.

Joseph, J., Liu, S.T., Jablonski, S.A., Yen, T.J., and Dasso, M. 2004. The RanGAP1-RanBP2 complex is essential for microtubule-kinetochore interactions in vivo. Curr. Biol. 14: 611–617.

Joseph, J., Tan, S.H., Karpova, T.S., McNally, J.G., and Dasso, M. 2002. SUMO-1 targets RanGAP1 to kinetochores and mitotic spindles. J. Cell Biol. 156: 595–602.

Kallio, M.J., Beardmore, V.A., Weinstein, J., and Gorbsky, G.J. 2002. Rapid microtubule-independent dynamics of Cdc20 at kinetochores and centrosomes in mammalian cells. J. Cell Biol. 158: 841–847.

Kallio, M., Eriksson, J.E., and Gorbsky, G.J. 2000. Differences in spindle association of the mitotic checkpoint protein Mad2 in mammalian spermatogenesis and oogenesis. Dev. Biol. 225: 112–123.

Kang, Y.H., Park, J.E., Yu, L.R., Soung, N.K., Yun, S.M., Bang, J.K., Seong, Y.S., Yu, H., Garfield, S., Veenstra, T.D., and Lee, K.S. 2006. Self-regulated Plk1 recruitment to kinetochores by the Plk1-PBIP1 interaction is critical for proper chromosome segregation. Mol. Cell 24: 409–422.

Kaplan, K.B., Hyman A.A., and Sorger, P.K. 1997. Regulating the yeast kinetochore by ubiquitin-dependent degradation and Skp1p-mediated phosphorylation. Cell 91: 491–500.

Kaplan, K.B., Burds, A.A., Swedlow, J.R., Bekir, S.S., Sorger, P.K., and Nathke, I.S. 2001. A role for the adenomatous polyposis coli protein in chromosome segregation. Nat. Cell Biol. 3: 429–432.

Kapoor, M., Montes de Oca Luna, R., Liu, G., Lozano, G., Cummings, C., Mancini, M., Ouspenski, I., Brinkley, B.R., and May, G.S. 1998. The cenpB gene is not essential in mice. Chromosoma 107: 570–576.

Katis, V.L., Galova, M., Rabitsch, K.P., Gregan, J., and Nasmyth, K. 2004. Maintenance of cohesin at centromeres after meiosis I in budding yeast requires a kinetochore-associated protein related to MEI-S332. Curr. Biol. 14: 560–572.

Kawashima, S.A., Tsukahara, T., Langegger, M., Hauf, S., Kitajima, T.S., and Watanabe, Y. 2007. Shugoshin enables tension-generating attachment of kinetochores by loading Aurora to centromeres. Genes Dev. 21: 420–435.

Keith, K.C., and Fitzgerald-Hayes, M. 2000. CSE4 genetically interacts with the *Saccharomyces cerevisiae* centromere DNA elements *CDE I* and *CDE II* but not *CDE III*. Implications for the path of the centromere DNA around a Cse4p variant nucleosome. Genetics 156: 973–981.

Kerrebrock, A.W., Miyazaki, W.Y., Birnby, D., and Orr-Weaver, T.L. 1992. The *Drosophila* mei-S332 gene promotes sister-chromatid cohesion in meiosis following kinetochore differentiation. Genetics 130: 827–841.

Kerres, A., Jakopec, V., and Fleig, U. 2007. The conserved Spc7 protein is required for spindle integrity and links kinetochore complexes in fission yeast. Mol. Biol. Cell 18: 2441–2454.

Kiburz, B.M., Reynolds, D.B., Megee P.C., Marston, A.L., Lee, B.H., Lee, T.I., Levine, S.S., Young R.A., and Amon, A. 2005. The core centromere and Sgo1 establish a 50-kb cohesin-protected domain around centromeres during meiosis I. Genes Dev. 19: 3017–3030.

Kilmartin, J.V. 1994. Genetic and biochemical approaches to spindle function and chromosome segregation in eukaryotic microorganisms. Curr. Opin. Cell Biol. 6: 50–54.

King, J.M., Hays, T.S., and Nicklas, R.B. 2000. Dynein is a transient kinetochore component whose binding is regulated by microtubule attachment, not tension. J. Cell Biol. 151: 739-748.

Kingwell, B., and Rattner, J.B. 1987. Mammalian kinetochore/centromere composition: a 50 kDa antigen is present in the mammalian kinetochore/centromere. Chromosoma 95: 403–407.

Kipling, D., and Warburton, P.E. 1997. Centromeres, CENP-B and Tigger too. Trends Genet. 13:141–145.

Kitajima, T.S., Yokobayashi, S., Yamamoto, M., and Watanabe, Y. 2003. Distinct cohesin complexes organize meiotic chromosome domains. Science 300: 1152–1155.

Kitajima, T.S., Kawashima, S.A., and Watanabe, Y. 2004. The conserved kinetochore protein shugoshin protects centromeric cohesion during meiosis. Nature 427: 510–517.

Kitajima, T.S., Sakuno, T., Ishiguro, K., Iemura, S.,Natsume, T., Kawashima, S.A., and Watanabe, Y. 2006. Shugoshin collaborates with protein phosphatase 2A to protect cohesin. Nature 441: 46–52.

Kitagawa, K., Skowyra, D., Elledge, S.J., Harper, J.W., and Hieter, P. 1999. SGT1 encodes an essential component of the yeast kinetochore assembly pathway and a novel subunit of the SCF ubiquitin ligase complex. Mol. Cell 4: 21–33.

Kiyomitsu, T., Obuse, C., and Yanagida, M. 2007. Human Blinkin/AF15q14 is required for chromosome alignment and the mitotic checkpoint through direct interaction with Bub1 and BubR1. Dev. Cell: 13: 1–15.

Klein, F., Mahr, P., Galova, M., Buonomo, S.B., Michaelis, C., Nairz, K., and Nasmyth, K. 1999. A central role for cohesins in sister chromatid cohesion, formation of axial elements, and recombination during yeast meiosis. Cell 98: 91–103.

Kline, S.L., Cheeseman, I.M., Hori, T., Fukagawa, T., and Desai, A. 2006. The human Mis12 complex is required for kinetochore assembly and proper chromosome segregation. J. Cell Biol. 173: 9–17.

Kniola, B., O'Toole, E., McIntosh, J. R., Mellone, B., Allshire, R., Mengarelli, S., Hultenby, K., and Ekwall, K. 2001. The domain structure of centromeres is conserved from fission yeast to humans. Mol. Biol. Cell 12: 2767–2775.

Koch, J. 2000. Neocentromeres and alpha satellite: a proposed structural code for functional human centromere DNA. Hum. Mol. Genet. 9: 149–154.

Kopski, K.M., and Huffaker, T.C. 1997. Suppressors of the *ndc10–2* mutation: a role for the ubiquitin system in *Saccharomyces cerevisiae* kinetochore function. Genetics 147: 409–420.

Koshland, D., Kent, J.C., and Hartwell, L.H. 1985. Genetic analysis of the mitotic transmission of minichromosomes. Cell 40: 393–403.

Koshland, D.H., Rutledge, L., Fitzgerald-Hayes, M., and Hartwell, L.H. 1987. A genetic analysis of dicentric minichromosomes in *Saccharomyces cerevisiae*. Cell 48: 801–812.

Kouprina, N., Kirillov, A., Kroll, E., Koryabin, M., Shestopalov, B., Bannikov, V., Zakharyev, V., and Larionov, V. 1993a. Identification and cloning of the *CHL4* gene controlling chromosome segregation in yeast. Genetics 135: 327–341.

Kouprina, N., Tsouladse, A., Koryabin, M., Hieter, P., Spencer, F., and Larionov, V. 1993b. Identification and mapping of *CHL* genes controlling mitotic chromosome transmission in yeast. Yeast 9: 11–19.

Kroll, E.S., Hyland, K.M., Hieter, P., and Li, J.J. 1996. Establishing genetic interactions by a synthetic dosage lethality phenotype. Genetics 143: 95–102.

Kwon, M.S., Hori, T., Okada, M., and Fukagawa, T. 2007. CENP-C is involved in chromosome segregation, mitotic checkpoint function, and kinetochore assembly. Mol. Biol. Cell 18: 2155–2168.

Le, M.H., Duricka, D., and Karpen, G.H. 1995. Islands of complex DNA are widespread in *Drosophila* centric heterochromatin. Genetics 141: 283–303.

Lechner, J. 1994. A zinc finger protein, essential for chromosome segregation, constitutes a putative DNA binding subunit of the *Saccharomyces cerevisiae* kinetochore complex, Cbf3. The EMBO J. 13: 5203–5211.

Lechner, J., and Carbon, J. 1991. A 240 kd multisubunit protein complex, CBF3, is a major component of the budding yeast centromere. Cell 64: 717–725.

Lee, Y.T., Jacob, J., Michowski, W., Nowotny, M., Kuznicki, J., and Chazin, W.J. 2004. Human Sgt1 binds HSP90 through the CHORD-Sgt1 domain and not the tetratricopeptide repeat domain. J Biol Chem. 279: 16511–16517.

Lee, J., Miyano, T., Dai, Y., Wooding, P., Yen, T.J., and Moor, R.M. 2000. Specific regulation of CENP-E and kinetochores during meiosis I/meiosis II transition in pig oocytes. Mol. Reprod. Dev. 56: 51–62.

Liao, H., Winkfein, R.J., Mack, G., Rattner, J.B., and Yen, T.J. 1995. CENP-F is a protein of the nuclear matrix that assembles onto kinetochores at late G2 and is rapidly degraded after mitosis. J. Cell Biol. 130: 507–518.

Lingelbach, L.B., and Kaplan, K.B. 2004. The interaction between Sgt1p and Skp1p is regulated by HSP90 chaperones and is required for proper CBF3 assembly. Mol. Cell. Biol. 24: 8938–8950.

Liu, X., McLeod, I., Anderson, S., Yates, J.R. 3rd, and He, X. 2005. Molecular analysis of kinetochore architecture in fission yeast. EMBO J. 24: 2919–2930.

Liu, S.T., Hittle, J.C., Jablonski, S.A., Campbell, M.S., Yoda, K., and Yen, T.J. 2003. Human CENP-I specifies localization of CENP-F, MAD1 and MAD2 to kinetochores and is essential for mitosis. Nat. Cell Biol. 5: 341–345.

Liu, S.T., Rattner, J.B., Jablonski, S.A., and Yen, T.J. 2006. Mapping the assembly pathways that specify formation of the trilaminar kinetochore plates in human cells. J. Cell Biol. 175: 41–53.

Lo, A.W., Craig, J.M., Saffery, R., Kalitsis, P., Irvine, D.V., Earle, E., Magliano, D.J., and Choo, K.H. 2001a. A 330 kb CENP-A binding domain and altered replication timing at a human neocentromere. EMBO J. 20: 2087–2096.

Lo, A.W., Magliano, D.J., Sibson, M.C., Kalitsis, P., Craig, J.M., and Choo, K.H. 2001b. A novel chromatin immunoprecipitation and array (CIA) analysis identifies a 460-kb CENP-A- binding neocentromere DNA. Genome Res. 11: 448–457.

Loïodice, I., Alves, A., Rabut, G., Van Overbeek, M., Ellenberg, J., Sibarita, J. B., and Doye, V. 2004. The entire Nup107-160 complex, including three new members, is targeted as one entity to kinetochores in mitosis. Mol. Biol. Cell 15: 3333–3344.

Luger, K., Mader, A.W., Richmond, R.K., Sargent. D.F., and Richmond, T.J. 1997. Crystal structure of the nucleosome core particle at 2.8 A resolution. Nature 389: 251–260.

Lutzmann, M., Kunze, R., Buerer, A., Aebi, U., and Hurt, E. 2002. Modular self-assembly of a Y-shaped multiprotein complex from seven nucleoporins. EMBO J. 21: 387–397.

Ma, W., Hou, Y., Sun, Q-Y, Sun, X.-F., and Wang, W.-H. 2003. Localization of centromere proteins and their association with chromosomes and microtubules during meiotic maturation in pig oocytes. Reproduction 126: 731–738.

Maddox, P.S., Oegema, K., Desai, A., and Cheeseman, I.M. 2004. "Holo"er than thou: chromosome segregation and kinetochore function in *C. elegans*. Chromosome Res. 12: 641–653.

Maddox, P.S., Hyndman, F., Monen, J., Oegema, K., and Desai, A. 2007. Functional genomics identifies a Myb domain-containing protein family required for assembly of CENP-A chromatin. J. Cell Biol. 176: 757–763.

Maiato, H., DeLuca. J., Salmon, E.D., and Earnshaw, W.C. 2004. The dynamic kinetochore-microtubule interface. J. Cell Sci. 117: 5461–5477.

Malik, H.S., and Henikoff, S. 2002. Conflict begets complexity: the evolution of centromeres. Curr. Opin. Genet. Dev. 12: 711–718.

Malik, H.S., Vermaak, D., and Henikoff, S. 2002. Recurrent evolution of DNA-binding motifs in the *Drosophila* centromeric histone. Proc. Natl. Acad. Sci. U.S.A. 99: 1449–1454.

Marahrens, Y., and Stillman, B. 1992. A yeast chromosomal origin of DNA replication defined by multiple functional elements. Science *255*, 817–823.

Marston, A.L., Tham, W.H., Shah, H., and Amon, A. 2004. A genome-wide screen identifies genes required for centromeric cohesion. Science 303: 1367–1370.

Masumoto, H., Masukata, H., Muro, Y., Nozaki, N., and Okazaki, T. 1989. A human centromere antigen (CENP-B) interacts with a short specific sequence in alphoid DNA, a human centromeric satellite. J. Cell. Biol. 109: 1963–1973.

McEwen, B.F., Arena, J.T., Frank, J., and Rieder, C.L. 1993. Structure of the colcemid treated PtK1 kinetochore outer plate as determined by high voltage electron microscopic tomography. J. Cell Biol. 120: 301–312.

McEwen, B.F., Hsieh, C.E., Mattheyses, A.L., and Rieder, C.L. 1998. A new look at kinetochore structure in vertebrate somatic cells using high pressure freezing and freeze substitution. Chromosoma 107: 366–375.

McGrew, J.T., Xiao, Z., and Fitzgerald-Hayes, M. 1989. *Saccharomyces cerevisiae* mutants defective in chromosome segregation. Yeast 5: 271–284.

Measday, V., Baetz, K., Guzzo, J., Yuen, K., Kwok, T., Sheikh, B., Ding, H., Ueta, R., Hoac, T., Cheng, B., Pot, I., Tong, A., Yamaguchi-Iwai, Y., Boone, C., Hieter, P., and Andrews. B.

2005. Systematic yeast synthetic lethal and synthetic dosage lethal screens identify genes required for chromosome segregation. Proc. Natl. Acad. Sci. U.S.A. 102: 13956–13961.

Measday, V., Hailey, D.W., Pot, I., Givan, S.A., Hyland, K.M., Cagney, G., Fields, S., Davis, T.N., and Hieter, P. 2002. Ctf3p, the Mis6 budding yeast homolog, interacts with Mcm22p and Mcm16p at the yeast outer kinetochore. Genes Dev. 16: 101–113.

Meeks-Wagner, D., Wood, J.S., Garvik, B., and Hartwell, L.H. 1986. Isolation of two genes that affect mitotic chromosome transmission in *S. cerevisiae*. Cell 44: 53–63.

Mellone, B.G., and Allshire, R.C. 2003. Stretching it: putting the CEN(P-A) in centromere. Curr. Opin. Genet. Dev. 13: 191–198.

Mellone, B., Erhardt, S., and Karpen, G.H. 2006. The ABCs of centromeres. Nat. Cell Biol. 8:427–429.

Mellor, J., Jiang, W., Funk, M., Rathjen, J., Barnes, C.A., Hinz, T., Hegemann, J.H., and Philippsen, P. 1990. CPF1, a yeast protein which functions in centromeres and promoters. EMBO J. 9: 4017–4026.

Meluh, P.B., and Koshland, D. 1995. Evidence that the MIF2 gene of *Saccharomyces cerevisiae* encodes a centromere protein with homology to the mammalian centromere protein CENP- C. Mol. Biol. Cell 6: 793–807.

Meluh, P.B., Yang, P., Glowczewski, L., Koshland, D., Smith, M.M. 1998. Cse4p is a component of the core centromere of *Saccharomyces cerevisiae*. Cell 94: 607–613.

Meraldi, P., Draviam, V.M., Sorger, P.K. 2004. Timing and checkpoints in the regulation of mitotic progression. Dev. Cell 7: 45–60.

Meraldi, P., McAinsh, A.D., Rheinbay, E., and Sorger, P.K. 2006. Phylogenetic and structural analysis of centromeric DNA and kinetochore proteins. Genome Biology 7: R23.

Merry, D.W., Pathak, S., Hsu, T.C., and Brinkley, B.R. 1985. Anti-kinetochore antibodies: use as probes for inactive centromeres. Am. J. Hum. Genet. 37: 425–430.

Mikami, Y., Hori, T., Kimura, H., and Fukagawa, T. 2005. The functional region of CENP-H interacts with the Nuf2 complex that localizes to centromere during mitosis. Mol. Cell. Biol. 25: 1958–1970.

Minoshima, Y., Hori, T., Okada, M., Kimura, H., Haraguchi, T., Hiraoka, Y., Bao, Y.C., Kawashima, T., Kitamura, T., and Fukagawa, T. 2005. The constitutive centromere component CENP-50 is required for recovery from spindle damage. Mol. Cell. Biol. 25: 10315–10328.

Miranda, J.J., De Wulf, P., Sorger, P.K., and Harrison, S.C. 2005. The yeast DASH complex forms closed rings on microtubules. Nat. Struct. Mol. Biol. 12: 138–143.

Miranda, J.J., King, D.S., and Harrison, S.C. 2007. Protein arms in the kinetochore-microtubule interface of the yeast DASH complex. Mol. Biol. Cell. 18: 2503–2510.

Mizuguchi, G., Xiao, H., Wisniewski, J., Smith, M.M., and Wu, C. 2007. Nonhistone Scm3 and histones CenH3-H4 assemble the core of centromere-specific nucleosomes. Cell 129: 1153–1164.

Monje-Casas, F., Prabhu, V.R., Lee, B.H., Boselli, M., and Amon, A. 2007. Kinetochore orientation during meiosis is controlled by Aurora B and the monopolin complex. Cell 128: 477–490.

Monen, J., Maddox, P.S., Hyndman, F., Oegema, K., and Desai, A. 2005. Differential role of CENP-A in the segregation of holocentric *C. elegans* chromosomes during meiosis and mitosis. Nat. Cell Biol. 7: 1248–1255.

Moore, W., Zhang, C, and Clarke, P.R. 2002. Targeting of RCC1 to chromosomes is required for proper spindle assembly in human cells. Curr. Biol. 12: 1442–1447.

Moroi, Y., Hartman, A.L., Nakane, P.K., and Tan, E.M. 1981. Distribution of kinetochore (centromere) antigen in mammalian cell nuclei. J. Cell Biol. 90: 254–259.

Moroi, Y., Peebles, C., Fritzler, M.J., Steigerwald, J., and Tan, E.M. 1980. Autoantibody to centromere (kinetochore) in scleroderma sera. Proc. Natl. Acad. Sci. U.S.A. 77: 1627–1631.

Murphy, T.D., and Karpen, G.H. 1995. Localization of centromere function in a *Drosophila* minichromosome. Cell 82: 599–609.

Mythreye, K., and Bloom, K.S. 2003. Differential kinetochore protein requirements for establishment versus propagation of centromere activity in *Saccharomyces cerevisiae*. J. Cell Biol. 160: 833–843.

Nakagawa, H., Lee, J.K., Hurwitz, J., Allshire, R.C., Nakayama, J., Grewal, S.I., Tanaka, K., and Murakami, Y. 2002. Fission yeast CENP-B homologs nucleate centromeric heterochromatin by promoting heterochromatin-specific histone tail modifications. Genes Dev. 16: 1766–1778.

Nasmyth, K. 2005. How might cohesin hold sister chromatids together? Philos. Trans. R. Soc. Lond. B Biol. Sci. 360: 483–496.

Nasmyth, K., and Haering, C.H. 2005. The structure and function of SMC and kleisin complexes. Annu. Rev. Biochem. 74: 595–648.

Nekrasov, V.S., Smith, M.A., Peak-Chew, S., and Kilmartin, J.V. 2003. Interactions between centromere complexes in *Saccharomyces cerevisiae*. Mol. Biol. Cell 14: 4931–4946.

Niikura, Y., Ohta, S., Vandenbeldt, K.J., Abdulle, R., McEwen, B.F., and Kitagawa, K. 2006. 17- AAG, an Hsp90 inhibitor, causes kinetochore defects: a novel mechanism by which 17-AAG inhibits cell proliferation. Oncogene 25: 4133–4146.

Nishihashi, A., Haraguchi, T., Hiraoka, Y., Ikemura, T., Regnier, V., Dodson, H., Earnshaw, W.C., and Fukagawa, T. 2002. CENP-I is essential for centromere function in vertebrate cells. Dev. Cell 2: 463–476.

Nishino, M., Kurasawa, Y., Evans, R., Lin, S.H., Brinkley, B.R., Yu-Lee, L.Y. 2006. NudC is required for Plk1 targeting to the kinetochore and chromosome congression. Curr. Biol. 16: 1414–1421.

Obuse, C., Iwasaki, O., Kiyomitsu, T., Goshima, G., Toyoda, Y., and Yanagida, M. 2004a. A conserved Mis12 centromere complex is linked to heterochromatic HP1 and outer kinetochore protein Zwint-1. Nature Cell Biol. 6: 1135–1141.

Obuse, C., Yang, H., Nozaki, N., Goto, S., Okazaki, T., and Yoda, K. 2004b. Proteomics analysis of the centromere complex from HeLa interphase cells: UV-damaged DNA binding protein 1 (DDB-1) is a component of the CEN-complex, while BMI-1 is transiently co-localized with the centromeric region in interphase. Genes Cells 9: 105–120.

Ochs, R.L., and Press, R.I. 1992. Centromere autoantigens are associated with the nucleolus. Exp. Cell Res. 200: 339–350.

Oegema, K., Desai, A., Rybina, S., Kirkham, M., and Hyman, A.A. 2001. Functional analysis of kinetochore assembly in *Caenorhabditis elegans*. J. Cell. Biol. 153: 1209–1226.

Ohkura, H., Adachi, Y., Kinoshita, N., Niwa, O., Toda, T., and Yanagida, M. 1988. Cold-sensitive and caffeine-supersensitive mutants of the *Schizosaccharomyces pombe dis* genes implicated in sister chromatid separation during mitosis. EMBO J. 7: 1465–1473.

Ohtsubo, M., Okazaki, H., and Nishimoto, T. 1989. The RCC1 protein, a regulator for the onset of chromosome condensation locates in the nucleus and binds to DNA. J. Cell Biol. 109: 1389–1397.

Ohzeki, J., Nakano, M., Okada, T., and Masumoto, H. 2002. CENP-B box is required for *de novo* centromere chromatin assembly on human alphoid DNA. J. Cell Biol. 159: 765–775.

Okada, M., Cheeseman, I.M., Hori, T., Okawa, K., McLeod, I.X., Yates, J.R. III, Desai, A., and Fukagawa, T. 2006. The CENP-H-I complex is required for the efficient incorporation of newly synthesized CENP-A into centromeres. Nat. Cell Biol. 8: 446–457.

Okada, T., Ohzeki, J., Nakano, M., Yoda, K., Brinkley, W.R., Larionov, V., and Masumoto, H. 2007. CENP-B controls centromere formation depending on the chromatin context. Cell 131: 1287–1300.

Okamoto, Y., Nakano, M., Ohzeki, J., Larionov, V., and Masumoto, H. 2007. A minimal CENP- A core is required for nucleation and maintenance of a functional human centromere. EMBO J. 26: 1279–1291.

Orjalo, A.V., Arnaoutov, A., Shen, Z., Boyarchuk, Y., Zeitlin, S.G., Fontoura, B., Briggs, B., Dasso, M., and Forbes, D.J. 2006. The Nup107-160 nucleoporin complex is required for correct bipolar spindle assembly. Mol. Biol. Cell 17: 3806–3818.

Ortiz, J., Stemmann, O., Rank, S., and Lechner, J. 1999. A putative protein complex consisting of Ctf19, Mcm21, and Okp1 represents a missing link in the budding yeast kinetochore. Genes Dev. 13: 1140–1155.

Palmer, D.K., O'Day K., Trong, H.L., Charbonneau, H., and Margolis, R.L. 1991. Purification of the centromere-specific protein CENP-A and demonstration that it is a distinctive histone. Proc. Natl. Acad. Sci. U.S.A. 88: 3734–3738.

Pasierbek, P., Jantsch, M., Melcher, M., Schleiffer, A., Schweizer, D., and Loidl, J. 2001. A *Caenorhabditis elegans* cohesion protein with functions in meiotic chromosome pairing and disjunction. Genes Dev. 15: 1349–1360.

Perez-Castro, A.V., Shamanski, F.L., Meneses, J.J., Lovato, T.L., Vogel, K.G., Moyzis, R.K., and Pedersen, R. 1998. Centromeric Protein B null mice are viable with no apparent abnormalities. Dev. Biol. 201: 135–143.

Peterson, J. B., and Ris, H. 1976. Electron-microscopic study of the spindle and chromosome movement in the yeast *Saccharomyces cerevisiae*. J. Cell Sci. 22: 219–242.

Petronczki, M., Matos, J., Mori, S., Gregan, J., Bogdanova, A., Schwickart, M., Mechtler, K., Shirahige, K., Zachariae, W., and Nasmyth, K. 2006. Monopolar attachment of sister kinetochores at meiosis I requires casein kinase 1. Cell 126: 1049–1064.

Pidoux, A.L., and Allshire, R.C. 2004. Kinetochore and heterochromatin domains of the fission yeast centromere. Chromosome Res. 12: 521–534.

Pietrasanta, L.I., Thrower, D., Hsieh, W., Rao, S., Stemmann, O., Lechner, J., Carbon, J., and Hansma, H. 1999. Probing the *Saccharomyces cerevisiae* centromeric DNA (CEN DNA)- binding factor 3 (CBF3) kinetochore complex by using atomic force microscopy. Proc. Natl. Acad. Sci. U.S.A. 96: 3757–3762.

Pimpinelli, S., and Goday, C. 1989. Unusual kinetochores and chromatin diminution in Parascaris. Trends Genet. 5: 310–315.

Pinsky, B.A., Kotwaliwale, C.V., Tatsutani, S.Y., Breed, C.A., and Biggins, S. 2006. Glc7/protein phosphatase 1 regulatory subunits can oppose the Ipl1/aurora protein kinase by redistributing Glc7. Mol. Cell. Biol. 26: 2648–2660.

Pluta, A. F., Cooke, C. A., and Earnshaw, W.C. 1990. Structure of the human centromere at metaphase. Trends Biol. Sci. 15: 181–185.

Pluta, A.F., and Earnshaw, W.C. 1996. Specific interaction between human kinetochore protein CENP-C and a nucleolar transcriptional regulator. J. Biol. Chem. 271: 18767–18774.

Pouwels, J., Kukkonen, A.M., Lan, W., Daum, J.R., Gorbsky, G.J., Stukenberg, T., and Kallio, M.J. 2007. Shugoshin 1 plays a central role in kinetochore assembly and is required for kinetochore targeting of Plk1. Cell Cycle 6: 1579–1585.

Prieto, I., Suja, J.A., Pezzi, N., Kremer, L., Martinez, A.C., Rufes, J.S., and Barbero, J.L. 2001. Mammalian STAG3 is a cohesin specific to sister chromatid arms in meiosis I. Nat. Cell Biol. 3: 761–766.

Przewloka, M.R., Zhang, W., Costa, P., Archambault, V., D'Avino, P.P., Lilley, K.S., Laue, E.D., McAinsh, A.D., and Glover, D.M. 2007. Molecular analysis of core kinetochore composition and assembly in *Drosophila melanogaster*. PLoS ONE 2: e478.

Purvis, A., and Singleton, M.R. 2008. Insights into kinetochore-DNA interactions from the structure of Cep3Delta. EMBO Rep. 9: 56–62.

Rabitsch, K.P., Gregan, J., Schleiffer, A., Javerzat, J.P., Eisenhaber, F., and Nasmyth, K. 2004. Two fission yeast homologs of *Drosophila* Mei-S332 are required for chromosome segregation during meiosis I and II. Curr. Biol. 14: 287–301.

Rabitsch, K.P., Petronczki, M., Javerzat, J.P., Genier, S., Chwalla, B., Schleiffer, A., Tanaka, T.U., and Nasmyth. K. 2003. Kinetochore recruitment of two nucleolar proteins is required for homolog segregation in meiosis I. Dev. Cell 4: 535–534.

Rasala, B.A., Orjalo, A.V., Shen, Z., Briggs, S., and Forbes, D.J. 2006. ELYS is a dual nucleoporin/kinetochore protein required for nuclear pore assembly and proper cell division. Proc Natl. Acad. Sci. U.S.A. 103: 17801–17806.

Ray-Gallet, D., Quivy, J.P., Scamps, C., Martini, E.M., Lipinski, M., and Almouzni, G. 2002. HIRA is critical for a nucleosome assembly pathway independent of DNA synthesis. Mol. Cell 9: 1091–1100.

Resnick, T.D., Satinover, D.L., MacIsaac, F., Stukenberg, P.T., Earnshaw, W.C., Orr-Weaver, T.L., and Carmena, M. 2006. INCENP and Aurora B promote meiotic sister chromatid cohesion through localization of the Shugoshin MEI-S332 in *Drosophila*. Dev. Cell 11: 57 68.

Regnier, V., Vagnarelli, P., Fukagawa, T., Zerjal, T., Burns, E., Trouche, D., Earnshaw, W., and Brown, W. 2005. CENP-A is required for accurate chromosome segregation and sustained kinetochore association of BubR1. Mol. Cell. Biol. 25: 3967–3981.

Riedel, C.G., Katis, V.L., Katou, Y., Mori, S., Itoh, T., Helmhart, W., Galova, M., Petronczki, M., Gregan, J., Cetin, B., Mudrak, I., Ogris, E., Mechtler, K., Pelletier, L., Buchholz, F., Shirahige, K., and Nasmyth, K. 2006. Protein phosphatase 2A protects centromeric sister chromatid cohesion during meiosis I. Nature 441: 53–61.

Rieder, C.L., Alexander, S.P., and Rupp, G. 1990. Kinetochores are transported poleward along a single astral microtubule during chromosome attachment to the spindle in newt lung cells. J. Cell Biol. 110: 81–95.

Ris, H., and Witt, P.L. 1981. Structure of the mammalian kinetochore. Chromosoma 82: 153–170.

Roos, U.P. 1973. Light and electron microscopy of rat kangaroo cells in mitosis. II. Kinetochore structure and function. Chromosoma 41: 195–220.

Saitoh, H., Tomkiel, J.E., Cooke, C.A., Ratrie, H.R., Maurer, M., Rothfield, N.F., and Earnshaw, W.C. 1992. CENP-C, an autoantigen in scleroderma, is a component of the human inner kinetochore plate. Cell 70: 115–125.

Salina, D., Enarson, P., Rattner, J.B., and Burke, B. 2003. Nup358 integrates nuclear envelope breakdown with kinetochore assembly. J. Cell Biol. 162: 991–1001.

Sandall, S., Severin, F., McLeod, I.X., Yates, J.R. 3rd, Oegema, K., Hyman, A., and Desai, A. 2006. A Bir1-Sli15 complex connects centromeres to microtubules and is required to sense kinetochore tension. Cell 127: 1179–1191.

Scharfenberger, M., Ortiz, J., Grau, N., Janke, C., Schiebel, E., and Lechner, J. 2003. Nsl1p is essential for the establishment of bipolarity and the localization of the Dam-Duo complex. EMBO J. 22: 6584–6597.

Schatten, G., Simerly, C., Palmer, D.K., Margolis, R.L., Maul, G., Andrews, B.S., and Schatten, H. 1988. Kinetochore appearance during meiosis, fertilization and mitosis in mouse oocytes and zygotes. Chromosoma 96: 341–352.

Schuh, M., Lechner, C.F., and Heidmann, S. 2007. Incorporation of *Drosophila* CID/CENP-A and CENP-C into centromeres during early embryonic anaphase. Curr. Biol. 17: 237–243.

Schuyler, S.C., and Pellman, D. 2002. Analysis of the size and shape of protein subcomplexes from yeast. Methods Enzymol. 351: 150–168.

Shah, J.V., Botvinick, E., Bonday, Z., Furnari, F., Berns, M., and Cleveland, D.W. 2004. Dynamics of centromere and kinetochore proteins: implications for checkpoint signaling and silencing. Curr. Biol. 14: 942–952.

Sharp, J.A., Franco, A.A., Osley, M.A., and Kaufman, P.D. 2002. Chromatin assembly factor I and Hir proteins contribute to building functional kinetochores in *S. cerevisiae*. Genes Dev. 16: 85–100.

Shelby, R.D., Monier, K., and Sullivan, K.F. 2000. Chromatin assembly at kinetochores is uncoupled from DNA replication. J. Cell Biol. 151: 1113–1118.

Shelby, R.D., Vafa, O., and Sullivan, K.F. 1997. Assembly of CENP-A into centromeric chromatin requires a cooperative array of nucleosomal DNA contact sites. J. Cell Biol. 136: 501–513.

Siniossoglou, S., Lutzmann, M., Santos-Rosa, H., Leonard, K., Mueller, S., Aebi, U., and Hurt, E. 2000. Structure and assembly of the Nup84p complex. J. Cell Biol. 149: 41–54.

Siniossoglou, S., Wimmer, C., Rieger, M., Doye, V., Tekotte, H., Weise, C., Emig, S., Segref, A., and Hurt, E.C. 1996. A novel complex of nucleoporins, which includes Sec13p and a Sec13p homolog, is essential for normal nuclear pores. Cell 84: 265–275.

Smith, M.M., Yang, P., Santisteban, M.S., Boone, P.W., Goldstein, A.T., and Megee, P.C. 1996. A novel histone H4 mutant defective in nuclear division and mitotic chromosome transmission. Mol. Cell. Biol. 16: 1017–1026.

Spencer, F., Gerring, S.L., Connelly, C., and Hieter P. 1990. Mitotic chromosome transmission fidelity mutants in *Saccharomyces cerevisiae*. Genetics 124: 237–249.

Steensgaard, P., Garre, M., Muradore, I., Transidico, P., Nigg, E.A., Kitagawa, K., Earnshaw, W.C., Faretta, M., and Musacchio, A. 2004. Sgt1 is required for human kinetochore assembly. EMBO Rep. 5: 626–631.

Stemmann, O., Neidig, A., Kocher, T., Wilm, M., and Lechner, J. 2002. Hsp90 enables Ctf13p/Skp1p to nucleate the budding yeast kinetochore. Proc. Natl. Acad. Sci. U.S.A. 99: 8585–8590.

Steuer, E.R., Wordeman, L., Schroer, T.A., and Sheetz, M.P. 1990. Localization of cytoplasmic dynein to mitotic spindles and kinetochores. Nature 345: 266–268.

Stoler S., Keith, K.C., Curnick, K.E., and Fitzgerald-Hayes, M.A. 1995. Mutation in CSE4, an essential gene encoding a novel chromatin-associated protein in yeast, causes chromosome nondisjunction and cell cycle arrest at mitosis. Genes Dev. 9: 573–586.

Stoler, S., Rogers, K., Weitze, S., Morey, L., Fitzgerald-Hayes, M., and Baker, R.E. 2007. Scm3, an essential *Saccharomyces cerevisiae* centromere protein required for G2/M progression and Cse4 localization. Proc. Natl. Acad. Sci. U.S.A. 104: 10571–10576.

Strunnikov, A.V., Kingsbury, J., and Koshland, D. 1995. CEP3 encodes a centromere protein of *Saccharomyces cerevisiae*. J. Cell Biol. 128: 749–760.

Sugata, N., Munekata, E., and Todokoro, K. 1999. Characterization of a novel kinetochore protein, CENP-H. J. Biol. Chem. 274: 27343–27346.

Sun, X., Le, H.D., Wahlstrom, J.M., and Karpen, G.H. 2003. Sequence analysis of a functional *Drosophila* centromere. Genome Res. 13: 182–194.

Sun, X., Wahlstrom, J., and Karpen, G. 1997. Molecular structure of a functional *Drosophila* centromere. Cell 91: 1007–1019.

Tagami, H., Ray-Gallet, D., Almouzni, G., and Nakatani, Y. 2004. Histone H3.1 and H3.3 complexes mediate nucleosome assembly pathways dependent or independent of DNA synthesis. Cell 116: 51–61.

Takahashi, K., Chen, E.S., and Yanagida, M. 2000. Requirement of Mis6 centromere connector for localizing a CENP-A-like protein in fission yeast. Science 288: 2215–2219.

Takahashi, K., Takayama, Y., Masuda, F., Kobayashi, Y., and Saitoh, S. 2005. Two distinct pathways responsible for the loading of CENP-A to centromeres in the fission yeast cell cycle. Philos. Trans. R. Soc. Lond. B Biol. Sci. 360: 595–606.

Takahashi, K., Yamada, H., and Yanagida, M. 1994. Fission yeast minichromosome loss mutants mis cause lethal aneuploidy and replication abnormality. Mol. Biol. Cell 5: 1145–1158.

Tanaka, K., Mukae, N., Dewar, H., van Breugel, M., James, E.K., Prescott, A.R., Antony, C., and Tanaka, T.U. 2005. Molecular mechanisms of kinetochore capture by spindle microtubules. Nature 434: 987–994.

Tang, Z., Shu, H., Qi, W., Mahmood, N.A., Mumby, M.C., and Yu, H. 2006. PP2A is required for centromeric localization of Sgo1 and proper chromosome segregation. Dev. Cell 10: 575–585.

Teixeira, M. T., Siniossoglou, S., Podtelejnikov, S., Benichou, J. C., Mann, M., Dujon, B., Hurt, E., and Fabre, E. 1997. Two functionally distinct domains generated by in vivo cleavage of Nup145p: a novel biogenesis pathway for nucleoporins. EMBO J. 16: 5086–5097.

Tirnauer, J.S., Canman, J.C., Salmon, E.D., and Mitchison, T.J. 2002. EB1 targets to kinetochores with attached, polymerizing microtubules. Mol. Biol. Cell 13: 4308–4316.

Toda, T., Umesono, K., Hirata, A., and Yanagida, M. 1983. Cold-sensitive nuclear division arrest mutants of the fission yeast *Schizosaccharomyces pombe*. J. Mol. Biol. 168: 251–270.

Tomkiel, J.E., Cooke, C.A., Saitoh, H., Bernat, R.L., and Earnshaw, W.C. 1994. CENP-C is required for maintaining proper kinetochore size and for a timely transition to anaphase. J. Cell Biol. 125: 531–545.

Valdivia, M.M., and Brinkley, B.R. 1985. Fractionation and initial characterization of the kinetochore from mammalian metaphase chromosomes. J. Cell Biol. 101: 1124–1134.

Van Hooser, A.A., Ouspenski, I.I., Gregson, H.C., Starr, D.A., Yen, T.J., Goldberg, M.L., Yokomori, K., Earnshaw, W.C., Sullivan, K.F., and Brinkley, B.R. 2001. Specification of kinetochore-forming chromatin by the histone H3 variant CENP-A. J. Cell Sci. 114: 3529–3542.

Vanoosthuyse, V., Prykhozhij, S., and Hardwick, K.G. 2007. Shugoshin 2 regulates localization of the chromosomal passenger proteins in fission yeast mitosis. Mol. Biol. Cell 18: 1657–1669.

Vaur, S., Cubizolles, F., Plane, G., Genier, S., Rabitsch, P.K., Gregan, J., Nasmyth, K., Vanoosthuyse, V., Hardwick, K.G., and Javerzat, J.P. 2005. Control of Shugoshin function during fission-yeast meiosis. Curr. Biol. 15: 2263–2270.

Verreault, A., Kaufman, P.D., Kobayashi, R., and Stillman, B. 1996. Nucleosome assembly by a complex of CAF-1 and acetylated histones H3/H4. Cell 87: 95–104.

Villasante, A., Abad, J.P., and Méndez-Lago, M. 2007. Centromeres were derived from telomeres during the evolution of the eukaryotic chromosome. Proc. Natl. Acad. Sci. U. S.A. 104: 10542–10547.

Warburton, P.E., Cooke, C.A., Bourassa, S., Vafa, O., Sullivan, B.A., Stetten, G., Gimelli, G., Warburton, D., Tyler-Smith, C., Sullivan, K.F., Poirier, G.G., and Earnshaw, W.C. 1997. Immunolocalization of CENP-A suggests a distinct nucleosome structure at the inner kinetochore plate of active centromeres. Curr. Biol. 7: 901–904.

Watanabe, Y., and Nurse, P. 1999. Cohesin Rec8 is required for reductional chromosome segregation at meiosis. Nature 400: 461–464.

Wei, R.R., Sorger, P.K., and Harrison, S.C. 2005. Molecular organization of the Ndc80 complex, an essential kinetochore component. Proc. Natl. Acad. Sci. U.S.A. 102: 5363–5367.

Wei, R.R., Al-Bassam, J., and Harrison, S.C. 2007. The Ndc80/HEC1 complex is a contact point for kinetochore-microtubule attachment. Nat. Struct. Mol. Biol. 14: 54–59.

Weiner, E.S., Hildebrandt, S., Senecal, J.-L., Daniels, L., Noell, S., Joyal, F., Roussin, A., Earnshaw, W., and Rothfield, N.F. 1991. Prognostic significance of anticentromere antibodies and anti-topoisomerase I antibodies in Raynaud's disease: A prospective study. Arthritis Rheum. 34: 68–77.

Westermann, S., Cheeseman, I.M., Anderson, S., Yates, J.R. III, Drubin, D.G., and Barnes, G. 2003. Architecture of the budding yeast kinetochore reveals a conserved molecular core. J. Cell Biol. 163: 215–222.

Westermann, S., Avila-Sakar, A., Wang, H.W., Niederstrasser, H., Wong, J., Drubin, D.G., Nogales, E., and Barnes, G. 2005. Formation of a dynamic kinetochore-microtubule interface through assembly of the Dam1 ring complex. Mol. Cell 17: 277–290.

Westermann, S., Drubin, D.G., and Barnes, G. 2007. Structures and functions of yeast kinetochore complexes. Annu. Rev. Biochem. 76: 563–591.

Wigge, P.A., and Kilmartin, J.V. 2001. The Ndc80p complex from *Saccharomyces cerevisiae* contains conserved centromere components and has a function in chromosome segregation. J. Cell Biol., 152: 349–360.

Winey, M., Mamay, C. L., O'Toole, E. T., Mastronarde, D. N., Giddings, T. H., Jr., McDonald, K. L., and McIntosh, J. R. 1995. Three-dimensional ultrastructural analysis of the *Saccharomyces cerevisiae* mitotic spindle. J. Cell Biol. 129: 1601–1615.

Winkler, A.A., Korstanje, R., Zonneveld, B.J., Hooykaas, P.J., and Steensma, H.Y. 2000. Isolation and characterization of KIUBP2, a ubiquitin hydrolase gene of *Kluyveromyces lactis* that can suppress a ts-mutation in CBF2, a gene encoding a centromeric protein of *Saccharomyces cerevisiae*. Curr. Genet. 38: 17–22.

Wojcik, E., Basto, R., Serr, M., Scaerou, F., Karess, R., and Hays, T. 2001. Kinetochore dynein: its dynamics and role in the transport of the Rough deal checkpoint protein. Nat. Cell Biol. 3: 1001–1007.

Wong, J., Nakajima, Y., Westermann, S., Shang, C., Kang, J.S., Goodner, C., Houshmand, P., Fields, S., Chan, C.S., Drubin, D., Barnes, G., and Hazbun, T. 2007. A protein interaction map of the mitotic spindle. Mol. Biol. Cell 18: 3800–3809.

Xu, H., Beasley, M.D., Warren, W. D., van der Horst, G.T., and McKay, M.J. 2005. Absence of mouse REC8 cohesin promotes synapsis of sister chromatids in meiosis. Dev. Cell 8: 949–961.

Yao, X., Anderson, K.L., and Cleveland, D.W. 1997. The microtubule-dependent motor centromere-associated protein E (CENP-E) is an integral component of kinetochore corona fibers that link centromeres to spindle microtubules. J. Cell Biol. 139: 435–447.

Yoda, K., Ando, S., Morishita S., Houmura, S., Hashimoto, K., Takeyassu, K, and Okazaki T. 2000. Human Centromere Protein A (CENP-A) can replace histone H3 in nucleosome reconstitution in Vitro. Proc. Natl. Acad. Sci. U.S.A 97: 7266–7271.

Yokobayashi, S., and Watanabe, Y. 2005. The kinetochore protein Moa1 enables cohesion-mediated monopolar attachment at meiosis I. Cell 123: 803–817.

Yoon, H.J., and Carbon, J. 1999. Participation of Bir1p, a member of the inhibitor of apoptosis family, in yeast chromosome segregation events. Proc. Natl. Acad. Sci. U.S.A. 96: 13208–13213.

Zinkowski, R.P., Meyne, J., and Brinkley, B.R. 1991. The centromere-kinetochore complex: a repeat subunit model. J. Cell Biol. 113: 1091–1110.

Zuccolo, M., Alves, A., Galy, V., Bolhy, S., Formstecher, E., Racine, V., Sibarita, J.B., Fukagawa, T., Shiekhattar, R., Yen, T., and Doye, V. 2007. The human Nup107–160 nuclear pore subcomplex contributes to proper kinetochore functions. EMBO J. 26: 1853–1864.

# Chapter 7
# Evolution of Centromeres and Kinetochores: A Two-Part Fugue

**Paul B. Talbert, Joshua J. Bayes, and Steven Henikoff**

Kinetochores are essential and universal features of eukaryotic chromosomes with a conserved set of functions. They are complex protein structures that are normally found at a unique location on each chromosome known as a centromere, which is often visible as a 'primary constriction' in a metaphase chromosome. Kinetochores serve to attach chromosomes to spindle microtubules in mitosis and meiosis in order to accomplish orderly chromosome segregation. They help to recruit cohesins and work in partnership with them to hold sister chromatids together locally in order to generate spindle tension (Eckert et al. 2007). They also carry out the spindle assembly checkpoint pathway, which assures that all kinetochores are attached to spindle fibers before commencing anaphase (Musacchio and Salmon 2007).

These highly conserved kinetochore functions might lead to the expectation that kinetochore proteins and the centromeres on which they assemble would be well conserved, like the microtubules to which they attach, but this is not the case. Instead, the size and DNA composition of centromeres vary considerably among eukaryotes, even between closely related species (Lee et al. 2005). This is enigmatic considering that the centromere is under strong selective constraints to permit chromosome segregation, and it suggests that organisms have 'reinvented' their centromeres over and over. Similarly, kinetochores are so varied that it has taken over two decades of work to gain an idea of their broadly conserved components (Meraldi et al. 2006). As if in an evolutionary fugue, the perpetual inventions of centromeres have been the subject for the contrapuntal voices of rapidly responding kinetochore proteins. Here we will examine the conservation and variation of centromeres and kinetochores, and consider the evolutionary forces that may shape them. We will emphasize the role of their interface in centromeric chromatin, and make particular reference to the centromere drive model, which offers hypotheses about how centromeres and kinetochores have become so varied.

---

P.B. Talbert (✉)
Howard Hughes Medical Institute and Fred Hutchinson Cancer Research Center, 1100 Fairview Avenue N, Seattle, Washington, 98109-1024, U.S.A.
e-mail: ptalbert@fhcrc.org

P. De Wulf, W.C. Earnshaw (eds.), *The Kinetochore*,
DOI 10.1007/978-0-387-69076-6_7, © Springer Science+Business Media, LLC 2009

## 7.1 Centromeric Nucleosomes as Kinetochore Subunits

Variation in kinetochores begins with their sizes, which vary in accordance with the size of the centromeres upon which they assemble. These in turn range from ~125 bp 'point' centromeres of the budding yeast *Saccharomyces cerevisiae* to 'regional' centromeres of most animals and flowering plants (Pluta et al. 1995) to 'holocentric' centromeres of some plants and invertebrates that assemble a kinetochore along the entire length of each chromosome (Buchwitz et al. 1999; Nagaki et al. 2005). Such extreme size variation is possible because kinetochores are composed of subunits. McClintock provided early evidence that kinetochores are divisible (McClintock 1938). Fragmentation and stretching of intact mammalian kinetochores led to a model for kinetochore structure in which repeated subunits fold together to form the kinetochore plate (Zinkowski et al. 1991), suggesting how kinetochores of such varied sizes could have a common underlying structure. A chromatin basis for the repeat subunit model of the kinetochore was suggested by the early finding that human kinetochores contain a centromere-specific variant of histone H3, known as CENP-A (Earnshaw and Rothfield 1985; Palmer et al. 1987), which replaces conventional H3 in centromeric nucleosomes (Yoda et al. 2000). Extensive investigations have shown CENP-A to be a 'foundation protein' necessary to assemble most other kinetochore components (Amor et al. 2004).

A centromere-specific H3 variant appears to be a kinetochore component of nearly all eukaryotes (Malik and Henikoff 2003). This makes the monophyletic origin of these H3 variants (and their orthology with CENP-A) an attractive hypothesis, and indeed monophyly of fungal centromere-specific H3 variants has been inferred (Baker and Rogers 2006). Phylogenetic analyses considering divergent eukaryotic groups, however, have been unable to exclude polyphyletic origins of centromeric H3 variants within the histone H3 family (Malik and Henikoff 2003; Dawson et al. 2007). Because of this ambiguity, the term 'CenH3' was introduced to refer to the functional class of centromere-specific H3s, regardless of whether or not they are shown to be orthologs of CENP-A (Talbert et al. 2002), and this term has become common especially among plant kinetochore investigators (Zhong et al. 2002; Nagaki et al. 2004; Maruyama et al. 2007). We take the monophyletic origin of CenH3s to be the most parsimonious working hypothesis and here treat CENP-A and CenH3 as equivalent terms, while pointing out possible exceptions to this equivalency. Where there are species-specific names for CENP-A/CenH3, we indicate these in parentheses on first occurrence.

Recently, the in vivo form of the CENP-A (Cid) nucleosome of the fruit fly *Drosophila melanogaster* has been shown to be a heterotypic tetramer of Cid/H4/H2B/H2A, differing fundamentally from the $(H3/H4/H2B/H2A)_2$ octameric structure of conventional H3 packaging nucleosomes (Dalal et al. 2007b). Native Cid nucleosomes have only half the height of octamers, and wrap no more than 120 bp of DNA rather than the ~150 bp wrapped by conventional octameric nucleosomes. The linker DNA between Cid nucleosomes is longer than that between octameric nucleosomes and appears to resist condensation into tightly packed chromatin under physiological conditions. Although the structure of CENP-A/CenH3 nucleosomes in other organisms is not yet known, there is evidence to suggest that they are also likely to be heterotypic tetramers (Dalal et al. 2007a), termed 'hemisomes' (Lavelle and Prunell 2007) to succinctly express their structural relationship to standard nucleosomes.

How a hemisome contributes to the formation of kinetochores remains to be investigated. Kinetochores are inherently asymmetric structures, connecting microtubules to DNA, and the inherent asymmetry of hemisomes, which differ from octameric nucleosomes in having nonequivalent faces, may be relevant to the assembly and orientation of the kinetochore (Dalal et al. 2007a). Similarly, the greater spacing of Cid hemisomes might accommodate the construction of the large multiprotein kinetochore.

## 7.2 The Epigenetic and Genetic Nature of Centromeres

Overexpressed Cid can induce ectopic kinetochores to form on previously noncentromeric sequences (Heun et al. 2006), which illustrates not only the foundational role of CENP-A/CenH3 nucleosomes in kinetochore assembly, but also the epigenetic nature of the centromere/kinetochore complex. A more dramatic illustration comes from 'neocentromeres', the functional kinetochores described in humans, flies, and barley that assemble on DNA sequences that lack any sequence similarity to the normal centromeres (Amor and Choo 2002; Nasuda et al. 2005). These observations suggest that kinetochores can assemble on any DNA sequence, but in nature they normally prefer specific centromere sequences at locations that can be maintained for tens of millions of years (Schueler et al. 2005; Obado et al. 2007).

The processes that maintain kinetochores at particular centromere sites are not well understood and may differ among organisms. Factors that are likely to be important are the primary sequence of the histone fold domain of CENP-A/CenH3s (Shelby et al. 1997; Keith et al. 1999; Vermaak et al. 2002; Morey et al. 2004), the histone chaperone RbAp48 (Hayashi et al. 2004; Furuyama et al. 2006), proteasome-mediated degradation of mislocalized CenH3 (Collins et al. 2004; Moreno-Moreno et al. 2006) and assorted 'upstream' kinetochore components and/or 'loading' factors for CENP-A/CenH3 that differ among organisms (Meluh and Koshland 1997; Takahashi et al. 2000; Camahort et al. 2007; Maddox et al. 2007; Stoler et al. 2007).

Despite the lack of specific DNA sequence requirements shown by human neocentromeres, native human centromere sequence is a critical variable in assembling functional kinetochores on human artificial chromosomes (Masumoto et al. 2004), and is likely to be of importance in the efficiency of assembling natural kinetochores. Since centromeres are under stringent selection at every cell division, subtle interactions between centromeres and CenH3 nucleosomes are likely to have consequences over time despite the absence of specific sequence requirements for kinetochore formation. In addition, other kinetochore proteins are known in some organisms that have greater DNA sequence specificity than CenH3 nucleosomes (Ng and Carbon 1987). The relative importance of genetics and epigenetics in centromere specification therefore differs among different organisms and among different structural classes of centromeres.

## 7.3 Point Centromeres in Budding Yeast

The budding yeast *Saccharomyces cerevisiae* and its relatives possess simple, 'point' centromeres (Pluta et al. 1995). Centromeres on each budding yeast chromosome have a consensus sequence that is only ~125 (bp), composed of three distinct regions - centromere DNA element (CDE)I, CDEII, and CDEIII. Whereas CDEII is conserved only in AT-richness and length, regions CDEI and CDEIII are strictly conserved as they serve as binding sites for yeast-specific kinetochore proteins (Clarke 1990) that are essential for recruiting a single CenH3 (Cse4)-containing nucleosome onto each centromere (Meluh and Koshland 1997; Furuyama and Biggins 2007).

Point centromeres are restricted to near relatives of budding yeast and were derived from more complex centromeres found in other yeasts and filamentous fungi that branch more basally in a phylogenetic tree (Baker and Rogers 2006; Meraldi et al. 2006). Cse4s of yeast with point centromeres share a conserved N-terminal 'END' domain that is thought to aid in efficient assembly of Cse4 nucleosomes at the point centromeres (Morey et al. 2004). The CDEI and CDEIII sequence elements and the proteins that bind them are conserved among these yeasts (Meraldi et al. 2006), but point centromeres are not interchangeable between species (Heus et al. 1994; Kitada et al. 1997). This indicates a genetic specification of point centromeres and co-evolution of their kinetochore proteins. The sequence-specific interaction between point centromeres and kinetochore proteins may serve as an example of 'optimization' of a stable evolutionary strategy.

## 7.4 Short Centromeres in Unicellular Eukaryotes

Some unicellular eukaryotes have 'short' centromeres that are typically a few kilobases long, tend to be AT-rich, and are located in somewhat larger gene-free regions. For example, the yeast *Candida albicans* has AT-rich (65%) centromeres of 3–4.5 kb that each has a unique sequence and that are found in gene-free regions of 4–18 kb (Sanyal et al. 2004). In *C. albicans*, short centromeres appear to function largely epigenetically, since a stable chromosome truncation with a 3 kb centromere bearing Cse4 nucleosomes, when extracted and re-introduced as naked DNA, fails to be stably inherited (Baum et al. 2006). Despite the lack of specific sequence requirements, the positions of these centromeres have been maintained over millions of years (Mishra et al. 2007).

The topoisomerase inhibitor etoposide has been used to map centromeres in the apicomplexan malarial parasite *Plasmodium falciparum* and in trypanosomes. In *Plasmodium,* centromeres are found in 6–12 kb gene-free regions that include ~2.5 kb of DNA with a 97% AT content (Kelly et al. 2006). Centromeres of *Trypanosoma* megabase-sized chromosomes are found in 11–16 kb 'strand-switch' regions between directional clusters of genes and are composed of degenerate retroelements that are GC-rich in *T. cruzi* or interspersed with AT-rich repetitive sequences in *T. brucei* (Obado et al. 2007).

*T. brucei* also has ~100 minichromosomes (30–150 kb) and 1–7 intermediate chromosomes (200–700 kb) in which no centromeres could be mapped using etoposide. While the megabase-sized chromosomes form end-on attachments with microtubules at the spindle periphery, the smaller chromosomes form lateral attachments to the spindle and move rapidly to the poles on the central spindle fibers (Ersfeld and Gull 1997). These chromosomes are composed of a central core of 177 bp tandem repeats, variable-length genetic regions, and telomeric repeats, all arranged in a palindrome around a central inversion point (Wickstead et al. 2004). Involvement of the 177 bp repeat in centromere function seems likely based on its similarity to centromeric satellite repeats of multicellular eukaryotes, but there is presently no evidence to support or refute this idea.

The occurrence of short AT-rich centromeres in at least three anciently diverged eukaryotic kingdoms raises the possibility that short centromeres are the ancestral form of centromere organization, but unfortunately, very little is known about this class of centromeres. With an increasing number of sequenced genomes from unicellular eukaryotes available, additional centromeres in this class are likely to be characterized in the near future.

## 7.5 Regional Centromeres

Another type of distantly related yeast, *Schizosaccharomyces pombe* (fission yeast), has a centromere structure that suggests an intermediate step between short centromeres and the well-studied class of 'regional' centromeres of multicellular eukaryotes (Pluta et al. 1995), which are distinguished by their large size and by being embedded in heterochromatin. In contrast to other yeasts, *S. pombe* has centromeric regions occupying 35–110 kb (Pidoux and Allshire 2004). These centromeres are more complex, consisting of a 4–7 kb nonrepetitive core domain (*cnt*) that is reminiscent of short centromeres, and that is flanked by domains of 'innermost' (*imr*) repeats and the outer region domain (*otr*; Clarke 1990; Pidoux and Allshire 2004). CenH3 (Cnp1) nucleosomes are found over the core domain and adjacent parts of the inner repeats (Takahashi et al. 2000), while the flanking regions assemble heterochromatin and mediate cohesion (Bernard et al. 2001).

Regional centromeres of multicellular eukaryotes differ from the centromeres of *S. pombe* in their greater size, which can range from hundreds to thousands of kilobases, and in the highly repetitive sequences bound by CENP-A/CenH3. Some regional centromeres in fungi, plants, and animals are composed of tandem arrays of very short repeats (3–60 bp) interspersed with transposons (Cambareri et al. 1998; Sun et al. 2003; Houben et al. 2007). In the fruitfly *Drosophila melanogaster*, the centromere of a stably inherited minichromosome is a complex organization of two simple 5 bp repeats, AATAT and TTCTC, and retrotransposons (Sun et al. 2003). The centromeric sequences of the native chromosomes in *D. melanogaster* are largely unknown, but the chromosomal locations of repeat families suggest that there are different repeats at each centromere (Lohe et al. 1993).

Most flowering plants and many animals have regional centromeres that are composed of highly repetitive AT-rich satellite DNA that is frequently 150–180 bp in repeat length (Choo 1997; Schueler et al. 2001; Cheng et al. 2002; Zhong et al. 2002). In humans and throughout the primate phylogeny, these satellite repeats are the 171 bp $\alpha$-satellite sequences (Willard 1991). In plants, the satellite monomers include the 155 bp CentO repeat of rice (*Oryza sativa*), the 156 bp CentC repeat of maize (*Zea mays*), and the 178 bp pAL1 ('180 bp') repeat of *Arabidopsis thaliana* (Murata et al. 1994; Ananiev et al. 1998; Cheng et al. 2002). While primate centromeres are characteristically free of transposon insertions within the 'core' $\alpha$-satellite repeats (Willard 1991; Schueler et al. 2001), the *crwydryn* family of Ty3-gypsy centromere-specific retrotransposons colonized the centromeres of grasses prior to their divergence and has been retained in most species (Langdon et al. 2000; Cheng et al. 2002; Zhong et al. 2002).

CENP-A/CenH3-containing nucleosomes are found on only a portion of large satellite arrays (Schueler et al. 2001; Jin et al. 2004; Shibata and Murata 2004; Houben et al. 2007). Chromatin domains of CENP-A nucleosomes are interspersed along the chromosome with domains of H3-containing nucleosomes. Despite this interspersion, CENP-A and H3 nucleosomes occupy different spatial domains suggesting a complex folding to form the kinetochore plate (Zinkowski et al. 1991; Blower et al. 2002). The 150–180 bp size of centromeric satellites suggests that they may serve to position nucleosomes, which might aid in folding centromeric chromatin into a functional kinetochore (Henikoff et al. 2001; Dalal et al. 2007a). CENP-A/Cid and H3 chromatin domains are dynamic. Their boundaries are not defined by the underlying sequence, and they can spread in response to dosage changes (Blower et al. 2002; Lam et al. 2006; Okamoto et al. 2007) The size of Cid domains in *Drosophila*, $\sim$15–40 kb, roughly corresponds with the minimum length of 30–70 kb of $\alpha$-satellite array that will multimerize into a functional human artificial chromosome, suggesting a minimum length necessary to maintain an epigenetic CENP-A chromatin state (Okamoto et al. 2007).

### *7.5.1 Centromeric Satellite Dynamics*

In contrast to the conservation of sequences in point centromeres, extensive characterization of satellites in animals and plants reveals rapid changes in satellite structure and composition. An example of how the overall structure of the centromeric repeats can change is provided by the $\alpha$-satellite sequences of primates. The 171 bp $\alpha$-satellites are arranged in highly homogenous chromosome-specific higher order repeat arrays in humans and many great apes, flanked by monomeric repeats (Willard 1991; Schueler et al. 2001; Schueler and Sullivan 2006). Only monomeric $\alpha$-satellites constitute the centromeres of lower primates, suggesting that the higher order arrays evolved at a later point in primate evolution (Schueler et al. 2001, 2005). Indeed the age of the homogenous $\alpha$-satellite in the human X pericentric region is very recent as estimated by the presence of a human-specific LINE1 insertion event (Schueler et al. 2001). These studies provide good evidence that monomeric $\alpha$-satellites, now relegated to pericentric regions in humans, represent an ancestral organization of the centromeric sequence that has been replaced by higher order repeats in humans and apes.

Studies of centromeric sequences in plants (rice, maize, and *Arabdopsis*) have illustrated just how labile satellite sequences can be (Ma et al. 2007b). For example, in comparisons between *O. sativa* and a wild rice species *O. brachyantha* (~7–9 million years diverged), it was found that *O. brachyantha* has replaced the CentO satellite found in *O. sativa* with another satellite sequence, referred to as CentO-F (Lee et al. 2005). In populations of *Arabdopsis thaliana*, a new variant of the 180 bp satellite was found to have spread in one particular ecotype (Ito et al. 2007). While the fixation of new centromeric satellite sequences can be quite spectacular, variation in satellite array size can also be dramatic. In humans the size of the predominant $\alpha$-satellite array on each chromosome varies more than 15-fold (Choo 1997). In rice, CentO arrays were estimated to range from 65 kb to 2 Mb on different chromosomes (Cheng et al. 2002), and in maize large differences in CentC array sizes were found both between different chromosomes and between different lines (Kato et al. 2004). These expansions or contractions of centromeric repeats are therefore occurring over only a very brief evolutionary time.

The expansions/contractions of satellites and their homogenization over time can be modeled by a mechanism of unequal crossover (Smith 1976). Since exchange occurs primarily through intra-chromosomal exchange versus inter-chromosomal exchange (Hall et al. 2005), this model explains the concerted evolution of satellite sequences on homologous chromosomes and high levels of unequal exchange directly observable in human and plant centromeres

(Rudd et al. 2006; Ma et al. 2007b). In higher order $\alpha$-satellite, recombination events occur preferentially in a 20–25 bp region of the 171 bp repeat that probably represents the linker region between phased nucleosomes. The frequent occurrence of these recombination events near the 17 bp sequence bound by the $\alpha$–satellite-binding protein CENP-B suggests a role for this protein in positioning these events, perhaps by blocking strand migration (Warburton et al. 1993).

The foundational role of CENP-A/CenH3 in kinetochore formation suggests that overall centromere size should be related to the size of the kinetochore and to the number of kinetochore microtubules bound, but this is only partially true. Comparison of a sampling of eukaryotes (Bloom 1993) shows that the average amount of genomic DNA 'cargo' per microtubule is fairly constant within a 50-fold range ($1–50 \times 10^6$ bp/ microtubule) over a broad range of genome sizes, with the number of microtubules per chromosome varying from 1 (yeasts, *Chlamydomonas*) up to ~30 (mammals), or exceptionally to 120 (blood lily). The region over which CENP-A/CenH3 assembles, however, varies from 125 bp to ~1 Mb, or nearly $10^4$-fold, and the underlying satellite arrays can be even larger. This 'CEN-value paradox' not only implies that only a tiny fraction of a large kinetochore actually binds microtubules, but suggests that some process works to favor seemingly unnecessarily large centromeres/ kinetochores.

## 7.6 Centromere Drive

The centromere drive model (Henikoff et al. 2001) proposes that this process is the 'selfish' behavior of centromeric DNA that can gain a transmission advantage in a meiosis that is asymmetric with respect to the fate of the meiotic products. Most commonly, asymmetric meiosis results in only one of four meiotic products being viable, as in the female meiosis of animals and some diatoms that produces a single egg. An analogous asymmetric meiosis in the megasporocytes of most seed plants produces a single megaspore that gives rise to the egg nucleus and embryo sac. Asymmetric meiosis also occurs in both parent cells during conjugation in ciliates. In all these cases a 'selfish' chromosome element has the opportunity to gain a transmission advantage when heterozygous with other alleles if it can increase the probability that it is included in the single nucleus that is propagated into the next generation (meiotic drive).

### *7.6.1 Classical Neocentromeres of Maize*

In the asymmetric meioses of both animals and seed plants, the viable egg or megaspore is determined by its position at one particular end of the bipolar spindle and resultant tetrad. Rhoades proposed the first model that used position on the meiotic spindle to explain ‘preferential segregation’ of the ‘classical neocentromeres’ of maize (Rhoades 1952; Dawe and Hiatt 2004). Unlike the true maize centromeres or human neocentromeres, the classical neocentromeres of maize (and other plants) take the form of heterochromatic knobs located distally on the chromosome arms. In plants heterozygous for one or more knobs, a crossover between a knob and its linked centromere results in the sister knobs moving to opposite spindle poles during anaphase I (Fig. 7.1). In the presence of the large chromosome 10 knob Ab10, all knobs become able to interact laterally with microtubules of the meiotic spindle and move rapidly to the poles (Yu et al. 1997), reminiscent of the behavior of trypanosome minichromosomes (Ersfeld and Gull 1997). In the megasporocyte, this rapid poleward movement pulls knob-bearing chromatids preferentially into polar positions where they remain through anaphase II and become incorporated into the two outside cells of the linear meiotic tetrad. Since one of these outside cells, the basal cell, becomes the surviving megaspore, knobs are preferentially transmitted to the next generation. In the microsporocyte, which gives rise to pollen, the meiotic tetrad is tetragonal and all four products survive, so no preferential transmission results (Rhoades 1952).

Maize knobs are composed of one or both of two classes of tandem satellite DNA, a 180-bp repeat and a 350-bp repeat known as TR-1 (Dawe and Hiatt 2004). The knob Ab10 contains, in addition to these satellites, novel chromosome segments that are essential for conferring neocentromere activity on the two classes of knob repeats (Hiatt and Dawe 2003). The proteins that confer this activity are unknown. It is considered likely that knob movement along spindle fibers is mediated by kinesins, but knobs do not assemble the inner kinetochore proteins CenH3 or CENP-C or the spindle checkpoint protein MAD2 (Yu et al. 1997; Dawe and Hiatt 2004).

The core elements of Ab10, including the 180-bp repeats, may have been introduced into maize from another genus and spread to new chromosomal locations because they confer a transmission advantage. The TR-1 repeats probably originated in a more recent episode of satellite expansion (Dawe and Hiatt 2004).

### *7.6.2 Karyotype Evolution*

The centromere drive model postulates that centromere variants can achieve a nonrandom segregation in asymmetric meiosis similar to that of maize knobs. Since centromeres do not cross over, their position on the spindle and inclusion in the egg are determined by the orientation of homologs at the first meiotic division (Fig. 7.1). If a centromere variant is preferentially oriented toward the pole that becomes the egg, it will increase its frequency in the population.

Evidence that such a preferential orientation is possible comes from Robertsonian translocations, the most common type of chromosome rearrangement in mammals. These translocations can be viewed as centromere fusions

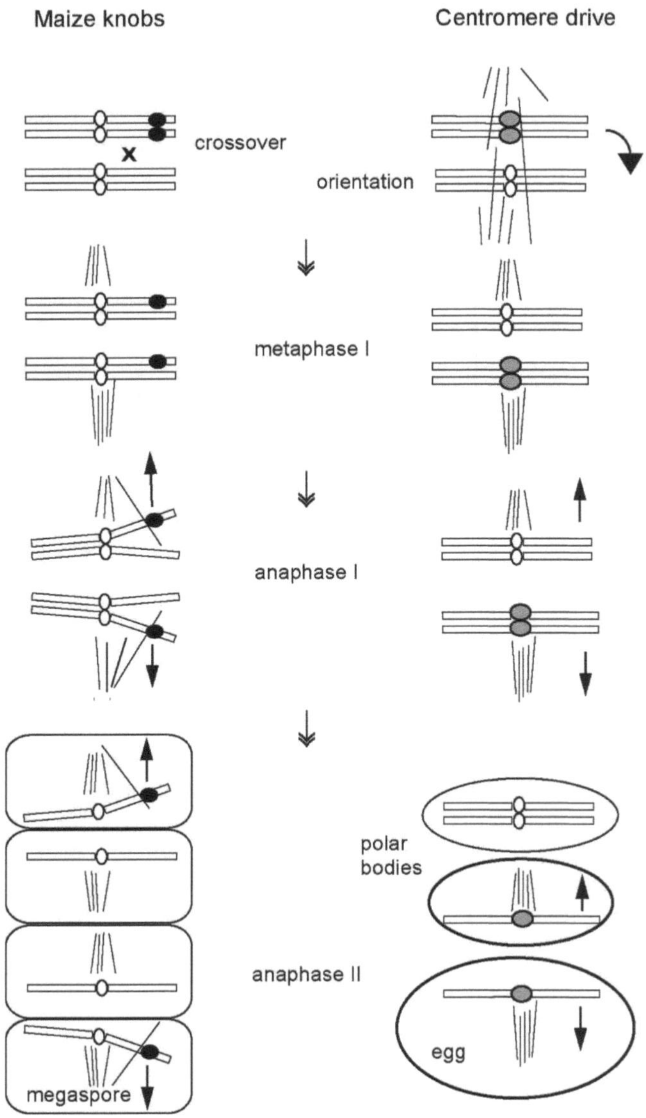


**Fig. 7.1** Comparison of classical neocentromere drive and centromere drive models. (*Left*) Classical neocentromeres or knobs (*black*) recombine so that sister knobs move rapidly to the poles in meiosis I, where they maintain their outside position in meiosis II. The basal cell becomes the megaspore. (*Right*) Selfish centromeres (*grey*) orient themselves toward the egg pole at meiosis I and are preferentially included in the egg at meiosis II

between two acrocentric chromosomes to form a metacentric chromosome with a reduction in centromere number. Human females preferentially transmit Robertsonian translocations over normal acrocentrics (Pardo-Manuel de Villena and Sapienza 2001c), whereas the reverse is true in female mice (Underkoffler et al.

2005). To explain this, Pardo-Manuel de Villena and Sapienza proposed a functional asymmetry to the oocyte meiotic spindle, with the polar body pole 'capturing' more centromeres in humans and the egg pole capturing more in mice (Pardo-Manuel de Villena and Sapienza 2001b). Consistent with this, the X chromosome is preferentially included in eggs from XO female mice (LeMaire-Adkins and Hunt 2000).

Analysis of karyotypes of 1,170 mammals revealed a strong bias toward predominantly acrocentric or predominantly metacentric chromosomes rather than a random mixture, suggesting that karyotype evolution is driven by nonrandom segregation of Robertsonian translocations, and that spindle asymmetry for centromere capture is general and frequently reversed in mammalian evolution (Pardo-Manuel de Villena and Sapienza 2001a). Other types of chromosome fusions would also be favored by an egg pole favoring fewer centromeres, such as the remarkable tandem centromere–telomere fusions seen in muntjac deer, in which the chromosome number varies from $2n = 46$ to $2n = 6/7$ between closely related species (Huang et al. 2006). Consistent with this view that one pole captures more centromeres, B chromosomes (the nonessential, usually heterochromatic 'selfish' chromosomes found in some individuals of a species) were found to be more frequently associated with acrocentric than metacentric chromosomes in mammals (Palestis et al. 2004), suggesting that capture of more centromeres by the egg pole provides opportunities for accumulation of B chromosomes. Similar associations of acrocentric and B chromosomes were found intraspecifically in 53 populations of the grasshopper *Dichroplus pratensis* polymorphic for Bs and Robertsonian translocations (Bidau and Marti 2004). One factor that might lead an egg pole to switch from capturing more centromeres to capturing fewer centromeres could therefore be selection to reduce the burden of B chromosomes. The reverse switch to capturing more centromeres might simply result from selection for increased efficiency of centromere capture.

### *7.6.3 The Spindle in Female Meiosis*

Preferential orientation of centromeres also appears to occur in birds, in which sex is determined by Z-W sex chromosomes: males are ZZ and females are ZW. The sex ratio of a brood can be maternally influenced, probably by influencing the orientation of the Z-W sex chromosome pair on the meiotic spindle (Sheldon 1999). This suggests that different individual centromeres, not just different centromere numbers, can have differential interactions with the spindle and exhibit nonrandom segregation. How could this work? The answer is unclear, but there are several unusual features of female meiotic spindles that might provide a basis for nonrandom chromosome orientation.

In mitosis, microtubules emanating from the centrosomal aster attach to kinetochores by a 'search-and-capture' process. Although this process was originally imagined as an efficient random search (Holy and Leibler 1994), more recent considerations suggest that the search must be biased toward chromatin, perhaps by the RanGTP gradient around chromosomes (Wollman et al. 2005). In addition, kinetochores can 'grow' kinetochore fibers by plus end microtubule polymerization at the kinetochore. These fibers are then captured by and integrated into the centrosomal spindle. It is proposed that these kinetochore-directed fibers are polymerized from short microtubules initiated around the chromatin that are regulated by RanGTP (Maiato et al. 2004).

In contrast, in female meiosis there are no centrosomes. Despite the highly conserved function of the meiotic spindle, there is variation among animal species in the details of meiosis I spindle formation following germinal vescicle (nuclear envelope) breakdown. Common features include spindle nucleation from small asters (Skold et al. 2005; Dumont et al. 2007), from monopolar asters, or from a transient microtubule array (Gard 1992); formation of a RanGTP gradient and growth of randomly oriented microtubules around the chromosomes (Cao et al. 2005; Dumont et al. 2007); self-organization of microtubules into a barrel-shaped spindle through the action of motor proteins (Skold et al. 2005); migration of the chromosomes and spindle to the oocyte cortex, which can interact with one pole of the spindle (Lutz et al. 1988; Gard 1992, 1993; Verlhac et al. 2000); and rotation of the spindle to extrude the first polar body (Gard 1992; Albertson and Thomson 1993). These processes of spindle nucleation and organization, chromosome and spindle migration, cortex attachment, and rotation could affect spindle asymmetry and provide cues for biased bivalent orientation. In mice, kinetochores interact laterally with microtubules as the chromosomes oscillate about the plate for several hours before the formation of stable end-on microtubule–kinetochore attachments and congression to the equatorial plate late in prometaphase I (Brunet et al. 1999). The orientation of bivalents might be influenced by this oscillation.

Kinetochore size and microtubule number are predicted to have large effects on time-to-capture in mitosis (Wollman et al. 2005), and might be important for determining a biased orientation of competing centromeres on the asymmetric spindle. Differences in the number of microtubules at the egg and polar body poles might determine the preference for more or fewer centromeres that seems to drive karyotype evolution, and differences in the sizes of kinetochores might influence the orientation of competing centromere variants on the spindle. Satellite expansions are likely to change the size of the kinetochores assembled on them, and expansion of particular satellite variants may change the efficiency of kinetochore assembly. As new centromere variants continually arise, some of them will sweep through the population because of their orientation advantage in female meiosis. Repeated expansions will lead to the accumulation of satellites, with older repeats diverging by mutation as they are replaced by the most recent sweep. Thus both the rapid changes in satellite arrays and the location of the functional centromere in the most recent satellite array (Schueler et al. 2001) are explicable in the centromere drive model as a consequence of the 'selfish' ability of satellites to use the kinetochore to achieve preferential transmission in asymmetric meiosis.

### 7.6.4 Centromere Repositioning

Comparisons of marker order among X chromosomes of primates provided unequivocal evidence that centromeres can emerge in new locations during evolution independent of chromosome rearrangements, and can acquire new satellites at the new centromere position and lose satellites at the old position (Ventura et al. 2001). Numerous examples of such centromere repositioning have now been documented in both mammals (Ventura et al. 2003, 2004, 2007; Ferreri et al. 2005; Cardone et al. 2006) and birds (Kasai et al. 2003). Centromere repositioning has been interpreted as the formation of a neocentromere on euchromatic sequence, followed by invasion and repopulation of the new centromere by satellite DNA (Ventura et al. 2001). Indeed, *centromere 8 (Cen8)* from rice, which has only ~65 kb of the centromeric *CentO* satellites in a CenH3-binding region that extends over 750 kb and includes active genes (Nagaki et al. 2004), has been interpreted to represent an intermediate phase in the development from a neocentromere to a typical regional centromere populated only with *CentO* satellites and transposons. Recently *Cen8* has been shown to lie within a ~1 Mb inversion that is not present in the wild rice species *O. brachyantha* and *O. officinalis*. These species have centromeres in the orthologous but uninverted chromosomal region (Lee et al. 2005; Ma et al. 2007a), indicating that rice *Cen8* be not may a neocentromere, but more probably derives from recently a rearranged centromere that retained only a fragment of its original satellite array. Whether derived from a fragmentary centromere or a neocentromere, centromere drive would be expected to favor *CentO* retention and expansion at *Cen8* over time, so that eventually it will resemble other rice centromeres with large satellite arrays.

'Hotspots' of human neocentromere formation have been observed at the cytological level, suggesting that some regions are more prone to neocentromere formation than others (Amor and Choo 2002). Some human neocentromere hotspots also appear to be sites for centromere repositioning during mammalian evolution (Ventura et al. 2004). Detailed analysis of three neocentromere hotspots found that in each case the neocentromeres were found on unrelated single-copy sequences distributed over 4–10 Mb, arguing strongly against a sequence basis for neocentromere formation, and indicating just how broad a region is represented by a cytological 'hotspot' (Alonso et al. 2003; Ventura et al. 2003; Cardone et al. 2006). Are there actually common centromere-promoting properties over such a large region?

Investigation of a hotspot at 15q24-26 found that the region contained segmental duplications (duplicons) distributed over a 16.5 Mb region around an ancestral centromere at 15q25 present in old world monkeys. The association of neocentromeres from this region with rearrangement/duplication boundaries on the chromosomes that carry them suggests that neocentromere formation can result from repair of rearrangements. Duplicons around centromeres may increase rearrangements through unequal crossover and to lead to 're-use' of pericentric regions as centromeres and neocentromeres (Ventura et al. 2003). Association of centromeres with rearrangements is supported by the finding of a centromeric retrotransposon at all centromeres in the tammar wallaby (*Macropus eugenii*) and at the positions of centromeres and of breaks of synteny in the karyotypes of other marsupials (Ferreri et al. 2005). Similarly, in Arabidopsis and related genera in the Brassicaceae, 24 of the 52 synteny breaks between karyotypes were found at centromeric regions, while 20 more were found at terminal positions rich in repetitive sequences (Lysak et al. 2006).

Investigation of a second neocentromere hotspot at 13q32 found no evidence of duplicons or rearrangements, but found that three neocentromeres from this site and three others were all in gene-poor regions or in a 2.5 Mb 'gene desert' (Alonso et al. 2003). Similarly, two repositioned centromeres at a hotspot at 13q21 mapped in a 3.9 Mb region devoid of genes (Cardone et al. 2006), and another repositioned centromere was found to lie within a 778 kb gene desert (Ventura et al. 2007). Together these studies strongly suggest that the absence of gene activity may be important in centromere establishment or maintenance. Although centromeres and neocentromeres can be compatible with gene expression (Saffery et al. 2003; Nagaki et al. 2004), neocentromeres have less CENP-A than established centromeres (Alonso et al. 2003; Irvine et al. 2004), and transcription results in nucleosome replacement (Schwartz and Ahmad 2005) that might destabilize neocentromeres. CENP-A/CenH3 nucleosomes can be normally incorporated into euchromatic regions (Talbert et al. 2002), and transcription may be important for evicting them so that they can be degraded by proteasomes (Collins et al. 2004; Moreno-Moreno et al. 2006). Regions of low gene activity may therefore favor retention of enough CENP-A/CenH3 for a neocentromere to become established.

## 7.7 Evolution of Kinetochore Components

Sequence divergence in kinetochore proteins has slowed recognition of universal structural components of the kinetochore, but substantial progress in identifying conserved proteins and subcomplexes has been made in recent years, particularly in budding yeast. The extensive dissection of the budding yeast kinetochore has yielded over 70 kinetochore proteins that have now been genetically, biochemically, and biophysically partitioned into a number of complexes and groupings. Those yeast inner kinetochore proteins that are known or predicted to bind DNA include Cse4/CenH3, Mif2/CENP-C, and the CBF3 complex. A central kinetochore of 'linker complexes' includes

Spc105/KNL1, the Ctf19 or COMA complex, the Mtw1 or MIND complex, and the Ndc80 complex (Westermann et al. 2007). The Ndc80 complex has been shown to bind and bundle microtubules in vitro (Cheeseman et al. 2006)). Spc105/KNL1 also binds microtubules in vitro, and may be stabilized by the MIND complex. Outer kinetochore proteins at the microtubule interface include microtubule plus end-tracking proteins or +TIPs, motor proteins, and the Dam1 or DASH complex, which forms a ring around the microtubule and is necessary to maintain bi-orientation and stabilize microtubule attachment. There are also mitotic checkpoint proteins, chromosomal passenger proteins that are necessary to establish bi-orientation, and others (Westermann et al. 2007).

### *7.7.1 Conservation of Kinetochore Components*

Meraldi and coworkers (Meraldi et al. 2006) set out to determine which of these several sets of proteins are unique to yeasts with point centromeres and which are parts of a conserved molecular core of the kinetochore that is common to all eukaryotes. They searched fungal genomes for homologs of 55 yeast kinetochore proteins, and found that 11 of these proteins were apparently unique to yeasts with point centromeres, including the CBF3 complex that binds to CDEIII, two subunits of the COMA complex, and several other proteins that require COMA for kinetochore localization. They also concluded that certain proteins such as Skp1 of the CBF3 complex and the transcription factor Cbf1 are most likely conserved in non-yeast eukaryotes for their independent non-kinetochore functions.

Searching additional nonfungal eukaryotic genomes revealed that the DASH complex is confined to fungi, while several proteins of the linker layer, including spc105/KNL1, are present in animals but were not detected in plants or apicomplexans. Of the 55 proteins examined, 34 putative orthologs were identified in animals, 21 in plants, 11 in apicomplexans, and 8 in the diplomonad *Giardia lamblia*. Allowing that some homologs may be too diverged to detect and that several major clades of eukaryotes remain unsampled, the results suggest that the ancestral eukaryotic kinetochore probably contained CENP-A/CenH3, CENP-C, the MIND complex (probably lacking Dsn1), the Ndc80 complex (without Spc24), the +TIPs Stu2/XMAP215 and Bim1 (which, however, also have nonkinetochore functions), an Aurora kinase and perhaps other chromosomal passenger proteins, and spindle assembly checkpoint proteins. In addition, homologs of the Mcm21 protein of the COMA complex were found in plants, animals and fungi, but have not so far been detected in apicomplexans or *Giardia*. In general, conservation between putative orthologs is restricted to short sequence blocks (Meraldi et al. 2006).

### 7.7.2 CENP-A/CenH3 Conservation

One of the most conserved kinetochore proteins is CENP-A/CenH3. CenH3s resemble conventional packaging H3s in having an N-terminal tail and a histone fold domain (HFD) made up of four alpha helices (designated N, 1, 2 and 3) that are separated by loops (Luger et al. 1997). The N-terminal tails of CenH3s are completely divergent between distant taxa and are usually longer than the tails of packaging H3s (Henikoff et al. 2001). Similarly, the C-termini of CenH3s are typically longer than the C-termini of packaging H3s (residues E133-A135), and show little conservation.

In contrast, the HFDs of CenH3s typically share ~50% amino acid identity with those of packaging H3s (Fig. 7.2), but are usually longer than other H3s in the loop 1 region that separates helices 1 and 2 of the HFD (Malik and Henikoff 2003). CenH3s replace the region from the end of helix 1 through most of loop 1 (corresponding to H3 residues A75 through L82) with a variety of amino acids (Fig. 7.2). This variable region is terminated by a conserved arginine that contacts DNA in H3 (R83). Adjacent to this DNA contact point, F84 of the H3 amino acid sequence is rather consistently replaced with W in CenH3s. Despite the conservation of this W, however, replacement of W with F did not affect the localization of CENP-A or the viability of Cse4 (Shelby et al. 1997; Keith et al. 1999). F84 in loop 1 of H3 projects into loop 2 of H4 (Fig. 7.2, bottom), and its replacement with W together with a longer loop 1 in CenH3 may be important for the conformation of H4. In H3 nucleosomes, Helix 2 of H4 is bent where it contacts H3 loop1, and the longer loop 1 and projecting W in CenH3s has been proposed to straighten H4 helix 2 and stabilize the 4-helix bundle it forms with H2B, favoring the formation of a stable hemisome (Dalal et al. 2007a).

Other residues in CenH3s are notable for their lack of conservation. V46 is absolutely conserved in H3s, but can apparently be replaced by almost any amino acid in CenH3s. V46 is constrained to fit between the two DNA gyres in H3 nucleosomes (Luger et al. 1997), but in a CenH3 hemisome, this constraint does not exist (Fig. 7.2). Residues S87 and P121 in H3s also show little or no conservation in CenH3s. In animals, S87 is essential to specify the CAF-1-mediated replication-coupled assembly pathway of H3 rather than the HirA-mediated replication-independent assembly pathway of H3.3, which has A at this position instead (Ahmad and Henikoff 2002). This amino acid position also differentiates pairs of packaging H3 variants in other eukaryotes (Malik and Henikoff 2003). The lack of conservation at this position in CenH3s may serve to preclude their entry into either of these packaging histone pathways. Significantly, of the 11 CenH3s represented in the alignment in Fig. 7.2, only that of the red alga *C. merolae* has the same amino acid at this position as its corresponding packaging histones. P121 lies at the loop2/helix 3 junction. Given the unique conformational constraints imposed by proline, substitutions of this residue in CenH3 might affect the angle and spacing between helix 3 and helix 2, which form the dimerization interface in H3 octamers. A change in the orientation of

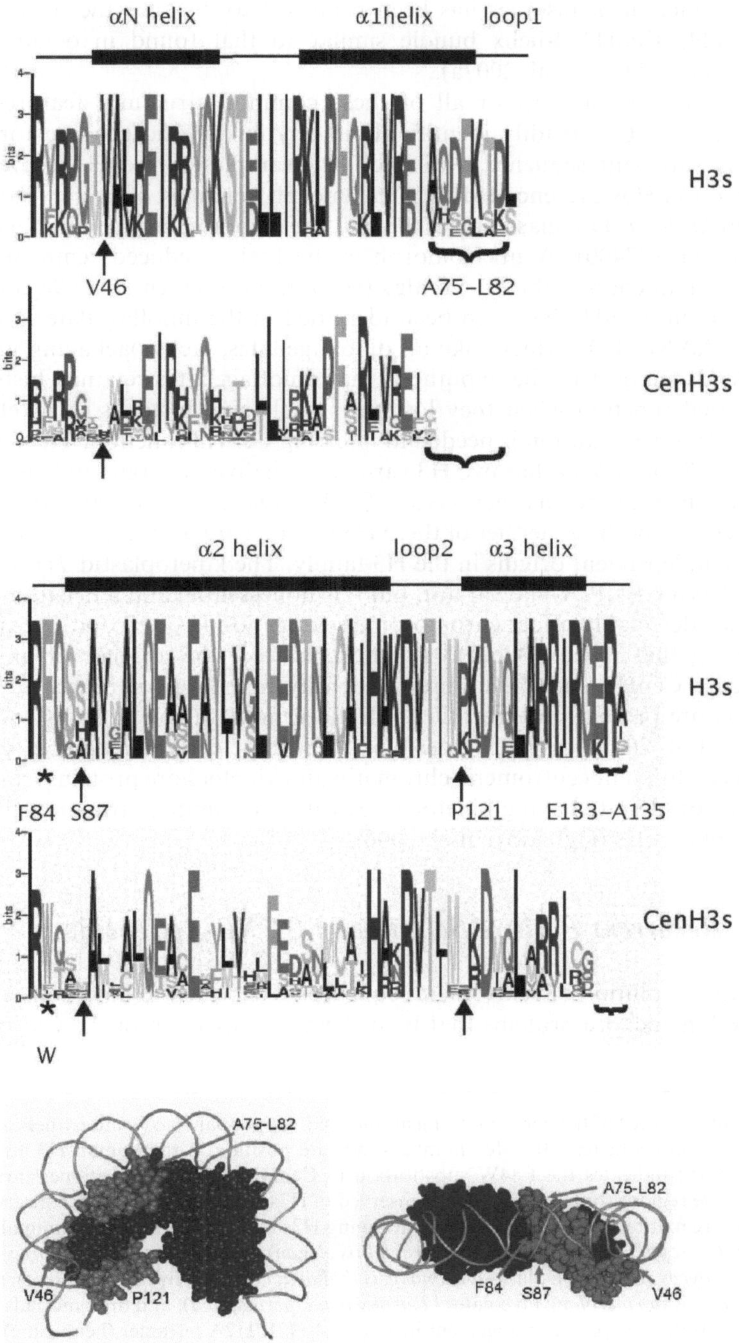


**Fig. 7.2** CenH3 conservation. (*Top*) Alignment of histone fold domains of CenH3s and packaging H3s in Logos format, in which the height of each letter reflects the conservation

this interface in hemisomes has been proposed to disfavor the formation of the CenH3–CenH3 4-helix bundle similar to that found in octameric H3 nucleosomes (Dalal et al. 2007a).

The presence of most or all of these common structural features make candidate CenH3s readily identifiable in any eukaryotic genome for which there is sufficient sequence available. As examples of CenH3 universality, putative CenH3s are encoded in nucleomorph genomes of the cryptomonad *Guillardia theta* (Douglas et al. 2001) and the chlorarachniophyte *Bigelowiella natans* (ABA27400). A nucleomorph is the highly reduced remnant of the nucleus of an endosymbiotic red alga (*Guillardia*) or green alga (*Bigelowiella*). An apparent CenH3 has even been identified in the dinoflagellate *Pyrocystis lunula* (AAN85430), which, like all dinoflagellates, lacks packaging histones. This lends support to the hypothesis that dinoflagellates retained histones of specialized function when they lost their packaging histones (Hackett et al. 2005). However, caution is needed in inferring CenH3 function. The diplomonad *Giardia intestinalis* has two H3 variants with divergent tails and longer loop 1s, but only one of them functions as a CenH3 (Dawson et al. 2007). It is unclear whether one, both, or neither of these H3 variants are orthologous to CENP-A or have independent origins in the H3 family. The kinetoplastid *Trypanosoma brucei* has a CENP-A-like variant, but it is nonessential and a deletion mutant does not detectably affect chromosome segregation (Lowell and Cross 2004), indicating that it lacks CenH3 function. Indeed, the absence of any other candidate CenH3 and the failure to identify homologs of nearly all known kinetochore proteins in three kinetoplastid genomes (Lowell and Cross 2004; Obado et al. 2007) raises the possibility that these early diverging eukaryotes may have different centromeric chromatin and kinetochore proteins from other eukaryotes, despite having kinetochores with a trilaminar structure similar to other eukaryotes (Ogbadoyi et al. 2000).

### *7.7.3 Recurrent Positive Selection in CENP-A/CenH3s*

The rapid evolution of centromere sequences is expected to have consequences for the kinetochore proteins that bind them. Comparison of the Cid genes of

**Fig. 7.2** (continued) of the corresponding amino acid. Upper panel: αN and α1 helices. Lower panel: α2 and α3 helices. Residue numbers indicate position in the human H3 amino acid sequence. (*) indicates the F84W substitution in CenH3s. Individual residues (arrows) or unalignable regions (brackets) that are conserved in H3s but show little or no conservation in CenH3s are marked. Sequences of both packaging H3s and CenH3s are from animals (*Homo sapiens, Drosophila melanogaster, Caenorhabditis elegan*); fungi (*Schizosaccharomyces pombe, Saccharomyces cerevisiae*); plants (*Arabidopsis thaliana, Oryza sativa*; *Luzula nivea*); red algae (*Cyanidioschyzon merolae*); alveolates (*Tetrahymena thermophila*); and diplomonads (*Giardia lamblia*). (Bottom) Spatial arrangement in a H3/H4/H2B/H2A tetramer (hemisome) of select H3 amino acids that are altered in CenH3s. Histone tails are not shown. Amber: DNA double helix. Green: H3; Blue: H2A, H2B, and H4. Magenta: selected H3 residues as labeled. Upper panel: 'front' view. Lower panel: 'side' view. (*See* Color Insert)

*Drosophila melanogaster* and *D. simulans* demonstrated that both the tail and the loop 1 region of the HFD are under positive (adaptive) selection (Malik and Henikoff 2001). The nearly invariant arginine at the edge of loop1 (R83) is predicted to contact DNA, based on the structure of H3 nucleosomes (Luger et al. 1997), suggesting that the positive selection in nearby amino acids might involve adaptation to centromeric DNA sequences. Adaptation would probably result from changes in nucleosome shape rather than through additional loop 1 - DNA contacts, since most of loop 1 interacts with helix 2 and loop 2 of H4. Loop 1 is essential for targeting CENP-A/CenH3s to centromeres (Shelby et al. 1997; Keith et al. 1999; Vermaak et al. 2002), and changes in shape adjacent to the DNA contact at R83 may be important in positioning CENP-A/CenH3 hemisomes on satellite sequences. Further analysis of additional *Drosophila* species revealed that the tails of some species encode minor groove binding motifs, suggesting that the tail also contacts DNA (Malik et al. 2002). Parallel results were obtained in *Arabidopsis thaliana* and its relatives, with both the tail and the loop1 region being subject to significant positive selection in some species (Talbert et al. 2002; Cooper and Henikoff 2004). Examination of two additional *Arabidopsis* species duplicated for CenH3 found little or no evidence for positive selection (Kawabe et al. 2006). However, the detection of positive selection in diverse eukaryotic lineages suggests continuing adaptation of the tail and loop 1 to a genetic conflict general to plants and animals.

Genetic conflict is implicit in the centromere drive model, which views satellite behavior as 'selfish' because it may result in deleterious consequences for the organism as a whole. The ability of a satellite expansion to assemble a larger kinetochore that is preferentially transmitted in the asymmetric meiosis of females might also lead to an imbalance between heterozygous centromeres in the symmetric meiosis of males, which is subject to more stringent control of segregation than female meiosis (Hunt and Hassold 2002). Unequal kinetochore tension on competing centromere variants might lead to nondisjunction, checkpoint arrest, or chromosome loss and consequent reduction in male fertility. Indeed, the preferential transmission of Robertsonian translocations in female meiosis of mammals is accompanied by a reduction in male fertility (Ogur et al. 2006). Reduced male fertility as a consequence of centromere drive creates a selective pressure for suppressors of centromere drive. Any mutations that result in a restoration of male fertility, presumably by restoring parity of centromere alleles in male meiosis, will therefore be selected. The cycle of centromere drive and suppression is expected to be a recurrent event in seed plants and animals, in which centromeres must be transmitted through both asymmetric female and symmetric male meioses.

One mechanism to restore male fertility would be to alter the DNA-binding interface of the kinetochore to neutralize the ability of driving centromere sequences to assemble kinetochores that result in unequal tension and distorted transmission. The positive selection seen in and around the DNA-contacting portions of CenH3s is proposed to reflect alterations in affinities for centromere variants that result in suppression of centromere drive (Malik and Henikoff

2001; Henikoff et al. 2001). In principal, either an increase in affinity for the nondriving allele(s) or a decrease in affinity for the driving allele(s) could restore parity, but selection may favor reducing affinity to new variants to avoid specification of the centromere sequence by driving satellites (Dawe and Henikoff 2006).

This model of centromere drive suppression predicts that CENP-A/CenH3s will be a common target of episodes of positive selection for drive suppression, but will not necessarily be a constant target or the only target, since other kinetochore proteins such as CENP-C also bind centromeric DNA. Indeed, mammalian CENP-A was found to be under negative (purifying) selection, as were CenH3s from maize, sorghum (*Sorghum bicolor*), and sugarcane (*Saccharum officinarum*). However, positive selection consistent with suppression of centromere drive was found in CENP-C in these lineages (Talbert et al. 2004).

### 7.7.4 CENP-C

CENP-C is a large kinetochore 'foundation protein' (Amor et al. 2004) dependent on CENP-A/CenH3 for centromere incorporation (Moore and Roth 2001; Oegema et al. 2001) and necessary for the assembly of the outer kinetochore plate (Tomkiel et al. 1994). It regulates the association of several outer kinetochore proteins (Liu et al. 2006). It is poorly conserved in sequence but not in phylogenomic distribution: it is recognized in diverse eukaryotic lineages by a conserved 24 amino acid motif (CENPC motif; Brown 1995; Talbert et al. 2004). The central region of human CENP-C and the C-terminus containing the CENPC motif have both been shown to bind $\alpha$-satellite in vivo (Politi et al. 2002; Trazzi et al. 2002), and have also recently been shown to specifically bind $\alpha$-satellite RNA, which is an integral component of the kinetochore (Wong et al. 2007). Centromeric RNAs were first found to be kinetochore components in maize (Topp et al. 2004). The role of centromeric RNAs is unclear, but RNase treatment led to significant loss of CENP-C at the kinetochore, and restoration of $\alpha$-satellite RNA was able to partially rescue CENP-C localization, suggesting that $\alpha$-satellite RNA is required for kinetochore function (Wong et al. 2007).

The central DNA/RNA-binding region of mammalian CENP-C was found to have undergone recurrent positive selection. Nearly half of the residues in the protein were predicted to be under positive selection by the PAML program, with the most highly significant sites in the DNA/RNA binding region. A similar result was obtained for CENP-Cs from grasses. Evidence for positive selection was found in the central region of CENP-Cs from all flowering plants and animals examined (Talbert et al. 2004). In contrast, the conserved C-terminus is under negative selection in nearly all lineages. Both the central and C-terminal regions confer centromere targeting in humans (Trazzi et al.

2002), but may do so in different ways. The central region recognizes the rapidly evolving centromere DNA and RNA directly (Politi et al. 2002; Wong et al. 2007). The C-terminus may do likewise, but the CENPC motif in the C-terminus has been proposed to recognize a conserved feature of CENP-A/CenH3 nucleosomes (Talbert et al. 2004). This proposed role of the CENPC motif is supported by the observation that the conserved C-terminus is found near CENP-A (Cid), while the N-terminus is near the Mis12 and Ndc80 complexes (Schittenhelm et al. 2007).

The centromere drive model predicts that positive selection of CENP-C will be found in seed plants and animals, but not in organisms with only symmetric meiosis. Indeed, both CenH3 (Cse4) and CENP-C (Mif2) were found to be under negative selection throughout their length in yeasts (Talbert et al. 2004), and CenH3s appear to be under negative selection in all fungi (Baker and Rogers 2006). Evidence strongly suggestive of negative selection in CENP-A/CenH3 and CENP-C genes is also seen (Table 7.1) in comparison of the rates of synonymous substitution ($K_S$) and nonsynonymous substitution ($K_A$), between other pairs of species with symmetric meiosis, where $K_A/K_S < 1$ is suggestive of negative (purifying) selection. Sliding window analysis of CenH3 or CENP-C genes from malarial parasites (*Plasmodium*), prasinophyte algae (*Ostreococcus*), ascomycete fungi (*Aspergillus*) and oomycete plant pathogens (*Phytophthora*) found all sites in these comparisons have $K_A/K_S < 1$, suggestive of purifying selection. Oomycetes are heterokonts that produce one or more eggs (oospheres) from a multinucleate oogonium. In *Phytophthora*, meiosis occurs simultaneously in 8–9 nuclei and appears to be symmetric. Subsequently one or two nuclei central in the oogonium become oosphere(s) while the others contribute to the periplasm that surrounds the oosphere (Sansome and Brasier 1973). The selection of nuclei to become oospheres does not appear to depend on chromosomal orientation on the meiotic spindles, but the mechanism of selection is unknown.

**Table 7.1** Pairwise estimates of synonymous ($K_S$) and nonsynonymous ($K_A$) substitution rates between CenH3 and CENP-C genes in species with symmetric meiosis or in which meiosis is not described (*Ostreococcus*). $K_A < K_S$ is evidence for negative selection. Estimates are from K-estimator (Comeron 1999). *Ostreococcus* and *Phytophthora* sequences are from the Joint Genome Institute (www.jgi.doe.gov)

| Gene | Species pair | Sequences | $K_A$ | $K_S$ |
|---|---|---|---|---|
| CenH3 | *Plasmodium falciparum* | XM_001350032 | 0.15 | 0.87 |
| | *P.yoelli* | XM_724470 | | |
| CenH3 | *Ostreococcus lucimarinus* | XM_001415867 | 0.02 | 0.78 |
| | *O. tauri* | Chr_01.0001: 612030-612914 | | |
| CENP-C | *Aspergillus nidulans*; | XM_657627 | 0.27 | 1.13 |
| | *A. terreus* | XM_001209513 | | |
| CENP-C | *Phytophthora sojae*; | estExt_fgenesh1_pg.C_46005 | 0.10 | 0.94 |
| | *P. ramorum* | fgenesh1_pg.C_scaffold_12000061 | | |

## 7.7.5 Ndc80/Hec1

Is positive selection observed in all kinetochore proteins in seed plants and animals? The centromere drive model hypothesizes that positive selection results from recurrent selection to restore kinetochore parity in symmetric male meiosis, and it is unlikely that all structural components of kinetochores are capable of affecting parity while maintaining function. Analysis of selection in most kinetochore proteins is nonexistent, but Ndc80/Hec1 is an example of a kinetochore protein that appears to be under negative selection throughout its length in all lineages that have been examined (Table 7.2). Its conserved function in binding microtubules is expected to result in optimized binding that is maintained by strong negative selection.

## 7.7.6 Nup 107–160 Complex

In vertebrates several components of the nuclear pore relocate to kinetochores during mitosis (Loiodice et al. 2004). The Nup107–160 complex accumulates on unattached kinetochores and on the spindle poles and proximal spindle fibers during mitosis in HeLa cells, similar to the checkpoint proteins Mad1, Mad2, Bub1, and Cdc20. In *Xenopus* egg extracts, it is found throughout the spindle and is required for stable spindle assembly, with depletion of the complex leading to lack of spindles or faint spindles with fewer microtubules (Orjalo et al. 2006). Recurrent adaptive evolution has been found in four Nup107–160 components (Nup75, Nup96, Nup107, Nup133) and two interacting proteins (Nup98 and Nup153) in *Drosophila* (Presgraves and Stephan 2007). Could the Nup107–160 complex be suppressing centromere drive by regulating the asymmetric spindle in female meiosis? Although the idea is intriguing, the

**Table 7.2** Pairwise estimates of synonymous ($K_S$) and nonsynonymous ($K_A$) substitution rates between Ndc80/Hec1 genes in animals, fungi, plants, and apicomplexans. Estimates are from K-estimator (Comeron 1999). *D. yakuba* sequence is CM000162: 7786557-7788776

| Species pair | Sequence | $K_A$ | $K_S$ |
|---|---|---|---|
| *Drosophila melanogaster* | NM_132674.2 | 0.05 | 0.33 |
| *D. yakuba* | CM000162 | | |
| *Xenopus laevis* | BC070725 | 0.02 | 0.32 |
| *X. tropicalis* | NM_001011162 | | |
| *Rattus norvegicus* | XM_001055564 | 0.05 | 0.21 |
| *Mus musculus* | NP_075783 | | |
| *Aspergillus nidulans* | XM_657481 | 0.08 | 2.09 |
| *A. fumigatus* | XM_749508 | | |
| *Arabidopsis thaliana* | NM_115320 | 0.05 | 0.19 |
| *Olimarabidopsis pumila* | AB299833 | | |
| *Plasmodium falicparum* | XM_961055.1 | 0.15 | 1.24 |
| *P. yoelli* | XM_724704.1 | | |

Nup107–160 complex has not been shown to localize at kinetochores in *Drosophila* and there are other plausible explanations for positive selection in the Nup107–160 complex, notably the possibility that it results from suppressing the well-documented segregation distortion (SD) system found in *D. melanogaster*, which involves the mislocalization of the nuclear transport regulator RanGAP (Presgraves and Stephan 2007). Additional investigation may reveal whether suppression of centromere drive plays a role in the evolution of nuclear pore proteins that also function in kinetochore and spindle regulation.

## 7.8 Holocentric Chromosomes

Most eukaryotes have monocentric chromosomes; their kinetochores occupy a single defined location on each chromosome that is generally visible as a primary constriction. However, in several eukaryotes the entire chromosome functions as a kinetochore. Such chromosomes, which lack visible constrictions, are said to be polycentric, holocentric, holokinetic, or to have diffuse centromeres. Holocentric chromosomes have been described in nematodes, in the plant families Cyperaceae and Juncaceae (sedges and rushes), in several orders of insects, and in other arthropods (White 1973). Investigation at the molecular level has occurred in the nematode *Caenorhabditis elegans* and to a lesser extent in other nematodes and in the snowy wood-rush *Luzula nivea*.

In C. elegans CENP-A, also know as holocentric centromere protein 3 (HCP-3), and other kinetochore proteins localize along the entire length of the chromosome and form a typical trilaminar kinetochore at mitosis (Braselton 1981; Albertson and Thomson 1982; Buchwitz et al. 1999; Oegema et al. 2001). Both chromosome condensation (Stear and Roth 2002; Chan et al. 2004) and a kinesin-mediated polar ejection force (Powers et al. 2004) have been implicated in maintaining a relatively rigid mitotic chromosome structure in which sister kinetochores are forced to directly face opposite poles, reducing the problem of kinetochore attachment to both poles (merotelic attachment).

Meiosis presents unique problems because homologs are held together in meiosis I by chiasmata, which provide the tension between homologs necessary to orient the bivalent for reductional division. To resolve chiasmata while separating homologs, monocentric chromosomes release cohesion from the chromosome arms, including the chiasmata, in anaphase I and then release cohesion from sister kinetochores in anaphase II. If the holocentric kinetochore were assembled all along a chromosome in meiosis as it is in mitosis, there would be no distinction between centromere and arm: every recombination event would occur within the kinetochore, and a recombinant chromatid would become attached to opposite spindle poles on the either side of the chiasma. Thus in order to prevent merotelic attachment, kinetic activity in metaphase/anaphase I apparently must be limited to a subregion of the chromosome that did not recombine in prophase I. In addition, a two-step release of cohesion must occur that resolves chiasmata and segregates chromosomes.

### 7.8.1 *Caenorhabditis*

In nematodes this problem has been solved by inventing an alternative method of directing kinetic activity. In prophase I, *Caenorhabditis* chromosomes are ordinarily limited to a single crossover forming a chiasma near one end of the chromosome, which separates the chromosomes into long and short arms (Albertson et al. 1997). Chromosome remodeling by condensins and differential loss of synaptonemal complex proteins from the long arm produce a highly condensed cruciform bivalent (Chan et al. 2004; Nabeshima et al. 2005) that has the appearance of an end-to-end association of homologs by light microscopy, as has been observed in several other holocentric organisms (White 1973). CENP-A (HCP-3) is absent in early prophase, but becomes assembled on chromosomes during late pachytene/diplotene as the chromosomes are desynapsing and remodeling (Monen et al. 2005). The timing raises the possibility that HCP-3 is deposited or modified differentially on the two arms.

At metaphase I, the bivalent aligns axially on the spindle, with its long arms parallel to the spindle fibers and perpendicular to the short arms and metaphase plate (Fig. 7.3). No typical trilaminar kinetochore plate is visible in meiosis by electron microscopy (Albertson and Thomson 1993; Albertson et al. 1997); instead, a kinetochore-like structure encircles the chromosome (Howe et al. 2001; Monen et al. 2005) The chromatid ends farther from the chiasma lead at anaphase I, separating the long arms of the bivalent reductionally and the short arms equationally (Albertson and Thomson 1993; Albertson et al. 1997). In meiosis II, kinetic activity shifts to the opposite ends of the chromatids, separating the short arms reductionally while the long arms separate equationally (Albertson and Thomson 1993). Thus, the ends of the chromosome appear to function much as kinetochores do on monocentric chromosomes, except that either end of the chromosome is capable of leading at the first division, while the opposite end leads at the second.

Does centromere drive affect holocentric chromosomes? In monocentric animals and seed plants, tandem satellite repeats around the centromere may occupy tens of megabases, and the absence of such regions in *Caenorhabditis* leads to the speculation that its unusual ability to form a kinetochore on any genomic sequence and even on extrachromosomal sequences (Collet and Westerman 1984) might have evolved to suppress the accumulation of repeated sequences by centromere drive. The recurrent conflict between genetic and epigenetic specification of centromeres (Dawe and Henikoff 2006) would be expected to be focused at *Caenorhabditis* chromosome ends because of their role in leading meiotic chromosome segregation. Some repetitive sequences such as the ~180 bp Cele14 element are nonrandomly distributed toward the chromosome ends, and this distribution has been suggested to reflect involvement in centromere function (Surzycki and Belknap 2000).

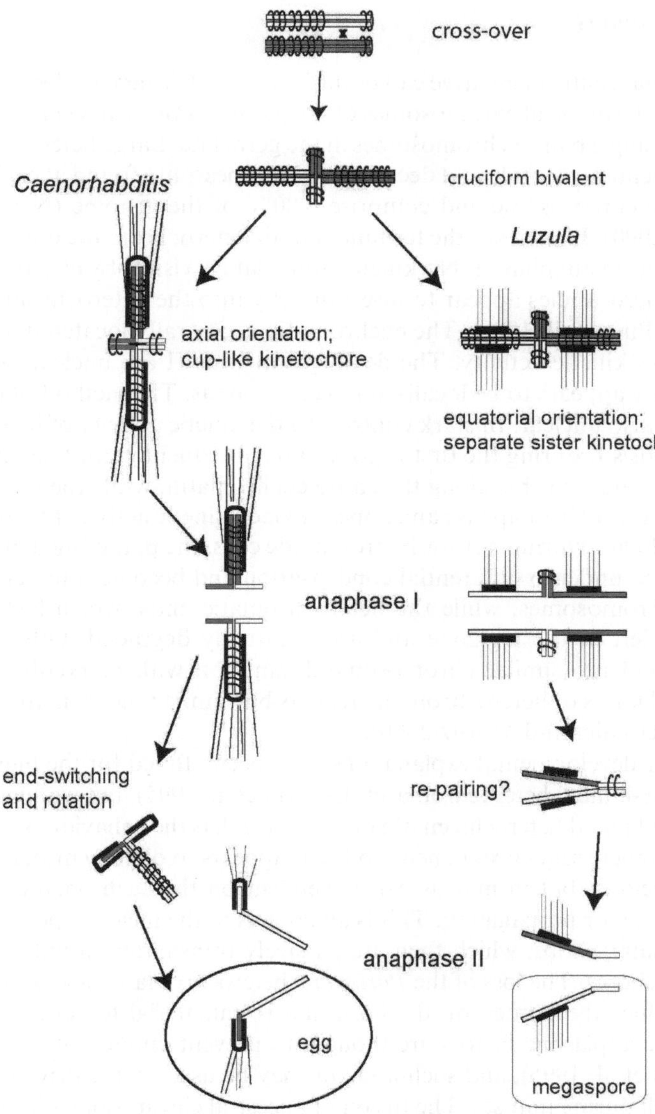


**Fig. 7.3** Models of meiosis of holocentric chromosomes in *Caenorhabditis* (*left*) and *Luzula* (*right*). A cross-over is re-modeled into a cruciform bivalent with long and short arms. (*Left*) Microtubules attach to a cup-like kinetochore on an axially oriented *Caenorhabditis* chromosome, and cohesion releases from the short arms in meiosis I. The kinetochore reforms in meiosis II over the short arms and cohesion releases from the long arms. (*Right*) Separate sister kinetochores form on equatorially oriented *Luzula* chromosomes, and cohesion releases from the long arms first so that most of the chromosome divides equationally in meiosis I and reductionally in meiosis II (inverted meiosis). Cohesion on the short arm and telomere association may 're-pair' homologs for meiosis II

### 7.8.2 *Parascaris*

The notion that centromere drive can occur in holocentric nematodes is strongly suggested by the unusual chromosomes of the parasite *Parascaris univalens*. This species has a single pair of chromosomes in the germ line. Large heterochromatic domains of pentanucleotide and decanucleotide repeats are found at each end of the germ-line chromosome and comprise ~80% of the genome (Niedermaier and Moritz 2000). In meiosis, the termini of both heterochromatic domains lead the chromatids in anaphase I. No kinetochore plate is visible by electron microscopy and microtubules appear to insert directly into the heterochromatin tips (Goday and Pimpinelli 1989). The euchromatin is centrally located in the bivalent and lacks kinetic activity. The details of meiosis II are unclear, although kinetic activity appears to be localized to certain areas. The method of chiasma resolution is also unclear. In stark contrast to the kinetic activity of heterochromatin in meiosis I, during the first zygotic and subsequent germ-line mitoses, a kinetochore plate is visible along the entire euchromatin, while the heterochromatic domains trail in anaphase and appear to lack kinetic activity (Niedermaier and Moritz 2000). During mitosis in presomatic cells, the pair of medial euchromatic domains undergo differential condensation and become fragmented into dozens of chromosomes, while the heterochromatic ends, which lack kinetic activity, are left in the midzone and are eventually degraded. Other ascarid nematodes undergo similar chromosome diminution with markedly different amounts and kinds of heterochromatic repeats but similar euchromatic complements (Niedermaier and Moritz 2000).

Cellular or developmental explanations have been offered for the highly variable germ-line-limited heterochromatin (Esteban et al. 1995), but we suggest that the germ-line-limited heterochromatin of *Parascaris* has the behavior expected for driving centromere sequences in nematodes. It appears to directly mediate kinetic activity in meiosis, but in mitosis it is dependent on the euchromatic holocentric kinetochore for propagation. This is analogous to the meiosis-specific kinetic activity of maize knobs, which then are passively transmitted in mitosis by the maize kinetochores. The loss of the *Parascaris* heterochromatic blocks in somatic cells may reduce the burden on the host, and is immaterial to germ-line drive. Prelocalized cytoplasmic factors are thought to prevent diminution in the germ line (Esteban et al. 1995), and such a factor may be used by the drive system to maintain itself during mitoses. The fusion of ancestral chromosomes into a single germ-line chromosome has been proposed to reduce nondisjunction (Niedermaier and Moritz 2000), but another beneficial effect of these fusions is to limit the number of chromosome ends at which driving sequences can accumulate.

### 7.8.3 *Luzula*

A different kind of chromosome 'diminution' is the phenomenon of agmatoploidy in wood-rushes of the genus *Luzula*. As in *Caenorhabditis*, *Luzula* chromosomes lack a primary constriction and have a CenH3-containing kinetochore structure along the length of each mitotic chromosome (Nagaki et al. 2005). The base chromosome number for this genus appears to be $2n = 12$, with six pairs of similar size, designated 'AL' chromosomes. However, some species have 24 'BL' chromosomes that are approximately half the size of the common AL chromosomes, and still others have 48 'CL' chromosomes about one quarter the size of the AL chromosomes. These karyotypes that appear to originate by fragmentation of the AL chromosomes are designated 'agmatoploid'. Some species or isolates have intermediate 'aneuploid' chromosome numbers, such as $2n = 14$ (10 AL + 4 BL; Nordenskiold 1951). Karyotype variability exists not only between species but also between individuals of a single species (Kuta et al. 2004). Chromosome fragmentation into an agmatoploid series with intraspecific variability suggests that much of each chromosome retains kinetic activity in both mitosis and meiosis, and this is supported by the segregation of irradiated chromosome fragments (Nordenskiold 1963).

In contrast to the cup-like meiotic kinetochore of nematodes, however, a separate kinetochore plate is found along each *Luzula* sister chromatid in meiosis (Braselton 1981). The extent of the kinetochore and how recombinant chromatids avoid merotelic attachment are unknown. Meiotic bivalents in *Luzula* microsporocytes form one or two nonterminalized chiasmata, and align equatorially on the metaphase plate (Nordenskiold 1962), in contrast to the axial orientation of nematode bivalents (Fig. 7.3). *Luzula* appears to solve the problem of chiasma resolution by undergoing an 'inverted meiosis' that releases most sister chromatid cohesion in anaphase I, similar to the release of arm cohesion in monocentric chromosomes (Nordenskiold 1962, 1963). The distinct sister kinetochores at this stage are necessary for the equational division of chromatids, but would be superfluous for directing reductional division of the chromatid ends as in *Caenorhabditis*. In late telophase I and interkinesis, the homologous chromatids appear to associate at their ends to form rings that then separate reductionally in meiosis II. Nordenskiöld (Nordenskiold 1962) interpreted this as the release of chiasmata followed by re-pairing of homologs. A similar association of ends leading to re-pairing of homologs has been described for the holocentric chromosomes of female mealybugs (*Planococcus citri*), which also undergo inverted meiosis. In this case, association of ends and re-pairing were suggested to be mediated by the cohesion between the sisters distal to a subterminal chiasma (Bongiorni et al. 2004), perhaps aided by telomere pairing.

If inverted meiosis in microsporocytes is indicative of meiosis in megasporocytes, there are implications for centromere drive. Competition between centromere alleles for inclusion in the megaspore would occur at meiosis II, not meiosis I, but of more consequence is that satellite expansion almost anywhere on the chromosome might contribute to centromere drive. This might explain the unusual dispersed distribution of *Luzula* satellite repeats, which can comprise over 20% of the genome in many species (Collet and Westerman 1987). In *L. nivea*, a 178 bp satellite related to the centromeric satellites of rice and maize has been identified and found to reside in at least five large clusters interspersed along each chromosome (Haizel et al. 2005). As in rice *cen8* (Nagaki et al. 2004), kinetochore activity is probably not restricted to the satellite sequences, but their distribution may approximate the extent of the meiotic kinetochore.

Chromosome fusions and fissions are characteristic of *Luzula* species and of the related sedges. The tendency of *Luzula* species to have 'euploid' (all AL or all BL, etc.) rather than 'aneuploid' chromosome numbers (Nordenskiold 1951) is reminiscent of the tendency of mammalian karyotypes to be predominantly metacentric or predominantly acrocentric, and might be similarly explained in terms of larger or smaller numbers of kinetochores being favored for inclusion in the surviving megaspore (Pardo-Manuel de Villena and Sapienza 2001a).

## 7.9 Origin of the Kinetochore?

Interest in the origin of eukaryotes is widespread, but surprisingly little attention has been given to the origin of the kinetochore, one of the defining features of eukaryotes. Margulis and colleagues (Dolan et al. 2002) have long argued for the origin of the microtubule cytoskeleton from a prokaryote endosymbiont, with its subsequent adaptation for use in mitosis. Although invoking a different ancestry for microtubules in an RNA-based 'chronocyte', Hartman and Fedorov also place the development of the microtubule cytoskeleton before its deployment in mitosis (Hartman and Fedorov 2002), but neither group has speculated on the origin of the kinetochore. Villasante and colleagues have recently proposed a hypothesis for the origin of centromeres from telomeres (Villasante et al. 2007). Their scheme involves the origin of linear chromosomes and telomeres by breakage of a circular genome with a prokaryote-like partitioning system, and subsequent origin of centromeres from subtelomeric repeats. The ancestral microtubule cytoskeleton is proposed to have begun interactions with the protocentromere by transport of ribonuclear protein complexes such as RNAi machinery to the repeats. While this scenario identifies possible steps necessary in a transition from a presumably ancestral prokaryote-like genome and partitioning system to a mitotic system, we find the emphasis on centromere origin, rather than kinetochore origin, to be misplaced, since centromeres seem to be able to originate at any sequence.

The heterotetrameric Cid hemisome suggests that centromeric nucleosomes may have a common origin with tetrameric packaging nucleosomes in archaea, with the current standard octameric packaging nucleosomes being a later innovation (Dalal et al. 2007a). At the same time, the apparent lack of a functional CenH3 in trypanosomes, whether a derived or ancestral condition, suggests flexibility in the connection of the outer kinetochore to chromatin. Thus we concur with previous suggestions that the microtubular cytoskeleton probably developed prior to its function in the modern kinetochore, although whether it aided in chromosome segregation at an earlier stage through more general interactions with chromatin remains a possibility. Among the earliest components of kinetochores, besides microtubules and motor proteins, are likely to have been an ancestral microtubule-binding Ndc80 complex, which in modern kinetochores appears to be connected to DNA via the Mis12 complex and CENP-C (Liu et al. 2006; Westermann et al. 2007). We suggest that these or analogous proteins formed the primitive kinetochore, with CenH3 nucleosomes evolving to help direct chromatin-binding to a specialized region. Further investigation of early branching eukaryotes may reveal clues to the steps in early kinetochore evolution.

## 7.10 Conclusion

The diversity of centromeres and kinetochores is based on the conserved subunit of centromeric chromatin, the CENP-A/CenH3 nucleosome. The structure of the *Drosophila* Cid nucleosome is a Cid/H4/H2B/H2A tetramer, which might underlie the asymmetric function and open conformation of the centromere/kinetochore complex. Centromeres are fundamentally epigenetic, being determined by the location of CENP-A/CenH3 nucleosomes, and the importance of centromeric DNA in determining the location of CenH3 nucleosomes varies among eukaryotes. Although DNA sequence is not determinative for centromere function in many eukaryotes, nearly all centromeres are found in DNA regions of low gene activity that are relatively AT-rich, suggesting that these properties favor the formation of centromeric chromatin. The centromere drive model provides a unifying hypothesis to explain the rapid evolution of centromeres and the kinetochore proteins that interact with them in organisms with asymmetric meiosis, and the relative stasis of centromeres and kinetochores in yeasts and other organisms with only symmetric meiosis. Centromere drive and its suppression also help to explain many aspects of karyotype evolution including fusions and fissions, centromere repositioning, accumulation of heterochromatin, and perhaps the occurrence of holocentric chromosomes and oddities such as chromatin diminution. Variations in the centromere/kinetochore complex in early branching eukaryotes are a largely unexplored avenue toward further understanding of the nature and development through time of this defining element of eukaryote biology. Progress in understanding the structures and roles of centromeric chromatin, centromeric RNAs, kinetochore proteins, and the spindles they act upon will likely clarify and sharpen our understanding of the forces that shape their evolution.

## References

Ahmad, K. and Henikoff, S. 2002. The histone variant H3.3 marks active chromatin by replication-independent nucleosome assembly. Mol Cell 9: 1191–200.

Albertson, D.G., Rose, A.M. and Villeneuve, A.M. 1997. *C. elegans* II: Chromosome Organization, Mitosis and Meiosis.

Albertson, D.G. and Thomson, J.N. 1982. The kinetochores of *Caenorhabditis elegans*. Chromosoma 86: 409–428.

Albertson, D.G. and Thomson, J.N. 1993. Segregation of holocentric chromosomes at meiosis in the nematode, *Caenorhabditis elegans*. Chromosome Res 1: 15–26.

Alonso, A., Mahmood, R., Li, S., et al. 2003. Genomic microarray analysis reveals distinct locations for the CENP-A binding domains in three human chromosome 13q32 neocentromeres. Hum Mol Genet 12: 2711–21.

Amor, D.J. and Choo, K.H. 2002. Neocentromeres: role in human disease, evolution, and centromere study. Am J Hum Genet 71: 695–714.

Amor, D.J., Kalitsis, P., Sumer, H., et al. 2004. Building the centromere: from foundation proteins to 3D organization. Trends Cell Biol 14: 359–368.

Ananiev, E.V., Phillips, R.L. and Rines, H.W. 1998. Chromosome-specific molecular organization of maize (*Zea mays* L.) centromeric regions. Proc Natl Acad Sci USA 95: 13073–8.

Baker, R.E. and Rogers, K. 2006. Phylogenetic analysis of fungal centromere H3 proteins. Genetics 174: 1481–1492.

Baum, M., Sanyal, K., Mishra, P.K., et al. 2006. Formation of functional centromeric chromatin is specified epigenetically in Candida albicans. Proc Natl Acad Sci USA 103: 14877–14882.

Bernard, P., Maure, J.F., Partridge, J.F., et al. 2001. Requirement of heterochromatin for cohesion at centromeres. Science 294: 2539–2542.

Bidau, C.J. and Marti, D.A. 2004. B chromosomes and Robertsonian fusions of *Dichroplus pratensis* (Acrididae): intraspecific support for the centromeric drive theory. Cytogenet Genome Res 106: 347–350.

Bloom, K. 1993. The centromere frontier: kinetochore components, microtubule-based motility, and the CEN-value paradox. Cell 73: 621–624.

Blower, M.D., Sullivan, B.A. and Karpen, G.H. 2002. Conserved organization of centromeric chromatin in flies and humans. Dev Cell 2: 319–330.

Bongiorni, S., Fiorenzo, P., Pippoletti, D., et al. 2004. Inverted meiosis and meiotic drive in mealybugs. Chromosoma 112: 331–341.

Braselton, J.P. 1981. The ultrastructure of meiotic kinetochores of *Luzula*. Chromosoma 82: 143–151.

Brown, M.T. 1995. Sequence similarities between the yeast chromosome segregation protein Mif2 and the mammalian centromere protein CENP-C. Gene 160: 111–116.

Brunet, S., Maria, A.S., Guillaud, P., et al. 1999. Kinetochore fibers are not involved in the formation of the first meiotic spindle in mouse oocytes, but control the exit from the first meiotic M phase. J Cell Biol 146: 1–12.

Buchwitz, B.J., Ahmad, K., Moore, L.L., et al. 1999. A histone-H3-like protein in *C. elegans*. Nature 401: 547–548.

Camahort, R., Li, B., Florens, L., et al. 2007. Scm3 is essential to recruit the histone h3 variant cse4 to centromeres and to maintain a functional kinetochore. Mol. Cell 26: 853–865.

Cambareri, E.B., Aisner, R. and Carbon, J. 1998. Structure of the chromosome VII centromere region in *Neurospora crassa*: degenerate transposons and simple repeats. Mol Cell Biol 18: 5465–5477.

Cao, Y.K., Zhong, Z.S., Chen, D.Y., et al. 2005. Cell cycle-dependent localization and possible roles of the small GTPase Ran in mouse oocyte maturation, fertilization and early cleavage. Reproduction 130: 431–440.

Cardone, M.F., Alonso, A., Pazienza, M., et al. 2006. Independent centromere formation in a capricious, gene-free domain of chromosome 13q21 in Old World monkeys and pigs. Genome Biol 7: R91.

Chan, R.C., Severson, A.F. and Meyer, B.J. 2004. Condensin restructures chromosomes in preparation for meiotic divisions. J Cell Biol 167: 613–625.

Cheeseman, I.M., Chappie, J.S., Wilson-Kubalek, E.M., et al. 2006. The conserved KMN network constitutes the core microtubule-binding site of the kinetochore. Cell 127: 983–997.

Cheng, Z., Dong, F., Langdon, T., et al. 2002. Functional rice centromeres are marked by a satellite repeat and a centromere-specific retrotransposon. Plant Cell 14: 1691–704.

Choo, K.H. 1997. The Centromere. Oxford: Oxford University Press.

Clarke, L. 1990. Centromeres of budding and fission yeasts. Trends Genet 6: 150–4.

Collet, C. and Westerman, M. 1984. Interspersed distribution patterns of C-bands and satellite DNA in the holocentric chromosomes of *Luzula flaccida*. Genetica 63: 175–179.

Collet, C. and Westerman, M. 1987. Interspecies comparison of the highly-repeated DNA of Australasian *Luzula* (Juncaceae). Genetica 74: 95–103.

Collins, K.A., Furuyama, S. and Biggins, S. 2004. Proteolysis contributes to the exclusive centromere localization of the yeast Cse4/CENP-A histone H3 variant. Curr Biol 14: 1968–1972.

Cooper, J.L. and Henikoff, S. 2004. Adaptive evolution of the histone fold domain in centromeric histones. Mol Biol Evol 21: 1712–8.

Dalal, Y., Furuyama, T., Vermaak, D., et al. 2007a. Structure, dynamics, and evolution of centromeric nucleosomes. Proc Natl Acad Sci USA 104: 15974–15981.

Dalal, Y., Wang, H., Lindsay, S., et al. 2007b. Tetrameric Structure of Centromeric Nucleosomes in Interphase Drosophila Cells. PLoS Biol 5: e218.

Dawe, R.K. and Henikoff, S. 2006. Centromeres put epigenetics in the driver's seat. Trends Biochem Sci 31: 662–669.

Dawe, R.K. and Hiatt, E.N. 2004. Plant neocentromeres: fast, focused, and driven. Chromosome Res 12: 655–669.

Dawson, S.C., Sagolla, M.S. and Cande, W.Z. 2007. The cenH3 histone variant defines centromeres in *Giardia intestinalis*. Chromosoma 116: 175–184.

Dolan, M.F., Melnitsky, H., Margulis, L., et al. 2002. Motility proteins and the origin of the nucleus. Anat Rec 268: 290–301.

Douglas, S., Zauner, S., Fraunholz, M., et al. 2001. The highly reduced genome of an enslaved algal nucleus. Nature 410: 1091–1096.

Dumont, J., Petri, S., Pellegrin, F., et al. 2007. A centriole- and RanGTP-independent spindle assembly pathway in meiosis I of vertebrate oocytes. J Cell Biol 176: 295–305.

Earnshaw, W.C. and Rothfield, N. 1985. Identification of a family of human centromere proteins using autoimmune sera from patients with scleroderma. Chromosoma 91: 313–321.

Eckert, C.A., Gravdahl, D.J. and Megee, P.C. 2007. The enhancement of pericentromeric cohesin association by conserved kinetochore components promotes high-fidelity chromosome segregation and is sensitive to microtubule-based tension. Genes Dev 21: 278–291.

Ersfeld, K. and Gull, K. 1997. Partitioning of large and minichromosomes in *Trypanosoma brucei*. Science 276: 611–614.

Esteban, M.R., Giovinazzo, G. and Goday, C. 1995. Chromatin diminution is strictly correlated to somatic cell behavior in early development of the nematode *Parascaris univalens*. J Cell Sci 108(Pt 6): 2393–2404.

Ferreri, G.C., Liscinsky, D.M., Mack, J.A., et al. 2005. Retention of latent centromeres in the Mammalian genome. J Hered 96: 217–224.

Furuyama, S. and Biggins, S. 2007. Centromere identity is specified by a single centromeric nucleosome in budding yeast. Proc Natl Acad Sci USA 104: 14706–14711.

Furuyama, T., Dalal, Y. and Henikoff, S. 2006. Chaperone-mediated assembly of centromeric chromatin in vitro. Proc Natl Acad Sci USA 103: 6172–6177.

Gard, D.L. 1992. Microtubule organization during maturation of *Xenopus* oocytes: assembly and rotation of the meiotic spindles. Dev Biol 151: 516–530.

Gard, D.L. 1993. Ectopic spindle assembly during maturation of *Xenopus* oocytes: evidence for functional polarization of the oocyte cortex. Dev Biol 159: 298–310.

Goday, C. and Pimpinelli, S. 1989. Centromere organization in meiotic chromosomes of *Parascaris univalens*. Chromosoma 98: 160–166.

Hackett, J.D., Scheetz, T.E., Yoon, H.S., et al. 2005. Insights into a dinoflagellate genome through expressed sequence tag analysis. BMC Genomics 6: 80.

Haizel, T., Lim, Y.K., Leitch, A.R., et al. 2005. Molecular analysis of holocentric centromeres of *Luzula* species. Cytogenet. Genome Res 109: 134–143.

Hall, S.E., Luo, S., Hall, A.E., et al. 2005. Differential rates of local and global homogenization in centromere satellites from *Arabidopsis* relatives. Genetics 170: 1913–1927.

Hartman, H. and Fedorov, A. 2002. The origin of the eukaryotic cell: a genomic investigation. Proc Natl Acad Sci USA 99: 1420–1425.

Hayashi, T., Fujita, Y., Iwasaki, O., et al. 2004. Mis16 and Mis18 are required for CENP-A loading and histone deacetylation at centromeres. Cell 118: 715–729.

Henikoff, S., Ahmad, K. and Malik, H.S. 2001. The centromere paradox: stable inheritance with rapidly evolving DNA. Science 293: 1098–1102.

Heun, P., Erhardt, S., Blower, M.D., et al. 2006. Mislocalization of the Drosophila centromere-specific histone CID promotes formation of functional ectopic kinetochores. Dev Cell 10: 303–315.

Heus, J.J., Zonneveld, B.J., Steensma, H.Y., et al. 1994. Mutational analysis of centromeric DNA elements of Kluyveromyces lactis and their role in determining the species specificity of the highly homologous centromeres from K. lactis and Saccharomyces cerevisiae. Mol Gen Genet 243: 325–333.

Hiatt, E.N. and Dawe, R.K. 2003. Four loci on abnormal chromosome 10 contribute to meiotic drive in maize. Genetics 164: 699–709.

Holy, T.E. and Leibler, S. 1994. Dynamic instability of microtubules as an efficient way to search in space. Proc Natl Acad Sci USA 91: 5682–5685.

Houben, A., Schroeder-Reiter, E., Nagaki, K., et al. 2007. CENH3 interacts with the centromeric retrotransposon cereba and GC-rich satellites and locates to centromeric substructures in barley. Chromosoma 116: 275–283.

Howe, M., McDonald, K.L., Albertson, D.G., et al. 2001. HIM-10 is required for kinetochore structure and function on *Caenorhabditis elegans* holocentric chromosomes. J Cell Biol 153: 1227–1238.

Huang, L., Wang, J., Nie, W., et al. 2006. Tandem chromosome fusions in karyotypic evolution of *Muntiacus*: evidence from *M. feae* and *M. gongshanensis*. Chromosome Res 14: 637–647.

Hunt, P.A. and Hassold, T.J. 2002. Sex matters in meiosis. Science 296: 2181–2183.

Irvine, D.V., Amor, D.J., Perry, J., et al. 2004. Chromosome size and origin as determinants of the level of CENP-A incorporation into human centromeres. Chromosome Res 12: 805–815.

Ito, H., Miura, A., Takashima, K., et al. 2007. Ecotype-specific and chromosome-specific expansion of variant centromeric satellites in *Arabidopsis thaliana*. Mol Genet Genomics 277: 23–30.

Jin, W., Melo, J.R., Nagaki, K., et al. 2004. Maize centromeres: organization and functional adaptation in the genetic background of oat. Plant Cell 16: 571–581.

Kasai, F., Garcia, C., Arruga, M.V., et al. 2003. Chromosome homology between chicken (*Gallus gallus domesticus*) and the red-legged partridge (*Alectoris rufa*); evidence of the occurrence of a neocentromere during evolution. Cytogenet Genome Res 102: 326–330.

Kato, A., Lamb, J.C. and Birchler, J.A. 2004. Chromosome painting using repetitive DNA sequences as probes for somatic chromosome identification in maize. Proc Natl Acad Sci USA 101: 13554–13559.

Kawabe, A., Nasuda, S. and Charlesworth, D. 2006. Duplication of centromeric histone H3 (HTR12) gene in *Arabidopsis halleri* and *A. lyrata*, plant species with multiple centromeric satellite sequences. Genetics 174: 2021–2032.

Keith, K.C., Baker, R.E., Chen, Y., et al. 1999. Analysis of primary structural determinants that distinguish the centromere-specific function of histone variant Cse4p from histone H3. Mol Cell Biol 19: 6130–6139.

Kelly, J.M., McRobert, L. and Baker, D.A. 2006. Evidence on the chromosomal location of centromeric DNA in *Plasmodium falciparum* from etoposide-mediated topoisomerase-II cleavage. Proc Natl Acad Sci USA 103: 6706–6711.

Kitada, K., Yamaguchi, E., Hamada, K., et al. 1997. Structural analysis of a *Candida glabrata* centromere and its functional homology to the *Saccharomyces cerevisiae* centromere. Curr Genet 31: 122–127.

Kuta, E., Bohanec, B., Dubas, E., et al. 2004. Chromosome and nuclear DNA study on *Luzula* – a genus with holokinetic chromosomes. Genome 47: 246–256.

Lam, A.L., Boivin, C.D., Bonney, C.F., et al. 2006. Human centromeric chromatin is a dynamic chromosomal domain that can spread over noncentromeric DNA. Proc Natl Acad Sci USA 103: 4186–4191.

Langdon, T., Seago, C., Mende, M., et al. 2000. Retrotransposon evolution in diverse plant genomes. Genetics 156: 313–325.

Lavelle, C. and Prunell, A. 2007. Chromatin polymorphism and the nucleosome superfamily: a genealogy. Cell Cycle 6: 2113–2119.

Lee, H.R., Zhang, W., Langdon, T., et al. 2005. Chromatin immunoprecipitation cloning reveals rapid evolutionary patterns of centromeric DNA in *Oryza* species. Proc Natl Acad Sci USA 102: 11793–11798.

LeMaire-Adkins, R. and Hunt, P.A. 2000. Nonrandom segregation of the mouse univalent X chromosome: evidence of spindle-mediated meiotic drive. Genetics 156: 775–783.

Liu, S.T., Rattner, J.B., Jablonski, S.A., et al. 2006. Mapping the assembly pathways that specify formation of the trilaminar kinetochore plates in human cells. J Cell Biol 175: 41–53.

Lohe, A.R., Hilliker, A.J. and Roberts, P.A. 1993. Mapping simple repeated DNA sequences in heterochromatin of *Drosophila melanogaster*. Genetics 134: 1149–74.

Loiodice, I., Alves, A., Rabut, G., et al. 2004. The entire Nup107–160 complex, including three new members, is targeted as one entity to kinetochores in mitosis. Mol Biol Cell 15: 3333–3344.

Lowell, J.E. and Cross, G.A. 2004. A variant histone H3 is enriched at telomeres in *Trypanosoma brucei*. J Cell Sci 117: 5937–5947.

Luger, K., Mader, A.W., Richmond, R.K., et al. 1997. Crystal structure of the nucleosome core particle at 2.8 A resolution. Nature 389: 251–260.

Lutz, D.A., Hamaguchi, Y. and Inoue, S. 1988. Micromanipulation studies of the asymmetric positioning of the maturation spindle in *Chaetopterus sp.* oocytes: I. Anchorage of the spindle to the cortex and migration of a displaced spindle. Cell Motil Cytoskeleton 11: 83–96.

Lysak, M.A., Berr, A., Pecinka, A., et al. 2006. Mechanisms of chromosome number reduction in Arabidopsis thaliana and related Brassicaceae species. Proc Natl Acad Sci USA 103: 5224–5229.

Ma, J., Wing, R.A., Bennetzen, J.L., et al. 2007a. Evolutionary History and Positional Shift of a Rice Centromere. Genetics Jul 29: [epub ahead of print].

Ma, J., Wing, R.A., Bennetzen, J.L., et al. 2007b. Plant centromere organization: a dynamic structure with conserved functions. Trends Genet 23: 134–9.

Maddox, P.S., Hyndman, F., Monen, J., et al. 2007. Functional genomics identifies a Myb domain-containing protein family required for assembly of CENP-A chromatin. J Cell Biol 176: 757–763.

Maiato, H., Rieder, C.L. and Khodjakov, A. 2004. Kinetochore-driven formation of kinetochore fibers contributes to spindle assembly during animal mitosis. J Cell Biol 167: 831–840.
Malik, H.S. and Henikoff, S. 2001. Adaptive evolution of Cid, a centromere-specific histone in *Drosophila*. Genetics 157: 1293–1298.
Malik, H.S. and Henikoff, S. 2003. Phylogenomics of the nucleosome. Nat Struct Biol 10: 882–891.
Malik, H.S., Vermaak, D. and Henikoff, S. 2002. Recurrent evolution of DNA-binding motifs in the *Drosophila* centromeric histone. Proc Natl Acad Sci USA 99: 1449–1454.
Maruyama, S., Kuroiwa, H., Miyagishima, S.Y., et al. 2007. Centromere dynamics in the primitive red alga Cyanidioschyzon merolae. Plant J 49: 1122–1129.
Masumoto, H., Nakano, M. and Ohzeki, J. 2004. The role of CENP-B and alpha-satellite DNA: *de novo* assembly and epigenetic maintenance of human centromeres. Chromosome Res 12: 543–556.
McClintock, B. 1938. The Production of Homozygous Deficient Tissues with Mutant Characteristics by Means of the Aberrant Mitotic Behavior of Ring-Shaped Chromosomes. Genetics 23: 315–376.
Meluh, P.B. and Koshland, D. 1997. Budding yeast centromere composition and assembly as revealed by in vivo cross-linking. Genes Dev 11: 3401–3412.
Meraldi, P., McAinsh, A.D., Rheinbay, E., et al. 2006. Phylogenetic and structural analysis of centromeric DNA and kinetochore proteins. Genome Biol 7: R23.
Mishra, P.K., Baum, M. and Carbon, J. 2007. Centromere size and position in Candida albicans are evolutionarily conserved independent of DNA sequence heterogeneity. Mol Genet Genomics 278: 455–465.
Monen, J., Maddox, P.S., Hyndman, F., et al. 2005. Differential role of CENP-A in the segregation of holocentric *C. elegans* chromosomes during meiosis and mitosis. Nat Cell Biol 7: 1248–1255.
Moore, L.L. and Roth, M.B. 2001. HCP-4, a CENP-C-like protein in Caenorhabditis elegans, is required for resolution of sister centromeres. J Cell Biol 153: 1199–1208.
Moreno-Moreno, O., Torras-Llort, M. and Azorin, F. 2006. Proteolysis restricts localization of CID, the centromere-specific histone H3 variant of *Drosophila*, to centromeres. Nucleic Acids Res 34: 6247–6255.
Morey, L., Barnes, K., Chen, Y., et al. 2004. The histone fold domain of Cse4 is sufficient for CEN targeting and propagation of active centromeres in budding yeast. Eukaryot. Cell 3: 1533–1543.
Murata, M., Ogura, Y. and Motoyoshi, F. 1994. Centromeric repetitive sequences in *Arabidopsis thaliana*. Jpn J Genet 69: 361–370.
Musacchio, A. and Salmon, E.D. 2007. The spindle-assembly checkpoint in space and time. Nat Rev Mol Cell Biol 8: 379–393.
Nabeshima, K., Villeneuve, A.M. and Colaiacovo, M.P. 2005. Crossing over is coupled to late meiotic prophase bivalent differentiation through asymmetric disassembly of the SC. J Cell Biol 168: 683–689.
Nagaki, K., Cheng, Z., Ouyang, S., et al. 2004. Sequencing of a rice centromere uncovers active genes. Nat Genet 36: 138–145.
Nagaki, K., Kashihara, K. and Murata, M. 2005. Visualization of diffuse centromeres with centromere-specific histone H3 in the holocentric plant *Luzula nivea*. Plant Cell 17: 1886–1893.
Nasuda, S., Hudakova, S., Schubert, I., et al. 2005. Stable barley chromosomes without centromeric repeats. Proc Natl Acad Sci USA 102: 9842–9847.
Ng, R. and Carbon, J. 1987. Mutational and in vitro protein-binding studies on centromere DNA from *Saccharomyces cerevisiae*. Mol Cell Biol 7: 4522–34.
Niedermaier, J. and Moritz, K.B. 2000. Organization and dynamics of satellite and telomere DNAs in *Ascaris*: implications for formation and programmed breakdown of compound chromosomes. Chromosoma 109: 439–452.

Nordenskiold, H. 1951. Cyto-taxonomical studies in the genus *Luzula* I. Somatic chromosomes and chromosome numbers. Hereditas 37: 325–355.
Nordenskiold, H. 1962. Studies of meiosis in *Luzula purpurea*. Hereditas 48: 503–519.
Nordenskiold, H. 1963. A study of meiosis in the progeny of X-irradiated *Luzula purpurea*. Hereditas 49: 33–47.
Obado, S.O., Bot, C., Nilsson, D., et al. 2007. Repetitive DNA is associated with centromeric domains in *Trypanosoma brucei* but not *Trypanosoma cruzi*. Genome Biol 8: R37.
Oegema, K., Desai, A., Rybina, S., et al. 2001. Functional analysis of kinetochore assembly in Caenorhabditis elegans. J Cell Biol 153: 1209–1226.
Ogbadoyi, E., Ersfeld, K., Robinson, D., et al. 2000. Architecture of the *Trypanosoma brucei* nucleus during interphase and mitosis. Chromosoma 108: 501–513.
Ogur, G., Van Assche, E., Vegetti, W., et al. 2006. Chromosomal segregation in spermatozoa of 14 Robertsonian translocation carriers. Mol Hum Reprod 12: 209–215.
Okamoto, Y., Nakano, M., Ohzeki, J., et al. 2007. A minimal CENP-A core is required for nucleation and maintenance of a functional human centromere. EMBO J 26: 1279–1291.
Orjalo, A.V., Arnaoutov, A., Shen, Z., et al. 2006. The Nup107–160 nucleoporin complex is required for correct bipolar spindle assembly. Mol Biol Cell 17: 3806–3818.
Palestis, B.G., Burt, A., Jones, R.N., et al. 2004. B chromosomes are more frequent in mammals with acrocentric karyotypes: support for the theory of centromeric drive. Proc Biol Sci 271(Suppl 3): S22–4.
Palmer, D.K., O'Day, K., Wener, M.H., et al. 1987. A 17-kD centromere protein (CENP-A) copurifies with nucleosome core particles and with histones. J Cell Biol 104: 805–815.
Pardo-Manuel de Villena, F. and Sapienza, C. 2001a. Female meiosis drives karyotypic evolution in mammals. Genetics 159: 1179–1189.
Pardo-Manuel de Villena, F. and Sapienza, C. 2001b. Nonrandom segregation during meiosis: the unfairness of females. Mamm. Genome 12: 331–339.
Pardo-Manuel de Villena, F. and Sapienza, C. 2001c. Transmission ratio distortion in offspring of heterozygous female carriers of Robertsonian translocations. Hum Genet 108: 31–36.
Pidoux, A.L. and Allshire, R.C. 2004. Kinetochore and heterochromatin domains of the fission yeast centromere. Chromosome Res 12: 521–34.
Pluta, A.F., Mackay, A.M., Ainsztein, A.M., et al. 1995. The centromere: hub of chromosomal activities. Science 270: 1591–1594.
Politi, V., Perini, G., Trazzi, S., et al. 2002. CENP-C binds the alpha-satellite DNA in vivo at specific centromere domains. J Cell Sci 115: 2317–2327.
Powers, J., Rose, D.J., Saunders, A., et al. 2004. Loss of KLP-19 polar ejection force causes misorientation and missegregation of holocentric chromosomes. J Cell Biol 166: 991–1001.
Presgraves, D.C. and Stephan, W. 2007. Pervasive adaptive evolution among interactors of the *Drosophila* hybrid inviability gene, Nup96. Mol Biol Evol 24: 306–314.
Rhoades, M.M. 1952. Preferential segregation in maize. In Heterosis, ed, J.W. Gowen, pp. 66–80. Ames. IA: Iowa State College Press.
Rudd, M.K., Wray, G.A. and Willard, H.F. 2006. The evolutionary dynamics of alpha-satellite. Genome Res 16: 88–96.
Saffery, R., Sumer, H., Hassan, S., et al. 2003. Transcription within a functional human centromere. Mol Cell 12: 509–516.
Sansome, E. and Brasier, C.M. 1973. Diploidy and chromosomal structure in *Phytophthora infestans*. Nature 241: 344–345.
Sanyal, K., Baum, M. and Carbon, J. 2004. Centromeric DNA sequences in the pathogenic yeast *Candida albicans* are all different and unique. Proc Natl Acad Sci USA 101: 11374–11379.
Schittenhelm, R.B., Heeger, S., Althoff, F., et al. 2007. Spatial organization of a ubiquitous eukaryotic kinetochore protein network in *Drosophila* chromosomes. Chromosoma 116: 385–402.

Schueler, M.G., Dunn, J.M., Bird, C.P., et al. 2005. Progressive proximal expansion of the primate X chromosome centromere. Proc Natl Acad Sci USA102: 10563–10568.
Schueler, M.G., Higgins, A.W., Rudd, M.K., et al. 2001. Genomic and genetic definition of a functional human centromere. Science 294: 109–115.
Schueler, M.G. and Sullivan, B.A. 2006. Structural and functional dynamics of human centromeric chromatin. Annu Rev Genomics Hum Genet 7: 301–13.
Schwartz, B.E. and Ahmad, K. 2005. Transcriptional activation triggers deposition and removal of the histone variant H3.3. Genes Dev. 19: 804–814.
Shelby, R.D., Vafa, O. and Sullivan, K.F. 1997. Assembly of CENP-A into centromeric chromatin requires a cooperative array of nucleosomal DNA contact sites. J Cell Biol 136: 501–513.
Sheldon, B.C. 1999. Sex allocation: At the females' whim. Curr Biol 9: R487–9.
Shibata, F. and Murata, M. 2004. Differential localization of the centromere-specific proteins in the major centromeric satellite of *Arabidopsis thaliana*. J Cell Sci 117: 2963–2970.
Skold, H.N., Komma, D.J. and Endow, S.A. 2005. Assembly pathway of the anastral *Drosophila* oocyte meiosis I spindle. J Cell Sci 118: 1745–1755.
Smith, G.P. 1976. Evolution of repeated DNA sequences by unequal crossover. Science 191: 528–535.
Stear, J.H. and Roth, M.B. 2002. Characterization of HCP-6, a *C. elegans* protein required to prevent chromosome twisting and merotelic attachment. Genes Dev 16: 1498–1508.
Stoler, S., Rogers, K., Weitze, S., et al. 2007. Scm3, an essential Saccharomyces cerevisiae centromere protein required for G2/M progression and Cse4 localization. Proc Natl Acad Sci USA. 104: 10571–10576.
Sun, X., Le, H.D., Wahlstrom, J.M., et al. 2003. Sequence analysis of a functional *Drosophila* centromere. Genome Res 13: 182–194.
Surzycki, S.A. and Belknap, W.R. 2000. Repetitive-DNA elements are similarly distributed on *Caenorhabditis elegans* autosomes. Proc Natl Acad Sci USA 97: 245–249.
Takahashi, K., Chen, E.S. and Yanagida, M. 2000. Requirement of Mis6 centromere connector for localizing a CENP-A-like protein in fission yeast. Science 288: 2215–2219.
Talbert, P.B., Bryson, T.D. and Henikoff, S. 2004. Adaptive evolution of centromere proteins in plants and animals. J Biol 3: 18.
Talbert, P.B., Masuelli, R., Tyagi, A.P., et al. 2002. Centromeric localization and adaptive evolution of an *Arabidopsis* histone H3 variant. Plant Cell 14: 1053–1066.
Tomkiel, J., Cooke, C.A., Saitoh, H., et al. 1994. CENP-C is required for maintaining proper kinetochore size and for a timely transition to anaphase. J Cell Biol 125: 531–545.
Topp, C.N., Zhong, C.X. and Dawe, R.K. 2004. Centromere-encoded RNAs are integral components of the maize kinetochore. Proc Natl Acad Sci USA 101: 15986–91.
Trazzi, S., Bernardoni, R., Diolaiti, D., et al. 2002. In vivo functional dissection of human inner kinetochore protein CENP-C. J Struct Biol 140: 39–48.
Underkoffler, L.A., Mitchell, L.E., Abdulali, Z.S., et al. 2005. Transmission ratio distortion in offspring of mouse heterozygous carriers of a (7.18) Robertsonian translocation. Genetics 169: 843–848.
Ventura, M., Antonacci, F., Cardone, M.F., et al. 2007. Evolutionary formation of new centromeres in macaque. Science 316: 243–246.
Ventura, M., Archidiacono, N. and Rocchi, M. 2001. Centromere emergence in evolution. Genome Res 11: 595–599.
Ventura, M., Mudge, J.M., Palumbo, V., et al. 2003. Neocentromeres in 15q24-26 map to duplicons which flanked an ancestral centromere in 15q25. Genome Res 13: 2059–2068.
Ventura, M., Weigl, S., Carbone, L., et al. 2004. Recurrent sites for new centromere seeding. Genome Res 14: 1696–703.
Verlhac, M.H., Lefebvre, C., Guillaud, P., et al. 2000. Asymmetric division in mouse oocytes: with or without Mos. Curr Biol 10: 1303–1306.

Vermaak, D., Hayden, H.S. and Henikoff, S. 2002. Centromere targeting element within the histone fold domain of Cid. Mol Cell Biol 22: 7553–61.

Villasante, A., Abad, J.P. and Mendez-Lago, M. 2007. Centromeres were derived from telomeres during the evolution of the eukaryotic chromosome. Proc Natl Acad Sci USA 104: 10542–10547.

Warburton, P.E., Waye, J.S. and Willard, H.F. 1993. Nonrandom localization of recombination events in human alpha satellite repeat unit variants: implications for higher-order structural characteristics within centromeric heterochromatin. Mol Cell Biol 13: 6520–6529.

Westermann, S., Drubin, D.G. and Barnes, G. 2007. Structures and functions of yeast kinetochore complexes. Annu Rev Biochem 76: 563–591.

White, M.J.D. 1973. Animal Cytology and Evolution. 3rd edn. London: Cambridge Universtiy Press.

Wickstead, B., Ersfeld, K. and Gull, K. 2004. The small chromosomes of *Trypanosoma brucei* involved in antigenic variation are constructed around repetitive palindromes. Genome Res 14: 1014–1024.

Willard, H.F. 1991. Evolution of alpha satellite. Curr Opin Genet Dev 1: 509–14.

Wollman, R., Cytrynbaum, E.N., Jones, J.T., et al. 2005. Efficient chromosome capture requires a bias in the 'search-and-capture' process during mitotic-spindle assembly. Curr Biol 15: 828–832.

Wong, L.H., Brettingham-Moore, K.H., Chan, L., et al. 2007. Centromere RNA is a key component for the assembly of nucleoproteins at the nucleolus and centromere. Genome Res 17: 1146–1160

Yoda, K., Ando, S., Morishita, S., et al. 2000. Human centromere protein A (CENP-A) can replace histone H3 in nucleosome reconstitution in vitro. Proc Natl Acad Sci USA 97: 7266–7271.

Yu, H.G., Hiatt, E.N., Chan, A., et al. 1997. Neocentromere-mediated chromosome movement in maize. J Cell Biol 139: 831–840.

Zhong, C.X., Marshall, J.B., Topp, C., et al. 2002. Centromeric retroelements and satellites interact with maize kinetochore protein CENH3. Plant Cell 14: 2825–36.

Zinkowski, R.P., Meyne, J. and Brinkley, B.R. 1991. The centromere-kinetochore complex: a repeat subunit model. J Cell Biol 113: 1091–1110.

# Chapter 8
# Mitotic Spindle Assembly Mechanisms

**Rebecca Heald and Claire E. Walczak**

**Abstract** The mitotic spindle consists of dynamic microtubules and many associated factors that form an antiparallel, bipolar array. Duplicated chromosomes are attached to microtubules of the spindle and then are physically separated by the spindle to opposite ends of the dividing cell. Spindles vary in their morphology and assembly pathway depending on the cell type and organism, but common underlying mechanisms derive from the dynamics of the microtubules and microtubule-based motor proteins, and the activities of chromosomes themselves. In this chapter, we describe the multiple mechanisms that promote assembly of the dynamic mitotic spindle.

## 8.1 Introduction

The goal of mitosis is to segregate chromosomes to two daughter cells, a feat accomplished by a dynamic macromolecular machine called the mitotic spindle. The spindle consists of microtubule fibers and many associated factors that organize to form a bipolar array during prometaphase. Duplicated and tethered sister chromatids attach by their kinetochores to microtubules emanating from opposite spindle poles, then separate and are transported to opposite ends of the spindle during anaphase. In cells possessing duplicated focal microtubule organizing centers (a centrosome in animal cells, or spindle pole body in yeast), the two nucleation centers define the position of the spindle poles. To ensure proper chromosome segregation, redundant mechanisms are utilized to assemble and organize the antiparallel microtubule array, with multiple pathways functioning in concert. By interacting with the cell cortex, spindle microtubules also function

R. Heald (✉)
Molecular and Cell Biology Department, University of California, Berkeley, CA 94720-3200, U.S.A.
e-mail: bheald@berkeley.edu

C.E. Walczak (✉)
Medical Sciences Program, Indiana University, Myers Hall 262, 915 East 3rd Street, Bloomington, IN 47405
e-mail: cwalczak@indiana.edu

P. De Wulf, W.C. Earnshaw (eds.), *The Kinetochore*,
DOI 10.1007/978-0-387-69076-6_8, © Springer Science+Business Media, LLC 2009

to define the position of the cleavage plane that divides the cell into two at cytokinesis (Eggert et al. 2006). In this review, we summarize what is known about the tactics and major players of spindle assembly, highlighting some of the factors identified that promote this essential and fascinating process.

## 8.2 Intrinsic Properties of Microtubules Facilitate Spindle Assembly and Function

*Microtubule polarity:* The dynamic nature of the mitotic spindle derives from the intrinsic properties of microtubules themselves. Microtubules assemble from tubulin heterodimers consisting of alpha and beta tubulin. Tubulin subunits arrange head to tail to form protofilaments, in which there are typically 13 protofilaments associated laterally to form a hollow tube (Nogales et al. 1999). The end with the terminal alpha tubulin subunit is the minus end, which grows more slowly than the plus end in vitro (Mitchison 1993), and is often temporarily capped or embedded in a microtubule organizing center in cells (Dammermann et al. 2003). The polarity of the tubulin dimer gives rise to a polarity along the length of the microtubule lattice. Microtubule motor proteins including cytoplasmic dynein and the kinesin family of motors recognize this polarity, and different motors move along the microtubule towards the plus or minus end. Microtubule-based motors play crucial roles in organizing spindle microtubules through cross-linking and sliding, by transporting activities along spindle microtubules, and by directly regulating their dynamics.

*Microtubule dynamics:* Microtubules grow and shrink by addition or loss of tubulin dimers at their ends, with individual microtubules switching stochastically between phases of growth and shrinkage, a property known as dynamic instability (Mitchison and Kirschner 1984). Dynamic instability results from GTP hydrolysis within the beta tubulin subunit that occurs upon assembly and destabilizes the lattice by promoting a conformational change (Desai and Mitchison 1997). Under conditions that favor polymerization, a cap of GTP-tubulin subunits is maintained at the growing end that holds the microtubule together (Drechsel and Kirschner 1994; Schek et al. 2007). However, if GTP hydrolysis catches up with subunit addition and the terminal subunits are converted to GDP-tubulin, the microtubule depolymerizes (Nogales 1999; Fig. 8.1a). Five major parameters describe microtubule behavior in a population: the rate of nucleation, the rate of growth, the rate of shrinkage, and the transition frequencies from growing to shrinking (a catastrophe) and shrinking to growing (a rescue; Walker et al. 1988; Desai and Mitchison 1997). Purified microtubules undergo dynamic instability in vitro, but microtubules inside cells grow more rapidly and undergo transitions more frequently, increasing their turnover rate dramatically, due to the presence of many microtubule- and tubulin-associated factors (Cassimeris 1993; Kinoshita et al. 2001). Cellular concentrations of tubulin ($\sim$15 $\mu$M) do not polymerize in pure protein solutions in the absence of pre-formed nucleation sites such as centrosomes. Thus,

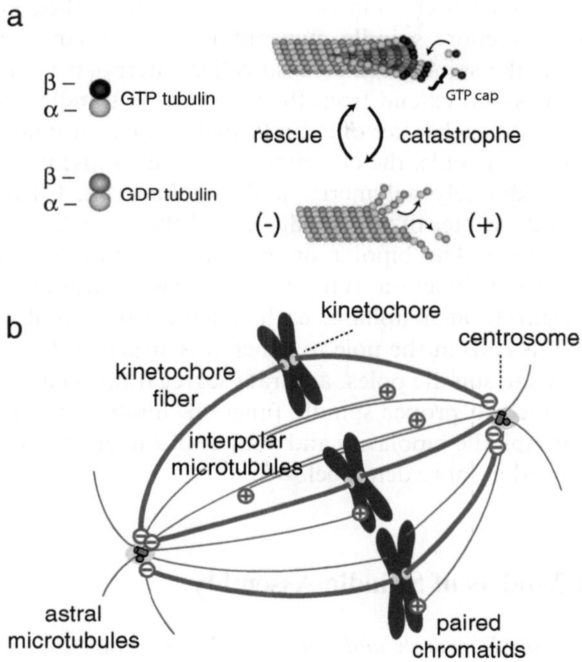


**Fig. 8.1** Cartoon illustrating the key components of the mitotic spindle. (**a**) The primary component of spindles, microtubules, constitute $\alpha/\beta$ tubulin heterodimers arranged in a head-to-tail configuration with a terminal $\beta$ subunit at the more dynamic plus (+) end of the polymer. Microtubules grow and shrink by addition and loss of tubulin dimers at their ends, with stochastic transitions from growth to shrinkage (catastrophe), and shrinkage to growth (rescue). The dynamic instability of microtubules results from GTP hydrolysis within the $\beta$ subunit. (**b**) Spindle microtubules are organized in an antiparallel array, with their minus ends located at the poles and their plus ends extending outward. Microtubules extend either to the kinetochores of paired sister chromatids (kinetochore or K-fiber bundles), to the central spindle where they form an overlapping array (interpolar microtubules), or away from the spindle towards the cell cortex (astral microtubules)

specialized nucleating activities determine where and when microtubules assemble within the cell and generate polarized arrays, while many regulatory factors positively or negatively stimulate microtubule dynamic behavior.

## 8.3 Structural Organization of the Mitotic Microtubule Array

Dynamic microtubules are major components of the mitotic spindle, and are organized with their minus ends at the spindle pole and their plus ends extending toward the cell cortex or toward the spindle equator. In vertebrates, spindle microtubules can be divided into three different classes: the bundles that attach the kinetochores of each chromosome to the spindle, called kinetochore or K-fibers; non-kinetochore spindle microtubules that emanate from each of the poles towards the spindle center, also called interpolar microtubules; and astral microtubules that extend from the poles of the spindle towards the cell cortex (Fig. 8.1b). As each sister of a duplicated pair of chromatids attaches to microtubules from one pole, the chromosome begins to oscillate as its K-fiber microtubules coordinately polymerize and depolymerize. Chromosomes can then congress to the center of the spindle, called the metaphase plate, where it continues to oscillate. The bipolar organization of microtubules within the spindle is critical for its function. When the sister chromatids of each duplicated chromosome separate at anaphase, each kinetochore is pulled towards the microtubule minus ends at the pole to which it is attached. Because sisters are attached to opposite spindle poles, accurate segregation is ensured. Due to its primary importance to proper spindle function, multiple mechanisms are in place to promote spindle bipolarity and antiparallel microtubule organization, which are discussed in more detail below.

## 8.4 Current Models of Spindle Assembly

*Centrosome-mediated "search and capture":* Historically, the centrosome has dominated models of spindle assembly due to its ability to define microtubule nucleation sites and generate polarized arrays. Astral arrays of microtubules emanating from centrosomes were among the first organelles observed in cells by light microscopy, and it was appreciated that centrosomes play a key role in determining the number of spindle poles, with each centrosome organizing a half-spindle. In vertebrate cells, the centrosome consists of a pair of centrioles surrounded by amorphous pericentriolar material (PCM), where microtubule nucleation and organization takes place (Doxsey et al. 2005). By the onset of mitosis, the centrosome has duplicated and recruited additional PCM, dramatically increasing its microtubule nucleation capacity (Kuriyama and Borisy 1981). The major nucleating element is the gamma tubulin ring complex ($\gamma$-TURC), named for its tubulin variant arranged in a lock-washer configuration that is thought to template microtubule growth (Zheng et al. 1995). Gamma tubulin recruitment to the centrosome increases approximately threefold at the onset of mitosis (Khodjakov and Rieder 1999), consistent with the increased nucleation capacity of mitotic centrosomes. The presence of two centrosomes functioning to generate dynamic microtubules led to the "search and capture" model of spindle assembly, in which bipolarity stems from the ability of each duplicated sister kinetochore to capture microtubules from the two nascent spindle poles (Kirschner and Mitchison 1986; McIntosh et al. 2002; Fig. 8.2a). Given their integral role in spindle formation, the assembly, duplication, and function of centrosomes are very active areas of investigation and the subject of excellent recent reviews (Doxsey et al. 2005; Tsou and Stearns 2006; Azimzadeh and Bornens 2007).

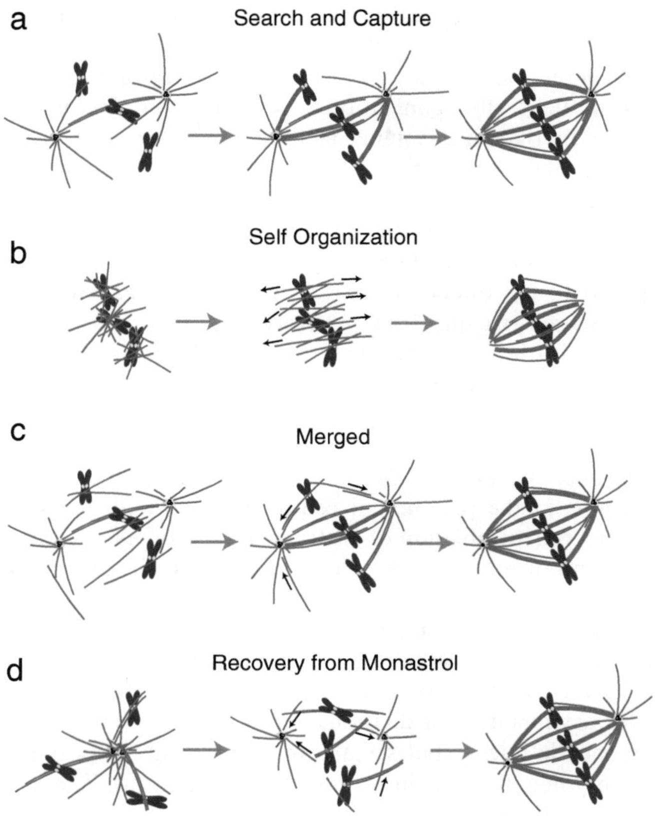


**Fig. 8.2** Models of Spindle Assembly. (**a**) Search and Capture. A bi-oriented chromosome and a newly captured mono-oriented chromosome are depicted. In this model, one sister kinetochore attaches to a centrosomally nucleated microtubule following a random encounter driven by cycles of microtubule growth and shrinkage. The chromosome rapidly translocates towards the minus end of the microtubule and its associated pole, often by sliding along the side of the microtubule, movement that is thought to be mediated by cytoplasmic dynein. The kinetochore attachment converts to an end on association with a microtubule bundle (K-fiber). The unattached kinetochore is subsequently contacted by a microtubule emanating from the opposite pole, and the chromosome begins to move toward the metaphase plate (congression). The leading kinetochore is defined as the kinetochore oriented towards the direction of movement, and the lagging kinetochore is trailing. The chromosome eventually congresses to the metaphase plate and will oscillate there. (**b**) Microtubule Self-Organization. In spindle self-assembly, microtubules are nucleated in the vicinity of chromatin, and a subset may be captured and bundled into short K-fibers. Microtubules are sorted and organized by molecular motor proteins into an antiparallel array as they continue to grow. Microtubule bundling generates a bipolar axis, and the spindle poles extend and become focused. (**c**) Merged model of spindle assembly. Peripheral microtubules or those emanating from chromosomes are captured and incorporated into the centrosome nucleated array to generate the spindle. (**d**) Recovery of spindle bipolarity following monastrol washout revealed that kinteochore fibers are captured by centrosomal nucleated microtubules and transported toward the microtubule minus ends

*Motor-dependent spindle "self-organization":* Since the identification of families of kinesins in the spindle, the phenomenon of motor-dependent organization has been revealed as another mechanism promoting spindle bipolarity, termed microtubule "self-organization" (Karsenti and Vernos 2001). This model was first invoked in systems lacking centrosomes, primarily female meiotic cells and plant cells, in which spindle microtubules do not emanate from duplicated focal nucleation sites, but from chromosomes or from the nuclear envelope. In this type of spindle assembly pathway, microtubules coalesce into bundles and extend spindle poles away from chromosomes. Therefore, as opposed to the outside-in search and capture pathway mediated by centrosomes, the self-organization pathway appears to be inside-out. This spindle self-organization requires that microtubule nucleation is favored in the vicinity of chromosomes, and that chromosomally nucleated microtubules are then organized and sorted into the bipolar array by microtubule-based motor proteins (Fig. 8.2b).

*Merged model of spindle assembly:* It is now recognized that the self-assembly pathway functions in the presence of centrosomes as well. Spindles still form following laser or genetic ablation of centrosomes (Khodjakov et al. 2000; Basto et al. 2006), and motor proteins are required to preserve bipolar spindle morphology even in the presence of centrosomes (Sawin et al. 1992; Walczak et al. 1998; Kapoor et al. 2000). Thus, centrosomes are neither necessary nor sufficient for spindle assembly under many situations, but by generating astral microtubules that interact with the cell cortex, centrosomes play key roles in defining spindle orientation critical for asymmetric cell divisions, and also provide a kinetic advantage to generating spindle microtubules that can be especially important during the rapid early embryonic divisions in several model organisms. The merged model postulates that spindle microtubules consist of two interacting populations, one that is centrosome-nucleated, and one that is self-organized (Fig. 8.2c). Under some conditions, different sources of spindle microtubules are apparent: for example in *Drosophila* spermatocytes (Rebollo et al. 2004), in *Drosophila* S2 cells (Maiato et al. 2004), and at low concentrations of nocodazole following spindle microtubule depolymerization (Tulu et al. 2006). But how do microtubules generated in the vicinity of the chromosomes become incorporated in the spindle? Insight into this problem was provided by use of the small molecule Kinesin-5 inhibitor monastrol, which arrests spindles in a monopolar configuration with the duplicated centrosomes in the center and microtubules extending outwards to form a rosette structure with the chromosomes at the periphery (Mayer et al. 1999). When cells were arrested in monastrol, K-fibers extending from chromosomes were frequently observed that were not connected to the centrosomes. Upon drug washout, these fibers were incorporated into the spindle by the sliding of their distal ends toward centrosomes by a mechanism requiring the activity of NuMA (Khodjakov et al. 2003), likely in a complex with the minus end-directed motor cytoplasmic dynein (Merdes et al. 1996). Therefore, in addition to kinetochores themselves, kinetochore-associated microtubules can be captured by centrosomal arrays to build the spindle in a motor-dependent manner (Khodjakov et al. 2003; Fig. 8.2d).

## 8.5 Microtubule Dynamics Affect Spindle Assembly and Bipolarity

*Microtubule dynamics are cell cycle regulated.* At the onset of mitosis in vertebrate cells, the interphase microtubule array is dismantled as the microtubule catastrophe frequency increases, and the rescue frequency decreases (Belmont et al. 1990; Gliksman et al. 1993; Rusan et al. 2001). Microtubule polymer levels drop transiently, then recover as the spindle forms (Zhai et al. 1996). However, mitotic microtubules are more dynamic than those of the interphase array, with a lifetime of about 30 s (Saxton et al. 1984). In vertebrates, the different populations of spindle microtubules have varying lengths and dynamics (McDonald et al. 1992; Mastronarde et al. 1993). Overall the kinetochore microtubules are longer and more stable than spindle microtubules, with a half-life of about 3 min. This increased stability may be important to maintain proper attachment of chromosomes to the spindle and also to maintain spindle organization. Intricate regulation of microtubule dynamics during mitosis is critical for spindle assembly. Although our knowledge of the molecules contributing to spindle microtubule dynamic behavior is incomplete, there is strong evidence that both stabilizing Microtubule Associated Proteins (MAPs) and destabilizing activities of the Kinesin-13 microtubule depolymerases are important players.

*MAPs of the XMAP215 family play central roles in generating microtubule dynamics:* A key regulatory MAP originally identified in *Xenopus* (Gard and Kirschner 1987), XMAP215 (termed CH-TOG in humans) and its orthologs in yeast (Stu2p), *Drosophila* (Msps), and C. elegans (Zyg-9) has been shown to be critical for spindle assembly, primarily by controlling the dynamics of microtubules (Cullen et al. 1999; Tournebize et al. 2000; Cassimeris and Morabito 2004; Holmfeldt et al. 2004; Brittle and Ohkura 2005; Srayko et al. 2005). Its activity is also important for chromosome alignment and anaphase spindle elongation (Kosco et al. 2001; Severin et al. 2001). In addition to a role in controlling microtubule dynamics, XMAP215/Stu2p may also function at the centrosome/spindle pole body to promote microtubule nucleation (Popov et al. 2002) or microtubule anchorage (Usui et al. 2003), a function that is likely modulated by its interaction with spindle pole TACC proteins (Lee et al. 2001; Bellanger and Gonczy 2003; Gergely et al. 2003; Srayko et al. 2003). The emerging model is that XMAP215 can funnel tubulin subunits or oligomers to growing microtubule plus ends by itself or in a complex with other MAPs (Srayko et al. 2003; Al-Bassam et al. 2006; Kerssemakers et al. 2006; Slep and Vale 2007; Kosco et al. 2001; van Breugel et al. 2003; Huang and Huffaker 2006; Wolyniak et al. 2006).

*MT plus-tip proteins are critical for regulation of plus end-microtubule dynamics during mitosis:* In addition to conventional MAPs, there exists a specialized class of MAPs known as the plus end-tracking proteins or plus-tip proteins (Howard and Hyman 2003; Lansbergen and Akhmanova 2006). These proteins track the growing plus end of microtubules both during interphase and mitosis. In mitosis, particular plus-tip proteins, such as APC, CLIP-170, and CLASP have been shown to be important for chromosome alignment (Dujardin et al. 1998; Maiato et al. 2002; Green and Kaplan 2003; Maiato et al. 2003; Green et al. 2005). In particular CLASP activity is required for regulating the plus end dynamics of kinetochore microtubules (Maiato et al. 2005). Despite the existence of these proteins in nearly all eukaryotic cells, it is not known how their activities are spatially controlled during mitosis nor do we understand the detailed mechanism by which they regulate plus end microtubule dynamics.

*MAP activity is opposed by Kinesin-13 microtubule depolymerizing enzymes.* In several systems, microtubule stabilizing activity of spindle MAPs is counteracted by members of the Kinesin-13 family of depolymerases, which regulate polymer levels both globally and at specific sites within the spindle, often at the kinetochore or centrosome (Tournebize et al. 2000; Holmfeldt et al. 2002; Noetzel et al. 2005; Niethammer et al. 2007). One model is that the MAPs bind to the microtubules, which subsequently can block access of the depolymerases to the microtubule polymer. However, it is unlikely to be that simple because of the complex spatial and temporal regulation of MAP and Kinesin-13 activities. Immunodepletion or knockdown of MCAK or related Kinesin-13 proteins results in an abnormal increase in microtubule polymer that can be associated with monopolar spindles (Walczak et al. 1996; Kline-Smith and Walczak 2002; Goshima and Vale 2003; Cassimeris and Morabito 2004; Rogers et al. 2004; Stout et al. 2006). Interestingly, the large microtubule arrays that form after depletion of the Kinesin-13 MCAK could be restored to their normal size by supplementing purified wild-type MCAK, but not truncated versions or the closely related Kinesin-13 Kif2A, even though equivalent amounts of microtubule destabilization activity were added (Ems-McClung et al. 2007; Ohi et al. 2007). These studies highlight the complex regulation of microtubule dynamics that occur during spindle assembly, but leave open the important question of how this is achieved.

The opposing activities of XMAP215 and Kinesin-13 MCAK on microtubules have also been demonstrated by depleting the proteins individually and in combination from *Xenopus* egg extracts (Tournebize et al. 2000). Furthermore, their importance is highlighted by the observation that incubation of XMAP215, MCAK, and purified tubulin is sufficient to reconstitute physiological microtubule dynamics in vitro (Kinoshita et al. 2001). A discrete interaction network among these factors is emerging that appears to be differentially regulated during interphase and mitosis (Niethammer et al. 2007).

*Microtubule dynamics affect spindle bipolarity:* In addition to disrupting microtubule polymer levels, interference with members of the Kinesin-13 family in multiple systems results in an increased percentage of monopolar spindles (Walczak et al. 1996; Kline-Smith and Walczak 2002; Goshima and Vale 2003; Gaetz and Kapoor 2004; Ganem and Compton 2004; Rogers et al. 2004; Zhu et al. 2005). In vertebrate cells, the monopolar spindles caused by knockdown of Kif2A can be rescued by co-depletion of a different member of the Kinesin-13 family (MCAK) or by agents that disrupt microtubule dynamics, such as the drug nocodazole (Ganem and Compton 2004), indicating that balanced microtubule dynamics are important for bipolarity. Spindle bipolarity can also be disrupted by depletion of CH-TOG or addition of paclitaxel to cells (Holmfeldt et al. 2004). It is currently unclear how CH-TOG contributes to multipolarity but it may be mediated in part by its interaction with TACC at the centrosome (Lee et al. 2001; Bellanger and Gonczy 2003; Gergely et al. 2003).

*Op18 is important in local regulation of microtubule dynamics:* Another important microtubule destabilizing activity in cells is the small heat-stable protein, Stathmin/Op18. Op18 or oncoprotein 18 is highly overexpressed in a number of tumors and a mutated form of Op18 has been identified in adenocarcinomas (Belmont et al. 1996; Cassimeris 2002; Misek et al. 2002; Holmfeldt et al. 2006). Its importance in regulating microtubule dynamics was revealed upon isolation of Op18 as a microtubule catastrophe-promoting activity from *Xenopus* egg extracts (Belmont et al. 1996; Belmont and Mitchison 1996), and subsequently it has been found to control microtubule dynamics in a number of different cell types (Cassimeris 2002). Key to Op18 function is its regulation by phosphorylation (Larsson et al. 1997; Kuntziger et al. 2001; Cassimeris 2002). In particular its activity appears to be regulated locally in the vicinity of chromatin and may be important in chromatin-mediated spindle formation (Andersen et al. 1997; Budde et al. 2001; Gadea and Ruderman 2006; Kelly et al. 2007).

## 8.6 Microtubule-Based Motors Are Critical for Spindle Organization

Microtubule motors play diverse roles in spindle assembly and function. Whereas the microtubule depolymerizing Kinesin-13 family, discussed above, affects spindle microtubule dynamics, most other spindle kinesins act primarily through microtubule cross-linking and sliding. Kinesin families functioning in the spindle are summarized in Table 8.1, and several different spindle activities of motor proteins are schematized in Fig. 8.3.

*Kinesin-5 family members are critical for spindle bipolarity and can drive poleward microtubule flux:* The roles of microtubule motors in spindle assembly were first elucidated in genetic studies in which kinesin superfamily members were identified as gene products mutated in strains with chromosome segregation defects (Enos and Morris 1990; Hagan and Yanagida 1990; McDonald and Goldstein 1990; McDonald, Stewart and Goldstein 1990; Meluh and Rose 1990; Wilson et al. 1992; Heck et al. 1993; Hoyt et al. 1993). From this early work it became apparent that members of the Kinesin-5 and Kinesin-14 families were critical for spindle assembly in nearly every organism examined (Walczak and Mitchison 1996; Sharp et al. 2000). The Kinesin-5 family members, including Cin8/Kip1 in yeast, Klp61F in *Drosophila*, and Eg5 in vertebrates are

**Table 8.1** Molecular Motor Functioning in Spindle Assembly and Function. The kinesin superfamily, its direction of movement along microtubules, localization, and proposed functions. Highly conserved motors are shown in bold

| Kinesin superfamily | Protein names-organism | Directionality | Localization | Proposed function |
|---|---|---|---|---|
| Kinesin-4 | Xklp1 – *Xenopus*<br>Klp3A – *Drosophila* | Plus end | Chromosome arms, spindle midzone | Chromosome positioning, chromosome MT attachment (Vernos et al. 1995; Antonio et al. 2000; Funabiki and Murray 2000; Levesque and Compton 2001; Tokai-Nishizumi et al. 2005) Coupling MT sliding to spindle elongation (Brust-Mascher et al. 2004) |
| Kinesin-5 | Eg5 – vertebrate<br>BimC yeast | Plus end | Spindle, enriched at poles | Spindle bipolarity (Sawin et al. 1992; Mayer et al. 1999), MT flux (Miyamoto et al. 2004; Shirasu-Hiza et al. 2004) |
| Kinesin-6 | MKLP-1 – vertebrate | Plus end | Spindle midzone | Central spindle organization, MT bundling, cytokinesis (Matuliene and Kuriyama 2002; Zhu et al. 2005) |
| Kinesin-7 | CENP-E – human | Plus end | Kinetochores | Chromosome congression, kinetochore-MT attachment (Schaar et al. 1997; Wood et al. 1997; Yao et al. 1997; Putkey et al. 2002; Kapoor et al. 2006) |
| Kinesin-8 | Kif18A – human<br>Klp67A – *Drosophila*<br>Kip3 – yeast | Plus-end, depolymerase | Kinetochores, spindle | Chromosome congression (Goshima and Vale 2003; Savoian et al. 2004; Zhu et al. 2005; Mayr et al. 2007; Stumpff et al. 2008) |
| Kinesin-10 | Kid – vertebrate | Plus end | Chromosome arms | Chromosome positioning, chromosome oscillations (Antonio et al. 2000; Funabiki and Murray 2000; Levesque and Compton 2001; Tokai-Nishizumi et al. 2005) |
| Kinesin-12 | Xklp2 – *Xenopus* | Plus end | Spindle pole | Spindle pole organization (Wittmann et al. 2000) |

**Table 8.1** (continued)

| Kinesin superfamily | Protein names-organism | Directionality | Localization | Proposed function |
|---|---|---|---|---|
| Kinesin-13 | MCAK – vertebrate<br>Kif2A – vertebrate<br>Klp59C – *Drosophila*<br>Klp10A – *Drosophila* | Depolymerase | Centromere, kinetochore, spindle poles | Error correction (Kline-Smith et al. 2004), chromosome segregation (Maney et al. 1998) Poleward MT flux (Gaetz and Kapoor 2004; Ganem et al. 2005) |
| Kinesin-14 | XCKT2 – *Xenopus*<br>Ncd – *Drosophila*<br>Kar3 – yeast | Minus-end, can act as depolymerase | Spindle pole | Spindle pole focusing and promotion of spindle bipolarity (Walczak et al. 1997) |

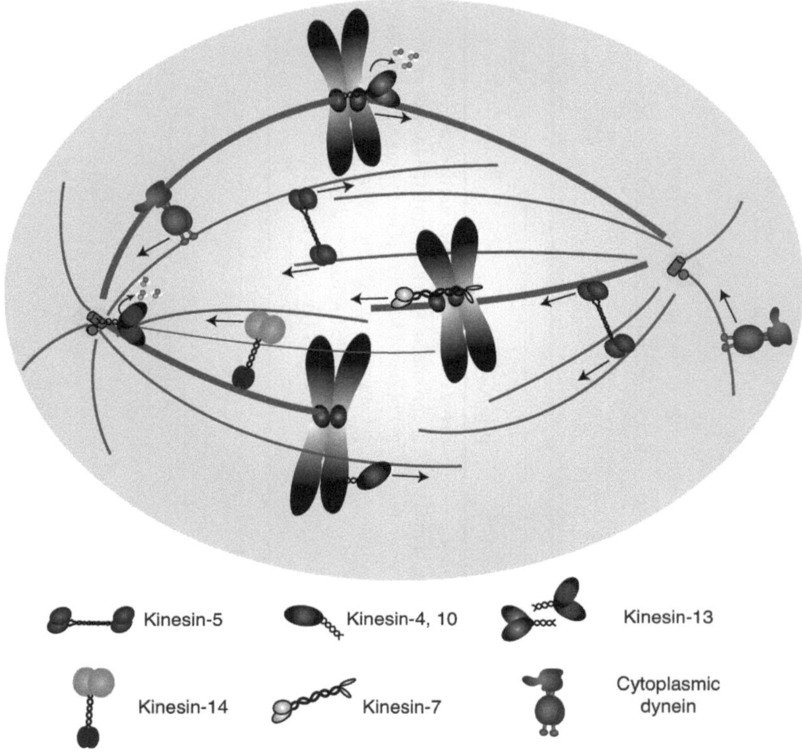


**Fig. 8.3** Motor function in the mitotic spindle. Various classes of motor proteins are illustrated, and their functions are described in Table 8.1. Tetrameric Kinesin-5 motors cross-link and can move along two microtubules in a parallel or antiparallel orientation towards their plus ends. Kinesin-14 can also cross-link microtubules via its minus end-directed motor at one end and its ATP-independent microtubule binding domain at the other end. Kinesin-4 and -10 are plus end-directed, chromatin-associated motors. Kinesin-7 is at the kinetochore and moves toward the microtubule plus end. Kinesin-13s are microtubule-depolymerizing enzymes that function at both the kinetochore and spindle poles. Cytoplasmic dynein can contribute to focusing of spindle poles, or pull on astral microtubules at the cortex

essential for establishing bipolarity, with a notable exception in *C. elegans* where they appear to act as a brake to control spindle pole separation (Saunders et al. 2007). A homotetramer, Kinesin-5 has a bipolar structure with two motor domains at each end (Kashina et al. 1996; Sharp et al. 1999). Kinesin-5 proteins can bind to and cross-link both parallel and antiparallel microtubules, but its activity is thought to be most critical on the antiparallel microtubules (Sharp et al. 1999; Sharp et al. 2000; Sharp, Rogers and Scholey 2000; Kapitein et al. 2005).

A by-product of forces in the spindle is a steady poleward flux of tubulin subunits within microtubules in higher eukaryotes, which occurs as tubulin subunits incorporate at microtubule plus ends and coordinately dissociate from minus ends within each half spindle. Particularly prominent in embryonic systems such as *Xenopus* egg extracts and *Drosophila* embryos, ~2 μm/min flux is driven by the Kinesin-5 –dependent sliding of antiparallel microtubules toward their minus ends (Brust-Mascher et al. 2004; Miyamoto et al. 2004). However, inhibition of Kinesin-5 in mammalian cultured cells (PtK1) resulted in only a minor reduction in the rate of kinetochore fiber flux from ~0.7 to ~0.5 μm/min, the same rate measured in monopolar spindles that lack antiparallel microtubules (Cameron et al. 2006). In systems with slower flux, a polar "pulling-in" mechanism utilizing a Kinesin-13 localized at kinetochore fiber minus ends may be the major contributor to poleward flux (Rogers et al. 2005). In addition, factors concentrating at microtubule plus ends, such as the CLASP protein, are thought to promote microtubule attachment and polymerization at the kinetochore (Maiato et al. 2005). Interestingly, blocking flux does not prevent chromosome alignment and segregation in cultured cells (Ganem et al. 2005), and its potential functional roles remain unclear.

*Kinesin-14 opposes the activity of Kinesin-5 and is involved in spindle pole formation:* The early genetic studies also revealed a unique balance of forces among motors, as mutation in members of the Kinesin-5 family could be rescued by corresponding mutations in members of the Kinesin-14 family (Hoyt et al. 1993; O'Connell et al. 1993). Kinesin-5 proteins likely provide an outward force on the spindle that is opposed by an inward-directed force generated by Kinesin-14 members. In *S. cerevisiae*, the inward-directed force by the Kinesin-14 Kar3p is diminished at anaphase onset to allow spindle elongation (Saunders et al. 1997). This balance of forces between Kinesin-5 and Kinesin-14 family members exists in multiple systems, suggesting that it is a fundamental aspect of spindle organization (Mountain et al. 1999; Sharp et al. 2000; Cytrynbaum et al. 2003).

Kinesin-14 family members are also important for focusing microtubule minus ends together at the spindle pole (Walczak et al. 1997; Walczak et al. 1998; Mountain et al. 1999), especially during female meiosis in the absence of centrosomes (McDonald et al. 1990; Nelson and Szauter 1992; Endow et al. 1994). In vertebrates this activity may be redundant with the action of cytoplasmic dynein, which is the major minus-end focusing activity in these systems (Verde et al. 1991; Merdes et al. 1996; Walczak et al. 1998).

*A minimal set of kinesins is required for spindle assembly in yeast:* In addition to the ubiquitous function of Kinesin-5 and Kinesin-14 family members, nearly all organisms also express one or more microtubule-destabilizing kinesins, again highlighting the importance of proper regulation of spindle dynamics. The minimal requirements for kinesin function during mitosis were perhaps best defined by genetic studies in *S. cerevesiae*, in which cells that are mutant for all of the kinesins except for a single member of the Kinesin-5 family member can progress through mitosis if they possess a temperature sensitive mutation in a member of the Kinesin-14 family (Cottingham et al. 1999), which has recently been shown to act as a microtubule depolymerase (Sproul et al. 2005). This requirement for the Kinesin-14 family member could be partially suppressed by incubation with low levels of benomyl (Cottingham et al. 1999), showing that microtubule sliding and dynamics are the most basic elements of motor function in the spindle.

*Conserved kinesins function in chromosome congression.* The kinetochore-associated Kinesin-8, Kif18A/Klp67A possesses both conventional microtubule-based motility and a plus end depolymerization activity, and was originally shown to be essential for chromosome congression to the metaphase plate, likely due to a role promoting chromosome motility (Goshima and Vale 2003; Savoian et al. 2004; Zhu et al. 2005; Mayr et al. 2007). However it has been shown recently that a major role of Kif18A is to control chromosome oscillations by suppressing chromosome movement (Stumpff et al. 2008), highlighting the complex activities at kinetochores of vertebrate chromosomes. While budding yeast has a single Kinesin-8 family member (Kip3), it is still controversial as to whether or not it plays a specific role at kinetochores (Gupta et al. 2006; Tytell and Sorger 2006; Varga et al. 2006). Mutants in *kip3* cause defects in spindle positioning, which may be due to its specific regulation of cortical microtubule dynamics (DeZwaan et al. 1997; Cottingham et al. 1999; Gupta et al. 2006; Varga et al. 2006). In addition, loss of Kip3p causes defects in synchronous poleward chromosome movement (Straight et al. 1998; Tytell and Sorger 2006). In contrast to budding yeast, studies in fission yeast clearly reveal a role for Kinesin-8 family members in chromosome motility (West et al. 2001; Garcia et al. 2002; Garcia et al. 2002; West et al. 2002). More work is needed to elucidate whether these family members play a single conserved role in all organisms or if they have adapted specialized roles within individual organisms.

*Vertebrate cells have specialized kinesin functions:*Compared to the simple budding yeast spindle, in which each kinetochore is attached to a single microtubule, vertebrate cells possess much more complex spindles and kinetochores and have evolved additional mitotic kinesins, such as Kinesin-7 (CENP-E), which plays a role in chromosome congression. CENP-E is localized to mitotic kinetochores, and its loss causes an increase in unattached and mono-oriented chromosomes near spindle poles (Yen et al. 1992; Schaar et al. 1997; Wood et al. 1997; Yao et al. 1997; Putkey et al. 2002) associated with a diminished number of microtubules attached to kinetochores (McEwen et al. 2001). Recent studies have identified a unique role for CENP-E in the congression of mono-oriented chromosomes to the spindle equator (Kapoor et al. 2006). CENP-E is proposed to drive the movement of chromosomes to the metaphase plate by utilizing its plus end motor activity to slide the unattached kinetochore along microtubules of an adjacent kinetochore fiber. This was the first study to suggest that duplicated chromosomes with two sister kinetochores can congress to the metaphase plate without actually being bi-oriented (attached to both spindle poles) on the spindle.

Other chromosomally bound kinesins play important roles in chromosome movement. The chromosome-associated Kinesin-4 (Kif4 or Xklp1) is important to establish a tight metaphase plate, and loss of this protein results in misalignment of chromosomes that may be due to an improper association of microtubule plus ends with chromosome arms (Vernos et al. 1995; Walczak et al. 1998). More recently, Kif4A has been shown to be involved in chromosome condensation, which may indirectly affect its interactions with microtubules (Mazumdar et al. 2004). However the role of Kif4 may be more complicated as other studies suggest loss of Kif4 does not affect condensation or chromosome alignment but instead results in defects in abscission and ultimate cell death (Samejima et al. 2008). In addition to Kinesin-4, the Kinesin-10 protein Kid is also necessary for proper chromosome alignment (Antonio et al. 2000; Funabiki and Murray 2000; Tokai-Nishizumi et al. 2005; Zhu et al. 2005). Kid is a DNA-binding kinesin that may provide a "polar wind" by moving chromosome arms towards spindle microtubule plus ends at the equator (Brouhard and Hunt 2005), and is also necessary for proper chromosome oscillations on the spindle (Levesque and Compton 2001). This is an interesting observation because it was originally thought that chromosome oscillations were driven primarily by forces at the kinetochore and not by forces on the chromosome arms (Kapoor and Compton 2002). Together these studies highlight the complex roles that the chromosome arms and their associated proteins play in mitosis.

Multiple other kinesins are important for proper spindle structure during anaphase and cytokinesis. For example members of the Kinesin-6 family, including MKLP1 and 2 are critical to establish proper central spindle formation by cross-linking antiparallel microtubules in this region (Kuriyama et al. 2002; Matuliene and Kuriyama 2002; Zhu et al. 2005). Loss of this activity leads to disorganized spindle midzones and ultimately failure in cytokinesis. The Kinesin-3 Kif14 has also been implicated in ensuring proper cytokinesis (Gruneberg et al. 2006), but its mechanism remains unclear.

*Cytoplasmic dynein plays multiple roles in mitosis:* In addition to the many kinesin proteins, cytoplasmic dynein also functions in multiple aspects of mitosis. Required early in mitosis to ensure proper centrosome separation, dynein may act at the cell cortex (Vaisberg et al. 1993) or at the nuclear envelope (Gonczy et al., 1999). Dynein activity is also critical for the rapid pole-directed movement of chromosomes early in mitosis (Yang et al. 2007) and for chromosome congression (Echeverri et al. 1996). During late mitosis, it is still controversial whether dynein contributes to poleward chromosome movements during anaphase A (Savoian et al. 2000; Sharp et al. 2000; Yang et al. 2007). In addition, dynein operates at spindle poles, where it promotes focusing of microtubule minus ends (Verde et al. 1991; Heald et al. 1996; Merdes et al. 1996; Gaglio et al. 1997; Heald et al. 1997) and delivers microtubule depolymerizing kinesins and other mitotic motors to the spindle poles (Gaetz and Kapoor 2004). The above studies highlight the numerous functions that have been attributed to dynein action on the mitotic spindle. Because dynein is proposed to act in so many places during mitosis, it is difficult to decipher whether a given phenotype is the result of a direct action of dynein at that site or whether it is a downstream consequence of disrupting other dynein functions. These problems have been overcome in part by the identification of multiple dynein-interacting proteins that target it to distinct locations (Echeverri et al. 1996; Scaerou et al. 2001; Yan et al. 2003; Li et al. 2005; Liang et al. 2007; Stehman et al. 2007). By disrupting dynein interacting proteins, researchers have begun to dissect the various roles of dynein in mitosis.

Dynein is clearly required for multiple aspects of spindle function in vertebrate cells, however it is not present in some higher plant cells, which lack focal centrosomes (King 2002; Wickstead and Gull 2007). These plant cells have likely evolved other minus-end directed motors for microtubule cross-linking functions, but it is unclear how the known kinetochore-based functions of dynein are bypassed in this system. The function of dynein also varies between the two yeasts. Early studies in budding yeast were somewhat surprising because dynein was not found to be essential in spindle function but rather in spindle positioning between the mother and daughter cells (Eshel et al. 1993; Li et al. 1993; Yeh et al. 2000). The current model is that dynein mediates proper interactions between microtubules and the cell cortex through interaction with plus-tip proteins such as Kar9p and Bim1p (Miller et al. 1999; Miller et al. 2000; Lew and Burke 2003; Grava et al. 2006), regulating cortical microtubule dynamics and thus nuclear and spindle positioning (Carminati and Stearns 1997; Shaw et al. 1997). In contrast to budding yeast, fission yeast dynein may be more akin to vertebrate dynein in that it is thought to be required for spindle pole focusing, chromosome movement, and the spindle checkpoint (Grishchuk et al. 2007). Overall the above data highlight the numerous functions that dynein plays to make critical contributions to proper spindle assembly, function, and checkpoint signaling.

Together all of the above studies show that motor activities and microtubule dynamics generate a balance of forces defining steady state spindle pole position during metaphase and that act both independently and coordinately to promote spindle assembly (Fig. 8.3). Ultimately, motor activity is also needed, either directly or indirectly, to ensure that the ultimate action of the mitotic spindle, chromosome segregation, is carried out in a timely and accurate fashion to distribute the genetic material to the two daughter cells.

## 8.7 Non-microtubule Structures in the Spindle

The mitotic spindle is defined as the composite of the microtubules, chromosomes, and associated factors. One conceptual problem in thinking about spindle function arises from the rapid turnover of microtubules within the spindle itself. The process of mitosis lasts approximately 1 h in vertebrate cells, and the short half-life of spindle microtubules means that they turn over many times during the event. Given their rates of movement, some motors could perhaps travel the length of the spindle on kinetochore microtubules (half-life $\sim$3 min), but not on spindle microtubules (half-life $\sim$20 s) because they are depolymerizing so frequently. Due to the dynamic turnover of mitotic microtubules and the forces that are generated, it has long been proposed that a non-microtubule "spindle matrix" exists to provide a framework on which motors and other proteins could interact with spindle microtubules (Pickett-Heaps et al. 1982; Pickett-Heaps 1986; Scholey et al. 2001). Furthermore, a matrix would provide a stable, non-dynamic substrate that could be utilized to help assemble and stabilize the mitotic spindle.

Despite the appeal of the matrix model, a precise molecular definition of its components has been elusive. Several candidate matrix proteins have been identified in the past decade, including *Drosophila* skeletor, which is found in a spindle shaped structure that appears before the mitotic spindle forms and persists in cells treated with drugs that depolymerize microtubules, two properties that one would envision for a spindle matrix (Walker et al. 2000). Skeletor is in a complex with chromator and megator, both of which exhibit similar localization patterns (Qi et al. 2004; Rath et al. 2004). However enthusiasm has been dampened by the apparent non-conserved nature of these proteins, which have yet to be identified in other eukaryotic organisms. The Kinesin-5 Eg5 has also been proposed as a spindle matrix protein, as it appears to be static in the spindle relative to tubulin, another feature essential to define a matrix (Kapoor and Mitchison 2001). Recently, a non-protein component, poly ADP ribose (PAR), was identified as a major component of a spindle matrix (Chang et al. 2004). PAR is a large negatively charged moiety that is added to proteins by a specific PAR polymerase (PARP). One such PARP was found localized on the spindle and its activity is required for proper spindle assembly (Chang et al. 2005). A major substrate of PARP is the spindle pole protein NuMA, itself a matrix candidate, which provides a potential model for how PARPs might modify different spindle components to help stabilize them even in the absence of microtubules (Chang et al. 2005; Compton 2005).

A third promising matrix candidate is the intermediate filament protein lamin B, which is better known for its role in the lamina underlying the nuclear membrane during interphase (Hayes 2006; Tsai et al. 2006; Zheng and Tsai 2006). A lamin B matrix can form independent of microtubules but depends on the RanGTP system, discussed below. Importantly, the lamin B matrix associates with several known spindle assembly factors, including Eg5, XMAP215, and NuMA (Tsai et al. 2006). Together these studies highlight potential molecular players that may constitute the spindle matrix, but it is not yet understood how these putative matrix components interact and assemble with temporal and spatial precision within the cell.

## 8.8 The Spindle Pole

Spindle poles maintain the integrity of the spindle, which is under strain due to the forces of chromosome alignment and segregation. In mitosis, proper chromosome movement requires anchorage of microtubules at the poles (Gordon et al. 2001), and defects such as supernumerary poles or unfocused microtubule minus ends can lead to the formation of micronuclei due to chromosome loss from the spindle (Fant et al. 2004). Like the spindle itself, the spindle pole is a dynamic structure that can form by centrosome-directed or self-organization pathways (Heald et al. 1997; Khodjakov and Rieder 2001). In both cases, microtubule minus ends are focused together by the same components, including motor proteins that play key roles in pole organization, creating a dynamic assembly that accommodates the rapid turnover of spindle microtubules due to dynamic instability, as well as poleward flux. In the presence of a centrosome, nucleated microtubules are transiently anchored at the pre-existing focus of microtubule minus ends. In self-organized spindles lacking centrosomes, the minus ends are sorted and focused into a *de novo* pole. Regardless, many of the same mechanisms are at work. Centrosomes define the position of the spindle poles, but are distinct structures, whose association with the poles are maintained by motor-dependent microtubule cross-linking (Heald et al. 1997; Merdes and Cleveland 1997).

*Components of spindle poles:* Three major classes of proteins as well as their regulatory factors are thought to contribute to spindle pole morphogenesis. These include microtubule minus end-associated MAPs, both plus and minus end-directed motor proteins, and microtubule destabilizing factors. A set of MAPs concentrates at the poles where they serve as cross-linkers. The best characterized of these is NuMA, a large coiled-coil protein found in vertebrates that localizes in a crescent-shaped pattern at the poles, and whose depletion causes loss of spindle pole focusing (Compton et al. 1992; Compton and Cleveland 1993). NuMA interacts with cytoplasmic dynein in *Xenopus* egg extracts, and is transported to microtubule minus ends where it remains associated (Merdes et al. 1996, 2000). Upon cross-linking to a second microtubule, the minus end-associated dynein–NuMA complex could move towards its minus end, thereby focusing the minus ends together. Such dynein/NuMA-dependent focusing activity has been visualized in *Xenopus* egg extracts by adding polarity-marked, fluorescent microtubule seeds to spindles, which can be seen translocating poleward with their minus ends leading (Heald et al. 1996). This dynein-dependent cross-linking activity helps to maintain association of the centrosomally nucleated microtubules with spindle pole microtubules (Heald et al. 1997; Goshima et al. 2005). Other MAP families contributing to spindle pole morphology include the TACC proteins, which perform a variety of functions supporting both centrosome-dependent microtubule nucleation and centrosome–independent spindle pole assembly, in part by targeting other crucial MAPs, such as XMAP215/CH-TOG (Theurkauf 2001; Gergely et al. 2003; O'Brien et al. 2005), and Asp, which unlike NuMA is expressed in many different animal phyla, where it is thought to play a similar role in pole focusing (Fant et al. 2004).

Depending on the system, minus-end directed kinesin-like motors also function in microtubule minus end focusing, notably the Kinesin-14 proteins Ncd in *Drosophila* and HSET in humans (Endow et al. 1994; Mountain et al. 1999). Importantly, opposing plus end-directed motor activities are essential to establish pole morphology. Pole/aster structures stimulated in mitotic HeLa extracts by taxol in the absence of Eg5 displayed an enlarged central core (Gaglio et al. 1996; Gaglio et al. 1997; Mountain et al. 1999). In vitro motility assays have demonstrated that Kinesin-5 antagonizes Kinesin-14 activity (Tao et al. 2006), suggesting that a balance of minus and plus end-directed motility is required to establish spindle poles.

In addition to MAPs and motor proteins, microtubule-destabilizing proteins of the Kinesin-13 family are important components of spindle poles whose inhibition leads to increased microtubule density and spindle length (Gaetz and Kapoor 2004; Goshima et al. 2005; Morales-Mulia and Scholey 2005). Recently, microtubule severing factors have also been implicated (Zhang et al. 2007). These findings are consistent with the requirement for depolymerization of microtubule minus ends at poles to drive poleward flux and to maintain constant spindle length by balancing plus-end growth. As noted above, the multiple Kinesin-13s found in vertebrate spindles perform distinct functions. Interestingly, depletion of Kinesin-13 Kif2b can rescue spindle formation in NuMA depleted cells (Manning et al. 2007). These results once again highlight the complex interactions among factors affecting microtubule organization and stability that act synergistically or antagonistically.

## 8.9 The Role of Chromosomes: Biochemical Signals

Both self-organization and search and capture spindle assembly models posit that the chromosomes act as a stabilizing force. By favoring the growth of microtubules in their vicinity, kinetochore-spindle microtubule interactions are facilitated. There appear to be at least two kinds of biochemical activity generated by chromatin-associated enzymes. These include RCC1, the exchange factor that loads the small GTPase Ran with GTP, and mitotic kinases such as Aurora B that phosphorylate chromosome and spindle substrates.

*Chromosome-generated gradient of RanGTP:* RanGTP is recognized to play important roles throughout the cell cycle as a marker for the genome due to the chromatin localization of its nucleotide exchange factor RCC1 and the cytoplasmic localization of its GTPase activating protein (RanGAP). These distinct localizations give rise to an enrichment of RanGTP in the interphase nucleus, and surrounding the mitotic chromosomes (Kalab et al. 2002). Ran provides directionality to nuclear transport during interphase, promotes spindle assembly during mitosis, as well as nuclear envelope and pore assembly (Hetzer et al. 2002). All of the known functions of Ran are mediated through its interaction with transport receptors of the importin $\beta$ superfamily. Nuclear cargoes and a number of spindle assembly factors (SAFs) bind to importin $\beta$, and are released upon RanGTP binding, either in the interphase nucleus or around mitotic chromosomes. The emerging model is that RanGTP sets off a positive feedback loop to promote spindle assembly (O'Connell and Khodjakov 2007). By stimulating microtubule polymerization, RanGTP also stimulates microtubule binding of additional importin cargoes, many of which are in complexes. Ran-regulated SAFs can be divided into two categories—those that have been biochemically determined to interact directly with importins as cargo molecules, and those that function further downstream and are regulated by or in complexes with the cargoes. Altogether, proteins in the mitotic Ran pathway now number in the double digits, and it seems likely that many more remain to be discovered.

*Mitotic cargoes of importins regulated by Ran:* Significant insight into Ran regulation of spindle function came from the discovery that a number of nuclear-localized spindle factors are regulated by RanGTP and importins, including NuMA and TPX2 (Gruss et al. 2001; Nachury et al. 2001; Wiese et al. 2001). NuMA may be activated by RanGTP near the chromosomes, but then delivered to the spindle poles by dynein where it functions in pole assembly (Merdes et al. 2000). Depletion of TPX2 also caused defects in spindle pole organization and centrosome-directed spindle assembly (Wittmann et al. 2000), and completely blocked microtubule growth in the absence of centrosomes (Gruss et al. 2001). This provided strong support for the idea that TPX2 is a key mediator of the chromatin-generated microtubule-stabilizing signal. This function appears to be conserved in somatic cells, since siRNA depletion of TPX2 caused defects in spindle organization (Garrett et al. 2002; Gruss et al. 2002). Perhaps the strongest evidence to demonstrate that TPX2 is functionally important for chromatin-mediated spindle assembly comes from studies by Tulu and colleagues in which they developed an assay to distinguish microtubule nucleation from centrosomes versus chromosomes in somatic cells (Tulu et al. 2006). They found that TPX2 knockdown completely abolished only chromosome-mediated microtubule nucleation in mammalian cells and had no effect on centrosomal microtubule nucleation. The ability of TPX2 to regulate microtubule nucleation in an importin-dependent manner can be reconstituted in vitro, suggesting that Ran and importins can act directly on TPX2 (Schatz et al. 2003).

NuMA and TPX2 are not the only MAPs that interact directly with importins. NuSAP is a microtubule bundling factor that likely contributes to microtubule–chromatin interactions (Ribbeck et al. 2006; Ribbeck et al. 2007) as it is found localized to chromosome-proximal microtubules and protects microtubules from depolymerization. Maskin is a TACC family member that stimulates microtubule growth at centrosomes (Gergely et al. 2003; Albee et al. 2006) and whose phosphorylation by the centrosomal kinase Aurora A is directly regulated by Ran-GTP and the importins (Kinoshita et al. 2005; Albee et al. 2006). A third Ran-regulated MAP is Rae1, an mRNA export factor previously identified in yeast (Brown et al. 1995). Interestingly, Rae1 exists in a large complex that requires RNA for its microtubule-stabilizing activity (Blower et al. 2005). However, not all mitotic importin cargoes that bind microtubules have obvious Ran-regulated functions. For example, the activity of Xnf7, a bundling MAP that contributes to spindle integrity, does not appear to be modulated directly by importin $\beta$ or by RanGTP (Maresca et al. 2005).

In addition to regulating microtubule dynamics, Ran also alters the activity of microtubule motor proteins to promote bipolar spindle organization (Carazo-Salas et al. 2001; Wilde et al. 2001). One spindle motor known to be directly regulated by Ran is the Kinesin-14 XCTK2, which interacts with importin $\alpha/\beta$ through an NLS in its tail domain that lies near a microtubule-binding site. Addition of importins to the XCTK2 tail in vitro inhibits binding to microtubules, and is relieved upon addition of Ran-GTP (Ems-McClung et al. 2004). XCTK2 enhances spindle assembly and works like dynein to focus microtubule minus ends, two activities that could clearly be regulated by the Ran pathway to promote spindle organization (Walczak et al. 1997; Walczak et al. 1998). The chromosomally localized Kinesin-10 Kid is also a likely target of Ran regulation. A domain of the protein containing NLS sequences interacts with importins in HeLa cell extracts, and importins inhibit Kid binding to microtubules, but not to DNA (Trieselmann et al. 2003). Kid is important for chromosome positioning (Antonio et al. 2000; Funabiki and Murray 2000; Levesque and Compton 2001; Tokai-Nishizumi et al. 2005), so perhaps RanGTP regulates its association with the plus ends of microtubules near chromosomes.

*Additional spindle assembly factors regulated by the Ran pathway:* One idea that has emerged from the studies of Ran-regulated spindle assembly factors is that many cargoes function in complexes, indicating that the Ran signaling pathway involves many components. One key Ran signaling pathway involves Aurora A and TPX2. RanGTP stimulates interaction between TPX2 (Tsai et al. 2003) and Aurora A kinase, which leads to activation of the kinase and TPX2 phosphorylation (Eyers et al. 2003; Tsai et al. 2003). While it is unclear what functional effect Aurora A phosphorylation has on TPX2, phosphorylation of other recognized substrates, including the Kinesin-5 Eg5 that promotes spindle bipolarity and TACC that promotes microtubule polymerization, are likely important factors in Ran-regulated spindle organization (Giet et al. 1999, 2002). Other known substrates include the oncogene HURP (Yu et al. 2005) and the tumor suppressor protein BRCA1 (Ouchi et al. 2004). Both of these proteins are found in large complexes that also contain other spindle regulators and importin cargoes, and evidence has recently emerged that they participate in the mitotic Ran pathway. HURP was isolated from *Xenopus* egg extracts together with TPX2, Aurora A, XMAP215, and Eg5 (Koffa et al. 2006). Interestingly, HURP is a MAP that localizes to K-fibers proximal to chromosomes, and is required for proper chromosome alignment (Koffa et al. 2006; Sillje et al. 2006; Wong and Fang 2006). Perhaps HURP is part of the Ran-signaling network that functions at kinetochores. The BRCA1/BARD1 heterodimer was found in a complex with TPX2, NuMA, and XRHAMM, another *Xenopus* protein functioning in Ran-regulated chromatin-driven microtubule polymerization and spindle pole assembly (Groen et al. 2004). Depletion of BRCA1/BARD1 from either egg extracts or cells caused both chromosome alignment and spindle pole defects (Joukov et al. 2006). Finally, RanGTP also plays a role in formation of a lamin B spindle matrix (Tsai et al. 2006). Eg5 and XMAP215 were found associated with the lamin B matrix, supporting the idea that this matrix could link together multiple SAFs. Altogether, these studies support the model that RanGTP functions in all aspects of spindle assembly, including microtubule nucleation, stabilization, and organization in the spindle. However, although binding to importin $\alpha/\beta$ has been demonstrated to alter some activities, the mechanisms by which the activities of these factors are regulated by RanGTP and importins are still poorly understood.

*Mitotic chromatin-associated kinases:* Numerous kinases are required for progression through mitosis, including Cyclin-dependent kinases, the Aurora, Polo, and NIMA families (Nigg 2001; O'Connell et al. 2003; Meraldi et al. 2004; van de Weerdt and Medema 2006). One kinase-driven, chromatin-mediated pathway of spindle assembly is driven by Aurora B, a component of the chromosomal passenger complex (CPC), which includes INCENP, survivin, and Dasra A/B/Borealin (Gassmann et al. 2004; Sampath et al. 2004). One target of the CPC is the Kinesin-13 MCAK, which is negatively regulated by Aurora B phosphorylation. This inhibits its interaction with chromatin, thereby promoting microtubule polymerization (Sampath et al. 2004; Zhang et al. 2007). Phosphorylation of Aurora B substrates including histone H3 and Op18/stathmin is suppressed by phosphatases in the cytoplasm, but CPC binding to chromatin induces substrate phosphorylation (Kelly et al. 2007). Interestingly, microtubules can also induce substrate phosphorylation. Since the Ran pathway promotes microtubule stabilization, it may contribute to the positive feedback loop by promoting CPC pathway activation.

In general, localized protein phosphorylation and dephosphorylation play important roles in many aspects of spindle assembly and function, but only a few pathways and downstream targets have been characterized. One identified substrate of Polo-like kinase, Kizuna, localizes to centrosomes and its phosphorylation on Thr 379 is required to maintain spindle pole integrity (Oshimori et al. 2006). Protein phosphatases, which counteract the activity of kinases, are also required for correct microtubule organization in the spindle. Regulation of phosphatase activity towards mitotic substrates is perhaps even more complicated, as a limited number of enzymes possess multiple regulatory and targeting subunits. In *C. elegans*, one PP2A-targeting complex called RSA has been identified that regulates both a Kinesin-13 and a TPX2-like MAP at the centrosome, maintaining microtubule stability (Schlaitz et al. 2007). Given that the majority of spindle proteins are regulated likely by phosphorylation, and that other post-translational modifications such as ubiquitination and sumoylation also exist (Kerscher et al. 2006), we are still at the tip of the iceberg with respect to understanding how spindle protein modifications contribute to mitosis.

## 8.10 The Role of Chromosomes: Microtubule Capture

In the classic search and capture model for spindle assembly, microtubules are thought to probe 3D space until they capture and stably attach to kinetochores on the spindle. A key component in this process is the Ndc80 complex, which is required for efficient kinetochore attachment in every model system analyzed (Chapter 10). For example, knockdown of the Nuf2 component of the Ndc80 complex results in a significant disruption of the kinetochore fiber microtubules (DeLuca et al. 2002, 2005). Using this approach it was found that kinetochore microtubule interactions are also important for spindle bipolarity. For example, the monopolar spindles caused by depletion of the Kinesin-13 Kif2a can be rescued by co-depletion of Nuf2, suggesting that spindle bipolarity is influenced by a stable microtubule–kinetochore interaction (Ganem and Compton 2004). In addition, spindle pole organization is dependent on proper integrity of the kinetochore-microtubules (Manning and Compton 2007). Thus although the bulk of the spindle is composed of non-kinetochore microtubules, and spindles can be assembled on chromatin in the absence of kinetochores, when kinetochore microtubules are present they influence the organization of microtubules both in the spindle and at the spindle pole.

## 8.11 Modern Approaches to Study Spindle Assembly

*Computational modeling approaches:* A valuable and complementary approach to study the interactions of microtubules, motors, and regulatory factors in the spindle is to analyze them through computational simulation (Mogilner et al. 2006). Since many spindle components are unknown, and there are a large number of parameters, the most informative theoretical studies have come from those focusing on the interactions of a few simple components. Based on experimental results in *Drosophila* S2 cells showing that Kinesin-14 Ncd localizes to the growing plus ends of spindle microtubules, computer modeling of K-fiber focusing suggested that Ncd could facilitate the capture and transport of kinetochore fibers along centrosomal microtubules (Goshima et al. 2005). Another recent example is one in which spindle organization via a "slide and cluster" mechanism was modeled (Burbank et al. 2007). Chromosome-nucleated microtubules slide outward due to the activity of a Kinesin-5 motor, and are clustered at their minus ends by a minus end-directed motor. Near the chromosomes, microtubules are pulled outward by both motors, but near the poles the clustering pulls microtubules inward and the plus and minus end motors become antagonistic to each other. This model generates a stable, bipolar spindle with focused poles, and importantly does not rely on unproven entities such as the spindle matrix (Burbank et al. 2007). Furthermore, this model has the potential to be readily tested in *Xenopus* egg extract spindles that possess many chromosome-nucleated microtubules. Visualization of the spindle microtubule lattice by low level incorporation of fluorescent tubulin subunits (speckle microscopy) revealed that outward microtubule sliding (flux) slowed near spindle poles only if the minus end-directed motility of dynein was active (Burbank et al. 2007). Thus, combining experimental and theoretical approaches has become a powerful tool to explore spindle assembly mechanisms.

*Large Scale Identification and Functional Analysis of Novel Spindle Components:* In the past, identification of proteins that play important roles in spindle assembly often involved genetic screens, primarily in yeast, *Drosophila*, and *C. elegans* that identified genes causing specific mitotic defects, which could then be cloned and characterized in a variety of systems. Alternative approaches have been to isolate factors based on their biochemical properties, such as association with microtubules or chromosomes, or by using antibodies that recognize proteins with specific localizations, for example at the centromere and kinetochore (Earnshaw and Rothfield 1985).

Recent technological developments have completely revolutionized how mitosis can be analyzed using large-scale proteomic and functional genomic approaches. Advances in mass spectrometry have allowed efficient identification of proteins cut out of gels as well as in complex mixtures found in isolated subcellular structures. The spindle itself as well as components including the centrosome (Andersen et al. 2003), mitotic chromosomes (Gassmann et al. 2005; Uchiyama et al. 2005), and the midbody (Skop et al. 2004), have been isolated and analyzed (Sauer et al. 2005). Perhaps not so surprising is that these studies have revealed nearly 1000 proteins that associate with the mitotic machinery. Many of these factors remain uncharacterized, highlighting the fact that our understanding of mitosis is still quite limited. Mass spectrometry has also been utilized to map mitotic phosphorylation sites, revealing 279 novel phosphorylation sites of known spindle proteins (Nousiainen et al. 2006). Such phosphoproteomic analysis is a key step towards understanding how many and which spindle proteins are regulated by the numerous kinases that act during mitosis (Ubersax et al. 2003; Meraldi et al. 2004; van de Weerdt and Medema 2006).

Proteomic studies are powerful in that they provide lists of potentially important mitotic factors, but they are limited in terms of functional information. Another large-scale approach has been to perform RNA interference screens to specifically and individually deplete a large set of individual proteins in cells, and identify those that cause particular mitotic defects. Such functional screens have been performed in *C. elegans* embryos (Fraser et al. 2000; Gonczy et al. 2000; Sonnichsen et al. 2005), *Drosophila* cell lines (Maiato et al. 2003) and human cell culture systems (Kittler and Buchholz 2005), and have identified factors with previously unappreciated mitotic functions.

Clever genetic screens in yeast have also been used to identify factors important for chromosome segregation. Synthetic lethal screens isolated 211 non-essential deletion mutants that were unable to tolerate defects in kinetochore function (Measday et al. 2005). A screen for factors that become essential upon increased sets of chromosomes (polyploidy) also highlighted factors affecting genomic stability and spindle function (Storchova et al. 2006).

The information obtained from proteomic, functional genomic, and large-scale genetic screens provides a valuable resource for those studying assembly and function of the mitotic spindle. However, these large lists should be viewed as a beginning and not an ending. To achieve a complete understanding, we need to know the temporal and spatial distribution of each protein, its interaction networks, and mechanism of function and regulation. Only then will we begin to fully grasp how the remarkable spindle machine operates at a molecular level to ensure accurate distribution of the genetic material to dividing cells.

**Acknowledgments** The authors would like to thank all members of their lab for constant discussions. We are especially grateful to Anne-Lore Schlaitz and Benjamin Freedman for critical reading of this manuscript. Research in the authors' labs was supported by grants from the NIH (R01GM057839, R01GM065232, and DP1OD000818 to RH and R01GM059618 to CEW), and the ACS (RSG CSM-106128 to CEW). The research in the lab of CEW was supported in part by the Indiana METACyt Initiative of Indiana University, funded in part through a major grant from the Lilly Endowment, Inc.

## References

Al-Bassam, J., van Breugel, M., Harrison, S. C. and Hyman, A. 2006. Stu2p binds tubulin and undergoes an open-to-closed conformational change. J Cell Biol 172: 1009–1022.

Albee, A. J., Tao, W. and Wiese, C. 2006. Phosphorylation of maskin by Aurora-A is regulated by RanGTP and importin beta. J Biol Chem 281: 38293–38301.

Andersen, J. S., Wilkinson, C. J., Mayor, T., Mortensen, P., Nigg, E. A. and Mann, M. 2003. Proteomic characterization of the human centrosome by protein correlation profiling. Nature 426: 570–574.

Andersen, S. S. L., Ashford, A. J., Tournebize, R., Gavet, O., A., S., Hyman, A. A. and Karsenti, E. 1997. Mitotic chromatin regulates phosphorylation of Stathmin/Op18. Nature 389: 640–643.

Antonio, C., Ferby, I., Wilhelm, H., Jones, M., Karsenti, E., Nebreda, A. R. and Vernos, I. 2000. Xkid, a chromokinesin required for chromosome alignment on the metaphase plate. Cell 102: 425–435.

Azimzadeh, J. and Bornens, M. 2007. Structure and duplication of the centrosome. J Cell Sci 120: 2139–2142.

Basto, R., Lau, J., Vinogradova, T., Gardiol, A., Woods, C. G., Khodjakov, A. and Raff, J. W. 2006. Flies without centrioles. Cell 125: 1375–1386.

Bellanger, J. M. and Gonczy, P. 2003. TAC-1 and ZYG-9 form a complex that promotes microtubule assembly in C. elegans embryos. Curr Biol 13: 1488–1498.

Belmont, L., Deacon, H. W. and Mitchison, T. J. 1996. Catastrophic revelations about Op18/stathmin. Trends Biochem Sci 21: 197–198.

Belmont, L. D., Hyman, A. A., Sawin, K. E. and Mitchison, T. J. 1990. Real-time visualization of cell cycle dependent changes in microtubule dynamics in cytoplasmic extracts. Cell 62: 579–589.

Belmont, L. D. and Mitchison, T. J. 1996. Identification of a protein that interacts with tubulin dimers and increases the catastrophe rate of microtubules. Cell 84: 623–631.

Blower, M. D., Nachury, M., Heald, R. and Weis, K. 2005. A Rae1-containing ribonucleoprotein complex is required for mitotic spindle assembly. Cell 121: 223–234.
Brittle, A. L. and Ohkura, H. 2005. Mini spindles, the XMAP215 homologue, suppresses pausing of interphase microtubules in Drosophila. Embo J 24: 1387–1396.
Brouhard, G. J. and Hunt, A. J. 2005. Microtubule movements on the arms of mitotic chromosomes: polar ejection forces quantified in vitro. Proc Natl Acad Sci USA 102: 13903–13908.
Brown, J. A., Bharathi, A., Ghosh, A., Whalen, W., Fitzgerald, E. and Dhar, R. 1995. A mutation in the Schizosaccharomyces pombe rae1 gene causes defects in poly(A)$^+$ RNA export and in the cytoskeleton. J Biol Chem 270: 7411–7419.
Brust-Mascher, I., Civelekoglu-Scholey, G., Kwon, M., Mogilner, A. and Scholey, J. M. 2004. Model for anaphase B: role of three mitotic motors in a switch from poleward flux to spindle elongation. Proc Natl Acad Sci USA 101: 15938–15943.
Budde, P. P., Kumagai, A., Dunphy, W. G. and Heald, R. 2001. Regulation of Op18 during spindle assembly in Xenopus egg extracts. J Cell Biol 153: 149–158.
Burbank, K. S., Mitchison, T. J. and Fisher, D. S. 2007. Slide-and-cluster models for spindle assembly. Curr Biol 17: 1373–1383.
Cameron, L. A., Yang, G., Cimini, D., Canman, J. C., Kisurina-Evgenieva, O., Khodjakov, A., Danuser, G. and Salmon, E. D. 2006. Kinesin 5-independent poleward flux of kinetochore microtubules in PtK1 cells. J Cell Biol 173: 173–179.
Carazo-Salas, R. E., Gruss, O. J., Mattaj, I. W. and Karsenti, E. 2001. Ran-GTP coordinates regulation of microtubule nucleation and dynamics during mitotic-spindle assembly. Nat Cell Biol 3: 228–234.
Carminati, J. L. and Stearns, T. 1997. Microtubules orient the mitotic spindle in yeast through dynein-dependent interactions with the cell cortex. J Cell Biol 138: 629–641.
Cassimeris, L. 1993. Regulation of microtubule dynamic instability. Cell Motil Cytoskeleton 26: 275–281.
Cassimeris, L. 2002. The oncoprotein 18/stathmin family of microtubule destabilizers. Curr Opin Cell Biol 14: 18–24.
Cassimeris, L. and Morabito, J. 2004. TOGp, the human homolog of XMAP215/Dis1, is required for centrosome integrity, spindle pole organization, and bipolar spindle assembly. Mol Biol Cell 15: 1580–1590.
Chang, P., Coughlin, M. and Mitchison, T. J. 2005. Tankyrase-1 polymerization of poly(-ADP-ribose) is required for spindle structure and function. Nat Cell Biol 7: 1133–1139.
Chang, P., Jacobson, M. K. and Mitchison, T. J. 2004. Poly(ADP-ribose) is required for spindle assembly and structure. Nature 432: 645–649.
Chang, W., Dynek, J. N. and Smith, S. 2005. NuMA is a major acceptor of poly(ADP-ribosyl)ation by tankyrase 1 in mitosis. Biochem J 391: 177–184.
Compton, D. A. 2005. Regulation of mitosis by poly(ADP-ribosyl)ation. Biochem J 391: e5–6.
Compton, D. A. and Cleveland, D. W. 1993. NuMA is required for the proper completion of mitosis. J Cell Biol 120: 947–957.
Compton, D. A., Szilak, I. and Cleveland, D. W. 1992. Primary structure of NuMA, an intranuclear protein that defines a novel pathway for segregation of proteins at mitosis. J Cell Biol 116: 1395–1408.
Cottingham, F. R., Gheber, L., Miller, D. L. and Hoyt, M. A. 1999. Novel roles for saccharomyces cerevisiae mitotic spindle motors. J Cell Biol 147: 335–350.
Cullen, C. F., Deak, P., Glover, D. M. and Ohkura, H. 1999. mini spindles: A gene encoding a conserved microtubule-associated protein required for the integrity of the mitotic spindle in Drosophila. J Cell Biol 146: 1005–1018.
Cytrynbaum, E. N., Scholey, J. M. and Mogilner, A. 2003. A force balance model of early spindle pole separation in Drosophila embryos. Biophys J 84: 757–769.
Dammermann, A., Desai, A. and Oegema, K. 2003. The minus end in sight. Curr Biol 13: R614–624.

DeLuca, J. G., Dong, Y., Hergert, P., Strauss, J., Hickey, J. M., Salmon, E. D. and McEwen, B. F. 2005. Hec1 and nuf2 are core components of the kinetochore outer plate essential for organizing microtubule attachment sites. Mol Biol Cell 16: 519–531.

DeLuca, J. G., Moree, B., Hickey, J. M., Kilmartin, J. V. and Salmon, E. D. 2002. hNuf2 inhibition blocks stable kinetochore-microtubule attachment and induces mitotic cell death in HeLa cells. J Cell Biol 159: 549–555.

Desai, A. and Mitchison, T. J. 1997. Microtubule polymerization dynamics. Annu Rev Cell Dev Biol 13: 83–117.

DeZwaan, T. M., Ellingson, E., Pellman, D. and Roof, D. M. 1997. Kinesin-related KIP3 of Saccharomyces cerevisiae is required for a distinct step in nuclear migration. J. Cell Biol. 138: 1023–1040.

Doxsey, S., McCollum, D. and Theurkauf, W. 2005. Centrosomes in cellular regulation. Annu Rev Cell Dev Biol 21: 411–434.

Drechsel, D. N. and Kirschner, M. W. 1994. The minimum GTP cap required to stabilize microtubules. Curr Biol 4: 1053–1061.

Dujardin, D., Wacker, U. I., Moreau, A., Schroer, T. A., Rickard, J. E. and De Mey, J. R. 1998. Evidence for a role of CLIP-170 in the establishment of metaphase chromosome alignment. J Cell Biol 141: 849–862.

Earnshaw, W. C. and Rothfield, N. 1985. Identification of a family of human centromere proteins using autoimmune sera from patients with scleroderma. Chromosoma 91: 313–321.

Echeverri, C. J., Paschal, B. M., Vaughan, K. T. and Vallee, R. B. 1996. Molecular characterization of the 50-kD subunit of dynactin reveals function for the complex in chromosome alignment and spindle organization during mitosis. J Cell Biol. 132: 617–633.

Eggert, U. S., Mitchison, T. J. and Field, C. M. 2006. Animal cytokinesis: from parts list to mechanisms. Annu Rev Biochem 75: 543–566.

Ems-McClung, S. C., Hertzer, K. M., Zhang, X., Miller, M. W. and Walczak, C. E. 2007. The interplay of the N- and C-terminal domains of MCAK control microtubule depolymerization activity and spindle assembly. Mol Biol Cell 18: 282–294.

Ems-McClung, S. C., Zheng, Y. and Walczak, C. E. 2004. Importin alpha/beta and Ran-GTP regulate XCTK2 microtubule binding through a bipartite nuclear localization signal. Mol Biol Cell 15: 46–57.

Endow, S. A., Chandra, R., Komma, D. J., Yamamoto, A. H. and Salmon, E. D. 1994. Mutants of the Drosophila ncd microtubule motor protein cause centrosomal and spindle pole defects in mitosis. J Cell Sci 107: 859–867.

Enos, A. P. and Morris, N. R. 1990. Mutation of a gene that encodes a kinesin-like protein blocks nuclear division in A. nidulans. Cell 60: 1019–1027.

Eshel, D., Urrestarazu, L. A., Vissers, S., Jauniaux, J.-C., van Vliet-Reedijk, J. C., Planta, R. J. and Gibbon, I. R. 1993. Cytoplasmic dynein is required for normal nuclear segregation in yeast. Proc Natl Acad Sci USA 90: 11172–11176.

Eyers, P. A., Erikson, E., Chen, L. G. and Maller, J. L. 2003. A novel mechanism for activation of the protein kinase Aurora A. Curr Biol 13: 691–697.

Fant, X., Merdes, A. and Haren, L. 2004. Cell and molecular biology of spindle poles and NuMA. Int Rev Cytol 238: 1–57.

Fraser, A. G., Kamath, R. S., Zipperlen, P., Martinez-Campos, M., Sohrmann, M. and Ahringer, J. 2000. Functional genomic analysis of C. elegans chromosome I by systematic RNA interference. Nature 408: 325–330.

Funabiki, H. and Murray, A. W. 2000. The Xenopus chromokinesin Xkid is essential for metaphase chromosome alignment and must be degraded to allow anaphase chromosome movement. Cell 102: 411–424.

Gadea, B. B. and Ruderman, J. V. 2006. Aurora B is required for mitotic chromatin-induced phosphorylation of Op18/Stathmin. Proc Natl Acad Sci USA 103: 4493–4498.

Gaetz, J. and Kapoor, T. M. 2004. Dynein/dynactin regulate metaphase spindle length by targeting depolymerizing activities to spindle poles. J Cell Biol 166: 465–471.

Gaglio, T., Dionne, M. A. and Compton, D. A. 1997. Mitotic spindle poles are organized by structural and motor proteins in addition to centrosomes. J Cell Biol 138: 1055–1066.
Gaglio, T., Saredi, A., Bingham, J. B., Hasbani, M. J., Gill, S. R., Schroer, T. A. and Compton, D. A. 1996. Opposing motor activities are required for the organization of the mammalian mitotic spindle pole. J Cell Biol 135: 399–414.
Ganem, N. J. and Compton, D. A. 2004. The KinI kinesin Kif2a is required for bipolar spindle assembly through a functional relationship with MCAK. J Cell Biol 166: 473–478.
Ganem, N. J., Upton, K. and Compton, D. A. 2005. Efficient mitosis in human cells lacking poleward microtubule flux. Curr Biol 15: 1827–1832.
Garcia, M. A., Koonrugsa, N. and Toda, T. 2002. Spindle-kinetochore attachment requires the combined action of Kin I-like Klp5/6 and Alp14/Dis1-MAPs in fission yeast. Embo J 21: 6015–6024.
Garcia, M. A., Koonrugsa, N. and Toda, T. 2002. Two kinesin-like Kin I family proteins in fission yeast regulate the establishment of metaphase and the onset of anaphase A. Curr Biol 12: 610–621.
Gard, D. L. and Kirschner, M. W. 1987. A microtubule-associated protein from Xenopus eggs that specifically promotes assembly at the plus-end. J Cell Biol 105: 2203–2215.
Garrett, S., Auer, K., Compton, D. A. and Kapoor, T. M. 2002. hTPX2 is required for normal spindle morphology and centrosome integrity during vertebrate cell division. Curr Biol 12: 2055–2059.
Gassmann, R., Carvalho, A., Henzing, A. J., Ruchaud, S., Hudson, D. F., Honda, R., Nigg, E. A., Gerloff, D. L. and Earnshaw, W. C. 2004. Borealin: a novel chromosomal passenger required for stability of the bipolar mitotic spindle. J Cell Biol 166: 179–191.
Gassmann, R., Henzing, A. J. and Earnshaw, W. C. 2005. Novel components of human mitotic chromosomes identified by proteomic analysis of the chromosome scaffold fraction. Chromosoma 113: 385–397.
Gergely, F., Draviam, V. M. and Raff, J. W. 2003. The ch-TOG/XMAP215 protein is essential for spindle pole organization in human somatic cells. Genes Dev 17: 336–341.
Giet, R., McLean, D., Descamps, S., Lee, M. J., Raff, J. W., Prigent, C. and Glover, D. M. 2002. Drosophila Aurora A kinase is required to localize D-TACC to centrosomes and to regulate astral microtubules. J Cell Biol 156: 437–451.
Giet, R., Uzbekov, R., Cubizolles, F., Le Guellec, K. and Prigent, C. 1999. The Xenopus laevis aurora-related protein kinase pEg2 associates with and phosphorylates the kinesin-related protein XlEg5. J Biol Chem 274: 15005–15013.
Gliksman, N. R., Skibbens, R. V. and Salmon, E. D. 1993. How the transition frequencies of microtubule dynamic instability (nucleation, catastrophe, and rescue) regulate microtubule dynamics in interphase and mitosis: analysis using a Monte Carlo computer simulation. Mol Biol Cell 4: 1035–1050.
Gönczy P, Pichler S, Kirkham M, Hyman AA. 1999. Cytoplasmic dynein is required for distinct aspects of MTOC positioning, including centrosome separation, in the one cell stage Caenorhabditis elegans embryo. J Cell Biol 4(147): 135–50.
Gonczy, P., et al. 2000. Functional genomic analysis of cell division in C. elegans using RNAi of genes on chromosome III. Nature 408: 331–336.
Gordon, M. B., Howard, L. and Compton, D. A. 2001. Chromosome movement in mitosis requires microtubule anchorage at spindle poles. J Cell Biol 152: 425–434.
Goshima, G., Nedelec, F. and Vale, R. D. 2005. Mechanisms for focusing mitotic spindle poles by minus end-directed motor proteins. J Cell Biol 171: 229–240.
Goshima, G. and Vale, R. D. 2003. The roles of microtubule-based motor proteins in mitosis: comprehensive RNAi analysis in the Drosophila S2 cell line. J Cell Biol 162: 1003–1016.
Goshima, G., Wollman, R., Stuurman, N., Scholey, J. M. and Vale, R. D. 2005. Length control of the metaphase spindle. Curr Biol 15: 1979–1988.

Grava, S., Schaerer, F., Faty, M., Philippsen, P. and Barral, Y. 2006. Asymmetric recruitment of dynein to spindle poles and microtubules promotes proper spindle orientation in yeast. Dev Cell 10: 425–439.

Green, R. A. and Kaplan, K. B. 2003. Chromosome instability in colorectal tumor cells is associated with defects in microtubule plus-end attachments caused by a dominant mutation in APC. J Cell Biol 163: 949–961.

Green, R. A., Wollman, R. and Kaplan, K. B. 2005. APC and EB1 function together in mitosis to regulate spindle dynamics and chromosome alignment. Mol Biol Cell 16: 4609–4622.

Grishchuk, E. L., Spiridonov, I. S. and McIntosh, J. R. 2007. Mitotic chromosome biorientation in fission yeast is enhanced by dynein and a minus-end-directed, kinesin-like protein. Mol Biol Cell 18: 2216–2225.

Groen, A. C., Cameron, L. A., Coughlin, M., Miyamoto, D. T., Mitchison, T. J. and Ohi, R. 2004. XRHAMM functions in ran-dependent microtubule nucleation and pole formation during anastral spindle assembly. Curr Biol 14: 1801–1811.

Gruneberg, U., Neef, R., Li, X., Chan, E. H., Chalamalasetty, R. B., Nigg, E. A. and Barr, F. A. 2006. KIF14 and citron kinase act together to promote efficient cytokinesis. J Cell Biol 172: 363–372.

Gruss, O. J., et al. 2001. Ran induces spindle assembly by reversing the inhibitory effect of importin alpha on TPX2 activity. Cell 104: 83–93.

Gruss, O. J., Wittmann, M., Yokoyama, H., Pepperkok, R., Kufer, T., Sillje, H., Karsenti, E., Mattaj, I. W. and Vernos, I. 2002. Chromosome-induced microtubule assembly mediated by TPX2 is required for spindle formation in HeLa cells. Nat Cell Biol 4: 871–879.

Gupta, M. L., Jr., Carvalho, P., Roof, D. M. and Pellman, D. 2006. Plus end-specific depolymerase activity of Kip3, a kinesin-8 protein, explains its role in positioning the yeast mitotic spindle. Nat Cell Biol 8: 913–923.

Hagan, I. and Yanagida, M. 1990. Novel potential mitotic motor protein encoded by the fission yeast $cut7^+$ gene. Nature 347: 563–566.

Hayes, S. 2006. Lamin in the spindle matrix. Nat Cell Biol 8: 550.

Heald, R., Tournebize, R., Blank, T., Sandaltzopoulos, R., Becker, P., Hyman, A. and Karsenti, E. 1996. Self-organization of microtubules into bipolar spindles around artificial chromosomes in Xenopus egg extracts. Nature 382: 420–425.

Heald, R., Tournebize, R., Habermann, A., Karsenti, E. and Hyman, A. 1997. Spindle assembly in Xenopus egg extracts: respective roles of centrosomes and microtubule self-organization. J Cell Biol 138: 615–628.

Heck, M. M. S., Pereira, A., Pesavento, P., Yannoni, Y., Spradling, A. C. and Goldstein, L. S. B. 1993. The kinesin-like protein KLP61F is essential for mitosis in Drosophila. J Cell Biol. 123: 665–679.

Hetzer, M., Gruss, O. J. and Mattaj, I. W. 2002. The Ran GTPase as a marker of chromosome position in spindle formation and nuclear envelope assembly. Nat Cell Biol 4: E177–184.

Holmfeldt, P., Brannstrom, K., Stenmark, S. and Gullberg, M. 2006. Aneugenic activity of Op18/stathmin is potentiated by the somatic Q18–>e mutation in leukemic cells. Mol Biol Cell 17: 2921–2930.

Holmfeldt, P., Brattsand, G. and Gullberg, M. 2002. MAP4 counteracts microtubule catastrophe promotion but not tubulin-sequestering activity in intact cells. Curr Biol 12: 1034–1039.

Holmfeldt, P., Stenmark, S. and Gullberg, M. 2004. Differential functional interplay of TOGp/XMAP215 and the KinI kinesin MCAK during interphase and mitosis. Embo J 23: 627–637.

Howard, J. and Hyman, A. A. 2003. Dynamics and mechanics of the microtubule plus end. Nature 422: 753–758.

Hoyt, M. A., He, L., Totis, L. and Saunders, W. S. 1993. Loss of function of *Saccharomyces cerevisiae* kinesin-related *CIN8* and *KIP1* is suppressed by *KAR3* motor domain mutations. Genetics 135: 35–44.

Huang, B. and Huffaker, T. C. 2006. Dynamic microtubules are essential for efficient chromosome capture and biorientation in S. cerevisiae. J Cell Biol 175: 17–23.

Joukov, V., Groen, A. C., Prokhorova, T., Gerson, R., White, E., Rodriguez, A., Walter, J. C. and Livingston, D. M. 2006. The BRCA1/BARD1 heterodimer modulates ran-dependent mitotic spindle assembly. Cell 127: 539–552.

Kalab, P., Weis, K. and Heald, R. 2002. Visualization of a Ran-GTP gradient in interphase and mitotic Xenopus egg extracts. Science 295: 2452–2456.

Kapitein, L. C., Peterman, E. J., Kwok, B. H., Kim, J. H., Kapoor, T. M. and Schmidt, C. F. 2005. The bipolar mitotic kinesin Eg5 moves on both microtubules that it crosslinks. Nature 435: 114–118.

Kapoor, T. M. and Compton, D. A. 2002. Searching for the middle ground: mechanisms of chromosome alignment during mitosis. J Cell Biol 157: 551–556.

Kapoor, T. M., Lampson, M. A., Hergert, P., Cameron, L., Cimini, D., Salmon, E. D., McEwen, B. F. and Khodjakov, A. 2006. Chromosomes can congress to the metaphase plate before biorientation. Science 311: 388–391.

Kapoor, T. M., Mayer, T. U., Coughlin, M. L. and Mitchison, T. J. 2000. Probing spindle assembly mechanisms with monastrol, a small molecule inhibitor of the mitotic kinesin, Eg5. J Cell Biol 150: 975–988.

Kapoor, T. M. and Mitchison, T. J. 2001. Eg5 is static in bipolar spindles relative to tubulin: evidence for a static spindle matrix. J Cell Biol 154: 1125–1133.

Karsenti, E. and Vernos, I. 2001. The mitotic spindle: a self-made machine. Science 294: 543–547.

Kashina, A. S., Baskin, R. J., Cole, D. G., Wedaman, K. P., Saxton, W. M. and Scholey, J. M. 1996. A bipolar kinesin. Nature 379: 270–272.

Kelly, A. E., Sampath, S. C., Maniar, T. A., Woo, E. M., Chait, B. T. and Funabiki, H. 2007. Chromosomal enrichment and activation of the aurora B pathway are coupled to spatially regulate spindle assembly. Dev Cell 12: 31–43.

Kerscher, O., Felberbaum, R. and Hochstrasser, M. 2006. Modification of proteins by ubiquitin and ubiquitin-like proteins. Annu Rev Cell Dev Biol 22: 159–180.

Kerssemakers, J. W., Munteanu, E. L., Laan, L., Noetzel, T. L., Janson, M. E. and Dogterom, M. 2006. Assembly dynamics of microtubules at molecular resolution. Nature 442: 709–712.

Khodjakov, A., Cole, R. W., Oakley, B. R. and Rieder, C. L. 2000. Centrosome-independent mitotic spindle formation in vertebrates. Curr Biol 10: 59–67.

Khodjakov, A., Copenagle, L., Gordon, M. B., Compton, D. A. and Kapoor, T. M. 2003. Minus-end capture of preformed kinetochore fibers contributes to spindle morphogenesis. J Cell Biol 160: 671–683.

Khodjakov, A. and Rieder, C. L. 1999. The sudden recruitment of gamma-tubulin to the centrosome at the onset of mitosis and its dynamic exchange throughout the cell cycle, do not require microtubules. J Cell Biol 146: 585–596.

Khodjakov, A. and Rieder, C. L. 2001. Centrosomes enhance the fidelity of cytokinesis in vertebrates and are required for cell cycle progression. J Cell Biol 153: 237–242.

King, S. M. 2002. Dyneins motor on in plants. Traffic 3: 930–931.

Kinoshita, K., Arnal, I., Desai, A., Drechsel, D. N. and Hyman, A. A. 2001. Reconstitution of physiological microtubule dynamics using purified components. Science 294: 1340–1343.

Kinoshita, K., Noetzel, T. L., Pelletier, L., Mechtler, K., Drechsel, D. N., Schwager, A., Lee, M., Raff, J. W. and Hyman, A. A. 2005. Aurora A phosphorylation of TACC3/maskin is required for centrosome-dependent microtubule assembly in mitosis. J Cell Biol 170: 1047–1055.

Kirschner, M. W. and Mitchison, T. J. 1986. Beyond self assembly: from microtubules to morphogenesis. Cell 45: 329–342.

Kittler, R. and Buchholz, F. 2005. Functional genomic analysis of cell division by endoribonuclease-prepared siRNAs. Cell Cycle 4: 564–567.

Kline-Smith, S. L., Khodjakov, A., Hergert, P. and Walczak, C. E. 2004. Depletion of centromeric MCAK leads to chromosome congression and segregation defects due to improper kinetochore attachments. Mol Biol Cell 15: 1146–1159.

Kline-Smith, S. L. and Walczak, C. E. 2002. The Microtubule-destabilizing Kinesin XKCM1 Regulates Microtubule Dynamic Instability in Cells. Mol Biol Cell 13: 2718–2731.

Koffa, M. D., Casanova, C. M., Santarella, R., Kocher, T., Wilm, M. and Mattaj, I. W. 2006. HURP is part of a Ran-dependent complex involved in spindle formation. Curr Biol 16: 743–754.

Kosco, K. A., Pearson, C. G., Maddox, P. S., Wang, P. J., Adams, I. R., Salmon, E. D., Bloom, K. and Huffaker, T. C. 2001. Control of microtubule dynamics by Stu2p is essential for spindle orientation and metaphase chromosome alignment in yeast. Mol Biol Cell 12: 2870–2880.

Kuntziger, T., Gavet, O., Manceau, V., Sobel, A. and Bornens, M. 2001. Stathmin/Op18 phosphorylation is regulated by microtubule assembly. Mol Biol Cell 12: 437–448.

Kuriyama, R. and Borisy, G. G. 1981. Microtubule-nucleating activity of centrosomes in chinese hamster ovary cells is independent of the centriole cycle but coupled to the mitotic cycle. J Cell Biol. 91: 822–826.

Kuriyama, R., Gustus, C., Terada, Y., Uetake, Y. and Matuliene, J. 2002. CHO1, a mammalian kinesin-like protein, interacts with F-actin and is involved in the terminal phase of cytokinesis. J Cell Biol 156: 783–790.

Lansbergen, G. and Akhmanova, A. 2006. Microtubule plus end: a hub of cellular activities. Traffic 7: 499–507.

Larsson, N., Marklund, U., Gradin, H. M., Brattsand, G. and Gullberg, M. 1997. Control of microtubule dynamics by oncoprotein 18: dissection of the regulatory role of multisite phosphorylation during mitosis. Mol Cell Biol 17: 5530–5539.

Lee, M. J., Gergely, F., Jeffers, K., Peak-Chew, S. Y. and Raff, J. W. 2001. Msps/XMAP215 interacts with the centrosomal protein D-TACC to regulate microtubule behaviour. Nat Cell Biol 3: 643–649.

Levesque, A. A. and Compton, D. A. 2001. The chromokinesin Kid is necessary for chromosome arm orientation and oscillation, but not congression, on mitotic spindles. J Cell Biol 154: 1135–1146.

Lew, D. J. and Burke, D. J. 2003. The spindle assembly and spindle position checkpoints. Annu Rev Genet 37: 251–282.

Li, J., Lee, W. L. and Cooper, J. A. 2005. NudEL targets dynein to microtubule ends through LIS1. Nat Cell Biol 7: 686–690.

Li, Y., Yeh, E., Hays, T. and Bloom, K. 1993. Disruption of mitotic spindle orientation in a yeast dynein mutant. Proc. Natl. Acad. Sci. USA 90: 10096–10100.

Liang, Y., et al. 2007. Nudel modulates kinetochore association and function of cytoplasmic dynein in M phase. Mol Biol Cell 18: 2656–2666.

Maiato, H., Fairley, E. A., Rieder, C. L., Swedlow, J. R., Sunkel, C. E. and Earnshaw, W. C. 2003. Human CLASP1 is an outer kinetochore component that regulates spindle microtubule dynamics. Cell 113: 891–904.

Maiato, H., Khodjakov, A. and Rieder, C. L. 2005. Drosophila CLASP is required for the incorporation of microtubule subunits into fluxing kinetochore fibres. Nat Cell Biol 7: 42–47.

Maiato, H., Khodjakov, A. and Rieder, C. L. 2005. MAST/Orbit is required for microtubule subunit incorporation into fluxing kinetochore fibers. Nat Cell Biol.

Maiato, H., Rieder, C. L. and Khodjakov, A. 2004. Kinetochore-driven formation of kinetochore fibers contributes to spindle assembly during animal mitosis. J Cell Biol 167: 831–840.

Maiato, H., Sampaio, P., Lemos, C. L., Findlay, J., Carmena, M., Earnshaw, W. C. and Sunkel, C. E. 2002. MAST/Orbit has a role in microtubule-kinetochore attachment and is essential for chromosome alignment and maintenance of spindle bipolarity. J Cell Biol 157: 749–760.

Maiato, H., Sunkel, C. E. and Earnshaw, W. C. 2003. Dissecting mitosis by RNAi in Drosophila tissue culture cells. Biol Proced Online 5: 153–161.

Maney, T., Hunter, A. W., Wagenbach, M. and Wordeman, L. 1998. Mitotic centromere-associated kinesin is important for anaphase chromosome segregation. J. Cell Biol 142: 787–801.

Manning, A. L. and Compton, D. A. 2007. Mechanisms of spindle-pole organization are influenced by kinetochore activity in mammalian cells. Curr Biol 17: 260–265.

Manning, A. L., Ganem, N. J., Bakhoum, S. F., Wagenbach, M., Wordeman, L. and Compton, D. A. 2007. The Kinesin-13 Proteins Kif2a, Kif2b, and Kif2c/MCAK Have Distinct Roles during Mitosis in Human Cells. Mol Biol Cell 18: 2970–2979.

Maresca, T. J., Niederstrasser, H., Weis, K. and Heald, R. 2005. Xnf7 contributes to spindle integrity through its microtubule-bundling activity. Curr Biol 15: 1755–1761.

Mastronarde, D. N., McDonald, K. L., Ding, R. and McIntosh, J. R. 1993. Interpolar spindle microtubules in PTK cells. J Cell Biol 123: 1475–1489.

Matuliene, J. and Kuriyama, R. 2002. Kinesin-like protein CHO1 is required for the formation of midbody matrix and the completion of cytokinesis in mammalian cells. Mol Biol Cell 13: 1832–1845.

Mayer, T. U., Kapoor, T. M., Haggarty, S. J., King, R. W., Schreiber, S. L. and Mitchison, T. J. 1999. Small molecule inhibitor of mitotic spindle bipolarity identified in a phenotype-based screen. Science 286: 971–974.

Mayr, M. I., Hummer, S., Bormann, J., Gruner, T., Adio, S., Woehlke, G. and Mayer, T. U. 2007. The human kinesin Kif18A is a motile microtubule depolymerase essential for chromosome congression. Curr Biol 17: 488–498.

Mazumdar, M., Sundareshan, S. and Misteli, T. 2004. Human chromokinesin KIF4A functions in chromosome condensation and segregation. J Cell Biol 166: 613–620.

McDonald, H. B. and Goldstein, L. S. 1990. Identification and characterization of a gene encoding a kinesin-like protein in Drosophila. Cell 61: 991–1000.

McDonald, H. B., Stewart, R. J. and Goldstein, L. S. B. 1990. The kinesin like ncd protein is a minus end directed microtubule motor. Cell 63: 1159–1165.

McDonald, K. L., O'Toole, E. T., Mastronarde, D. N. and McIntosh, J. R. 1992. Kinetochore microtubules in PTK cells. J Cell Biol 118: 369–383.

McEwen, B. F., Chan, G. K., Zubrowski, B., Savoian, M. S., Sauer, M. T. and Yen, T. J. 2001. CENP-E is essential for reliable bioriented spindle attachment, but chromosome alignment can be achieved via redundant mechanisms in mammalian cells. Mol Biol Cell 12: 2776–2789.

McIntosh, J. R., Grishchuk, E. L. and West, R. R. 2002. Chromosome-microtubule interactions during mitosis. Annu Rev Cell Dev Biol 18: 193–219.

Measday, V., et al. 2005. Systematic yeast synthetic lethal and synthetic dosage lethal screens identify genes required for chromosome segregation. Proc Natl Acad Sci U S A 102: 13956–13961.

Meluh, P. B. and Rose, M. D. 1990. KAR3, a kinesin-related gene required for yeast nuclear fusion. Cell 60: 1029–1041.

Meraldi, P., Honda, R. and Nigg, E. A. 2004. Aurora kinases link chromosome segregation and cell division to cancer susceptibility. Curr Opin Genet Dev 14: 29–36.

Merdes, A. and Cleveland, D. W. 1997. Pathways of spindle pole formation: different mechanisms; conserved components. J Cell Biol 138: 953–956.

Merdes, A., Heald, R., Samejima, K., Earnshaw, W. C. and Cleveland, D. W. 2000. Formation of spindle poles by dynein/dynactin-dependent transport of NuMA. J Cell Biol 149: 851–862.

Merdes, A., Ramyar, K., Vechio, J. D. and Cleveland, D. W. 1996. A complex of NuMA and cytoplasmic dynein is essential for mitotic spindle assembly. Cell 87: 447–458.

Miller, R. K., Cheng, S. C. and Rose, M. D. 2000. Bim1p/Yeb1p mediates the Kar9p-dependent cortical attachment of cytoplasmic microtubules. Mol Biol Cell 11: 2949–2959.

Miller, R. K., Matheos, D. and Rose, M. D. 1999. The cortical localization of the microtubule orientation protein, Kar9p, is dependent upon actin and proteins required for polarization. J Cell Biol 144: 963–975.

Misek, D. E., Chang, C. L., Kuick, R., Hinderer, R., Giordano, T. J., Beer, D. G. and Hanash, S. M. 2002. Transforming properties of a Q18–>E mutation of the microtubule regulator Op18. Cancer Cell 2: 217–228.

Mitchison, T. and Kirschner, M. 1984. Dynamic instability of microtubule growth. Nature 312: 237–242.

Mitchison, T. J. 1993. Localization of an exchangeable GTP binding site at the plus end of microtubules. Science 261: 1044–1047.

Miyamoto, D. T., Perlman, Z. E., Burbank, K. S., Groen, A. C. and Mitchison, T. J. 2004. The kinesin Eg5 drives poleward microtubule flux in Xenopus egg extract spindles. J Cell Biol. 167: 813–818.

Mogilner, A., Wollman, R., Civelekoglu-Scholey, G. and Scholey, J. 2006. Modeling mitosis. Trends Cell Biol 16: 88–96.

Morales-Mulia, S. and Scholey, J. M. 2005. Spindle pole organization in Drosophila S2 cells by dynein, abnormal spindle protein (Asp), and KLP10A. Mol Biol Cell 16: 3176–3186.

Mountain, V., Simerly, C., Howard, L., Ando, A., Schatten, G. and Compton, D. A. 1999. The kinesin-related protein, HSET, opposes the activity of Eg5 and cross-links microtubules in the mammalian mitotic spindle. J Cell Biol 147: 351–366.

Nachury, M. V., Maresca, T. J., Salmon, W. C., Waterman-Storer, C. M., Heald, R. and Weis, K. 2001. Importin beta is a mitotic target of the small GTPase Ran in spindle assembly. Cell 104: 95–106.

Nelson, C. R. and Szauter, P. 1992. Timing of mitotic chromosome loss caused by the ncd mutation of Drosophila melanogaster. Cell Motil Cytoskeleton 23: 34–44.

Niethammer, P., Kronja, I., Kandels-Lewis, S., Rybina, S., Bastiaens, P. and Karsenti, E. 2007. Discrete states of a protein interaction network govern interphase and mitotic microtubule dynamics. PLoS Biol 5: e29.

Nigg, E. A. 2001. Mitotic kinases as regulators of cell division and its checkpoints. Nat Rev Mol Cell Biol 2: 21–32.

Noetzel, T. L., Drechsel, D. N., Hyman, A. A. and Kinoshita, K. 2005. A comparison of the ability of XMAP215 and tau to inhibit the microtubule destabilizing activity of XKCM1. Philos Trans R Soc Lond B Biol Sci 360: 591–594.

Nogales, E. 1999. A structural view of microtubule dynamics. Cell Mol Life Sci 56: 133–142.

Nogales, E., Whittaker, M., Milligan, R. A. and Downing, K. H. 1999. High-resolution model of the microtubule. Cell 96: 79–88.

Nousiainen, M., Sillje, H. H., Sauer, G., Nigg, E. A. and Korner, R. 2006. Phosphoproteome analysis of the human mitotic spindle. Proc Natl Acad Sci USA 103: 5391–5396.

O'Brien, L. L., Albee, A. J., Liu, L., Tao, W., Dobrzyn, P., Lizarraga, S. B. and Wiese, C. 2005. The Xenopus TACC homologue, maskin, functions in mitotic spindle assembly. Mol Biol Cell 16: 2836–2847.

O'Connell, C. B. and Khodjakov, A. L. 2007. Cooperative mechanisms of mitotic spindle formation. J Cell Sci 120: 1717–1722.

O'Connell, M. J., Krien, M. J. and Hunter, T. 2003. Never say never. The NIMA-related protein kinases in mitotic control. Trends Cell Biol 13: 221–228.

O'Connell, M. J., Meluh, P. B., Rose, M. D. and Morris, N. R. 1993. Suppression of the bimC4 mitotic spindle defect by deletion of klpA, a gene encoding a KAR3-related kinesin-like protein in Aspergillus nidulans. J Cell Biol 120: 153–162.

Ohi, R., Burbank, K., Liu, Q. and Mitchison, T. J. 2007. Nonredundant functions of Kinesin-13s during meiotic spindle assembly. Curr Biol 17: 953–959.

Oshimori, N., Ohsugi, M. and Yamamoto, T. 2006. The Plk1 target Kizuna stabilizes mitotic centrosomes to ensure spindle bipolarity. Nat Cell Biol 8: 1095–1101.

Ouchi, M., et al. 2004. BRCA1 phosphorylation by Aurora-A in the regulation of G2 to M transition. J Biol Chem 279: 19643–19648.

Pickett-Heaps, J., McIntosh, J. R. and Porter, K. R. 1982. Rethinking mitosis. Cell 29: 729–744.

Pickett-Heaps, J. D. 1986. Mitotic mechanisms: an alternative view. Trends Biochem Sci 11: 504–507.

Popov, A. V., Severin, F. and Karsenti, E. 2002. XMAP215 is required for the microtubule-nucleating activity of centrosomes. Curr Biol 12: 1326–1330.

Putkey, F. R., Cramer, T., Morphew, M. K., Silk, A. D., Johnson, R. S., McIntosh, J. R. and Cleveland, D. W. 2002. Unstable kinetochore-microtubule capture and chromosomal instability following deletion of CENP-E. Dev Cell 3: 351–365.

Qi, H., et al. 2004. Megator, an essential coiled-coil protein that localizes to the putative spindle matrix during mitosis in Drosophila. Mol Biol Cell 15: 4854–4865.

Rath, U., Wang, D., Ding, Y., Xu, Y. Z., Qi, H., Blacketer, M. J., Girton, J., Johansen, J. and Johansen, K. M. 2004. Chromator, a novel and essential chromodomain protein interacts directly with the putative spindle matrix protein skeletor. J Cell Biochem 93: 1033–1047.

Rebollo, E., Llamazares, S., Reina, J. and Gonzalez, C. 2004. Contribution of noncentrosomal microtubules to spindle assembly in Drosophila spermatocytes. PLoS Biol 2: E8.

Ribbeck, K., et al. 2006. NuSAP, a mitotic RanGTP target that stabilizes and cross-links microtubules. Mol Biol Cell 17: 2646–2660.

Ribbeck, K., Raemaekers, T., Carmeliet, G. and Mattaj, I. W. 2007. A role for NuSAP in linking microtubules to mitotic chromosomes. Curr Biol 17: 230–236.

Rogers, G. C., Rogers, S. L., Schwimmer, T. A., Ems-McClung, S. C., Walczak, C. E., Vale, R. D., Scholey, J. M. and Sharp, D. J. 2004. Two mitotic kinesins cooperate to drive sister chromatid separation during anaphase. Nature 427: 364–370.

Rogers, G. C., Rogers, S. L. and Sharp, D. J. 2005. Spindle microtubules in flux. J Cell Sci 118: 1105–1116.

Rusan, N. M., Fagerstrom, C. J., Yvon, A. M. and Wadsworth, P. 2001. Cell cycle-dependent changes in microtubule dynamics in living cells expressing green fluorescent protein-alpha tubulin. Mol Biol Cell 12: 971–980.

Samejima, K., et al. 2008. A promoter-hijack strategy for conditional shutdown of multiply spliced essential cell cycle genes. Proc Natl Acad Sci USA 105: 2457–2462.

Sampath, S. C., Ohi, R., Leismann, O., Salic, A., Pozniakovski, A. and Funabiki, H. 2004. The chromosomal passenger complex is required for chromatin-induced microtubule stabilization and spindle assembly. Cell 118: 187–202.

Sauer, G., Korner, R., Hanisch, A., Ries, A., Nigg, E. A. and Sillje, H. H. 2005. Proteome analysis of the human mitotic spindle. Mol Cell Proteomics 4: 35–43.

Saunders, A. M., Powers, J., Strome, S. and Saxton, W. M. 2007. Kinesin-5 acts as a brake in anaphase spindle elongation. Curr Biol 17: R453–R454.

Saunders, W., Hornack, D., Lengyel, V. and Deng, C. 1997. The Saccharomyces cerevisiae kinesin-related motor Kar3p acts at preanaphase spindle poles to limit the number and length of cytoplasmic microtubules. J Cell Biol 137: 417–431.

Savoian, M. S., Gatt, M. K., Riparbelli, M. G., Callaini, G. and Glover, D. M. 2004. Drosophila Klp67A is required for proper chromosome congression and segregation during meiosis I. J Cell Sci 117: 3669–3677.

Savoian, M. S., Goldberg, M. L. and Rieder, C. L. 2000. The rate of poleward chromosome motion is attenuated in Drosophila zw10 and rod mutants. Nat Cell Biol 2: 948–952.

Sawin, K. E., LeGuellec, K., Philippe, M. and Mitchison, T. J. 1992. Mitotic spindle organization by a plus-end directed microtubule motor. Nature 359: 540–543.

Saxton, W. M., Stemple, D. L., Leslie, R. J., Salmon, E. D., Zavortink, M. and McIntosh, J. R. 1984. Tubulin dynamics in cultured mammalian cells. J Cell Biol. 99: 2175–2186.

Scaerou, F., Starr, D. A., Piano, F., Papoulas, O., Karess, R. E. and Goldberg, M. L. 2001. The ZW10 and Rough Deal checkpoint proteins function together in a large, evolutionarily conserved complex targeted to the kinetochore. J Cell Sci 114: 3103–3114.

Schaar, B. T., Chan, G. K. T., Maddox, P., Salmon, E. D. and Yen, T. J. 1997. CENP-E function at kinetochores is essential for chromosome alignment. J Cell Biol. 139: 1373–1382.

Schatz, C. A., Santarella, R., Hoenger, A., Karsenti, E., Mattaj, I. W., Gruss, O. J. and Carazo-Salas, R. E. 2003. Importin alpha-regulated nucleation of microtubules by TPX2. Embo J 22: 2060–2070.

Schek, H. T., 3rd, Gardner, M. K., Cheng, J., Odde, D. J. and Hunt, A. J. 2007. Microtubule assembly dynamics at the nanoscale. Curr Biol 17: 1445–1455.

Schlaitz, A. L., et al. 2007. The C. elegans RSA complex localizes protein phosphatase 2A to centrosomes and regulates mitotic spindle assembly. Cell 128: 115–127.

Scholey, J. M., Rogers, G. C. and Sharp, D. J. 2001. Mitosis, microtubules, and the matrix. J Cell Biol 154: 261–266.

Severin, F., Habermann, B., Huffaker, T. and Hyman, T. 2001. Stu2 promotes mitotic spindle elongation in anaphase. J Cell Biol 153: 435–442.

Sharp, D. J., Brown, H. M., Kwon, M., Rogers, G. C., Holland, G. and Scholey, J. M. 2000. Functional coordination of three mitotic motors in drosophila embryos. Mol Biol Cell 11: 241–253.

Sharp, D. J., McDonald, K. L., Brown, H. M., Matthies, H. J., Walczak, C., Vale, R. D., Mitchison, T. J. and Scholey, J. M. 1999. The bipolar kinesin, KLP61F, cross-links microtubules within interpolar microtubule bundles of Drosophila embryonic mitotic spindles. J Cell Biol 144: 125–138.

Sharp, D. J., Rogers, G. C. and Scholey, J. M. 2000. Cytoplasmic dynein is required for poleward chromosome movement during mitosis in Drosophila embryos. Nat Cell Biol 2: 922–930.

Sharp, D. J., Rogers, G. C. and Scholey, J. M. 2000. Microtubule motors in mitosis. Nature 407: 41–47.

Shaw, S. L., Yeh, E., Maddox, P., Salmon, E. D. and Bloom, K. 1997. Astral microtubule dynamics in yeast: a microtubule-based searching mechanism for spindle orientation and nuclear migration into the bud. J Cell Biol 139: 985–994.

Shirasu-Hiza, M., Perlman, Z. E., Wittmann, T., Karsenti, E. and Mitchison, T. J. 2004. Eg5 causes elongation of meiotic spindles when flux-associated microtubule depolymerization is blocked. Curr Biol 14: 1941–1945.

Sillje, H. H., Nagel, S., Korner, R. and Nigg, E. A. 2006. HURP is a Ran-importin beta-regulated protein that stabilizes kinetochore microtubules in the vicinity of chromosomes. Curr Biol 16: 731–742.

Skop, A. R., Liu, H., Yates, J., 3rd, Meyer, B. J. and Heald, R. 2004. Dissection of the mammalian midbody proteome reveals conserved cytokinesis mechanisms. Science 305: 61–66.

Slep, K. C. and Vale, R. D. 2007. Structural basis of microtubule plus end tracking by XMAP215, CLIP-170, and EB1. Mol Cell 27: 976–991.

Sonnichsen, B., et al. 2005. Full-genome RNAi profiling of early embryogenesis in Caenorhabditis elegans. Nature 434: 462–469.

Sproul, L. R., Anderson, D. J., Mackey, A. T., Saunders, W. S. and Gilbert, S. P. 2005. Cik1 targets the minus-end kinesin depolymerase kar3 to microtubule plus ends. Curr Biol 15: 1420–1427.

Srayko, M., Kaya, A., Stamford, J. and Hyman, A. A. 2005. Identification and characterization of factors required for microtubule growth and nucleation in the early C. elegans embryo. Dev Cell 9: 223–236.

Srayko, M., Quintin, S., Schwager, A. and Hyman, A. A. 2003. Caenorhabditis elegans TAC-1 and ZYG-9 form a complex that is essential for long astral and spindle microtubules. Curr Biol 13: 1506–1511.

Stehman, S. A., Chen, Y., McKenney, R. J. and Vallee, R. B. 2007. NudE and NudEL are required for mitotic progression and are involved in dynein recruitment to kinetochores. J Cell Biol.

Storchova, Z., Breneman, A., Cande, J., Dunn, J., Burbank, K., O'Toole, E. and Pellman, D. 2006. Genome-wide genetic analysis of polyploidy in yeast. Nature 443: 541–547.

Stout, J. R., Rizk, R. S., Kline, S. L. and Walczak, C. E. 2006. Deciphering protein function during mitosis in PtK cells using RNAi. BMC Cell Biol 7: 26.

Straight, A. F., Sedat, J. W. and Murray, A. W. 1998. Time-lapse microscopy reveals unique roles for kinesins during anaphase in budding yeast. J Cell Biol 143: 687–694.

Stumpff, J., von Dassow, G., Wagenbach, M., Asbury, C. and Wordeman, L. 2008. The kinesin-8 motor Kif18A suppresses kinetochore movements to control mitotic chromosome alignment. Dev Cell 14: 252–262.

Tao, L., Mogilner, A., Civelekoglu-Scholey, G., Wollman, R., Evans, J., Stahlberg, H. and Scholey, J. M. 2006. A homotetrameric kinesin-5, KLP61F, bundles microtubules and antagonizes Ncd in motility assays. Curr Biol 16: 2293–2302.

Theurkauf, W. E. 2001. TACCing down the spindle poles. Nat Cell Biol 3: E159–E161.

Tokai-Nishizumi, N., Ohsugi, M., Suzuki, E. and Yamamoto, T. 2005. The chromokinesin Kid is required for maintenance of proper metaphase spindle size. Mol Biol Cell 16: 5455–5463.

Tournebize, R., et al. 2000. Control of microtubule dynamics by the antagonistic activities of XMAP215 and XKCM1 in Xenopus egg extracts. Nat Cell Biol 2: 13–19.

Trieselmann, N., Armstrong, S., Rauw, J. and Wilde, A. 2003. Ran modulates spindle assembly by regulating a subset of TPX2 and Kid activities including Aurora A activation. J Cell Sci 116: 4791–4798.

Tsai, M. Y., Wang, S., Heidinger, J. M., Shumaker, D. K., Adam, S. A., Goldman, R. D. and Zheng, Y. 2006. A mitotic lamin B matrix induced by RanGTP required for spindle assembly. Science 311: 1887–1893.

Tsai, M. Y., Wiese, C., Cao, K., Martin, O., Donovan, P., Ruderman, J., Prigent, C. and Zheng, Y. 2003. A Ran signalling pathway mediated by the mitotic kinase Aurora A in spindle assembly. Nat Cell Biol 5: 242–248.

Tsou, M. F. and Stearns, T. 2006. Controlling centrosome number: licenses and blocks. Curr Opin Cell Biol 18: 74–78.

Tulu, U. S., Fagerstrom, C., Ferenz, N. P. and Wadsworth, P. 2006. Molecular requirements for kinetochore-associated microtubule formation in mammalian cells. Curr Biol 16: 536–541.

Tytell, J. D. and Sorger, P. K. 2006. Analysis of kinesin motor function at budding yeast kinetochores. J Cell Biol 172: 861–874.

Ubersax, J. A., Woodbury, E. L., Quang, P. N., Paraz, M., Blethrow, J. D., Shah, K., Shokat, K. M. and Morgan, D. O. 2003. Targets of the cyclin-dependent kinase Cdk1. Nature 425: 859–864.

Uchiyama, S., et al. 2005. Proteome analysis of human metaphase chromosomes. J Biol Chem 280: 16994–17004.

Usui, T., Maekawa, H., Pereira, G. and Schiebel, E. 2003. The XMAP215 homologue Stu2 at yeast spindle pole bodies regulates microtubule dynamics and anchorage. Embo J 22: 4779–4793.

Vaisberg, E. A., Koonce, M. P. and McIntosh, J. R. 1993. Cytoplasmic dynein plays a role in mammalian mitotic spindle formation. J Cell Biol 123: 849–858.

van Breugel, M., Drechsel, D. and Hyman, A. 2003. Stu2p, the budding yeast member of the conserved Dis1/XMAP215 family of microtubule-associated proteins is a plus end-binding microtubule destabilizer. J Cell Biol 161: 359–369.

van de Weerdt, B. C. and Medema, R. H. 2006. Polo-like kinases: a team in control of the division. Cell Cycle 5: 853–864.

Varga, V., Helenius, J., Tanaka, K., Hyman, A. A., Tanaka, T. U. and Howard, J. 2006. Yeast kinesin-8 depolymerizes microtubules in a length-dependent manner. Nat Cell Biol 8: 957–962.

Verde, F., Berrez, J.-M., Antony, C. and Karsenti, E. 1991. Taxol-induced microtubule asters in mitotic extracts of Xenopus eggs: requirement for phosphorylated factors and cytoplasmic dynein. J Cell Biol 112: 1177–1188.

Vernos, I., Raats, J., Hirano, T., Heasman, J., Karsenti, E. and Wylie, C. 1995. Xklp1, a chromosomal Xenopus kinesin-like protein essential for spindle organization and chromosome positioning. Cell 81: 117–127.

Walczak, C. E. and Mitchison, T. J. 1996. Kinesin-related proteins at mitotic spindle poles: function and regulation. Cell 85: 943–946.

Walczak, C. E., Mitchison, T. J. and Desai, A. 1996. XKCM1: a Xenopus kinesin-related protein that regulates microtubule dynamics during mitotic spindle assembly. Cell 84: 37–47.

Walczak, C. E., Verma, S. and Mitchison, T. J. 1997. XCTK2: A Kinesin-related protein that promotes mitotic spindle assembly in Xenopus laevis egg extracts. J Cell Biol. 136: 859–870.

Walczak, C. E., Vernos, I., Mitchison, T. J., Karsenti, E. and Heald, R. 1998. A model for the proposed roles of different microtubule-based motor proteins in establishing spindle bipolarity. Curr Biol 8: 903–913.

Walker, D. L., Wang, D., Jin, Y., Rath, U., Wang, Y., Johansen, J. and Johansen, K. M. 2000. Skeletor, a novel chromosomal protein that redistributes during mitosis provides evidence for the formation of a spindle matrix. J Cell Biol 151: 1401–1412.

Walker, R. A., O'Brien, E. T., Pryer, N. K., Soboeiro, M. F., Voter, W. A., Erickson, H. P. and Salmon, E. D. 1988. Dynamic instability of individual microtubules analyzed by video light microscopy: rate constants and transition frequencies. J Cell Biol 107: 1437–1448.

West, R. R., Malmstrom, T. and McIntosh, J. R. 2002. Kinesins *klp5*$^{+}$ and *klp6*$^{+}$ are required for normal chromosome movement in mitosis. J Cell Sci 115: 931–940.

West, R. R., Malmstrom, T., Troxell, C. L. and McIntosh, J. R. 2001. Two related kinesins, *klp5*$^{+}$ and *klp6*$^{+}$, foster microtubule disassembly and are required for meiosis in fission yeast. Mol Biol Cell 12: 3919–3932.

Wickstead, B. and Gull, K. 2007. Dyneins across eukaryotes: a comparative genomic analysis. Traffic 8: 1708–1721.

Wiese, C., Wilde, A., Moore, M. S., Adam, S. A., Merdes, A. and Zheng, Y. 2001. Role of importin-beta in coupling Ran to downstream targets in microtubule assembly. Science 291: 653–656.

Wilde, A., Lizarraga, S. B., Zhang, L., Wiese, C., Gliksman, N. R., Walczak, C. E. and Zheng, Y. 2001. Ran stimulates spindle assembly by altering microtubule dynamics and the balance of motor activities. Nat Cell Biol 3: 221–227.

Wilson, P. G., Heck, M. and Fuller, M. T. 1992. Monastral spindles are generated by mutations in urchin, a bimC homolog in Drosophila. Mol Biol Cell 3: 343a.

Wittmann, T., Wilm, M., Karsenti, E. and Vernos, I. 2000. TPX2, A novel xenopus MAP involved in spindle pole organization. J Cell Biol 149: 1405–1418.

Wolyniak, M. J., Blake-Hodek, K., Kosco, K., Hwang, E., You, L. and Huffaker, T. C. 2006. The regulation of microtubule dynamics in Saccharomyces cerevisiae by three interacting plus-end tracking proteins. Mol Biol Cell 17: 2789–2798.

Wong, J. and Fang, G. 2006. HURP controls spindle dynamics to promote proper inter-kinetochore tension and efficient kinetochore capture. J Cell Biol 173: 879–891.

Wood, K. W., Sakowicz, R., Goldstein, L. S. B. and Cleveland, D. W. 1997. CENP-E is a plus end-directed kinetochore motor required for metaphase chromosome alignment. Cell 91: 357–366.

Yan, X., Li, F., Liang, Y., Shen, Y., Zhao, X., Huang, Q. and Zhu, X. 2003. Human Nudel and NudE as regulators of cytoplasmic dynein in poleward protein transport along the mitotic spindle. Mol Cell Biol 23: 1239–1250.

Yang, Z., Tulu, U. S., Wadsworth, P. and Rieder, C. L. 2007. Kinetochore dynein is required for chromosome motion and congression independent of the spindle checkpoint. Curr Biol 17: 973–980.

Yao, X., Anderson, K. L. and Cleveland, D. W. 1997. The microtubule-dependent motor centromere-associated protein E (CENP-E) is an integral component of kinetochore corona fibers that link centromeres to spindle microtubules. J Cell Biol139: 435–447.

Yeh, E., Yang, C., Chin, E., Maddox, P., Salmon, E. D., Lew, D. J. and Bloom, K. 2000. Dynamic positioning of mitotic spindles in yeast: role of microtubule motors and cortical determinants. Mol Biol Cell 11: 3949–3961.

Yen, T. J., Li, G., Schaar, B. T., Szilak, I. and Cleveland, D. W. 1992. CENP-E is a putative kinetochore motor that accumulates just before mitosis. Nature 359: 536–539.

Yu, C. T., Hsu, J. M., Lee, Y. C., Tsou, A. P., Chou, C. K. and Huang, C. Y. 2005. Phosphorylation and stabilization of HURP by Aurora-A: implication of HURP as a transforming target of Aurora-A. Mol Cell Biol 25: 5789–5800.

Zhai, Y., Kronebusch, P. J., Simon, P. M. and Borisy, G. G. 1996. Microtubule dynamics at the $G_2$/M transition: Abrupt breakdown of cytoplasmic microtubules at nuclear envelope breakdown and implications for spindle morphogenesis. J Cell Biol 135: 201–214.

Zhang, D., Rogers, G. C., Buster, D. W. and Sharp, D. J. 2007. Three microtubule severing enzymes contribute to the "Pacman-flux" machinery that moves chromosomes. J Cell Biol 177: 231–242.

Zhang, X., Lan, W., Ems-McClung, S. C., Stukenberg, P. T. and Walczak, C. E. 2007. Aurora B Phosphorylates Multiple Sites on Mitotic Centromere-associated Kinesin to Spatially and Temporally Regulate Its Function. Mol Biol Cell 18: 3264–3276.

Zheng, Y. and Tsai, M. Y. 2006. The mitotic spindle matrix: a fibro-membranous lamin connection. Cell Cycle 5: 2345–2347.

Zheng, Y., Wong, M. L., Alberts, B. and Mitchison, T. 1995. Nucleation of microtubule assembly by a gamma-tubulin-containing ring complex. Nature 378: 578–583.

Zhu, C., Zhao, J., Bibikova, M., Leverson, J. D., Bossy-Wetzel, E., Fan, J. B., Abraham, R. T. and Jiang, W. 2005. Functional analysis of human microtubule-based motor proteins, the kinesins and dyneins, in mitosis/cytokinesis using RNA interference. Mol Biol Cell 16: 3187–3199.

# Chapter 9
# Kinetochore-Microtubule Interactions

**Lesley Clayton and Tomoyuki U. Tanaka**

## 9.1 Introduction

Kinetochores are the multiprotein macromolecular assemblies on chromatin that ensure the accurate and timely segregation of chromosomes at during mitosis. To achieve this, kinetochores must interact with the microtubules of the spindle and microtubule-associated proteins. The nature of the kinetochore–microtubule interaction varies during the stages of the mitotic cycle, starting with initial capture and progressing through bi-orientation and congression at prometaphase/metaphase, then finally separation of sister kinetochores/chromatids during anaphase. All the while during this process, kinetochores are able to signal their state of microtubule binding to the cell cycle control machinery. They are also able to influence microtubule dynamics in order to achieve chromosome segregation. Determining the structure and biochemistry of these various interactions continues to be a major objective of research in this field.

Much of the cell biology/cytology of cell division has originally been described in metazoan cells in culture, e.g. PtK cells, newt lung, mouse and human cell lines etc., as their relatively large size and flat growth characteristics in culture make them easy to work with. However recent advances in microscopy, particularly fluorescence techniques, have made it possible to visualise spindle components in living cells of both budding and fission yeast, and also *Drosophila* cells, all of which had previously proven too small to image successfully, but which have powerful advantages in terms of genetics and proteomics.

A great deal of research on kinetochore structure and function has been performed using the budding yeast *Saccharomyces cerevisiae* as an experimental system. Many of the proteins that make up the yeast kinetochore have counterparts in other organisms, including mammals (McAinsh et al. 2003). In addition, the yeast centromere DNA spans only about 130 bp (Hegemann and Fleig 1993), and a kinetochore interacts with a single microtubule in metaphase,

L. Clayton (✉)
College of Life Sciences, University of Dundee, Wellcome Trust Biocentre, Dow Street, Dundee DD1 5EH, U.K.
e-mail: l.clayton@dundee.ac.uk

P. De Wulf, W.C. Earnshaw (eds.), *The Kinetochore*,
DOI 10.1007/978-0-387-69076-6_9, © Springer Science+Business Media, LLC 2009 

compared with dozens in higher organisms (Winey et al. 1995; O'Toole et al. 1999). These features, coupled with the ease of genetic manipulation and the extensive proteomic information available have meant that much progress has been made in studying the yeast kinetochore.

This chapter will examine, and compare where possible, the various kinds of interactions that occur between kinetochores and microtubules at different times during mitosis, both in yeast and in higher organisms.

## 9.2 Kinetochore Capture of Microtubules

The initial capture of microtubules by kinetochores is the first critical step in preparing for the chromosome segregation that takes place in anaphase, and occurs rapidly either following nuclear envelope breakdown in metazoan cells, or after kinetochore reassembly during S-phase in budding yeast. Yeast chromosomes do not condense before segregation, remaining tethered to the spindle pole body (SPB) by microtubules throughout most of the cell cycle. They are, however, released from microtubules for a brief period during S-phase that coincides with replication of the centromere region of the chromosome, and subsequently rapidly re-capture microtubules (Kitamura et al. 2007). The yeast cell cycle also differs from that of metazoans in that the nuclear envelope (NE) does not break down before cell division. This is known as a closed mitosis, and the bipolar spindle is formed within the nucleus; the microtubules extending from a microtubule organizing centre (MTOC) embedded in the nuclear membrane, called an SPB.

Time-lapse microscopy of vertebrate cells such as newt lung cells has shown that kinetochores are initially captured by a lateral surface of a single microtubule extending from one of the spindle poles Fig. 9.1a; (Hayden et al. 1990; Rieder and Alexander 1990). This is also the case for *S. cerevisiae*. Although difficult to visualise in the normal cell cycle of living cells, because capture takes place during a very short period of S-phase and within close proximity of the SPBs, visualisation was recently achieved (Tanaka et al. 2005) by preventing activation of a single GFP-marked centromere enabling it to be displaced from the remainder of the centromeres already aligned on the spindle. After reactivating the centromere, allowing kinetochore assembly, capture by a single microtubule could be shown. As in vertebrate cells, capture took place at the lateral surface of the microtubule, followed by transport polewards along the microtubule. Kinetochore attachment to microtubule lateral surfaces has also been seen during pro-metaphase in diatom cells (Tippit et al. 1980; Pickett-Heaps 1991), indicating that this mechanism has a long evolutionary history, presumably its advantage being that the microtubule lattice provides a larger surface area for kinetochore capture which is sterically permissive, with productive contacts between kinetochores and microtubules formed over a range of angles and distances, compared with microtubule tips which require a 'bulls-eye' hit of a microtubule directly into the kinetochore. Figure 9.1 depicts kinetochore–microtubule interactions in prometaphase, metaphase and anaphase.

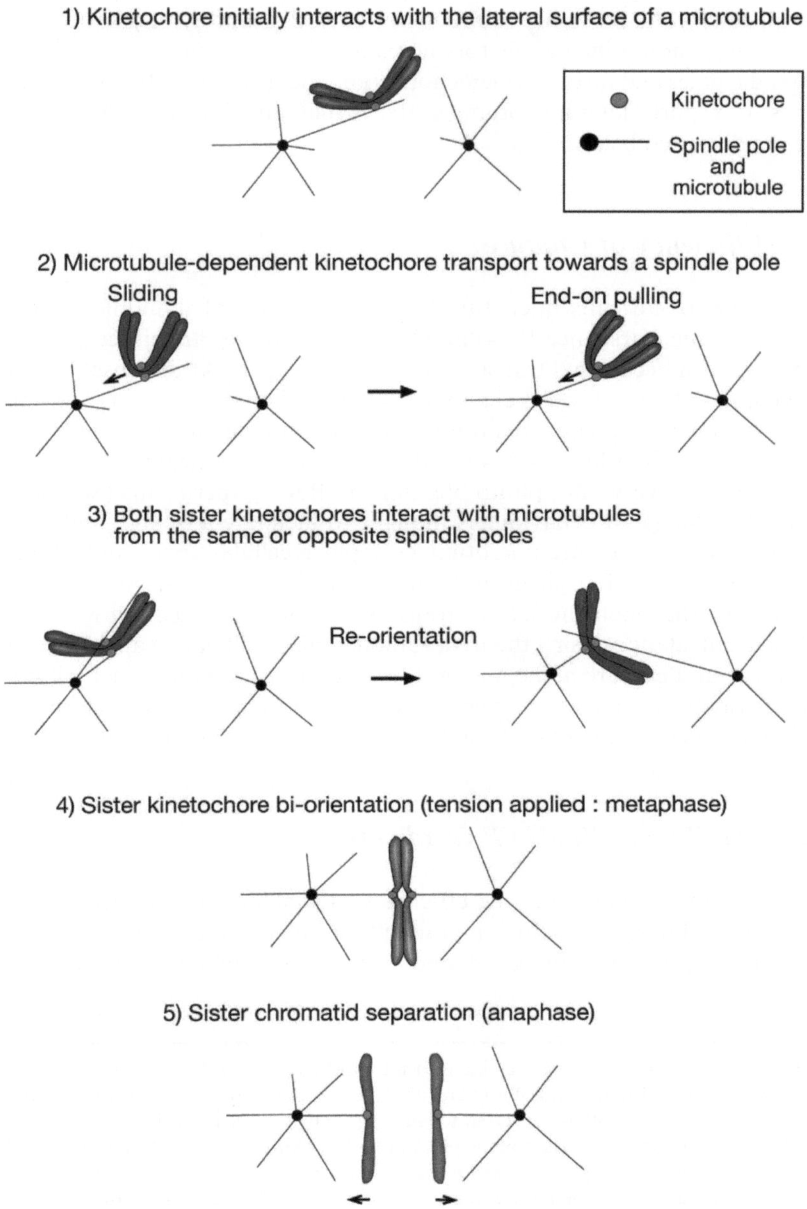


**Fig. 9.1 Kinetochore–microtubule interactions in prometaphase and metaphase (a)** Kinetochores initially interact with the lateral surface of a single microtubule extending from one of the spindle poles. This initial encounter with microtubules happens quickly following nuclear envelope breakdown (metazoan cells), or once kinetochore assembly is complete (budding yeast). **(b)** Once captured, kinetochores are transported along single microtubules towards the spindle pole. Transport can occur in two distinct ways; lateral sliding, in which centromeres

In all eukaryotic cells, the requirements for initial capture of microtubules by kinetochores should include factors necessary for microtubule extension from the spindle poles and distinct kinetochore components used for binding to microtubules, transport along microtubules and regulation of microtubule dynamics; these are discussed below.

### *9.2.1 Efficiency of Capture*

In most metazoan cells microtubules are in a state of dynamic instability (Mitchison and Kirschner 1984) in which microtubules are either growing or shrinking, with occasional transitions between the two. At the onset of mitosis the dynamics of microtubules changes, transitions between growing and shrinking become more frequent and more shorter microtubules are nucleated from the centrosome. The half-life of microtubules decreases dramatically as measured by fluorescence recovery after photo-bleaching (FRAP) experiments (Saxton et al. 1984). This change in behaviour is thought to facilitate a 'search and capture' process by which growing microtubules explore cellular space and ultimately bind a kinetochore (Kirschner and Mitchison 1986), thus becoming stabilised and allowing the kinetochore to be transported polewards (see below).

Recent calculations using theoretical modelling (Wollman et al. 2005) suggest that search and capture alone may not be efficient enough to attach all kinetochores in order to complete mitosis with the observed timing, particularly during rapid division cycles e.g. of early embryos such as *Xenopus* and *Drosophila.*

### *9.2.2 The Role of RanGTP Gradients*

Various means of increasing the efficiency of kinetochore capture have been proposed, including invoking a gradient of microtubule attracting signals centred upon chromatin and kinetochores. One such signal is the small GTPase

**Fig. 9.1** (continued) move along the side of a microtubule; and end-on pulling, in which the centromere is tethered to the end of a microtubule and is pulled polewards as the microtubule shrinks. Lateral sliding is often converted to end-on pulling but the opposite conversion is rare. (**c**) As kinetochores approach the spindle poles both sister kinetochores attach to microtubules extending from the same (*left*) or opposite (*right*) spindle poles. Mal-oriented kinetochore–microtubule interactions must be re-oriented to establish proper bi-orientation. (**d**) Sister kinetochore bi-orientation; cessation of re-orientation is thought to depend on the tension generated by microtubule pulling forces applied on kinetochores. The number of microtubules whose plus-ends attach to a single kinetochore increases during these processes in metazoan cells, whereas only a single microtubule is thought to attach to each kinetochore in budding yeast (the latter case is shown here for simplicity). (**e**) Once all kinetochores bi-orient on the spindle, cohesion between sister chromatids is removed, causing segregation of sister chromatids to opposite spindle poles during anaphase. Figure modified from (Tanaka et al. 2005)

Ran in its GTP (active) form. Ran, like other small Ras-like GTPases, undergoes switching between GTP- and GDP-bound forms; the transitions are governed by a guanine nucleotide exchange factor (GEF) called Regulator of Chromatin Condensation 1 (RCC1), and the GTPase activating factor RanGAP, which requires additional Ran binding proteins (RanBP1 and RanBP2) for GTPase activity (Kalab et al. 1999). A gradient of RanGTP centred on, and directed away from chromosomes is generated by the action of its exchange factor RCC1 which is associated with chromatin (Caudron et al. 2005). RanGTP stimulates the dissociation of spindle assembly-promoting factors such as the proteins NuMA and TPX2 from their complexes with Importin $\beta$ (Gruss and Vernos 2004)$\beta$ thus changing microtubule polymerisation dynamics in the vicinity of chromosomes, adding bias to the 'search and capture' process and possibly 'guiding' microtubules towards kinetochores and promoting spindle assembly (Carazo-Salas et al. 2001; Kalab et al. 2002). RanGTP may also function directly at the kinetochore. It has been shown in HeLa cells that RanGAP localisation is targeted to kinetochores by Sumoylation, and RanBP2 is also located at kinetochores during mitosis (Joseph et al. 2002). These two proteins appear to be targeted to kinetochores as a complex.

One can understand how generating such a directional microtubule assembly bias may be an advantage in large cells such as eggs and early embryos, however there is evidence that RanGTP may also have a generally conserved function. Mutations in yeast Ran and its GEF Prp20/Mtr1 result in reduced microtubule extension from a spindle pole. Microtubule extension is however unbiased in its direction with respect to kinetochores (Tanaka et al. 2005). Thus there appears to be no RanGTP gradient to provide a directional cue for microtubule polymerisation within the closed mitosis of yeast.

### *9.2.3 Kinetochore-Derived Microtubules*

Encounters between kinetochores and microtubules extending from spindle poles might also be facilitated by the generation of microtubules from the kinetochore itself, or from chromatin. Such kinetochore-derived microtubules were first reported in vitro, assembled onto HeLa cell mitotic chromosome kinetochores (Telzer et al. 1975), and subsequently in vivo in cells recovering from low temperature, or treatment with spindle poisons (Witt et al. 1980; De Brabander et al. 1981; Czaban and Forer 1985), but their physiological relevance was unknown. Recently however, such microtubule assembly has also been observed in unperturbed vertebrate and *Drosophila* cells (Khodjakov et al. 2003; Maiato et al. 2004). In these organisms, kinetochore-derived microtubules extend by addition of tubulin dimers at their plus ends, which are embedded in the kinetochore. These microtubules interact with pole-derived microtubules, which may serve to 'guide' both them and their associated kinetochores towards the spindle poles. The kinetochore- and chromatin-generated microtubules are frequently bundled by motor proteins and collected together to form a spindle pole ((Wittmann et al. 2001) for a review). This mechanism for kinetochore capture and spindle formation is common in cells that do not posses centrosomes, such as plant cells, early mammalian embryos and some *Drosophila* cells, but detailed observation and experimentation have shown that it can also play a role in cells that contain centrosomes (Khodjakov et al. 2000; Wadsworth and Khodjakov 2004). Kinetochore-derived microtubules have also recently been observed in *S. cerevisiae*, and they appear to associate along their length with spindle pole-derived microtubules (K. Tanaka, E.Kitamura and T.U. Tanaka, unpublished). Their contribution to kinetochore capture in normal cell cycles remains to be addressed.

### 9.2.4 Kinetochore Transport; Sliding vs. Pulling

After kinetochores have been captured by the lateral surface of microtubules (Fig. 9.1a) they are then conveyed towards a pole, i.e. towards a microtubule minus-end (Fig. 9.1b). This movement facilitates bi-orientation, as it locates kinetochores in close proximity to the spindle where microtubules extend from both spindle poles at a high density. Transport polewards initially takes place by sliding of the kinetochore along the microtubule lattice surface Fig. 9.1b; (Rieder and Alexander 1990; Tanaka et al. 2005). The ATP-driven motor proteins of the kinesin and dynein protein superfamilies are involved in this process.

In *S. cerevisiae* there are only six kinesin-family proteins and only one dynein heavy chain (Hildebrandt and Hoyt 2000). Dynein is an exclusively cytoplasmic protein and does not participate in kinetochore sliding within the closed mitosis of yeast. Systematic genetic analysis has shown that Kar3 (a kinesin 14 family member) is involved in poleward transport along the microtubule (Tanaka et al. 2005). Kar3 is the only minus-end directed kinesin with nuclear localisation in yeast; indeed Kar3 is also localised at kinetochores as well as at spindle poles (Tytell and Sorger 2006; Tanaka et al. 2007). In wild-type $KAR3^{+}$ cells, centromere sliding is directed progressively polewards, whereas in Kar3-deleted mutant cells, laterally attached centromeres move randomly by diffusion, indicating that Kar3 is the main, and probably sole, factor directing kinetochore sliding in yeast (Tanaka et al. 2007). Chromosomes do however reach the spindle poles in Kar3-deleted cells, demonstrating that other mechanisms are involved: detailed observation reveals that kinetochores are attached to the plus ends of microtubules, and are conveyed polewards as these microtubules shrink Fig. 9.1b; (Tanaka et al. 2007).

In yeast, lateral attachment of the kinetochore is converted to end-on attachment at the plus end of the microtubule as the chromosome nears the pole, but the opposite conversion is rare. End-on attachment is presumably more stable than lateral attachment and occurs in preparation for bi-orientation and chromosome separation in anaphase. Chromosomes continue to be pulled towards the pole once end-on attached (this process is denoted end-on pulling in the figure), and this mode of transport is more rapid than sliding. The Dam1 complex is presumably involved in tethering the kinetochore to the microtubule end during this conversion from lateral to end-on attachment, and certainly facilitates the progression of end-on pulling (Tanaka et al. 2007; see next section).

Dam1 appears to have an important role in the conversion of the free energy stored in the microtubule lattice to kinetochore pulling force *via* microtubule depolymerisation (Asbury et al. 2006; Westermann et al. 2006; Franck et al. 2007). By contrast, kinetochore sliding is driven by Kar3 motor activity that is dependent on ATP hydrolysis, thus consuming additional energy to achieve a less processive mode of transport.

Given the advantages in terms of energy and transport velocity of end-on pulling over sliding, why is kinetochore sliding conserved from yeast to vertebrates? Kinetochore sliding may have the following merits: firstly, to establish end-on pulling, kinetochores once bound to the microtubule lateral wall must wait until the microtubule shrinks and the plus end reaches the kinetochore. When kinetochores are already close to a spindle pole when captured, they may reach a spindle pole much earlier by sliding; secondly, a single microtubule plus end can presumably bind only one kinetochore by end-on attachment and pulling (Winey et al. 1995), whereas several kinetochores could be transported simultaneously by sliding (K. Tanaka and T.U. Tanaka, unpublished); thirdly, microtubule rescue (see below), which occurs during sliding but not end-on pulling may increase the chance that kinetochores further afield are captured by the same microtubule.

In vertebrate cells kinetochore sliding along the microtubule lattice is very rapid (11–14 $\mu m\ min^{-1}$ in newt lung cells, compared with 0.5–2.0 $\mu m\ min^{-1}$ in budding yeast), and dynein has been proposed to be involved, based on the speed of its mobility along microtubules in vitro and its localisation at unattached kinetochores in early mitosis (Starr et al. 1998; Pfarr et al. 1990; King et al. 2000). In human cells kinetochore-associated dynein has recently been shown directly to be responsible for the rapid poleward transport (Yang et al. 2007). It is clear that vertebrate cells make end-on attachments between microtubules and kinetochores by metaphase, when chromosomes are bi-oriented, however end-on pulling has not yet been unequivocally shown for poleward kinetochore transport in normal prometaphase.

Metazoan cells as yet have no reported Dam1 equivalent, but the +TIP protein MAST/ORBIT and the checkpoint protein Bub1 may be important for end-on attachment because their depletion causes lateral attachments to persist (Maiato et al. 2002; Meraldi and Sorger 2005). Alternatively in metazoans with multiple kinetochore microtubules, sliding is faster than end-on pulling and rapidly moves the kinetochore nearer to the pole where the chance of making end-on attachments is much higher, thus end-on pulling may not be as advantageous.

## 9.3 The Kinetochore–Microtubule Interface

The structure, assembly and functional organisation of kinetochores have been revealed by recent studies, which are detailed in chapter 10 of this book. The kinetochore consists of around 100 highly conserved proteins, and defects in many of these result in loss of chromosome–microtubule binding, mainly due to deleterious effects on the highly ordered organisation/assembly of the kinetochore, but what constitutes the direct kinetochore–microtubule interface where interactions take place? Recent studies have begun to pinpoint the components directly involved in microtubule association. The outer kinetochore components are those involved in interactions with microtubules; specifically the Ndc80/HEC1 complex, which comprise Ndc80/Hec1, Nuf2, Spc24 and Spc25. These four components assemble into a heterotetrameric rod structure, with globular heads at either end of a coiled-coil shaft (Wei et al. 2005). Biochemical reconstitution experiments and electron microscopy have shown that Ndc80-Nuf2 subunits of this complex interact directly with the microtubule lattice by decorating the microtubule walls in vitro (Cheeseman et al. 2006; Wei et al. 2007). The crystal structure of the head region of Ndc80/Hec1 reveals that it is folded into a calponin-homology (CH) domain similar to that found in the microtubule-binding region of the plus end-associated protein EB1 (Wei et al. 2007). Thus the Ndc80 complex seems to provide a contact point of kinetochores to the lateral surface of microtubules. Consistent with this notion, the Ndc80 complex is required for kinetochore association with the microtubule lateral surface in vivo in budding yeast; see Fig. 9.2, step 2 (Tanaka et al. 2005).

Then, which molecules provide the kinetochore–microtubule interface after kinetochores attach to the ends of microtubules? In budding yeast, the Dam1 (DASH/DD) complex seems to make this interface. The Dam1 complex localises along microtubules but does not co-localise with kinetochores, while they are associated with the microtubule lateral surface. However, the Dam1 complex has important roles in subsequent tethering of kinetochore at the plus ends of microtubules (Tanaka et al. 2007) Fig. 9.2, step 4; the Dam1 complex association with the Ndc80 complex is presumably involved in this tethering (Shang et al. 2003). The Dam1 complex oligomerises and forms a ring around microtubules in vitro (Westermann et al. 2005, 2006; Miranda et al. 2007), and there is evidence from live cell studies in yeast that it forms a similar structure in vivo (Tanaka et al. 2007). In this way the Dam1 complex may stably attach the microtubule end to the kinetochore and also harness the energy of depolymerising microtubules to generate pulling force in a minus-end direction along the microtubule, e.g. in end-on pulling of chromosomes during pro-metaphase and anaphase (Westermann et al. 2005; Asbury et al. 2006; Westermann et al. 2006; Miranda et al. 2007; Tanaka et al. 2007). Dam1 and another component of the complex, Spc34, are also phosphorylated by Ipl1 kinase in *S. cerevisiae* (Cheeseman et al. 2002; Shang et al. 2003), and thus involved in regulation of bi-orientation (see section below).

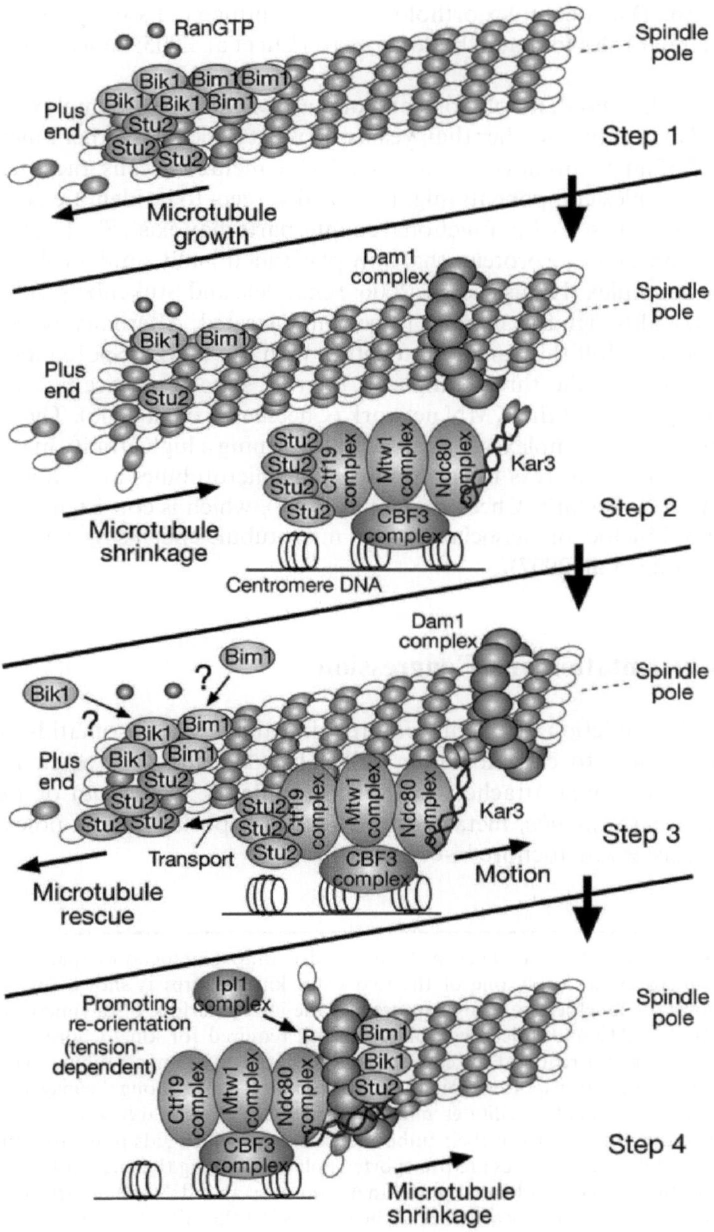


**Fig. 9.2 Kinetochore capture of microtubules in budding yeast** In Step 1, a microtubule extends from a spindle pole, facilitated by the microtubule plus-end-tracking proteins Stu2, Bim1 and Bik1 which localise at the growing plus-end of the microtubule, and the small GTPase Ran bound to GTP. In Step 2, the amount of Stu2, Bim1 and Bik1 protein at the microtubule plus-end decreases, and the microtubule depoymerises and shrinks. Stu2 and

Recently the Dam1 complex orthologue was identified in fission yeast and seems to play similar roles to that in budding yeast (Liu et al. 2005; Sanchez-Perez et al. 2005).

On the other hand, no obvious orthologues of the Dam1 complex have been identified in organisms other than yeasts, so what molecules tether kinetochores at the end of microtubules in metazoan cells? In metazoan cells, the homologues of Dam1 complex components might be too divergent to be identified by primary sequence comparisons, but functional counterparts may exist. Ska1/2, Cep57 and Bod1 are all vertebrate proteins that may play functionally similar roles to that of the Dam1 complex (Hanisch et al. 2006; Emanuele and Stukenberg 2007; Porter et al. 2007), although this remains to be demonstrated. Alternatively, the Ndc80 complex may fulfill this function, together with the KNL1/Spc105 and Mis12/Mtw1 complexes; the three complexes interact stoichiometrically and form a larger complex called the KMN network (Cheeseman et al. 2006). The combination of these three complexes is necessary for forming a high affinity microtubule-binding site in vitro. It is thus proposed that microtubules are embedded in a fibrous KMN network (Cheeseman et al. 2006), which is consistent with recent imaging of kinetochore association with microtubule ends using electron tomography (Dong et al. 2007).

## 9.4 Bi-Orientation and Congression

Successful completion of cell division requires that sister chromatids are segregated accurately to each daughter cell. This depends upon each replicated chromatid becoming attached to a microtubule (*S. cerevisiae*) or bundle of microtubules (*S. pombe,* metazoan cells) from opposite spindle poles. This is known as sister kinetochore bi-orientation.

**Fig. 9.2** (continued) Kar3 (a kinesin14) are loaded onto kinetochores that have not yet captured microtubules (only one of the two sister kinetochores is shown for simplicity). Step 3 shows the kinetochore being captured by the lateral surface of the microtubule. The CBF3, Mtw1, Ctf19 and Ndc80 complexes are all required for kinetochore capture. The Dam1 complex is not required for capture, and localises on the microtubule. After kinetochore capture Stu2 is transported from the captured kinetochores along the microtubule to its plus-end, which probably facilitates microtubule rescue. Presumably Bim1 and Bik1 are recruited to the plus-end from their unbound fractions, which aids in microtubule rescue. Once captured the kinetochores are transported polewards along the microtubule promoted by Kar3. In Step 4 Dam1 is loaded onto kinetochores in an Ndc80-dependent manner and the kinetochore subsequently attaches to the plus-end of the microtubule, stabilised in this end-on attachment by the Dam1 complex. End-on pulling of kinetochores towards the pole can occur as the microtubule shrinks, in pro/metaphase. Phosphorylation of the Dam1 and Ndc80 complex components (and possibly other kinetochore components) by the Ipl1 kinase complex facilitates turnover of kinetochore–microtubule attachments until bi-orientation generates tension, leading to separation of sister kinetochores. Figure modified from (Tanaka et al. 2005)

As discussed above, a single sister kinetochore often becomes attached to firstly one spindle pole. This is termed monotelic attachment or mono-orientation (Fig. 9.3). Subsequently its sister kinetochore becomes attached to the opposite pole and the chromosome becomes bi-oriented. During this process incorrect microtubule attachments may occur (Fig. 9.3): e.g. both sister kinetochores attached to the same pole (syntelic attachment), or one kinetochore attached to both poles (merotelic attachment). Correct, bi-oriented attachments are called amphitelic attachments.

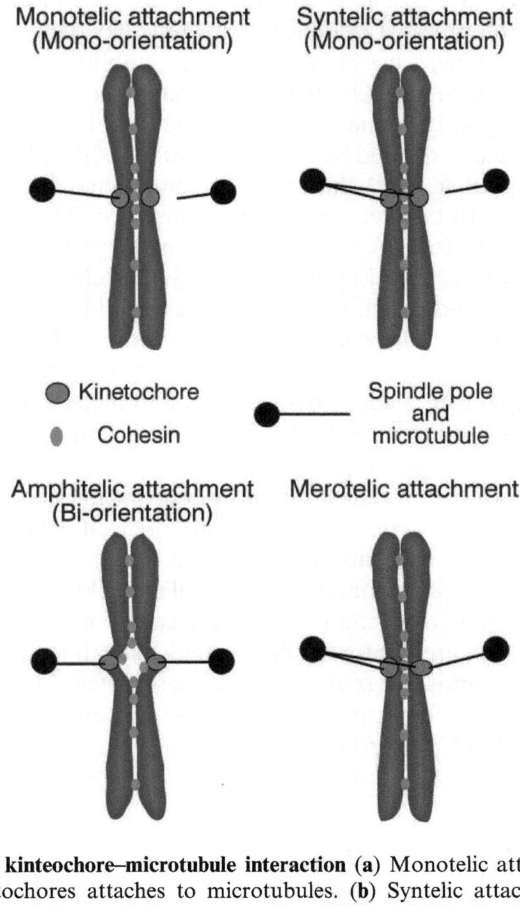


**Fig. 9.3 Modes of kinteochore–microtubule interaction** (**a**) Monotelic attachment; only one of the sister kinetochores attaches to microtubules. (**b**) Syntelic attachment: both sister kinetochores attach to microtubules extending from the same spindle pole. Both monotelic and syntelic interactions result in chromosomes being mono-oriented i.e. connected to only one spindle pole. (**c**) Amphitelic attachment: each sister kinetochore attaches to microtubules extending from opposite spindle poles. Chromosomes are thus bi-oriented. (**d**) Merotelic attachment: one sister kinetochore attaches simultaneously to microtubules extending from both spindle poles. Figure modified from (Tanaka et al. 2005)

Incorrect attachments are, to some degree, discouraged by the back-to-back geometry of sister kinetochores on mitotic chromosomes i.e. the kinetochores face in opposite directions, making binding microtubules from opposite poles more likely. However it is thought that such geometry is not always present on chromosomes during the first meiotic division in which two homologous kinetochores, each consisting of sister kinetochores fused together, must attach to microtubules that extend from opposite spindle poles. These homologous kinetochores can flexibly change their geometry because they are connected only by chiasmata formed on the chromosome arms, as there is no cohesion that directly links the two homologous centromeres. This flexibility precludes a key role of the geometry-dependent mechanism in facilitating bi-orientation. In this situation there is biophysical evidence to suggest that the presence of tension between the sister kinetochores is required to stabilise correct microtubule attachments (Nicklas and Koch 1969). The relative contribution of geometry and tension-sensing to the generation of bi-orientation has been debated and the problem has been addressed by taking advantage of the pliable experimental system of budding yeast. The behaviour of a constructed, unreplicated circular minichromosome with two centromeres was observed. This chromosome was designed to lack back-to-back kinetochore geometry but was able to generate tension if properly bi-oriented on the spindle. Such a chromosome was able to bi-orient efficiently, indicating that tension across kinetochore pairs is sufficient, and the geometry dispensable for bi-orientation (Dewar et al. 2004). However, this does not preclude a role for geometry in assisting bi-orientation, particularly in organisms such as fission yeast and metazoan cells where merotelic attachments are more likely due to the ability of kinetochores to bind more than one microtubule.

It has been shown that incorrect microtubule attachments do in fact occur in mitotic cells, and that, with the exception of merotelic attachments, they trigger the spindle checkpoint (or wait-anaphase signal), in order that they can be removed prior to anaphase onset. The spindle checkpoint is a surveillance mechanism that senses whether all chromosomes have been properly replicated and all kinetochores are captured, bi-oriented and aligned at metaphase in preparation for chromosome segregation. If these criteria are not fulfilled, the spindle assembly checkpoint prevents premature chromosome separation. This is covered in detail in Chapter 11 of this book.

### *9.4.1 Role of Tension in Bi-orientation*

In addition to proteins necessary for the kinetochore–microtubule attachment, other factors are necessary for ensuring sister kinetochore bi-orientation. The cohesin protein complex is a multiprotein complex required physically to bind sister chromatids together from replication until anaphase onset (Nasmyth 2005), and is necessary for the generation of tension between sister kinetochores when bi-orientation is established during mitosis (Tanaka et al. 2000). Bi-orientation is defective in both yeast and tissue culture cells in which cohesin is depleted (Tanaka et al. 2000; Sonoda et al. 2001). However tension and bi-orientation can be restored, at least partially, by inhibition of topoisomerase II which is required to decatenate sister chromatids after replication (Dewar et al. 2004; Vagnarelli et al. 2004). Thus cohesin seems to provide the physical connection necessary to generate tension that is crucial for establishment of bi-orientation in mitosis. During meiosis the homologous chromosomes are held together through chiasmata and it is these, rather than cohesin, that provide the physical connection responsible for generating tension between homologous kinetochore pairs.

In metazoan cells, although cohesin is removed from the chromosome arms early in mitosis, it is retained at the centromere region, thus maintaining sister chromatid association where spindle forces are strongest. This retention at the centromere crucially depends on the Shugoshin (Sgo) proteins. The founding member of this protein family was originally identified as MEI-S322 involved in *Drosophila* meiosis (Kerrebrock et al. 1995; Moore and Orr-Weaver 1998), and subsequently found to be conserved from yeast to vertebrates (Katis et al. 2004; Kitajima et al. 2004; Marston et al. 2004; Rabitsch et al. 2004) Sgo is required to protect cohesin at centromeres during mitosis (Kitajima et al. 2004; McGuinness et al. 2005) and this protection depends on its co-operation with PP2A phosphatase (Kitajima et al. 2006; Riedel et al. 2006). In yeast by contrast, cohesin is not removed from chromosome arms until anaphase onset, but Sgo is required in yeast for the tension-dependent spindle assembly checkpoint (Indjeian et al. 2005).

### *9.4.2 The Chromosomal Passenger Complex (CPC)*

The Chromosomal Passenger Complex (CPC) is comprised of the kinase Aurora B (Ipl1 in yeast), with the non-enzymic components INCENP (Sli15 in yeast), Survivin (Bir1 in yeast) and Borealin (Dasra B in Xenopus but with no known equivalent in yeast), see (Ruchaud et al. 2007) for a comprehensive review. This complex changes its distribution and plays roles at many stages of mitosis. In particular it has an essential role in the regulation of bi-orientation. Ipl1 is the only Aurora kinase in yeast, whereas metazoan cells express two or three. Yeast cell mutants for either Ipl1 or Sli15 show extensive defects in chromosome segregation (Chan and Botstein 1993; Biggins et al. 1999; He et al. 2001; Tanaka et al. 2002). In such cells, sister centromeres frequently segregate to a single spindle pole. This was found to be due to mono-orientation of the microtubule–kinetochore interactions, and the cell's inability to correct such mis-oriented attachments (Tanaka et al. 2002). The Ipl1–Sli15 complex thus promotes the re-orientation of incorrect kinetochore–microtubule attachments, until correct amphitelic connections are made, generating tension and thus signalling to the complex to cease re-orientation. The functions of this complex are conserved through evolution, as it has been shown that Aurora B is required to correct both syntelic and merotelic attachments in mammalian cells (Hauf et al. 2003; Lampson et al. 2004; Cimini 2007) and defects in Aurora B orthologues in fission yeast, worms and flies leads to impaired bi-orientation and chromosome segregation (Adams et al. 2000; Kaitna et al. 2000; Andrews et al. 2003; Carmena and Earnshaw 2003; Fig. 9.4).

Re-orientation of kinetochore–microtubule attachments, facilitated by Aurora B, may involve transient kinetochore detachment from microtubules, as is indeed found in metazoan cells (Hauf et al. 2003; Lampson et al. 2004). This notion is consistent with the observation that in yeast, reduced Ipl1 activity restores defective kinetochore–microtubule interactions (Pinsky et al. 2003). Such Aurora B-dependent kinetochore detachment from microtubules may be responsible for spindle checkpoint activation in the absence of tension on kinetochores (Biggins and Murray 2001; Kallio et al. 2002; Hauf et al. 2003), However there is recent evidence from yeast that Ipl1 can also activate the checkpoint independently of such detachment (King et al. 2007).

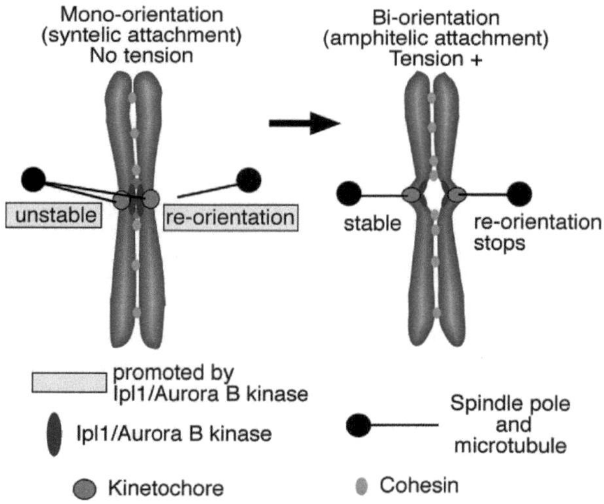


**Fig. 9.4 A model for sister kinetochore bi-orientation** The Ipl1–Sli5/Aurora B–INCENP kinase complex facilitates bi-orientation by promoting the re-orientation of kinetochore–spindle connections in a tension-dependent manner. Because syntelic attachment (*left*) does not generate tension on kinetochore-to-pole connections, the Ipl1–Sli5 complex promotes re-orientation of these connections by phosphorylating kinetochore components such as the Dam1 complex. When an amphitelic attachment is established (*right*), tension is applied on kinetochore-to-pole connections, and as a result the Ipl1–Sli5 complex stops promoting their re-orientation, which causes preferential selection of the amphitelic attachment. It is still unclear how tension leads to the cessation of re-orientation. Figure modified from (Tanaka et al. 2005)

How re-orientation is achieved and the signal(s) generated by tension to stop it are currently not known. It is reasonable to suppose that the Aurora B –INCENP kinase complex promotes re-orientation of kinetochore to pole connections by phosphorylating kinetochore components, because the complex localises at kinetochores from G1 until anaphase onset (He et al. 2001; Buvelot et al. 2003; Tanaka et al. 2005). Indeed it was found that yeast Dam1 and Spc34, two components of the Dam1 kinetochore complex, are phosphorylated by the Ipl1 kinase, and that their phosphorylation is vital to ensure sister kinetochore bi-orientation (Cheeseman et al. 2002). Perhaps the phosphorylation weakens the association between the Dam1 and Ndc80 kinetochore complexes, thereby facilitating the re-orientation of kinetochore–microtubule attachments (Cheeseman et al. 2002; Shang et al. 2003). In vertebrate cells the kinesin13 MCAK is a substrate for Aurora B kinase and this phosphorylation is required to ensure bi-orientation (Andrews et al. 2004; Lan et al. 2004; Ohi et al. 2004; Knowlton et al. 2006; Zhang et al. 2007). Ndc80 has also recently been shown to be an Aurora B substrate in vitro (Cheeseman et al. 2006). The N-terminus of Ndc80, contains Aurora B kinase consensus sites, and is required for high microtubule-binding affinity of the Ndc80/Nuf2 heterodimer. Indeed, the interaction between Ndc80 and microtubules can be significantly reduced in vitro by Aurora B-dependent phosphorylation (Cheeseman et al. 2006). Consistent with this result, microinjection of an antibody to the N-terminus of Ndc80/Hec1 increases improper orientation of kinetochore–microtubule interactions (DeLuca et al. 2006). It remains to be seen how these phosphorylations relate to the presence and/or absence of tension on kinetochores. In one model, it is proposed that the INCENP–Survivin (Sli15-Bir1) sub-complex of the CPC senses the tension and regulates the Aurora B kinase activity accordingly (Sandall et al. 2006).

Given the importance of Aurora B kinase, it seems reasonable that a counteracting phosphatase(s) may also be involved in regulating bi-orientation. The protein phosphatase PP1 and its orthologue Glc7in yeast are the phosphatases that antagonise Aurora B kinases (Francisco et al. 1994), and it appears that yeast Glc7 mutants show elevated rates of chromosome loss and spindle checkpoint activation, indicative of kinetochore detachment from microtubules (Sassoon et al. 1999; Hsu et al. 2000). In human cell lines the protein phosphatase PP1$\gamma$ localises to kinetochores in prometaphase and metaphase (Trinkle-Mulcahy et al. 2003), and thus may have a similar function.

### *9.4.3 Mps1 Kinase*

In addition to Aurora B kinase, Mps1 is another evolutionarily conserved protein kinase, required for the spindle assembly checkpoint and, in some organisms, for duplication of microtubule-organizing centres (Winey and Huneycutt 2002). Separately from these functions, however, Mps1 has an important role in chromosome segregation (Jones et al. 2005). In fact, Mps1 has a crucial role in establishing sister kinetochore bi-orientation on the mitotic spindle (Maure et al. 2007; Jelluma et al. 2008). Similarly to Aurora B, Mps1 promotes re-orientation of kinetochore–spindle pole connections and eliminates those that do not generate tension between sister kinetochores (Maure et al. 2007). In humans, Mps1 regulates Aurora B kinase activity by phosphorylating Borealin/Dasra B that forms a complex with Aurora B (Jelluma et al. 2008); this explains the role of Mps1 in bi-orientation. Budding yeast, however, does not have an orthologue of Borealin/Dasra B, and Mps1 and Aurora B may work in parallel pathways (Maure et al. CB 2007).

### *9.4.4 Congression*

Congression occurs in vertebrate cells where chromosomes move towards, and oscillate around the equator at the metaphase plate in preparation for segregation. It is not clear how this movement occurs. It has been thought that congression happens after a chromosome becomes bi-oriented, i.e. microtubules from the opposite pole bind the unattached sister kinetochore, and pull it back towards the centre of the developing spindle. However it has been shown that congression of a mono-oriented chromosome may also occur before bi-orientation. In this case the unoccupied sister kinetochore may bind to, and slide along the side of a microtubule attached to an already bi-oriented chromosome (Kapoor et al. 2006). The plus end-directed kinesin7 family motor protein CENP-E is responsible for this sliding.

## 9.5 Kinetochore Influence on Microtubule Dynamics

When kinetochores become bound to microtubules, it appears that they are able to exert influence on microtubule dynamics. For instance, in budding yeast, when a kinetochore becomes bound by the lateral surface of a microtubule emanating from a pole, that microtubule frequently switches from disassembly to assembly at its plus (distal) end; i.e. it becomes ‘rescued’. Lateral attachment of kinetochores to microtubules has the advantage of providing a large surface area for microtubule attachment compared with microtubule tips, however a kinetochore may potentially become detached by dropping off the microtubule if it persistently depolymerises, e.g. in yeast the rate of microtubule shrinkage exceeds the rate of poleward kinetochore sliding (Tanaka et al. 2005). The likelihood of kinetochore detachment is decreased by the ‘rescue’ of microtubules once a kinetochore is bound. This undoubtedly contributes to the efficiency of kinetochore capture in the early stages of M-phase (see above). In yeast, the +TIP protein Stu2 (XMAP215/TOG/Dis1) is one of the crucial mediators of this rescue. Stu2 localises at the plus end of microtubules but the amount decreases as microtubules shrink. It is also localised at un-captured kinetochores, and, once capture has occurred it is intermittently transported along the microtubule towards the plus end. Microtubule rescue appears to coincide with the arrival of Stu2 at the plus end (Tanaka et al. 2005). Stu2 and its orthologues are known to alter microtubule dynamics at the plus end in vitro (van Breugel et al. 2003), thus Stu2 seems to be responsible for the kinetochore-dependent microtubule rescue (Fig. 9.2, step 3).

Even after kinetochores are bound to the plus ends of microtubules, they are able to influence microtubule polymerisation and depolymerisation, for instance during congression when already bi-oriented chromosomes move toward the metaphase plate. Here, already attached microtubules must adjust their length by co-ordinated switches between polymerisation and depolymerisation at the kinetochore plus ends (Skibbens et al. 1993; Rieder and Salmon 1994; Gardner et al. 2005). It is thought that these kinetochore movements are based on the dynamic instability of microtubule plus ends, which is partly controlled by tension at the kinetochore. In mammalian cultured cells, low tension promotes switching to kinetochore microtubule polymerisation, and high tension promotes switching to kinetochore microtubule depolymerisation (Rieder and Salmon 1994; Maiato et al. 2004). Indeed, these microtubule dynamics have recently been recapitulated in vitro by application of tension to the microtubule-associated Dam1 complex in an optical trap (Franck, Powers et al 2007).

Budding yeast have no clear metaphase plate at which chromosomes align, but the kinetochores are localised and oscillate in a zone between the SPBs (Goshima and Yanagida 2000; He et al. 2000; Tanaka et al. 2000; Pearson et al. 2001). Stu2 is required for such metaphase oscillations of the kinetochores (He et al. 2001; Pearson et al. 2003) and, as in vertebrate cells, tension between microtubules and kinetochores is thought to regulate microtubule dynamics (Gardner et al. 2005).

Continual poleward flux is a feature of spindle microtubules throughout mitosis in metazoan cells and can be visualised by 'fluorescence speckle' microscopy in which a small proportion of tubulin subunits is fluorescently labelled and incorporated into the microtubule lattice; these can be tracked moving towards the spindle poles (Waterman-Storer et al. 1998; Maddox et al. 2002). Exactly how spindle microtubule dynamics are regulated at kinetochores and poles is not known, but the +TIP protein CLASP (MAST/Orbit and Stu1 in *Drosophila* & yeast respectively) is involved in maintaining dynamic behaviour of microtubules at kinetochores (Maiato et al. 2003), see (Kapoor and Compton 2002) for a more extensive review.

## 9.6 Anaphase

After successful capture and bi-orientation of all chromosomes and kinetochores the cell is able to progress into anaphase and segregate the chromosomes. Major poleward forces that achieve this segregation are generated in two ways;

by microtubule plus end disassembly at the kinetochore (Gorbsky et al. 1987; the so-called Pacman mechanism), and also by microtubule flux, which transports (or 'reels in') the whole microtubule with its attached kinetochore polewards by disassembly at the pole (minus-end; Mitchison and Salmon 2001; Rogers et al. 2005). The nature of the anaphase microtubule flux and its regulation may indeed differ from that in metaphase, where no net change in length of microtubules is apparent. The precise mechanistic details of either the Pacman or flux processes are unclear. The disassembly of the microtubule ends may be governed by numerous regulatory factors some of which may be members of the microtubule-depolymerising kinesins. The generation of force by disassembly at the kinetochore might be brought about by the outward curling of tubulin protofilaments from the plus end as it disassembles (Grishchuk et al. 2005; Efremov et al. 2007), e.g. pushing against a restraining collar around the microtubule at the kinetochore. In yeast this restraining collar may be the ring of the Dam1 complex, and thus moving it and its attached kinetochore/chromosome polewards (Westermann et al. 2006; Tanaka et al. 2007).

The relative contributions of microtubule flux and kinetochore-based disassembly to anaphase chromosome segregation differs, depending on the cell type. Microtubule flux plays a major role at anaphase in *Xenopus* egg extracts and *Drosophila* embryos (Maddox et al. 2002), whereas in vertebrate cells and yeast disassembly at the kinetochore may be the main force-generator, with flux contributing only 25–30% to poleward movement in vertebrates (Mitchison and Salmon 1992; Compton 2002).

In *S. cerevisiae*, poleward flux seems not to occur in either astral cytoplasmic or interpolar microtubules (Maddox et al. 2000), or in kinetochore microtubules (Maddox et al. 2000; Tanaka et al. 2005). Moreover FRAP experiments addressed overall dynamics of spindle microtubules in yeast and revealed their cell cycle-dependent regulation (Higuchi and Uhlmann 2005); an increase in dynamics at metaphase is followed by a decrease in turnover at anaphase. This change in anaphase requires the release of phosphatase Cdc14 from the nucleolus. This conserved phosphatase is also responsible for dephosphorylating INCENP at kinetochores, thereby redirecting the CPC complex from the kinetochore to the spindle midzone in anaphase (Pereira and Schiebel 2003). Removal of the CPC complex from kinetochores is presumably necessary to prevent further microtubule re-orientation during anaphase, when tension applied on kinetochores is much reduced relative to metaphase.

## 9.7 Conclusions and Perspectives

There is still much to discover about how kinetochores interact with microtubules during the different stages of cell division, and how those interactions are regulated. Central questions remaining to be answered are; how tension, or the lack of it, is sensed, enabling mal-oriented kinetochore–microtubule connections to be corrected; and what is the mechanism by which phosphorylation of kinetochore components by Aurora B/Ipl1 kinase leads to re-orientation. In addition, our understanding of the influence of kinetochores and associated proteins on microtubule assembly and disassembly is still limited. Doubtless many of these questions will be answered by the combined power of genetics, biochemistry and rapidly developing advances in cell biological methods.

**Acknowledgments** Work in the authors' laboratory was supported by Cancer Research UK, the Wellcome Trust, the Human Frontier Science Program, the Lister Research Institute Prize and the Association for International Cancer Research. T.U. Tanaka is a Senior Research Fellow of Cancer Research UK.

## References

Adams, R. R., S. P. Wheatley, et al. 2000. INCENP binds the Aurora-related kinase AIRK2 and is required to target it to chromosomes, the central spindle and cleavage furrow. Curr Biol 10: 1075–8.
Andrews, P. D., E. Knatko, et al. 2003. Mitotic mechanics: the auroras come into view. Curr Opin Cell Biol 15: 672–83.
Andrews, P. D., Y. Ovechkina, et al. 2004. Aurora B regulates MCAK at the mitotic centromere. Dev Cell 6: 253–68.
Asbury, C. L., D. R. Gestaut, et al. 2006. The Dam1 kinetochore complex harnesses microtubule dynamics to produce force and movement. Proc Natl Acad Sci USA 103: 9873–8.
Biggins, S. and A. W. Murray 2001. The budding yeast protein kinase Ipl1/Aurora allows the absence of tension to activate the spindle checkpoint. Genes Dev 15: 3118–29.
Biggins, S., F. F. Severin, et al. 1999. The conserved protein kinase Ipl1 regulates microtubule binding to kinetochores in budding yeast. Genes Dev 13: 532–44.
Buvelot, S., S. Y. Tatsutani, et al. 2003. The budding yeast Ipl1/Aurora protein kinase regulates mitotic spindle disassembly. J Cell Biol 160: 329–39.
Carazo-Salas, R. E., O. J. Gruss, et al. 2001. Ran-GTP coordinates regulation of microtubule nucleation and dynamics during mitotic-spindle assembly. Nat Cell Biol 3: 228–34.
Carmena, M. and W. C. Earnshaw 2003. The cellular geography of aurora kinases. Nat Rev Mol Cell Biol 4: 842–54.
Caudron, M., G. Bunt, et al. 2005. Spatial coordination of spindle assembly by chromosome-mediated signaling gradients. Science 309: 1373–6.
Chan, C. S. and D. Botstein 1993. Isolation and characterization of chromosome-gain and increase-in-ploidy mutants in yeast. Genetics 135: 677–91.
Cheeseman, I. M., S. Anderson, et al. 2002. Phospho-regulation of kinetochore-microtubule attachments by the Aurora kinase Ipl1p. Cell 111: 163–72.
Cheeseman, I. M., J. S. Chappie, et al. 2006. The conserved KMN network constitutes the core microtubule-binding site of the kinetochore. Cell 127: 983–97.
Cimini, D. 2007. Detection and correction of merotelic kinetochore orientationby Aurora B and its partners. Cell Cycle 6: 1558–64.
Compton, D. A. 2002. Chromosome segregation: pulling from the poles. Curr Biol 12: R651–3.
Czaban, B. B. and A. Forer 1985. The kinetic polarities of spindle microtubules in vivo, in crane-fly spermatocytes. I. Kinetochore microtubules that re-form after treatment with colcemid. J Cell Sci 79: 1–37.
De Brabander, M., G. Geuens, et al. 1981. Nucleated assembly of mitotic microtubules in living PTK2 cells after release from nocodazole treatment. Cell Motil 1: 469–83.

DeLuca, J. G., W. E. Gall, et al. 2006. Kinetochore microtubule dynamics and attachment stability are regulated by Hec1. Cell 127: 969–82.
Dewar, H., K. Tanaka, et al. 2004. Tension between two kinetochores suffices for their bi-orientation on the mitotic spindle. Nature 428: 93–7.
Dong, Y., K. J. Vanden Beldt, et al. 2007. The outer plate in vertebrate kinetochores is a flexible network with multiple microtubule interactions. Nat Cell Biol 9: 516–22.
Efremov, A., E. L. Grishchuk, et al. 2007. In search of an optimal ring to couple microtubule depolymerization to processive chromosome motions. Proc Natl Acad Sci U S A 104: 19017–22.
Emanuele, M. J. and P. T. Stukenberg 2007. Xenopus Cep57 is a novel kinetochore component involved in microtubule attachment. Cell 130: 893–905.
Francisco, L., W. Wang, et al. 1994. Type 1 protein phosphatase acts in opposition to IpL1 protein kinase in regulating yeast chromosome segregation. Mol Cell Biol 14: 4731–40.
Franck, A. D., A. F. Powers, et al. 2007. Tension applied through the Dam1 complex promotes microtubule elongation providing a direct mechanism for length control in mitosis. Nat Cell Biol 9: 832–7.
Gardner, M. K., C. G. Pearson, et al. 2005. Tension-dependent regulation of microtubule dynamics at kinetochores can explain metaphase congression in yeast. Mol Biol Cell 16: 3764–75.
Gorbsky, G. J., P. J. Sammak, et al. 1987. Chromosomes move poleward in anaphase along stationary microtubules that coordinately disassemble from their kinetochore ends. J Cell Biol 104: 9–18.
Goshima, G. and M. Yanagida 2000. Establishing biorientation occurs with precocious separation of the sister kinetochores, but not the arms, in the early spindle of budding yeast. Cell 100: 619–33.
Grishchuk, E. L., M. I. Molodtsov, et al. 2005. Force production by disassembling microtubules. Nature 438: 384–8.
Gruss, O. J. and I. Vernos 2004. The mechanism of spindle assembly: functions of Ran and its target TPX2. J Cell Biol 1667: 94–55.
Hanisch, A., H. H. Sillje, et al. 2006. Timely anaphase onset requires a novel spindle and kinetochore complex comprising Ska1 and Ska2. Embo J 25: 5504–15.
Hauf, S., R. W. Cole, et al. 2003. The small molecule Hesperadin reveals a role for Aurora B in correcting kinetochore-microtubule attachment and in maintaining the spindle assembly checkpoint. J Cell Biol 161: 281–94.
Hayden, J. H., S. S. Bowser, et al. 1990. Kinetochores capture astral microtubules during chromosome attachment to the mitotic spindle: direct visualization in live newt lung cells. J Cell Biol 111: 1039–45.
He, X., S. Asthana, et al. 2000. Transient sister chromatid separation and elastic deformation of chromosomes during mitosis in budding yeast. Cell 101: 763–75.
He, X., D. R. Rines, et al. 2001. Molecular analysis of kinetochore-microtubule attachment in budding yeast. Cell 106: 195–206.
Hegemann, J. H. and U. N. Fleig 1993. The centromere of budding yeast. Bioessays 15: 451–60.
Higuchi, T. and F. Uhlmann 2005. Stabilization of microtubule dynamics at anaphase onset promotes chromosome segregation. Nature 433: 171–6.
Hildebrandt, E. R. and M. A. Hoyt 2000. Mitotic motors in Saccharomyces cerevisiae. Biochim Biophys Acta 1496: 99–116.
Hsu, J. Y., Z. W. Sun, et al. 2000. Mitotic phosphorylation of histone H3 is governed by Ipl1/aurora kinase and Glc7/PP1 phosphatase in budding yeast and nematodes. Cell 102: 279–91.
Indjeian, V. B., B. M. Stern, et al. 2005. The centromeric protein Sgo1 is required to sense lack of tension on mitotic chromosomes. Science 307: 130–3.

Jelluma, N., A. B. Brenkman, et al. 2008. Mps1 phosphorylates Borealin to control Aurora B activity and chromosome alignment. Cell 132: 233–46.
Jones, M. H., B. J. Huneycutt, et al. 2005. Chemical genetics reveals a role for Mps1 kinase in kinetochore attachment during mitosis. Curr Biol 15: 160–5.
Joseph, J., S. H. Tan, et al. 2002. SUMO-1 targets RanGAP1 to kinetochores and mitotic spindles. J Cell Biol 156: 595–602.
Kaitna, S., M. Mendoza, et al. 2000. Incenp and an aurora-like kinase form a complex essential for chromosome segregation and efficient completion of cytokinesis. Curr Biol 10: 1172–81.
Kalab, P., R. T. Pu, et al. 1999. The ran GTPase regulates mitotic spindle assembly. Curr Biol 9: 481–4.
Kalab, P., K. Weis, et al. 2002. Visualization of a Ran-GTP gradient in interphase and mitotic Xenopus egg extracts. Science 295: 2452–6.
Kallio, M. J., M. L. McCleland, et al. 2002. Inhibition of aurora B kinase blocks chromosome segregation, overrides the spindle checkpoint, and perturbs microtubule dynamics in mitosis. Curr Biol 12: 900–5.
Kapoor, T. M. and D. A. Compton 2002. Searching for the middle ground: mechanisms of chromosome alignment during mitosis. J Cell Biol 157: 551–6.
Kapoor, T. M., M. A. Lampson, et al. 2006. Chromosomes can congress to the metaphase plate before biorientation. Science 311: 388–91.
Katis, V. L., M. Galova, et al. 2004. Maintenance of cohesin at centromeres after meiosis I in budding yeast requires a kinetochore-associated protein related to MEI-S332. Curr Biol 14: 560–72.
Kerrebrock, A. W., D. P. Moore, et al. 1995. Mei-S332, a Drosophila protein required for sister-chromatid cohesion, can localize to meiotic centromere regions. Cell 83: 247–56.
Khodjakov, A., R. W. Cole, et al. 2000. Centrosome-independent mitotic spindle formation in vertebrates. Curr Biol 10: 59–67.
Khodjakov, A., L. Copenagle, et al. 2003. Minus-end capture of preformed kinetochore fibers contributes to spindle morphogenesis. J Cell Biol 160: 671–83.
King, E. M., N. Rachidi, et al. 2007. Ipl1p-dependent phosphorylation of Mad3p is required for the spindle checkpoint response to lack of tension at kinetochores. Genes Dev 21: 1163–8.
King, J. M., T. S. Hays, et al. 2000. Dynein is a transient kinetochore component whose binding is regulated by microtubule attachment, not tension. J Cell Biol 151: 739–48.
Kirschner, M. and T. Mitchison 1986. Beyond self-assembly: from microtubules to morphogenesis. Cell 45: 329–42.
Kitajima, T. S., S. A. Kawashima, et al. 2004. The conserved kinetochore protein shugoshin protects centromeric cohesion during meiosis. Nature 427: 510–7.
Kitajima, T. S., T. Sakuno, et al. 2006. Shugoshin collaborates with protein phosphatase 2A to protect cohesin. Nature 441: 46–52.
Kitamura, E., K. Tanaka et al. 2007. Kinetochore microtubule interaction during S phase in Saccharomyces Cerevisiae. Genes Dev 21: 3319–3330.
Knowlton, A. L., W. Lan, et al. 2006. Aurora B is enriched at merotelic attachment sites, where it regulates MCAK. Curr Biol 1617: 17–10.
Lampson, M. A., K. Renduchitala, et al. 2004. Correcting improper chromosome-spindle attachments during cell division. Nat Cell Biol 6: 232–7.
Lan, W., X. Zhang, et al. 2004. Aurora B phosphorylates centromeric MCAK and regulates its localization and microtubule depolymerization activity. Curr Biol 14: 273–86.
Liu, X., I. McLeod, et al. 2005. Molecular analysis of kinetochore architecture in fission yeast. Embo J 24: 2919–30.
Maddox, P., A. Desai, et al. 2002. Poleward microtubule flux is a major component of spindle dynamics and anaphase a in mitotic Drosophila embryos. Curr Biol 12: 1670–4.
Maddox, P. S., K. S. Bloom, et al. 2000. The polarity and dynamics of microtubule assembly in the budding yeast Saccharomyces cerevisiae. Nat Cell Biol 2: 36–41.

Maiato, H., J. DeLuca, et al. 2004. The dynamic kinetochore-microtubule interface. J Cell Sci 117: 5461–77.
Maiato, H., E. A. Fairley, et al. 2003. Human CLASP1 is an outer kinetochore component that regulates spindle microtubule dynamics. Cell 113: 891–904.
Maiato, H., C. L. Rieder, et al. 2004. Kinetochore-driven formation of kinetochore fibers contributes to spindle assembly during animal mitosis. J Cell Biol 167: 831–40.
Maiato, H., P. Sampaio, et al. 2002. MAST/Orbit has a role in microtubule-kinetochore attachment and is essential for chromosome alignment and maintenance of spindle bipolarity. J Cell Biol 157: 749–60.
Marston, A. L., W. H. Tham, et al. 2004. A genome-wide screen identifies genes required for centromeric cohesion. Science 303: 1367–70.
Maure, J. F., E. Kitamura, et al. 2007. Mps1 kinase promotes sister-kinetochore bi-orientation by a tension-dependent mechanism. Curr Biol 17: 2175–82.
McAinsh, A. D., J. D. Tytell, et al. 2003. Structure, function, and regulation of budding yeast kinetochores. Annu Rev Cell Dev Biol 19: 519–39.
McGuinness, B. E., T. Hirota, et al. 2005. Shugoshin prevents dissociation of cohesin from centromeres during mitosis in vertebrate cells. PLoS Biol 3: e86.
Meraldi, P. and P. K. Sorger 2005. A dual role for Bub1 in the spindle checkpoint and chromosome congression. Embo J 24: 1621–33.
Miranda, J. J., D. S. King, et al. 2007. Protein arms in the kinetochore-microtubule interface of the yeast DASH complex. Mol Biol Cell 18: 2503–10.
Mitchison, T. and M. Kirschner 1984. Dynamic instability of microtubule growth. Nature 312: 237–42.
Mitchison, T. J. and E. D. Salmon 1992. Poleward kinetochore fiber movement occurs during both metaphase and anaphase-A in newt lung cell mitosis. J Cell Biol 119: 569–82.
Mitchison, T. J. and E. D. Salmon 2001. Mitosis: a history of division. Nat Cell Biol 3: E17–21.
Moore, D. P. and T. L. Orr-Weaver 1998. Chromosome segregation during meiosis: building an unambivalent bivalent. Curr Top Dev Biol 37: 263–99.
Nasmyth, K. 2005. How might cohesin hold sister chromatids together? Philos Trans R Soc Lond B Biol Sci 360: 483–96.
Nicklas, R. B. and C. A. Koch 1969. Chromosome micromanipulation. 3. Spindle fiber tension and the reorientation of mal-oriented chromosomes. J Cell Biol 43: 40–50.
O'Toole, E. T., M. Winey, et al. 1999. High-voltage electron tomography of spindle pole bodies and early mitotic spindles in the yeast Saccharomyces cerevisiae. Mol Biol Cell 10: 2017–31.
Ohi, R., T. Sapra, et al. 2004. Differentiation of cytoplasmic and meiotic spindle assembly MCAK functions by Aurora B-dependent phosphorylation. Mol Biol Cell 15: 2895–906.
Pearson, C. G., P. S. Maddox, et al. 2001. Budding yeast chromosome structure and dynamics during mitosis. J Cell Biol 152: 1255–66.
Pearson, C. G., P. S. Maddox, et al. 2003. Yeast kinetochores do not stabilize Stu2p-dependent spindle microtubule dynamics. Mol Biol Cell 14: 4181–95.
Pereira, G. and E. Schiebel 2003. Separase regulates INCENP-Aurora B anaphase spindle function through Cdc14. Science 302: 2120–4.
Pfarr, C. M., M. Coue, et al. 1990. Cytoplasmic dynein is localized to kinetochores during mitosis. Nature 345: 263–5.
Pickett-Heaps, J. D. 1991. Cell division in diatoms. Int. Rev.Cytol. 128: 63–107.
Pinsky, B. A., S. Y. Tatsutani, et al. 2003. An Mtw1 complex promotes kinetochore biorientation that is monitored by the Ipl1/Aurora protein kinase. Dev Cell 5: 735–45.
Porter, I. M., S. E. McClelland, et al. 2007. Bod1, a novel kinetochore protein required for chromosome biorientation. J Cell Biol 179: 187–97.
Rabitsch, K. P., J. Gregan, et al. 2004. Two fission yeast homologs of Drosophila Mei-S332 are required for chromosome segregation during meiosis I and II. Curr Biol 14: 287–301.

Riedel, C. G., V. L. Katis, et al. 2006. Protein phosphatase 2A protects centromeric sister chromatid cohesion during meiosis I. Nature 441: 53–61.

Rieder, C. L. and S. P. Alexander 1990. Kinetochores are transported poleward along a single astral microtubule during chromosome attachment to the spindle in newt lung cells. J Cell Biol 110: 81–95.

Rieder, C. L. and E. D. Salmon 1994. Motile kinetochores and polar ejection forces dictate chromosome position on the vertebrate mitotic spindle. J Cell Biol 124: 223–33.

Rogers, G. C., S. L. Rogers, et al. 2005. Spindle microtubules in flux. J Cell Sci 118: 1105–16.

Ruchaud, S., M. Carmena, et al. 2007. Chromosomal passengers: conducting cell division. Nat Rev Mol Cell Biol 8: 798–812.

Sanchez-Perez, I., S. J. Renwick, et al. 2005. The DASH complex and Klp5/Klp6 kinesin coordinate bipolar chromosome attachment in fission yeast. Embo J 24: 2931–43.

Sandall, S., F. Severin, et al. 2006. A Bir1-Sli15 complex connects centromeres to microtubules and is required to sense kinetochore tension. Cell 127: 1179–91.

Sassoon, I., F. F. Severin, et al. 1999. Regulation of Saccharomyces cerevisiae kinetochores by the type 1 phosphatase Glc7p. Genes Dev 13: 545–55.

Saxton, W. M., D. L. Stemple, et al. 1984. Tubulin dynamics in cultured mammalian cells. J Cell Biol 99: 2175–86.

Shang, C., T. R. Hazbun, et al. 2003. Kinetochore protein interactions and their regulation by the Aurora kinase Ipl1p. Mol Biol Cell 14: 3342–55.

Skibbens, R. V., V. P. Skeen, et al. 1993. Directional instability of kinetochore motility during chromosome congression and segregation in mitotic newt lung cells: a push-pull mechanism. J Cell Biol 122: 859–75.

Sonoda, E., T. Matsusaka, et al. 2001. Scc1/Rad21/Mcd1 is required for sister chromatid cohesion and kinetochore function in vertebrate cells. Dev Cell 1: 759–70.

Starr, D. A., B. C. Williams, et al. 1998. ZW10 helps recruit dynactin and dynein to the kinetochore. J Cell Biol 142: 763–74.

Tanaka, K., E. Kitamura, et al. 2007. Molecular mechanisms of microtubule-dependent kinetochore transport toward spindle poles. J Cell Biol. 178: 269–281.

Tanaka, K., N. Mukae, et al. 2005. Molecular mechanisms of kinetochore capture by spindle microtubules. Nature 434: 987–94.

Tanaka, T., J. Fuchs, et al. 2000. Cohesin ensures bipolar attachment of microtubules to sister centromeres and resists their precocious separation. Nat Cell Biol 2: 492–9.

Tanaka, T. U., N. Rachidi, et al. 2002. Evidence that the Ipl1-Sli15 (Aurora kinase-INCENP) complex promotes chromosome bi-orientation by altering kinetochore-spindle pole connections. Cell 108: 317–29.

Tanaka, T. U., M. J. Stark, et al. 2005. Kinetochore capture and bi-orientation on the mitotic spindle. Nat Rev Mol Cell Biol 6: 929–42.

Telzer, B. R., M. J. Moses, et al. 1975. Assembly of microtubules onto kinetochores of isolated mitotic chromosomes of HeLa cells. Proc Natl Acad Sci USA 72: 4023–7.

Tippit, D. H., J. D. Pickett-Heaps, et al. 1980. Cell division in two large pennate diatoms Hantzschia and Nitzschia III. A new proposal for kinetochore function during prometaphase. J Cell Biol 86: 402–16.

Tytell, J. D. and P. K. Sorger 2006. Analysis of kinesin motor function at budding yeast kinetochores. J Cell Biol 172: 861–74.

Vagnarelli, P., C. Morrison, et al. 2004. Analysis of Scc1-deficient cells defines a key metaphase role of vertebrate cohesin in linking sister kinetochores. EMBO Rep 5: 167–71.

van Breugel, M., D. Drechsel, et al. 2003. Stu2p, the budding yeast member of the conserved Dis1/XMAP215 family of microtubule-associated proteins is a plus end-binding microtubule destabilizer. J Cell Biol 161: 359–69.

Wadsworth, P. and A. Khodjakov 2004. E pluribus unum: towards a universal mechanism for spindle assembly. Trends Cell Biol 14: 413–9.

Waterman-Storer, C. M., A. Desai, et al. 1998. Fluorescent speckle microscopy, a method to visualize the dynamics of protein assemblies in living cells. Curr Biol 8: 1227–30.
Wei, R. R., J. Al-Bassam, et al. 2007. The Ndc80/HEC1 complex is a contact point for kinetochore-microtubule attachment. Nat Struct Mol Biol 14: 54–9.
Wei, R. R., P. K. Sorger, et al. 2005. Molecular organization of the Ndc80 complex, an essential kinetochore component. Proc Natl Acad Sci USA 102: 5363–7.
Westermann, S., A. Avila-Sakar, et al. 2005. Formation of a dynamic kinetochore- microtubule interface through assembly of the Dam1 ring complex. Mol Cell 17: 277–90.
Westermann, S., H. W. Wang, et al. 2006. The Dam1 kinetochore ring complex moves processively on depolymerizing microtubule ends. Nature 440: 565–9.
Winey, M. and B. J. Huneycutt 2002. Centrosomes and checkpoints: the MPS1 family of kinases. Oncogene 21: 6161–9.
Winey, M., C. L. Mamay, et al. 1995. Three-dimensional ultrastructural analysis of the Saccharomyces cerevisiae mitotic spindle. J Cell Biol 129: 1601–15.
Witt, P. L., H. Ris, et al. 1980. Origin of kinetochore microtubules in Chinese hamster ovary cells. Chromosoma 81: 483–505.
Wittmann, T., A. Hyman, et al. 2001. The spindle: a dynamic assembly of microtubules and motors. Nat Cell Biol 3: E28–34.
Wollman, R., E. N. Cytrynbaum, et al. 2005. Efficient chromosome capture requires a bias in the 'search-and-capture' process during mitotic-spindle assembly. Curr Biol 15: 828–32.
Yang, Z., U. S. Tulu, et al. 2007. Kinetochore dynein is required for chromosome motion and congression independent of the spindle checkpoint. Curr Biol 17: 973–80.
Zhang, X., W. Lan, et al. 2007. Aurora B Phosphorylates Multiple Sites on MCAK to Spatially and Temporally Regulate Its Function. Mol Biol Cell. 18: 3264–3276.

# Chapter 10
# Post-Translational Modifications that Regulate Kinetochore Activity

**Chitra V. Kotwaliwale and Sue Biggins**

## 10.1 Introduction

To ensure accurate chromosome segregation, the interactions between kinetochores and microtubules must be precisely regulated throughout every mitotic and meiotic cell cycle. Following genome replication, the kinetochores of duplicated chromosomes (sister chromatids) must make bioriented attachments to microtubules arising from opposite spindle poles to ensure that chromosomes are properly segregated at anaphase (see Chapters 2 and 10 for details). At metaphase, sister kinetochores come under tension due to the pulling forces of microtubules on linked sister chromatids. When the cell does not make proper kinetochore–microtubule attachments, a signal transduction system called the spindle checkpoint delays the metaphase to anaphase transition to give the cell time to correct the defects (for review, see Musacchio and Salmon 2007). Elucidating the mechanisms that control kinetochore function is daunting given the extensive number of kinetochore proteins that have recently been identified, as well as the dynamic changes that kinetochores and microtubules undergo throughout the cell cycle. We are now faced with the challenge of determining the precise molecular role that each kinetochore protein contributes to kinetochore structure and function. To add to this complexity, many kinetochore proteins are subject to additional controls, such as post-translational modifications, that further modulate kinetochore activity. Although these regulatory mechanisms are critical to ensure the fidelity of chromosome segregation, we are just beginning to identify and characterize the functions of modifications that occur on kinetochore proteins. Here, we review six types of post-translational modifications that have demonstrated roles in regulating kinetochore activity: phosphorylation, ubiquitylation, sumoylation, methylation, acetylation, and farnesylation.

Sue Biggins (✉)
Division of Basic Sciences,
Fred Hutchinson Cancer Research Center,
1100 Fairview Ave N,
Seattle, WA 98109
e-mail: sbiggins@fhcrc.org

P. De Wulf, W.C. Earnshaw (eds.), *The Kinetochore*,
DOI 10.1007/978-0-387-69076-6_10, 

After introducing the modifications and key modifying enzymes, we review their contributions to the assembly of centromeric chromatin and the regulation of microtubule–kinetochore interactions during mitosis. We refer the reader to Chapter 12 for a discussion of the role of modifications in the spindle checkpoint, and Chapter 14 for a review of kinetochore activity during meiosis.

## 10.2 The Post-Translational Modifications

### *10.2.1 Phosphorylation*

Phosphorylation is the most well-studied and commonly employed modification in living systems. Phosphorylation is catalyzed by kinases that transfer the $\gamma$-phosphate from ATP to hydroxyls of serines, threonines, or tyrosines within substrate proteins, while dephosphorylation is catalyzed by phosphatases. Given the widespread role of phosphoproteins in all basic cellular processes, such as cell division, differentiation and metabolism, the 518 kinases identified in humans constitute one of the largest gene families, making up $\sim$2% of the eukaryotic genome (Manning et al., 2002). The importance of phosphorylation in cell cycle regulation first became apparent with the discovery of the master regulator of the eukaryotic cell cycle, cyclin-dependent kinase (Cdk1; for review, see Bloom and Cross 2007). It is now clear that many kinases and phosphatases localize to the kinetochore and have critical roles in regulating kinetochore assembly and function. The negative charge introduced by phosphorylation can lead to dynamic changes in protein–protein interactions, alter the stability of substrate proteins, create binding sites for partner proteins, or change the catalytic activity of enzymes.

Protein kinases are divided into two main groups: those that phosphorylate serine/threonine residues (Ser/Thr kinases) and those that phosphorylate tyrosine residues (Tyr kinases). The residues that flank the phosphorylated residue typically contribute to kinase–substrate recognition and constitute a consensus site for a given kinase (http://scansite.mit.edu/). The consensus sites of several mitotic kinases have been mapped and have contributed significantly to the identification of their substrates. Like most modifications, phosphorylation is reversible and the balance between phosphorylation and dephosphorylation is important for protein function. The dephosphorylation of substrates is catalyzed by phosphatases that are classified into either Ser/Thr, Tyr, or dual-specificity (Ser/Thr and Tyr) phosphatases (for review, see Trinkle-Mulcahy and Lamond 2006). Despite mounting evidence that dephosphorylation also regulates the cell cycle, comparatively less is known about the specificity and activity of phosphatases in regulating kinetochore function than kinases.

### *10.2.2 Ubiquitylation and Sumoylation*

Modification with ubiquitin or ubiquitin-like proteins represents a unique situation in that the modification itself is a polypeptide. Ubiquitin-mediated proteolysis is a common mechanism to regulate protein function that is involved in both centromere specification and kinetochore assembly. Proteins are targeted for degradation by the 26S proteasome when four or more ubiquitin molecules are conjugated in a multi-step reaction (for reviews, see Weissman 2001; Glickman and Ciechanover 2002; Kerscher, Felberbaum and Hochstrasser 2006). Ubiquitin is initially synthesized as a fusion protein that is processed by cleavage at the carboxy-terminal glycine residue, resulting in ubiquitin monomers. A ubiquitin-activating enzyme (E1) forms a thiol-ester bond with ubiquitin in an ATP-dependent manner, and then a ubiquitin-conjugating enzyme (E2) accepts ubiquitin from the E1. The E2 transfers ubiquitin to the $\varepsilon$-amino group of a lysine residue within the substrate. This reaction is often catalyzed by an E3 ubiquitin protein ligase that is also primarily responsible for substrate specificity. The addition of ubiquitin monomers to the previously conjugated ubiquitin via a lysine 48 linkage within ubiquitin results in the formation of ubiquitin chains that are subsequently recognized by the 26S proteasome. While conjugation of a substrate via lysine 48 of ubiquitin results in degradation, one of the other lysines in ubiquitin, such as lysine 63, can also serve to covalently modify a substrate. However, this modification does not signal degradation and instead regulates protein interactions (for review, see Sun and Chen 2004). A consensus sequence for ubiquitylation has not been defined, and ubiquitin is usually conjugated to multiple lysines in substrate proteins. It is often difficult to experimentally prevent ubiquitylation of target proteins because alternative lysines can be employed when a preferred lysine is removed. Similar to phosphorylation, ubiquitylation is reversible and is catalyzed by de-ubiquitylating enzymes (DUBs), some of which have critical roles in cell cycle progression (Vong et al., 2005; Reddy et al., 2007; Stegmeier et al., 2007). The importance of DUBs in the cell cycle suggests that there is a careful balance between protein stability and protein destruction, but it is not known how this balance is attained.

Ubiquitylation can also alter protein function in ways that do not involve protein degradation. For example, histones and a number of proteins involved in endocytosis are mono-ubiquitylated (for review, see Sigismund, Polo and Di Fiore 2004). It has been postulated that mono-ubiquitylation of histones results in an "opening up" of chromatin structure. Despite the versatility of ubiquitylation, mono-ubiquitylation has not yet been implicated in kinetochore function.

Several other proteins similar to ubiquitin also post-translationally modify proteins (for reviews, see Johnson 2004; Welchman, Gordon and Mayer 2005). One of these modifications, the small ubiquitin-like molecule (SUMO), has important roles in chromosome segregation (for review, see Watts 2007). Similar to ubiquitin, SUMO is highly conserved. While invertebrates contain a single SUMO gene called *SMT3/pmt3*, three SUMO family members (SUMO-1, 2, and 3) have been identified in vertebrates, with SUMO-1 being the most closely related to the invertebrate SUMO gene. SUMO-2 and SUMO-3 are almost identical, so they are called SUMO-2/3. SUMO-1 localizes to the mitotic spindle and midzone, while SUMO-2/3 localize to centromeres and condensed chromosomes (Zhang et al., 2008). The sumoylation pathway is mechanistically similar to ubiquitylation, in that SUMO is synthesized as an inactive precursor, undergoes a C-terminal proteolytic cleavage, and is subsequently conjugated to substrates in multiple steps requiring distinct E1, E2, and E3 enzyme activities. Sumoylation is reversed by isopeptidases called Ubl (yeast) or SENP (mammals; for review, see Mukhopadhyay and Dasso 2007). In contrast to the ubiquitin reactions, there is only a single E2 that has been identified to date and a sumoylation consensus site has been mapped. Sumoylation most commonly occurs on lysine residues found in the consensus tetrapeptide motif ψKXE/D, where ψ is any large hydrophobic residue (Rodriguez, Dargemont and Hay 2001). Instead of leading to protein degradation, sumoylation has been implicated in altering protein localization and protein–protein interactions. SUMO has generated significant interest in the mitosis field following the recent identification of a number of sumoylated kinetochore proteins (Table 10.1).

### *10.2.3 Methylation and Acetylation*

Protein methylation and acetylation are two other key modifications with important roles in centromere specification and kinetochore function (for reviews, see Schueler and Sullivan 2006; An 2007). Methylation characteristically occurs on lysine or arginine residues. Lysines can be mono-, di- or trimethylated, while arginines can be mono- or dimethylated. In addition, arginine dimethylation can be symmetric or asymmetric. Protein methylation has been intensely studied, especially in the context of chromatin, and found to influence diverse processes such as gene regulation, signal transduction, and chromatin remodeling (for review, see Martin and Zhang 2005). Methylation does not change the overall charge of the substrate protein, so its effects are usually related to changes in the interaction of the substrate with other proteins. The enzymes responsible for methylation are called methyltransferases and they exhibit specificity toward the amino acid they methylate. In addition, methyltransferases often generate a defined methylation state such that a single lysine must be modified by more than one methyltransferase to become di- or trimethylated. The consensus motifs recognized by a few methyltransferases have been identified and appear to be specific to a given enzyme (Couture and Trievel 2006; Porras-Yakushi, Whitelegge and Clarke 2006). Given the large number of methyltransferases present in the genome, it is likely that the consensus motifs are highly divergent, thus contributing to substrate specificity. Although methylation was long thought to be a static modification, the recent identification of demethylases has demonstrated that it is also a dynamic protein modification (for review, see Klose and Zhang 2007). To date, only a single histone arginine demethylase has been identified (Chang et al., 2007). However, arginine residues can be covalently modified by deimination to produce citrulline in a reaction that can either remove or inhibit methylation (for review, see Klose and Zhang 2007).

**Table 10.1** Post-translational modifications of kinetochore proteins. Note that modifications to the key modifying enzymes are not included

| Substrate | Modification | Sites[a] | Enzyme | Organism | Function |
|---|---|---|---|---|---|
| Ask1 | Phosphorylation | Ser/Thr 26, 118, 134, 140, 200, 250 | Aurora B (Ser 200) | Budding yeast | Unknown (Cheeseman et al. 2002) |
| | Phosphorylation | Ser 216, 250 | Cdk1 | Budding yeast | Microtubule dynamics (Cheeseman et al. 2001; Li and Elledge 2003; Ubersax et al. 2003; Higuchi and Uhlmann 2005; Miranda, King and Harrison 2007) |
| Borealin/ DasraB | Phosphorylation | Thr 88, Thr 94, Thr 169, Thr 230 | Mps1 | Humans | Aurora B activation and kinetochore biorientation (Jelluma et al. 2008) |
| Bub1 | Phosphorylation | Ser 593, Thr 609 | Cdk1 | Humans, Fission Yeast | Plk1 recruitment to kinetochore (Yamaguchi, Decottignies and Nurse 2003; Qi, Tang and Yu 2006) |
| | Sumoylation | Unknown | Ubc9 | Humans | May regulate CENP-E kinetochore localization (Zhang et al. 2008) |
| | Phosphorylation | Not Determined | Plk1 | Humans | Plk1 recruitment to kinetochore (Qi, Tang and Yu 2006) |
| BubR1 | Phosphorylation | Ser 676 | Plk1 | Humans | Stabilization of kinetochore-microtubule attachments (Elowe et al. 2007) |
| | Phosphorylation | Not determined | Plk1 | *Xenopus* | 3F3/2 epitope formation (Ahonen et al. 2005; Wong and Fang 2005, 2007) |
| | Phosphorylation | Thr 620 Thr605 | Cdk1 | Humans *Xenopus* | Priming phosphorylation (Elowe et al. 2007; Wong and Fang 2007) |

**Table 10.1** (continued)

| Substrate | Modification | Sites[a] | Enzyme | Organism | Function |
|---|---|---|---|---|---|
| CENP-A | Phosphorylation | Ser 7 | Aurora A, Aurora B | Humans | Cytokinesis and kinetochore function (Zeitlin, Shelby and Sullivan 2001; Kunitoku et al. 2003) |
| | Phosphorylation | Ser50 | Unknown | Maize | Unknown (Zhang et al. 2005b) |
| | Ubiquitylation | Multiple Lys | Unknown | Budding yeast Drosophila | Centromere specification (Collins, Furuyama and Biggins 2004; Moreno-Moreno, Torras-Llort and Azorin 2006) |
| | Poly(ADP-ribosylation) | Unknown | PARP-1 | Mouse | Unknown (Saxena et al. 2002) |
| CENP-B | Phosphorylation | Ser 56, 65[b] | Unknown | Humans | Unknown (Nousiainen et al. 2006) |
| | Poly(ADP-ribosylation) | Unknown | PARP-1 | Mouse | Unknown (Saxena et al. 2002) |
| CENP-C/Mif2 | Phosphorylation | Ser 54, 325 | Aurora B | Budding yeast | Unknown (Westermann et al. 2003) |
| | Phosphorylation | Ser 177, 316, 333, 538[b] | Unknown | Humans | Unknown (Nousiainen et al. 2006) |
| | Sumoylation | Multiple Lys | Ubc9 | Humans | Unknown (Chung et al. 2004) |
| CENP-E | Phosphorylation | Ser 2567, 2570, 2601, 2616 | Cdk1 and MAP Kinase | Humans | Inhibits CENP-E microtubule binding (Liao, Li and Yen 1994; Zecevic et al. 1998) |
| | Phosphorylation | Thr 422, 1267, Ser 454, 611, 1211, 2601, 2613 and/or 2616[b] | Unknown | Humans | Unknown (Nousiainen et al. 2006) |

**Table 10.1** (continued)

| Substrate | Modification | Sites[a] | Enzyme | Organism | Function |
|---|---|---|---|---|---|
| | Phosphorylation | C-terminal tail | Cdk1, Mps1 | *Xenopus* | Activation of CENP-E motor activity (Espeut et al. 2008) |
| | Farnesylation | Unknown | Farnesyl transferase | Humans | Association with microtubules and kinetochores (Ashar et al. 2000; Schafer-Hales et al. 2007) |
| | Sumoylation | Unknown | Ubc9 | Humans | Unknown (Zhang et al. In Press) |
| CENP-F | Phosphorylation | Ser 821, 1324, 1747, 1750, 1988, 2513, 2996 and/or 3007, 3094, 3119, 3175 and/or 3179[b] | Unknown | Humans | Unknown (Nousiainen et al. 2006) |
| | Farnesylation | Unknown | Farnesyl transferase | Humans | CENP-F kinetochore localization and subsequent degradation (Ashar et al. 2000; Hussein and Taylor 2002; Schafer-Hales et al. 2007) |
| Cep3 | Phosphorylation | Ser 574 | Unknown | Budding yeast | Unknown (Westermann et al. 2003) |
| | Sumoylation | Unknown | Unknown | Budding yeast | Unknown (Montpetit et al. 2006) |
| Dam1 | Phosphorylation | Ser 20, 257, 265, 292 | Aurora B | Budding yeast | Proper microtubule attachments (Cheeseman et al. 2002; Li et al. 2002) |
| | Phosphorylation | Ser 13, 49, 217, 218, 221, 232 | Mps1 | Budding yeast | Kinetochore positioning (Shimogawa et al. 2006) |

**Table 10.1** (continued)

| Substrate | Modification | Sites[a] | Enzyme | Organism | Function |
|---|---|---|---|---|---|
| | Methylation | Lys 233 | Set1 | Budding yeast | Regulation of Aurora B phosphorylation (Zhang et al. 2005a) |
| Dis1 | Phosphorylation | Thr 279, Ser 293, 300, 551, 556, 590 | Cdk1 | Fission yeast | Accurate chromosome segregation (Aoki et al. 2006) |
| Dsn1 | Phosphorylation | Ser 250 | Aurora B | Budding yeast | Unknown (Westermann et al. 2003) |
| | Phosphorylation | Ser 28, 30[b] | Unknown | Humans | Unknown (Nousiainen et al. 2006) |
| INCENP/ Sli15* | Phosphorylation | Thr 848, Ser 850 | Aurora B | Xenopus | Aurora B activation (Sessa et al. 2005) |
| | Phosphorylation | Ser 598, 599 | Aurora B | *C. elegans* | Aurora B activation (Bishop and Schumacher 2002) |
| | Phosphorylation | Thr 893, Ser 894, 895 | Aurora B | Humans | Aurora B activation (Honda, Korner and Nigg 2003) |
| | Phosphorylation | Ser 59, 388 | Cdk1 | Humans | Plk1 localization (Goto et al. 2006) |
| | Sumoylation | Unknown | Ubc9 | Budding Yeast | Unknown (Wohlschlegel et al. 2004; Montpetit et al. 2006) |
| MCAK | Phosphorylation | Ser 92, 106, 108, 112, 186 | Aurora B | Humans | Inhibition of MCAK depolymerase activity (Andrews et al. 2004) |
| | Phosphorylation | Ser/Thr 95, 110 161, 162, 177, 196, 229, 253 | Aurora B | Xenopus | Inhibition of MCAK depolymerase activity (Lan et al. 2004; Ohi et al. 2004) |
| | Phosphorylation | N-terminus | Plk1 | *Xenopus* | Priming phosphorylation for Aurora B (Rosasco-Nitcher et al. 2008) |

**Table 10.1** (continued)

| Substrate | Modification | Sites[a] | Enzyme | Organism | Function |
|---|---|---|---|---|---|
| Mcm21 | Sumoylation | Unknown | Unknown | Budding yeast | Unknown (Montpetit et al. 2006) |
| Mis6/CENP-I | Phosphorylation | Ser 22[b] | Unknown | Humans | Unknown (Nousiainen et al. 2006) |
| Mis12/Mtw1 | Phosphorylation | Ser 190 or Thr 192, Ser 213 | Unknown | Fission Yeast | May regulate Mis12 binding to centromeres (Goshima et al. 2003) |
| Mis14/DC8 | Phosphorylation | Thr 242[b] | Unknown | Humans | Unknown (Nousiainen et al. 2006) |
| Ndc10 | Phosphorylation | Unknown | Aurora B | Budding yeast | Unknown (Biggins et al. 1999; Sassoon et al. 1999) |
| | Phosphorylation | Unknown | Mck1 | Budding yeast | Unknown (Jiang et al. 1995) |
| | Sumoylation | Lys 556, 651, 652, 779 | Nfi1, Siz1 | Budding yeast | Spindle stability (Montpetit et al. 2006) |
| Ndc80/Hec1 | Phosphorylation | Ser 100 | Aurora B | Budding yeast | Unknown (Cheeseman et al. 2002) |
| | Phosphorylation | Ser/Thr 5, 15, 49, 55, 69 | Aurora B | Humans | Inhibits microtubule attachments (DeLuca et al. 2006) |
| | Phosphorylation | Ser 55, 62, 76/77 | Unknown | Humans | Unknown (Nousiainen et al. 2006) |
| | Phosphorylation | Thr 8, Ser 18, 44, 51 | Aurora B | C. Elegans | Inhibits microtubule attachments (Cheeseman et al. 2006) |
| | Sumoylation | Lys 231 | Unknown | Budding yeast | Unknown (Montpetit et al. 2006) |
| | Phosphorylation | Ser 165 | Nek2 | *Aspergillus nidulans*Humans | Accurate chromosome segregation (Chen et al. 2002; Du et al. 2008) |

**Table 10.1** (continued)

| Substrate | Modification | Sites[a] | Enzyme | Organism | Function |
|---|---|---|---|---|---|
| NudC | Phosphorylation | Ser 274, 326 | Polo Kinase | Humans | Plk1 localization (Zhou et al. 2003; Nishino et al. 2006) |
| Nuf2 | Phosphorylation | Ser 247[b] | Unknown | Humans | Unknown (Nousiainen et al. 2006) |
| | Sumoylation | Unknown | Ubc9 | Humans | May regulate CENP-E localization (Zhang et al. In Press) |
| PBIP1 | Phosphorylation | Thr 78 | Polo Kinase | Humans | Plk1 localization (Kang et al. 2006) |
| PICH | Phosphorylation | Thr 1063 Unknown | Cdk1 Plk1 | Humans Humans | Plk1 priming phosphorylation Restricts PICH to kinetochores (Baumann et al. 2007) |
| RanGAP1 | Sumoylation | Lys 524 | Unknown | Humans | Localization to spindle and kinetochores (Joseph et al. 2002) |
| | Sumoylation | Lys 526 | Unknown | Mice | Localization to nuclear pore (Matunis, Wu and Blobel 1998) |
| Sgo1 | Phosphorylation | Ser 14, 5-7 | Nek2A | Humans | Chromosome congression (Fu et al. 2007) |
| Spc19 | Phosphorylation | Ser 107, 116 | Unknown | Budding yeast | Unknown (Cheeseman, Drubin and Barnes 2002) |
| Spc24 | Phosphorylation | Thr 130[b] | Unknown | Humans | Unknown (Nousiainen et al. 2006) |
| Spc34 | Phosphorylation | Thr 199 | Aurora B | Budding yeast | Accurate chromosome segregation (Cheeseman et al. 2002) |
| Survivin/Bir1 | Phosphorylation | Thr 735, 747 | Cdk1 | Budding yeast | Spindle stability (Widlund et al. 2006) |

**Table 10.1** (continued)

| Substrate | Modification | Sites[a] | Enzyme | Organism | Function |
|---|---|---|---|---|---|
| | Phosphorylation | Thr 117 | Aurora B | Humans | Survivin localization to kinetochore (Wheatley et al. 2004; Wheatley et al. 2007) |
| | Phosphorylation | Thr 34 | Cdk1 | Humans and mice | Survivin localization to the spindle (O'7;Connor et al. 2000; Nousiainen et al. 2006) |
| | Sumoylation | Unknown | Unknown | Budding yeast | Unknown (Zhou, Ryan and Zhou 2004; Montpetit et al. 2006) |
| | Ubiquitylation | Lys 23,\ 62, 78, 79 | Ufd1 | Humans | Survivin localization to centromere (Vong et al. 2005; Delacour-Larose et al. 2007) |
| | Ubiquitylation | Unknown | Unknown | *Xenopus* | Unknown (Ramadan et al. 2007) |
| TopoII | Sumoylation | Lys 1220, 1246, 1247, 1277, 1278 | Ubc9 | Budding yeast | Centromeric cohesion (Bachant et al. 2002) |
| ZW10 | Phosphorylation | Thr 437[b] | Unknown | Humans | Unknown (Nousiainen et al. 2006) |

[a] These sites reflect in vitro and/or in vivo modifications

[b] These sites are present in peptides, which were found to be phosphorylated in a proteomic analysis, and are therefore putative phosphosites (Nousiainen et al. 2006)

*A large number of INCENP phosphorylation sites have been identified in Humans and are not listed here (see (Nousiainen et al. 2006))

Similar to methylation, the role of acetylation in protein function has been most intensively studied in the context of chromatin structure. However, acetylation has also been identified on non-histone proteins involved in processes such as DNA replication and repair, sister chromatid cohesion, cellular signaling and cell motility, highlighting the widespread use of this modification (for review, see Glozak et al., 2005). $N^{\varepsilon}$-acetylation by acetyltranferases neutralizes the positive charge of the lysine side chain and impairs its ability to form hydrogen bonds. The change in charge due to acetylation can alter protein–DNA, protein–RNA, and protein–protein interactions, as well as change the subcellular location of a substrate. Acetyltransferases are often large complexes composed of a catalytic subunit that is regulated by other subunits in the complex. Acetylation is a reversible reaction that is governed by acetyltransferase and deacetylase activities. Deacetylases, extensively studied in the context of histone deacetylation, are often overexpressed or mutated in malignant cells (for review, see Gallinari et al., 2007), making them good targets for cancer therapy.

### *10.2.4 Farnesylation*

Farnesylation is a modification catalyzed by farnesyl transferases (FT) that covalently add a lipid moiety to cysteine residues in a C-terminal CAAX motif (for review, see Zhang and Casey 1996). Although most farnesylation events lead to membrane association of substrate proteins, the modification can also be important for protein–protein interactions. The treatment of cells with farnesyl transferase inhibitors (FTI) causes cells to arrest in G2/M (Moasser et al., 1998; Crespo et al., 2001), and FTI are promising cancer therapeutic agents because they also cause cancer cells to arrest in mitosis. The only kinetochore proteins known to be farnesylated to date are centromere proteins (CENP)-E and CENP-F (Ashar et al., 2000), and the effect of FTI could be due to the role of farnesylation in localizing these proteins to kinetochores (Hussein and Taylor 2002; Schafer-Hales et al., 2007).

### *10.2.5 The Dynamic Control of Modifications*

Because an assortment of post-translational modifications can occur on a given protein, dissecting the roles of each modification is complex because they do not exist in isolation. In addition, a diverse set of modifications can occur on lysine residues, so additional control can result from dynamic changes in the modification state of a single residue on a given protein. For example, acetylation of lysine can prevent the conjugation of other post-translational modifications at the same residue, therefore affecting the stability, localization, and function of the protein. Also, protein modifications on one residue can dramatically influence other modifications on the same substrate. For example, phosphorylation of a protein can promote its subsequent ubiquitylation and degradation. The addition of SUMO can prevent ubiquitylation and therefore stabilize a protein. Moreover, a single protein can be modified by a specific combination of modifications that can alter protein function. Like other cellular structures, the kinetochore is regulated by a complex set of protein modifications that dictate its assembly and cell cycle activity, and fully understanding the post-translational modifications that occur on kinetochore proteins may also reveal novel regulatory mechanisms.

## 10.3 The Regulatory Enzymes

A number of enzymes that modify kinetochore proteins and the underlying centromeric chromatin have been identified. Currently, protein kinases appear to be the major class of regulatory proteins that modulate kinetochore activity. Two conserved protein kinase families, Polo and Aurora, play particularly critical roles in regulating kinetochore assembly and microtubule attachment. Here, we introduce the major enzymes that are known to modify centromeric chromatin and kinetochore proteins.

### *10.3.1 The Kinases*

*Polo-like Kinases*: Polo-like kinases (Plks) comprise a family of structurally related Ser/Thr kinases that are characterized by the presence of an N-terminal catalytic domain and one or more C-terminal polo-box domain (PBD; for reviews, see Barr, Sillje and Nigg 2004; Lowery, Lim and Yaffe 2005). While the kinase domain is responsible for phosphorylating substrates, substrate recognition is carried out at least in part by the PBD, which serves as a phospho-serine/threonine binding module (Elia et al., 2003). The sequence recognized by the PBD, S-(pS/pT)-(P/X), corresponds to the Cdk1 consensus site. Many Plk substrates or docking partners are therefore primed for their interaction with Plk by prior phosphorylation by Cdk1 or other proline-directed kinases (Elia, Cantley, and Yaffe 2003). However, in some cases, Plk phosphorylates this site on the substrate to create a stable docking site (Kang et al., 2006). Besides, some Polo substrates do not contain the PBD binding site, so there are additional controls over the specificity of Polo binding to substrates that have yet to be identified. Although a precise Plk consensus site has not yet been identified, a hydrophobic residue at + 1 and an acidic residue at -2 appear to be optimal for phosphorylation (Nakajima et al., 2003). Two conserved kinases, Plkk1 and PKA, have been identified in *X. laevis* that can phosphorylate and activate Plk in vitro (Qian, Erikson and Maller 1998; Kelm et al., 2002). These kinases phosphorylate a conserved residue within the Plk1 catalytic domain T-loop, resulting in Plk1 activation. However, it is unclear whether these kinases are responsible for Plk1 activation in vivo. In addition, Plk is also regulated by cell cycle changes in transcription and ubiquitin-mediated proteolysis in various organisms.

While flies and yeast contain a single Plk ortholog, multiple family members are found in multicellular eukaryotes, with mammals containing four members (Plk1-4). The most extensively studied mammalian family member is Plk1 due to its similarity in function and localization to other eukaryotic Polo-like kinases, as well as its connection to cellular transformation. Like other essential kinases, Plk1 has multiple cell cycle functions, regulating centrosome duplication, mitotic entry, and cytokinesis (for review, see Barr, Sillje and Nigg 2004). Consistent with its various functions, Plk1 exhibits a dynamic localization pattern, especially during mitosis. It localizes to the spindle poles and kinetochores during prometaphase and metaphase, and then it relocalizes to the spindle midzone after the initiation of anaphase. The control over Plk1 localization to the kinetochore is complex and appears to be regulated by phosphorylation. During interphase, Plk1 phosphorylates a kinetochore protein called PBIP1 (polo-box-interacting protein 1), a homolog of the chicken protein CENP-50 (human CENP-U) that is required for normal chromosome segregation in DT40 cells (Kang et al., 2006). The phosphorylation of T78 on PBIP1 creates a Plk1 self-docking site, so Plk1 regulates its own recruitment to the kinetochore. The phosphorylation of PBIP1 by Plk1 is then thought to promote PBIP1 degradation later in mitosis, thus freeing the kinetochore-bound Plk1 to phosphorylate its other substrates. Plk1 localization has also been suggested to require Cdk1 phosphorylation of inner centromere protein (INCENP), a binding partner for the Aurora B kinase (see below), as well as the Bub1 kinetochore protein (Goto et al., 2006; Qi, Tang and Yu 2006). However, it is difficult to determine whether all of these modifications directly mediate Plk1 localization to the kinetochore, or whether their role in Plk1 localization is a secondary effect of other roles in kinetochore assembly.

Although the localization of Plk1 and its interaction with numerous kinetochore proteins indicates that it has important roles at the kinetochore, Plk1's precise contribution to kinetochore activity has been difficult to analyze due to its multiple cellular functions. Plk1 inhibition by RNA interference (RNAi) results in severe spindle defects (Sumara et al., 2004; van Vugt et al., 2004; Lenart et al., 2007), and experiments using small molecule inhibitors of Plk1 have shown that it is required to stabilize chromosome-associated microtubules (Lenart et al., 2007) and mediate spindle positioning in budding yeast (Snead et al., 2007). Plk1 also appears to be involved in monitoring the state of kinetochore attachment to microtubules. When sister kinetochores biorient, they come under tension because the pulling forces exerted by microtubules from opposite poles are opposed by the linkage between sister chromatids. Unaligned kinetochores lacking tension generate a Plk1-mediated phosphoepitope on BubR1 that is recognized by the 3F3/2 monoclonal antibody (Ahonen et al., 2005; Wong and Fang 2005, 2007), suggesting that Plk1 has a role in sensing tension at the kinetochore. Taken together, Plk1 appears to have multiple roles at the kinetochore.

*Aurora Kinases:* Another key family of conserved kinases involved in chromosome segregation are the Ser/Thr-directed Aurora protein kinases that have generated significant interest in recent years due to their amplification in many types of cancers (for review, see Ruchaud, Carmena, and Earnshaw 2007). In multicellular eukaryotes, the Aurora kinases are subdivided into three families (Aurora A, B, and C), while there is only a single Aurora kinase in budding and fission yeast. The Aurora kinases contain a divergent N-terminal tail and a conserved C-terminal catalytic domain and their activity peaks during mitosis. Using synthetic peptides as substrates, the Aurora A consensus site has been mapped to (R/K/N)(R)(X)(S/T)(B), where B is any hydrophobic residue except proline (Ferrari et al., 2005). Although a similar consensus site ((R/K)X(S/T)(S/L/V)) for Aurora B has been proposed based on in vitro phosphorylation data (Cheeseman et al., 2002), not all known Aurora sites match this consensus.

The Aurora protein kinases can be classified by their localization patterns and activators. Aurora A mainly localizes to centrosomes and spindles (Roghi et al., 1998) and has at least two known activators, the TPX-2 and Ajuba proteins (Eyers et al., 2003; Hirota et al., 2003; Tsai et al., 2003). However, it is not clear whether these activators are responsible for all of the regulation of Aurora A nor whether these activators exist in all organisms. In most organisms, the Aurora B kinase exists in a complex called the chromosomal passenger complex (CPC) that contains the INCENP, Survivin, and Dasra B/Borealin/Csc1 (for review, see Ruchaud, Carmena and Earnshaw 2007). INCENP is a potent activator of Aurora B kinase activity and other CPC members also appear to regulate Aurora B activity (Kang et al., 2001; Bishop and Schumacher 2002; Bolton et al., 2002; Honda, Korner and Nigg 2003; Sessa et al., 2005; Jelluma et al., 2008). Aurora B phosphorylates INCENP, leading to additional activation of Aurora B in a feedback loop (Bishop and Schumacher 2002; Honda, Korner, and Nigg 2003; Sessa et al., 2005; Kelly et al., 2007). In addition, chromatin can stimulate Aurora B in *Xenopus* egg extracts (Kelly et al., 2007), and the TD-60 protein and microtubules can synergistically activate Aurora B in vitro (Rosasco-Nitcher et al., 2008). In *C. elegans*, the Tousled-like kinase-1 protein is both a substrate and activator of Aurora B (Han et al., 2005). Aurora B activity can also be inhibited by its substrates, and this inhibition can be reversed when other kinases such as Plk1 phosphorylate the substrates (Rosasco-Nitcher et al., 2008). Aurora B is therefore subject to multiple levels of regulation in various organisms. Like Plk1, Aurora B exhibits a dynamic localization pattern during the cell cycle (for review, see Ruchaud, Carmena, and Earnshaw 2007). It is first detected along the entire chromosome, and then it concentrates at the inner centromere region from prometaphase through metaphase. At the onset of anaphase, Aurora B dissociates from kinetochores and binds to the central spindle, eventually accumulating at the spindle midzone at telophase. It is also on the equatorial cortex where the cleavage furrow will form from late anaphase to telophase. Aurora C is less characterized than Aurora A and B. Although it exhibits a localization pattern similar to Aurora B, Aurora C is mainly expressed in testes, with low levels of expression in a number of other tissues (Yan et al., 2005).

The localization pattern of the Aurora kinases closely reflects their functions. The major roles identified for Aurora A include centrosome maturation, centrosome separation, and bipolar spindle assembly, while Aurora C has undefined functions in specific tissues such as the testes (for reviews, see Vagnarelli and Earnshaw 2004; Ruchaud, Carmena and Earnshaw 2007). Aurora B has been implicated in many cell cycle processes, including chromosome condensation, chromosome segregation, sister chromatid cohesion, and cytokinesis. Of the Aurora family members, Aurora B is the major regulator of kinetochores where it appears to have two general functions. First, it is required for the localization of numerous proteins to the kinetochore in multicellular eukaryotes. Second, it appears to detect inappropriate kinetochore–microtubule attachments and facilitate their correction. A variety of data suggest that Aurora B detects the lack of tension on kinetochores that results from inappropriate microtubule–kinetochore attachments and subsequently destabilizes these attachments, giving cells a chance to make proper attachments (see Chapters 10 and 12 for more details).

*Additional kinases that regulate the kinetochore:* Several additional kinases regulate kinetochore function, although less is known about their specific roles at the kinetochore in comparison to the Plk1 and Aurora B kinases. The spindle checkpoint kinases, Mps1, Bub1, and BubR1, are enriched on kinetochores at prometaphase (Howell et al., 2004). The roles of these proteins in checkpoint function are reviewed in Chapter 12. A variety of data indicates that these kinases also regulate kinetochore function separately from the spindle checkpoint. Mps1 has a conserved role in kinetochore biorientation (Maure, Kitamura, and Tanaka 2007; Jelluma et al., 2008), and this function may be linked to activation of the Aurora B kinase via phosphorylation of the CPC subunit Borealin/DasraB in vertebrates (Jelluma et al., 2008). In addition, Mps1 was recently reported to aid in activation of the CENP-E motor protein in vitro (Espeut et al., 2008). The Bub1 kinase is required for kinetochore function and proper centromere localization of the Sgo1 protein, a regulator of sister chromatid cohesion (Johnson et al., 2004; Tang et al., 2004; Kitajima et al., 2005; Fernius and Hardwick 2007; Pouwels et al., 2007). The BubR1 checkpoint kinase also appears to have a constitutive role in kinetochore function because its downregulation in an unperturbed cell cycle causes severe chromosome misalignment (Lampson and Kapoor 2005).

There are also non-checkpoint kinases that are implicated in kinetochore activity. The Mitogen Activated protein (MAP) kinases, which are required for diverse processes such as differentiation and secretion, also localize to kinetochores and recruit spindle checkpoint proteins (Zhao and Chen 2006). Specifically, the phosphorylated form of MAP kinase, which is the activated form, is found at kinetochores (Shapiro et al., 1998; Zecevic et al., 1998). The major cyclin-dependent kinase, Cdk1, also regulates kinetochore activity. Cdk1 is responsible for generating the Plk1 priming phosphorylation site on a number of kinetochore substrates (Elia, Cantley, and Yaffe 2003). In addition, Cdk1 also regulates kinetochore-associated kinase complexes such as the CPC (Pereira and Schiebel 2003; Goto et al., 2006). It is likely that Cdk1 activity plays additional roles at the kinetochore that have not yet been elucidated due to the pleiotropic nature of knockdown experiments that make it difficult to distinguish primary versus secondary effects. Depletion of the conserved Haspin kinase prevents normal chromosome alignment at metaphase, suggesting a critical role in chromosome segregation (Dai et al., 2005). Haspin phosphorylates histone H3 at Thr3 and is required to maintain centromeric cohesion, but it is not yet known if H3 is the key substrate for this function (Dai, Sullivan and Higgins 2006). The NIMA (*n*ever in *m*itosis A)-related kinases (also called Nek), which have important roles in centrosome function, also contribute to kinetochore activity (Chen et al., 2002), although their precise roles are relatively unknown (for review, see O'Regan, Blot and Fry 2007).

### *10.3.2 The Phosphatases*

The reversal of phosphorylation also significantly influences protein function, and the dephosphorylation of key substrates has been proposed to be an important mechanism that regulates kinetochore activity. Kinetochores are regulated by the catalytic subunits of two Ser/Thr phosphatases that are involved in a diverse set of processes: PP1 (Protein Phosphatase 1), PP2A (Protein Phosphatase 2A) and the Cdc14 phosphatase (for reviews, see Trinkle-Mulcahy and Lamond 2006; Moorhead, Trinkle-Mulcahy and Ulke-Lemee 2007). While the substrate specificity of PP1 and PP2A is mainly determined by their associated regulatory subunits, the Cdc14 phosphatase reverses Cdk1-mediated phosphorylation.

Mammalian cells express three PP1 isoforms ($\alpha$, $\beta$, and $\gamma$) that exhibit distinct localization patterns (Andreassen et al., 1998; Trinkle-Mulcahy, Sleeman, and Lamond 2001). The $\alpha$ and $\gamma$ PP1 isoforms are nuclear and cytoplasmic during interphase and subsequently recruited to kinetochores as cells enter mitosis. At the metaphase to anaphase transition, PP1$\gamma$ relocalizes to the entire chromosome. Finally, during telophase, PP1$\alpha$ and $\gamma$ accumulate at the cell cortex and the midbody. A major function of PP1 in regulating kinetochore function is to antagonize Aurora B activity, by either dephosphorylating Aurora B substrates or by inhibiting Aurora B kinase activity, depending on the organism (Francisco and Chan 1994; Francisco, Wang and Chan 1994; Pinsky et al., 2006a). However, despite the identification of more than 50 regulatory subunits that regulate PP1 activity in diverse processes, the subunit that targets PP1 specifically to kinetochores to regulate its mitotic functions remains unknown. Although it is possible that PP1 does not require a targeting subunit for its mitotic functions, mutation of the hydrophobic patch that mediates binding to many of its regulatory subunits leads to the mislocalization of PP1 from kinetochores (Wu and Tatchell 2001).

It will therefore be critical to identify both the regulatory subunit and substrates of PP1 in the future to fully understand its role at the kinetochore.

Although PP1 is the most well-studied phosphatase in the context of kinetochore activity, two other phosphatases, PP2A and Cdc14, help to ensure accurate chromosome segregation (Tang et al., 2004). The PP2A core consists of the PP2A catalytic subunit (PP2A-C) and a scaffold protein (PP2A-A). This core dimer recruits one of four regulatory subunits (PP2A-B, PP2A-B', PP2A-B'' or PP2A-B''') to form oligomers with unique substrate specificities (for review, see Janssens and Goris 2001). PP2A is recruited to centromeres by the PP2A-B' regulatory subunit, where it functions to protect centromeric cohesion until the metaphase to anaphase transition (Kitajima et al., 2006; Riedel et al., 2006). In addition, PP2A has important roles in regulating exit from mitosis. It is therefore likely that kinetochore substrates that are dephosphorylated by PP2A will be identified in the future.

The Cdc14 phosphatase localizes to kinetochores in many organisms and appears to regulate chromosome segregation (for reviews, see Stegmeier and Amon 2004; Sullivan and Morgan 2007). Cdc14 is a proline-directed phosphatase that mediates the dephosphorylation of Cdk1 substrates to promote exit from mitosis. Cdc14 dephosphorylates a number of budding yeast kinetochore proteins to promote their localization to the mitotic spindle at anaphase (Pereira and Schiebel 2003; Bouck and Bloom 2005; Stoepel et al., 2005). In *S. pombe*, deletion of the Cdc14 homolog results in mono-oriented kinetochores and chromosome missegregation, suggesting that dephosphorylation of fission yeast kinetochore proteins is important for correcting errors in microtubule–kinetochore attachments (Trautmann, Rajagopalan, and McCollum 2004). Given the widespread phosphorylation by Cdk1, it will be critical to identify Cdc14 substrates at the kinetochore in *S. pombe* as well as other organisms.

### *10.3.3 Ubiquitin and SUMO Enzymes*

Ubiquitin-mediated proteolysis is the key regulatory step that initiates the metaphase to anaphase transition. The anaphase promoting complex (APC) is an E3 ubiquitin ligase that catalyzes the ubiquitylation and subsequent destruction of the anaphase inhibitor securin when it is activated by the Cdc20 protein (for review, see Thornton and Toczyski 2006). Components of the spindle checkpoint, which monitors the state of kinetochore–microtubule attachments and inhibits the cell cycle in the presence of aberrant attachments (for review, see Musacchio and Hardwick 2002 and Chapter 12), form inhibitory complexes with Cdc20 (for review, see Yu 2007). Strikingly, Cdc20 and a number of APC subunits are recruited to kinetochores during prometaphase when the spindle checkpoint is activated (Acquaviva et al., 2004; Melloy and Holloway 2004; Vigneron et al., 2004). Moreover, the kinetochore localization of the APC and

Cdc20 depends on the presence of the spindle checkpoint proteins (Acquaviva et al., 2004; Vigneron et al., 2004), raising intriguing questions about the spatial regulation of the APC as well as the functional consequences of the localization of this E3 ligase to kinetochores. It is possible that APC localization to the kinetochore promotes the specific turnover of kinetochore-bound substrates. Indeed, localized ubiquitylation of APC substrates has been previously reported in the context of centrosomes (Clute and Pines 1999). Given the importance of this enzyme in cell cycle progression, it will be critical to elucidate the functions of the kinetochore-bound APC.

Besides APC, an *S. pombe* ubiquitin ligase appears to be important for kinetochore activity through its regulation of heterochromatin (Hong et al., 2005; Jia, Kobayashi and Grewal 2005). Cul4 is a member of the cullin family of E3 ubiquitin ligases. Methylated H3-K9 promotes the formation of heterochromatin, which plays a role in centromere specification. Cul4 is required to localize Clr4, the enzyme responsible for the methylation of histone H3 at lysine 9, to heterochromatin (Jia, Kobayashi, and Grewal 2005; see below for more details). Indeed, *cul4* mutants are defective in H3K9 methylation and exhibit chromosome segregation defects. Although Cul4 is a ubiquitin ligase, the regulation of Clr4 does not appear to be via protein degradation. It has been proposed that the Cul4–Clr4 complex is recruited to DNA by a specificity factor that allows Cul4 to ubiquitylate histone H2B and Clr4 to promote methylation of H3K9 (Hong et al., 2005). It will be interesting to determine the functional interplay between H2B ubiquitylation and H3K9 methylation to establish epigenetic states like heterochromatin.

Deubiquitylating enzymes (DUBs) are involved in regulating kinetochore activity. In HeLa cells, the inhibition of a DUB called hFAM results in chromosome alignment defects and lagging chromosomes during anaphase. hFAM deubiquitylates the CPC protein survivin, which is thought to promote the dissociation of survivin from kinetochores (Vong et al., 2005). It has been proposed that survivin dissociation from kinetochores may aid in regulating Aurora B activity specifically at kinetochores. In addition, a DUB called Usp44 has recently been implicated in the spindle checkpoint (Reddy et al., 2007; Stegmeier et al., 2007). Deubiquitylation of the APC activator Cdc20 by Usp44 promotes its interaction with the spindle checkpoint protein Mad2, resulting in Cdc20 inhibition. In human cells, the Ubp-M enzyme deubiquitylates H2A in a reaction that promotes Aurora-B mediated phosphorylation of histone H3 (Joo et al., 2007). Consistent with this, depletion of Ubp-M leads to defects in mitotic progression.

To date, no specific enzyme required for SUMO modification has been identified at the kinetochore, although the budding yeast SUMO E3 enzymes Siz1 and Siz2 appear to regulate the pericentric pool of topoisomerase II (Takahashi et al., 2006). Given the number of kinetochore proteins modified by SUMO, it will be important to determine how SUMO modification of kinetochore proteins is regulated.

### *10.3.4 Methyltransferases and Acetyltransferases*

Similar to the rest of the genome, centromeric and pericentric chromatin are modulated by both methylation and acetylation. To date, two acetyltransferases have been directly implicated in chromosome segregation. The essential, multi-subunit NuA4 histone acetyltransferase (HAT) complex that is involved in many cellular processes (for review, see Doyon and Cote 2004) is known to acetylate the budding yeast H2A.Z histone variant and H4 (Allard et al., 1999; Galarneau et al., 2000; Keogh et al., 2006). Although an H2A.Z mutant that cannot be acetylated by NuA4 is defective in segregation (Keogh et al., 2006), it is not yet known whether NuA4 acetylates H2A.Z in other organisms, and nonacetylatable budding yeast histone H4 mutants also exhibit phenotypes consistent with a role in kinetochore function (Le Masson et al., 2003). It is therefore unclear whether the role of NuA4 in segregation is restricted to H2A.Z acetylation or whether there are additional acetyltransferase enzymes that affect chromosome segregation. In addition, it is not yet known whether NuA4 activity toward H2A.Z is regulated. The other acetyltransferase implicated in chromosome segregation is the budding yeast Gcn5 protein (Vernarecci et al., 2008). Gcn5 appears to localize to kinetochores and regulate centromeric chromatin structure, but the kinetochore target has not yet been identified.

Centromeric and pericentric chromatin generally contain hypoacetylated histone H3 and H4, and a number of experiments suggest that the inhibition of histone deacetylases (HDAC) alters kinetochore function (for review, see Ekwall 2007). To date, HDACs that are required to maintain the methylation status of H3-K9 have been identified (Yamada et al., 2005), but it is not known to what extent these HDACs regulate the histones at centromeres versus exerting their effects on segregation through changes in transcription elsewhere in the genome or by regulating the acetylation status of non-histone proteins. However, there is widespread promise that HDAC inhibitors will be useful as anti-cancer agents (for review, see Marchion and Munster 2007), so it will be critical to further elucidate the roles of HDACs in chromosome segregation.

Similar to HATs, methyltransferases have also been implicated in kinetochore function, mainly through their regulation of histones within centromeres and heterochromatin. Although 28 methyltransferases exist in humans (Allis et al., 2007), the hallmark of pericentric heterochromatin in all organisms is methylation of H3-K9 by the conserved Suv39h/clr4 histone methyltransferases (HMTases; for review, see Klose and Zhang 2007). These enzymes contain a SET domain that is a hallmark of many chromatin-interacting proteins, as well as a chromodomain that directs the association of the HMTases with chromatin. Recently, components of the RNAi machinery have been implicated in targeting these HMTases to heterochromatin (for review, see Klose and Zhang 2007). In mammals, there are two Suv39 HMTases, Suv39h1 and Suv39h2 that are specifically responsible for the trimethylation of H3 on K9 that recruits the heterochromatin protein HP1 (Rice et al., 2003). Interestingly, the Suv39h1 protein also binds to the outer region of the centromere from prometaphase to metaphase (Aagaard et al., 2000). This centromere localization is correlated with phosphorylation of Suv39h1, suggesting that there is a dynamic regulation of this HMTase at centromeres. However, the identity of the kinase and phosphatase that regulate Suv39h1, as well as the centromere-related function for the enzyme, have not yet been elucidated. The Suv39 HMTases are also involved in establishing the trimethylation of H4 on K20, another modification enriched in pericentric heterochromatin (Schotta et al., 2004). However, they do not directly methylate this residue and instead likely contribute to creating the binding site for two additional SET domain HMTases, Suv4-20h1 and Suv4-20h2, that modify this residue. In fission yeast, the HMTase Set2 methylates histone H3 on K36 to promote heterochromatin assembly in a pathway parallel to the Suv39/Clr4 pathway (Folco et al., 2008). Another enzyme involved in modifying H3 is the Polycomb group protein Enhancer of Zeste 2 (EZH2; for review, see Cao et al., 2002). EZH2, which functions in a transcription repressive complex, is required for the maintenance of the spatial patterning of the homeotic box genes during early embryonic development. EZH2 executes this function by methylating K27 of H3. While this modification is also found in pericentric regions, it is unclear whether EZH2 is the enzyme responsible for this modification. H3-K27 is demethylated by two related demethylases UTX and JMJD3, members of the Jumonji C family of proteins (Hong et al., 2007; Lan et al., 2007; Swigut and Wysocka 2007; Xiang et al., 2007; Smith et al., 2008). However, their role in modifying H3 at pericentric regions remains to be elucidated. In the future, it will be critical to determine how these enzymes are regulated to understand the contributions of heterochromatin modifications to centromere function.

The budding yeast Set1 methyltransferase appears to methylate the Dam1 kinetochore protein (Zhang et al., 2005a). Although Set1 is mainly known for its role in methylation of H3-K4, its effects on kinetochore function do not appear to be linked to this modification. Set1 and Dam1 co-immunoprecipitate, and the methylation of Dam1 depends on Set1 function. The identification of a methylated kinetochore protein highlights the importance of further studies directed at detecting and characterizing modifications such as methylation.

## 10.4 Centromere Specification

A specialized centromeric chromatin structure is essential to specify centromere identity and ensure the epigenetic propagation of kinetochores at a single chromosomal locus in all eukaryotic organisms (for review, see Dunleavy, Pidoux, and Allshire 2005 and Chapters 3 and 8 for additional details). The fundamental unit of chromatin is the nucleosome, $\sim$2 turns of DNA wrapped around an octamer containing two molecules of each core histone (H2A, H2B, H3 and H4; for review, see Luger 2003). Chromatin is subject to three major

**Table 10.2** Post-translational modifications associated with canonical and histone variants that reside in centromeric chromatin and pericentric heterochromatin

| Substrate | Modification | Sites[a] | Enzyme | Organism |
|---|---|---|---|---|
| Centromeric chromatin | | | | |
| CENP-A | Phosphorylation | Ser 7<br>Ser 50 | Aurora A + B<br>Aurora B | Humans<br>(Zeitlin, Shelby and Sullivan 2001; Kunitoku et al. 2003)<br>Maize<br>(Zhang et al. 2005b) |
| | Ubiquitylation | Not applicable | Unknown | Budding yeast<br>(Collins, Furuyama and Biggins 2004)<br>*Drosophila* (Moreno-Moreno, Torras-Llort and Azorin 2006) |
| H2A.Z | Acetylation | Lys 14 | NuA4 acetyl-transferase | Budding Yeast<br>(Keogh et al. 2006) |
| Histone H2A | Phosphorylation | Thr 119 | Aurora B | *Drosophila*<br>(Brittle et al. 2007) |
| Histone H3 | Dimethylation | Lys 4 | Likely Set1 | *Drosophila*, Humans<br>(Sullivan and Karpen 2004) |
| | Phosphorylation | Thr 3 | Haspin | Mammals<br>(Dai and Higgins 2005) |
| | Phosphorylation | Thr 11 | Likely Dlk1/ZIP | Mammals<br>(Preuss, Landsberg and Scheidtmann 2003; Polioudaki et al. 2004) |
| Pericentric Heterochromatin | | | | |
| Histone H3 | Dimethylation | Lys 9 | Clr4<br>Su(var)3-9 | Fission yeast<br>(Noma, Allis and Grewal 2001)<br>*Drosophila* (Schotta et al. 2002) |
| | Trimethylation | Lys 36 | Set2 | Fission yeast<br>(Chen et al. 2008) |
| | Trimethylation | Lys 9 | Suv39h1, Suv39h2 | Mammals<br>(Rea et al. 2000; Peters et al. 2003; Rice et al. 2003) |

**Table 10.2** (continued)

| Substrate | Modification | Sites[a] | Enzyme | Organism |
|---|---|---|---|---|
| | Monomethylation | Lys 27 | Ezh2 | Mammals (Cao et al. 2002; Peters et al. 2003) |
| | Phosphorylation | Ser 10 | Aurora B<br>NIMA | Most eukaryotes (Hendzel et al. 1997; Wei et al. 1998; Hsu et al. 2000; Petersen et al. 2001)<br>*Aspergillus nidulans* (De Souza et al. 2000) |
| | Phosphorylation | Ser 28 | Aurora B | Mammals (Goto et al. 2002) |
| Histone H3.3 | Phosphorylation | Ser 31 | Unknown | Mammals (Hake et al. 2005) |
| Histone H4 | Trimethylation | Lys 20 | Unknown | Mammals, *Drosophila* (Schotta et al. 2004) |

a These sites reflect in vivo modifications

forms of regulation that can alter gene expression and epigenetic states. First, there are chromatin remodeling complexes that contain ATPase subunits that can cause nucleosome sliding, histone replacement, or altered histone–DNA interactions (for reviews, see Tsukiyama 2002; Langst and Becker 2004). Second, chromatin function is modulated by the incorporation of histone variants, non-allelic forms of the major histones that can have significant differences in primary sequence (for review, see Polo and Almouzni 2006). Third, the histones are post-translationally modified, especially on the N-terminal tails (for review, see Kouzarides 2007). Here, we review the major modifications that are associated with the canonical and variant histones within centromeric and pericentric chromatin (Table 10.2). A number of chromatin remodeling complexes also regulate kinetochore function, and we refer the reader to (Ekwall 2007) for a review of these enzymes.

Like the rest of the genome, centromeres are packaged into chromatin composed of canonical histones as well as histone variants. The hallmark of centromeric chromatin is an essential histone H3 variant, CENP-A, that appears to replace histone H3 in nucleosomes (for reviews, see Dunleavy, Pidoux, and Allshire 2005; Carroll and Straight 2006). CENP-A is exclusively localized to centromeres and is not found in the surrounding heterochromatin. With the exception of budding yeast centromeres that only have a single CENP-A/Cse4 nucleosome (Meluh et al., 1998; Furuyama and Biggins 2007), the centromeres in most organisms contain both CENP-A and canonical H3 nucleosomes (Sullivan and Karpen 2004). In contrast, H2A.Z, a histone H2A variant, has been detected at both pericentric and centromeric chromatin (Rangasamy et al., 2003; Krogan et al., 2004; Rangasamy, Greaves and Tremethick 2004; Foltz et al., 2006; Greaves et al., 2007). Centromeres may also contain other histone variants. The MacroH2A variant was enriched when CENP-A nucleosomes were purified from HeLa cells (Foltz et al., 2006), but it is not yet clear if this reflects direct centromeric association.

### *10.4.1 Canonical Histone Modifications at the Centromere*

Centromeres have a canonical histone modification pattern that contains hallmarks of both heterochromatin and euchromatin. In contrast to heterochromatin that is enriched for H3-K9 methylation (for review, see Grewal and Jia 2007), the H3 within centromeres is dimethylated on K4, a modification associated with euchromatin (Sullivan and Karpen 2004). However, similar to heterochromatin, histones H3 and H4 within centromeric chromatin are hypoacetylated (Hayashi et al., 2004; Sullivan and Karpen 2004). Therefore, centromeres appear to have a unique combination of histone modifications that could contribute to centromere function and propagation. Although the role of H3-K4 dimethylation in centromere function has not yet been explored, histone acetylation is correlated with centromere activity (for review, see Ekwall 2007). The inhibition of HDAC activity by trichostatin A (TSA) leads to hyperacetylated centromeres and chromosome segregation defects in *S. pombe* and prolonged mitotic arrest in HeLa cells (Shin et al., 2003). In addition, acetylation appears to play a role in CENP-A localization (see below). Centromeric chromatin is also regulated by phosphorylation. Histone H3 is phosphorylated on T3 during mitosis by the Haspin kinase (Dai et al., 2005; Dai, Sullivan and Higgins 2006). The phosphorylation of H3-T3 can prime H3 for subsequent phosphorylation by Aurora B in vitro (Rosasco-Nitcher et al., 2008), but the direct function of this modification within centromeric chromatin in vivo is not yet known. Histone H3 is also phosphorylated by the Dlk kinase on Thr11 from prophase to early anaphase (Preuss, Landsberg and Scheidtmann 2003). Although this modification is enriched at centromeres during mitosis, its function in vivo is not yet clear.

### *10.4.2 The CENP-A Histone Variant*

CENP-A is essential for accurate chromosome segregation in every organism examined (for reviews, see Dunleavy, Pidoux, and Allshire 2005; Henikoff and Dalal 2005; Kamakaka and Biggins 2005; Carroll and Straight 2006). Because CENP-A is localized to all centromeres and active neocentromeres, it is an excellent candidate for the epigenetic mark that specifies the site of kinetochore formation (Warburton et al., 1997; Van Hooser et al., 2001; Amor and Choo 2002; Heun et al., 2006). It is therefore critical to determine how CENP-A is targeted to and exclusively maintained at centromeres.

To date, no specific post-translational modification of CENP-A has been identified that is required for its localization to centromeres. However, acetylation of CENP-A or another protein appears to be involved. Mutants in the *S. pombe mis16* and *mis18* genes lead to CENP-A/Cnp1 mislocalization from the centromere as well as an increase in acetylated histone H3 and H4 within the central core (Hayashi et al., 2004). In addition, the loss of Mis18 protein in human cells by RNAi prevents newly synthesized CENP-A from assembling at centromeres (Fujita et al., 2007). Strikingly, the inhibition of HDAC activity by TSA partially restores CENP-A localization to the centromere in cells lacking Mis18. In addition, the RbAp48 protein (homologous to Mis16), a conserved chromatin factor required for CENP-A localization in vivo and in vitro (Hayashi et al., 2004; Furuyama, Dalal and Henikoff 2006), interacts with the Hat1 histone acetyltransferase catalytic subunit (Verreault et al., 1996, 1998). Together, these data suggest that histone acetylation positively regulates CENP-A localization. However, HDACs can act on non-histone substrates (Kaiser and James 2004), and TSA can inhibit other enzymatic reactions (Chen et al., 2005), so it will be important to identify the TSA target that alters CENP-A deposition.

The ubiquitin-mediated proteolysis of CENP-A/Cse4/Cid is required to prevent it from localizing to euchromatin in budding yeast and flies (Collins, Furuyama, and Biggins 2004; Moreno-Moreno, Torras-Llort, and Azorin

2006). There are low levels of soluble CENP-A in other organisms, suggesting that this could be a widely conserved mode of CENP-A regulation (Furuyama and Henikoff 2006). Although a CENP-A/Cse4 mutant that is not ubiquitinated associates with euchromatin (Collins, Furuyama, and Biggins 2004), it is not yet known whether the chromatin-bound pool is specifically degraded. The identification and characterization of the specific proteolysis machinery that degrades CENP-A is critical to fully understanding how CENP-A localization is regulated.

CENP-A is also modified by phosphorylation. The human and maize CENP-A/CenH3 are phosphorylated on the N-terminal tail from prophase to the onset of anaphase (Zeitlin et al., 2001; Zhang et al., 2005b). Although histone H3 appears to be specifically phosphorylated by Aurora B on S10 (Hsu et al., 2000), human CENP-A is phosphorylated on S7 by both Aurora A and B (Zeitlin, Shelby, and Sullivan 2001; Kunitoku et al., 2003). The Aurora A-mediated phosphorylation occurs first and appears to prime CENP-A for subsequent phosphorylation by Aurora B. The phosphorylation of human CENP-A is involved in localizing Aurora B to inner centromeres, as well as ensuring that Aurora B, INCENP, and PP1 transfer from chromosomes to the spindle midzone at the onset of anaphase. Consistent with a role in kinetochore function, expression of a non-phosphorylatable CENP-A mutant leads to defects in chromosome misalignment and the recruitment of checkpoint proteins to kinetochores (Zeitlin, Shelby and Sullivan 2001). Because CENP-A tails are highly divergent and do not contain any conserved phosphorylation sites (Malik and Henikoff 2003), it is unclear whether CENP-A is phosphorylated in other organisms. The maize CENP-A is phosphorylated on S50 in a temporal pattern that is similar to the phosphorylation of S7 in human CENP-A, so it is possible that CENP-A phosphorylation occurs in all organisms in divergent regions of the tail (Zhang et al., 2005b).

### *10.4.3 The H2A.Z Histone Variant*

The other histone variant that is required for accurate chromosome segregation is H2A.Z (for review, see Raisner and Madhani 2006). H2A.Z accounts for $\sim$10% of the cellular histone H2A and it is distributed throughout the genome where it functions in processes such as gene silencing and transcriptional activation and repression (for review, see Kamakaka and Biggins 2005). H2A.Z also localizes to centromeres and pericentromeres and is required for chromosome stability (Carr et al., 1994; Rangasamy et al., 2003; Krogan et al., 2004; Rangasamy, Greaves and Tremethick 2004; Foltz et al., 2006; Greaves et al., 2007). Although the precise localization of H2A.Z to centromeres and pericentromeres varies between organisms and cell types, H2A.Z appears to have a direct role in chromosome segregation. A budding yeast allele of H2A.Z that is specifically defective in segregation but not its other functions was recently isolated (Keogh et al., 2006). Strikingly, this allele is a mutation in K14, one of the four known sites of acetylation in the budding yeast H2A.Z. In addition, mutants in the NuA4 histone acetyltransferase that modify K14 on H2A.Z also have chromosome segregation defects (Krogan et al., 2004). Although these data strongly suggest that acetylation of H2A.Z is required for its function in chromosome segregation, it is not yet known whether this modification is conserved nor whether it is restricted to the pool of H2A.Z that localizes to the centromeric region.

### 10.4.4 Heterochromatin Modifications

Centromeres are embedded in pericentric heterochromatin, which contributes to chromosome segregation and is subject to its own set of distinct histone modifications (for review, see Ekwall 2007). Similar to centromeres, the surrounding heterochromatin is characterized by hypoacetylated canonical histones. However, in contrast to centromeres, pericentric chromatin is characterized by dimethylation (in flies and fission yeast) or trimethylation (in mammalian cells) of H3-K9, monomethylation of mammalian H3-K27, and trimethylation of mammalian H4-K20 (Noma, Allis, and Grewal 2001; Peters et al., 2003; Rice et al., 2003; Schotta et al., 2004; Sullivan and Karpen 2004). Histone H3-K9 methylation recruits the HP1/Swi6 protein to establish and maintain heterochromatin boundaries (for review, see Sims, Nishioka, and Reinberg 2003) and the cohesin protein complex, which holds sister chromatids together until anaphase onset, in *S. pombe* (Bailis et al., 2003). These RNAi-directed properties of heterochromatin have recently been shown to be required to establish CENP-A chromatin in fission yeast (Folco et al., 2008). HP1 dissociation also appears to be regulated by one or more histone modifications. Histone H3-S10 phosphorylation initiates within pericentric chromatin during G2 and correlates with HP1 dissociation (Hendzel et al., 1997). Although it has been suggested that Aurora B modification of H3-S10 is sufficient to cause HP1 removal from chromatin (Fischle et al., 2005), these data are complicated by the finding that Aurora B activity is also required for the localization of the H3-K9 HMtase SUV39H1 (Terada 2006). In addition, factors such as the acetylation of H3-K14 may also be required for HP1 dissociation (Fass et al., 2002; Mateescu et al., 2004). The mammalian H3.3 histone variant also localizes to heterochromatin and is modified by phosphorylation of ser31 (Hake et al., 2005). In contrast to H3-S10 phosphorylation which begins during prophase and disappears throughout telophase, H3.3-S31 is modified during late prometaphase and metaphase by an unidentified kinase. Although H3.3-S31 is conserved, it is not known whether the phosphorylation is conserved and its functions have not yet been characterized.

Although post-translational modifications associated with centromeric histones have been identified, it is likely that there are unique modifications that have not yet been identified. Modifications of the canonical and variant histones could influence many other aspects of chromosome segregation. For example, histone modifications are often associated with boundary function and could therefore restrict CENP-A nucleosome formation to the centromere to ensure that it does not spread into euchromatin where ectopic kinetochores could form. Histone modifications could also alter the underlying structure of the centromeric chromatin to influence the timing of kinetochore assembly, as well as aspects of the formation and stability of the kinetochore complex. It is therefore critical to continue to identify and elucidate the functions of centromeric chromatin modifications.

## 10.5 The Regulation of Kinetochore–Microtubule Attachments

The kinetochore–microtubule interface mediates the attachment of chromosomes to the spindle microtubules and is therefore subject to precise regulation. Kinetochores must establish proper bioriented attachments to microtubules by metaphase, and then maintain these attachments throughout anaphase. When erroneous kinetochore–microtubule attachments are made, they must be detected and corrected prior to anaphase. Evidence suggests that the phosphorylation of proteins residing at the kinetochore–microtubule interface is a key mechanism that regulates this interaction. Multiple kinases are involved in detecting such defects in kinetochore–microtubule attachments, and the mechanism by which kinases regulate the spindle checkpoint is reviewed in Chapter 12. Here, we review the major modifications that occur on various kinetochore subcomplexes to regulate microtubule–kinetochore attachments (Table 10.2). We refer readers to Chapters 6, 7, and 10 for a more detailed description of the kinetochore subcomplexes and their corresponding functions.

### *10.5.1 The Ndc80 Complex*

The conserved Ndc80 complex, which is composed of the Ndc80, Nuf2, Spc24, and Spc25 proteins (for review, see Ciferri, Musacchio, and Petrovic 2007), has been proposed to be part of a larger network called KMN that mediates microtubule attachment (Cheeseman et al., 2006). Consistent with this, Ndc80 localizes to the outer-plate of the kinetochore, the region where microtubule plus ends terminate (Deluca et al., 2005; Du and Dawe 2007). Moreover, Ndc80 and Nuf2 form heterodimers that directly bind to microtubules in vitro (Wei, Al-Bassam and Harrison 2007), and the depletion or inactivation of Ndc80 results in unstable kinetochore microtubule interactions (for review, see Kline-Smith, Sandall, and Desai 2005). The Ndc80 complex is therefore an excellent candidate for controlling interactions between kinetochores and microtubules, and recent data suggest that phosphorylation of Ndc80 by the Aurora B kinase destabilizes kinetochore–microtubule attachments (Cheeseman et al., 2006; DeLuca et al., 2006). Although the N-terminus of Ndc80 is highly variable in size and sequence among organisms (Wei, Al-Bassam, and Harrison 2007), the human and *C. elegans* N-termini are phosphorylated by Aurora B in vitro (Cheeseman et al., 2006; DeLuca et al., 2006), and at least one residue in the budding yeast Ndc80 N-terminus is phosphorylated in vivo (Cheeseman et al., 2002). It is possible that the positive charge of the Ndc80 N-terminus favors binding to microtubules, so phosphorylation by Aurora B would reduce the positive charge of this domain, thus weakening kinetochore–microtubule interactions. Indeed, phosphorylation of the Ndc80 complex by Aurora B reduces its affinity for microtubules in vitro (Cheeseman et al., 2006). In addition, the elimination of the Aurora B phosphorylation sites in the human Ndc80 protein, or the steric blocking of phosphorylation due to microinjection of an antibody against the Ndc80 N-terminal globular domain in mitotic PtK1 cells, results in chromosome alignment defects and merotelic attachments (DeLuca et al., 2006). These phenotypes are reminiscent of Aurora B downregulation where inappropriate kinetochore–microtubule attachments are stabilized (Cimini et al., 2006). In addition, two hybrid interactions between the budding yeast Ndc80 protein and its interacting partners are altered by mutation of the Aurora B phosphorylation sites (Wong et al., 2007). Together, these experiments provide compelling evidence for the regulation of Ndc80 by Aurora B-mediated phosphorylation (Fig. 10.1). Future studies will therefore need to address the spatial and temporal regulation of the Ndc80 phosphorylation to fully understand the relative contribution of Ndc80 phosphorylation by Aurora B to ensuring proper kinetochore–microtubule attachments.

### *10.5.2 The Mtw1/Mis12 Complex*

Components of the conserved Mtw1/Mis12 complex (Mtw1/Mis12, Dsn1, Nnf1, and Nsl1 proteins) were first identified in a screen for genes required for the proper segregation of a minichromosome in fission yeast (Takahashi, Yamada and Yanagida 1994). Although the bacterially expressed Mis12 complex does not interact with microtubules in vitro, its presence leads to a synergistic enhancement of the microtubule binding activities of the Ndc80 complex and another kinetochore protein, KNL-1/Spc105 (Cheeseman et al., 2006). This observation led to the proposal that these proteins (KNL-1/Spc105, Mis12, Ndc80), which comprise the KMN network, exhibit an array of low-affinity binding sites that cooperate with other factors to create a dynamic kinetochore–spindle interface. One prediction of this model is that other microtubule binding components might also be subject to phosphoregulation, similar to Ndc80. Although the Dsn1 component of the budding yeast Mtw1/Mis12 complex is a phosphoprotein in vivo and can be phosphorylated by Aurora B in vitro (Westermann et al., 2003), the elimination of the predicted Aurora B

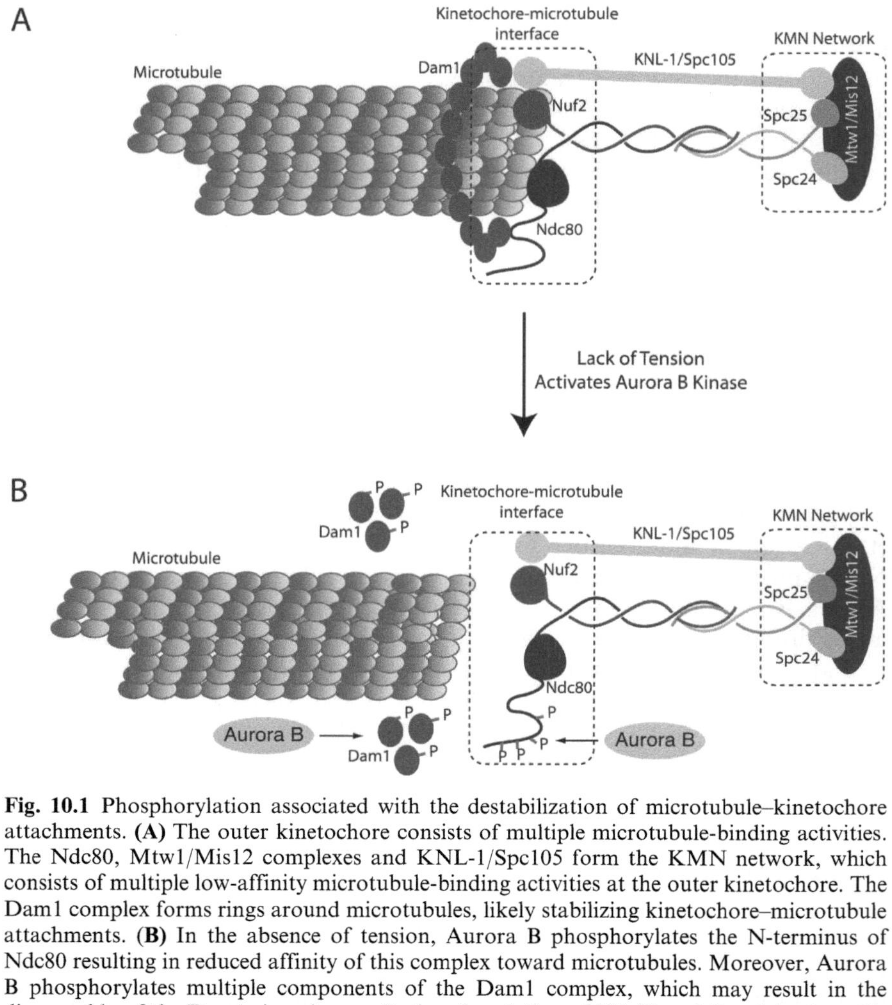


**Fig. 10.1** Phosphorylation associated with the destabilization of microtubule–kinetochore attachments. **(A)** The outer kinetochore consists of multiple microtubule-binding activities. The Ndc80, Mtw1/Mis12 complexes and KNL-1/Spc105 form the KMN network, which consists of multiple low-affinity microtubule-binding activities at the outer kinetochore. The Dam1 complex forms rings around microtubules, likely stabilizing kinetochore–microtubule attachments. **(B)** In the absence of tension, Aurora B phosphorylates the N-terminus of Ndc80 resulting in reduced affinity of this complex toward microtubules. Moreover, Aurora B phosphorylates multiple components of the Dam1 complex, which may result in the disassembly of the Dam1 ring. Aurora B phosphorylation of Ndc80 and Dam1 leads to the destabilization of the kinetochore–microtubule interactions giving the cell a chance to establish proper attachments

phosphorylation site does not result in a segregation defect. Dsn1 may be phosphorylated by Aurora B at additional, unidentified sites in vivo, or other components of the Mis12 complex may be subject to phosphoregulation. In fact, the fission yeast Mtw1/Mis12 protein is phosphorylated on at least two residues in its C-terminus in vivo (Goshima et al., 2003). Although the *S. pombe* Mtw1/Mis12 protein is regulated by the Gsk3 kinase and opposing Ppe1 phosphatase (Goshima et al., 2003), it is not yet known whether Mtw1/Mis12 phosphorylation is directly mediated by Gsk3 nor whether this is a conserved modification. It will be critical to map the in vivo phosphorylation sites on the subunits of the Mis12 complex as well as the KNL-1/Spc105 protein for a more complete understanding of the regulation of microtubule-binding activity at the kinetochore.

### 10.5.3 The Dam1/DASH/DDD Complex

The 10 subunit Dam1–DASH–DDD–kinetochore complex is found in budding and fission yeasts (for review, see Davis and Wordeman 2007). Dam1 is essential in budding yeast, though not in *S. pombe*. However, loss of function mutants exhibit spindle and chromosome segregation defects in both organisms. Recently, a *Xenopus* protein, xCep57, has been proposed to be a Dam1 homolog, suggesting it may be conserved throughout eukaryotes (Emanuele and Stukenberg 2007). Dam1 binds to microtubules in vitro and ultrastructural analysis has shown that the complex assembles into ring-shaped structures around microtubules (Miranda et al., 2005; Westermann et al., 2005). This led to the proposal that this complex could mediate the attachment between kinetochores and microtubules and remain bound while microtubules polymerized and depolymerized (Westermann et al., 2006). Similar to Ndc80 complex, the Dam1 complex is a target of the budding yeast Aurora B kinase in vivo (Cheeseman et al., 2002). However, it has been difficult to precisely determine the consequences of Aurora B phosphorylation of Dam1 because the elimination of numerous Aurora consensus phosphorylation sites is lethal, and could disrupt the protein structure or its interactions with partners (Cheeseman et al., 2002). Studies of an allele that was originally reported to mimic the phenotype of a mutant in the Aurora kinase showed that it does not truly phenocopy a lack of Aurora activity (Pinsky et al., 2006b). In addition, there is no apparent cell cycle regulation to Dam1 phosphorylation as would be expected if it were critical for achieving biorientation. Recent structural analyses on a phosphomimetic mutant of Dam1 suggest that although Aurora B phosphorylation does not alter the interaction of Dam1 with microtubules, the phosphorylation could decrease its ability to form rings around microtubules (Wang et al., 2007). In addition, two hybrid data using Dam1 phosphomimetic mutants suggests that phosphorylation weakens its interactions with other kinetochore proteins (Shang et al., 2003). Together, these data indicate that phosphorylation of Dam1 likely results in the destabilization of the interactions between microtubules and kinetochores (Fig. 10.1), but it will take additional analyses to determine the precise contribution of Aurora kinase-mediated phosphorylation of Dam1 in vivo. In addition, it will be important to determine how the phosphorylation of Dam1 cooperates with Aurora B-mediated regulation of the KMN proteins.

The Dam1 protein appears to be controlled by additional post-translational modifications, further complicating the ability to dissect the contributions of each specific modification. An antibody generated against a Dam1 peptide containing dimethylated K233 recognizes the Dam1 protein in a Set1 methyltransferase-dependent manner in vivo (Zhang et al., 2005a). The K233 residue is flanked by three conserved serine residues (S232, S234 and S235), and at least of one these serines appears to be phosphorylated by Aurora B in vitro. This led to the proposal that Dam1 methylation could block Aurora B from phosphorylating the neighboring serines. Consistent with this, Aurora B exhibits reduced activity against a Dam1 peptide containing dimethylated K233 in vitro, and inactivating the Set1 methyltransferase partially suppresses the chromosome segregation defect associated with loss of Aurora B function. Dam1 was also recently shown to be a target of the Mps1 kinase (Shimogawa et al., 2006), and the S232 residue that flanks K233 is phosphorylated by Mps1 in vitro. Although the functions of Mps1 phosphorylation of S232 have not been studied, Mps1 phosphorylates at least six residues in Dam1 in vitro, and the mutation of two of these sites (S218 or S221) leads to an altered arrangement of kinetochores in vivo. Unlike wild-type cells where the kinetochores co-localize with a microtubule plus end-binding protein (Bik1) near the spindle poles, the kinetochores cluster near the poles in Dam1S221F cells, but the plus ends are distributed throughout the spindle. This led to the interesting proposal that the kinetochores in these *dam1* mutant cells are able to biorient without making proper plus end attachments and that Mps1 phosphorylation of Dam1 promotes its association with microtubule plus ends. If this model is true, the viability of these mutants suggests that yeast cells have additional mechanisms to ensure biorientation and it will be critical to identify these pathways in the future. In addition, it will be important to determine which Dam1 residues are actually phosphorylated in vivo and how methylation affects Mps1 phosphorylation.

### *10.5.4 The Budding Yeast CBF3 Complex*

The budding yeast CBF3 complex (Ndc10, Cep3, Ctf13, Skp1) binds directly to centromeric DNA and is required to initiate the assembly of the kinetochore. Ndc10 was originally shown to be a substrate of the Aurora B kinase and PP1 phosphatase (Biggins et al., 1999; Sassoon et al., 1999). However, the phosphorylation sites in Ndc10 have not been identified. Recently, Ndc10 has been proposed to be part of the tension-sensing machinery that controls the activity of the Aurora B kinase (Sandall et al., 2006). Although it will be interesting to determine whether the Aurora B-mediated phosphorylation of Ndc10 is involved in controlling Aurora activity in response to tension defects, most eukaryotic kinetochores lack Ndc10 so this could not be a widespread mechanism to regulate Aurora B activity.

The CBF3 components Ndc10 and Cep3 are also modified by sumoylation, and elimination of the four SUMO consensus sites in Ndc10 results in chromosomal instability, consistent with a role in segregation (Yoon and Carbon 1999;

Montpetit et al., 2006). Although Ndc10 is sumoylated throughout the cell cycle, activation of the spindle checkpoint causes Ndc10 sumoylation to decrease to undetectable levels. However, Ndc10 sumoylation is not required for spindle checkpoint activation. One possibility is that sumoylation of CBF3 proteins is involved in response to defects in tension, a role that has not yet been tested. It is noteworthy that two components of the CPC that is required for tension-sensing, INCENP and survivin, are also modified by sumoylation (Table 10.1).

The CBF3 complex is also subject to ubiquitin-mediated proteolysis, which is catalyzed by the SCF E3 ligase. However, it is not clear whether this activity is related to kinetochore functions or other roles for CBF3 (Chapter 13). Taken together, it is clear that the CBF3 complex is subject to many types of post-translational modification, so it will be critical to determine the interplay between the numerous modifications that occur on CBF3.

### *10.5.5 Microtubule-Associated Proteins*

In addition to the kinetochore subcomplexes discussed above, a number of motor and microtubule-associated proteins (MAPs) localize to kinetochores in multicellular eukaryotes. These motors and MAPs also have important functions in microtubule attachment and force generation necessary for chromosome movement and are subject to a number of post-translational modifications.

*The Mitotic Centromere-Associated Kinesin (MCAK)*: Because chromosome segregation requires microtubule disassembly, the MCAK motor protein that catalyzes microtubule depolymerization has been of particular interest (Wordeman and Mitchison 1995; Walczak, Mitchison and Desai 1996; Desai et al., 1999; Moore and Wordeman 2004). Centromere-bound MCAK appears to be involved in correcting inappropriate kinetochore–microtubule attachments (Kline-Smith et al., 2004; Wordeman, Wagenbach and von Dassow 2007). Strikingly, one feature of MCAK is its dynamic co-localization with the Aurora B kinase at the inner centromere of unaligned chromosomes in prometaphase (Andrews et al., 2004; Lan et al., 2004; Knowlton, Lan and Stukenberg 2006), and data suggests that Aurora B regulates MCAK localization and activity on these kinetochores (Andrews et al., 2004; Lan et al., 2004; Ohi et al., 2004; Zhang et al., 2007). Aurora B phosphorylates the N-terminus of MCAK on at least six sites in vitro, and at least two of these sites are phosphorylated in an Aurora B-dependent manner in vivo (Andrews et al., 2004; Lan et al., 2004; Ohi et al., 2004). Although one might expect Aurora B to stimulate MCAK activity and thus destabilize inappropriate microtubule attachments, phosphorylation by Aurora actually inhibits the depolymerase activity of MCAK. One possibility is that Aurora B initially helps MCAK localize to centromeres, and then phosphorylates MCAK when proper attachments are made to prevent it from destabilizing them. Consistent with this, individual kinetochores appear to contain populations of MCAK that are both unphosphorylated and phosphorylated by Aurora B (Lan et al., 2004), and phospho-deficient and phospho-mimic mutants display different localization patterns at the kinetochore (Andrews et al., 2004). Recently, it was shown that the Plk1 kinase also phosphorylates MCAK, and this primes MCAK for Aurora B phosphorylation in vitro (Rosasco-Nitcher et al., 2008). Together, these data suggest that there is complex regulation of both MCAK localization and activity by Aurora B and Plk1 that is critical in ensuring proper kinetochore–microtubule attachments (Fig. 10.2).

*The Centromere Protein E (CENP-E)*: Another motor protein with important functions in kinetochore–microtubule attachments is the CENP-E that localizes to kinetochores specifically from prometaphase until metaphase (Yen et al., 1992). During anaphase, CENP-E dissociates from the kinetochores and localizes to the spindle midzone. The absence of CENP-E activity in *Xenopus*, *Drosophila*, and mice results in chromosome alignment defects, and ultrastructural analysis of primary mouse fibroblasts in which CENP-E had been deleted showed reduced numbers of microtubules interacting with kinetochores (Putkey et al., 2002). CENP-E is phosphorylated specifically in mitosis in vivo, though it is not currently clear whether phosphorylation alters the activity of kinetochore-bound CENP-E. Although it was previously reported that phosphorylation of the CENP-E C-terminus by Cdk1 inhibits its microtubule binding activity in vitro (Liao, Li and Yen 1994), recent studies indicate that phosphorylation of the CENP-E tail by either the Mps1 or Cdk1 kinases relieves autoinhibition of its motor activity in vitro (Espeut et al., 2008). MAPK can also phosphorylate CENP-E in vitro (Zecevic et al., 1998), so it will be important to determine the relative contributions of each kinase to the phosphorylation and activity of CENP-E in vivo.

CENP-E is also regulated by sumoylation (Zhang et al., 2008). When sumoylation at kinetochores is inhibited by the overexpression of the SUMO isopeptidase SENP-2, CENP-E no longer localizes to kinetochores and cells exhibit unaligned chromosomes. CENP-E contains a motif that binds to polymeric SUMO-2/3 chains, and this domain is also required for CENP-E localization to the kinetochore. Because the BubR1 and Nuf2 kinetochore proteins are modified by SUMO-2/3 and are required for CENP-E localization, they may be the critical SUMO-2/3 kinetochore substrates that regulate CENP-E localization. However, CENP-E is also directly modified by SUMO-2/3, so future studies will be needed to fully dissect the complex regulation of CENP-E localization by sumoylation.

*Nuclear Distribution Protein C (NudC)*: The NudC protein that localizes to the outer plate of the vertebrate kinetochore primarily in prometaphase, metaphase, and anaphase is a Plk1 target (Nishino et al., 2006). NudC has been implicated in kinetochore–microtubule attachments because depletion of NudC results in significant chromosome congression defects. Plk1 phosphorylates two highly conserved residues (S274 and S326) on NudC, and mutation

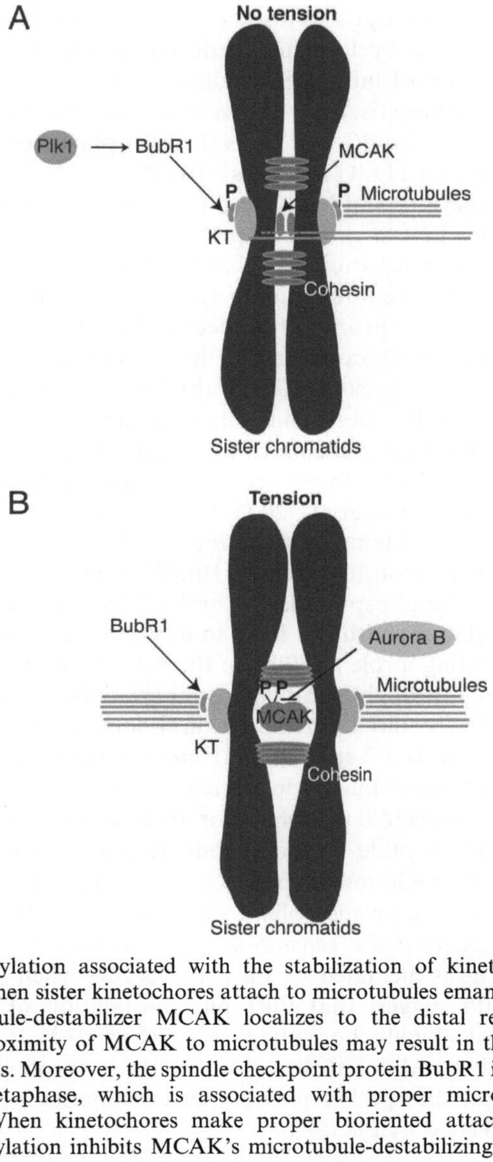


**Fig. 10.2** Phosphorylation associated with the stabilization of kinetochore–microtubule attachments. **(A)** When sister kinetochores attach to microtubules emanating from the same pole, the microtubule-destabilizer MCAK localizes to the distal regions of the inner centromere. The proximity of MCAK to microtubules may result in the destabilization of aberrant attachments. Moreover, the spindle checkpoint protein BubR1 is phosphorylated by Plk1 during prometaphase, which is associated with proper microtubule–kinetochore attachments. **(B)** When kinetochores make proper bioriented attachments, Aurora B-mediated phosphorylation inhibits MCAK's microtubule-destabilizing activity. Moreover, Aurora B phosphorylation results in the concentration of MCAK in the central region of the inner centromere, possibly titrating it away from microtubules. This results in the stabilization of kinetochore–microtubule attachments. In addition, BubR1 is no longer phosphorylated by Plk1, which also contributes to stable attachments

of S326 results in significant chromosome congression and alignment defects. Moreover, NudC-deficient cells exhibit reduced end-on kinetochore–microtubule attachments and Plk1 fails to enrich at the kinetochores. Strikingly, Plk1 localization to kinetochores is restored in cells that ectopically express wild-type NudC but not a mutant NudC that lacks the Plk1 phosphorylation sites. This led to the proposal that NudC interacts with Plk1 in prometaphase, and the subsequent phosphorylation of NudC results in a conformational change in the Plk1–NudC complex that targets these proteins to the outer plate of the kinetochore. NudC phosphorylation by Plk1 is also important for the CenpE localization to the kinetochore, leading to the idea that the localized Plk1 activity at the kinetochore promotes the accumulation of proteins like CENP-E to ensure proper kinetochore–microtubule attachments.

*BubR1*: Recent data suggests that the BubR1 protein, initially identified due to its role in the spindle checkpoint, also has an independent function in regulating microtubule attachments to kinetochores (Ditchfield et al., 2003; Lampson and Kapoor 2005). BubR1 is phosphorylated during mitosis in vivo by the Plk1 kinase (Ahonen et al., 2005; Wong and Fang 2005; Elowe et al., 2007; Matsumura, Toyoshima and Nishida 2007; Wong and Fang 2007). Similar to other Plk1 substrates, human BubR1 is primed for Plk1-mediated phosphorylation by the phosphorylation of T620 by Cdk1. Elimination of the priming phosphorylation in BubR1 leads to unstable microtubule–kinetochore attachments, suggesting a role for Plk1 in the regulation of microtubule–kinetochore attachments through BubR1 phosphorylation. In addition, the Cdk1 priming site (T605) is required for spindle checkpoint arrest and the formation of the tension-sensitive 3F3/2 epitope in *Xenopus* egg extracts (Wong and Fang 2007). Using quantitative mass spectrometry, residue S676 in BubR1 was identified as being threefold enriched for phosphorylation during mitosis. Strikingly, a phosphopeptide antibody generated against this site in BubR1 specifically stains kinetochores when they are forming proper microtubule–kinetochore attachments at prometaphase. Consistent with its identification as a kinase that can generate the tension-sensitive 3F3/2 epitope, BubR1-S676 kinetochore staining is mediated by Plk1 and occurs when kinetochores lack tension. These data suggest that Plk1 phosphorylation of BubR1 helps to stabilize microtubule–kinetochore attachments (Fig. 10.2). It will be important to determine whether additional sites in BubR1 are phosphorylated by Plk1 in vivo, as well as test the specific phenotypes of mutants in residues that are directly phosphorylated by Plk1.

#### 10.5.5.1 Plk1-Interacting Checkpoint "Helicase" (PICH)

PICH, a SNF2 class ATPase of the superfamily 2 helicases, is a human Plk1-PBD binding protein that is phosphorylated by Plk1 after being primed by Cdk1 phosphorylation (Baumann et al., 2007). PICH colocalizes with Plk1 on kinetochores from prometaphase until anaphase. In metaphase and anaphase, PICH also exhibits a unique localization pattern where it appears as short, thread-like structures that stretch between sister kinetochores of unaligned chromosomes. It has been proposed that these threads contain stretched chromatin and PICH localization to these threads might enable the protein to act as a tension sensor for centromeric chromatin. Consistent with this, cells depleted of PICH and treated with drugs that disrupt kinetochore–microtubule attachments fail to arrest in mitosis. The Plk1-mediated phosphorylation of PICH appears to restrict PICH to kinetochores where it could facilitate the correction of inappropriate microtubule–kinetochore attachments.

## 10.6 Summary and Perspectives

In summary, a vast array of dynamic post-translational modifications regulate eukaryotic kinetochore function. Although we have focused on the modifications that specify the site of kinetochore assembly and regulate the kinetochore–microtubule interface, many additional modifications control the localization of kinetochore proteins. In addition, the key modifying enzymes themselves are subject to a wide array of modifications that regulate their activity. Like the rest of the genome, the assembly and function of the underlying centromeric chromatin is subject to diverse modifications that exist in a unique pattern that specifies centromere identity and contributes to kinetochore function. The kinetochore itself is subject to a wealth of modifications that ensure that it assembles into a macromolecular complex that makes proper attachments to microtubules. Currently, the most common modification that has been shown to regulate the microtubule–kinetochore interface is phosphorylation by the conserved Plk1 and Aurora B kinases. In general, the activity of Aurora B is associated with the destabilization of inappropriate kinetochore–microtubule attachments, while Plk1 activity is linked to the stabilization of microtubule attachments. However, several additional kinases and modifying enzymes have recently been found to function at the kinetochore. Future studies will no doubt identify an even more complex array of modifications, and this will present an even greater challenge in trying to define the precise functions of each modification. Although significant effort has gone into identifying and characterizing kinetochore protein modifications, the mutants in modifiable residues rarely mimic the phenotypes of mutants in the modifying enzyme due to the large number of substrates for each enzyme as well as the redundancy that exists between modifications. It is also becoming increasingly clear that there is a complex interplay between modifications such that they cannot be studied in isolation. Many of the kinases regulate each other as well as structural components of the kinetochore, further complicating phenotypic analyses.

To fully understand the regulation of kinetochore activity in the future, it will be critical to identify and analyze the key modifications associated with specific states of microtubule–kinetochore attachment, such as kinetochore biorientation. The abundant sequence information that is now available should allow a more systematic identification of substrates for different enzymes. In addition, the continued development of sophisticated mass spectrometry techniques should help identify modifications, especially those that are dynamic or less abundant. Sensitive mass spectrometry should also help detect modifications, such as methylation and acetylation, that are not easily identified by changes in protein migration and/or abundance like phosphorylation and ubiquitylation. A precise understanding of the molecular basis of kinetochore activity will depend on elucidating the full complement of modifications that regulate kinetochores, a goal that will be aided by continued technological advances.

**Acknowledgments** We thank all past and present members of the Biggins lab for thoughtful discussions regarding the regulation of kinetochore activity by post-translational modifications. We also thank Bungo Akiyoshi, Suzanne Furuyama and Leigh Ann Higa for critical comments on the manuscript. We apologize to our colleagues whose work we omitted due to space considerations. C. V. K. was supported by a DOD Breast Cancer predoctoral fellowship. Work in the Biggins' laboratory is funded by NIH grants R01 GM078069 and R01 GM064386. S. B. is a Scholar of the Leukemia and Lymphoma Society.

## References

Aagaard, L., Schmid, M., Warburton, P., and Jenuwein, T. 2000. Mitotic phosphorylation of SUV39H1, a novel component of active centromeres, coincides with transient accumulation at mammalian centromeres. J Cell Sci 113 (Pt 5): 817–29.

Acquaviva, C., Herzog, F., Kraft, C., and Pines, J. 2004. The anaphase promoting complex/cyclosome is recruited to centromeres by the spindle assembly checkpoint. Nat Cell Biol 6: 892–8.

Ahonen, L. J., Kallio, M. J., Daum, J. R., Bolton, M., Manke, I. A., Yaffe, M. B., Stukenberg, P. T., and Gorbsky, G. J. 2005. Polo-like Kinase 1 Creates the Tension-Sensing 3F3/2 Phosphoepitope and Modulates the Association of Spindle-Checkpoint Proteins at Kinetochores. Current Biology 15: 1078–89.

Allard, S., Utley, R. T., Savard, J., Clarke, A., Grant, P., Brandl, C. J., Pillus, L., Workman, J. L., and Cote, J. 1999. NuA4, an essential transcription adaptor/histone H4 acetyltransferase complex containing Esa1p and the ATM-related cofactor Tra1p. Embo J 18: 5108–19.

Allis, C. D., Berger, S. L., Cote, J., Dent, S., Jenuwien, T., Kouzarides, T., Pillus, L., Reinberg, D., Shi, Y., Shiekhattar, R., Shilatifard, A., Workman, J., and Zhang, Y. 2007. New nomenclature for chromatin-modifying enzymes. Cell 131: 633–6.

Amor, D. J., and Choo, K. H. 2002. Neocentromeres: role in human disease, evolution, and centromere study. Am J Hum Genet 71: 695–714.

An, W. 2007. Histone acetylation and methylation: combinatorial players for transcriptional regulation. Subcell Biochem 41: 351–69.

Andreassen, P. R., Lacroix, F. B., Villa-Moruzzi, E., and Margolis, R. L. 1998. Differential subcellular localization of protein phosphatase-1 alpha, gamma1, and delta isoforms during both interphase and mitosis in mammalian cells. J Cell Biol 141: 1207–15.

Andrews, P. D., Ovechkina, Y., Morrice, N., Wagenbach, M., Duncan, K., Wordeman, L., and Swedlow, J. R. 2004. Aurora B regulates MCAK at the mitotic centromere. Dev Cell 6: 253–68.

Aoki, K., Nakaseko, Y., Kinoshita, K., Goshima, G., and Yanagida, M. 2006. CDC2 phosphorylation of the fission yeast dis1 ensures accurate chromosome segregation. Curr Biol 16: 1627–35.

Ashar, H. R., James, L., Gray, K., Carr, D., Black, S., Armstrong, L., Bishop, W. R., and Kirschmeier, P. 2000. Farnesyl transferase inhibitors block the farnesylation of CENP-E and CENP-F and alter the association of CENP-E with the microtubules. J Biol Chem 275: 30451–7.

Bachant, J., Alcasabas, A., Blat, Y., Kleckner, N., and Elledge, S. J. 2002. The SUMO-1 isopeptidase Smt4 is linked to centromeric cohesion through SUMO-1 modification of DNA topoisomerase II.Mol Cell 9: 1169–82.

Bailis, J. M., Bernard, P., Antonelli, R., Allshire, R. C., and Forsburg, S. L. 2003. Hsk1-Dfp1 is required for heterochromatin-mediated cohesion at centromeres. Nat Cell Biol 5: 1111–6.

Barr, F. A., Sillje, H. H., and Nigg, E. A. 2004. Polo-like kinases and the orchestration of cell division. Nat Rev Mol Cell Biol 5: 429–40.

Baumann, C., Korner, R., Hofmann, K., and Nigg, E. A. 2007. PICH, a centromere-associated SNF2 family ATPase, is regulated by Plk1 and required for the spindle checkpoint. Cell 128: 101–14.

Biggins, S., Severin, F. F., Bhalla, N., Sassoon, I., Hyman, A. A., and Murray, A. W. 1999. The conserved protein kinase Ipl1 regulates microtubule binding to kinetochores in budding yeast. Genes Dev 13: 532–44.

Bishop, J. D., and Schumacher, J. M. 2002. Phosphorylation of the carboxyl terminus of inner centromere protein (INCENP) by the Aurora B Kinase stimulates Aurora B kinase activity. J Biol Chem 277: 27577–80.

Bloom, J., and Cross, F. R. 2007. Multiple levels of cyclin specificity in cell-cycle control. Nat Rev Mol Cell Biol 8: 149–60.

Bolton, M. A., Lan, W., Powers, S. E., McCleland, M. L., Kuang, J., and Stukenberg, P. T. 2002. Aurora B kinase exists in a complex with survivin and INCENP and its kinase activity is stimulated by survivin binding and phosphorylation. Mol Biol Cell 13: 3064–77.

Bouck, D. C., and Bloom, K. S. 2005. The kinetochore protein Ndc10p is required for spindle stability and cytokinesis in yeast. Proc Natl Acad Sci USA 102: 5408–13.

Brittle, A. L., Nanba, Y., Ito, T., and Ohkura, H. 2007. Concerted action of Aurora B, Polo and NHK-1 kinases in centromere-specific histone 2A phosphorylation. Exp Cell Res 313: 2780–5.

Cao, R., Wang, L., Wang, H., Xia, L., Erdjument-Bromage, H., Tempst, P., Jones, R. S., and Zhang, Y. 2002. Role of histone H3 lysine 27 methylation in Polycomb-group silencing. Science 298: 1039–43.

Carr, A. M., Dorrington, S. M., Hindley, J., Phear, G. A., Aves, S. J., and Nurse, P. 1994. Analysis of a histone H2A variant from fission yeast: evidence for a role in chromosome stability. Mol Gen Genet 245: 628–35.

Carroll, C. W., and Straight, A. F. 2006. Centromere formation: from epigenetics to self-assembly. Trends Cell Biol 16: 70–8.

Chang, B., Chen, Y., Zhao, Y., and Bruick, R. K. 2007. JMJD6 is a histone arginine demethylase. Science 318: 444–7.

Cheeseman, I. M., Anderson, S., Jwa, M., Green, E. M., Kang, J., Yates, J. R., Chan, C. S., Drubin, D. G., and Barnes, G. 2002. Phospho-Regulation of Kinetochore-Microtubule Attachments by the Aurora Kinase Ipl1p. Cell 111: 163–172.

Cheeseman, I. M., Brew, C., Wolyniak, M., Desai, A., Anderson, S., Muster, N., Yates, J. R., Huffaker, T. C., Drubin, D. G., and Barnes, G. 2001. Implication of a novel multiprotein Dam1p complex in outer kinetochore function. J Cell Biol 155: 1137–45.

Cheeseman, I. M., Chappie, J. S., Wilson-Kubalek, E. M., and Desai, A. 2006. The conserved KMN network constitutes the core microtubule binding site of the kinetochore. Cell 127: 983–97.

Cheeseman, I. M., Drubin, D. G., and Barnes, G. 2002. Simple centromere, complex kinetochore: linking spindle microtubules and centromeric DNA in budding yeast. J Cell Biol 157: 199–203.

Chen, C. S., Weng, S. C., Tseng, P. H., and Lin, H. P. 2005. Histone acetylation-independent effect of histone deacetylase inhibitors on Akt through the reshuffling of protein phosphatase 1 complexes. J Biol Chem 280: 38879–87.
Chen, E. S., Zhang, K., Nicolas, E., Cam, H. P., Zofall, M., and Grewal, S. I. 2008. Cell cycle control of centromeric repeat transcription and heterochromatin assembly. Nature 451: 734–7.
Chen, Y., Riley, D. J., Zheng, L., Chen, P. L., and Lee, W. H. 2002. Phosphorylation of the mitotic regulator protein Hec1 by Nek2 kinase is essential for faithful chromosome segregation. J Biol Chem 277: 49408–16.
Chung, T. L., Hsiao, H. H., Yeh, Y. Y., Shia, H. L., Chen, Y. L., Liang, P. H., Wang, A. H., Khoo, K. H., and Shoei-Lung Li, S. 2004. In vitro modification of human centromere protein CENP-C fragments by small ubiquitin-like modifier (SUMO) protein: definitive identification of the modification sites by tandem mass spectrometry analysis of the isopeptides. J Biol Chem 279: 39653–62.
Ciferri, C., Musacchio, A., and Petrovic, A. 2007. The Ndc80 complex: hub of kinetochore activity. FEBS Lett 581: 2862–9.
Cimini, D., Wan, X., Hirel, C. B., and Salmon, E. D. 2006. Aurora kinase promotes turnover of kinetochore microtubules to reduce chromosome segregation errors. Curr Biol 16: 1711–8.
Clute, P., and Pines, J. 1999. Temporal and spatial control of cyclin B1 destruction in metaphase. Nat Cell Biol 1: 82–7.
Collins, K. A., Furuyama, S., and Biggins, S. 2004. Proteolysis contributes to the exclusive centromere localization of the yeast Cse4/CENP-A histone H3 variant. Curr Biol 14: 1968–72.
Couture, J. F., and Trievel, R. C. 2006. Histone-modifying enzymes: encrypting an enigmatic epigenetic code. Curr Opin Struct Biol 16: 753–60.
Crespo, N. C., Ohkanda, J., Yen, T. J., Hamilton, A. D., and Sebti, S. M. 2001. The farnesyltransferase inhibitor, FTI-2153, blocks bipolar spindle formation and chromosome alignment and causes prometaphase accumulation during mitosis of human lung cancer cells. J Biol Chem 276: 16161–7.
Dai, J., and Higgins, J. M. 2005. Haspin: a mitotic histone kinase required for metaphase chromosome alignment. Cell Cycle 4: 665–8.
Dai, J., Sullivan, B. A., and Higgins, J. M. 2006. Regulation of mitotic chromosome cohesion by Haspin and Aurora B. Dev Cell 11: 741–50.
Dai, J., Sultan, S., Taylor, S. S., and Higgins, J. M. 2005. The kinase haspin is required for mitotic histone H3 Thr 3 phosphorylation and normal metaphase chromosome alignment. Genes Dev 19: 472–88.
Davis, T. N., and Wordeman, L. 2007. Rings, bracelets, sleeves, and chevrons: new structures of kinetochore proteins. Trends Cell Biol 17: 377–82.
De Souza, C. P., Osmani, A. H., Wu, L. P., Spotts, J. L., and Osmani, S. A. 2000. Mitotic histone H3 phosphorylation by the NIMA kinase in Aspergillus nidulans. Cell 102: 293–302.
Delacour-Larose, M., Thi, M. N., Dimitrov, S., and Molla, A. 2007. Role of survivin phosphorylation by aurora B in mitosis. Cell Cycle 6: 1878–85.
Deluca, J. G., Dong, Y., Hergert, P., Strauss, J., Hickey, J. M., Salmon, E. D., and McEwen, B. F. 2005. Hec1 and Nuf2 are Core Components of the Kinetochore Outer Plate Essential for Organizing Microtubule Attachment Sites. Mol Biol Cell 16: 519–31.
DeLuca, J. G., Gall, W. E., Ciferri, C., Cimini, D., Musacchio, A., and Salmon, E. D. 2006. Kinetochore microtubule dynamics and attachment stability are regulated by Hec1. Cell 127: 969–82.
Desai, A., Verma, S., Mitchison, T. J., and Walczak, C. E. 1999. Kin I kinesins are microtubule-destabilizing enzymes. Cell 96: 69–78.
Ditchfield, C., Johnson, V. L., Tighe, A., Ellston, R., Haworth, C., Johnson, T., Mortlock, A., Keen, N., and Taylor, S. S. 2003. Aurora B couples chromosome alignment with anaphase by targeting BubR1, Mad2, and Cenp-E to kinetochores. J Cell Biol 161: 267–80.

Doyon, Y., and Cote, J. 2004. The highly conserved and multifunctional NuA4 HAT complex. Curr Opin Genet Dev 14: 147–54.

Du, J., Cai, X., Yao, J., Ding, X., Wu, Q., Pei, S., Jiang, K., Zhang, Y., Wang, W., Shi, Y., Lai, Y., Shen, J., Teng, M., Huang, H., Fei, Q., Reddy, E. S., Zhu, J., Jin, C., and Yao, X. 2008. The mitotic checkpoint kinase NEK2A regulates kinetochore microtubule attachment stability. Oncogene 27: 4107–14.

Du, Y., and Dawe, R. K. 2007. Maize NDC80 is a constitutive feature of the central kinetochore. Chromosome Res 15: 767–75.

Dunleavy, E., Pidoux, A., and Allshire, R. 2005. Centromeric chromatin makes its mark. Trends Biochem Sci 30: 172–5.

Ekwall, K. 2007. Epigenetic Control of Centromere Behavior. Annu Rev Genet 41: 63–81.

Elia, A. E., Cantley, L. C., and Yaffe, M. B. 2003. Proteomic screen finds pSer/pThr-binding domain localizing Plk1 to mitotic substrates. Science 299: 1228–31.

Elia, A. E., Rellos, P., Haire, L. F., Chao, J. W., Ivins, F. J., Hoepker, K., Mohammad, D., Cantley, L. C., Smerdon, S. J., and Yaffe, M. B. 2003. The molecular basis for phosphodependent substrate targeting and regulation of Plks by the Polo-box domain. Cell 115: 83–95.

Elowe, S., Hummer, S., Uldschmid, A., Li, X., and Nigg, E. A. 2007. Tension-sensitive Plk1 phosphorylation on BubR1 regulates the stability of kinetochore microtubule interactions. Genes Dev 21: 2205–19.

Emanuele, M. J., and Stukenberg, P. T. 2007. Xenopus cep57 is a novel kinetochore component involved in microtubule attachment. Cell 130: 893–905.

Espeut, J., Gaussen, A., Bieling, P., Morin, V., Prieto, S., Fesquet, D., Surry, T., and Abrieu, A. 2008. Phosphorylation relieves autoinhibition of the kinetochore motor CENP-E. Molecular Cell 29: 637–643.

Eyers, P. A., Erikson, E., Chen, L. G., and Maller, J. L. 2003. A novel mechanism for activation of the protein kinase Aurora A. Curr Biol 13: 691–7.

Fass, E., Shahar, S., Zhao, J., Zemach, A., Avivi, Y., and Grafi, G. 2002. Phosphorylation of histone h3 at serine 10 cannot account directly for the detachment of human heterochromatin protein 1gamma from mitotic chromosomes in plant cells. J Biol Chem 277: 30921–7.

Fernius, J., and Hardwick, K. G. 2007. Bub1 kinase targets Sgo1 to ensure efficient chromosome biorientation in budding yeast mitosis. PLoS Genet 3: e213.

Ferrari, S., Marin, O., Pagano, M. A., Meggio, F., Hess, D., El-Shemerly, M., Krystyniak, A., and Pinna, L. A. 2005. Aurora-A site specificity: a study with synthetic peptide substrates. Biochem J 390: 293–302.

Fischle, W., Tseng, B. S., Dormann, H. L., Ueberheide, B. M., Garcia, B. A., Shabanowitz, J., Hunt, D. F., Funabiki, H., and Allis, C. D. 2005. Regulation of HP1-chromatin binding by histone H3 methylation and phosphorylation. Nature 438: 1116–22.

Folco, H. D., Pidoux, A. L., Urano, T., and Allshire, R. C. 2008. Heterochromatin and RNAi are required to establish CENP-A chromatin at centromeres. Science 319: 94–7.

Foltz, D. R., Jansen, L. E., Black, B. E., Bailey, A. O., Yates, J. R., 3rd, and Cleveland, D. W. 2006. The human CENP-A centromeric nucleosome-associated complex. Nat Cell Biol 8: 458–69.

Francisco, L., and Chan, C. S. 1994. Regulation of yeast chromosome segregation by Ipl1 protein kinase and type 1 protein phosphatase. Cell Mol Biol Res 40: 207–13.

Francisco, L., Wang, W., and Chan, C. S. 1994. Type 1 protein phosphatase acts in opposition to Ipl1 protein kinase in regulating yeast chromosome segregation. Mol Cell Biol 14: 4731–40.

Fu, G., Ding, X., Yuan, K., Aikhionbare, F., Yao, J., Cai, X., Jiang, K., and Yao, X. 2007. Phosphorylation of human Sgo1 by NEK2A is essential for chromosome congression in mitosis. Cell Res 17: 608–18.

Fujita, Y., Hayashi, T., Kiyomitsu, T., Toyoda, Y., Kokubu, A., Obuse, C., and Yanagida, M. 2007. Priming of centromere for CENP-A recruitment by human hMis18alpha, hMis18beta, and M18BP1. Dev Cell 12: 17–30.

Furuyama, S., and Biggins, S. 2007. Centromere identity is specified by a single centromeric nucleosome in budding yeast. Proc Natl Acad Sci USA 104: 14706–11.
Furuyama, T., Dalal, Y., and Henikoff, S. 2006. Chaperone-mediated assembly of centromeric chromatin in vitro. Proc Natl Acad Sci USA 103: 6172–7.
Furuyama, T., and Henikoff, S. 2006. Biotin-Tag Affinity Purification of a Centromeric Nucleosome Assembly Complex. Cell Cycle 5: 1269–74.
Galarneau, L., Nourani, A., Boudreault, A. A., Zhang, Y., Heliot, L., Allard, S., Savard, J., Lane, W. S., Stillman, D. J., and Cote, J. 2000. Multiple links between the NuA4 histone acetyltransferase complex and epigenetic control of transcription. Mol Cell 5: 927–37.
Gallinari, P., Di Marco, S., Jones, P., Pallaoro, M., and Steinkuhler, C. 2007. HDACs, histone deacetylation and gene transcription: from molecular biology to cancer therapeutics. Cell Res 17: 195–211.
Glickman, M. H., and Ciechanover, A. 2002. The ubiquitin-proteasome proteolytic pathway: destruction for the sake of construction. Physiol Rev 82: 373–428.
Glozak, M. A., Sengupta, N., Zhang, X., and Seto, E. 2005. Acetylation and deacetylation of non-histone proteins. Gene 363: 15–23.
Goshima, G., Iwasaki, O., Obuse, C., and Yanagida, M. 2003. The role of Ppe1/PP6 phosphatase for equal chromosome segregation in fission yeast kinetochore. Embo J 22: 2752–2763.
Goto, H., Kiyono, T., Tomono, Y., Kawajiri, A., Urano, T., Furukawa, K., Nigg, E. A., and Inagaki, M. 2006. Complex formation of Plk1 and INCENP required for metaphase-anaphase transition. Nat Cell Biol 8: 180–7.
Goto, H., Yasui, Y., Nigg, E. A., and Inagaki, M. 2002. Aurora-B phosphorylates Histone H3 at serine28 with regard to the mitotic chromosome condensation. Genes Cells 7: 11–7.
Greaves, I. K., Rangasamy, D., Ridgway, P., and Tremethick, D. J. 2007. H2A.Z contributes to the unique 3D structure of the centromere. Proc Natl Acad Sci USA 104: 525–30.
Grewal, S. I., and Jia, S. 2007. Heterochromatin revisited. Nat Rev Genet 8: 35–46.
Hake, S. B., Garcia, B. A., Kauer, M., Baker, S. P., Shabanowitz, J., Hunt, D. F., and Allis, C. D. 2005. Serine 31 phosphorylation of histone variant H3.3 is specific to regions bordering centromeres in metaphase chromosomes. Proc Natl Acad Sci USA 102: 6344–9.
Han, Z., Riefler, G. M., Saam, J. R., Mango, S. E., and Schumacher, J. M. 2005. The C. elegans Tousled-like kinase contributes to chromosome segregation as a substrate and regulator of the Aurora B kinase. Curr Biol 15: 894–904.
Hayashi, T., Fujita, Y., Iwasaki, O., Adachi, Y., Takahashi, K., and Yanagida, M. 2004. Mis16 and Mis18 are required for CENP-A loading and histone deacetylation at centromeres. Cell 118: 715–29.
Hendzel, M. J., Wei, Y., Mancini, M. A., Van Hooser, A., Ranalli, T., Brinkley, B. R., Bazett-Jones, D. P., and Allis, C. D. 1997. Mitosis-specific phosphorylation of histone H3 initiates primarily within pericentromeric heterochromatin during G2 and spreads in an ordered fashion coincident with mitotic chromosome condensation. Chromosoma 106: 348–60.
Henikoff, S., and Dalal, Y. 2005. Centromeric chromatin: what makes it unique? Curr Opin Genet Dev 15: 177–84.
Heun, P., Erhardt, S., Blower, M. D., Weiss, S., Skora, A. D., and Karpen, G. H. 2006. Mislocalization of the Drosophila Centromere-Specific Histone CID Promotes Formation of Functional Ectopic Kinetochores. Dev Cell 10: 303–15.
Higuchi, T., and Uhlmann, F. 2005. Stabilization of microtubule dynamics at anaphase onset promotes chromosome segregation. Nature 433: 171–6.
Hirota, T., Kunitoku, N., Sasayama, T., Marumoto, T., Zhang, D., Nitta, M., Hatakeyama, K., and Saya, H. 2003. Aurora-A and an interacting activator, the LIM protein Ajuba, are required for mitotic commitment in human cells. Cell 114: 585–98.
Honda, R., Korner, R., and Nigg, E. A. 2003. Exploring the functional interactions between Aurora B, INCENP, and survivin in mitosis. Mol Biol Cell 14: 3325–41.

Hong, E. J., Villen, J., Gerace, E. L., Gygi, S. P., and Moazed, D. 2005. A cullin E3 ubiquitin ligase complex associates with Rik1 and the Clr4 histone H3-K9 methyltransferase and is required for RNAi-mediated heterochromatin formation. RNA Biol 2: 106–11.

Hong, S., Cho, Y. W., Yu, L. R., Yu, H., Veenstra, T. D., and Ge, K. 2007. Identification of JmjC domain-containing UTX and JMJD3 as histone H3 lysine 27 demethylases. Proc Natl Acad Sci USA 104: 18439–44.

Howell, B. J., Moree, B., Farrar, E. M., Stewart, S., Fang, G., and Salmon, E. D. 2004. Spindle checkpoint protein dynamics at kinetochores in living cells. Curr Biol 14: 953–64.

Hsu, J. Y., Sun, Z. W., Li, X., Reuben, M., Tatchell, K., Bishop, D. K., Grushcow, J. M., Brame, C. J., Caldwell, J. A., Hunt, D. F., Lin, R., Smith, M. M., and Allis, C. D. 2000. Mitotic phosphorylation of histone H3 is governed by Ipl1/aurora kinase and Glc7/PP1 phosphatase in budding yeast and nematodes. Cell 102: 279–91.

Hussein, D., and Taylor, S. S. 2002. Farnesylation of Cenp-F is required for G2/M progression and degradation after mitosis. J Cell Sci 115: 3403–14.

Janssens, V., and Goris, J. 2001. Protein phosphatase 2A: a highly regulated family of serine/threonine phosphatases implicated in cell growth and signalling. Biochem J 353: 417–39.

Jelluma, N., Brenkman, A. B., van den Broek, N. J., Cruijsen, C. W., van Osch, M. H., Lens, S. M., Medema, R. H., and Kops, G. J. 2008. Mps1 phosphorylates Borealin to control Aurora B activity and chromosome alignment. Cell 132: 233–46.

Jia, S., Kobayashi, R., and Grewal, S. I. 2005. Ubiquitin ligase component Cul4 associates with Clr4 histone methyltransferase to assemble heterochromatin. Nat Cell Biol 7: 1007–13.

Jiang, W., Lim, M. Y., Yoon, H. J., Thorner, J., Martin, G. S., and Carbon, J. 1995. Overexpression of the yeast MCK1 protein kinase suppresses conditional mutations in centromere-binding protein genes CBF2 and CBF5. Mol Gen Genet 246: 360–6.

Johnson, E. S. 2004. Protein modification by SUMO. Annu Rev Biochem 73: 355–82.

Johnson, V. L., Scott, M. I., Holt, S. V., Hussein, D., and Taylor, S. S. 2004. Bub1 is required for kinetochore localization of BubR1, Cenp-E, Cenp-F and Mad2, and chromosome congression. J Cell Sci 117: 1577–89.

Joo, H. Y., Zhai, L., Yang, C., Nie, S., Erdjument-Bromage, H., Tempst, P., Chang, C., and Wang, H. 2007. Regulation of cell cycle progression and gene expression by H2A deubiquitination. Nature 449: 1068–72.

Joseph, J., Tan, S. H., Karpova, T. S., McNally, J. G., and Dasso, M. 2002. SUMO-1 targets RanGAP1 to kinetochores and mitotic spindles. J Cell Biol 156: 595–602.

Kaiser, C., and James, S. R. 2004. Acetylation of insulin receptor substrate-1 is permissive for tyrosine phosphorylation. BMC Biol 2: 23.

Kamakaka, R. T., and Biggins, S. 2005. Histone variants: deviants? Genes Dev 19: 295–310.

Kang, J., Cheeseman, I. M., Kallstrom, G., Velmurugan, S., Barnes, G., and Chan, C. S. 2001. Functional cooperation of Dam1, Ipl1, and the inner centromere protein (INCENP)-related protein Sli15 during chromosome segregation. J Cell Biol 155: 763–74.

Kang, Y. H., Park, J. E., Yu, L. R., Soung, N. K., Yun, S. M., Bang, J. K., Seong, Y. S., Yu, H., Garfield, S., Veenstra, T. D., and Lee, K. S. 2006. Self-regulated Plk1 recruitment to kinetochores by the Plk1-PBIP1 interaction is critical for proper chromosome segregation. Mol Cell 24: 409–22.

Kelly, A. E., Sampath, S. C., Maniar, T. A., Woo, E. M., Chait, B. T., and Funabiki, H. 2007. Chromosomal enrichment and activation of the aurora B pathway are coupled to spatially regulate spindle assembly. Dev Cell 12: 31–43.

Kelm, O., Wind, M., Lehmann, W. D., and Nigg, E. A. 2002. Cell cycle-regulated phosphorylation of the Xenopus polo-like kinase Plx1. J Biol Chem 277: 25247–56.

Keogh, M. C., Mennella, T. A., Sawa, C., Berthelet, S., Krogan, N. J., Wolek, A., Podolny, V., Carpenter, L. R., Greenblatt, J. F., Baetz, K., and Buratowski, S. 2006. The Saccharomyces cerevisiae histone H2A variant Htz1 is acetylated by NuA4. Genes Dev 20: 660–5.

Kerscher, O., Felberbaum, R., and Hochstrasser, M. 2006. Modification of Proteins by Ubiquitin and Ubiquitin-Like Proteins. Annu Rev Cell Dev Biol 22: 159–80.

Kitajima, T. S., Hauf, S., Ohsugi, M., Yamamoto, T., and Watanabe, Y. 2005. Human Bub1 defines the persistent cohesion site along the mitotic chromosome by affecting Shugoshin localization. Curr Biol 15: 353–9.

Kitajima, T. S., Sakuno, T., Ishiguro, K., Iemura, S., Natsume, T., Kawashima, S. A., and Watanabe, Y. 2006. Shugoshin collaborates with protein phosphatase 2A to protect cohesin. Nature 441: 46–52.

Kline-Smith, S. L., Khodjakov, A., Hergert, P., and Walczak, C. E. 2004. Depletion of centromeric MCAK leads to chromosome congression and segregation defects due to improper kinetochore attachments. Mol Biol Cell 15: 1146–59.

Kline-Smith, S. L., Sandall, S., and Desai, A. 2005. Kinetochore-spindle microtubule interactions during mitosis. Curr Opin Cell Biol 17: 35–46.

Klose, R. J., and Zhang, Y. 2007. Regulation of histone methylation by demethylimination and demethylation. Nat Rev Mol Cell Biol 8: 307–18.

Knowlton, A. L., Lan, W., and Stukenberg, P. T. 2006. Aurora B is enriched at merotelic attachment sites, where it regulates MCAK. Curr Biol 16: 1705–10.

Kouzarides, T. 2007. Chromatin modifications and their function. Cell 128: 693–705.

Krogan, N. J., Baetz, K., Keogh, M. C., Datta, N., Sawa, C., Kwok, T. C., Thompson, N. J., Davey, M. G., Pootoolal, J., Hughes, T. R., Emili, A., Buratowski, S., Hieter, P., and Greenblatt, J. F. 2004. Regulation of chromosome stability by the histone H2A variant Htz1, the Swr1 chromatin remodeling complex, and the histone acetyltransferase NuA4. Proc Natl Acad Sci USA 101: 13513–8.

Kunitoku, N., Sasayama, T., Marumoto, T., Zhang, D., Honda, S., Kobayashi, O., Hatakeyama, K., Ushio, Y., Saya, H., and Hirota, T. 2003. CENP-A phosphorylation by Aurora-A in prophase is required for enrichment of Aurora-B at inner centromeres and for kinetochore function. Dev Cell 5: 853–64.

Lampson, M. A., and Kapoor, T. M. 2005. The human mitotic checkpoint protein BubR1 regulates chromosome-spindle attachments. Nat Cell Biol 7: 93–8.

Lan, F., Bayliss, P. E., Rinn, J. L., Whetstine, J. R., Wang, J. K., Chen, S., Iwase, S., Alpatov, R., Issaeva, I., Canaani, E., Roberts, T. M., Chang, H. Y., and Shi, Y. 2007. A histone H3 lysine 27 demethylase regulates animal posterior development. Nature 449: 689–94.

Lan, W., Zhang, X., Kline-Smith, S. L., Rosasco, S. E., Barrett-Wilt, G. A., Shabanowitz, J., Hunt, D. F., Walczak, C. E., and Stukenberg, P. T. 2004. Aurora B phosphorylates centromeric MCAK and regulates its localization and microtubule depolymerization activity. Curr Biol 14: 273–86.

Langst, G., and Becker, P. B. 2004. Nucleosome remodeling: one mechanism, many phenomena? Biochim Biophys Acta 1677: 58–63.

Le Masson, I., Yu, D. Y., Jensen, K., Chevalier, A., Courbeyrette, R., Boulard, Y., Smith, M. M., and Mann, C. 2003. Yaf9, a novel NuA4 histone acetyltransferase subunit, is required for the cellular response to spindle stress in yeast. Mol Cell Biol 23: 6086–102.

Lenart, P., Petronczki, M., Steegmaier, M., Di Fiore, B., Lipp, J. J., Hoffmann, M., Rettig, W. J., Kraut, N., and Peters, J. M. 2007. The small-molecule inhibitor BI 2536 reveals novel insights into mitotic roles of polo-like kinase 1. Curr Biol 17: 304–15.

Li, Y., Bachant, J., Alcasabas, A. A., Wang, Y., Qin, J., and Elledge, S. J. 2002. The mitotic spindle is required for loading of the DASH complex onto the kinetochore. Genes Dev. 16:183–97.

Li, Y., and Elledge, S. J. 2003. The DASH complex component Ask1 is a cell cycle-regulated Cdk substrate in Saccharomyces cerevisiae. Cell Cycle 2: 143–8.

Liao, H., Li, G., and Yen, T. J. 1994. Mitotic regulation of microtubule cross-linking activity of CENP-E kinetochore protein. Science 265: 394–8.

Lowery, D. M., Lim, D., and Yaffe, M. B. 2005. Structure and function of Polo-like kinases. Oncogene 24: 248–59.

Luger, K. 2003. Structure and dynamic behavior of nucleosomes. Curr Opin Genet Dev 13: 127–35.

Malik, H. S., and Henikoff, S. 2003. Phylogenomics of the nucleosome. Nat Struct Biol 10: 882–91.
Manning, G., Whyte, D. B., Martinez, R., Hunter, T., and Sudarsanam, S. 2002. The protein kinase complement of the human genome. Science 298: 1912–34.
Marchion, D., and Munster, P. 2007. Development of histone deacetylase inhibitors for cancer treatment. Expert Rev Anticancer Ther 7: 583–98.
Martin, C., and Zhang, Y. 2005. The diverse functions of histone lysine methylation. Nat Rev Mol Cell Biol 6: 838–49.
Mateescu, B., England, P., Halgand, F., Yaniv, M., and Muchardt, C. 2004. Tethering of HP1 proteins to chromatin is relieved by phosphoacetylation of histone H3. EMBO Rep 5: 490–6.
Matsumura, S., Toyoshima, F., and Nishida, E. 2007. Polo-like kinase 1 facilitates chromosome alignment during prometaphase through BubR1. J Biol Chem 282: 15217–27.
Matunis, M. J., Wu, J., and Blobel, G. 1998. SUMO-1 modification and its role in targeting the Ran GTPase-activating protein, RanGAP1, to the nuclear pore complex. J Cell Biol 140: 499–509.
Maure, J. F., Kitamura, E., and Tanaka, T. U. 2007. Mps1 kinase promotes sister-kinetochore bi-orientation by a tension-dependent mechanism. Curr Biol 17: 2175–82.
Melloy, P. G., and Holloway, S. L. 2004. Changes in the localization of the Saccharomyces cerevisiae anaphase-promoting complex upon microtubule depolymerization and spindle checkpoint activation. Genetics 167: 1079–94.
Meluh, P. B., Yang, P., Glowczewski, L., Koshland, D., and Smith, M. M. 1998. Cse4p is a component of the core centromere of Saccharomyces cerevisiae. Cell 94: 607–13.
Miranda, J. J., De Wulf, P., Sorger, P. K., and Harrison, S. C. 2005. The yeast DASH complex forms closed rings on microtubules. Nat Struct Mol Biol 12: 138–43.
Miranda, J. J., King, D. S., and Harrison, S. C. 2007. Protein arms in the kinetochore-microtubule interface of the yeast DASH complex. Mol Biol Cell 18: 2503–10.
Moasser, M. M., Sepp-Lorenzino, L., Kohl, N. E., Oliff, A., Balog, A., Su, D. S., Danishefsky, S. J., and Rosen, N. 1998. Farnesyl transferase inhibitors cause enhanced mitotic sensitivity to taxol and epothilones. Proc Natl Acad Sci USA 95: 1369–74.
Montpetit, B., Hazbun, T. R., Fields, S., and Hieter, P. 2006. Sumoylation of the budding yeast kinetochore protein Ndc10 is required for Ndc10 spindle localization and regulation of anaphase spindle elongation. J Cell Biol 174: 653–63.
Moore, A., and Wordeman, L. 2004. The mechanism, function and regulation of depolymerizing kinesins during mitosis. Trends Cell Biol 14: 537–46.
Moorhead, G. B., Trinkle-Mulcahy, L., and Ulke-Lemee, A. 2007. Emerging roles of nuclear protein phosphatases. Nat Rev Mol Cell Biol 8: 234–44.
Moreno-Moreno, O., Torras-Llort, M., and Azorin, F. 2006. Proteolysis restricts localization of CID, the centromere-specific histone H3 variant of Drosophila, to centromeres. Nucleic Acids Res 34: 6247–55.
Mukhopadhyay, D., and Dasso, M. 2007. Modification in reverse: the SUMO proteases. Trends Biochem Sci 32: 286–95.
Musacchio, A., and Hardwick, K. G. 2002. The spindle checkpoint: structural insights into dynamic signalling. Nat Rev Mol Cell Biol 3: 731–41.
Musacchio, A., and Salmon, E. D. 2007. The spindle-assembly checkpoint in space and time. Nat Rev Mol Cell Biol 8: 379–93.
Nakajima, H., Toyoshima-Morimoto, F., Taniguchi, E., and Nishida, E. 2003. Identification of a consensus motif for Plk (Polo-like kinase) phosphorylation reveals Myt1 as a Plk1 substrate. J Biol Chem 278: 25277–80.
Nishino, M., Kurasawa, Y., Evans, R., Lin, S. H., Brinkley, B. R., and Yu-Lee, L. Y. 2006. NudC is required for Plk1 targeting to the kinetochore and chromosome congression. Curr Biol 16: 1414–21.
Noma, K., Allis, C. D., and Grewal, S. I. 2001. Transitions in distinct histone H3 methylation patterns at the heterochromatin domain boundaries. Science 293: 1150–5.

Nousiainen, M., Sillje, H. H., Sauer, G., Nigg, E. A., and Korner, R. 2006. Phosphoproteome analysis of the human mitotic spindle. Proc Natl Acad Sci USA 103: 5391–6.

O'Connor, D. S., Grossman, D., Plescia, J., Li, F., Zhang, H., Villa, A., Tognin, S., Marchisio, P. C., and Altieri, D. C. 2000. Regulation of apoptosis at cell division by p34cdc2 phosphorylation of survivin. Proc Natl Acad Sci USA 97: 13103–7.

O'Regan, L., Blot, J., and Fry, A. M. 2007. Mitotic regulation by NIMA-related kinases. Cell Div 2: 25.

Ohi, R., Sapra, T., Howard, J., and Mitchison, T. J. 2004. Differentiation of cytoplasmic and meiotic spindle assembly MCAK functions by Aurora B-dependent phosphorylation. Mol Biol Cell 15: 2895–906.

Pereira, G., and Schiebel, E. 2003. Separase regulates INCENP-Aurora B anaphase spindle function through Cdc14. Science 302: 2120–4.

Peters, A. H., Kubicek, S., Mechtler, K., O'Sullivan, R. J., Derijck, A. A., Perez-Burgos, L., Kohlmaier, A., Opravil, S., Tachibana, M., Shinkai, Y., Martens, J. H., and Jenuwein, T. 2003. Partitioning and plasticity of repressive histone methylation states in mammalian chromatin. Mol Cell 12: 1577–89.

Petersen, J., Paris, J., Willer, M., Philippe, M., and Hagan, I. M. 2001. The S. pombe aurora-related kinase Ark1 associates with mitotic structures in a stage dependent manner and is required for chromosome segregation. J Cell Sci 114: 4371–84.

Pinsky, B. A., Kotwaliwale, C. V., Tatsutani, S. Y., Breed, C. A., and Biggins, S. 2006a. Glc7/Protein Phosphatase-1 Regulatory Subunits can oppose the Ipl1/Aurora Protein Kinase by redistributing Glc7. Mol Cell Biol 26: 2648–80.

Pinsky, B. A., Kung, C., Shokat, K. M., and Biggins, S. 2006b. The Ipl1-Aurora protein kinase activates the spindle checkpoint by creating unattached kinetochores. Nat Cell Biol 8: 78–83.

Polioudaki, H., Markaki, Y., Kourmouli, N., Dialynas, G., Theodoropoulos, P. A., Singh, P. B., and Georgatos, S. D. 2004. Mitotic phosphorylation of histone H3 at threonine 3. FEBS Lett 560: 39–44.

Polo, S. E., and Almouzni, G. 2006. Chromatin assembly: a basic recipe with various flavours. Curr Opin Genet Dev 16: 104–11.

Porras-Yakushi, T. R., Whitelegge, J. P., and Clarke, S. 2006. A novel SET domain methyltransferase in yeast: Rkm2-dependent trimethylation of ribosomal protein L12ab at lysine 10. J Biol Chem 281: 35835–45.

Pouwels, J., Kukkonen, A. M., Lan, W., Daum, J. R., Gorbsky, G. J., Stukenberg, T., and Kallio, M. J. 2007. Shugoshin 1 plays a central role in kinetochore assembly and is required for kinetochore targeting of Plk1. Cell Cycle 6: 1579–85.

Preuss, U., Landsberg, G., and Scheidtmann, K. H. 2003. Novel mitosis-specific phosphorylation of histone H3 at Thr11 mediated by Dlk/ZIP kinase. Nucleic Acids Res 31: 878–85.

Putkey, F. R., Cramer, T., Morphew, M. K., Silk, A. D., Johnson, R. S., McIntosh, J. R., and Cleveland, D. W. 2002. Unstable kinetochore-microtubule capture and chromosomal instability following deletion of CENP-E. Dev Cell 3: 351–65.

Qi, W., Tang, Z., and Yu, H. 2006. Phosphorylation- and polo-box-dependent binding of Plk1 to Bub1 is required for the kinetochore localization of Plk1. Mol Biol Cell 17: 3705–16.

Qian, Y. W., Erikson, E., and Maller, J. L. 1998. Purification and cloning of a protein kinase that phosphorylates and activates the polo-like kinase Plx1. Science 282: 1701–4.

Raisner, R. M., and Madhani, H. D. 2006. Patterning chromatin: form and function for H2A. Z variant nucleosomes. Curr Opin Genet Dev 16: 119–24.

Ramadan, K., Bruderer, R., Spiga, F. M., Popp, O., Baur, T., Gotta, M., and Meyer, H. H. 2007. Cdc48/p97 promotes reformation of the nucleus by extracting the kinase Aurora B from chromatin. Nature 450: 1258–62.

Rangasamy, D., Berven, L., Ridgway, P., and Tremethick, D. J. 2003. Pericentric heterochromatin becomes enriched with H2A.Z during early mammalian development. Embo J 22: 1599–607.

Rangasamy, D., Greaves, I., and Tremethick, D. J. 2004. RNA interference demonstrates a novel role for H2A.Z in chromosome segregation. Nat Struct Mol Biol 11: 650–5.

Rea, S., Eisenhaber, F., O'Carroll, D., Strahl, B. D., Sun, Z. W., Schmid, M., Opravil, S., Mechtler, K., Ponting, C. P., Allis, C. D., and Jenuwein, T. 2000. Regulation of chromatin structure by site-specific histone H3 methyltransferases. Nature 406: 593–9.

Reddy, S. K., Rape, M., Margansky, W. A., and Kirschner, M. W. 2007. Ubiquitination by the anaphase-promoting complex drives spindle checkpoint inactivation. Nature 446: 921–5.

Rice, J. C., Briggs, S. D., Ueberheide, B., Barber, C. M., Shabanowitz, J., Hunt, D. F., Shinkai, Y., and Allis, C. D. 2003. Histone methyltransferases direct different degrees of methylation to define distinct chromatin domains. Mol Cell 12: 1591–8.

Riedel, C. G., Katis, V. L., Katou, Y., Mori, S., Itoh, T., Helmhart, W., Galova, M., Petronczki, M., Gregan, J., Cetin, B., Mudrak, I., Ogris, E., Mechtler, K., Pelletier, L., Buchholz, F., Shirahige, K., and Nasmyth, K. 2006. Protein phosphatase 2A protects centromeric sister chromatid cohesion during meiosis I. Nature 441: 53–61.

Rodriguez, M. S., Dargemont, C., and Hay, R. T. 2001. SUMO-1 conjugation in vivo requires both a consensus modification motif and nuclear targeting. J Biol Chem 276: 12654–9.

Roghi, C., Giet, R., Uzbekov, R., Morin, N., Chartrain, I., Le Guellec, R., Couturier, A., Doree, M., Philippe, M., and Prigent, C. 1998. The Xenopus protein kinase pEg2 associates with the centrosome in a cell cycle-dependent manner, binds to the spindle microtubules and is involved in bipolar mitotic spindle assembly. J Cell Sci 111: 557–72.

Rosasco-Nitcher, S. E., Lan, W., Khorasanizadeh, S., and Stukenberg, P. T. 2008. Centromeric Aurora-B activation requires TD-60, microtubules, and substrate priming phosphorylation. Science 319: 469–72.

Ruchaud, S., Carmena, M., and Earnshaw, W. C. 2007. Chromosomal passengers: conducting cell division. Nat Rev Mol Cell Biol 8: 798–812.

Sandall, S., Severin, F., McLeod, I., Yates, J. R., Oegema, K., Hyman, A., and Desai, A. 2006. A complex of Bir1-Sli15 (Survivin-INCENP) connects centromeres to microtubules and is the likely tension sensor controlling Aurora B activation. Cell 127: 1179–91.

Sassoon, I., Severin, F. F., Andrews, P. D., Taba, M. R., Kaplan, K. B., Ashford, A. J., Stark, M. J., Sorger, P. K., and Hyman, A. A. 1999. Regulation of Saccharomyces cerevisiae kinetochores by the type 1 phosphatase Glc7p. Genes Dev 13: 545–55.

Saxena, A., Saffery, R., Wong, L. H., Kalitsis, P., and Choo, K. H. 2002. Centromere proteins Cenpa, Cenpb, and Bub3 interact with poly(ADP-ribose) polymerase-1 protein and are poly(ADP-ribosyl)ated. J Biol Chem 277: 26921–6.

Schafer-Hales, K., Iaconelli, J., Snyder, J. P., Prussia, A., Nettles, J. H., El-Naggar, A., Khuri, F. R., Giannakakou, P., and Marcus, A. I. 2007. Farnesyl transferase inhibitors impair chromosomal maintenance in cell lines and human tumors by compromising CENP-E and CENP-F function. Mol Cancer Ther 6: 1317–28.

Schotta, G., Ebert, A., Krauss, V., Fischer, A., Hoffmann, J., Rea, S., Jenuwein, T., Dorn, R., and Reuter, G. 2002. Central role of Drosophila SU(VAR)3–9 in histone H3-K9 methylation and heterochromatic gene silencing. Embo J 21: 1121–31.

Schotta, G., Lachner, M., Sarma, K., Ebert, A., Sengupta, R., Reuter, G., Reinberg, D., and Jenuwein, T. 2004. A silencing pathway to induce H3-K9 and H4-K20 trimethylation at constitutive heterochromatin. Genes Dev 18: 1251–62.

Schueler, M. G., and Sullivan, B. A. 2006. Structural and functional dynamics of human centromeric chromatin. Annu Rev Genomics Hum Genet 7: 301–13.

Sessa, F., Mapelli, M., Ciferri, C., Tarricone, C., Areces, L. B., Schneider, T. R., Stukenberg, P. T., and Musacchio, A. 2005. Mechanism of Aurora B activation by INCENP and inhibition by hesperadin. Mol Cell 18: 379–91.

Shang, C., Hazbun, T. R., Cheeseman, I. M., Aranda, J., Fields, S., Drubin, D. G., and Barnes, G. 2003. Kinetochore protein interactions and their regulation by the Aurora kinase Ipl1p. Mol Biol Cell 14: 3342–55.

Shapiro, P. S., Vaisberg, E., Hunt, A. J., Tolwinski, N. S., Whalen, A. M., McIntosh, J. R., and Ahn, N. G. 1998. Activation of the MKK/ERK pathway during somatic cell mitosis: direct interactions of active ERK with kinetochores and regulation of the mitotic 3F3/2 phosphoantigen. J Cell Biol 142: 1533–45.

Shimogawa, M. M., Graczyk, B., Gardner, M. K., Francis, S. E., White, E. A., Ess, M., Molk, J. N., Ruse, C., Niessen, S., Yates, J. R., 3rd, Muller, E. G., Bloom, K., Odde, D. J., and Davis, T. N. 2006. Mps1 phosphorylation of Dam1 couples kinetochores to microtubule plus ends at metaphase. Curr Biol 16: 1489–501.

Shin, H. J., Baek, K. H., Jeon, A. H., Kim, S. J., Jang, K. L., Sung, Y. C., Kim, C. M., and Lee, C. W. 2003. Inhibition of histone deacetylase activity increases chromosomal instability by the aberrant regulation of mitotic checkpoint activation. Oncogene 22: 3853–8.

Sigismund, S., Polo, S., and Di Fiore, P. P. 2004. Signaling through monoubiquitination. Curr Top Microbiol Immunol 286: 149–85.

Sims, R. J., 3rd, Nishioka, K., and Reinberg, D. 2003. Histone lysine methylation: a signature for chromatin function. Trends Genet 19: 629–39.

Smith, E. R., Lee, M. G., Winter, B., Droz, N. M., Eissenberg, J. C., Shiekhattar, R., and Shilatifard, A. 2008. Drosophila UTX is a histone H3 Lys27 demethylase that colocalizes with the elongating form of RNA polymerase II. Mol Cell Biol 28: 1041–6.

Snead, J. L., Sullivan, M., Lowery, D. M., Cohen, M. S., Zhang, C., Randle, D. H., Taunton, J., Yaffe, M. B., Morgan, D. O., and Shokat, K. M. 2007. A coupled chemical-genetic and bioinformatic approach to Polo-like kinase pathway exploration. Chem Biol 14: 1261–72.

Stegmeier, F., and Amon, A. 2004. Closing mitosis: the functions of the Cdc14 phosphatase and its regulation. Annu Rev Genet 38: 203–32.

Stegmeier, F., Rape, M., Draviam, V. M., Nalepa, G., Sowa, M. E., Ang, X. L., McDonald, E. R., 3rd, Li, M. Z., Hannon, G. J., Sorger, P. K., Kirschner, M. W., Harper, J. W., and Elledge, S. J. 2007. Anaphase initiation is regulated by antagonistic ubiquitination and deubiquitination activities. Nature 446: 876–81.

Stoepel, J., Ottey, M. A., Kurischko, C., Hieter, P., and Luca, F. C. 2005. The mitotic exit network Mob1p-Dbf2p kinase complex localizes to the nucleus and regulates passenger protein localization. Mol Biol Cell 16: 5465–79.

Sullivan, B. A., and Karpen, G. H. 2004. Centromeric chromatin exhibits a histone modification pattern that is distinct from both euchromatin and heterochromatin. Nat Struct Mol Biol 11: 1076–83.

Sullivan, M., and Morgan, D. O. 2007. Finishing mitosis, one step at a time. Nat Rev Mol Cell Biol 8: 894–903.

Sumara, I., Gimenez-Abian, J. F., Gerlich, D., Hirota, T., Kraft, C., de la Torre, C., Ellenberg, J., and Peters, J. M. 2004. Roles of polo-like kinase 1 in the assembly of functional mitotic spindles. Curr Biol 14: 1712–22.

Sun, L., and Chen, Z. J. 2004. The novel functions of ubiquitination in signaling. Curr Opin Cell Biol 16: 119–26.

Swigut, T., and Wysocka, J. 2007. H3K27 demethylases, at long last. Cell 131: 29–32.

Takahashi, K., Yamada, H., and Yanagida, M. 1994. Fission yeast minichromosome loss mutants mis cause lethal and replication abnormality. Mol Biol Cell 5: 1145–58.

Takahashi, Y., Yong-Gonzalez, V., Kikuchi, Y., and Strunnikov, A. 2006. SIZ1/SIZ2 control of chromosome transmission fidelity is mediated by the sumoylation of topoisomerase II. Genetics 172: 783–94.

Tang, Z., Sun, Y., Harley, S. E., Zou, H., and Yu, H. 2004. Human Bub1 protects centromeric sister-chromatid cohesion through Shugoshin during mitosis. Proc Natl Acad Sci USA 101: 18012–7.

Terada, Y. 2006. Aurora-B/AIM-1 regulates the dynamic behavior of HP1alpha at the G2-M transition. Mol Biol Cell 17: 3232–41.

Thornton, B. R., and Toczyski, D. P. 2006. Precise destruction: an emerging picture of the APC. Genes Dev 20: 3069–78.

Trautmann, S., Rajagopalan, S., and McCollum, D. 2004. The S. pombe Cdc14-like phosphatase Clp1p regulates chromosome biorientation and interacts with Aurora kinase. Dev Cell 7: 755–62.

Trinkle-Mulcahy, L., and Lamond, A. I. 2006. Mitotic phosphatases: no longer silent partners. Curr Opin Cell Biol 18: 623–31.

Trinkle-Mulcahy, L., Sleeman, J. E., and Lamond, A. I. 2001. Dynamic targeting of protein phosphatase 1 within the nuclei of living mammalian cells. J Cell Sci 114: 4219–28.

Tsai, M. Y., Wiese, C., Cao, K., Martin, O., Donovan, P., Ruderman, J., Prigent, C., and Zheng, Y. 2003. A Ran signalling pathway mediated by the mitotic kinase Aurora A in spindle assembly. Nat Cell Biol 5: 242–8.

Tsukiyama, T. 2002. The in vivo functions of ATP-dependent chromatin-remodelling factors. Nat Rev Mol Cell Biol 3: 422–9.

Ubersax, J. A., Woodbury, E. L., Quang, P. N., Paraz, M., Blethrow, J. D., Shah, K., Shokat, K. M., and Morgan, D. O. 2003. Targets of the cyclin-dependent kinase Cdk1. Nature 425: 859–64.

Vagnarelli, P., and Earnshaw, W. C. 2004. Chromosomal passengers: the four-dimensional regulation of mitotic events. Chromosoma 113: 211–22.

Van Hooser, A. A., Ouspenski, II, Gregson, H. C., Starr, D. A., Yen, T. J., Goldberg, M. L., Yokomori, K., Earnshaw, W. C., Sullivan, K. F., and Brinkley, B. R. 2001. Specification of kinetochore-forming chromatin by the histone H3 variant CENP-A. J Cell Sci 114: 3529–42.

van Vugt, M. A., van de Weerdt, B. C., Vader, G., Janssen, H., Calafat, J., Klompmaker, R., Wolthuis, R. M., and Medema, R. H. 2004. Polo-like kinase-1 is required for bipolar spindle formation but is dispensable for anaphase promoting complex/Cdc20 activation and initiation of cytokinesis. J Biol Chem 279: 36841–54.

Vernarecci, S., Ornaghi, P., Bagu, A., Cundari, E., Ballario, P., and Filetici, P. 2008. Gcn5p plays an important role in centromere kinetochore function in budding yeast. Mol Cell Biol 28: 988–96.

Verreault, A., Kaufman, P. D., Kobayashi, R., and Stillman, B. 1996. Nucleosome assembly by a complex of CAF-1 and acetylated histones H3/H4. Cell 87: 95–104.

Verreault, A., Kaufman, P. D., Kobayashi, R., and Stillman, B. 1998. Nucleosomal DNA regulates the core-histone-binding subunit of the human Hat1 acetyltransferase. Curr Biol 8: 96–108.

Vigneron, S., Prieto, S., Bernis, C., Labbe, J. C., Castro, A., and Lorca, T. 2004. Kinetochore localization of spindle checkpoint proteins: who controls whom? Mol Biol Cell 15: 4584–96.

Vong, Q. P., Cao, K., Li, H. Y., Iglesias, P. A., and Zheng, Y. 2005. Chromosome alignment and segregation regulated by ubiquitination of survivin. Science 310: 1499–504.

Walczak, C. E., Mitchison, T. J., and Desai, A. 1996. XKCM1: a Xenopus kinesin-related protein that regulates microtubule dynamics during mitotic spindle assembly. Cell 84: 37–47.

Wang, H. W., Ramey, V. H., Westermann, S., Leschziner, A. E., Welburn, J. P., Nakajima, Y., Drubin, D. G., Barnes, G., and Nogales, E. 2007. Architecture of the Dam1 kinetochore ring complex and implications for microtubule-driven assembly and force-coupling mechanisms. Nat Struct Mol Biol 14: 721–6.

Warburton, P. E., Cooke, C. A., Bourassa, S., Vafa, O., Sullivan, B. A., Stetten, G., Gimelli, G., Warburton, D., Tyler-Smith, C., Sullivan, K. F., Poirier, G. G., and Earnshaw, W. C. 1997. Immunolocalization of CENP-A suggests a distinct nucleosome structure at the inner kinetochore plate of active centromeres. Curr Biol 7: 901–4.

Watts, F. Z. 2007. The role of SUMO in chromosome segregation. Chromosoma 116: 15–20.

Wei, R. R., Al-Bassam, J., and Harrison, S. C. 2007. The Ndc80/HEC1 complex is a contact point for kinetochore-microtubule attachment. Nat Struct Mol Biol 14: 54–9.

Wei, Y., Mizzen, C. A., Cook, R. G., Gorovsky, M. A., and Allis, C. D. 1998. Phosphorylation of histone H3 at serine 10 is correlated with chromosome condensation during mitosis and meiosis in Tetrahymena. Proc Natl Acad Sci U S A 95: 7480–4.

Weissman, A. M. 2001. Themes and variations on ubiquitylation. Nat Rev Mol Cell Biol 2: 169–78.

Welchman, R. L., Gordon, C., and Mayer, R. J. 2005. Ubiquitin and ubiquitin-like proteins as multifunctional signals. Nat Rev Mol Cell Biol 6: 599–609.

Westermann, S., Avila-Sakar, A., Wang, H. W., Niederstrasser, H., Wong, J., Drubin, D. G., Nogales, E., and Barnes, G. 2005. Formation of a dynamic kinetochore- microtubule interface through assembly of the Dam1 ring complex. Mol Cell 17: 277–90.

Westermann, S., Cheeseman, I. M., Anderson, S., Yates, J. R., 3rd, Drubin, D. G., and Barnes, G. 2003. Architecture of the budding yeast kinetochore reveals a conserved molecular core. J Cell Biol 163: 215–22.

Westermann, S., Wang, H. W., Avila-Sakar, A., Drubin, D. G., Nogales, E., and Barnes, G. 2006. The Dam1 kinetochore ring complex moves processively on depolymerizing microtubule ends. Nature 440: 565–9.

Wheatley, S. P., Barrett, R. M., Andrews, P. D., Medema, R. H., Morley, S. J., Swedlow, J. R., and Lens, S. M. 2007. Phosphorylation by aurora-B negatively regulates survivin function during mitosis. Cell Cycle 6: 1220–30.

Wheatley, S. P., Henzing, A. J., Dodson, H., Khaled, W., and Earnshaw, W. C. 2004. Aurora-B phosphorylation in vitro identifies a residue of survivin that is essential for its localization and binding to inner centromere protein (INCENP) in vivo. J Biol Chem 279: 5655–60.

Widlund, P. O., Lyssand, J. S., Anderson, S., Niessen, S., Yates, J. R., 3rd, and Davis, T. N. 2006. Phosphorylation of the chromosomal passenger protein Bir1 is required for localization of Ndc10 to the spindle during anaphase and full spindle elongation. Mol Biol Cell 17: 1065–74.

Wohlschlegel, J. A., Johnson, E. S., Reed, S. I., and Yates, J. R., 3rd. 2004. Global analysis of protein sumoylation in Saccharomyces cerevisiae. J Biol Chem 279: 45662–8.

Wong, J., Nakajima, Y., Westermann, S., Shang, C., Kang, J. S., Goodner, C., Houshmand, P., Fields, S., Chan, C. S., Drubin, D., Barnes, G., and Hazbun, T. 2007. A Protein Interaction Map of the Mitotic Spindle. Mol Biol Cell 18: 3800–9.

Wong, O. K., and Fang, G. 2005. Plx1 is the 3F3/2 kinase responsible for targeting spindle checkpoint proteins to kinetochores. J Cell Biol 170: 709–19.

Wong, O. K., and Fang, G. 2007. Cdk1 phosphorylation of BubR1 controls spindle checkpoint arrest and Plk1-mediated formation of the 3F3/2 epitope. J Cell Biol 179: 611–7.

Wordeman, L., and Mitchison, T. J. 1995. Identification and partial characterization of mitotic centromere-associated kinesin, a kinesin-related protein that associates with centromeres during mitosis. J Cell Biol 128: 95–104.

Wordeman, L., Wagenbach, M., and von Dassow, G. 2007. MCAK facilitates chromosome movement by promoting kinetochore microtubule turnover. J Cell Biol 179: 869–79.

Wu, X., and Tatchell, K. 2001. Mutations in yeast protein phosphatase type 1 that affect targeting subunit binding. Biochemistry 40: 7410–20.

Xiang, Y., Zhu, Z., Han, G., Lin, H., Xu, L., and Chen, C. D. 2007. JMJD3 is a histone H3K27 demethylase. Cell Res 17: 850–7.

Yamada, T., Fischle, W., Sugiyama, T., Allis, C. D., and Grewal, S. I. 2005. The nucleation and maintenance of heterochromatin by a histone deacetylase in fission yeast. Mol Cell 20: 173–85.

Yamaguchi, S., Decottignies, A., and Nurse, P. 2003. Function of Cdc2p-dependent Bub1p phosphorylation and Bub1p kinase activity in the mitotic and meiotic spindle checkpoint. Embo J 22: 1075–87.

Yan, X., Cao, L., Li, Q., Wu, Y., Zhang, H., Saiyin, H., Liu, X., Zhang, X., Shi, Q., and Yu, L. 2005. Aurora C is directly associated with Survivin and required for cytokinesis. Genes Cells 10: 617–26.

Yen, T. J., Li, G., Schaar, B. T., Szilak, I., and Cleveland, D. W. 1992. CENP-E is a putative kinetochore motor that accumulates just before mitosis. Nature 359: 536–9.

Yoon, H. J., and Carbon, J. 1999. Participation of Bir1p, a member of the inhibitor of apoptosis family, in yeast chromosome segregation events. Proc Natl Acad Sci USA 96: 13208–13.

Yu, H. 2007. Cdc20: a WD40 activator for a cell cycle degradation machine. Mol Cell 27: 3–16.

Zecevic, M., Catling, A. D., Eblen, S. T., Renzi, L., Hittle, J. C., Yen, T. J., Gorbsky, G. J., and Weber, M. J. 1998. Active MAP kinase in mitosis: localization at kinetochores and association with the motor protein CENP-E. J Cell Biol 142: 1547–58.

Zeitlin, S. G., Barber, C. M., Allis, C. D., and Sullivan, K. F. 2001. Differential regulation of CENP-A and histone H3 phosphorylation in G2/M. J Cell Sci 114: 653–61.

Zeitlin, S. G., Shelby, R. D., and Sullivan, K. F. 2001. CENP-A is phosphorylated by Aurora B kinase and plays an unexpected role in completion of cytokinesis. J Cell Biol 155: 1147–57.

Zhang, F. L., and Casey, P. J. 1996. Protein prenylation: molecular mechanisms and functional consequences. Annu Rev Biochem 65: 241–69.

Zhang, K., Lin, W., Latham, J. A., Riefler, G. M., Schumacher, J. M., Chan, C., Tatchell, K., Hawke, D. H., Kobayashi, R., and Dent, S. Y. 2005a. The Set1 methyltransferase opposes Ipl1 aurora kinase functions in chromosome segregation. Cell 122: 723–34.

Zhang, X., Lan, W., Ems-McClung, S. C., Stukenberg, P. T., and Walczak, C. E. 2007. Aurora B Phosphorylates Multiple Sites on Mitotic Centromere-associated Kinesin to Spatially and Temporally Regulate Its Function. Mol Biol Cell 18: 3264–76.

Zhang, X., Li, X., Marshall, J. B., Zhong, C. X., and Dawe, R. K. 2005b. Phosphoserines on maize CENTROMERIC HISTONE H3 and histone H3 demarcate the centromere and pericentromere during chromosome segregation. Plant Cell 17: 572–83.

Zhang, X.-D., Goeres, J., Zhang, H., Yen, T. J., Porter, A. C. G., and Matunis, M. J. 2008. SUMO-2/3 modification and binding regulate the association of CENP-E with kinetochores and progression through mitosis. Molecular Cell 29: 729–41.

Zhao, Y., and Chen, R. H. 2006. Mps1 phosphorylation by MAP kinase is required for kinetochore localization of spindle-checkpoint proteins. Curr Biol 16: 1764–9.

Zhou, T., Aumais, J. P., Liu, X., Yu-Lee, L. Y., and Erikson, R. L. 2003. A role for Plk1 phosphorylation of NudC in cytokinesis. Dev Cell 5: 127–38.

Zhou, W., Ryan, J. J., and Zhou, H. 2004. Global analyses of sumoylated proteins in Saccharomyces cerevisiae. Induction of protein sumoylation by cellular stresses. J Biol Chem 279: 32262–8.

# Chapter 11
# The Role of the Kinetochore in Spindle Checkpoint Signaling

**P. Todd Stukenberg and Daniel J. Burke**

## 11.1 Background

It was observed, with the advent of live cell imaging in the 1950s, that cells did not enter into anaphase until the last chromosomes arrived at the metaphase plate suggesting intricate regulation between chromosome movements and cell cycle progression (Carlson, 1956; Bajer and Mole-Bajer, 1961). Spermatocytes of praying mantids provided a dramatic demonstration of this intricate regulation (Callan and Jacobs, 1957). Male mantids have $X_1X_2Y$ sex determination, a result of an ancient event that split the X chromosome in two. During meiosis the $X_1X_2Y$ trivalent must disjoin to produce an $X_1X_2$–containing gamete and a Y-containing gamete. However, there is chromosome misalignment in 10% of meiotic divisions producing an X-Y bivalent and one X chromosome that is unaligned at metaphase. Interestingly, cells with the unaligned X chromosome remain arrested at metaphase of meiosis I. Callan and Jacobs (1957) proposed that cells sense the single unaligned chromosome and inhibit the onset of anaphase. Zirkle (1970) confirmed this experimentally by using an ultraviolet microbeam to dislodge chromosomes from metaphase spindles and dramatically alter the onset of anaphase. He suggested that the arrival of the last kinetochore to the metaphase plate was the critical signal for the onset of anaphase.

The 1970s marked the beginning of the genetic analysis of the cell cycle in budding yeast. Leland Hartwell and his students isolated the temperature sensitive *cdc* mutants that could not complete the cell cycle at the restrictive temperature (Hartwell, 1978). Despite the large collection of mutants, there were a relatively small number of phenotypes; most mutants arrested as either unbudded cells or large budded cells with undivided nuclei. Epistasis experiments suggested a substrate–product model to explain how the cell cycle was ordered (Hartwell, 1978). The model was conceptually simple. DNA replication

P.T. Stukenberg (✉)
Department of Biochemistry and Molecular Genetics, University of Virginia Medical Center, Charlottesville VA 22908-0733, U.S.A.
e-mail: djb6t@virginia.edu, pts7h@virginia.edu

P. De Wulf, W.C. Earnshaw (eds.), *The Kinetochore*,
DOI 10.1007/978-0-387-69076-6_11, 

preceded mitosis because a substrate for mitosis (for example centromeres) had to be synthesized during S phase. Mitosis would not happen until S phase was completed if centromeres were the very last DNA sequences to be replicated. The precursor–product model broke down over time with new observations, new mutants, and molecular cloning of cell division cycle (*CDC*) genes. For example, careful timing experiments showed that centromeres replicated at the beginning of S phase (McCarroll and Fangman, 1988). Furthermore, *cdc17* cells arrest after S phase, prior to mitosis however the gene was identified through molecular cloning as the alpha-subunit of DNA polymerase (Lucchini et al., 1990). *cdc17* mutants should have arrested in early S phase if the precursor–product model were correct. However, *cdc17* cells have reduced amounts of DNA polymerase, replicate their DNA poorly, and arrest in mitosis. These and many other observations made it clear that the simple precursor–product model was insufficient to explain how order was achieved in the cell cycle. Yet the epistasis experiments correctly described the order-of-function for *CDC* genes. How could the observations be reconciled? The answer was both simple and brilliant. The epistasis experiments revealed the order-of-function but said nothing about the nature of the dependencies. There was born the concept of checkpoints. Hartwell and Weinert envisioned a series of regulatory systems (checkpoints) that imposed order on the cell cycle by preventing key transitions from occurring until previous steps were properly completed (Hartwell and Weinert, 1989). The concept not only explained the epistasis data but nicely explained the paucity of *cdc* phenotypes indicative of key steps in the cell cycle that were under checkpoint regulation.

Hartwell and Weinert introduced the concept of checkpoints and showed that cells did not progress into mitosis in the presence of DNA damage (Weinert and Hartwell, 1988; Hartwell and Weinert, 1989). They also proposed that there was a similar checkpoint that prevented the onset of anaphase in response to misaligned chromosomes. A spindle checkpoint could explain the early observations that cells delay anaphase until the last chromosomes reach the metaphase plate or when chromosomes are improperly aligned. Li and Murray (1991) and Hoyt et al., (1991) executed the first screens in budding yeast to identify spindle checkpoint mutants that could not arrest in mitosis in response to the benzimidazole drugs benomyl and nocodazole. They identified the *MAD* (mitotic arrest deficient) and *BUB* (budding uninhibited by benzimidazole) genes. Subsequent work identified Mps1 as an additional component of the checkpoint (Weiss and Winey, 1996). Paralogs of the yeast proteins have been identified from many organisms showing that the spindle checkpoint is evolutionarily conserved from yeast to humans (Reviewed in: Cleveland et al., 2003; Musacchio and Salmon, 2007; Lew and Burke, 2003). Interestingly, the checkpoint proteins localize to kinetochores of unaligned chromosomes in higher cells suggesting a relationship between the kinetochore and checkpoint signaling (Cleveland et al., 2003; Musacchio and Salmon, 2007). There are a number of proteins that have been identified in other organisms that are required for the spindle checkpoint but are absent in yeast suggesting a more intricate

**Table 11.1** Spindle Checkpoint Proteins

| Protein | SP | Function and Major Domains | Complex |
|---|---|---|---|
| APC Inhibitors | | | |
| Cdc20 | Con. | APC activator, Mad2 binding domain | MCC |
| Mad2 | Con. | Cdc20 inhibitor | MCC |
| BubR1,Mad3 | Con. | Mad3 homology, CENP-E activated kinase, binding, TPR-containing | MCC |
| Bub3 | Con. | WD40 Repeat | MCC |
| Mad2 Regulator | | | |
| Mad1 | Con. | Mad2 binding, coiled-coil | Mad1/ Mad2 |
| Dynein Regulators | | | |
| ZW-10 | F,V | | RZZ |
| Rod | F,V | | RZZ |
| Zwilch | F,V | | RZZ |
| CENP-E | F,V | MT binding, Kinesin motor | BubR1 |
| KMN | | | |
| Ndc80 complex | Con. | MT binding, Mad1 binding, coiled coil | KMN |
| MIND/Mis12 complex | Con | | KMN |
| KLN-1/Spc105/ Blinkin | Con | Bub1, BubR1 binding, MT binding | KMN |
| Kinases | | | |
| Mps1 | Con | Kinase, Mad3 homology | |
| MAP kinase | V | Kinase | |
| Tao1 | V | Kinase | |
| Bub1 | Con. | Kinase, Mad3 homology, binding | Mad3,Bub3 |
| NEK2A | Con. | Centrosome regulator, kinase | Mad1 |
| Inner Centromere | | | |
| AuroraB/Ipl1 | Con | Kinase | CPC |
| INCENP/Sli15 | Con. | Kinase activator, MT binding | CPC |
| Borealin | V | Chromatin binding | CPC |
| Survibin/Bir1 | Con. | Centromere targeting | CPC |
| Sgo1 | Con. | PP2A binding | PP2a (PP2A?)) |
| PICH | V | Swi/Snf ATPase | |

Abbreviations: Con. Conserved, F Fruit Flies (Drosophila), MT Microtubules,PP2a Phosphatase type 2A SP Species, V Vertebrate

regulatory process in higher eukaryotic cells. The majority of these proteins also localize to kinetochores in mitosis. There are conserved motifs in many of the proteins that suggest a molecular function and the proteins are summarized in Table 11.1. Interestingly, seven of the genes are predicted to encode protein kinases suggesting that protein phosphorylation is important in checkpoint signaling.

The spindle checkpoint regulates the metaphase to anaphase transition by controlling the ubiquitin-dependent proteolysis of securin, called Pds1 in yeast (Yamamoto et al., 1996). Securin is bound to separase, a protease which cleaves the cohesin proteins that bind sister chromatids together and active separase is the trigger for anaphase (Nasmyth, 2002). Securin is a substrate of the 26S proteasome and is regulated by an E3 ubiquitin ligase called the anaphase promoting complex (APC) and the substrate specificity factor Cdc20 which targets Pds1 for ubiquitylation and destruction (Nasmyth, 2002). The spindle checkpoint inhibits the activity of Cdc20 so that securin is stable and separase remains inactive (Hwang et al., 1998).

## 11.2 The Role of the Kinetochore in Spindle Checkpoint Signaling

McIntosh proposed that the spindle checkpoint was active as long as there was a lack of tension between sister kinetochores at mitosis (McIntosh, 1991). There were two important implications of the McIntosh model. The first was that kinetochores were the domain on chromosomes that regulated mitotic progression through the checkpoint and the second was that tension was the event that was monitored. Goh and Kilmartin (1993) provided the first evidence for a role of the kinetochore in the spindle checkpoint. They identified Ndc10 as a kinetochore protein and showed that temperature sensitive *ndc10-1* cells lost kinetochore function and abrogated the spindle checkpoint. Rieder et al., (1995) provided supporting evidence from higher cells by ablating kinetochores using a laser microbeam and abrogating the checkpoint (Rieder et al., 1995). Li and Nicklas (1995) tested the tension model directly using mantid spermatocytes with misaligned X chromosomes. They used a microneedle to apply tension to the misaligned chromosome and drove arrested cells into anaphase. This showed that the lack of tension on meiotic chromosomes can arrest the onset of anaphase. They used an antibody, 3F3/2, that recognizes a phospho-specific epitope on kinetochores to show that 3F3/2 staining on kinetochores correlated with the lack of tension (Nicklas et al., 1995; Waters et al., 1999). 3F3/2 stained mantid and mammalian kinetochores on unaligned chromosomes suggesting a link between protein phosphorylation, tension, and the spindle checkpoint (Nicklas et al., 1995). A role for tension is complicated by the fact that microtubule binding to kinetochores is also tension-sensitive (Ault and Nicklas, 1989). An alternative model for the spindle checkpoint is that the kinetochore monitors the lack of microtubule occupancy, and the effect of tension is indirect (Cleveland et al., 2003). It is currently unclear which model is correct or if they are both correct and the spindle checkpoint monitors both tension and occupancy in the kinetochore.

An early model for the role of the kinetochore in the checkpoint incorporated two important observations. The first was that checkpoint proteins localize to unattached kinetochores and the second derived from an in vitro assay for APC regulation. Tetrameric recombinant mitotic arrest deficient protein 2 (Mad2) inhibits the ubiquitin ligase activity of the APC in vitro (Fang et al., 1998). However, Mad2 does not exist as a tetramer in vivo and it was assumed that the tetramer of the recombinant protein assumed a conformation similar to one that Mad2 assumes after it associates with kinetochores, referred to as Mad2*. The proposed role of the kinetochore was to act as a platform to convert Mad2 to Mad2* and it was Mad2* that inhibited the APC. The model was enhanced when fluorescence recovery after photobleaching (FRAP) data showed that Mad2, and other checkpoint proteins, cycled rapidly through the kinetochore (Howell et al., 2004; Kallio et al., 2002a; Vink et al., 2006). The Mad2* model was simple and appealing although it did not account for any of the other checkpoint proteins. Additional work identified a more potent inhibitor of the APC called the mitotic checkpoint complex (MCC) that is a complex of BubR1 (Mad3 in yeast), Bub3, Mad2, and Cdc20 (Sudakin and Yen, 2004; Hardwick et al., 2000; Fang, 2002; Fraschini et al., 2001). The model evolved so that checkpoint proteins dynamically associate with kinetochores of unaligned chromosomes and are assembled into MCC (Cleveland et al., 2003; Musacchio and Salmon, 2007). The key step is believed to be the formation of Mad2–Cdc20 complexes. Mad1 and Cdc20 share a sequence that binds Mad2 and there is a conformational change of Mad2 induced upon binding (Luo et al., 2000, 2002; Musacchio and Hardwick, 2002; Sironi et al., 2002). An elegant model for conformational change in Mad2, catalyzed by Mad1 has been proposed and this induced change in conformation of Mad2 is proposed as the key step in checkpoint signaling. The model has been elegantly presented in a recent review and will not be discussed further here (Musacchio and Salmon, 2007). We will review the data that maps checkpoint activity within the kinetochore and we will propose a complementary model where the spindle checkpoint is intimately linked to microtubule binding sites within the kinetochore. Our model proposes that the kinetochore coordinates signaling by checkpoint protein kinases.

## 11.3 Mapping the Spindle Checkpoint Within the Kinetochore in Yeast

The yeast kinetochore is a complex macromolecular structure made up of more than 65 proteins that are assembled onto centromere DNA in a hierarchical manner (De Wulf et al., 2003; McAinsh et al., 2003; Westermann et al., 2003; Cheeseman et al., 2002). Most of the proteins are assembled into sub-complexes that have been well characterized. To understand how the spindle checkpoint signal is generated it is important to determine which kinetochore proteins participate and this is accomplished by finding kinetochore mutants that are checkpoint deficient. About half the yeast genes that encode kinetochore proteins are nonessential and deletion mutants eliminate the proteins from kinetochores. Deletion mutants of all nonessential kinetochore genes are checkpoint proficient and arrest in mitosis in response to nocodazole (McCleland et al., 2003; Gardner et al., 2001). Assaying the essential genes requires conditional (usually temperature sensitive) mutants and is more complex because mutants often retain some residual gene activity at the restrictive temperature but are below the threshold required for viability. This is an important consideration for assaying the spindle checkpoint because a single chromosome is capable of eliciting a response. Therefore, mutants may appear checkpoint proficient if they retain at least one active kinetochore, but be inviable because one or more kinetochores are non-functional. Despite this potential limitation, three kinetochore proteins were identified as components of the spindle checkpoint (Goh and Kilmartin, 1993; Tavormina and Burke, 1998; Janke et al., 2001). Temperature sensitive mutations in *NDC10*, *SPC24,* and *SPC25* abrogate the checkpoint at the restrictive temperature. Ndc10 is a member of the chromosome binding factor 3 (CBF3) complex of proteins and temperature sensitive mutations in other members are checkpoint proficient (McCleland et al., 2003; Gardner et al., 2001). Similarly, Spc24 and Spc25 are members of the Ndc80 complex of proteins and mutations in the other two genes that encode members of the complex are checkpoint proficient (Gillett et al., 2004, McCleland et al., 2003). This underscores the need for having true loss of function mutations that completely eliminate the protein and mimics the phenotype that results from a complete deletion of the gene.

Temperature sensitive "degrons" are epitopes that target fusion proteins for destruction by the 26S proteasome in a temperature-dependent fashion (Dohmen and Varshavsky, 2005). Degrons have been used to generate temperature sensitive kinetochore mutants to assay the essential genes for checkpoint activity (McCleland et al., 2003; Gardner et al., 2001). The result is that seven different kinetochore proteins were identified as required for the checkpoint. The CBF3 complex forms the "core" of the kinetochore and degron alleles of CBF3 genes eliminate the checkpoint. In the absence of CBF3, the kinetochore does not assemble and therefore the effect of the mutants on checkpoint activity is likely to be indirect (Gardner et al., 2001). Mutants defective for the Ndc80 complex are unable to attach chromosomes to the spindle and are also checkpoint defective suggesting that this could be a molecular link between spindle attachment and checkpoint activity in the kinetochore (McCleland et al., 2003; He et al., 2001). Mutations that eliminate essential proteins from other kinetochore complexes Cse4 and Scm1 (centromeric nucleosome), Mif2, Okp1 (COMA), Mtw1 (MIND), Duo1 (Dam1 complex) and Stu2 are checkpoint proficient (Brown et al., 1993; Gardner et al., 2001; Cheeseman et al., 2001; Kosco et al., 2001; Mizuguchi et al., 2007; Stoler et al., 1995, 2007). Degron alleles of *SPC105* have not been tested. However, temperature sensitive *spc105* mutants (*spc105-4*, *spc105-15*) have been isolated (Nekrasov et al., 2003). Under permissive conditions, the Spc105-4 and Spc105-15 cells do not arrest in the cell cycle when the checkpoint is activated (De Wulf and Sorger, pers. comm.). Thus involvement of the Spc105 complex in checkpoint seems conserved from budding yeast to humans. These data suggest that the Ndc80 and Spc105 complexes play a major role in orchestrating the spindle checkpoint within the kinetochore in yeast and there is an intimate relationship between kinetochore–microtubule binding and checkpoint signaling. The Ndc80 complex is also required for spindle checkpoint signaling in higher cells as described below. This relationship between kinetochore–microtubule binding and checkpoint signaling is strengthened by the observation that Mps1, a checkpoint kinase and a kinetochore protein is implicated in regulating kinetochore–microtubule attachments (Jones et al., 2005). In addition, the *C. elegans* homolog of Spc105 (KNL-1) binds microtubules in vitro which further supports a molecular link between kinetochore–microtubule attachments and checkpoint signaling (Nekrasov et al., 2003; Cheeseman et al., 2006).

Yeast spindle checkpoint proteins only associate with kinetochores that are detached from microtubules and do not associate with kinetochores that lack tension (Gillett et al., 2004). The concentrations of nocodazole used in yeast experiments to initiate checkpoint signals (typically 15 μg /ml) do not cause the dissolution of the spindle and complete loss of kinetochore microtubules as seen in vertebrate cells. Yeast spindles collapse when cells are grown in the presence of nocodazole (Jacobs et al., 1988). Most kinetochores remain attached to the collapsed spindle; sister kinetochores lack tension and surprisingly, spindle checkpoint proteins are not localized to these kinetochores (Gillett et al., 2004). A few kinetochores detach and spindle checkpoint proteins localize robustly. Moreover, mutations that eliminate sister chromatid cohesion result in sister kinetochores that lack tension and spindle checkpoint proteins are not localized to these kinetochores (Gillett et al., 2004). Thus the lack of occupancy triggers binding of checkpoint proteins to kinetochores but low tension does not. These observations question whether there is a role for tension in the spindle checkpoint in yeast. Mutations that eliminate the Ipl1 kinase, the Aurora B homolog in yeast, cannot properly establish bipolar orientation of chromosomes (Tanaka et al., 2002). The kinetochores lack tension and cells progress through mitosis with unaligned chromosomes suggesting that the checkpoint is inactive and that Ipl1 is required for the checkpoint. Surprisingly, temperature sensitive *ipl1* mutants arrest in response to nocodazole suggesting that the checkpoint is intact (Biggins and Murray, 2001). One possibility to reconcile these observations is that there are two branches to the checkpoint, one for tension and the other for occupancy and that Ipl1 is required for the tension branch of the checkpoint. Nocodazole should trigger both branches since chromosomes detach and most kinetochores in the collapsed spindle are not under tension. *ipl1* cells would arrest in nocodazole if Ipl1 were required only for the tension branch of the checkpoint. This was tested using mutations that eliminate Cdc6 resulting in cells that progress through the cell cycle in the absence of DNA replication. Since there are no sister chromatids, there is no tension across kinetochores. Cdc6 cells delay mitosis in a Mad2-dependent fashion suggesting that the lack of tension activates the checkpoint (Stern and Murray, 2001). Loss of Ipl1 abrogates the delay suggesting that Ipl1 is required for the tension branch of the checkpoint (Biggins and Murray, 2001). However, the interpretation is complicated because Ipl1 detaches chromosomes from the spindle as part of the error correction mechanism for misaligned chromosomes that lack tension (Dewar et al., 2004; Tanaka et al., 2002). The Mad2-dependent delay in *cdc6* cells could be strictly due to occupancy at the kinetochore because Ipl1 must be detaching microtubules from kinetochores. The delay would be abrogated in an *ipl1* mutant. The role of Ipl1 in error correction makes it difficult to differentiate a role in tension from occupancy strictly from analyzing *ipl1* mutants (Pinsky et al., 2006). However, the existence of the tension checkpoint in yeast is strengthened by identification of two other mutants, *sgo1* and *skp1*, both of which are required for the tension checkpoint (Indjeian et al., 2005; Kitagawa et al., 2003). Sgo1 is a kinetochore protein that is implicated in sensing tension and generating the tension-dependent spindle checkpoint signal (Indjeian et al., 2005). Skp1 is a member of the CBF3 complex, but a *skp1-AA* allele is a separation of function mutant that retains the essential function (kinetochore assembly) and lacks the tension checkpoint function (Kitagawa et al., 2003). In addition, mutations eliminating an Ipl1-dependent phosphorylation of Mad3 eliminate the tension branch of the spindle checkpoint (King et al., 2007). Together, these data strongly suggest that Ipl1-dependent phosphorylation of Mad3 in response to a lack of tension activates the spindle checkpoint and requires the two kinetochore proteins Sgo1 and Skp1.

## 11.4 Early Lessons from Metazoan Systems

Characterization of vertebrate homologs of the spindle checkpoint genes demonstrated that the pathway is evolutionarily conserved. The first gene cloned was human Mad2 (Li and Benezra, 1996). Subsequently, homologs of Bub1 and Bub3 were identified and all were shown to have conserved roles in checkpoint signaling (Taylor and McKeon, 1997; Taylor et al., 1998). A surprising finding was that Mad3 in vertebrates contained a Bub1 related kinase domain on its C-terminus and was therefore renamed BubR1 (Taylor et al., 1998). Deletion analysis demonstrated that Bub3 bound a similar region on both Bub1 and BubR1 and that this domain is required to bind kinetochores. Thus, a role of Bub3 is to target the Bub kinases to kinetochores (Taylor et al., 1998). Mad2, Bub1, BubR1, and Bub3 were all shown to be required to arrest cells in response to nocodazole (Taylor and McKeon, 1997; Taylor et al., 1998; Li and Benezra, 1996). These experiments also established that each of these proteins localize to unaligned kinetochores, an important correlation that links the signaling proteins to the source of the signal. Kinetochore localization of the proteins has become a cornerstone of all models, however only a small percentage of some checkpoint proteins localize to kinetochores and localization correlates with checkpoint signaling. It has never been formally proven that these events are required to generate a signal.

It may be a mistake to consider Bub1, BubR1, Bub3, and Mps1 as checkpoint proteins that localize to kinetochores. It has subsequently become clear that these core spindle checkpoint proteins have additional mitotic functions. Thus an emerging picture is that they are kinetochore proteins that have roles in checkpoint signaling. This is not simply a semantic difference but is critical in interpreting the loss of function phenotypes. For example, Bub1 is a regulator of both kinetochore assembly and the maintenance of cohesion. It is required to localize Sgo1 to inner centromeres and chromosomes lacking Bub1 eventually lose cohesion and cannot be aligned (Kitajima et al., 2005; Kitajima et al., 2004; Meraldi and Sorger, 2005). Bub1 is also important in *Xenopus* for chromosomal passenger complex (CPC) localization that includes Aurora B (Ipl1) INCENP, Survivin, and Dasra/Borealin (Boyarchuk et al., 2007). Mislocalization of the CPC leads to a variety of other defects. These concerns highlight the need to generate separation of function mutants that lack checkpoint signaling but properly perform other functions.

Important insights to the spindle checkpoint have come from *Drosophila* genetics. A complex of three proteins, abbreviated as RZZ contains Rough Deal (Rod), Zeste-White 10 (ZW10) and Zwilch and is required for the spindle checkpoint (Chan et al., 2000; Basto et al., 2000). There are homologs of all three proteins in higher cells (they have not yet been identified in yeast) and their checkpoint functions are conserved. In addition, RZZ interacts with the microtubule motor dynein and its associated regulator dynactin and this interaction is believed to be important for silencing the checkpoint, as described below (Starr et al., 1998). Finally, mutations that eliminate Mad2 in *Drosophila* are viable suggesting that the checkpoint in flies, as in yeast, is nonessential under laboratory conditions (Buffin et al., 2007).

A system to assay spindle checkpoint function in *Xenopus* extracts has been used extensively to characterize checkpoint signaling (Minshull et al., 1994). *Xenopus* Mad2 (Xmad2) was identified as the human Mad2 ortholog (Chen et al., 1996). X*mad2* was purified from egg extracts in a tight complex with Xmad1 (Chen et al., 1998). Extracts treated with inactivating antibodies to either protein were unable to generate a spindle checkpoint signal. Subsequently, requirements of Bub1, BubR1, and Mps1 were confirmed by immunodepletion from *Xenopus* extracts and all of the proteins were shown to localize to kinetochores during prometaphase and to unaligned kinetochores (Sharp-Baker and Chen, 2001; Chen, 2002; Abrieu et al., 2001). Early studies also demonstrated that the aforementioned oligomeric form of Mad2 (Mad2*) inhibited anaphase in the absence of kinetochores, which suggests that Mad2 is downstream of the kinetochore in the signaling pathway (Fang et al., 1998). Thus from these early experiments it became clear that the spindle checkpoint components were conserved from yeast to humans and localized to unaligned kinetochores. Moreover they established the concept that kinetochores ultimately regulate Mad2 to inhibit APC.

There are complex dependencies of spindle checkpoint proteins for kinetochore localization. These dependencies have been extensively explored in both egg extracts and tissue culture cells and are summarized in Fig. 11.1A. In frogs there is a mutual dependency for Bub1, BubR1, Bub3 and Mps1 (Sharp-Baker and Chen, 2001; Chen, 2002; Abrieu et al., 2001). All of these proteins are required for Mad1 recruitment, which is required to localize Mad2 (Murray et al., 1999). The dependencies are different in tissue culture cells as assayed by siRNA knockdown. Mps1 is recruited independently of the Bubs and Bub1 independently of BubR1 (Stucke et al., 2002, 2004; Liu et al., 2003). This difference may be due to the degree of knockdown or may indicate an important role of Bub1 to assemble the inner centromere and recruit the CPC in *Xenopus* extracts (Boyarchuk et al., 2007). In *Xenopus* extracts none of the outer kinetochore proteins assemble when the CPC is depleted (Emanuele et al., 2005). Although this prevents checkpoint signaling one cannot discern if this is caused by poor kinetochore assembly or a direct role of the CPC in the generating the signal (Kallio et al., 2002b). Despite potential differences in the dependencies in *diverse* experimental systems, it is clear from Fig. 11.1 that most arrows ultimately point to Mad1 suggesting that Mad1 recruitment to the kinetochore is a critical event in checkpoint signaling. We incorporate this observation as an important component of the model we present below.

## 11.5 Roles of Protein Kinases in Checkpoint Signaling

Numerous protein kinases localize to unattached kinetochores suggesting that the kinetochore is a signal transduction platform. Although it is debated, it appears that the kinase domains of both Bub1 and BubR1 strengthen the checkpoint signal (Chen, 2002; Sharp-Baker and Chen, 2001; Chan et al., 1999; Mao et al., 2003; Robbins et al., 2005; Tang et al., 2004). Mps1 is recruited to signaling kinetochores independently of Mad1 (Stucke et al., 2004; Abrieu et al., 2001; Stucke et al., 2002). Localization of Mps1 requires the Ndc80 complex, but no direct interactions between kinetochore proteins and Mps1 have been identified (Martin-Lluesma et al., 2002; Stucke et al., 2004).

Early *Xenopus* experiments suggested a role for p38 MAP Kinase in the spindle checkpoint (Minshull et al., 1994). Antibodies that recognize activated MAP kinase localize to kinetochores in tissue culture cells (Zecevic et al., 1998). Recent papers have nicely identified key targets of MAP kinase signaling. In *Xenopus* extracts MAP kinase phosphorylates two residues on Cdc20 that are required for checkpoint signaling (Chung and Chen, 2003). Mutation of MAP kinase sites on Bub1 produced a weakened checkpoint signal that is similar to that produced by a mutation in Bub1 that renders the kinase inactive (Chen, 2004). In addition, p38 MAP kinase also controls the kinetochore localization of Mps1 (Zhao and Chen, 2006). Mutation of a MAP kinase site on Mps1 to a residue incapable of phosphorylation perturbs its kinetochore localization and blocks checkpoint signaling, but the kinase is still active. This is the best

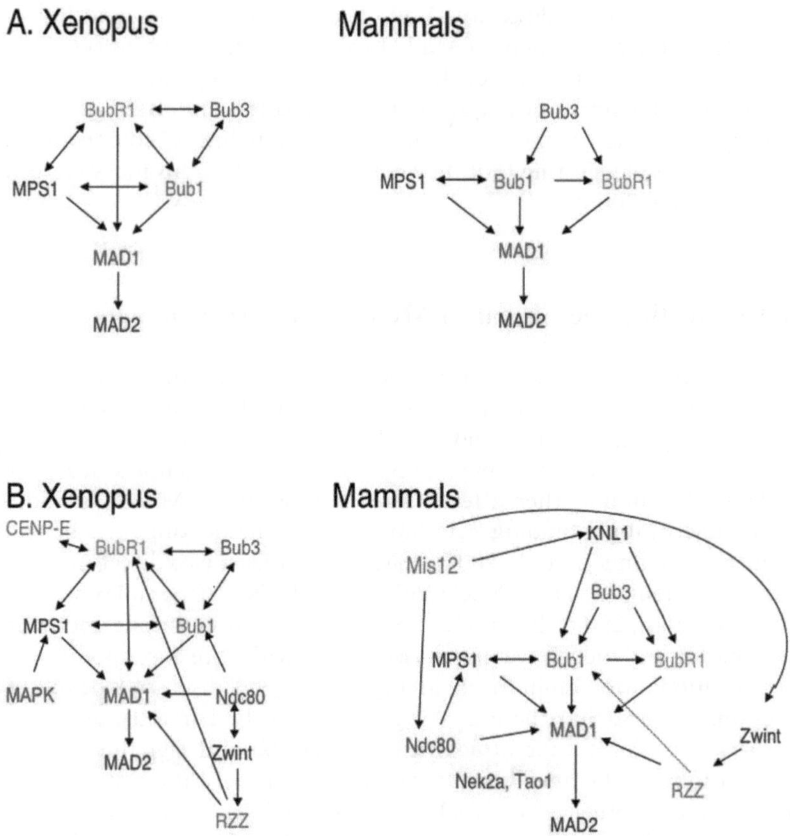


**Fig. 11.1** (**A**) Dependencies for kinetochore localization of conserved spindle checkpoint proteins in extracts of *Xenopus laevis* and tissue culture systems. Arrows indicates a strict dependency with the arrow head pointing to the dependent protein. Dual arrowheads represent interdependencies for kinetochore localization. Colors of proteins are conserved in all subsequent figures. (**B**) Dependency relationships of checkpoint and kinetochore proteins. The dependency relationships for kinetochore proteins is superimposed upon the checkpoint proteins (Fig. 11.1A). The arrows have the same meanings

evidence that localization of Mps1 to kinetochores is important for generating the signal.

Many kinases are required for the step between the recruitment of Mad1 and the localization of Mad2 suggesting that this is a highly regulated event. CDK1 priming phosphorylation on Bub1 targets Plk1 to kinetochores, which allows Mad2 and BubR1 localization (Qi et al., 2006). Although Plk1 has never been shown to have a checkpoint role, it has been strongly argued that it plays a role that has been masked by either poor knockdown or redundant activities (Qi et al., 2006). Nek2A binds Mad1 and in the absence of Nek2A Mad1 binds kinetochores and Mad2 does not (Lou et al., 2004). Tao1 is another kinase responsible for the recruitment of Mad2 to kinetochores (Draviam et al., 2007). Why so many kinases are required, how the kinases are regulated and what their substrates are remain critical questions. Both Bub1 and BubR1 are highly phosphorylated when isolated on checkpoint signaling chromosomes from *Xenopus* extracts, suggesting that they may be a target of this mass recruitment of kinases (Chen, 2002).

## 11.6 Connecting the Signal to Microtubule Attachments

The kinetochore has at least four key protein complexes that directly interact with microtubules to coordinate kinetochore-based movements of chromosomes on the spindle. Dynein and CENP-E are microtubule motors that play specific roles during chromosome segregation. KMN (a complex of AF15Q14/KNL-1/Spc105/Blinkin, (hereafter referred to as KNL-1), Mis12 and Ndc80) is a critical microtubule-binding interface in the kinetochore for microtubule attachments (Emanuele et al., 2007). KMN has at least two direct microtubule interacting proteins the Hec1/Ndc80 subunit of the Ndc80 complex and KNL-1 (Cheeseman et al., 2006). Each of these proteins has been implicated in checkpoint signaling and most have direct association with core checkpoint proteins. The associations are summarized in Fig. 11.1B and discussed below. These interactions suggest that there is a relationship in higher cells, as there is in yeast, between microtubule attachment and checkpoint signaling. Moreover these microtubule attachment complexes are not strongly associated in solution but only come together at kinetochores (Emanuele et al., 2005). We propose that this assembly mechanism is important for the checkpoint which assures that checkpoint signals are produced at kinetochores.

Current models suggest that the Ndc80 complex is the key to microtubule interaction that directs depolymerization-coupled movement of chromosomes. All four members of the Ndc80 complex are required for congression of chromosomes to the metaphase plate and anaphase segregation of sister chromatids (DeLuca et al., 2002; Martin-Lluesma et al., 2002; McCleland et al., 2003, 2004). The N-terminus of Hec1 has an unstructured tail followed by a globular head with limited homology to the microtubule plus end-binding protein EB1 (Wei et al., 2007). The N-terminus of Hec1 binds microtubules with a Kd of $\sim$3 μM. Ndc80 has been localized to the outer kinetochore, and siRNA of any of the four subunits of the complex leads to poor kinetochore assembly as assayed by immunofluorescence of other kinetochore proteins and by electron microscopy (DeLuca et al., 2005; McCleland et al., 2003, 2004).

The Ndc80 complex is implicated in spindle checkpoint signaling in vertebrates; however it is important to remember that the complex is required for kinetochore assembly and so a direct role in checkpoint signaling is difficult to prove. *Xenopus* extracts depleted of the Ndc80 complex have no spindle checkpoint, and they cannot localize any of the checkpoint proteins including the RZZ complex. However if kinetochores are first assembled and then antibodies to the Ndc80 component are added, the extracts cannot signal but both the RZZ and dynactin proteins remain at kinetochores (McCleland et al., 2003). Similar findings were obtained after injecting antibodies against each of the four subunits into cells (McCleland et al., 2003; McCleland et al., 2004). These are the strongest data that vertebrate Ndc80 has a direct role in checkpoint signaling. siRNA of Hec1 from HeLa cells resulted in a prometaphase arrest with low levels of Mad1 and Mps1 at kinetochores (Martin-Lluesma et al., 2002). It appears that the arrest is produced by residual Hec1, as later reports showed that cells lost the checkpoint if greater than 98% of Nuf2 is eliminated (Meraldi et al., 2004).

The mechanism is still unclear but one can draw an exciting model from preliminary data. A two-hybrid interaction between the first coiled-coil domain of Hec1 and Mad1 has been reported, and the Ndc80 complex is required for kinetochore assembly of Mad1, Mad2, Mps1, and RZZ (Martin-Lluesma et al., 2002; McCleland et al., 2003, 2004; Martin-Lluesma et al., 2002; Stucke et al., 2004). Bub1 and BubR1 still localize after knockdown of Ndc80. This suggests that Ndc80 is a platform for Mad1 and perhaps Mps1 binding, but this needs additional experimentation. The clear role of Ndc80 in kinetochore–microtubule attachment and the implication in checkpoint signaling suggests that recognition of microtubule binding is as simple as a mutually exclusive binding site on the Ndc80 complex for microtubules and Mad1. Testing this model is an important area for future experimentation.

Biochemical data suggest that the Ndc80 complex is associated with the Mis12/MIND complex and the KNL-1 protein at the kinetochore (Cheeseman et al., 2004; Goshima et al., 2003). This supercomplex (KMN) may be the scaffold that brings the checkpoint proteins together. HeLa cells fail to arrest in nocodazole after siRNA of the Mis12 proteins (McAinsh et al., 2006). KNL-1 is required to generate checkpoint signals and, significantly, is the platform to direct Bub1 and BubR1 to kinetochores (Kiyomitsu et al., 2007). This elegant study shows Bub1 and BubR1 to have TPR domains that interact with the KNL-1 N-terminus. Point mutants of the Bubs that knockout TPR function cannot localize to kinetochore or generate checkpoint signals. Similarly loss of the N-and middle domains of KNL-1 prevents checkpoint signaling and mislocalizes the Bub proteins. KNL-1 is a large and poorly studied protein with roles in kinetochore assembly, microtubule binding, and checkpoint signaling. Further characterization of this critical protein is an important area of future research.

RZZ has two roles at kinetochores. It is required to regulate Dynein binding to kinetochores and generate the checkpoint signal. In *Xenopus*, RZZ is required for Bub1 binding which is highly disruptive to kinetochore architecture (Kops et al., 2005). However, Bub1 and BubR1 properly localize in human cells after RZZ knockdown, but Mad1 and Mad2 do not (Wang et al., 2004; Kops et al., 2005; Chan et al., 2000). Chromosomes align and undergo anaphase movements suggesting that the KMN is intact. These data suggest that Bub1 and Ndc80 are not sufficient for Mad1 binding in vertebrates and there is an additional requirement for RZZ. The Zwint-1 protein appears to be a link between the KMN complex and RZZ. Furthermore, Kre28 in budding yeast may be the Zwint-1 homolog (Sorger and De Wulf, Pers. Comm.). Zwint-1 is required for the checkpoint in higher cells presumably to recruit RZZ (Kops et al., 2005).

Dynein has been implicated in silencing the checkpoint after microtubule attachment. Aligned chromosomes with normal numbers of bound microtubules retained checkpoint proteins after inhibition of Dynein/Dynactin activity (Howell et al., 2001). Dynein carries the checkpoint protein complexes away from kinetochores after microtubule attachment to help silence the checkpoint (Howell et al., 2001; Siller et al., 2005; Basto et al., 2004). Although this model is likely to be correct it is important to be careful in interpreting these results because it is impossible to separate roles of dynein in properly attaching microtubules, which would generate a spindle checkpoint signal, from roles in turning off a checkpoint signal after proper attachment.

The plus end kinesin CENP-E has also been shown to be important for spindle checkpoint signaling. An interaction between CENP-E and BubR1 was first demonstrated in human cells (Chan et al., 1998). Elegant experiments in *Xenopus* established that CENP-E binds to BubR1 only under checkpoint signaling conditions and this activates kinase activity (Mao et al., 2003). When CENP-E binds to microtubules it no longer activates BubR1 (Mao et al., 2005). This suggests that microtubule attachment blocks the activation of BubR1 kinase activity and thus inhibits signaling. The role of CENP-E in generating a spindle checkpoint signal in tissue culture cells and normal mouse cells is less clear since a mouse deletion mutant is alive (although slightly compromised) and in both systems the checkpoint appears to be largely intact in the absence of CENP-E (Putkey et al., 2002; Weaver et al., 2003).

## 11.7 A Model for Kinetochore Regulation of Occupancy Checkpoint Signaling

We propose that KMN acts as a platform for the spindle checkpoint signal with independent binding sites for checkpoint proteins. KNL-1 binds Bub1 and BubR1 proteins, Ndc80 recruits Mad1 and Mps1 and Mis12/Zwint recruits RZZ. The evolutionarily conserved regulated event is Mad1 recruitment (Fig. 11.2A). Mad1 has at least three independent interactions with kinetochore proteins, Bub1 via KNL-1, Ndc80 and a member of RZZ. These proteins only interact at kinetochores, thus ensuring that the checkpoint signal can only be produced there. We propose that Mad1 acts as a kinase scaffold, similar to Ste5 which acts in the MAP kinase signaling pathway in the yeast pheromone response (Elion, 2001). Mad1 binds strongly to cyclin B1-Cdk1 (Mark Jackman and Jonathon Pines, pers. comm.) and thus could recruit two kinases to the

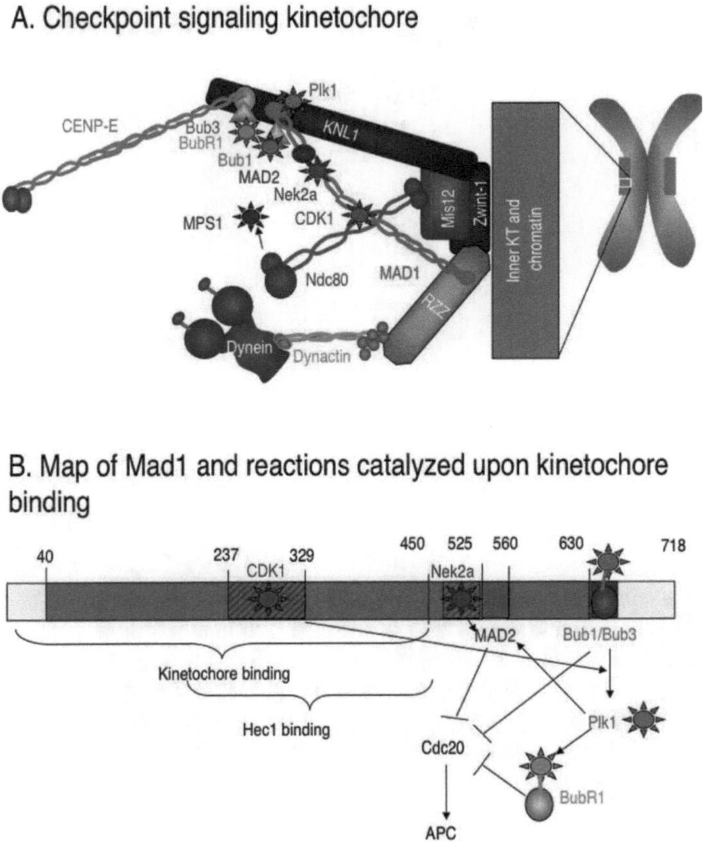


**Fig. 11.2** (**A**) Model of KMN as a scaffold for generating a spindle checkpoint complex. Possible organization of checkpoint proteins in an unattached kinetochore. Interactions are inferred if proteins have been identified in complexes, although the interactions may not be direct. The sun shapes indicate kinase domains. No physical interaction between Mps1 and a kinetochore protein has been demonstrated but Ndc80 is required (arrow). Tao1 kinase associates with kinetochores but the molecular nature of the interaction is unknown and therefore it has been excluded from the diagram. (**B**) Map of Mad1 and reactions catalyzed upon kinetochore binding. The bar represents the domain structure of Mad1. The potential regulation of Cdc20 is indicated below Mad1, see text for details

kinetochore Cdc2/Cdk1 and Nek2A, and a key substrate Mad2 as depicted in Fig. 11.2B (Lou et al., 2004). Once at the kinetochore, Mad1 would complete a signaling network that could include five additional kinases that are independently localized to the kinetochore including Bub1, BubR1, Mps1, Plk1, and Tao1 to generate the signal (Fig. 11.2A). Each kinetochore in vertebrates may have 900 KMN complexes allowing a large amount of signal to be generated from a single unattached kinetochore (Emanuele et al., 2005).

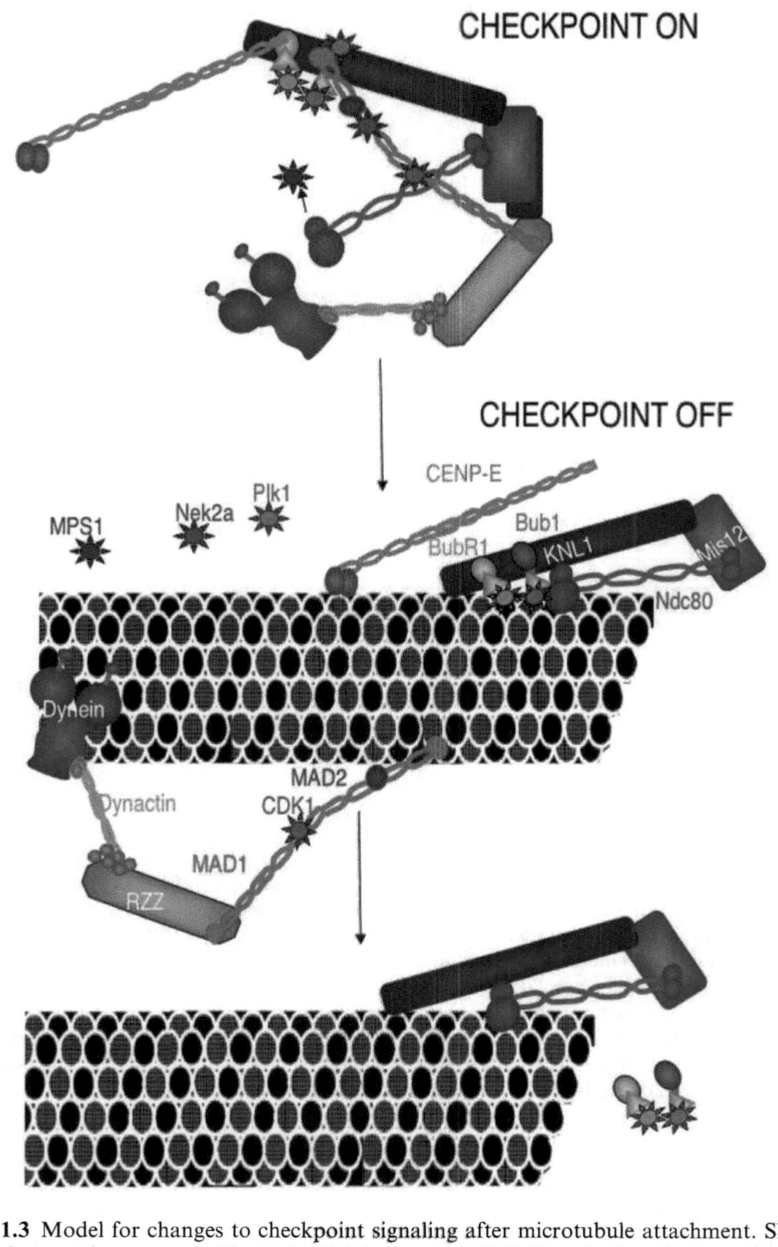


**Fig. 11.3** Model for changes to checkpoint signaling after microtubule attachment. Shapes and colors are the same as in Fig. 11.2. KMN cannot bind Mad1 and microtubules at the same time, which is shown as a conformational change upon microtubule binding. This leads to disassembly of many of the checkpoint proteins although the Bub proteins can still associate with kinetochores after microtubule binding. See text for details

There is a tight inverse correlation of Mad1 localization and microtubule binding, which suggests that Mad1 association is the microtubule-regulated step. Moreover many of the proteins that directly bind microtubules in the kinetochore have weak interactions with Mad1 including Ndc80 and RZZ. There is a complex of Bub1/Bub3 and Mad1 in budding yeast suggesting that there is an indirect association between Mad1 and KNL-1 through Bub1. Thus a simple model for the occupancy checkpoint is that KMN binds either microtubules or Mad1 but cannot bind both simultaneously (Fig. 11.3). The numerous requirements to localize Mad1 to kinetochores would also ensure that the signal is only generated at unattached kinetochores. We suggest that Mad1 recruitment initiates three independent pathways that inhibit Cdc20 (Fig. 11.2B). Mad1 is a long coiled-coil protein that interacts tightly with Mad2 and recruits it to kinetochores. Bub1 kinase phosphorylates and inhibits Cdc20 directly. Moreover, CDK1 priming phosphorylation on Bub1 targets Polo kinase to kinetochores; an event that allows Mad2 and BubR1 localization thus concentrating all of the components of MCC (Qi et al., 2006).

Microtubule attachment would block the generation of the signal at various points. First we propose that Ndc80 and KNL-1 cannot simultaneously bind microtubules and Mad1. Note that the model predicts that Mad1 at kinetochores would need to be dynamic to allow microtubule binding. Microtubule binding would also block the CENP-E/BubR1 interaction, shut off kinase activity, and allow Dynein to carry RZZ, Mad1, and Mad2 away from KMN.

## 11.8 The Tension Checkpoint and Roles of the CPC

Bipolar attachment of spindle microtubules to sister kinetochores pulls sister kinetochores apart and thus produces "tension" that could be sensed at either the inner centromere region or at the kinetochore–microtubule interface. Cells arrest with aligned chromosomes in response to low doses of taxol or vinblastin, suggesting that kinetochores are attached to microtubules. However, the distance between sister kinetochore is reduced suggesting that tension is not generated (Waters et al., 1998). Interestingly there is a biochemical difference; Bub1 and BubR1 localize to aligned kinetochores under low tension, but Mad1 and Mad2 do not (Waters et al., 1998; Skoufias et al., 2001). Such studies support the model, mentioned previously, that kinetochores generate distinct signals by either the lack of microtubule attachment or under low tension. Another difference is a requirement for Aurora B kinase activity (which is part of the CPC). Cells in taxol cannot arrest in the presence of small molecule inhibitors of Aurora B, or after siRNA knockdown of survivin but cells in nocodazole are able to arrest for several hours (Hauf et al., 2003; Ditchfield et al., 2003). There may be a stoichiometric role for the CPC in the occupancy checkpoint as checkpoint signals cannot be generated after antibody injections to Aurora B or siRNA against Aurora B (Kallio et al., 2002b; Ditchfield et al.,

2003; Carvalho et al., 2003). Aurora B has many functions so a concern is that its effects may be indirect, however a recent elegant study suggests direct roles of the CPC in checkpoint signaling. After siRNA knockdown of INCENP and replacement with a mutant that lacks a coiled-coil domain most functions of the CPC were rescued but these cells were still deficient in the tension branch of the checkpoint (Vader et al., 2007). This study focuses future experiments on the tension branch of signaling on this critical domain. Moreover it argues that the role of the CPC is not simply to release microtubules and stimulate the occupancy signal. An exciting new protein involved in tension checkpoint signaling is PICH, a SWI/SNF type ATPase that localizes to stretched chromatin in anaphase and is required for checkpoint signaling. (Baumann et al., 2007).

Polo kinase has also been implicated in tension signaling. The monoclonal antibody 3F3/2 recognizes a phosphorylation event on kinetochores that are not under tension (Nicklas et al., 1995). Polo generates the 3F3/2 epitope arguing that it is activated in response to tension (Ahonen et al., 2005; Wong and Fang, 2005). How the tension signal is generated is still mysterious and models are highly speculative. However, it is interesting that Aurora B and Mps1 are both activated by microtubules (Rosasco-Nitcher et al., 2008, Stucke et al., 2004). Moreover Polo generates priming phosphorylations on substrates to activate Aurora B (Rosasco-Nitcher et al., 2008). Thus kinetochores under low tension could act as a scaffold to concentrate microtubules, Polo, Aurora B, Bub1, BubR1, and Mps1 to initiate a signal transduction cascade.

## 11.9 Summary and Future Directions

Analysis of the spindle checkpoint from yeast to humans has shown that a great deal of the checkpoint is evolutionarily conserved and that the checkpoint is biochemically more complex in higher cells. A prevailing theme for the role of the kinetochore is that the spindle checkpoint is intimately related to the kinetochore proteins that mediate microtubule binding. The prevailing model for the role of the kinetochore is that it catalyzes the assembly of Mad2–Cdc20 complexes, especially MCC. This emphasis is mostly historical and can be traced to the discovery of Mad2* and the innumerable studies showing a correlation between checkpoint activation and checkpoint protein localization to kinetochores. It is clear in yeast that Mad2–Cdc20 and MCC are essential for the checkpoint (Hardwick et al., 2000; Fraschini et al., 2001; Poddar et al., 2005). However, regulated assembly of Mad2–Cdc20 complexes have nothing to do with checkpoint signaling. Mad2–Cdc20 complexes form in mitosis independently of the checkpoint and independently of the kinetochore (Fraschini et al., 2001; Poddar et al., 2005). MCC assembly is also kinetochore independent in human cells (Sudakin and Yen, 2004). It is possible that the kinetochore is the site of regulated assembly of some other yet to be appreciated checkpoint protein complexes. We propose a different model; that spindle checkpoint signaling is mediated in large part by protein kinases. We propose that the kinetochore is the site of assembly of a Mad1-dependent scaffold that orchestrates a plethora of checkpoint kinases to inhibit mitosis in response to either a lack of tension or a lack of occupancy. We envision that the occupancy checkpoint is a simple two-state switch. When microtubules are absent the Mad1 scaffold assembles on KMN-RZZ and the checkpoint is on. When microtubules bind, RZZ is removed, the scaffold is displaced, and the checkpoint is off. The model needs only a slight modification for yeast where RZZ is not present. One consequence of the model is that it explains why checkpoint proteins have a short half life in kinetochores. If they associated tightly to kinetochores then microtubules would be precluded from binding. Our model is consistent with a large number of observations but almost certainly wrong in specific details. This reflects the fact that we have not identified all of the proteins in the spindle checkpoint or in the kinetochore. We are only beginning to understand how the checkpoint is organized in the kinetochore. The model makes predictions, for example about the nature of separation-of-function mutants that should be possible. It highlights the importance of understanding how the checkpoint kinases are localized and regulated. It also highlights the importance of identifying substrates and how phosphorylation alters their activity.

Our model is less developed for the tension branch of the checkpoint which reflects our need to study this important branch more carefully. Our poorer understanding of the tension branch of the checkpoint is partly due to a lack of tools in mammalian cells that would match those in yeast such as cdc6 mutants and cohesin mutants. The role of the kinetochore in the tension checkpoint appears to involve the inner centromere and is distinct from KMN, RZZ and the microtubule binding proteins. We believe this is an important hint to the mechanism of the tension checkpoint that needs to be fully explored. It is clear that there is much to be done but we are hopeful that a clearer understanding of the role of the kinetochore in all aspects of the spindle checkpoint is imminent.

**Acknowledgments** We thank Sue Biggins and the members of the Stukenberg and Burke labs for helpful discussions and comments on the manuscript. We also thank Mitsuhiro Yanagida, Mark Jackman, and Jonathon Pines for communicating results prior to publication. We apologize to those colleagues whose work was not cited due to space limitations.

## References

Abrieu, A., Magnaghi-Jaulin, L., Kahana, J.A., Peter, M., Castro, A., Vigneron, S., Lorca, T., Cleveland, D.W., and Labbe, J.C. (2001). Mps1 is a kinetochore-associated kinase essential for the vertebrate mitotic checkpoint. Cell 106, 83–93.

Ahonen, L.J., Kallio, M.J., Daum, J.R., Bolton, M., Manke, I.A., Yaffe, M.B., Stukenberg, P.T., and Gorbsky, G.J. (2005). Polo-like kinase 1 creates the tension-sensing 3F3/2 phosphoepitope and modulates the association of spindle-checkpoint proteins at kinetochores. Curr Biol 15, 1078–89.

Ault, J.G. and Nicklas, R.B. (1989). Tension, microtubule rearrangements, and the proper distribution of chromosomes in mitosis. Chromosoma 98, 33–9.

Bajer, A. and Mole-Bajer, J. (1961). UV microbeam irradiation of chromosomes during mitosis in endosperm. Exp Cell Res 25, 251–267.

Basto, R., Gomes, R., and Karess, R.E. (2000). Rough deal and Zw10 are required for the metaphase checkpoint in *Drosophila*. Nat Cell Biol 2, 939–43.

Basto, R., Scaerou, F., Mische, S., Wojcik, E., Lefebvre, C., Gomes, R., Hays, T., and Karess, R. (2004). In vivo dynamics of the rough deal checkpoint protein during *Drosophila* mitosis. Curr Biol 14, 56–61.

Baumann, C., Korner, R., Hofmann, K., and Nigg, E.A. (2007). PICH, a centromere-associated SNF2 family ATPase, is regulated by Plk1 and required for the spindle checkpoint. Cell 128, 101–114.

Biggins, S. and Murray, A.W. (2001). The budding yeast protein kinase Ipl1/Aurora allows the absence of tension to activate the spindle checkpoint. Genes Dev 15, 3118–3129.

Boyarchuk, Y., Salic, A., Dasso, M., and Arnaoutov, A. (2007). Bub1 is essential for assembly of the functional inner centromere. J Cell Biol 176, 919–28.

Brown, M.T., Goetsch, L., and Hartwell, L.H. (1993). MIF2 is required for mitotic spindle integrity during anaphase spindle elongation in Saccharomyces cerevisiae. J Cell Biol 123(2), 387–403.

Buffin, E., Emre, D., and Karess, R.E. (2007). Flies without a spindle checkpoint. Nat Cell Biol 9, 565–72.

Callan, H. G and Jacobs, P. A. (1957). The meiotic process in Mantis Religiosa L. males. J. Genetics 200, 200–217.

Carlson, J.G. (1956). On the mitotic movements of chromosomes. Science 124, 203–206.

Carvalho, A., Carmena, M., Sambade, C., Earnshaw, W.C., and Wheatley, S.P. (2003). Survivin is required for stable checkpoint activation in taxol-treated HeLa cells. J Cell Sci 116, 2987–2998.

Chan, G.K., Jablonski, S.A., Starr, D.A., Goldberg, M.L., and Yen, T.J. (2000). Human Zw10 and ROD are mitotic checkpoint proteins that bind to kinetochores. Nat Cell Biol 2, 944–7.

Chan, G.K., Jablonski, S.A., Sudakin, V., Hittle, J.C., and Yen, T.J. (1999). Human BUBR1 is a mitotic checkpoint kinase that monitors CENP-E functions at kinetochores and binds the cyclosome/APC. J Cell Biol 146, 941–54.

Chan, G.K., Schaar, B.T., and Yen, T.J. (1998). Characterization of the kinetochore binding domain of CENP-E reveals interactions with the kinetochore proteins CENP-F and hBUBR1. J Cell Biol 143, 49–63.

Cheeseman, I.M., Chappie, J.S., Wilson-Kubalek, E.M., and Desai, A. (2006). The conserved KMN network constitutes the core microtubule-binding site of the kinetochore. Cell 127, 983–97.

Cheeseman, I.M., Drubin, D.G., and Barnes, G. (2002). Simple centromere, complex kinetochore: linking spindle microtubules and centromeric DNA in budding yeast. J Cell Biol 157, 199–203.

Cheeseman, I.M., Enquist-Newman, M., Muller-Reichert, T., Drubin, D.G., and Barnes, G. (2001). Mitotic spindle integrity and kinetochore function linked by the Duo1p/Dam1p complex. J Cell Biol 152, 197–212.

Cheeseman, I.M., Niessen, S., Anderson, S., Hyndman, F., Yates, J.R.3., Oegema, K., and Desai, A. (2004). A conserved protein network controls assembly of the outer kinetochore and its ability to sustain tension. Genes Dev 18, 2255–68.

Chen, R.H. (2002). BubR1 is essential for kinetochore localization of other spindle checkpoint proteins and its phosphorylation requires Mad1. J Cell Biol 158, 487–96.

Chen, R.H. (2004). Phosphorylation and activation of Bub1 on unattached chromosomes facilitate the spindle checkpoint. EMBO J 23, 3113–3121.

Chen, R.H., Shevchenko, A., Mann, M., and Murray, A.W. (1998). Spindle checkpoint protein Xmad1 recruits Xmad2 to unattached kinetochores. J Cell Biol 143, 283–95.

Chen, R.H., Waters, J.C., Salmon, E.D., and Murray, A.W. (1996). Association of spindle assembly checkpoint component XMAD2 with unattached kinetochores. Science 274, 242–6.

Chung, E. and Chen, R.H. (2003). Phosphorylation of Cdc20 is required for its inhibition by the spindle checkpoint. Nat Cell Biol 5, 748–753.

Cleveland, D.W., Mao, Y., and Sullivan, K.F. (2003). Centromeres and kinetochores: from epigenetics to mitotic checkpoint signaling. Cell 112, 407–421.

De Wulf, P., McAinsh, A.D., and Sorger, P.K. (2003). Hierarchical assembly of the budding yeast kinetochore from multiple subcomplexes. Genes Dev 17, 2902–2921.

DeLuca, J.G., Dong, Y., Hergert, P., Strauss, J., Hickey, J.M., Salmon, E.D., and McEwen, B.F. (2005). Hec1 and nuf2 are core components of the kinetochore outer plate essential for organizing microtubule attachment sites. Mol Biol Cell 16, 519–31.

DeLuca, J.G., Moree, B., Hickey, J.M., Kilmartin, J.V., and Salmon, E.D. (2002). hNuf2 inhibition blocks stable kinetochore-microtubule attachment and induces mitotic cell death in HeLa cells. J Cell Biol 159, 549–55.

Dewar, H., Tanaka, K., Nasmyth, K., and Tanaka, T.U. (2004). Tension between two kinetochores suffices for their bi-orientation on the mitotic spindle. Nature 428, 93–97.

Ditchfield, C., Johnson, V.L., Tighe, A., Ellston, R., Haworth, C., Johnson, T., Mortlock, A., Keen, N., and Taylor, S.S. (2003). Aurora B couples chromosome alignment with anaphase by targeting BubR1, Mad2, and Cenp-E to kinetochores. J Cell Biol 161, 267–80.

Dohmen, R.J. and Varshavsky, A. (2005). Heat-inducible degron and the making of conditional mutants. Methods Enzymol. 399, 799–822.

Draviam, V.M., Stegmeier, F., Nalepa, G., Sowa, M.E., Chen, J., Liang, A., Hannon, G.J., Sorger, P.K., Harper, J.W., and Elledge, S.J. (2007). A functional genomic screen identifies a role for TAO1 kinase in spindle-checkpoint signalling. Nat Cell Biol 9, 556–64.

Elion, E.A. (2001). The Ste5p scaffold. J. Cell Sci. 114, 3967–3978.

Emanuele, M., Burke, D.J., and Stukenberg, P.T. (2007). A Hec of a microtubule attachment. Nat Struct Mol Biol 14, 11–3.

Emanuele, M.J., McCleland, M.L., Satinover, D.L., and Stukenberg, P.T. (2005). Measuring the Stoichiometry and Physical Interactions between Components Elucidates the Architecture of the Vertebrate Kinetochore. Mol Biol Cell 16, 4882–4892.

Fang, G. (2002). Checkpoint protein BubR1 acts synergistically with Mad2 to inhibit anaphase-promoting complex. Mol Biol Cell 13, 755–66.

Fang, G., Yu, H., and Kirschner, M.W. (1998). The checkpoint protein MAD2 and the mitotic regulator CDC20 form a ternary complex with the anaphase-promoting complex to control anaphase initiation. Genes Dev 12, 1871–83.

Fraschini, R., Beretta, A., Sironi, L., Musacchio, A., Lucchini, G., and Piatti, S. (2001). Bub3 interaction with Mad2, Mad3 and Cdc20 is mediated by WD40 repeats and does not require intact kinetochores. EMBO J. 20, 6648–6659.

Gardner, R.D., Poddar, A., Yellman, C., Tavormina, P.A., Monteagudo, M.C., and Burke, D.J. (2001). The spindle checkpoint of the yeast Saccharomyces cerevisiae requires kinetochore function and maps to the CBF3 domain. Genetics 157, 1493–1502.

Gillett, E.S., Espelin, C.W., and Sorger, P.K. (2004). Spindle checkpoint proteins and chromosome-microtubule attachment in budding yeast. J Cell Biol 164, 535–546.

Goh, P.Y. and Kilmartin, J.V. (1993). NDC10: a gene involved in chromosome segregation in Saccharomyces cerevisiae. J Cell Biol 121, 503–12.

Goshima, G., Kiyomitsu, T., Yoda, K., and Yanagida, M. (2003). Human centromere chromatin protein hMis12, essential for equal segregation, is independent of CENP-A loading pathway. J Cell Biol 160, 25–39.

Hardwick, K.G., Johnston, R.C., Smith, D.L., and Murray, A.W. (2000). MAD3 encodes a novel component of the spindle checkpoint which interacts with Bub3p, Cdc20p, and Mad2p. J. Cell Biol. 148, 871–882.

Hartwell, L.H. (1978). Cell division from a genetic perspective. J Cell Biol 77, 627–637.

Hartwell, L.H. and Weinert, T.A. (1989). Checkpoints: controls that ensure the order of cell cycle events. Science 246, 629–634.

Hauf, S., Cole, R.W., LaTerra, S., Zimmer, C., Schnapp, G., Walter, R., Heckel, A., van Meel, J., Rieder, C.L., and Peters, J.M. (2003). The small molecule Hesperadin reveals a role for Aurora B in correcting kinetochore-microtubule attachment and in maintaining the spindle assembly checkpoint. J Cell Biol 161, 281–94.

He, X., Rines, D.R., Espelin, C.W., and Sorger, P.K. (2001). Molecular analysis of kinetochore-microtubule attachment in budding yeast. Cell 106, 195–206.

Howell, B.J., McEwen, B.F., Canman, J.C., Hoffman, D.B., Farrar, E.M., Rieder, C.L., and Salmon, E.D. (2001). Cytoplasmic dynein/dynactin drives kinetochore protein transport to the spindle poles and has a role in mitotic spindle checkpoint inactivation. J Cell Biol 155, 1159–72.

Howell, B.J., Moree, B., Farrar, E.M., Stewart, S., Fang, G., and Salmon, E.D. (2004). Spindle checkpoint protein dynamics at kinetochores in living cells. Curr Biol 14, 953–964.

Hoyt, M.A., Totis, L., and Roberts, B.T. (1991). S. cerevisiae genes required for cell cycle arrest in response to loss of microtubule function. Cell 66, 507–517.

Hwang, L.H., Lau, L.F., Smith, D.L., Mistrot, C.A., Hardwick, K.G., Hwang, E.S., Amon, A., and Murray, A.W. (1998). Budding yeast Cdc20: a target of the spindle checkpoint. Science 279, 1041–1044.

Indjeian, V.B., Stern, B.M., and Murray, A.W. (2005). The centromeric protein Sgo1 is required to sense lack of tension on mitotic chromosomes. Science 307, 130–133.

Jacobs, C.W., Adams, A.E., Szaniszlo, P.J., and Pringle, J.R. (1988). Functions of microtubules in the Saccharomyces cerevisiae cell cycle. J Cell Biol 107, 1409–1426.

Janke, C., Ortiz, J., Lechner, J., Shevchenko, A., Shevchenko, A., Magiera, M.M., Schramm, C., and Schiebel, E. (2001). The budding yeast proteins Spc24p and Spc25p interact with Ndc80p and Nuf2p at the kinetochore and are important for kinetochore clustering and checkpoint control. EMBO J 20, 777–791.

Jones, M.H., Huneycutt, B.J., Pearson, C.G., Zhang, C., Morgan, G., Shokat, K., Bloom, K., and Winey, M. (2005). Chemical genetics reveals a role for Mps1 kinase in kinetochore attachment during mitosis. Curr Biol 15, 160–165.

Kallio, M.J., Beardmore, V.A., Weinstein, J., and Gorbsky, G.J. (2002a). Rapid microtubule-independent dynamics of Cdc20 at kinetochores and centrosomes in mammalian cells. J Cell Biol 158, 841–847.

Kallio, M.J., McCleland, M.L., Stukenberg, P.T., and Gorbsky, G.J. (2002b). Inhibition of aurora B kinase blocks chromosome segregation, overrides the spindle checkpoint, and perturbs microtubule dynamics in mitosis. Curr Biol 12, 900–5.

King, E.M., Rachidi, N., Morrice, N., Hardwick, K.G., and Stark, M.J. (2007). Ipl1p-dependent phosphorylation of Mad3p is required for the spindle checkpoint response to lack of tension at kinetochores. Genes Dev 21(10), 1163–8.

Kitagawa, K., Abdulle, R., Bansal, P.K., Cagney, G., Fields, S., and Hieter, P. (2003). Requirement of Skp1-Bub1 interaction for kinetochore-mediated activation of the spindle checkpoint. Mol Cell 11, 1201–1213.

Kitajima, T.S., Hauf, S., Ohsugi, M., Yamamoto, T., and Watanabe, Y. (2005). Human Bub1 defines the persistent cohesion site along the mitotic chromosome by affecting Shugoshin localization. Curr Biol 15, 353–9.

Kitajima, T.S., Kawashima, S.A., and Watanabe, Y. (2004). The conserved kinetochore protein shugoshin protects centromeric cohesion during meiosis. Nature 427, 510–7.

Kiyomitsu, T., Obuse, C., and Yanagida, M. (2007). Human Blinkin/AF15q14 is required for chromosome alignment and the mitotic checkpoint through direct interaction with Bub1 and BubR1. Dev Cell 13(5), 663–76.

Kops, G.J., Kim, Y., Weaver, B.A., Mao, Y., McLeod, I., Yates, J.R.3., Tagaya, M., and Cleveland, D.W. (2005). ZW10 links mitotic checkpoint signaling to the structural kinetochore. J Cell Biol 169, 49–60.

Kosco, K.A., Pearson, C.G., Maddox, P.S., Wang, P.J., Adams, I.R., Salmon, E.D., Bloom, K., and Huffaker, T.C. (2001). Control of microtubule dynamics by Stu2p is essential for spindle orientation and metaphase chromosome alignment in yeast. Mol Biol Cell 12, 2870–2880.

Lew, D.J. and Burke, D.J. (2003). The spindle assembly and spindle position checkpoints. Annu Rev Genet 37, 251–282.

Li, R. and Murray, A.W. (1991). Feedback control of mitosis in budding yeast. Cell 66, 519–531.

Li, X. and Nicklas, R.B. (1995). Mitotic forces control a cell-cycle checkpoint. Nature 373, 630–632.

Li, Y. and Benezra, R. (1996). Identification of a human mitotic checkpoint gene: hsMAD2. Science 274, 246–8.

Liu, S.T., Chan, G.K., Hittle, J.C., Fujii, G., Lees, E., and Yen, T.J. (2003). Human MPS1 kinase is required for mitotic arrest induced by the loss of CENP-E from kinetochores. Mol Biol Cell 14, 1638–51.

Lou, Y., Yao, J., Zereshki, A., Dou, Z., Ahmed, K., Wang, H., Hu, J., Wang, Y., and Yao, X. (2004). NEK2A interacts with MAD1 and possibly functions as a novel integrator of the spindle checkpoint signaling. J Biol Chem 279, 20049–57.

Lucchini, G., Falconi, M.M., Pizzagalli, A., Aguilera, A., Klein, H.L., and Plevani, P. (1990). Nucleotide sequence and characterization of temperature-sensitive pol1 mutants of Saccharomyces cerevisiae. Gene 90, 99–104.

Luo, X., Fang, G., Coldiron, M., Lin, Y., Yu, H., Kirschner, M.W., and Wagner, G. (2000). Structure of the Mad2 spindle assembly checkpoint protein and its interaction with Cdc20. Nat Struct. Biol. 7, 224–229.

Luo, X., Tang, Z., Rizo, J., and Yu, H. (2002). The Mad2 spindle checkpoint protein undergoes similar major conformational changes upon binding to either Mad1 or Cdc20. Mol Cell 9, 59–71.

Mao, Y., Abrieu, A., and Cleveland, D.W. (2003). Activating and silencing the mitotic checkpoint through CENP-E-dependent activation/inactivation of BubR1. Cell 114, 87–98.

Mao, Y., Desai, A., and Cleveland, D.W. (2005). Microtubule capture by CENP-E silences BubR1-dependent mitotic checkpoint signaling. J Cell Biol 170, 873–80.

Martin-Lluesma, S., Stucke, V.M., and Nigg, E.A. (2002). Role of Hec1 in spindle checkpoint signaling and kinetochore recruitment of Mad1/Mad2. Science 297, 2267–70.

McAinsh, A.D., Meraldi, P., Draviam, V.M., Toso, A., and Sorger, P.K. (2006). The human kinetochore proteins Nnf1R and Mcm21R are required for accurate chromosome segregation. Embo J 25, 4033–49.

McAinsh, A.D., Tytell, J.D., and Sorger, P.K. (2003). Structure, function, and regulation of budding yeast kinetochores. Annu Rev Cell Dev Biol 19, 519–539.

McCarroll, R.M. and Fangman, W.L. (1988). Time of replication of yeast centromeres and telomeres. Cell 54, 505–513.

McCleland, M.L., Gardner, R.D., Kallio, M.J., Daum, J.R., Gorbsky, G.J., Burke, D.J., and Stukenberg, P.T. (2003). The highly conserved Ndc80 complex is required for kinetochore assembly, chromosome congression, and spindle checkpoint activity. Genes Dev 17, 101–114.

McCleland, M.L., Kallio, M.J., Barrett-Wilt, G.A., Kestner, C.A., Shabanowitz, J., Hunt, D. F., Gorbsky, G.J., and Stukenberg, P.T. (2004). The vertebrate Ndc80 complex contains Spc24 and Spc25 homologs, which are required to establish and maintain kinetochore-microtubule attachment. Curr Biol 14, 131–7.

McIntosh, J.R. (1991). Structural and mechanical control of mitotic progression. Cold Spring Harb Symp Quant Biol 56, 613–619.

Meraldi, P., Draviam, V.M., and Sorger, P.K. (2004). Timing and checkpoints in the regulation of mitotic progression. Dev Cell 7, 45–60.

Meraldi, P. and Sorger, P.K. (2005). A dual role for Bub1 in the spindle checkpoint and chromosome congression. Embo J 24, 1621–33.

Minshull, J., Sun, H., Tonks, N.K., and Murray, A.W. (1994). A MAP kinase-dependent spindle assembly checkpoint in *Xenopus* egg extracts. Cell 79, 475–86.

Mizuguchi, G., Xiao, H., Wisniewski, J., Smith, M.M., and Wu, C. (2007). Nonhistone Scm3 and histones CenH3-H4 assemble the core of centromere-specific nucleosomes. Cell 129(6), 1153–64.

Murray, D., Mirzayans, R., and Chen, R.H. (1999). Spindle checkpoint protein Xmad1 recruits Xmad2 to unattached kinetochores. Br J Cancer 81, 959–65.

Musacchio, A. and Hardwick, K.G. (2002). The spindle checkpoint: structural insights into dynamic signalling. Nat Rev Mol Cell Biol 3, 731–741.

Musacchio, A. and Salmon, E.D. (2007). The spindle-assembly checkpoint in space and time. Nat Rev Mol Cell Biol 8, 379–393.

Nasmyth, K. (2002). Segregating sister genomes: the molecular biology of chromosome separation. Science 297, 559–565.

Nekrasov, V.S., Smith, M.A., Peak-Chew, S., and Kilmartin, J.V. (2003). Interactions between centromere complexes in Saccharomyces cerevisiae. Mol Biol Cell 14, 4931–4946.

Nicklas, R.B., Ward, S.C., and Gorbsky, G.J. (1995). Kinetochore chemistry is sensitive to tension and may link mitotic forces to a cell cycle checkpoint. J Cell Biol 130, 929–939.

Pinsky, B.A., Kung, C., Shokat, K.M., and Biggins, S. (2006). The Ipl1-Aurora protein kinase activates the spindle checkpoint by creating unattached kinetochores. Nat Cell Biol 8, 78–83.

Poddar, A., Stukenberg, P.T., and Burke, D.J. (2005). Two complexes of spindle checkpoint proteins containing Cdc20 and Mad2 assemble during mitosis independently of the kinetochore in Saccharomyces cerevisiae. Eukaryot Cell 4, 867–878.

Putkey, F.R., Cramer, T., Morphew, M.K., Silk, A.D., Johnson, R.S., McIntosh, J.R., and Cleveland, D.W. (2002). Unstable kinetochore-microtubule capture and chromosomal instability following deletion of CENP-E. Dev Cell 3, 351–65.

Qi, W., Tang, Z., and Yu, H. (2006). Phosphorylation- and polo-box-dependent binding of Plk1 to Bub1 is required for the kinetochore localization of Plk1. Mol Biol Cell 17, 3705–16.

Rieder, C.L., Cole, R.W., Khodjakov, A., and Sluder, G. (1995). The checkpoint delaying anaphase in response to chromosome monoorientation is mediated by an inhibitory signal produced by unattached kinetochores. J Cell Biol 130, 941–948.

Robbins, A.R., Jablonski, S.A., Yen, T.J., Yoda, K., Robey, R., Bates, S.E., and Sackett, D.L. (2005). Inhibitors of histone deacetylases alter kinetochore assembly by disrupting pericentromeric heterochromatin. Cell Cycle 4, 717–26.

Rosasco-Nitcher, S.E., Lan, W., Khorasanizadeh, S., Stukenberg, P.T. (2008) Centromeric Aurora-B activation requires TD-60, microtubules, and substrate priming phosphorylation. Science 319(5862), 469–7

Sharp-Baker, H. and Chen, R.H. (2001). Spindle checkpoint protein Bub1 is required for kinetochore localization of Mad1, Mad2, Bub3, and CENP-E, independently of its kinase activity. J Cell Biol 153, 1239–50.

Siller, K.H., Serr, M., Steward, R., Hays, T.S., and Doe, C.Q. (2005). Live imaging of *Drosophila* brain neuroblasts reveals a role for Lis1/dynactin in spindle assembly and mitotic checkpoint control. Mol Biol Cell 16, 5127–40.

Sironi, L., Mapelli, M., Knapp, S., De Antoni, A., Jeang, K.T., and Musacchio, A. (2002). Crystal structure of the tetrameric Mad1-Mad2 core complex: implications of a 'safety belt' binding mechanism for the spindle checkpoint. EMBO J 21, 2496–2506.

Skoufias, D.A., Andreassen, P.R., Lacroix, F.B., Wilson, L., and Margolis, R.L. (2001). Mammalian mad2 and bub1/bubR1 recognize distinct spindle-attachment and kinetochore-tension checkpoints. Proc Natl Acad Sci USA 98, 4492–7.

Starr, D.A., Williams, B.C., Hays, T.S., and Goldberg, M.L. (1998). ZW10 helps recruit dynactin and dynein to the kinetochore. J Cell Biol 142, 763–74.

Stern, B.M. and Murray, A.W. (2001). Lack of tension at kinetochores activates the spindle checkpoint in budding yeast. Curr Biol 11, 1462–1467.

Stoler, S., Keith, K.C., Curnick, K.E., and Fitzgerald-Hayes, M. (1995). A mutation in CSE4, an essential gene encoding a novel chromatin-associated protein in yeast, causes chromosome nondisjunction and cell cycle arrest at mitosis. Genes Dev 9(5), 573–86.

Stoler, S., Rogers, K., Weitze, S., Morey, L., Fitzgerald-Hayes, M., Baker, R.E. (2007). Scm3, an essential Saccharomyces cerevisiae centromere protein required for G2/M progression and Cse4 localization. Proc Natl Acad Sci U S A 104(25), 10571–6 (Epub 2007 Jun 4).

Stucke, V.M., Baumann, C., and Nigg, E.A. (2004). Kinetochore localization and microtubule interaction of the human spindle checkpoint kinase Mps1. Chromosoma 113, 1–15.

Stucke, V.M., Sillje, H.H., Arnaud, L., and Nigg, E.A. (2002). Human Mpsl kinase is required for the spindle assembly checkpoint but not for centrosome duplication. Embo J 21, 1723–32.

Sudakin, V. and Yen, T.J. (2004). Purification of the mitotic checkpoint complex, an inhibitor of the APC/C from HeLa cells. Methods Mol Biol 281, 199–212.

Tanaka, T.U., Rachidi, N., Janke, C., Pereira, G., Galova, M., Schiebel, E., Stark, M.J., and Nasmyth, K. (2002). Evidence that the Ipl1-Sli15 (Aurora kinase-INCENP) complex promotes chromosome bi-orientation by altering kinetochore-spindle pole connections. Cell 108, 317–329.

Tang, Z., Shu, H., Oncel, D., Chen, S., and Yu, H. (2004). Phosphorylation of Cdc20 by Bub1 provides a catalytic mechanism for APC/C inhibition by the spindle checkpoint. Mol. Cell 16, 387–397.

Tavormina, P.A. and Burke, D.J. (1998). Cell cycle arrest in cdc20 mutants of Saccharomyces cerevisiae is independent of Ndc10p and kinetochore function but requires a subset of spindle checkpoint genes. Genetics 148, 1701–1713.

Taylor, S.S., Ha, E., and McKeon, F. (1998). The human homologue of Bub3 is required for kinetochore localization of Bub1 and a Mad3/Bub1-related protein kinase. J Cell Biol 142, 1–11.

Taylor, S.S. and McKeon, F. (1997). Kinetochore localization of murine Bub1 is required for normal mitotic timing and checkpoint response to spindle damage. Cell 89, 727–35.

Vader, G., Cruijsen, C.W., van Harn, T., Vromans, M.J., Medema, R.H., Lens, S.M. (2007). The chromosomal passenger complex controls spindle checkpoint function independent from its role in correcting microtubule kinetochore interactions. Mol Biol Cell 18(11), 4553–64 (Epub 2007 Aug 15).

Vink, M., Simonetta, M., Transidico, P., Ferrari, K., Mapelli, M., De Antoni, A., Massimiliano, L., Ciliberto, A., Faretta, M., Salmon, E.D., and Musacchio, A. (2006). In vitro FRAP identifies the minimal requirements for Mad2 kinetochore dynamics. Curr Biol 16, 755–766.

Wang, H., Hu, X., Ding, X., Dou, Z., Yang, Z., Shaw, A.W., Teng, M., Cleveland, D.W., Goldberg, M.L., Niu, L., and Yao, X. (2004). Human Zwint-1 specifies localization of Zeste White 10 to kinetochores and is essential for mitotic checkpoint signaling. J Biol Chem 279, 54590–8.

Waters, J.C., Chen, R.H., Murray, A.W., Gorbsky, G.J., Salmon, E.D., and Nicklas, R.B. (1999). Mad2 binding by phosphorylated kinetochores links error detection and checkpoint action in mitosis. Curr Biol 9, 649–652.

Waters, J.C., Chen, R.H., Murray, A.W., and Salmon, E.D. (1998). Localization of Mad2 to kinetochores depends on microtubule attachment, not tension. J Cell Biol 141, 1181–91.

Weaver, B.A., Bonday, Z.Q., Putkey, F.R., Kops, G.J., Silk, A.D., and Cleveland, D.W. (2003). Centromere-associated protein-E is essential for the mammalian mitotic checkpoint to prevent aneuploidy due to single chromosome loss. J Cell Biol 162, 551–63.

Wei, R.R., Al-Bassam, J., and Harrison, S.C. (2007). The Ndc80/HEC1 complex is a contact point for kinetochore-microtubule attachment. Nat Struct Mol Biol 14, 54–9.

Weinert, T.A. and Hartwell, L.H. (1988). The RAD9 gene controls the cell cycle response to DNA damage in Saccharomyces cerevisiae. Science 241, 317–322.

Weiss, E. and Winey, M. (1996). The Saccharomyces cerevisiae spindle pole body duplication gene MPS1 is part of a mitotic checkpoint. J Cell Biol 132, 111–123.

Westermann, S., Cheeseman, I.M., Anderson, S., Yates, J.R., III, Drubin, D.G., and Barnes, G. (2003). Architecture of the budding yeast kinetochore reveals a conserved molecular core. J Cell Biol 163, 215–222.

Wong, O.K. and Fang, G. (2005). Plx1 is the 3F3/2 kinase responsible for targeting spindle checkpoint proteins to kinetochores. J Cell Biol 170, 709–19.

Yamamoto, A., Guacci, V., and Koshland, D. (1996). Pds1p, an inhibitor of anaphase in budding yeast, plays a critical role in the APC and checkpoint pathway(s). J Cell Biol 133, 99–110.

Zecevic, M., Catling, A.D., Eblen, S.T., Renzi, L., Hittle, J.C., Yen, T.J., Gorbsky, G.J., and Weber, M.J. (1998). Active MAP kinase in mitosis: localization at kinetochores and association with the motor protein CENP-E. J Cell Biol 142, 1547–58.

Zhao, Y. and Chen, R.H. (2006). Mps1 phosphorylation by MAP kinase is required for kinetochore localization of spindle-checkpoint proteins. Curr Biol 16, 1764–9.

Zirkle, R.E. (1970). Ultraviolet-microbeam irradiation of newt-cell cytoplasm: spindle destruction, false anaphase, and delay of true anaphase. Radiat Res 41, 516–537.

# Chapter 12
# Kinetochore Regulation of Anaphase and Cytokinesis

**Scott Thomas and Kenneth B. Kaplan**

**Abstract** Since the initial discovery of kinetochore antigens, a plethora of kinetochore-associated proteins have been identified. Their roles in connecting chromosomes to the mitotic spindle have been the subject of intense investigation. However, a surprising number of kinetochore proteins perform non-centromeric functions during mitosis. This class of kinetochore proteins has been best characterized through studies of the so-called "chromosomal passengers," proteins that associate with kinetochores at the start of mitosis and then redistribute to the anaphase spindle. The conserved behavior of chromosomal passengers suggests that redistribution of kinetochore-associated proteins is a commonly used strategy for cells to temporally and spatially orchestrate mitotic events. The activities of chromosomal passengers are closely linked to cell cycle regulation, placing them in a position to transmit regulatory changes to the cell division machinery. As cells enter mitosis, chromosomal passengers alter chromatin organization. At kinetochores, they ensure that sister chromatids form proper attachments with the mitotic spindle. During anaphase, they organize spindle structures to direct the cytokinetic machinery. In this chapter, we will discuss the expanding role for chromosomal passengers in regulating anaphase events and how the redistribution of other kinetochore-associated proteins might contribute to the orderly progression of mitosis.

## 12.1 Introduction

Chromosomal passenger proteins associate with mitotic chromatin and kinetochores early in mitosis and then redistribute to the mitotic spindle in anaphase—hence their description as "passengers" riding on chromosomes to their anaphase spindle destination. This chapter will highlight new insights into chromosomal passengers from studies in budding yeast and integrate these observations with

Kenneth B. Kaplan (✉)
Department of Molecular and Cellular Biology, University of California, Davis, One Shields Ave., Davis, CA 95616, U.S.A.
e-mail: kbkaplan@ucdavis.edu

P. De Wulf, W.C. Earnshaw (eds.), *The Kinetochore*, 
DOI 10.1007/978-0-387-69076-6_12, 

the classic studies of chromosomal passengers in other systems. We will discuss how a surprisingly large number of kinetochore-associated proteins may assemble into multiple regulatory complexes that control discrete anaphase events. Finally, we will present a model that describes how chromosomal passengers link the regulators of mitosis to the orderly mechanical events of cell division.

### *12.1.1 Regulation of Anaphase Events*

In considering how kinetochores influence events in anaphase, it is useful to first review the events of anaphase. The process of cell division represents a well-orchestrated series of molecular and biochemical events that result in the mechanical segregation of chromosomes and other cellular organelles. Although the transformation of an interphase cell to a mitotic cell requires the large-scale modification of proteins by cyclin-dependent kinases (CDKs), it is the orderly reversal of these transient modifications that contributes to the progression of chromosome segregation, cytokinesis, and the return of cells to G1 (i.e., mitotic exit). Cells use at least two mechanisms to reverse CDK–cyclin modifications in anaphase: (i) multiple ubiquitin-modifying enzymes, including the anaphase promoting complex (APC), contribute to the destruction of mitotic cyclins as well as CDK–cyclin substrates and (ii) cellular phosphatases reverse mitotic phosphorylations as CDK–cyclin activity decreases in anaphase. Although there are many details that are not understood, current evidence suggests that the degradation of mitotic proteins via the APC is tightly coupled to the dephosphorylation of CDK-substrates. Here, we will briefly discuss these pathways as they relate to chromosomal passenger proteins but a more complete discussion of mitotic exit is available in several excellent reviews (Morgan, 1999; Sullivan and Morgan, 2007).

Multiple mitotic CDK–cyclin complexes contribute to the orderly assembly of the mitotic apparatus (e.g., chromosome condensation, nuclear envelope breakdown, spindle formation) by modifying substrates in specific subcellular compartments. For example, CDK1-cyclin B is targeted to the nucleus only during late prophase to modify nuclear lamins and induce nuclear envelope breakdown. A similar spatial and temporal regulation of CDK–cyclin complexes and their substrates is likely to govern the orderly events associated with anaphase. A dramatic example of this type of regulation involves the coordinated destruction of securin (Pds1 in budding yeast) and B-type cyclins by the $APC^{cdc20}$ ubiquitin-ligase (Cohen-Fix and Koshland, 1999; Shirayama et al., 1999; Tinker-Kulberg and Morgan, 1999). Securin inhibits the activity of separase (Esp1 in yeast), the protease that releases sister-chromatid cohesion, and its destruction allows sisters to be untethered. To ensure the efficient segregation of sisters, cyclin B must also be degraded so that Aurora B, which destabilizes kinetochore–microtubule attachments in metaphase, can be redistributed from kinetochores to the anaphase spindle (Parry et al., 2003; Parry

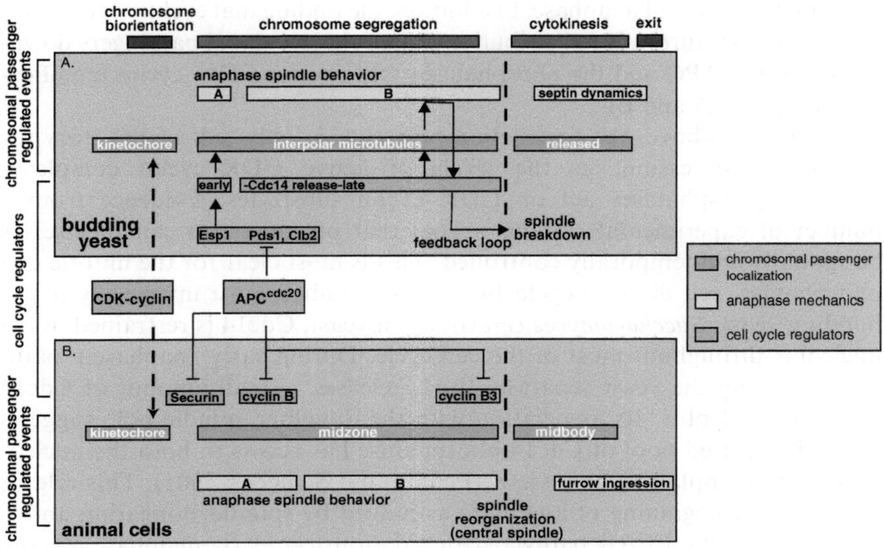


**Fig. 12.1** The relationship between cell cycle regulation, chromosomal passenger localization and cell division. The mitotic cyclin-dependent kinase (CDK) complexes and the anaphase promoting complex ($APC^{cdc20}$) are common master regulators of anaphase progression in both budding yeast (A) and animal cells (B). The cell cycle regulators are denoted by green bars, the localization changes associated with chromosomal passengers by blue bars and the mechanical events of anaphase by yellow bars. (**A**) In budding yeast, $APC^{cdc20}$ works via Esp1 (Separase) to control the spatial and temporal release of the Cdc14 protein phosphatase. Together with cyclin destruction, Cdc14 coordinates the events of anaphase, in part by relocalizing chromosomal passengers to the anaphase spindle. The arrows and bar-lines indicate genetic or biochemical interactions between the indicated gene products or anaphase events. The arrows labeled feedback loop suggest a regulatory scheme where Cdc14 release is required for proper localization of chromosomal passengers to the anaphase spindle; chromosomal passengers modulate the rate of spindle elongation which in turn feeds back to change the activity of Cdc14 in regulating spindle breakdown. (**B**) In animal cells, the coordination of anaphase events is driven mainly by the coordinated destruction of mitotic cyclins. As in budding yeast, a reversal of mitotic modifications allows for the change in chromosomal passenger localization and the coordination of anaphase events

and O'Farrell, 2001). On the other hand, the degradation of cyclin B3 is delayed in anaphase relative to the degradation of securin and cyclin B, as its destruction is required for spindle breakdown (Fig. 12.1B). Thus, by carefully controlling when each substrate is modified with ubiquitin, the APC ensures that sister-chromatid cohesion is released even as kinetochore–microtubule attachments and the anaphase spindle are stabilized. Consequently, forces at kinetochores can efficiently segregate chromosomes to opposite ends of the cell. While not all of the relevant APC substrates are known, the temporal control of securin and cyclin destruction provides insight into how this process can contribute to the coordinated events of anaphase. In addition, the finding that cyclin degradation is required for Aurora B redistribution places chromosomal passengers downstream of the APC and the phosphatases that reverse CDK–cyclin modifications (Fig. 12.1A and B).

As implied above, the destruction of cyclins is only half of the story for anaphase progression; as the levels of active CDK–cyclin complexes decrease, phosphatases act on CDK–cyclin substrates. Evidence from a number of experimental systems argues that phosphatases can themselves be spatially and temporally controlled. This is most clear for the mitotic exit phosphatase, cell division cycle 14 (Cdc14), studied most intensively in the budding yeast, *Saccharomyces cerevisiae*. In yeast, Cdc14 is restrained in the nucleolus throughout most of the cell cycle. During early anaphase a pathway involving the yeast separase, Esp1, releases a small amount of Cdc14 from the nucleolus. Its association with the daughter spindle pole suggests that this released pool of Cdc14 phosphatase has access to both the nucleoplasm and cytoplasm (for review, (Pereira and Schiebel, 2001). This release occurs at the beginning of anaphase as judged by spindle elongation and is referred to as the FEAR pathway (for Cdc-*f*ourteen-*e*arly *a*naphase *r*elease; Dumitrescu and Saunders, 2002). The release of the remaining nucleolar Cdc14 is under control of a complex signaling pathway, known as the *m*itotic *e*xit *n*etwork (MEN), that restrains completion of anaphase until the nucleus and thus the segregating chromosomes have been positioned at the bud neck for proper segregation to mother and daughter cells (Bardin and Amon, 2001).

In yeast, it is clear that the spatial and temporal control of the Cdc14 phosphatase directly affects the function of chromosomal passengers in anaphase (Fig. 12.1A). The *in*ner *cen*tromeric *p*rotein (INCENP) homolog, Sli15, is phosphorylated early in mitosis and then dephosphorylated at the start of anaphase in a Cdc14-dependent manner (Pereira and Schiebel, 2003). As we will discuss in detail below, dephosphorylation of Sli15 appears to be required for redistributing chromosomal passengers during anaphase and in this way Sli15 phosphorylation can regulate both kinetochore–microtubule attachment as well as anaphase spindle elongation. It is less clear if a similar regulatory phosphatase dominates the timing of anaphase in animal cells. CDC14A in mammals and Xenopus is required for proper centrosome behavior, chromosome congression, and has been implicated in the late cytokinetic event of abscission (Kaiser et al., 2002). Interestingly, nuclear export is also required for the action of CDC14A in regulating mitotic processes, suggesting that as in yeast, the spatial regulation of this phosphatase is important in metazoans (Mailand et al., 2002). Clearly, other phosphatases play an important role in anaphase but their precise impact on chromosomal passengers and anaphase progression has not been fully elucidated (Chiroli et al., 2007; Dobbelaere et al., 2003; Queralt et al., 2006; Tang and Wang, 2006; Trinkle-Mulcahy and Lamond, 2006; Vagnarelli et al., 2006).

### *12.1.2 Chromosomal Passengers and Mitosis*

Here, we will briefly outline the discovery of chromosomal passengers and how the ensuing body of research surrounding their function has contributed to our view of chromosome segregation and mitosis. We will first consider the conserved core components of the chromosomal passenger complex (CPC). The founding member of this core complex, INCENP, was discovered after generating monoclonal antibodies to chromosomal proteins (Cooke et al., 1987). INCENP was found to dynamically localize to various structures during mitosis. INCENP first localizes to chromosome arms in prometaphase and then concentrates between sister centromeres in metaphase. Upon anaphase onset INCENP redistributes to the central spindle, to a pericortical band at the cleavage site, and to the spindle midbody structure. Subsequent work identified the INCENP binding partner and kinase, Aurora B and the associated Survivin protein as part of a well-conserved core of chromosomal passenger proteins (more below and reviewed in: (Adams et al., 2001; Vagnarelli and Earnshaw, 2004)). It was their dynamic localization that first led to the prescient model that chromosomal passenger proteins coordinate chromosome behavior by "assembling a functional spindle and positioning the cleavage furrow" (Earnshaw and Bernat, 1991). These observations also heralded a paradigm shift from considering chromosomes as passive cargo (Earnshaw and Bernat, 1991) to our current view of chromosomes as active regulators of the mitotic process.

### *12.1.3 Chromosomal Passengers Regulate Multiple Mitotic Events*

One implication from the dynamic localization of proteins like INCENP is that these proteins have multiple functions in mitosis. Experiments using different model systems have confirmed this view. The first, and perhaps most dramatic evidence, comes from the analysis of dominant negative INCENP mutants that displace the endogenous protein and cause defects in chromosome congression, chromosome segregation, and in the completion of cytokinesis (Mackay et al., 1998). Biochemical studies (see below) and the conservation of INCENP, Aurora B, and Survivin from metazoans to yeast, argues that these proteins have evolved to regulate mechanistically diverse mitoses. Indeed, inhibition studies in multiple experimental systems give rise to remarkably consistent defects in both metaphase and anaphase.

The conserved mitotic functions of chromosomal passengers raise the possibility that there is a common biochemical mechanism by which the chromosomal passenger complexes affect so many mitotic events. One possibility supported by structural studies of chromosomal passenger fragments is that the same stable core complex regulates multiple mitotic events (Jeyaprakash et al., 2007), perhaps by delivering the Aurora B kinase to different substrates. A second possibility is that there are multiple chromosomal passenger

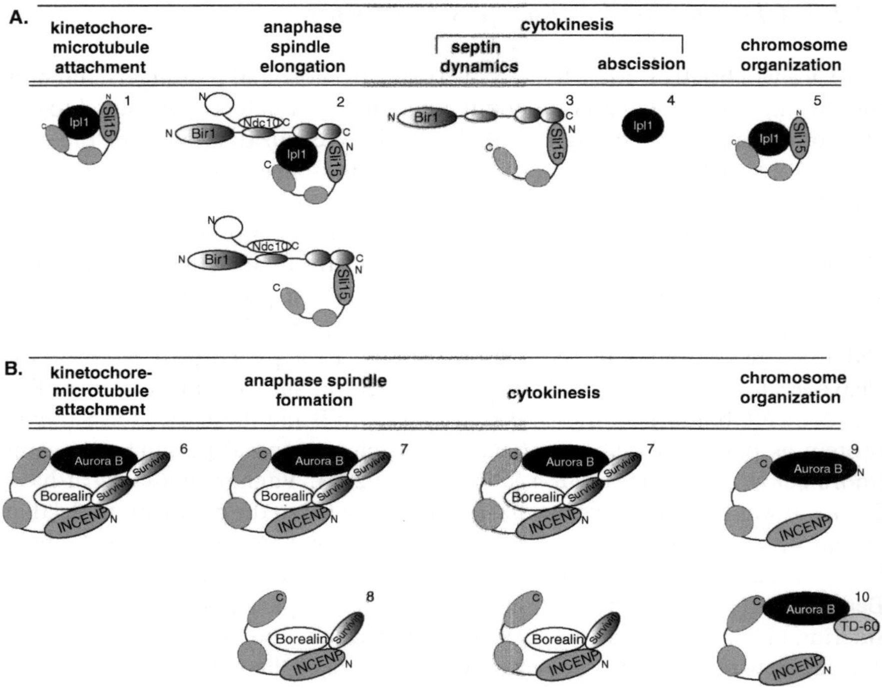


**Fig. 12.2** Chromosomal passenger "complexology": Chromosomal passenger complexes are cartooned based on data from studies in budding yeast (A) and in animal cells (B). The cartoons represent possible complexes and are based on both genetic and biochemical findings. The assignment of functional roles for each complex is based on inhibitor or mutant studies. The number notations to the right of each complex refer to the following supporting references: 1, (Biggins and Murray, 2001; Chan and Botstein, 1993); 2, Rozelle, Hansen and Kaplan; unpublished observations; 3, (Thomas and Kaplan, 2007); 4, (Norden et al., 2006); 5, (Hsu et al., 2000); 6, (Jelluma et al., 2008; Ruchaud et al., 2007); 7, (Cao et al., 2006; Eckley et al., 1997; Gassmann et al., 2004; Yang et al., 2004) and reviewed in (Adams et al., 2001; Ruchaud et al., 2007); 8, (Bourhis et al., 2007; Jeyaprakash et al., 2007); 9, (Adams et al., 2001; Gassmann et al., 2004; Klein et al., 2006); 10, (Rosasco-Nitcher et al., 2008) and as discussed in (Ruchaud et al., 2007)

complexes that each carry out distinct functions (Fig. 12.2). In this scenario, it may be that each sub-complex is sensitive to the overall levels of all the subunits, possibly because the sub-complexes exchange with each other. This latter possibility is supported by biochemical studies in mammalian cells, which show biochemically distinct sub-complexes (Gassmann et al., 2004; Klein et al., 2006). In addition, studies in yeast show that reducing the levels of one chromosomal passenger sub-complex reduces the levels of another functionally distinct sub-complex, suggesting that sub-complexes share or exchange subunits (Sandall et al., 2006; Thomas and Kaplan, 2007). Distinguishing between these possibilities will require more discriminating chromosome passenger mutants (e.g., separation of function alleles) and more attention to the full range of anaphase defects in mutants.

What are the events in anaphase that require temporal regulation? Considering this question from the perspective of the mitotic apparatus, it is important to maintain the integrity of the anaphase spindle to ensure that chromosomes are completely segregated prior to the induction of the cytokinetic furrow. However, a full answer to this question is dependent on the extent to which we understand anaphase events. Recent findings in yeast highlight the complexities of anaphase events affected by chromosomal passengers and suggest there is still much to learn. In the context of anaphase, our current understanding implicates chromosomal passengers, either directly or indirectly, in regulating chromosome condensation, septins, anillins, microtubules, actin as well as membrane deposition machinery (in yeast, cell wall deposition as well; (Glotzer, 2005; Lavoie et al., 2004; Norden et al., 2006; Thomas and Kaplan, 2007)). Remarkably, the precise interactions that link chromosomal passengers to these anaphase events are not understood. Early cytological studies have suggested that chromosomal passengers may affect many anaphase events through their interaction with the anaphase spindle midzone. For example, inhibition of Aurora B or INCENP in metazoans prevents proper organization of the central spindle, a structure required for the completion of cytokinesis (Severson et al., 2000). Studies in yeast have begun to shed more light on the specific anaphase targets of chromosomal passengers, although it remains an open question as to how well conserved these pathways are between yeast and animal cells. Ipl1, the Aurora B homolog in yeast, and the cytokinetic anillins are involved in inhibiting abscission in the presence of an intact spindle (Norden et al., 2006). Chromosomal passengers in budding yeast have also been implicated in the proper timing of septin filament dynamics. In budding yeast, septins form a ring thought to act as a diffusion barrier in order to concentrate the cytokinetic machinery at the bud neck (Dobbelaere and Barral, 2004). Late in telophase, septin rings become dynamic and disassemble after cytokinesis is complete. Mutants in chromosomal passengers prevent this disassembly, indicating that there is a link between chromosomal passengers and regulators of septin dynamics (Gillis et al., 2005; Thomas and Kaplan, 2007). Changes in chromosome organization are another important aspect of anaphase that may require careful temporal control. For example, ensuring that chromosomes decondense (i.e., return to their interphase state) only after anaphase is complete is likely to be important to prevent DNA damage that could result from the forces acting on chromosomes during mitosis. Even more specialized changes in chromosome organization may be under tight temporal regulation; in yeast chromosome XII contains the rDNA repeats and is the last part of the genome to be segregated. It is believed that this segregation is driven by chromosome condensation rather than by the mitotic spindle (i.e., compaction-generated rather than spindle-generated forces; for review (Strunnikov, 2005; Torres-Rosell et al., 2005)). Interestingly, the condensation-driven segregation of rDNA requires both Cdc14 and Ipl1 and therefore places chromosomal passengers in a position to coordinate spindle elongation with the changes in chromatin necessary for completion of chromosome segregation. These examples are striking in that they consistently place chromosomal passengers downstream of APC and phosphatases to coordinate cell cycle regulators with the mechanical events of cell division (Fig. 12.1).

## 12.2 Catalog of Chromosomal Passenger Complexes

The literature on kinetochore proteins that play non-centromeric roles in anaphase has mainly focused on the well-characterized chromosomal passengers: INCENP, Aurora B, and Survivin and the less conserved Borealin and TD-60. However, recent studies, especially in budding yeast, have identified a surprisingly large number of kinetochore proteins that appear to function in anaphase independent of their role at kinetochores. One can divide kinetochore proteins that redistribute during mitosis into two general classes: (i) the classically characterized "core" chromosomal passenger proteins that localize to pre-anaphase chromosomes and kinetochores and then redistribute to the spindle during anaphase; this class appears to have very *distinct* activities in metaphase and anaphase and (ii), non-core chromosomal passengers that localize and function both at anaphase kinetochores and spindles; in some cases these proteins may interact with the core CPC. We will generally refer to both classes as chromosomal passengers and make distinctions between core and non-core chromosomal passengers where appropriate.

### *12.2.1 Core Chromosomal Passengers*

*INCENP–Aurora B–Survivin*: Historically, studies of kinetochore proteins that redistribute during anaphase have focused primarily on the chromosomal passengers. While their number varies between yeasts and metazoans, the conserved core forms a complex consisting of INCENP, Aurora B, and Survivin (Sli15, Ipl1 and Bir1 in budding yeast; for review, see Ruchaud et al., 2007; Vagnarelli and Earnshaw, 2004). The interaction between INCENP and Aurora B enhances the enzymatic activity of the Aurora B kinase and targets the kinase to specific mitotic structures (i.e., kinetochores and anaphase spindles, (Adams et al., 2000; Bishop and Schumacher, 2002). INCENP–Aurora B can assemble on its own (although this sub-complex may not be functional (Jeyaprakash et al., 2007)) or in a complex that also contains Survivin and Borealin ((Gassmann et al., 2004); see below and Fig. 12.2B). In animal cells, INCENP, Aurora B, Borealin, and Survivin depend on each other for localization and for their spindle related functions; however, this dependence is less clear for their association with chromosomes (Bourhis et al., 2007). In contrast, the yeast homologs appear to form a hierarchical complex whose localization to the anaphase spindle depends solely on the INCENP homolog, Sli15. However, even in yeast, biochemical and genetic findings indicate that there are distinct chromosomal passenger complexes that contribute differentially to anaphase progression (Fig. 12.2B; Thomas and Kaplan, 2007).

*TD-60, CSC-1, and Borealin*: Three additional chromosomal passengers have been identified in metazoans: *t*elophase *d*isc (TD)-60; Borealin, and *c*hromosome *s*egregation and *c*ytokinesis defective (CSC)-1. Cloning of human TD-60 showed it to have homology to a GTP exchange factor (GEF) and bind to the small G protein, Rac1, although its GEF activity remains unconfirmed (Mollinari et al., 2003). Studies in Xenopous extracts show that TD-60 is important for the localization and activation of Aurora B at centromeres in mitotic cells, arguing that TD-60 may have multiple biochemical activities (Rosasco-Nitcher et al., 2008). Borealin was discovered in HeLa cells, where it was shown to be important for the localization of Aurora B and is an Aurora B substrate (Gassmann et al., 2004). While not required for the enzymatic activity of Aurora B, structural studies indicate that Borealin is an obligate component of the core CPC. Taken together with its ability to bind DNA, it has been suggested that Borealin-containing chromosomal passenger complexes target the mitotic reorganization of chromosomes (Klein et al., 2006). CSC-1 has weak sequence homology to Borealin but has not been identified outside of *C. elegans*, suggesting it may be a nematode specific chromosomal passenger protein (Romano et al., 2003).

Although the polo kinase qualifies as a chromosomal passenger based on its localization pattern in animal cells and its interaction with INCENP, it regulates a multitude of mitotic targets and therefore defies simple categorization (Arnaud et al., 1998; Goto et al., 2006). Polo, however, does provide an important case in point, demonstrating that chromosomal passengers carry out their mitotic functions in the context of complex regulatory networks. Therefore, to fully understand the role of chromosomal passengers in anaphase, future efforts will need to consider how multiple regulatory pathways influence the targets of chromosomal passengers.

### *12.2.2 Non-Core Chromosomal Passengers (The CBF3 and DASH Complexes, and CENP-F)*

*Centromere-bound complexes*: Perhaps one of the more surprising kinetochore protein complexes to redistribute during anaphase is the budding yeast *c*entromere *b*inding *f*actor-*3* (CBF3) complex. Originally purified using its centromere (*CEN*)-DNA binding activity and later shown by genetics to be critical for kinetochore formation, CBF3 consists of three core proteins (Ctf13p, Cep3p, and Ndc10p) and associated assembly factors (Skp1p, Sgt1p, and Hsp90; for review, (McAinsh et al., 2003)). The observation that a *CEN*-DNA binding complex also associates with the anaphase spindle independent of chromosomes is a striking example of how kinetochore complexes can have two completely different functions during mitosis. Biochemical analysis shows that the Ndc10 subunit of CBF3 binds to Bir1 and is recruited to the anaphase spindle when Bir1 interacts with Sli15 (Fig. 12.2B and (Bouck and Bloom, 2005; Thomas and Kaplan, 2007; Widlund et al., 2006)). One distinguishing feature of this complex is that it does not leave kinetochores in anaphase, giving rise to kinetochore bound and anaphase spindle bound populations of CBF3 (see micrographs in Fig. 12.3C)—hence its classification as a non-core chromosomal passenger. It is noteworthy that CBF3 requires the help of an Hsp90–Sgt1 chaperone complex to assemble and bind to *CEN*-DNA

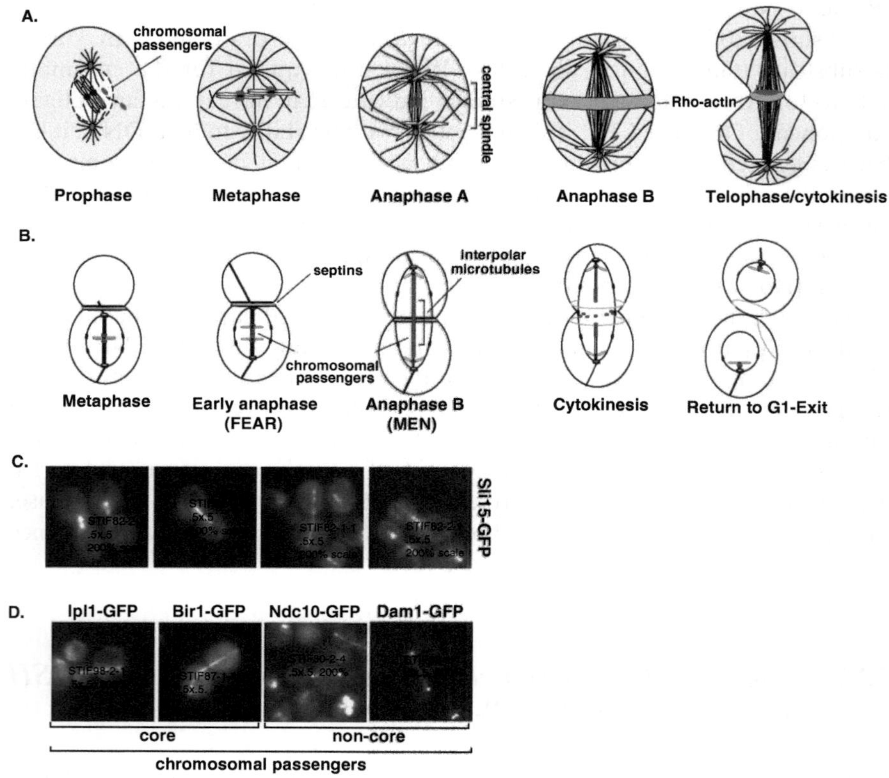


**Fig. 12.3** Chromosomal passengers during anaphase in animal cells and budding yeast. Chromosomal passenger localization is shown by the red ovals and related to each stage of mitosis for (**A**) animal cells and (**B**) budding yeast. (**C**) Examples of budding yeast expressing Sli15-GFP; images represent examples from different cells that show the temporal sequence of localization from metaphase to late telophase. (**D**) Representative images from budding yeast expressing the indicated chromosomal passenger protein fused to GFP. Note that the grey-scale values for the Dam1-GFP image were adjusted to highlight the anaphase spindle localization. (*See* Color Insert)

(Lingelbach and Kaplan, 2004; Stemmann et al., 2002). Mutants that interfere with the Hsp90–Sgt1-mediated assembly of CBF3 also give rise to anaphase defects, arguing that Hsp90–Sgt1 is required for both the kinetochore and anaphase roles of CBF3 (Gillis et al., 2005). This relationship raises the interesting possibility that Hsp90 is required to provide biochemical "plasticity" for CBF3 complexes, perhaps aiding their transition from metaphase to anaphase functions. This same Hsp90–Sgt1 chaperone complex is also required for proper kinetochore assembly in human cells (Niikura et al., 2006; Steensgaard et al., 2004). This observation together with the finding that Survivin is an Hsp90 client (Fortugno et al., 2003), argues that there is a conserved requirement for Hsp90 in mediating the multiple functions of core and non-core chromosomal passengers that redistribute in anaphase. While a function for centromeric binding proteins in animal cell anaphase is less clear, it is intriguing that the centromeric specific histone H3 variant, CENP-A, is a substrate of Aurora B and remains phosphorylated in anaphase (Zeitlin et al., 2001). Antibodies that detect the phospho-specific form of CENP-A have shown that the modified CENP-A goes to the midbody in anaphase. Mutants in the Aurora B phosphorylation sites in CENP-A dominantly interfere with cytokinesis (Li et al., 2007; Zeitlin et al., 2001). Together, these findings raise the possibility that the link between centromeric binding factors and anaphase regulation may be conserved between yeast and animal cells.

*Microtubule-associated passengers*: Although not traditionally discussed as "passengers," many microtubule-associated proteins undergo anaphase redistribution and in some cases, are directly implicated in core chromosomal passenger function. As these proteins can be found at both kinetochores and on the spindle in anaphase, they are classified as non-core chromosomal passengers. In some cases, these proteins presumably change localization to regulate different classes of microtubules during mitosis, first at the kinetochore and then at the anaphase spindle. In addition, some of these proteins may be downstream targets of the core CPC. In general, less is known about how these proteins influence anaphase events but they are important to consider in a discussion of kinetochore proteins that contribute to an orderly anaphase.

In yeast, the proper linkage between kinetochores and microtubules requires the multiprotein complex named, DASH (also known as the Dam1 complex). Consisting of ten components, this complex of proteins forms ring-like structures around the microtubule filament in vitro (Miranda et al., 2005, 2007; Westermann et al., 2006). Remarkably, the components of this complex also decorate the anaphase spindle while remaining at kinetochores throughout the completion of mitosis (Fig. 12.3D); in this sense, the DASH complex resembles the behavior of the CBF3 complex during mitosis. Mutations in the genes encoding for the subunits of the DASH complex result in kinetochore–microtubule attachment defects as well as loss of anaphase spindle integrity (Jones et al., 2001). Although little is known about the DASH complex in anaphase, it is possible that its role in binding to microtubules is important for organizing overlapping microtubules at the spindle midzone. An equivalent of the DASH complex has not been found in animal cells and it may be restricted to yeasts to compensate for the small number of kinetochore microtubules and overlapping midzone microtubules in anaphase.

In animal cells, recent work has shown that the kinetochore protein NudEL has a role in recruiting the microtubule-associated proteins Lis1, CLIP170 and the dynein–dynactin motor complex to kinetochores (Li et al., 2005; Stehman et al., 2007; Yan et al., 2003). Although studied in less detail in anaphase, NudEL is present at the midbody in telophase and the implication is that it recruits the same complex of proteins to the midbody as it does to the kinetochore. Interestingly, the outer kinetochore protein CENP-F has recently been shown to link the NudEL complex to kinetochores in metaphase. CENP-F is also found at the spindle midbody in anaphase (Rattner et al., 1993). While the functional relationship between CENP-F and NudEL in anaphase has not been directly studied, studies in *Drosophila* S2 cells show that dynein–dynactin is deposited at microtubule plus ends early in anaphase and is required to recruit other chromosomal passengers as well as the centralspindlin complex to the overlapping midzone microtubules (Delcros et al., 2006). Thus, dynein–dynactin, possibly together with CENP-F and NudEL, appears to regulate both kinetochore microtubules in metaphase and spindle midzone microtubules in anaphase.

In a number of systems, microtubule +TIPs (+ end *t*rack*i*ng *p*rotein) are found associated with kinetochores before anaphase and later at the plus ends of microtubules of the anaphase spindle. Although this class of proteins also acts outside of mitosis to regulate microtubule dynamics, they also appear to be involved in the ability of chromosomal passengers to regulate anaphase events. These include Bim1/EB1, adenomatous polyposis coli (APC), Stu2/XMAP215/Dis1 and Bik1/Clip170 (for review; (Akhmanova and Hoogenraad, 2005; Carvalho et al., 2003)). The fact that three of these +TIPs (Bim1, Stu2, and Bik1) interact in yeast argues that they function together (Wolyniak et al., 2006). Although best known for their role in regulating microtubule dynamics, it appears that +TIPs can also affect the "activity" of microtubule plus ends in mitosis. For example, inhibition of EB1 can reduce the efficiency of kinetochore–microtubule attachment and block checkpoint signals from unattached kinetochores (Draviam et al., 2006; Green et al., 2005). Similarly, during anaphase EB1 and other +TIPs are important for the activity of astral microtubule plus ends to regulate the cell cortex and induce a cytokinetic furrow (Rogers et al., 2004; Strickland et al., 2005). Most relevant to this discussion is the observation that at least some +TIPs appear to be downstream of the core chromosomal passenger complexes; EB1 dynamics require Survivin while Aurora B is required to maintain the dynamic state of astral microtubules necessary for proper cytokinesis (Miyauchi et al., 2007; Rosa et al., 2006). Although a full understanding of these relationships will require more analysis, it is possible that chromosomal passengers regulate microtubules in part through +TIPs.

## 12.3 Kinetochore and Chromosomal Passengers Regulation of Anaphase Mechanics

Our understanding of how the redistribution of chromosomal passengers regulates anaphase is incomplete. The large numbers of kinetochore-associated proteins that redistribute during anaphase point to a network that acts downstream of cell cycle regulators to spatially and temporally coordinate the anaphase cell division machinery (Fig. 12.3). In this section, we will summarize which anaphase event this network affects and how its design might contribute to an orderly anaphase.

### *12.3.1 Chromosome Segregation: Metaphase – Anaphase A*

As cells transition from metaphase to anaphase, sister cohesion is lost and kinetochores mediate the poleward movement of chromosomes in anaphase A. The metaphase activity of the Aurora B complex that detaches kinetochores from microtubules to ensure proper bivalent kinetochore–microtubule attachments must be turned off to preserve kinetochore–microtubule attachments in anaphase (DeLuca et al., 2006). In animal cells, this important transition has been separately reported to be under the control of the kinesin 6 family member, Mklp2 (Gruneberg et al., 2004) as well as the Cul3-based E-3 ubiquitin ligase (Sumara et al., 2007), which is required for the targeted removal of centromeric Aurora B (and Survivin but not of INCENP). It is not yet clear how the kinesin and E-3 activities are coordinated. One possibility is that the motor activity of Mklp2 is required to bring the E-3 ligase to the kinetochore to modify Aurora B as cells enter anaphase. It is likely that these two pathways lay downstream of cyclin degradation. In *Drosophila*, expression of stable cyclin B mutants prevents Aurora B from leaving kinetochores and give rise to lagging chromosomes in anaphase (i.e., unattached from spindle; (Parry et al., 2003)). Interestingly, lagging chromosomes were not reported when either Mklp2 or the Cul3 ubiquitin ligase is inhibited, suggesting that failing to degrade cyclin B affects more than just the removal of chromosomal passengers from kinetochores. In budding yeast, a similar loss of chromosomal passenger complexes from kinetochores takes place at the beginning of anaphase, as Ipl1 (Aurora B), Sli15 (INCENP) and Bir1 (Survivin) disappear from kinetochores at the start of anaphase (Fig. 12.3B–D). While the removal mechanism for Sli15 complexes in yeast is unclear, the reversal of mitotic phosphorylations on Sli15 by the Cdc14 phosphatase is required to localize complexes to the anaphase spindle (Pereira and Schiebel, 2003) where they are maintained by the action of other MEN regulators (Stoepel et al., 2005). Mutations that prevent phosphorylation of Sli15 in budding yeast cause an increase in chromosome missegregation indicating that premature spindle association compromises the fidelity of anaphase (Pereira and Schiebel, 2003). Together, these findings highlight the importance of the timely removal and spindle association of chromosomal passengers in all eukaryotes.

### *12.3.2 Spindle Integrity – Anaphase B*

Evidence from localization and functional studies suggest that chromosomal passengers can regulate anaphase spindle microtubules. The anaphase spindle is composed of multiple sets of microtubules that perform distinct roles. Microtubules that overlap at the midzone and contact the cell cortex are important for the sliding and pulling forces necessary for spindle elongation (for review, Scholey et al., 2003). In metazoans, anaphase microtubules also contact the cell cortex and serve a regulatory role by spatially directing actin polymerization. In particular, microtubules that assemble between segregating chromosomes, and are bundled together to form the midbody, play an important role in allowing the completion of the cytokinetic furrow (Fig. 12.3A). In contrast, yeasts have relatively simple midzones that are compromised of relatively few overlapping microtubules and cytoplasmic microtubules (Fig. 12.3B). Although the yeast anaphase spindle does not determine the plane of cell division, anaphase microtubules and chromosomal passengers are similarly important in ensuring the correct timing of cytokinesis.

Several observations directly implicate chromosomal passengers in the regulation of anaphase spindle behavior in eukaryotes. First, a large number of kinetochore-associated proteins redistribute to midzone spindle structures and inhibition of chromosomal passengers results in organizational changes of midzone microtubules. It is worth drawing a functional distinction between the effect of chromosomal passengers and microtubule cross-linking proteins (i.e., MAPs and motors). For example, loss of the microtubule cross-linking protein Ase1 in budding yeast causes the spindle to breakdown at approximately half of its normal length (Pellman et al., 1995; Schuyler et al., 2003). On the other hand, loss of chromosomal passenger proteins (as discussed below) appears to affect the timing of anaphase spindle elongation rather than its stability per se. Their effect on elongation and timing further argues that chromosomal passengers act upstream of the MAPs and motors that are associated with the anaphase spindle.

Although the spindle midzone and midbody appear disorganized after inhibition of chromosomal passengers in animal cells, this relatively mild phenotype offers few insights into the mechanism of microtubule regulation. Recent work in budding yeast offers the potential for a more detailed understanding; mutations in chromosomal passengers affect the final size of the anaphase spindle. *ipl1-321* results in longer spindles whereas the non-core chromosomal passenger mutant, *ndc10-1*(a CBF3 subunit), gives rise to shorter spindles (Bouck and Bloom, 2005; Buvelot et al., 2003; Widlund et al., 2006). Live cell imaging of chromosomal passenger mutants reveals their role in modulating the rate of anaphase B spindle elongation rather than the stability of the spindle per se

(Rozelle, Hansen and Kaplan, unpublished data). Interestingly, mutants in different chromosomal passengers can either increase or decrease the rate of spindle elongation. Alterations in spindle elongation rates may help to explain the different lengths of spindles observed at fixed intervals following entry of cells into mitosis. Importantly, changes in the rate of spindle elongation are compensated by changes in the duration of anaphase, suggesting that there is a feedback loop between spindle length and mitotic exit (Rozelle, Hansen and Kaplan, unpublished data; see arrows in Fig. 12.1A).

In animal cells, there are several intriguing connections between chromosomal passengers and microtubule regulators that offer insight into how these pathways may be connected. For example, in human cells the Aurora B phosphorylation of Mklp1, a kinesin-6 family member, depends on the transport of Aurora B to the spindle midzone via another kinesin, Mklp2 (Neef et al., 2006). Thus, motor-dependent delivery of chromosomal passengers may trigger changes in the activity of kinesins important for organizing midzone microtubules. Another substrate of Aurora B, OP18 normally promotes microtubule depolymerization due to its sequestration of free tubulin. Aurora B phosphorylation inhibits the activity of OP18, thus stabilizing microtubules near chromosomes (Gadea and Ruderman, 2006). It is unclear if Aurora B similarly regulates OP18 during anaphase but such regulation could influence the stability of midzone microtubules. Finally, in budding yeast the DASH complex is a target of Ipl1. As mentioned earlier, the Ipl1 negatively modulates kinetochore–microtubule attachments. The fact that *DAM1* and *DUO1* (encoding DASH subunits) mutants affect anaphase spindle length (Cheeseman et al., 2001) and the observation that DASH associates with the anaphase spindle raises the possibility that Ipl1 could influence anaphase spindles through modification of DASH subunits. Consistent with this type of regulatory mechanism, in vitro studies show that Ipl1 negatively affects the assembly of the DASH complex (Miranda et al., 2007). While these connections are intriguing, it remains unclear whether spindle-associated kinetochore proteins directly or indirectly regulate microtubules. Addressing this issue will require direct observations of spindle dynamics combined with the genetic dissection of the anaphase spindle-associated components.

### 12.3.3 Cytokinesis

Numerous experiments have established that chromosomal passengers are critical for cytokinesis. While details of their precise actions during cytokinesis are less clear, attention has focused on the substrates of the Aurora B kinase. Perhaps the most compelling anaphase substrates are Mklp1/Zen4 (the kinesin 6 family member (Neef et al., 2006)) and MgcRacGAP/Cyk4 (Minoshima et al., 2003). Together, these proteins form a complex known as centralspindlin (Mishima et al., 2002; Pavicic-Kaltenbrunner et al., 2007). Phosphorylation of Mklp1 is required to target the complex to the central spindle and modification of MgcRacGAP is believed to convert it to a GAP (*G*TPase *a*ctivating *p*rotein) for Rho, a small G protein critical for induction of the cytokinetic furrow. Inhibition of other core chromosomal passengers (i.e., INCENP and Survivin) also inhibits cytokinesis. A simple interpretation of the these observations is that loss of the INCENP or Survivin prevents Aurora B from acting at the proper place and time (Vagnarelli and Earnshaw, 2004), although this idea has not been rigorously tested.

While a model that invokes the regulation of a Rho GAP by chromosomal passengers is appealing, it is almost certainly the tip of the iceberg. Recent findings in budding yeast indicate that chromosomal passengers impact multiple aspects of cytokinesis. In the so-called "no-cut" pathway, Ipl1 (Aurora B) was shown to play a role in delaying abscission and completion of cytokinesis until the end of anaphase B (Norden et al., 2006), thus preventing the "cut" or chromosome breakage phenotype associated with inappropriate furrow ingression (Fig. 12.4). Genetic analysis suggests that Ipl1 (Aurora B) targets the anillin pathway that is conserved in metazoans. It is not known how Ipl1 targets pathways necessary for abscission nor whether other chromosomal passengers are important. Nonetheless, the connection between Ipl1, anillins, and abscission reinforces the idea that chromosomal passengers target multiple anaphase events. The coordination between changes in the spindle midzone and abscission in yeast may be analogous to the reorganization that occurs at the spindle midzone during abscission in metazoans. In future work, it will be important to analyze the relationships between chromosomal passengers, midzone microtubules, and the mechanisms underlying abscission in animal cells.

In addition to their involvement in abscission, chromosomal passengers have also been linked to regulating septin dynamics. In budding yeast, there are five related septin proteins that assemble filaments at the site of bud emergence and that are required to determine the plane of cell division in the subsequent mitosis (Figs. 12.3B and 12.4). The septin proteins bind and hydrolyze GTP but do not obviously have mechano-chemical properties (Longtine and Bi, 2003). Rather, septins have been proposed to act as a diffusion barrier trapping proteins important for cytokinesis at the bud neck (Dobbelaere and Barral, 2004). The completion of cytokinesis triggers the disassembly of the mother cell septin ring, allowing it to reform in an axial position to begin the next bud cycle (in haploid yeast; see diagrams in Fig. 12.3B). Premature septin disassembly would eliminate the diffusion barrier and prevent the proper completion of cytokinesis. Mutants that interfere with the Bir1–Sli15 interaction lead to a hyper-stable septin ring in the mother cell, suggesting that this complex normally induces the disassembly of septins at the end of anaphase (Thomas and Kaplan, 2007). Interestingly, mutations in Ipl1 have no effect on septin disassembly, reinforcing the idea that distinct chromosomal passenger complexes are required to regulate specific anaphase events. To date, there is no evidence that chromosomal passengers directly contact septins and therefore it is likely that they regulate septin modifiers. It will be important to elucidate the details of this pathway as it is a clear example of a chromosomal passenger mediated, Aurora

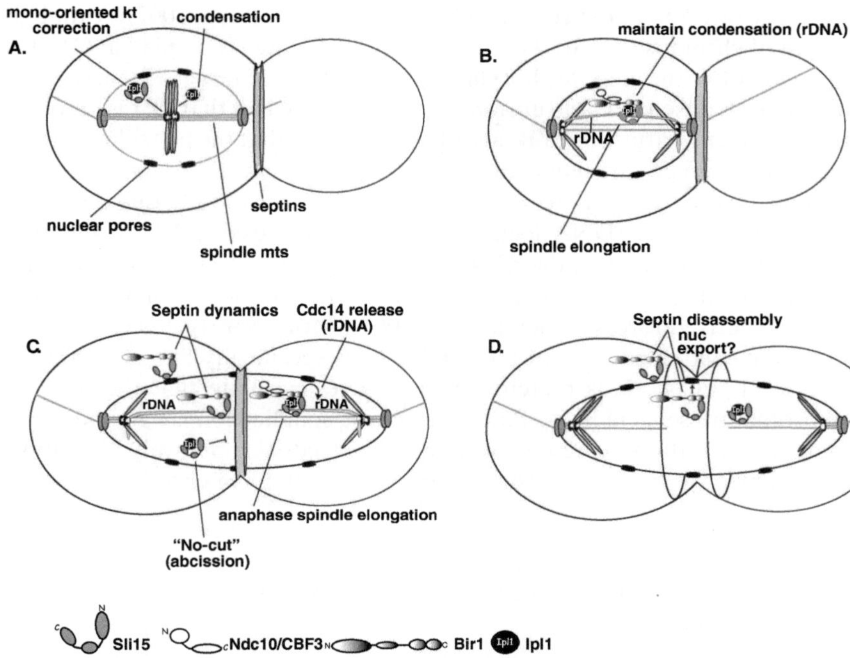


**Fig. 12.4** Model of chromosomal passenger coordination of anaphase mechanics in budding yeast. (**A**) In metaphase cells, the Sli15–Ipl1 complex regulates kinetochore (kt)–microtubule (mt) attachments. (**B**) In early anaphase, all chromosomal passengers associate with the elongating anaphase spindle. (**C**) An Ipl1–Sli15–Bir1 complex regulates anaphase spindle elongation and, Ipl1, maintains rDNA condensation. The rate of spindle elongation may depend on the segregation of rDNA and on feedback from chromosomal passengers. (**D**) Spindle breakdown releases chromosomal passengers from interpolar microtubules, freeing them to mediate changes in septin dynamics necessary for cytokinesis and the return of cells to G1. The activity of chromosomal passengers may affect pathways that originate in the nucleus or may be transported from the nucleus to the cytosol

B-independent regulation of anaphase. The connections between the chromosomal passengers, the abscission machinery, and septin dynamics underscore the variety of pathways that are coordinated to ensure an orderly anaphase.

*Chromatin (decondensation)*: As mentioned, a dramatic series of changes occur at the end of mitosis to allow chromosomes to return to their G1 state of organization. The control of chromosome organization during cell division is complicated and chromosomal passengers have been implicated in both establishing and maintaining chromosome condensation. This type of regulation may be particularly important to ensure that segregation of specialized chromatin is coordinated with spindle elongation. Aurora B modifies both condensin I and the histone-H3 subunit of nucleosomes and therefore contributes in multiple ways to the compaction of mitotic chromosomes (Crosio et al., 2002;

Giet and Glover, 2001; Lipp et al., 2007; Shannon and Salmon, 2002). Although the return of chromosomes to their non-compacted state is complex in G1, it is likely to involve the dephosphorylation of Aurora B targets mediated by protein phosphatase 1 (PP1; Murnion et al., 2001). The fact that PP1 is reported to interact directly with Aurora B raises the possibility that the phosphorylation state of chromatin components are tightly coupled to chromosomal passenger localization and/or activity. As mentioned, a more specialized issue relates to the segregation of the rDNA locus located on chromosome XII in budding yeast (Fig. 12.4). A region of the rDNA repeats segregates last during anaphase and its movement to the poles depends on chromatin condensation rather than just spindle-derived forces (Machin et al., 2005). Moreover, Ipl1 is required to maintain the condensed state of the rDNA locus specifically during anaphase (Lavoie et al., 2004). It is therefore tempting to speculate that the roles of chromosomal passengers in regulating anaphase spindle elongation and chromosome condensation represent an important feedback pathway that links spindle behavior to the segregation of specialized regions of chromosomes.

## 12.4 A Model of Chromosomal Passenger Coordination of Anaphase

To place the proteins and anaphase events reviewed here into a broader context, we will review the key elements of the chromosomal passenger model. First, our review shows that there are a large number of kinetochore-associated proteins that have anaphase-specific roles. As delineated in a number of experimental systems, the core CPC has a wide range of targets that regulate spindle organization and cytokinesis. In addition, the number of non-core chromosomal passengers that are found at both kinetochores and the anaphase spindle suggests that their activities are required at both the kinetochore–microtubule and midzone microtubule interfaces. The redistribution of kinetochore passengers are downstream of cyclin destruction and the reversal of mitotic phosphorylations (Fig. 12.1). This relationship allows chromosomal passengers to serve as an important link between the temporal information that controls mitotic progression (i.e., cyclin degradation and phosphatase activity) and the spatial events that accompany proper anaphase progression.

Despite the complexities that arise when comparing mitoses from different experimental systems, we believe that chromosomal passengers control a conserved set of anaphase events. Using data drawn from multiple model systems, we can propose a plausible sequence of events that link chromosomal passengers to anaphase progression (Fig. 12.4). The events in anaphase begin with cyclin destruction and/or Cdc14 release from the nucleolus, providing a temporal cue for anaphase initiation. The latter is translated spatially by the relocalization of chromosomal passengers from centromeres to the spindle. Chromosomal passengers then act to modulate spindle elongation and maintain the condensed state of rDNA (and possibly other chromosome regions). We speculate that these activities ensure that genomes are fully segregated prior to spindle breakdown. Our unpublished work suggests that chromosomal passengers operate in a feedback loop with the mitotic exit network to allow spatial information, possibly spindle length or DNA position, to modulate anaphase duration and ensure complete chromosome segregation. The anaphase spindle interacts with chromosomal passengers that induce abscission and inhibit septin dynamics. After spindle breakdown, chromosomal passengers are released, allowing septin rings to disassemble at the end of mitosis. The spatial information associated with spindle breakdown is hence linked to the proper timing of cytokinetic induction. In animal cells, a similar relationship between formation of the spindle midzone and activation of the cytokinetic machinery connects the completion of anaphase B with furrow induction. The novel insights into pathways controlled by chromosomal passengers, especially those described in budding yeast, provide a unique opportunity to identify anaphase targets of chromosomal passengers. Progress will undoubtedly take advantage of genetically tractable systems and advanced imaging approaches to begin to shed light on the details that underlie the long-accepted paradigm that chromosomal passengers coordinate mitotic progression to ensure mitotic fidelity.

## References


Adams, R.R., M. Carmena, and W.C. Earnshaw. 2001. Chromosomal passengers and the (aurora) ABCs of mitosis. *Trends Cell Biol.* 11:49–54.

Adams, R.R., S.P. Wheatley, A.M. Gouldsworthy, S.E. Kandels-Lewis, M. Carmena, C. Smythe, D.L. Gerloff, and W.C. Earnshaw. 2000. INCENP binds the Aurora-related kinase AIRK2 and is required to target it to chromosomes, the central spindle and cleavage furrow. *Curr Biol.* 10:1075–8.

Akhmanova, A., and C.C. Hoogenraad. 2005. Microtubule plus-end-tracking proteins: mechanisms and functions. *Curr Opin Cell Biol.* 17:47–54.

Arnaud, L., J. Pines, and E.A. Nigg. 1998. GFP tagging reveals human Polo-like kinase 1 at the kinetochore/centromere region of mitotic chromosomes. *Chromosoma.* 107:424–9.

Bardin, A.J., and A. Amon. 2001. Men and sin: what's the difference? *Nat Rev Mol Cell Biol.* 2:815–26.

Biggins, S., and A.W. Murray. 2001. The budding yeast protein kinase Ipl1/Aurora allows the absence of tension to activate the spindle checkpoint. *Genes Dev.* 15:3118–29.

Bishop, J.D., and J.M. Schumacher. 2002. Phosphorylation of the carboxyl terminus of inner centromere protein (INCENP) by the Aurora B Kinase stimulates Aurora B kinase activity. *J Biol Chem.* 277:27577–80.

Bouck, D.C., and K.S. Bloom. 2005. The kinetochore protein Ndc10p is required for spindle stability and cytokinesis in yeast. *Proc Natl Acad Sci USA.* 102:5408–13.

Bourhis, E., S.G. Hymowitz, and A.G. Cochran. 2007. The mitotic regulator Survivin binds as a monomer to its functional interactor Borealin. *J Biol Chem.* 282:35018–23.

Buvelot, S., S.Y. Tatsutani, D. Vermaak, and S. Biggins. 2003. The budding yeast Ipl1/Aurora protein kinase regulates mitotic spindle disassembly. *J Cell Biol.* 160:329–39.

Cao, L., X. Yan, Y. Wu, H. Hu, Q. Li, T. Zhou, S. Jiang, and L. Yu. 2006. Survivin mutant (Surv-DD70, 71AA) disrupts the interaction of Survivin with Aurora B and causes multinucleation in HeLa cells. *Biochem Biophys Res Commun.* 346:400–7.

Carvalho, P., J.S. Tirnauer, and D. Pellman. 2003. Surfing on microtubule ends. *Trends Cell Biol.* 13:229–37.

Chan, C.S., and D. Botstein. 1993. Isolation and characterization of chromosome-gain and increase-in-ploidy mutants in yeast. *Genetics.* 135:677–91.

Cheeseman, I.M., M. Enquist-Newman, T. Muller-Reichert, D.G. Drubin, and G. Barnes. 2001. Mitotic spindle integrity and kinetochore function linked by the Duo1p/Dam1p complex. *J Cell Biol.* 152:197–212.

Chiroli, E., V. Rossio, G. Lucchini, and S. Piatti. 2007. The budding yeast PP2ACdc55 protein phosphatase prevents the onset of anaphase in response to morphogenetic defects. *J Cell Biol.* 177:599–611.

Cohen-Fix, O., and D. Koshland. 1999. Pds1p of budding yeast has dual roles: inhibition of anaphase initiation and regulation of mitotic exit. *Genes Dev.* 13:1950–9.

Cooke, C.A., M.M. Heck, and W.C. Earnshaw. 1987. The inner centromere protein (INCENP) antigens: movement from inner centromere to midbody during mitosis. *J Cell Biol.* 105:2053–67.

Crosio, C., G.M. Fimia, R. Loury, M. Kimura, Y. Okano, H. Zhou, S. Sen, C.D. Allis, and P. Sassone-Corsi. 2002. Mitotic phosphorylation of histone H3: spatio-temporal regulation by mammalian Aurora kinases. *Mol Cell Biol.* 22:874–85.

Delcros, J.G., C. Prigent, and R. Giet. 2006. Dynactin targets Pavarotti-KLP to the central spindle during anaphase and facilitates cytokinesis in Drosophila S2 cells. *J Cell Sci.* 119:4431–41.

DeLuca, J.G., W.E. Gall, C. Ciferri, D. Cimini, A. Musacchio, and E.D. Salmon. 2006. Kinetochore microtubule dynamics and attachment stability are regulated by Hec1. *Cell.* 127:969–82.

Dobbelaere, J., and Y. Barral. 2004. Spatial coordination of cytokinetic events by compartmentalization of the cell cortex. *Science.* 305:393–6.

Dobbelaere, J., M.S. Gentry, R.L. Hallberg, and Y. Barral. 2003. Phosphorylation-dependent regulation of septin dynamics during the cell cycle. *Dev Cell.* 4:345–57.

Draviam, V.M., I. Shapiro, B. Aldridge, and P.K. Sorger. 2006. Misorientation and reduced stretching of aligned sister kinetochores promote chromosome missegregation in EB1- or APC-depleted cells. *Embo J.* 25:2814–27.

Dumitrescu, T.P., and W.S. Saunders. 2002. The FEAR Before MEN: networks of mitotic exit. *Cell Cycle.* 1:304–7.

Earnshaw, W.C., and R.L. Bernat. 1991. Chromosomal passengers: toward an integrated view of mitosis. *Chromosoma.* 100:139–46.

Eckley, D.M., A.M. Ainsztein, A.M. Mackay, I.G. Goldberg, and W.C. Earnshaw. 1997. Chromosomal proteins and cytokinesis: patterns of cleavage furrow formation and inner centromere protein positioning in mitotic heterokaryons and mid-anaphase cells. *J Cell Biol.* 136:1169–83.

Fortugno, P., E. Beltrami, J. Plescia, J. Fontana, D. Pradhan, P.C. Marchisio, W.C. Sessa, and D.C. Altieri. 2003. Regulation of survivin function by Hsp90. *Proc Natl Acad Sci USA.* 100:13791–6.

Gadea, B.B., and J.V. Ruderman. 2006. Aurora B is required for mitotic chromatin-induced phosphorylation of Op18/Stathmin. *Proc Natl Acad Sci USA.* 103:4493–8.

Gassmann, R., A. Carvalho, A.J. Henzing, S. Ruchaud, D.F. Hudson, R. Honda, E.A. Nigg, D.L. Gerloff, and W.C. Earnshaw. 2004. Borealin: a novel chromosomal passenger required for stability of the bipolar mitotic spindle. *J Cell Biol.* 166:179–91.

Giet, R., and D.M. Glover. 2001. Drosophila aurora B kinase is required for histone H3 phosphorylation and condensin recruitment during chromosome condensation and to organize the central spindle during cytokinesis. *J Cell Biol.* 152:669–82.

Gillis, A.N., S. Thomas, S.D. Hansen, and K.B. Kaplan. 2005. A novel role for the CBF3 kinetochore-scaffold complex in regulating septin dynamics and cytokinesis. *J Cell Biol.* 171:773–84.

Glotzer, M. 2005. The molecular requirements for cytokinesis. *Science*. 307:1735–9.
Goto, H., T. Kiyono, Y. Tomono, A. Kawajiri, T. Urano, K. Furukawa, E.A. Nigg, and M. Inagaki. 2006. Complex formation of Plk1 and INCENP required for metaphase-anaphase transition. *Nat Cell Biol*. 8:180–7.
Green, R.A., R. Wollman, and K.B. Kaplan. 2005. APC and EB1 Function Together in Mitosis to Regulate Spindle Dynamics and Chromosome Alignment. *Mol Biol Cell*. 16:4609–22.
Gruneberg, U., R. Neef, R. Honda, E.A. Nigg, and F.A. Barr. 2004. Relocation of Aurora B from centromeres to the central spindle at the metaphase to anaphase transition requires MKlp2. *J Cell Biol*. 166:167–72.
Hsu, J.Y., Z.W. Sun, X. Li, M. Reuben, K. Tatchell, D.K. Bishop, J.M. Grushcow, C.J. Brame, J.A. Caldwell, D.F. Hunt, R. Lin, M.M. Smith, and C.D. Allis. 2000. Mitotic phosphorylation of histone H3 is governed by Ipl1/aurora kinase and Glc7/PP1 phosphatase in budding yeast and nematodes. *Cell*. 102:279–91.
Jelluma, N., A.B. Brenkman, N.J. van den Broek, C.W. Cruijsen, M.H. van Osch, S.M. Lens, R.H. Medema, and G.J. Kops. 2008. Mps1 phosphorylates Borealin to control Aurora B activity and chromosome alignment. *Cell*. 132:233–46.
Jeyaprakash, A.A., U.R. Klein, D. Lindner, J. Ebert, E.A. Nigg, and E. Conti. 2007. Structure of a Survivin-Borealin-INCENP core complex reveals how chromosomal passengers travel together. *Cell*. 131:271–85.
Jones, M.H., X. He, T.H. Giddings, and M. Winey. 2001. Yeast Dam1p has a role at the kinetochore in assembly of the mitotic spindle. *Proc Natl Acad Sci USA*. 98:13675–80.
Kaiser, B.K., Z.A. Zimmerman, H. Charbonneau, and P.K. Jackson. 2002. Disruption of centrosome structure, chromosome segregation, and cytokinesis by misexpression of human Cdc14A phosphatase. *Mol Biol Cell*. 13:2289–300.
Klein, U.R., E.A. Nigg, and U. Gruneberg. 2006. Centromere targeting of the chromosomal passenger complex requires a ternary subcomplex of Borealin, Survivin, and the N-terminal domain of INCENP. *Mol Biol Cell*. 17:2547–58.
Lavoie, B.D., E. Hogan, and D. Koshland. 2004. In vivo requirements for rDNA chromosome condensation reveal two cell-cycle-regulated pathways for mitotic chromosome folding. *Genes Dev*. 18:76–87.
Li, D., R. Liu, L. Song, H. Zhou, J. Chen, and X. Huang. 2007. The special location of p-H3 and p-CENP-A on heterochromatin during mitosis in MCF-7. *Mol Biol Rep*. 10.1007/s11033-007-9136-9
Li, J., W.L. Lee, and J.A. Cooper. 2005. NudEL targets dynein to microtubule ends through LIS1. *Nat Cell Biol*. 7:686–90.
Lingelbach, L.B., and K.B. Kaplan. 2004. The Interaction between Sgt1p and Skp1p Is Regulated by HSP90 Chaperones and Is Required for Proper CBF3 Assembly. *Mol Cell Biol*. 24:8938–50.
Lipp, J.J., T. Hirota, I. Poser, and J.M. Peters. 2007. Aurora B controls the association of condensin I but not condensin II with mitotic chromosomes. *J Cell Sci*. 120:1245–55.
Longtine, M.S., and E. Bi. 2003. Regulation of septin organization and function in yeast. *Trends Cell Biol*. 13:403–9.
Machin, F., J. Torres-Rosell, A. Jarmuz, and L. Aragon. 2005. Spindle-independent condensation-mediated segregation of yeast ribosomal DNA in late anaphase. *J Cell Biol*. 168:209–19.
Mackay, A.M., A.M. Ainsztein, D.M. Eckley, and W.C. Earnshaw. 1998. A dominant mutant of inner centromere protein (INCENP), a chromosomal protein, disrupts prometaphase congression and cytokinesis. *J Cell Biol*. 140:991–1002.
Mailand, N., C. Lukas, B.K. Kaiser, P.K. Jackson, J. Bartek, and J. Lukas. 2002. Deregulated human Cdc14A phosphatase disrupts centrosome separation and chromosome segregation. *Nat Cell Biol*. 4:317–22.
McAinsh, A.D., J.D. Tytell, and P.K. Sorger. 2003. Structure, function, and regulation of budding yeast kinetochores. *Annu Rev Cell Dev Biol*. 19:519–39.

Minoshima, Y., T. Kawashima, K. Hirose, Y. Tonozuka, A. Kawajiri, Y.C. Bao, X. Deng, M. Tatsuka, S. Narumiya, W.S. May, Jr., T. Nosaka, K. Semba, T. Inoue, T. Satoh, M. Inagaki, and T. Kitamura. 2003. Phosphorylation by aurora B converts MgcRacGAP to a RhoGAP during cytokinesis. *Dev Cell.* 4:549–60.

Miranda, J.J., P. De Wulf, P.K. Sorger, and S.C. Harrison. 2005. The yeast DASH complex forms closed rings on microtubules. *Nat Struct Mol Biol.* 12:138–43.

Miranda, J.J., D.S. King, and S.C. Harrison. 2007. Protein arms in the kinetochore-microtubule interface of the yeast DASH complex. *Mol Biol Cell.* 18:2503–10.

Mishima, M., S. Kaitna, and M. Glotzer. 2002. Central spindle assembly and cytokinesis require a kinesin-like protein/RhoGAP complex with microtubule bundling activity. *Dev Cell.* 2:41–54.

Miyauchi, K., X. Zhu, C. Foong, H. Hosoya, and M. Murata-Hori. 2007. Aurora B kinase activity is required to prevent polar cortical ingression during cytokinesis. *Cell Cycle.* 6:2549–53.

Mollinari, C., C. Reynaud, S. Martineau-Thuillier, S. Monier, S. Kieffer, J. Garin, P.R. Andreassen, A. Boulet, B. Goud, J.P. Kleman, and R.L. Margolis. 2003. The mammalian passenger protein TD-60 is an RCC1 family member with an essential role in prometaphase to metaphase progression. *Dev Cell.* 5:295–307.

Morgan, D.O. 1999. Regulation of the APC and the exit from mitosis. *Nat Cell Biol.* 1:E47–53.

Murnion, M.E., R.R. Adams, D.M. Callister, C.D. Allis, W.C. Earnshaw, and J.R. Swedlow. 2001. Chromatin-associated protein phosphatase 1 regulates aurora-B and histone H3 phosphorylation. *J Biol Chem.* 276:26656–65.

Neef, R., U.R. Klein, R. Kopajtich, and F.A. Barr. 2006. Cooperation between mitotic kinesins controls the late stages of cytokinesis. *Curr Biol.* 16:301–7.

Niikura, Y., S. Ohta, K.J. Vandenbeldt, R. Abdulle, B.F. McEwen, and K. Kitagawa. 2006. 17-AAG, an Hsp90 inhibitor, causes kinetochore defects: a novel mechanism by which 17-AAG inhibits cell proliferation. *Oncogene.* 25:4133–46.

Norden, C., M. Mendoza, J. Dobbelaere, C.V. Kotwaliwale, S. Biggins, and Y. Barral. 2006. The NoCut pathway links completion of cytokinesis to spindle midzone function to prevent chromosome breakage. *Cell.* 125:85–98.

Parry, D.H., G.R. Hickson, and P.H. O'Farrell. 2003. Cyclin B destruction triggers changes in kinetochore behavior essential for successful anaphase. *Curr Biol.* 13:647–53.

Parry, D.H., and P.H. O'Farrell. 2001. The schedule of destruction of three mitotic cyclins can dictate the timing of events during exit from mitosis. *Curr Biol.* 11:671–83.

Pavicic-Kaltenbrunner, V., M. Mishima, and M. Glotzer. 2007. Cooperative assembly of CYK-4/MgcRacGAP and ZEN-4/MKLP1 to form the centralspindlin complex. *Mol Biol Cell.* 18:4992–5003.

Pellman, D., M. Bagget, Y.H. Tu, G.R. Fink, and H. Tu. 1995. Two microtubule-associated proteins required for anaphase spindle movement in Saccharomyces cerevisiae. *J Cell Biol.* 130:1373–85.

Pereira, G., and E. Schiebel. 2001. The role of the yeast spindle pole body and the mammalian centrosome in regulating late mitotic events. *Curr Opin Cell Biol.* 13:762–9.

Pereira, G., and E. Schiebel. 2003. Separase regulates INCENP-Aurora B anaphase spindle function through Cdc14. *Science.* 302:2120–4.

Queralt, E., C. Lehane, B. Novak, and F. Uhlmann. 2006. Downregulation of PP2A(Cdc55) phosphatase by separase initiates mitotic exit in budding yeast. *Cell.* 125:719–32.

Rattner, J.B., A. Rao, M.J. Fritzler, D.W. Valencia, and T.J. Yen. 1993. CENP-F is a .ca 400 kDa kinetochore protein that exhibits a cell-cycle dependent localization. *Cell Motil Cytoskeleton.* 26:214–26.

Rogers, S.L., U. Wiedemann, U. Hacker, C. Turck, and R.D. Vale. 2004. Drosophila RhoGEF2 associates with microtubule plus ends in an EB1-dependent manner. *Curr Biol.* 14:1827–33.

Romano, A., A. Guse, I. Krascenicova, H. Schnabel, R. Schnabel, and M. Glotzer. 2003. CSC-1: a subunit of the Aurora B kinase complex that binds to the survivin-like protein BIR-1 and the incenp-like protein ICP-1. *J Cell Biol.* 161:229–36.

Rosa, J., P. Canovas, A. Islam, D.C. Altieri, and S.J. Doxsey. 2006. Survivin modulates microtubule dynamics and nucleation throughout the cell cycle. *Mol Biol Cell.* 17:1483–93.

Rosasco-Nitcher, S.E., W. Lan, S. Khorasanizadeh, and P.T. Stukenberg. 2008. Centromeric Aurora-B activation requires TD-60, microtubules, and substrate priming phosphorylation. *Science.* 319:469–72.

Ruchaud, S., M. Carmena, and W.C. Earnshaw. 2007. Chromosomal passengers: conducting cell division. *Nat Rev Mol Cell Biol.* 8:798–812.

Sandall, S., F. Severin, I.X. McLeod, J.R. Yates, 3rd, K. Oegema, A. Hyman, and A. Desai. 2006. A Bir1-Sli15 complex connects centromeres to microtubules and is required to sense kinetochore tension. *Cell.* 127:1179–91.

Scholey, J.M., I. Brust-Mascher, and A. Mogilner. 2003. Cell division. *Nature.* 422:746–52.

Schuyler, S.C., J.Y. Liu, and D. Pellman. 2003. The molecular function of Ase1p: evidence for a MAP-dependent midzone-specific spindle matrix. Microtubule-associated proteins. *J Cell Biol.* 160:517–28.

Severson, A.F., D.R. Hamill, J.C. Carter, J. Schumacher, and B. Bowerman. 2000. The aurora-related kinase AIR-2 recruits ZEN-4/CeMKLP1 to the mitotic spindle at metaphase and is required for cytokinesis. *Curr Biol.* 10:1162–71.

Shannon, K.B., and E.D. Salmon. 2002. Chromosome dynamics: new light on Aurora B kinase function. *Curr Biol.* 12:R458–60.

Shirayama, M., A. Toth, M. Galova, and K. Nasmyth. 1999. APC(Cdc20) promotes exit from mitosis by destroying the anaphase inhibitor Pds1 and cyclin Clb5 [see comments]. *Nature.* 402:203–7.

Steensgaard, P., M. Garre, I. Muradore, P. Transidico, E.A. Nigg, K. Kitagawa, W.C. Earnshaw, M. Faretta, and A. Musacchio. 2004. Sgt1 is required for human kinetochore assembly. *EMBO Rep.* 5:626–31.

Stehman, S.A., Y. Chen, R.J. McKenney, and R.B. Vallee. 2007. NudE and NudEL are required for mitotic progression and are involved in dynein recruitment to kinetochores. *J Cell Biol.* 178:583–94.

Stemmann, O., A. Neidig, T. Kocher, M. Wilm, and J. Lechner. 2002. Hsp90 enables Ctf13p/Skp1p to nucleate the budding yeast kinetochore. *Proc Natl Acad Sci USA.* 99:8585–90.

Stoepel, J., M.A. Ottey, C. Kurischko, P. Hieter, and F.C. Luca. 2005. The mitotic exit network Mob1p-Dbf2p kinase complex localizes to the nucleus and regulates passenger protein localization. *Mol Biol Cell.* 16:5465–79.

Strickland, L.I., Y. Wen, G.G. Gundersen, and D.R. Burgess. 2005. Interaction between EB1 and p150glued is required for anaphase astral microtubule elongation and stimulation of cytokinesis. *Curr Biol.* 15:2249–55.

Strunnikov, A.V. 2005. A case of selfish nucleolar segregation. *Cell Cycle.* 4:113–7.

Sullivan, M., and D.O. Morgan. 2007. Finishing mitosis, one step at a time. *Nat Rev Mol Cell Biol.* 8:894–903.

Sumara, I., M. Quadroni, C. Frei, M.H. Olma, G. Sumara, R. Ricci, and M. Peter. 2007. A Cul3-based E3 ligase removes Aurora B from mitotic chromosomes, regulating mitotic progression and completion of cytokinesis in human cells. *Dev Cell.* 12:887–900.

Tang, X., and Y. Wang. 2006. Pds1/Esp1-dependent and -independent sister chromatid separation in mutants defective for protein phosphatase 2A. *Proc Natl Acad Sci USA.* 103:16290–5.

Thomas, S., and K.B. Kaplan. 2007. A Bir1p Sli15p kinetochore passenger complex regulates septin organization during anaphase. *Mol Biol Cell.* 18:3820–34.

Tinker-Kulberg, R.L., and D.O. Morgan. 1999. Pds1 and Esp1 control both anaphase and mitotic exit in normal cells and after DNA damage. *Genes Dev.* 13:1936–49.

Torres-Rosell, J., F. Machin, and L. Aragon. 2005. Cdc14 and the temporal coordination between mitotic exit and chromosome segregation. *Cell Cycle*. 4:109–12.

Trinkle-Mulcahy, L., and A.I. Lamond. 2006. Mitotic phosphatases: no longer silent partners. *Curr Opin Cell Biol*. 18:623–31.

Vagnarelli, P., and W.C. Earnshaw. 2004. Chromosomal passengers: the four-dimensional regulation of mitotic events. *Chromosoma*. 113:211–22.

Vagnarelli, P., D.F. Hudson, S.A. Ribeiro, L. Trinkle-Mulcahy, J.M. Spence, F. Lai, C.J. Farr, A.I. Lamond, and W.C. Earnshaw. 2006. Condensin and Repo-Man-PP1 co-operate in the regulation of chromosome architecture during mitosis. *Nat Cell Biol*. 8:1133–42.

Westermann, S., H.W. Wang, A. Avila-Sakar, D.G. Drubin, E. Nogales, and G. Barnes. 2006. The Dam1 kinetochore ring complex moves processively on depolymerizing microtubule ends. *Nature*. 440:565–9.

Widlund, P.O., J.S. Lyssand, S. Anderson, S. Niessen, J.R. Yates, 3rd, and T.N. Davis. 2006. Phosphorylation of the chromosomal passenger protein Bir1 is required for localization of Ndc10 to the spindle during anaphase and full spindle elongation. *Mol Biol Cell*. 17:1065–74.

Wolyniak, M.J., K. Blake-Hodek, K. Kosco, E. Hwang, L. You, and T.C. Huffaker. 2006. The regulation of microtubule dynamics in Saccharomyces cerevisiae by three interacting plus-end tracking proteins. *Mol Biol Cell*. 17:2789–98.

Yan, X., F. Li, Y. Liang, Y. Shen, X. Zhao, Q. Huang, and X. Zhu. 2003. Human Nudel and NudE as regulators of cytoplasmic dynein in poleward protein transport along the mitotic spindle. *Mol Cell Biol*. 23:1239–50.

Yang, D., A. Welm, and J.M. Bishop. 2004. Cell division and cell survival in the absence of survivin. *Proc Natl Acad Sci USA*. 101:15100–5.

Zeitlin, S.G., R.D. Shelby, and K.F. Sullivan. 2001. CENP-A is phosphorylated by Aurora B kinase and plays an unexpected role in completion of cytokinesis. *J Cell Biol*. 155:1147–57.

# Chapter 13
# Roles of Centromeres and Kinetochores in Meiosis

**Adele L. Marston**

Meiosis is the cell division process by which haploid gametes are produced from a diploid progenitor cell. Reduction of the genome by half requires that DNA replication is followed not by one nuclear division, as in mitosis, but by two consecutive divisions. The sorting and segregation of chromosomes during these two nuclear divisions is tightly controlled, thereby ensuring that each of the gametes inherits a complete haploid set of chromosomes. Errors in chromosome segregation during meiosis generate gametes with too few or too many chromosomes, a condition known as aneuploidy, which is associated with birth defects and infertility (Hassold and Hunt, 2001). This chapter reviews our current understanding of the role the centromere and kinetochore play in bringing about the specialized segregation of chromosomes during meiosis.

## 13.1 Overview of Meiosis and the Role of the Kinetochore

Centromere is the name given to the chromosomal DNA onto which the large macromolecular complex, known as the kinetochore, will assemble and mediate the attachment of chromosomes to microtubules. The DNA region immediately flanking the centromere is known as the pericentromere. This chapter focuses on the modifications made to centromeres, pericentromeres and kinetochores that allow them to carry out their meiosis-specific functions, but refers to general principles where relevant.

A diploid cell enters meiosis with two homologous sets of chromosomes, of which one is paternally derived and the other maternally derived. During DNA replication, an identical copy of both sets of chromosomes is synthesized and the newly duplicated 'sister chromatids' are held together by cohesion (Fig. 13.1). For the accurate segregation of chromosomes during the ensuing

A.L. Marston (✉)
The Wellcome Trust Centre for Cell Biology, University of Edinburgh, School of Biological Sciences, Michael Swann Building, Mayfield Road, Edinburgh, EH9 3JR, U.K.
e-mail: adele.marston@ed.ac.uk

P. De Wulf, W.C. Earnshaw (eds.), *The Kinetochore*,
DOI 10.1007/978-0-387-69076-6_13, 

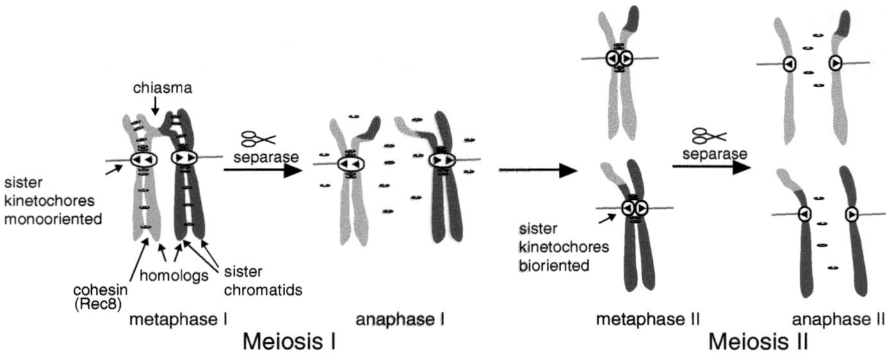


**Fig. 13.1 Chromosome segregation during meiosis.** Schematic diagram showing the key events directing chromosome segregation during meiosis. For details see text

meiotic divisions, linkages must also be established between the homologous chromosomes. In most organisms, meiotic recombination generates crossovers, which will become chromatin bridges known as chiasmata. Chiasmata provide inter-homologue linkages due to the cohesion between exchanged sister chromatid arms. This complex of four chromosomes, called a bivalent (two homologous pairs of sister chromatids), is the template upon which the chromosome segregation machinery will act.

During the first nuclear division, meiosis I, a segregation event that is unique to meiosis occurs because homologous chromosomes are separated away from each other but sister chromatids stay together. The second nuclear division, meiosis II, separates the sister chromatids to opposite poles, as in mitosis. In addition to linkages between homologues (usually chiasmata), two major modifications to the segregation machinery must be made to bring about this specialized segregation pattern (Fig. 13.1). Firstly, the direction in which kinetochores attach to microtubules changes during meiosis. During meiosis I, kinetochores of sister chromatids (sister kinetochores) attach to microtubules emanating from the same spindle pole (mono-orientation, also called co-orientation). This is in contrast to mitosis and meiosis II where sister kinetochores attach to microtubules emanating from opposite spindle poles (bi-orientation). The second major modification is that, in contrast to mitosis where all cohesion is lost in a single nuclear division, cohesion loss is spread out over two nuclear divisions. Cohesion on chromosome arms is lost during meiosis I and this triggers the segregation of homologues to opposite poles. Cohesion around centromeric regions is, however, maintained until meiosis II. Loss of this centromeric cohesion during meiosis II triggers the segregation of sister chromatids to opposite poles. Events at the kinetochore direct both the mono-orientation of kinetochores and the retention of centromeric cohesion and are considered in detail in the following sections. The correct attachment of chromosomes to the spindle is monitored by a kinetochore-driven surveillance mechanism known as the spindle checkpoint and our current understanding of how this is modified for homologue segregation during meiosis I is reviewed. Finally, the role of the centromere in the pairing of homologous chromosomes, which is a prerequisite for the establishment of inter-homologue linkages during meiosis I, is discussed.

## 13.2 Centromeres and Cohesin

During meiosis, linkages (cohesion) between centromeres must be preserved during meiosis I to ensure the accurate segregation of chromosomes during meiosis II. This section reviews what is currently known about the mechanism by which this is achieved.

### *13.2.1 Cohesin Regulation in Mitosis*

During both mitosis and meiosis, sister chromatids must be held together from the time of their synthesis until their separation in anaphase. This cohesion between sister chromatids is generated by a protein complex known as cohesin (reviewed in Nasmyth, 2005). Cohesin is a four-subunit complex that is composed of two SMC proteins, Smc1 and Smc3, together with the Scc3/SA/STAG protein and an $\alpha$-kleisin (Scc1/Mcd1/Rad21) subunit (reviewed in Nasmyth and Haering, 2005). How cohesin holds sister chromatids together is a matter of debate (Milutinovich and Koshland, 2003; Huang et al., 2005; Losada and Hirano, 2005; Nasmyth, 2005). One attractive proposal is the 'embrace' model, in which the cohesin complex forms a ring that encapsulates the two DNA molecules (Haering and Nasmyth, 2003). However, alternative models, such as the 'snap' model in which sister DNA molecules are bound individually by cohesin molecules that oligomerize, have not been ruled out (Milutinovich and Koshland, 2003). Furthermore, cohesin at the centromere may adopt a configuration that allows it to link two regions of DNA on the same molecule, rather than sisters (Yeh et al., 2008).

Cohesin opposes the force exerted by microtubules on bi-oriented sister kinetochores and facilitates their alignment on the mitotic spindle at metaphase (Fig. 13.2). Once proper bipolar attachment is achieved, a protease known as separase, cleaves the Scc1/Mcd/Rad21 subunit of cohesin, thereby triggering the movement of sister chromatids to opposite poles (Fig. 13.2). Prior activation of separase is prevented by the binding of an inhibitor, known as securin (Ciosk et al., 2000). Destruction of securin occurs following its ubiquitination by the anaphase-promoting complex (APC), coupled to an activator, Cdc20. Securin destruction releases active separase and initiates chromosome segregation (reviewed in Yu, 2007). In mammalian cells, Cdk1-cyclin B also complexes with, and inhibits, separase prior to the destruction of cyclin B by the APC (Stemmann et al., 2001; Gorr et al., 2005). A surveillance mechanism known as

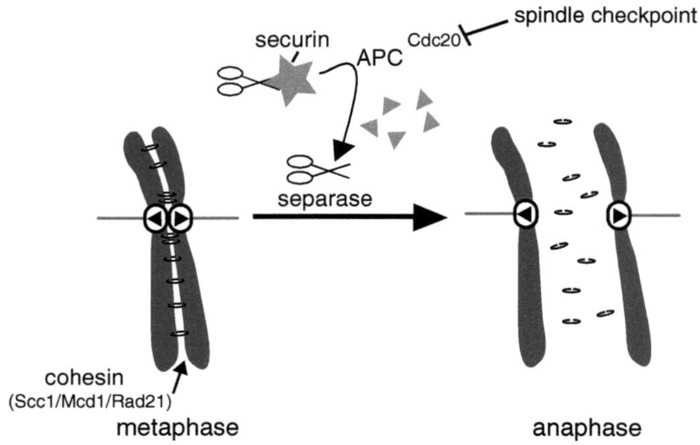


**Fig. 13.2 Chromosome segregation during mitosis.** The movement of sister chromatids to opposite poles is triggered by the cleavage of the Scc1/Mcd1/Rad21 cohesin subunit by separase. Upon correct bipolar attachment of sister chromatids, the spindle checkpoint relieves its inhibition of the $APC^{Cdc20}$, which ubiquitinates securin, thereby targeting it for destruction and consequently liberating separase. Note that in mammalian cells, the bulk of cohesin is removed from chromosome arms prior to metaphase but chromosome segregation is triggered by separase-dependent cleavage of the residual cohesin. Reviewed in Nasmyth (2001)

the spindle assembly checkpoint (SAC) prevents APC activation, and thereby separase activation and anaphase onset until all kinetochores are correctly attached to the spindle (reviewed in Musacchio and Salmon, 2007).

In budding and fission yeast mitosis, all cohesin is cleaved simultaneously along the length of chromosomes upon separase activation. This is in contrast to mammalian cells where the majority of cohesin is removed from chromosome arms prior to metaphase in a non-proteolytic pathway that is separase-independent but which requires the Polo-like kinase (Plk1) and Aurora-B kinase (the so-called 'prophase pathway' (Losada et al., 2002; Sumara et al., 2002). The residual cohesin that resides particularly around centromeres in mammalian cells is, however, sufficient to hold sister chromatids together until it is cleaved upon separase activation, initiating chromosome segregation (Waizenegger et al., 2000; Hauf et al., 2001). Therefore, separase-dependent cleavage of cohesin is the trigger for chromosome segregation during mitosis.

## *13.2.2 Composition of Meiotic Cohesin*

Cohesin loss must be spatially regulated to allow the consecutive segregation of homologues and sister chromatids during meiosis I and meiosis II, respectively (Fig. 13.1). Cohesin loss from chromosome arms during meiosis I abolishes the linkages between homologues (chiasmata) and triggers their poleward movement. In contrast, cohesins must be retained in the vicinity of centromeres during meiosis I to ensure that sister chromatids are held together until meiosis II.

A conserved feature of meiotic cohesin is the replacement of the Scc1/Mcd1/Rad21 subunit by its meiosis-specific homologue, Rec8 (Klein et al., 1999; Watanabe and Nurse, 1999; Pasierbek et al., 2001; Xu et al., 2005). At least in mammalian meiosis and grasshoppers, Rec8-containing cohesin co-exists alongside its Scc1/Mcd1/Rad21-containing counterpart (Prieto et al., 2002; Parra et al., 2004; Xu et al., 2004; Gomez et al., 2007). Cytological studies in budding yeast, fission yeast, *C. elegans* and mammals have demonstrated that Rec8 is localized along the length of chromosomes and is lost from chromosome arms during meiosis I but retained at centromeres until meiosis II (Klein et al., 1999; Watanabe and Nurse, 1999; Pasierbek et al., 2001; Eijpe et al., 2003; Lee et al., 2003, 2006). In some organisms the composition of cohesin complexes on chromosome arms and centromeres differs during meiosis. In fission yeast meiosis, the mitotic Scc3-like Psc3 subunit is found in cohesin complexes around centromeres but is replaced on chromosome arms with a meiosis-specific variant, Rec11 (Kitajima et al., 2003b). Similarly, in mammals, the Scc3-like STAG3 is meiosis-specific and localized specifically on chromosome arms (Prieto et al., 2001). This specialization of arm cohesin seems to facilitate meiotic recombination, but could also contribute to the differential timing with which arm and centromeric cohesins are lost in meiosis.

### *13.2.3 Separase is the Trigger for Chromosome Segregation in Meiosis I and Meiosis II*

One way in which the step-wise loss of cohesion in meiosis could be achieved is to destroy linkages between homologues through the non-proteolytic removal of cohesin at meiosis I and limit separase activation until meiosis II. However, cleavage of Rec8 by separase was shown to be necessary for the disjunction of homologues during meiosis I in both budding and fission yeast. Homologue disjunction was blocked either in the absence of separase or in the presence of a version of Rec8 with the cleavage sites mutated (Buonomo et al., 2000; Kitajima et al., 2003a). Similarly, the resolution of chiasmata has been shown to require separase and the APC in *C. elegans* (Siomos et al., 2001; Davis et al., 2002) and mouse oocytes (Herbert et al., 2003; Terret et al., 2003; Kudo et al., 2006).

A requirement for separase during meiosis II was shown by some ingenious experiments using fission yeast. Watanabe and colleagues took advantage of the different composition of cohesin complexes at centromeres and on chromosome arms during meiosis and engineered yeast with non-cleavable cohesin at centromeres and lacking any cohesin on chromosome arms (Kitajima et al., 2003a). Homologue segregation was not blocked at meiosis I, but sister chromatids failed to disjoin at meiosis II. Consistent with a need to activate separase twice in meiosis, its inhibitor, securin, accumulates during both meiosis I and meiosis II and is destroyed at the onset of anaphase in both divisions in budding yeast, mouse and pig oocytes (Salah and Nasmyth, 2000; Herbert et al., 2003; Huo et al., 2006).

Despite the definitive evidence that separase activity is required for homologue disjunction in organisms ranging from yeast to mammals, other mechanisms of cohesin removal could contribute to chiasmata resolution. In budding yeast, a non-proteolytic and Polo kinase-dependent mechanism of cohesin removal, akin to the prophase pathway that operates during mammalian mitosis, has been reported to operate during the early stages of meiosis (Yu and Koshland, 2005). In addition, depletion of either the APC or securin from *Xenopus* extracts did not prevent meiosis I chromosome segregation, indicating that other, separase-independent, mechanisms might remove cohesins in this organism during meiosis I (Peter et al., 2001; Taieb et al., 2001). However, separase is activated during meiosis I and two waves of securin accumulation and destruction have been observed in *Xenopus*, suggesting that separase-dependent cleavage of cohesin is likely to contribute to homologue segregation in this organism too (Fan et al., 2006).

### *13.2.4 Protectors of Centromeric Cohesion*

Because separase is activated during both meiosis I and meiosis II (Fig. 13.1), a mechanism must exist to make cohesins in the vicinity of the centromere resistant to separase activity during meiosis I. The kinetochore is well-placed to provide such a function, but what are the factors that mediate the protection of centromeric cohesion in this region? An insight into potential players in this process came with the isolation of a *Drosophila* mutant, *mei-S332*, in which sister chromatids were randomly segregated at meiosis II, although the segregation of homologues at meiosis I was not dramatically perturbed (Kerrebrock et al., 1992). This is exactly the outcome that would be predicted if all cohesion were lost in meiosis I. Further experiments indicated that MEI-S332 localizes around centromeres but disappears from these chromosomal regions during meiosis II at the time at which centromeric cohesion is lost (Kerrebrock et al., 1995; Moore et al., 1998). These findings implicated MEI-S322 as having a direct role in the maintenance of centromeric cohesion, however, homologues of MEI-S322 in other organisms went undetected until genetic screens in budding and fission yeast identified a coiled-coil protein, that confers protection to centromeric cohesion in these organisms (Katis et al., 2004a; Kitajima et al., 2004; Marston et al., 2004; Rabitsch et al., 2004). These proteins turn out to be distant homologues of MEI-S332 that are conserved in the eukaryotes and which are collectively called Shugoshins ('guardian spirit' in Japanese), (Kitajima et al., 2004; Rabitsch et al., 2004).

Budding yeast and *Drosophila* each have a single Shugoshin, called Sgo1 and MEI-S332, respectively. However, fission yeast and mammals each have two Shugoshins, Sgo1 and Sgo2. In addition to a role in cohesin protection, Shugoshin family members are involved in promoting and monitoring the correct bipolar attachment of chromosomes (Section 13.4.1). The two functions of cohesion protection and kinetochore bi-orientation are conferred within the single budding yeast Shugoshin, Sgo1 (Katis et al., 2004a; Marston et al., 2004; Indjeian et al., 2005; Kiburz et al., 2008). In fission yeast, Sgo1 functions in cohesion protection during meiosis, whereas Sgo2 ensures the generation of correct bipolar attachments (Kitajima et al., 2004; Rabitsch et al., 2004; Kawashima et al., 2007; Vanoosthuyse et al., 2007). Shugoshins also function in cohesion protection in mitosis in mammals (Section 13.2.6), with Sgo1 likely to be predominant and Sgo2 having a similar, although controversial role (Salic et al., 2004; Tang et al., 2004; Kitajima et al., 2005; McGuinness et al., 2005; Huang et al., 2007). Centromeric cohesion protection in mouse oocytes has recently been shown to require mammalian Sgo2, but not Sgo1 (Lee et al., 2008). Furthermore, recently, a splice variant of mammalian Sgo1, sSgo1, has been found to play a role in centriole cohesion during mitosis (Wang et al., 2008).

### 13.2.5 Protection of Centromeric Cohesion During Meiosis

Consistent with a role for Sgo1 in cohesion protection, like *Drosophila* MEI-S332, chromosomes are segregated randomly during meiosis II in *sgo1*Δ mutants of both fission and budding yeast and Rec8 is not retained at centromeric regions during meiosis I (Katis et al., 2004a; Kitajima et al., 2004; Marston et al., 2004; Rabitsch et al., 2004). Sgo1 function in meiosis is also conserved in plants because a mutation in the maize *zmsgo1* gene results in a failure to maintain centromeric cohesion beyond meiosis I (Hamant et al., 2005). Similarly, premature separation of sister centromeres was observed in mouse oocytes depleted of *Sgo2* (Lee et al., 2008).

An elegant experiment performed in fission yeast demonstrated that Sgo1 ensures maintenance of centromeric cohesion during meiosis I by preventing cohesin cleavage. Rabitsch et al. (2004) used a version of cohesin that has its separase recognition sites mutated and therefore cannot be cleaved. Using the *rec11* mutant, which lacks arm cohesion, the separase-resistant cohesin was assembled only at centromeres. Importantly, sister chromatid separation was blocked by the separase-resistant centromeric cohesion, even in cells lacking Sgo1. This clearly shows that Sgo1 is a factor that protects cohesin during meiosis I by counteracting separase activity in the vicinity of the centromere, but how is this achieved mechanistically? A clue came from the finding that Sgo1 forms a complex with a specific form of (PP2A) in both budding and fission yeast during meiosis (Kitajima et al., 2006; Riedel et al., 2006). PP2A is a heterotrimeric complex consisting of catalytic (C), scaffold (A) and one of four types of regulatory (B) subunit (called B, B', B'' or B'''), the PP2A-B' form of which was found to associate with Sgo1 (Kitajima et al., 2006; Riedel et al., 2006). Like Sgo1, PP2A-B', is required for the protection of centromeric cohesion during meiosis I and localizes to centromeres (Kitajima et al., 2006; Riedel et al., 2006). The centromere localization of PP2A-B' was found to depend on yeast Sgo1, but not *vice versa*. Similarly, Sgo2 is required for centromeric PP2A localization in mouse oocytes (Lee et al., 2008). A model in which Shugoshin protects cohesin by recruiting PP2A-B' to centromeres is supported by the finding that ectopic localization of the fission yeast B' subunit, Par1, to chromosomes is sufficient to prevent loss of cohesion at these sites even in the absence of Sgo1 (Kitajima et al., 2006; Riedel et al., 2006).

The observation that the protection of centromeric cohesion requires PP2A's catalytic subunit (Kitajima et al., 2006; Riedel et al., 2006) predicts that PP2A-B' inhibits cohesin cleavage by removing phosphates from at least one key substrate. Several observations made the meiotic $\alpha$-kleisin subunit, Rec8, a good candidate for a PP2A-B' substrate. In both budding and fission yeast, replacement of the mitotic Scc1/Mcd1/Rad21 subunit of cohesin with the meiotic Rec8 subunit is essential for the maintenance of centromeric cohesion until meiosis II, suggesting that a property of Rec8 that is not shared with Scc1/Mcd1/Rad21 allows cohesin to be protected (Toth et al., 2000; Yokobayashi et al., 2003). The specific requirement for Rec8 in yeast might be because Rec8 and Scc1/Mcd1 differ in their ability to be cleaved by separase in the absence of phosphorylation by the Polo-like kinase, Cdc5. In budding yeast mitosis, phosphorylation of Scc1/Mcd1 by, Cdc5 facilitates its cleavage, but is not absolutely required (Alexandru et al., 2001). In meiosis, however, Rec8 cleavage is completely blocked in the absence of Cdc5 (Clyne et al., 2003; Lee and Amon, 2003). Rec8 phosphorylation is reduced both in budding yeast cells where Cdc5 has been depleted (Lee and Amon, 2003) and in fission yeast where PP2A-B' has been targeted along the length of chromosomes (Riedel et al., 2006). The importance of Rec8 phosphorylation for its timely cleavage at meiosis I was demonstrated in budding yeast using the *rec8-17A* allele in which 17 in vivo mapped phosphorylation sites (14 of them Cdc5-dependent) had been mutated (Brar et al., 2006). Rec8 phosphorylation might be generally excluded from the centromere because a phospho-specific antibody raised to one of the Cdc5-independent sites stained chromosome arms but was largely excluded from the centromeric regions (Brar et al., 2006). Together, these findings suggest that Polo kinase plays a role in cohesin cleavage during meiosis I, which is at least in part mediated through phosphorylation of Rec8. In this model, Sgo1-dependent recruitment of PP2A-B' to centromeres opposes Polo kinase activity, thereby ensuring that Rec8 in this region is unphosphorylated and refractory to cleavage by separase.

### *13.2.6 Protection of Centromeric Cohesion During Mammalian Mitosis*

Shugoshins act not only to protect centromeric cohesins from cleavage by separase during meiosis I, but are also found at centromeres during mammalian mitosis from prophase until the onset of anaphase. Of the two mammalian Shugoshin proteins, Sgo1 is important to prevent the removal of centromeric cohesin at the hands of the prophase pathway during mitosis (Salic et al., 2004; Tang et al., 2004; Kitajima et al., 2005; McGuinness et al., 2005). Mammalian Sgo2 has also been reported to have a similar, although controversial, role (Kitajima et al., 2005; Huang et al., 2007). The removal of cohesin by the prophase pathway is thought to occur not via phosphorylation of the $\alpha$-kleisin subunit of cohesin, but as a result of the phosphorylation of the Scc3-SA subunit, probably by Polo (Hauf et al., 2005). In mammalian cells, Sgo1 appears to function in the maintenance of centromeric cohesion in mitosis by preventing phosphorylation of the Scc3-SA subunit, since the cohesion defect seen in Sgo1 depleted cells can be rescued by a version of Scc3-SA with its mitotic phosphorylation sites mutated (McGuinness et al., 2005). Again, Sgo1 and Sgo2 are likely to act at least partially through the PP2A-B' phosphatase because Sgo1 and Sgo2 associate with PP2A-B' in mammalian mitosis and PP2A-B' is required to prevent precocious separation of sister centromeres in these cells (Kitajima et al., 2006; Tang et al., 2006). Furthermore, Sgo1 immunoprecipitates containing PP2A-B' can dephosphorylate Scc3-SA that had been previously phosphorylated by Polo in vitro (Kitajima et al., 2006).

Several observations suggest that Sgo1 functions in cohesion protection in ways other than influencing the state of cohesin phosphorylation through recruitment of PP2A-B'. First, in contrast to the situation in budding yeast meiosis, PP2A can localize to centromeres in the absence of Sgo1 in human cells, but it is unable to protect centromeric cohesion (Kitajima et al., 2006; Tang et al., 2006). Second, in fission yeast, co-overexpression of Rec8 and Sgo1 in mitosis causes a block to cohesin cleavage that is independent of PP2A (Kitajima et al., 2006). Third, depletion of Polo in mammalian mitosis cannot rescue the precocious separation of sister centromeres in Sgo1-deficient cells, but it is able to do so in PP2A-B'-deficient cells (McGuinness et al., 2005; Tang et al., 2006). Similarly, preventing cohesin phosphorylation in budding yeast meiosis, either by Cdc5 depletion or using the phospho-site mutant *rec8-A17* blocks cohesin cleavage only in the presence of Sgo1 (Brar et al., 2006). These findings suggest that Sgo1 has a PP2A-independent function in the protection of centromeric cohesion during both yeast meiosis and mammalian mitosis. Interestingly, a phospho-site *Scc3-SA* mutant rescues sister centromere separation caused by depletion of Sgo1 in mammalian cells, raising the possibility that, in addition to antagonising Polo activity in concert with PP2A-B', Sgo1 also counteracts other kinases to prevent cohesin removal. Indeed, removal of arm cohesin in both *C. elegans* meiosis and mammalian mitosis requires Aurora B kinase and the contribution of other kinases has not been ruled out (Rogers et al., 2002; McGuinness et al., 2005). Together, these findings indicate that Sgo1 acts at least partially independently of PP2A-B' to prevent loss of centromeric cohesion, but the mechanism by which it does this is currently unclear.

### 13.2.7 Establishment of a Specialized Domain of Cohesin Around the Centromere

How is the domain of cohesin around the centromere earmarked for protection during meiosis I? A similar mechanism must also exist to set up a domain of protected cohesin during mammalian mitosis. How this occurs is poorly understood, but there are hints that establishment of the protector could be coupled to dedicated mechanisms of cohesin establishment around the centromere. In most eukaryotes, centromeres are buried in heterochromatin and one important function of the pericentric heterochromatin could be to attract cohesin to this region (Bernard et al., 2001b; Nonaka et al., 2002; Kitajima et al., 2003b). In fission yeast, Rec8 is localized at both the core centromere and the pericentromeric region (Watanabe et al., 2001; Fig. 13.3). The heterochromatic nature of the pericentromere is important to attract cohesion to the pericentromere (Bernard et al., 2001b; Nonaka et al., 2002; Kitajima et al., 2003b). In meiosis, recruitment of Rec8–Psc3 complexes to the pericentric region, but not to the central core, of the fission yeast centromere depends on two factors that are required for heterochromatin formation in the pericentromere, Clr4 and Swi6. This suggests that cohesin complexes localize to core centromeres and pericentromeric regions through different mechanisms (Kitajima et al., 2003b). Furthermore, Rec8 in the central core and pericentromere appears to have different functions during meiosis in fission yeast. Importantly, Clr4 and Swi6 are required for maintenance of cohesion beyond meiosis I, indicating that the cohesin that resides in the pericentric region must be protected during meiosis I (Kitajima et al., 2003b). Accordingly, Sgo1 resides within these pericentromeric regions where cohesins must be preserved (Kitajima et al., 2004; Rabitsch et al., 2004), although recently it has additionally been found in the central core (Riedel et al., 2006). In contrast, cohesins in the central core are dispensable for the protection of centromeric cohesion but instead are essential for the mono-orientation of sister kinetochores (Yokobayashi et al., 2003; Yokobayashi and Watanabe, 2005). A factor called Moa1 is important for cohesin establishment in the central core region (Fig. 13.3; Yokobayashi and Watanabe, 2005; see below).

Notably, budding yeast lacks pericentric heterochromatin, but analysis of cohesin distribution along both mitotic and meiotic chromosomes has revealed that the highest levels of cohesin are found in pericentric regions (Blat and Kleckner, 1999; Laloraya et al., 2000; Glynn et al., 2004; Lengronne et al., 2004; Weber et al., 2004; Kiburz et al., 2005). In addition to narrowly defined 'cohesin-associated regions' (CARs) on chromosome arms, cohesin is enriched in an ~50 kb domain surrounding the centromere, although the centromere itself is comprised of only ~120 bp (Glynn et al., 2004; Lengronne et al., 2004; Weber

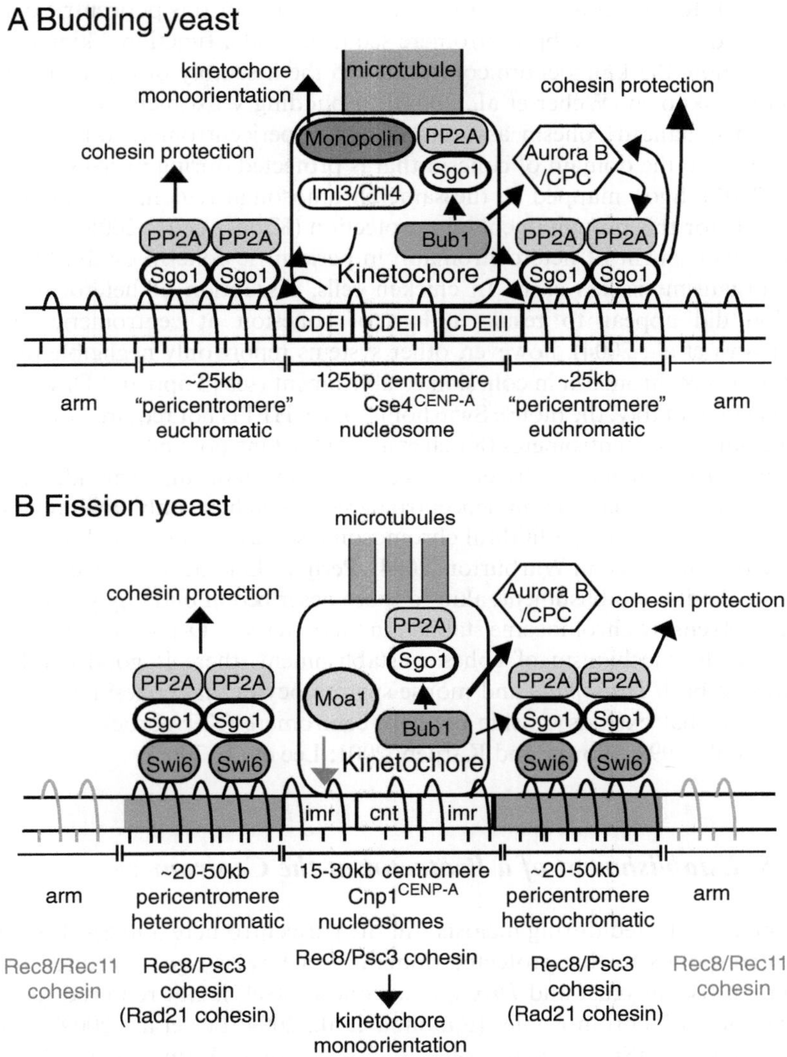


**Fig. 13.3 Model of yeast meiotic centromeres**. A schematic diagram showing the functional organization of the budding (A) and fission (B) yeast centromeres. The locations of regulators of cohesin protection and kinetochore mono-orientation and known localization dependencies are shown. (**A**) Budding yeast have point centromeres and lack pericentric heterochromatin. Nevertheless, the existence of a functional equivalent is suggested because cohesin and Sgo1 are highly enriched in ~50 kb region around the centromere where cohesin is protected. (**B**) In fission yeast, the pericentric heterochromatin directs cohesin binding to this region. Centromeric, Moa1-dependent cohesin is important for kinetochore mono-orientation during meiosis I. The composition of cohesin complexes in the centromere/pericentromere and chromosome arms is distinct. For details see text

et al., 2004; Kiburz et al., 2005; Fig. 13.3). Assembly of this pericentric cohesin domain requires the 120 bp centromere sequence and a functional kinetochore, although how the kinetochore contributes to the assembly of this cohesin-rich region is unknown (Weber et al., 2004). In budding yeast meiosis, the ~50 kb region of enhanced cohesin association in the pericentromere was shown to correspond to the domain of cohesin that is protected during meiosis I (Kiburz et al., 2005). Sgo1 mapped to the same chromosomal region, in support of a direct role for this protein in cohesin protection (Kiburz et al., 2005).

The involvement of heterochromatin in pericentric cohesin establishment in other organisms is less clear. In chicken cells, disruption of heterochromatin function did appear to result in loss of cohesion at centromeric regions (Fukagawa et al., 2004), however, other systems report only negligible requirement of heterochromatin in cohesion establishment (see (Topp and Dawe, 2006) for review). Notably, the mouse Swi6 homologue, HP1 is not required for cohesin localization at pericentromeres (Koch et al., 2008). One possibility is that cohesin assembly in this region is driven by redundant mechanisms. This idea is supported by the fact that human neocentromeres, which have only small regions of heterochromatin, support faithful chromosome segregation during both mitosis and meiosis (reviewed in Warburton, 2004). Perhaps kinetochore-driven mechanisms of cohesion establishment, akin to those described in budding yeast (Weber et al., 2004) ensure chromosome stability in this situation. Despite the uncertainty regarding the mechanism of cohesin establishment, there is good cytological evidence in both *Drosophila* and mouse spermatocytes that cohesin is protected in a region that extends outside the core centromere (Kerrebrock et al., 1995; Moore et al., 1998; Blower and Karpen, 2001; Lee et al., 2006).

### 13.2.8 Establishment of a Protector at the Centromere

Cohesin is protected during meiosis I in the pericentromere and Sgo1 localization corresponds to this protected domain, but how is Sgo1 recruited to this region? In fission yeast and *Drosophila*, cohesin itself is not required for Sgo1 association with chromosomes (Kitajima et al., 2004; Lee et al., 2004b). However, in maize, ZmSgo1 localization is Rec8-dependent (Hamant et al., 2005). In budding yeast, Sgo1 localization was found to be partially dependent on cohesin because Rec8 is required for Sgo1 association with the pericentromere but not the core centromere (Kiburz et al., 2005; Fig. 13.3). Similarly, in the absence of either Iml3 or Chl4, two kinetochore proteins that are required for the maintenance of centromeric cohesion during meiosis (Marston et al., 2004), Sgo1 associates normally with the core centromere, but is reduced at pericentromeres (Kiburz et al., 2005). Perhaps these results reflect the existence of two separate pathways for Sgo1/MEI-S332 localization to the core centromere and pericentromere. In addition to the protector proteins Sgo1 and PP2A-B', several other factors have been found to regulate centromeric cohesion, some of which influence Sgo1/PP2A-B' localization at kinetochores. Chief among these are the major mitotic and meiotic kinases, Bub1, Aurora B and Polo. Several studies have examined the inter-dependencies between Sgo1, PP2A and these kinases. The outcome of these studies has revealed a different hierarchy of assembly at the kinetochore, between yeast meiosis (Fig. 13.3) and mammalian mitosis.

The spindle checkpoint protein, Bub1, which is localized at kinetochores, plays a universal role in mediating Sgo1, Sgo2 and PP2A-B' localization, suggesting that Bub1 is a master regulator of protector assembly. In fission yeast, budding yeast and mouse oocytes, cells lacking Bub1 prematurely lose sister centromere cohesion during meiosis (Bernard et al., 2001a; Kitajima et al., 2004, 2005; Kiburz et al., 2005). This can be explained by the failure to localize Sgo1 in the absence of Bub1 (Tang et al., 2004; Kiburz et al., 2005; Kitajima et al., 2005; Riedel et al., 2006; Boyarchuk et al., 2007). Interestingly, Sgo1 was displaced onto chromosome arms in mammalian and budding yeast cells lacking Bub1, suggesting that Sgo1 does not require Bub1 to localize to chromatin, but only for its concentration at centromeres (Kitajima et al., 2005; Kueng et al., 2006; Riedel et al., 2006). Mutations in the catalytic site of the predicted kinase domain in Bub1 in both budding and fission yeast also prevent proper Sgo1 localization to the centromere, suggesting that the kinase activity of Bub1 is important for its Sgo1-localizing activity (Kitajima et al., 2004; Rabitsch et al., 2004; Vaur et al., 2005; Fernius and Hardwick, 2007). Similarly, Bub1 is required for Sgo2 localization in mammalian mitosis (Huang et al., 2007) and for PP2A-B' localization to centromeres in fission yeast meiosis and mammalian mitosis (Riedel et al., 2006; Tang et al., 2006). Furthermore, Bub1 is required for the kinetochore localization of another regulator of centromeric cohesion, Aurora B (see below), in budding and fission yeast meiosis (Hauf et al., 2007; Yu and Koshland, 2007), *Xenopus* extracts and mammalian mitosis (Boyarchuk et al., 2007; Pouwels et al., 2007). This is, however, controversial because other studies have found Aurora B to be localized normally in mammalian cells lacking Bub1 (Johnson et al., 2004; Meraldi and Sorger, 2005). Nevertheless, these findings implicate Bub1 as a regulatory scaffold that coordinates the assembly of protector proteins at the centromere.

The Aurora B kinase is a constituent of the 'chromosome passenger complex' (CPC), which also contains inner centromere protein (INCENP), Survivin and Borealin (see (Ruchaud et al., 2007) for review). The CPC, which regulates many aspects of mitosis and meiosis, characteristically dissociates from the inner centromere and relocates to the spindle midzone at the onset of anaphase (Ruchaud et al., 2007). During meiosis, however, the CPC components, INCENP and Aurora B were found to be maintained on centromeres during anaphase I in mouse spermatocytes (Parra et al., 2003) and similar localization patterns have been reported for INCENP and Aurora B in *Drosophila* and budding yeast meiosis, respectively (Resnick et al., 2006; Monje-Casas et al., 2007; Yu and Koshland, 2007). Consistent with their retention on centromeres during anaphase I, *Drosophila* INCENP and budding yeast Aurora B (Ipl1) are required for the maintenance of centromeric cohesion during meiosis (Resnick et al., 2006; Monje-Casas et al., 2007; Yu and Koshland, 2007). In both

*Drosophila* and budding yeast meiosis, the important role of the CPC is likely the establishment of the protector complex at centromeres, although the mechanism could differ in the two systems.

In *Drosophila,* INCENP is required for MEI-S332 localization at centromeres, both during mitosis in cultured cells and in male meiosis, but no effect on INCENP localization was observed in *mei-S332* mutant flies (Resnick et al., 2006). A direct role for the CPC in recruitment of MEI-S332 to centromeres was suggested by the observations that MEI-S332 and INCENP physically interact and that MEI-S332 can be phosphorylated by Aurora B in vitro (Resnick et al., 2006). Furthermore, a version of MEI-S332 in which one of the three potential consensus sites for Aurora B is mutated does not associate properly with centromeres. The interaction between CPC components and the localization dependence of Sgo1 and Sgo2 on CPCs is conserved in *Xenopus* egg extracts and mammalian mitosis (Boyarchuk et al., 2007; Kawashima et al., 2007). Depletion of survivin or Aurora B by RNA interference (RNAi) in human mitotic cells caused Sgo1 to be displaced from the centromeres and, instead to be weakly localized along the length of chromosomes, similar to the situation reported in Bub1-depleted cells (Kitajima et al., 2005; Dai et al., 2006; Kueng et al., 2006; Riedel et al., 2006; Boyarchuk et al., 2007; Kawashima et al., 2007; Pouwels et al., 2007). Conversely, depletion of Sgo1 by RNAi did not prevent Aurora B or Survivin localizing to the centromere in these systems (Boyarchuk et al., 2007; Kawashima et al., 2007). These observations suggest that Sgo1 requires neither the CPC nor Bub1 to associate with chromosomes, but that they are required to concentrate Sgo1 in the centromeric region. Furthermore, the increased arm cohesion that has been observed in CPC-deficient cells, could be explained due to ectopic protection misplaced Sgo1, since depletion of both Sgo1 and Aurora B allows sister chromatids to separate (McGuinness et al., 2005; Dai et al., 2006).

Although important in protector establishment in yeast meiosis, the role of the CPC appears to be divergent. In budding yeast, Sgo1 localization was only slightly perturbed during meiosis in Ipl1-deficient cells. Similarly, in fission yeast meiosis, Sgo1 localization is only marginally altered in cells with impaired survivin (Bir1) function (Kawashima et al., 2007; Yu and Koshland, 2007). These observations raise the possibility that the CPC does not protect meiotic cohesin by localizing Sgo1 in these organisms. Instead, at least in budding yeast, retention of PP2A-B' at centromeres could be the important activity of the CPC in cohesion protection. Although the PP2A-B' subunit, Rts1 localizes to centromeres normally early in meiosis, it is not maintained at the onset of anaphase I (Yu and Koshland, 2007). Contrary to what has been found in other systems, Ipl1 localization was reported to be at least partially dependent on Sgo1, suggesting that an Ipl1-dependent mechanism could contribute to the Sgo1-mediated maintenance of PP2A-B' at kinetochores during anaphase (Yu and Koshland, 2007), although another study found that Ipl1 localized normally in the absence of Sgo1 (Kiburz et al., 2008).

Another important difference between protector establishment in meiosis and mammalian mitosis is the inter-dependence of Sgo1 and PP2A. In yeast meiosis,

PP2A-B' depends on Sgo1 for its localization at kinetochores, but the opposite is not true (Kitajima et al., 2006; Riedel et al., 2006). Similarly, in mouse oocytes, Sgo2 is required for PP2A-B' localization (Lee et al., 2008). In contrast, during mammalian mitosis, Sgo1 requires PP2A-B' to be recruited onto kinetochores, but not the reverse (Kitajima et al., 2006; Tang et al., 2006). Instead, Sgo2, recruits PP2A-B' to kinetochores (Kitajima et al., 2006). In conclusion, Bub1 appears to act as a master organizer of cohesin protection through collaboration of Sgo1, PP2A and Aurora B. In particular, the inter-dependence of Sgo1 and PP2A, and of Sgo1 and Aurora B, appears to be reversed in meiosis compared to mitosis in the systems so far examined. How the interplay between Sgo1/PP2A, Aurora B/CPC and Polo kinase could lead to cohesin protection is summarized in the speculative model shown in Fig. 13.4.

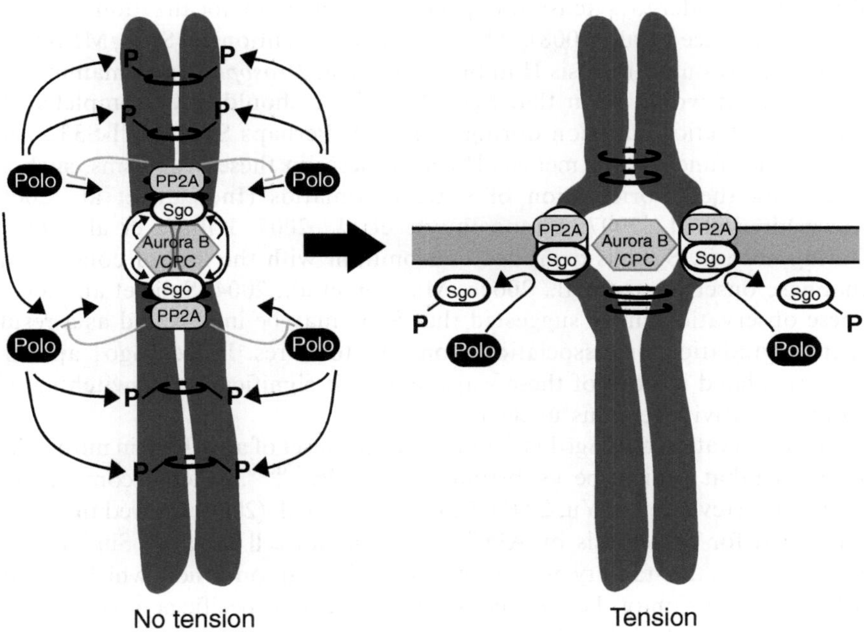


**Fig. 13.4 Generalized model for the mechanism of cohesin protection at the centromere**. In the absence of tension across sister kinetochores, Aurora B/ chromosome passenger complex (CPC) ensures that the Sgo/ protein phosphatase 2A (PP2A) protector complex associates stably with centromeres. This allows PP2A to antagonise Polo kinase-dependent phosphorylation of cohesin and Shugoshin in the vicinity, thereby ensuring their retention at centromeres. Polo-dependent phosphorylation of cohesin on chromosome arms, however, contributes to its release. Upon bipolar attachment of microtubules, sister kinetochores come under tension and the Sgo/PP2A protector complex is spatially separated from both cohesin and Aurora B/CPC. This frees cohesin from protection and concomitantly renders Sgo susceptible to phosphorylation from Polo, leading to the dissociation of the protector from kinetochores. Note that this model is both generalized and speculative. In particular, the details of the interplay between key factors differs depending on the organism studied (see text)

### *13.2.9 Switching Off the Protector*

The notion that the association of Sgo1 with kinetochores is important for its cohesion-protection activity at the centromeres predicts that Sgo1 must be cleared from this region or otherwise inactivated to allow sister chromatids to separate at the appropriate time. How might this occur? In some organisms, Sgo1 dissociates from kinetochores either prior to meiosis II and the need to lose centromeric cohesion as in fission yeast and maize (Kitajima et al., 2004; Rabitsch et al., 2004; Hamant et al., 2005). However, in budding yeast and *Drosophila* Sgo1/MEI-S332 dissociate from kinetochores during anaphase II, concomitant with centromeric cohesin loss (Moore et al., 1998; Katis et al., 2004a; Marston et al., 2004). In mammalian oocytes and spermatocytes, Sgo2 is also present at centromeres during meiosis II, although it undergoes tension-dependent changes in localization (Gomez et al., 2007; Lee et al., 2008). The reason for retention of Sgo1/MEI-S322 at centromeres until meiosis II in budding yeast, *Drosophila* and mammals is unclear, as it would seem that Sgo1/MEI-S332 should have completed its cohesin-protection function during meiosis I. Perhaps Sgo1/MEI-S332 performs other functions at meiosis II centromeres in these organisms, such as promoting the bi-orientation of sister chromatids (Indjeian et al., 2005; Kawashima et al., 2007; Vanoosthuyse et al., 2007; Kiburz et al., 2008). During mitosis, Sgo1 is degraded concomitant with the loss of cohesion at anaphase onset (Katis et al., 2004a; Marston et al., 2004; Salic et al., 2004). These observations have suggested that Sgo1 may be inactivated as a result of its degradation or dissociation from kinetochores. Indeed Sgo1 appears to be regulated in both of these ways, but their significance in switching off protector activity remains unclear.

The observation that Sgo1 is degraded at the onset of anaphase in mitosis led to the idea that it might be a substrate for the $APC^{Cdc20}$, which becomes active at this time (reviewed in Yu, 2007). Indeed, Salic et al. (2004) showed that Sgo1 is targeted for proteolysis by $APC^{Cdc20}$ in human cell extracts. Similarly, in budding yeast, the stability of Sgo1 in meiosis depends on Mnd2, which inhibits $APC^{Ama1}$, a version of the APC coupled to a meiosis-specific activator, Ama1 (Oelschlaegel et al., 2005; Penkner et al., 2005). Other pathways are likely to inactivate Sgo1, however, because the segregation of sister chromatids at meiosis II is not prevented in *ama1*Δ mutants, even though Sgo1 is stabilized (Penkner et al., 2005). In addition, work in fission yeast has also indicated that degradation of Sgo1 might not be a prerequisite for chromosome segregation during meiosis II. Altering the 3' UTR of the *sgo1* gene stabilized Sgo1 on kinetochores during meiosis II, but this did not block sister centromere separation (Rabitsch et al., 2004).

Another candidate for a regulator of Sgo1 in budding yeast is Spo13, which is present only in meiosis I (Katis et al., 2004b; Lee et al., 2004a). Spo13 is required for the maintenance of centromeric cohesion and the control of kinetochore orientation in meiosis (Katis et al., 2004b; Lee et al., 2004a; see below). In *spo13Δ* mutants, Rec8 at anaphase I centromeres is not completely lost, but is severely diminished, suggesting that centromeric cohesin is not properly protected (Klein et al., 1999; Katis et al., 2004b; Lee et al., 2004a). This could be explained because Sgo1 binding at the centromeres is dramatically reduced in *spo13Δ* cells (Kiburz et al., 2005). Consistent with such a role, Spo13 is enriched at centromeres during meiosis (Katis et al., 2004b; Lee et al., 2004a). However, overexpression of *SPO13* in mitosis causes a block to cohesin cleavage that does not require *SGO1* (Lee et al., 2002, 2004a; Shonn et al., 2002). This implies that Spo13 can prevent precocious loss of centromeric cohesion independently of Sgo1, although the involvement of the spindle checkpoint was not ruled out. Indeed, the requirement for Spo13 could be rather indirect as it is also an inhibitor of the APC, which could explain how Spo13 can control multiple key meiosis I events (Katis et al., 2004b). However, Sgo1 is stable in *spo13Δ* mutants (Kiburz et al., 2005), implying that the function of Spo13 in the maintenance of centromeric cohesion is not to shield Sgo1 from the APC.

In *Drosophila*, MEI-S332 dissociation from kinetochores is regulated by phosphorylation by Polo kinase (Clarke et al., 2005). In mammalian cells, Polo is also required for Sgo1 dissociation and, remarkably, Sgo1 localization in PP2A-depleted cells is restored by co-depletion of Polo (Tang et al., 2006). Mutation of the Polo-dependent phosphorylation sites in MEI-S332 prevents its dissociation from kinetochores, but surprisingly, does not prevent sister chromatid separation at meiosis II, suggesting that dissociation of Sgo1 from kinetochores is also not required for the inactivation of the protector (Clarke et al., 2005). An attractive hypothesis is that redundant mechanisms inactivate the protector complex. Interestingly, cohesin may lose its competence to be protected after a critical point in the cell cycle. Work in budding yeast has suggested that the meiotic cohesin, Rec8, is highly dependent on phosphorylation to be susceptible to cleavage in meiosis I, but less so in meiosis II (Brar et al., 2006). The difference in the mode of kinetochore–microtubule attachment between meiosis I and meiosis II may also influence the ability of Shugoshins to protect cohesin. In mammalian cells, immunolocalization studies have found that Sgo2 co-localizes with cohesins in the inner centromere during meiosis I when sister kinetochores are mono-oriented and therefore not under tension (Gomez et al., 2007; Lee et al., 2008). During meiosis II, bi-orientation of sister kinetochores generates tension, which correlates with Sgo2 redistribution to sites close to the kinetochores and away from cohesin (Gomez et al., 2007; Huang et al., 2007; Lee et al., 2008). These findings suggest a model in which tension at meiosis II kinetochores pulls Sgo2 away from cohesin, resulting in its deprotection (Gomez et al., 2007; Lee et al., 2008; Fig. 13.4).

## 13.3 Mono-Orientation of Kinetochores

During mitosis and meiosis II, kinetochores of sister chromatids (sister kinetochores) attach to microtubules from opposite poles. This arrangement, known as bi-orientation, or amphitelic attachment (Fig. 13.5) leads to the separation of sister chromatids to opposite poles upon cohesin destruction. During meiosis I, however, a unique segregation event occurs in which homologous chromosomes are segregated to opposite poles and sister chromatids, rather than being separated, co-segregate (Fig. 13.5). To accomplish this, sister kinetochores must be directed to microtubules emanating from the same pole, called mono-orientation or monopolar attachment. This requires some important modifications to the kinetochore, the molecular basis of which are just beginning to be understood.

### *13.3.1 Monopolar Attachment is Achieved by Modification of the Kinetochore*

Micromanipulation experiments using grasshopper spermatocytes demonstrated that mono-orientation is a property of the kinetochore rather than the spindle or cytoplasmic factors. A pair of homologues from grasshopper cells in meiosis I segregate in a meiosis I-like manner when transplanted on a meiosis II spindle (Paliulis and Nicklas, 2000). Cytological experiments have indicated that the geometry of sister kinetochores changes during meiosis and that sister kinetochores behave as a single microtubule-binding unit during meiosis I (reviewed in Moore and Orr-Weaver, 1998). Visualization of sister kinetochores of *Drosophila* and mouse spermatocytes by electron microscopy has suggested that they are fused at the time of microtubule attachment in meiosis I, but differentiate into two separate kinetochores in later stages of meiosis I (Goldstein, 1981; Parra et al., 2004). The fusion, or side-to-side arrangement, of sister kinetochores is thought to facilitate their monopolar attachment in meiosis I by providing steric constraints, just as a back-to-back geometry is believed to promote the bi-orientation of sister kinetochores during mitosis and meiosis II (reviewed in Hauf and Watanabe, 2004).

### *13.3.2 Monopolin Achieves Monopolar Attachment in Budding Yeast*

Recently, studies in yeast have begun to uncover the molecular mechanism of mono-orientation in meiosis I. In budding yeast, a 'monopolin' complex assembles at kinetochores and directs monopolar attachment. To date, monopolin is known to have four subunits, a meiosis-specific protein, Mam1, the Casein kinase 1 $\delta/\epsilon$ orthologue (CK1 $\delta/\epsilon$) Hrr25 along with two proteins, Csm1 and

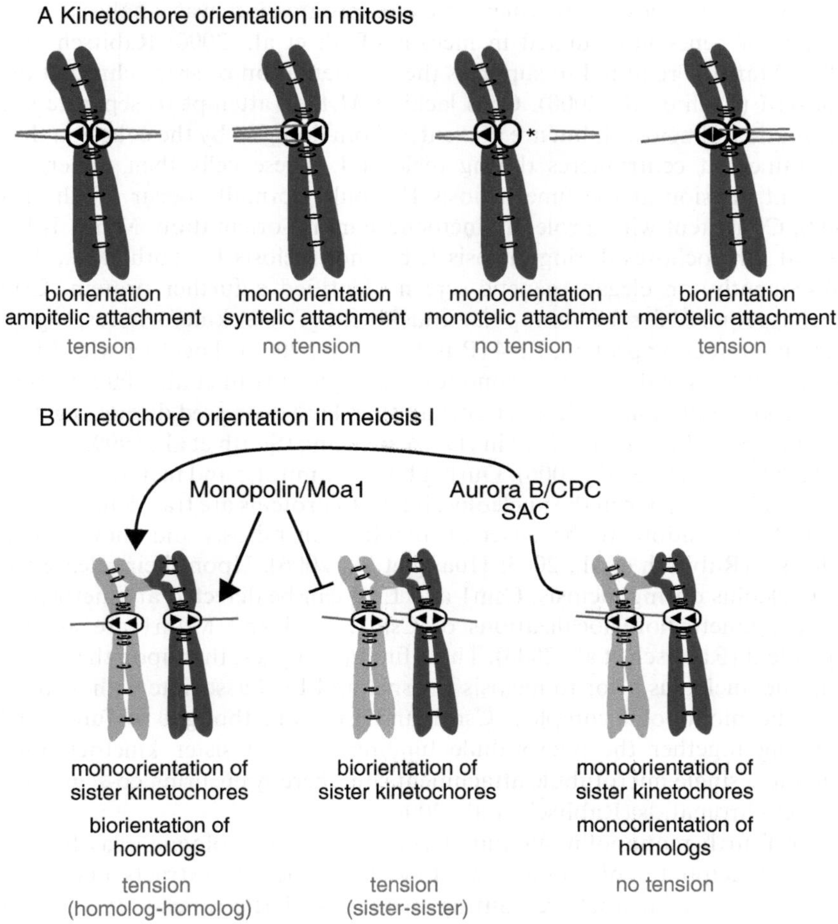


**Fig. 13.5** Modes of kinetochore–microtubule attachment during mitosis (A) and meiosis I (B). (**A**) Syntelic and monotelic attachment are prevented in mitosis by the spindle checkpoint and Aurora B, which sense unattached kinetochores and the lack of tension. In fission yeast, Pcs1 and Mde4 prevent merotelic attachment, possibly by clamping together adjacent microtubule binding sites. Note that merotelic attachment is not permitted in budding yeast since there is a single microtubule binding site per kinetochore (Winey et al., 1995). (**B**) During meiosis I, sister kinetochores are mono-oriented and homologues are bi-oriented. Linkages between homologues (usually chiasmata) allow tension to be established between them. Sister kinetochore bi-orientation is prevented during meiosis I by monopolin in budding yeast and Moa1-dependent assembly of cohesin at the centromere in fission yeast. Mono-orientation of homologues is monitored and corrected by Aurora B in budding yeast

Lrs4, that generally reside in the nucleolus, but which promote mono-orientation upon their release. Mam1 (*mo*nopolar microtubule *a*ttachment during *m*eiosis I) was the first monopolin subunit to be identified through a functional genomics study where meiotic progression was analysed in the absence of genes upregulated in meiosis (Toth et al., 2000; Rabitsch et al., 2003). Mam1 is required to suppress the bi-orientation of sister chromatids at meiosis I (Toth et al., 2000). Cells lacking *MAM1* attempt to separate sister chromatids at meiosis I, but are prevented from doing so by the cohesion that is maintained at centromeres during meiosis I. These cells then undergo an aberrant division at the time meiosis II would normally occur (Toth et al., 2000). Consistent with a role in kinetochore mono-orientation, Mam1 is localized at kinetochores during meiosis I, but not meiosis II (Toth et al., 2000). Subsequently, an elegant genetic screen identified a further protein, Csm1, which is required for the monopolar attachment of sister kinetochores, together with an interacting protein, Lrs4 (Rabitsch et al., 2003). Together with Mam1, Csm1 and Lrs4 make up the 'monopolin' complex (Toth et al., 2000; Rabitsch et al., 2003), but, unlike Mam1, Csm1 and Lrs4 are expressed during vegetative growth where they are involved in rDNA silencing (Smith et al., 1999; Rabitsch et al., 2003; Huang et al., 2006). During both the mitotic and meiotic cell cycles, Csm1 and Lrs4 reside in the nucleolus, but both proteins are transiently released for a short window at the onset of anaphase in mitosis and shortly before meiosis I (Rabitsch et al., 2003; Huang et al., 2006). Upon their release from the nucleolus during meiosis, Csm1 and Lrs4 can be detected at kinetochores, and the kinetochore localizations of Csm1, Lrs4 and Mam1 are all interdependent (Rabitsch et al., 2003). These findings suggest that upon their release from the nucleolus prior to meiosis I, Csm1 and Lrs4 associate with Mam1 to form the monopolin complex. Csm1 and Lrs4 are thought to function by clamping together the microtubule binding sites on sister kinetochores to generate a single microtubule attachment site, thereby ensuring co-segregation of sister chromatids (Rabitsch et al., 2004).

The fourth monopolin subunit, Hrr25, a CK1 orthologue, was found in immunoprecipitates of Mam1 and Lrs4 from meiotic extracts and is also associated with kinetochores during metaphase I (Petronczki et al., 2006). The CK1 family of protein kinases have been implicated in diverse cellular functions (Knippschild et al., 2005). Consistent with the fact that Hrr25 is the sole budding yeast CK1 that is not membrane associated, deletion of *HRR25* is very deleterious to cell growth and progression through meiosis (Petronczki et al., 2006). However, two alleles of *HRR25* have been generated which specifically abolish kinetochore mono-orientation and exhibit phenotypes very similar to that of cells lacking *MAM1*, *CSM1* or *LRS4* (Petronczki et al., 2006). The *hrr25-zo* allele produces a protein that is incapable of interacting with Mam1 and was used to demonstrate that the Mam1–Hrr25 interaction is necessary for the association of Mam1 with kinetochores (Petronczki et al., 2006). The *hrr25-as1* allele is sensitive to chemical inhibition by an adenine analogue that can bind in the enlarged cleft of the mutant protein and inhibit its kinase activity (Bishop et al., 2000; Petronczki et al., 2006). Analysis of this mutant showed that CK1 activity is not required for recruitment of Mam1 to kinetochores, although it is essential for the mono-orientation of sister kinetochores in budding yeast (Petronczki et al., 2006). What then is the relevant substrate for CK1 in mono-orientation? Both Mam1 and Rec8 were found to depend on CK1 for normal levels of phosphorylation, suggesting that CK1 could promote sister kinetochore mono-orientation by phosphorylating one or both of these proteins.

The finding that Rec8 phosphorylation depends on Hrr25 is significant as Rec8 is important for kinetochore orientation in fission yeast, *Arabidopsis* and maize (Watanabe and Nurse, 1999; Yu and Dawe, 2000; Chelysheva et al., 2005). However, in budding yeast, there does not seem to be a specific requirement for the meiotic cohesin in mono-orientation of sister chromatids because replacement of Rec8 with the mitotic cohesin, Scc1/Mcd1, permits mono-orientation, although centromeric cohesion is not protected (Toth et al., 2000). Hrr25 was found to phosphorylate Rec8, but not Scc1, suggesting that cohesin might not be the critical substrate of Hrr25 in specifying mono-orientation (Petronczki et al., 2006). Remarkably, in the absence of either cohesin, mono-orientation is still established at meiosis I, provided that monopolin is present, indicating that monopolin can hold sister kinetochores together at meiosis I independently of cohesin (Monje-Casas et al., 2007). Therefore, although Hrr25-dependent phosphorylation of Rec8 could contribute to kinetochore mono-orientation, other critical targets (e.g. Mam1 (Petronczki et al., 2006)) must exist.

The association of monopolin with kinetochores directs their monopolar attachment, but how is monopolin controlled to ensure it accumulates at kinetochores only during meiosis I? A key event in monopolin formation is the release of Csm1 and Lrs4 from the nucleolus. During meiosis, the gene encoding budding yeast Polo kinase, *CDC5*, is required for the transient release of Csm1 and Lrs4 that occurs shortly before the first meiotic division (Clyne et al., 2003). Consistent with this, cells depleted for Cdc5 fail to establish monopolar attachment and fail to accumulate Mam1 at kinetochores (Clyne et al., 2003; Lee and Amon, 2003). Phosphorylation of both Mam1 and Lrs4 is dependent on Cdc5, suggesting that Cdc5 could promote mono-orientation through phosphorylating multiple targets (Clyne et al., 2003; Katis et al., 2004b). A second mechanism to ensure monopolin is functional only in meiosis I, is to limit *MAM1* expression to early meiosis (Toth et al., 2000). Mam1 may be the only critical subunit of monopolin that is meiosis-specific, since mono-orientation of sister kinetochores can be induced in mitosis by *MAM1* overexpression, provided that the gene encoding Polo kinase, *CDC5*, is also overexpressed (Monje-Casas et al., 2007). The need for *CDC5* overexpression is due to the fact that in mitosis, Csm1 and Lrs4 are not released from the nucleolus until anaphase (Toth et al., 2000; Rabitsch et al., 2003; Huang et al., 2006) but artificial release of Lrs4, and likely also Csm1, can be induced by *CDC5* overexpression (Monje-Casas et al., 2007). The fact that the timing of Csm1/Lrs4 nucleolar release differs in mitosis and meiosis, despite the fact that Polo kinase is present in both kinds of division, raise the possibility that specialized mechanisms regulate Polo in meiosis. A candidate for such a regulator is the meiosis I-specific Spo13 that regulates multiple meiotic events, including monopolin function (see above). In *spo13Δ* cells, Mam1 is not stably retained at kinetochores (Katis et al., 2004b; Lee et al., 2004a). Furthermore, phosphorylation of Lrs4 is reduced in *spo13Δ* cells, as well as *CDC5*-depleted cells (Katis et al., 2004b). This raises the possibility that Spo13 promotes phosphorylation of Lrs4 by Cdc5. In the artificially induced mono-orientation caused by co-overexpression of *MAM1* and *CDC5* in mitosis, simultaneous overexpression of *SPO13* was found not to enhance mono-orientation (Monje-Casas et al., 2007). This finding suggests that Spo13 might play a regulatory, rather than direct, role in monopolin function.

### *13.3.3 Cohesin is Required for Monopolar Attachment in Other Organisms*

Is the monopolin complex conserved in other organisms? CK1s are highly conserved and at least in fission yeast, a role in mono-orientation is likely. Mutation of both fission yeast CK1$\delta/\epsilon$ homologues, *hhp1* and *hhp2*, results in defective meiosis I chromosome segregation that would be consistent with bi-orientation of sister chromatids (Petronczki et al., 2006). Aside from CK1, however, orthologues of other monopolin subunits have been identified only in fungi (Rabitsch et al., 2003). Fission yeast possesses orthologues of both Csm1 and Lrs4, called Pcs1 and Mde4, respectively, however, neither of these proteins are required for mono-orientation during meiosis I (Rabitsch et al., 2003; Gregan et al., 2007). Instead, Pcs1 and Mde4 contribute to chromosome segregation during mitosis and meiosis II (Rabitsch et al., 2003; Gregan et al., 2007). Nevertheless, Csm1/Lrs4 and Pcs1/Mde4 complexes have been proposed to perform the same physiological function: clamping together adjacent microtubule binding sites (Rabitsch et al., 2003; Gregan et al., 2007). In budding yeast mitosis, a single microtubule binds per sister kinetochore, so that merotely (attachment of a single chromatid to microtubules from opposite poles) is not possible (Winey et al., 1995). In fission yeast, however, each kinetochore binds between two and four microtubules (Ding et al., 1993), so a mechanism is required to ensure that binding sites on the same sister kinetochore are occupied by microtubules from the same spindle pole body, thereby preventing merotely. Pcs1 and Mde4 are likely to provide this function (Gregan et al., 2007). In the absence of Pcs1 or Mde4, single lagging chromatids are observed in mitosis, characteristic of merotelic attachment and indicative of a failure to clamp together adjacent microtubule binding sites (Gregan et al., 2007). Csm1 and Lrs4 are likely to carry out a similar function in budding yeast meiosis by clamping together the microtubule binding sites of the sister chromatids (Rabitsch et al., 2003; Gregan et al., 2007). However, it is unlikely that both sister kinetochores can bind microtubules, as, interestingly, Winey et al. (2005) obtained evidence that a pair of sister chromatids are attached to a single microtubule in meiosis I. This suggests that in addition to being clamped, sister kinetochores are either fused, or one of the kinetochores is silenced. Presumably the other monopolin subunits collaborate with Csm1 and Lrs4 to reduce the number of microtubule binding sites.

If Pcs1/Mde4 are not required for meiosis I, what does promote mono-orientation of kinetochores in fission yeast meiosis I? In contrast to the situation in budding yeast (Toth et al., 2000), in fission yeast, the meiotic cohesin, Rec8 plays a role in mono-orientation that is not fulfilled if it is replaced by its mitotic counterpart, Rad21 (Yokobayashi et al., 2003). Rec8 is localized both to the central core and the pericentromere of the fission yeast centromere (Watanabe et al., 2001), whereas substitutive Rad21, which cannot confer mono-orientation, localizes only to the pericentromere in meiosis. This suggests that cohesin complexes in the central core are important for the mono-orientation of sister kinetochores (Yokobayashi et al., 2003; Fig. 13.3B). Recently, a meiosis-specific factor, Moa1 (*mo*nopolar *a*ttachment), was identified that functions together with Rec8 to establish kinetochore mono-orientation (Yokobayashi and Watanabe, 2005). Moa1 localizes to the central core of the fission yeast centromere until the onset of anaphase in meiosis I and appears to be required for the proper establishment of cohesion, specifically in this region (Yokobayashi and Watanabe, 2005). Consistently, Moa1 interacts with Rec8-containing cohesin complexes, but not Rad21-containing cohesin complexes (Yokobayashi and Watanabe, 2005). In the absence of Moa1, mono-orientation of sister kinetochores is disrupted, but the protection of centromeric cohesion is conserved. These results confirm that cohesins in the central core bring about the monopolar attachment of sister kinetochores in fission yeast. However, other factors must be involved because expression of Rec8 in mitosis is not sufficient to direct monopolar attachment (Yokobayashi et al., 2003). The function of Rec8 in monopolar attachment appears to be conserved in maize and *Arabidopsis* (Chelysheva et al., 2005; Hamant et al., 2005), but how cohesin brings about the mono-orientation of sister kinetochores is unknown. It is thought that centromeric cohesin alters the geometry of sister kinetochores, promoting their attachment to microtubules from the same pole (reviewed in Hauf and Watanabe, 2004).

## 13.4 Bi-Orientation of Homologues

In addition to a mechanism that ensures the monopolar attachment of sister kinetochores, a means must exist to ensure that homologous sister chromatid pairs are segregated to opposite poles (Fig. 13.5). This bi-orientation of homologous chromosomes in meiosis I is regulated by the SAC, which is best understood in mitosis (reviewed in Musacchio and Salmon, 2007).

### 13.4.1 Spindle Assembly Checkpoint in Mitosis

Briefly, the SAC monitors the correct bipolar attachment of sister chromatids to the mitotic spindle. Ultimately, the SAC inhibits anaphase onset by preventing $APC^{Cdc20}$ activation, and thereby cohesin cleavage, until all sister chromatids are bi-oriented on microtubules. The SAC requires the BUB and MAD proteins (BUB1, BUBR1/Mad3, MAD1, MAD2, MAD3) that are enriched on unattached kinetochores. How the signal is transmitted from unattached kinetochores to the $APC^{Cdc20}$ is not clear. In addition to kinetochore–microtubule attachment, a mechanism exists to monitor tension between sister kinetochores and this requires Aurora B kinase and other CPC components. In budding yeast, in the absence of tension between sister kinetochores, as would arise if they were attached to microtubules from the same pole (syntelic attachment, Fig. 13.5), Aurora B/Ipl1 severs faulty kinetochore–microtubule connections (Tanaka et al., 2002). The resulting unattached kinetochores activate the SAC and prevent anaphase onset (Pinsky et al., 2006). Therefore, the response to lack of tension requires both the core SAC components (Mad and Bub proteins) and the CPC/Aurora B. Interestingly, the budding yeast cohesin-protection protein, Sgo1, is also required to sense the lack of tension that results from the aberrant attachment of sister kinetochores to microtubules from the same pole (Indjeian et al., 2005). Such a function for Shugoshin family members is likely to be conserved. In addition to the Sgo1 protein that functions in cohesion protection, fission yeast have a second Shugoshin homologue, called Sgo2 that does not share this role (Kitajima et al., 2004; Rabitsch et al., 2004). Recently, two studies have shown that Sgo2 promotes the bi-orientation of sister chromatids in fission yeast mitosis (Kawashima et al., 2007; Vanoosthuyse et al., 2007). Similarly, mammalian Sgo2 is required to correct defective microtubule–kinetochore attachments (Huang et al., 2007). Fission yeast mutants lacking Sgo2 exhibit defects in the localization of CPCs, which likely explains their impaired ability to bi-orient sister chromatids (Kawashima et al., 2007; Vanoosthuyse et al., 2007). In contrast, CPCs are localized normally in Sgo2-depleted mammalian cells and the failure to detect aberrant kinetochore–microtubule attachments could instead be attributed to mislocalization of the error-correcting microtubule depolymerase, MCAK (Huang et al., 2007).

### 13.4.2 The Spindle Checkpoint is Required in Meiosis

In meiosis I, the mono-orientation of sister kinetochores would mean that they are not under tension, which implies that the tension-sensing apparatus must somehow be silenced at sister kinetochores. Instead, linkages between homologues, such as chiasmata, would allow tension to be generated by the bipolar attachment of homologous chromosomes that could be detected by the tension-sensing apparatus in meiosis I (Fig. 13.5). Does the SAC/Aurora B also monitor kinetochore attachment and bi-orientation of homologues during meiosis I and, if so, how is it programmed to allow homologue–homologue tension but not sister–sister tension?

In micromanipulation experiments in grasshopper spermatocytes, mal-aligned pairs of homologous chromosomes re-align on the meiosis I spindle through a trial and error mechanism, making both inappropriate connections, which are unstable, and proper bipolar connections, which are stable (Nicklas, 1967). This revealed the existence of an error-correction mechanism and further micro-manipulation experiments demonstrated the importance of tension in this process. A pair of homologues that were attached to the same pole were stabilized by using micromanipulation to provide an opposing force (Nicklas and Koch, 1969). The involvement of a checkpoint monitoring tension was demonstrated by the finding that artificial tension allows cell cycle progression (Li and Nicklas, 1995). More recent experiments using insect spermatocytes have found that, as in mitosis, kinetochore–microtubule attachment seems to be monitored in addition to tension during meiosis I (Nicklas et al., 2001).

The SAC has also been shown to be required for proper meiosis in budding yeast (Shonn et al., 2000; Shonn et al., 2003) *C. elegans* (Stein et al., 2007) and mouse (Homer et al., 2005). Mad2-depleted mouse oocytes enter anaphase I prematurely and exhibit a high incidence of chromosome mis-segregation at meiosis I (Homer et al., 2005). In budding yeast cells lacking Mad2, meiosis I non-disjunction (segregation of both homologues to the same pole) occurred frequently (Shonn et al., 2000). Mad2 is required to respond both to spindle perturbation and tension defects in meiosis I. Incubation of wild-type cells with a microtubule-depolymerizing drug causes them to arrest in metaphase I with high levels of securin (Pds1 in budding yeast), indicating that inhibition of the $APC^{Cdc20}$ has occurred. In contrast, cells lacking Mad2 fail to arrest in the presence of these drugs and Pds1 is not stabilized (Shonn et al., 2000). To assess the role of tension between homologues during meiosis I in spindle checkpoint function, cells defective in the initiation of meiotic recombination were examined. Meiotic recombination is completely abolished in the *spo11*Δ mutant and therefore no linkages between homologues are produced and this prohibits tension being generated between homologues. Similar to kinetochore attachment errors, the lack of tension in the *spo11*Δ mutant prevented Pds1 degradation, but only in the presence of Mad2 (Shonn et al., 2000). Surprisingly, the absence of another spindle checkpoint protein, Mad3, does not lead to meiosis I non-disjunction, although it is also important for arresting the cell in response to lack of microtubules or tension (Shonn et al., 2003). The difference in requirement for Mad2 and Mad3 appears to stem from the fact that in addition to delaying the cell cycle, Mad2 plays a role in reorienting homologous chromosomes that are incorrectly attached to the meiosis I spindle (Shonn et al., 2003). The functional distinction between Mad2 and Mad3 was substantiated by a study using a yeast strain in which a single yeast chromosome fails to recombine. Surprisingly, this 'non-exchange' chromosome segregates homologues to opposite poles remarkably efficiently in meiosis, in a manner that is highly dependent on Mad3, but less so on Mad2 (and Mad1; Cheslock et al., 2005). Further experiments revealed that Mad3 acts as a cell cycle timer, functioning in every meiosis to delay spindle assembly, whereas Mad1 and Mad2 delay chromosome segregation only in response to defective homologue alignment (Cheslock et al., 2005). It is unclear why non-exchange chromosomes are so dependent on a delay in spindle assembly for their accurate segregation. It is worthwhile to note that whereas Mad2 is required to prevent non-disjunction during meiosis I, it has little role during an unperturbed mitosis in budding yeast (Hardwick et al., 1999). Perhaps during mitosis, the back-to-back geometry of sister kinetochores increases the probability that they will be captured from opposite poles in the first instance, thereby decreasing their dependence on reorientation mechanisms.

The mechanism of correcting faulty attachments in meiosis I appears to work via the CPC, as in mitosis, because Ipl1, the budding yeast Aurora B kinase, is also required for the bi-orientation of homologues (Monje-Casas et al., 2007) and fission yeast Aurora B and survivin (Bir1) are required to prevent non-disjunction of homologues during meiosis I (Hauf et al., 2007; Kawashima et al., 2007). The requirement for Shugoshin in tension-sensing in meiosis I differs between budding and fission yeast (Kawashima et al., 2007; Kiburz et al., 2008). Fission yeast Sgo2 is required for tension-sensing at meiosis I, probably through its role in localizing the CPC (Kawashima et al., 2007), but budding yeast Sgo1 plays only a minor role in this process (Kiburz et al., 2008). A key question is what ensures that the kinetochore–microtubule severing activity of Aurora B is directed only towards homologues that lack tension and not sister chromatids? Monopolin is a crucial factor in ensuring that tension is only generated when homologues are bi-oriented and it likely accomplishes this role by providing a link between sister kinetochores that is independent of cohesins (Monje-Casas et al., 2007). Presumably, sister kinetochores linked by monopolin escape recognition as separate entities by the SAC and Aurora B.

## 13.5 Roles of Centromeres in Meiotic Prophase

In addition to well-defined roles in cohesion regulation and specifying the mode of attachment of kinetochores to microtubules, centromeres have a poorly understood role in the pairing of homologous chromosomes in preparation for segregation at meiosis I. As described above, linkages between homologues are essential for their accurate segregation during meiosis I, and chiasmata most frequently provide these linkages. However, in some organisms, homologues lacking chiasmata do not segregate entirely randomly, indicating the existence of other mechanisms to segregate these chromosomes. Most notably, the fourth chromosome does not undergo exchange in *Drosophila* oocytes, but is still segregated efficiently (Dernburg et al., 1996; Karpen et al., 1996). An exchange-independent mechanism of homologue segregation was also demonstrated to be approximately 90% effective in budding yeast using a strain that was engineered to contain a single non-exchange pair of chromosomes (Kemp et al., 2004). Pairing of homologous centromeres has been observed during meiotic prophase in *Drosophila*, budding yeast, fission yeast and plants (Dernburg et al., 1996; Karpen et al., 1996; Martinez-Perez et al., 2001; Ding et al., 2004; Kemp et al., 2004). In *Drosophila*, the pairing of centromeric heterochromatin of homologous chromosomes during prophase plays an important role in the segregation of achiasmate chromosomes (Dernburg et al., 1996; Karpen et al., 1996); however, in budding yeast, centromere-pairing is independent of surrounding sequences (Kemp et al., 2004; Tsubouchi and Roeder, 2005). Recently, Zip1, a component of the synaptonemal complex (SC), was found to be important for the sequence-independent coupling of centromeres (Tsubouchi and Roeder, 2005). Initially, non-homologous centromeres pair, but these interactions switch to homologous interactions in a manner that is dependent on the Spo11 endonuclease, which is responsible for the initiation of meiotic recombination (Tsubouchi and Roeder, 2005). These findings led to a model in which sequence-independent centromere coupling initiates homologue pairing, which is stabilized in a homology-dependent manner. Centromere pairing may be important not only for the segregation of non-exchange chromosomes, but also for the proper assembly of the SC and, consequently, the segregation of exchange chromosomes. An attractive hypothesis is that centromere pairing facilitates homology search on chromosome arms by restricting them in space.

In fission yeast, during meiotic prophase, centromeres become detached from the spindle pole body/centrosome and telomeres attach instead. This arrangement is known as the bouquet and the associated 'horsetail' movement is thought to facilitate homologue pairing by facilitating interactions (reviewed in Scherthan, 2006). The bouquet arrangement has been found to occur in meiotic prophase in many organisms, but it has been most intensively studied in fission yeast, where the NMS (Ndc80-Mis12-Spc7) group of kinetochore proteins are known to dissociate from the centromere, leading to its detachment from the spindle pole body (Asakawa et al., 2005; Hayashi et al., 2006). These findings suggest that the kinetochore could influence homologue pairing in several ways.

## 13.6 The Meiotic Kinetochore and Disease

Human meiosis is astonishingly error-prone. It is estimated that at least 5% of all human zygotes are aneuploid, accounting for stillbirths, miscarriages and birth defects, such as Down's syndrome (Hassold and Hunt, 2001). How do these aneuploidies arise and are kinetochore defects to blame? Since embryonic monosomy (one copy of a chromosome) is nearly always lethal at an early stage of foetal development, most studies have examined cases of trisomy (three copies of a chromosome), which frequently allows at least some development. An extra copy of some chromosomes is viable, but associated with developmental disorders, for example, trisomy chromosome 21, which causes Down's syndrome and has been most intensely studied. A key question in understanding how aneuploidy in humans has arisen is to learn whether it results from the non-disjunction of chromosomes at meiosis I, or at meiosis II. In humans, the origin of the extra chromatid can be inferred by the examination of polymorphisms near the centromere of the trisomic chromosome. This kind of analysis has revealed that aneuploid gametes arise most frequently due to maternal non-disjunction at meiosis I (MacDonald et al., 1994; Hassold et al., 1996; Lamb et al., 1996). In model organisms, the generation of crossovers during meiotic recombination is essential for the proper disjunction of homologues at meiosis I (Ross-Macdonald and Roeder, 1994; Sym and Roeder, 1994; Koehler et al., 1996; Kouznetsova et al., 2007). Similarly, in mammalian germ cells, failure to crossover and form a chiasma is associated with meiosis I non-disjunction (MacDonald et al., 1994; Hassold et al., 1996; Lamb et al., 1996). In mouse oocytes that lack a structural component of the synaptonemal complex, a proteinacious structure that assembles between homologues and facilitates recombination, some homologues lack chiasmata (Kouznetsova et al., 2007). These achiasmate homologues fail to bi-orient, but sister chromatids bi-orient instead and silence the SAC, resulting in the generation of aneuploid gametes (Kouznetsova et al., 2007). Another less common, but still significant, cause of aneuploidy in humans is meiosis II non-disjunction (Lamb et al., 1996; Hassold and Hunt, 2001). Although the mis-segregation of chromosomes occurs at meiosis II, the error originated during recombination in meiosis I. Studies in yeast have demonstrated that placement of crossovers too close to the centromere results in the premature separation of sister chromatids at meiosis I, and consequently meiosis II non-disjunction (Sears et al., 1995; Ross et al., 1996; Rockmill et al., 2006). These findings have been borne out in aneuploid *Drosophila* and human gametes, where recombination in the centromere and pericentromere is associated with meiosis II non-disjunction (Koehler et al., 1996; Lamb et al., 1996, 1997). The disruption of centromeric cohesion by centromere-proximal chiasmata is probably the source of the non-disjunction, since cohesins appear to be excluded from crossover sites in cytological studies (Eijpe et al., 2003; Parra et al., 2004). Therefore, mechanisms must exist to control both the number and position of crossovers during meiosis. The centromere plays an important role in controlling the placement of crossovers because it suppresses crrosovers in its vicinity (Lambie and Roeder, 1986; Lambie and Roeder, 1988; Rockmill et al., 2006). How the centromere accomplishes this, and the role of the specialized domain of pericentric cohesin in this process will be important questions for the future.

The incidence of aneuploidy in gametes increases dramatically with maternal age (Risch et al., 1986; Morton et al., 1988). This has been attributed to the fact that in females, meiosis I is protracted over decades, since it begins in the foetal ovary and is not completed until the time of ovulation. In the interim period, oocytes are arrested in prophase I having already undergone recombination and generated crossovers. With increasing maternal age, trisomic oocytes are not so tightly associated with recombination defects (Lamb et al., 2005). These findings are consistent with a 'two-step' model for the genesis of aneuploidy in the female germ line (Lamb et al., 2005). In the first step, unfavourably placed chiasmata lead to an increased risk of non-disjunction, but can be tolerated, provided that the chromosome segregation machinery is otherwise functioning optimally. In the second step, damage occurs to the segregation machinery, the likelihood of which increases with maternal age, leading to an increased chance that unfavourably linked homologues will be mis-segregated. One attractive candidate that could be susceptible to degradation over time is cohesin, which must hold sister chromatids together throughout the protracted arrest. This idea has been corroborated because oocytes from SMC1β-deficient mice show an increase in aneuploidy with age (Hodges et al., 2005). Given the evidence from studies on model organisms, cohesin defects are likely to be just one example of the errors that could give rise to aneuploid gametes. In this light, it is worthwhile bearing in mind that studies in humans are limited to those oocytes that show some development because gametes with the most severe defects are lost at an early stage. Given the central role of the meiotic kinetochore in setting up a specialized pattern of chromosome segregation, it seems very likely that many of its constituents will turn out to be significant in human disease.

**Acknowledgments** I am grateful to Angelika Amon, Kevin Hardwick, Hiro Ohkura and Alison Pidoux for helpful comments on the manuscript and to the Wellcome Trust for funding. I apologise to those colleagues whose work I was unable to cite directly due to space restrictions.

## References

Alexandru, G., F. Uhlmann, K. Mechtler, M. A. Poupart and K. Nasmyth 2001. Phosphorylation of the cohesin subunit Scc1 by Polo/Cdc5 kinase regulates sister chromatid separation in yeast. Cell 105: 459–72.

Asakawa, H., F. Hayashi, T. Haraguchi and Y. Hiraoka 2005. Dissociation of the Nuf2-Ndc80 complex releases centromeres from the spindle-pole body during meiotic prophase in fission yeast. Mol Biol Cell 16: 2325–38.

Bernard, P., F. F. Maure and J. P. Javerzat 2001a. Fission yeast Bub1 is essential in setting up the meiotic pattern of chromosome segregation. Nat Cell Biol 3: 522–6.

Bernard, P., F. F. Maure, J. F. Partridge, S. Genier, J. P. Javerzat and R. C. Allshire 2001b. Requirement of heterochromatin for cohesion at centromeres. Science 294: 2539–42.

Bishop, A. C., J. A. Ubersax, D. T. Petsch, D. P. Matheos, N. S. Gray, J. Blethrow, E. Shimizu, J. Z. Tsien, P. G. Schultz, M. D. Rose, J. L. Wood, D. O. Morgan and K. M. Shokat 2000. A chemical switch for inhibitor-sensitive alleles of any protein kinase. Nature 407: 395–401.

Blat, Y. and N. Kleckner 1999. Cohesins bind to preferential sites along yeast chromosome III, with differential regulation along arms versus the centric region. Cell 98: 249–59.

Blower, M. D. and G. H. Karpen 2001. The role of Drosophila CID in kinetochore formation, cell-cycle progression and heterochromatin interactions. Nat Cell Biol 3: 730–9.

Boyarchuk, Y., F. Salic, M. Dasso and A. Arnaoutov 2007. Bub1 is essential for assembly of the functional inner centromere. J Cell Biol 176: 919–28.

Brar, G. A., B. M. Kiburz, Y. Zhang, J. E. Kim, F. White and A. Amon 2006. Rec8 phosphorylation and recombination promote the step-wise loss of cohesins in meiosis. Nature 441: 532–6.

Buonomo, S. B., R. K. Clyne, J. Fuchs, J. Loidl, F. Uhlmann and K. Nasmyth 2000. Disjunction of homologous chromosomes in meiosis I depends on proteolytic cleavage of the meiotic cohesin Rec8 by separin. Cell 103: 387–98.

Chelysheva, L., F. Diallo, D. Vezon, G. Gendrot, N. Vrielynck, K. Belcram, N. Rocques, A. Marquez-Lema, A. M. Bhatt, C. Horlow, R. Mercier, C. Mezard and M. Grelon 2005. AtREC8 and AtSCC3 are essential to the monopolar orientation of the kinetochores during meiosis. J Cell Sci 118: 4621–32.

Cheslock, P. S., B. J. Kemp, R. M. Boumil and D. S. Dawson 2005. The roles of MAD1, MAD2 and MAD3 in meiotic progression and the segregation of nonexchange chromosomes. Nat Genet 37: 756–60.

Ciosk, R., F. Shirayama, A. Shevchenko, T. Tanaka, A. Toth, A. Shevchenko and K. Nasmyth 2000. Cohesin's binding to chromosomes depends on a separate complex consisting of Scc2 and Scc4 proteins. Mol Cell 5: 243–54.

Clarke, A. S., T. T. Tang, D. L. Ooi and T. L. Orr-Weaver 2005. POLO kinase regulates the Drosophila centromere cohesion protein MEI-S332. Dev Cell 8: 53–64.

Clyne, R. K., V. L. Katis, L. Jessop, K. R. Benjamin, I. Herskowitz, M. Lichten and K. Nasmyth 2003. Polo-like kinase Cdc5 promotes chiasmata formation and cosegregation of sister centromeres at meiosis I. Nat Cell Biol 5: 480–5.

Dai, J., F. A. Sullivan and J. M. Higgins 2006. Regulation of mitotic chromosome cohesion by Haspin and Aurora B. Dev Cell 11: 741–50.

Davis, E. S., L. Wille, B. A. Chestnut, P. L. Sadler, D. C. Shakes and A. Golden 2002. Multiple subunits of the Caenorhabditis elegans anaphase-promoting complex are required for chromosome segregation during meiosis I. Genetics 160: 805–13.

Dernburg, A. F., J. W. Sedat and R. S. Hawley 1996. Direct evidence of a role for heterochromatin in meiotic chromosome segregation. Cell 86: 135–46.

Ding, D. Q., A. Yamamoto, T. Haraguchi and Y. Hiraoka 2004. Dynamics of homologous chromosome pairing during meiotic prophase in fission yeast. Dev Cell 6: 329–41.

Ding, R., F. L. McDonald and J. R. McIntosh 1993. Three-dimensional reconstruction and analysis of mitotic spindles from the yeast, Schizosaccharomyces pombe. J Cell Biol 120: 141–51.

Eijpe, M., F. Offenberg, R. Jessberger, E. Revenkova and C. Heyting 2003. Meiotic cohesin REC8 marks the axial elements of rat synaptonemal complexes before cohesins SMC1beta and SMC 3. J Cell Biol 160: 657–70.

Fan, H. Y., Q. Y. Sun and H. Zou 2006. Regulation of Separase in meiosis: Separase is activated at the metaphase I-II transition in Xenopus oocytes during meiosis. Cell Cycle 5: 198–204.

Fernius, J. and K. G. Hardwick 2007. Bub1 kinase targets Sgo1 to ensure efficient chromsoome biorientation in budding yeast mitosis. PLoS Genet 3: e213.

Fukagawa, T., F. Nogami, M. Yoshikawa, M. Ikeno, T. Okazaki, Y. Takami, T. Nakayama and M. Oshimura 2004. Dicer is essential for formation of the heterochromatin structure in vertebrate cells. Nat Cell Biol 6: 784–91.

Glynn, E. F., P. C. Megee, H. G. Yu, C. Mistrot, E. Unal, D. E. Koshland, J. L. DeRisi and J. L. Gerton 2004. Genome-wide mapping of the cohesin complex in the yeast Saccharomyces cerevisiae. PLoS Biol 2: E259.

Goldstein, L. S. 1981. Kinetochore structure and its role in chromosome orientation during the first meiotic division in male D. melanogaster. Cell 25: 591–602.

Gomez, R., F. Valdeolmillos, M. T. Parra, A. Viera, C. Carreiro, F. Roncal, J. S. Rufas, J. L. Barbero and J. A. Suja 2007. Mammalian SGO2 appears at the inner centromere domain and redistributes depending on tension across centromeres during meiosis II and mitosis. EMBO Rep.

Gorr, I. H., D. Boos and O. Stemmann 2005. Mutual inhibition of separase and Cdk1 by two-step complex formation. Mol Cell 19: 135–41.

Gregan, J., F. G. Riedel, A. L. Pidoux, Y. Katou, C. Rumpf, A. Schleiffer, S. E. Kearsey, K. Shirahige, R. C. Allshire and K. Nasmyth 2007. The kinetochore proteins pcs1 and mde4 and heterochromatin are required to prevent merotelic orientation. Curr Biol 17: 1190–200.

Haering, C. H. and K. Nasmyth 2003. Building and breaking bridges between sister chromatids. Bioessays 25: 1178–91.

Hamant, O., F. Golubovskaya, R. Meeley, E. Fiume, L. Timofejeva, A. Schleiffer, K. Nasmyth and W. Z. Cande 2005. A REC8-dependent plant Shugoshin is required for maintenance of centromeric cohesion during meiosis and has no mitotic functions. Curr Biol 15: 948–54.

Hardwick, K. G., R. Li, C. Mistrot, R. H. Chen, P. Dann, A. Rudner and A. W. Murray 1999. Lesions in many different spindle components activate the spindle checkpoint in the budding yeast Saccharomyces cerevisiae. Genetics 152: 509–18.

Hassold, T., F. Abruzzo, K. Adkins, D. Griffin, M. Merrill, E. Millie, D. Saker, J. Shen and M. Zaragoza 1996. Human aneuploidy: incidence, origin, and etiology. Environ Mol Mutagen 28: 167–75.

Hassold, T. and P. Hunt 2001. To err (meiotically) is human: the genesis of human aneuploidy. Nat Rev Genet 2: 280–91.

Hauf, S., F. Biswas, M. Langegger, S. A. Kawashima, T. Tsukahara and Y. Watanabe 2007. Aurora controls sister kinetochore mono-orientation and homolog bi-orientation in meiosis-I. Embo J 26: 4475–86.

Hauf, S., F. Roitinger, B. Koch, C. M. Dittrich, K. Mechtler and J. M. Peters 2005. Dissociation of cohesin from chromosome arms and loss of arm cohesion during early mitosis depends on phosphorylation of SA2. PLoS Biol 3: e69.

Hauf, S., F. C. Waizenegger and J. M. Peters 2001. Cohesin cleavage by separase required for anaphase and cytokinesis in human cells. Science 293: 1320–3.

Hauf, S. and Y. Watanabe 2004. Kinetochore orientation in mitosis and meiosis. Cell 119: 317–27.

Hayashi, A., F. Asakawa, T. Haraguchi and Y. Hiraoka 2006. Reconstruction of the kinetochore during meiosis in fission yeast Schizosaccharomyces pombe. Mol Biol Cell 17: 5173–84.

Herbert, M., F. Levasseur, H. Homer, K. Yallop, A. Murdoch and A. McDougall 2003. Homologue disjunction in mouse oocytes requires proteolysis of securin and cyclin B1. Nat Cell Biol 5: 1023–5.

Hodges, C. A., E. Revenkova, R. Jessberger, T. J. Hassold and P. A. Hunt 2005. SMC1beta-deficient female mice provide evidence that cohesins are a missing link in age-related nondisjunction. Nat Genet 37: 1351–5.

Homer, H. A., A. McDougall, M. Levasseur, K. Yallop, A. P. Murdoch and M. Herbert 2005. Mad2 prevents aneuploidy and premature proteolysis of cyclin B and securin during meiosis I in mouse oocytes. Genes Dev 19: 202–7.

Huang, C. E., M. Milutinovich and D. Koshland 2005. Rings, bracelet or snaps: fashionable alternatives for Smc complexes. Philos Trans R Soc Lond B Biol Sci 360: 537–42.

Huang, H., F. Feng, J. Famulski, J. B. Rattner, S. T. Liu, G. D. Kao, R. Muschel, G. K. Chan and T. J. Yen 2007. Tripin/hSgo2 recruits MCAK to the inner centromere to correct defective kinetochore attachments. J Cell Biol 177: 413–24.

Huang, J., F. L. Brito, J. Villen, S. P. Gygi, A. Amon and D. Moazed 2006. Inhibition of homologous recombination by a cohesin-associated clamp complex recruited to the rDNA recombination enhancer. Genes Dev 20: 2887–901.

Huo, L. J., Z. S. Zhong, C. G. Liang, Q. Wang, S. Yin, J. S. Ai, L. Z. Yu, D. Y. Chen, H. Schatten and Q. Y. Sun 2006. Degradation of securin in mouse and pig oocytes is dependent on ubiquitin-proteasome pathway and is required for proteolysis of the cohesion subunit, Rec8, at the metaphase-to-anaphase transition. Front Biosci 11: 2193–202.

Indjeian, V. B., B. M. Stern and A. W. Murray 2005. The centromeric protein Sgo1 is required to sense lack of tension on mitotic chromosomes. Science 307: 130–3.

Johnson, V. L., M. I. Scott, S. V. Holt, D. Hussein and S. S. Taylor 2004. Bub1 is required for kinetochore localization of BubR1, Cenp-E, Cenp-F and Mad2, and chromosome congression. J Cell Sci 117: 1577–89.

Karpen, G. H., M. H. Le and H. Le 1996. Centric heterochromatin and the efficiency of achiasmate disjunction in Drosophila female meiosis. Science 273: 118–22.

Katis, V. L., M. Galova, K. P. Rabitsch, J. Gregan and K. Nasmyth 2004a. Maintenance of cohesin at centromeres after meiosis I in budding yeast requires a kinetochore-associated protein related to MEI-S332. Curr Biol 14: 560–72.

Katis, V. L., J. Matos, S. Mori, K. Shirahige, W. Zachariae and K. Nasmyth 2004b. Spo13 facilitates monopolin recruitment to kinetochores and regulates maintenance of centromeric cohesion during yeast meiosis. Curr Biol 14: 2183–96.

Kawashima, S. A., T. Tsukahara, M. Langegger, S. Hauf, T. S. Kitajima and Y. Watanabe 2007. Shugoshin enables tension-generating attachment of kinetochores by loading Aurora to centromeres. Genes Dev 21: 420–35.

Kemp, B., F. M. Boumil, M. N. Stewart and D. S. Dawson 2004. A role for centromere pairing in meiotic chromosome segregation. Genes Dev 18: 1946–51.

Kerrebrock, A. W., W. Y. Miyazaki, D. Birnby and T. L. Orr-Weaver 1992. The Drosophila mei-S332 gene promotes sister-chromatid cohesion in meiosis following kinetochore differentiation. Genetics 130: 827–41.

Kerrebrock, A. W., D. P. Moore, J. S. Wu and T. L. Orr-Weaver 1995. Mei-S332, a Drosophila protein required for sister-chromatid cohesion, can localize to meiotic centromere regions. Cell 83: 247–56.

Kiburz, B. M., A. Amon and A. L. Marston 2008. Shugoshin Promotes Sister Kinetochore Biorientation in Saccharomyces cerevisiae. Mol Biol Cell 19: 1199–209.

Kiburz, B. M., D. B. Reynolds, P. C. Megee, A. L. Marston, B. H. Lee, T. I. Lee, S. S. Levine, R. A. Young and A. Amon 2005. The core centromere and Sgo1 establish a 50-kb cohesin-protected domain around centromeres during meiosis I. Genes Dev 19: 3017–30.

Kitajima, T. S., S. Hauf, M. Ohsugi, T. Yamamoto and Y. Watanabe 2005. Human Bub1 defines the persistent cohesion site along the mitotic chromosome by affecting Shugoshin localization. Curr Biol 15: 353–9.

Kitajima, T. S., S. A. Kawashima and Y. Watanabe 2004. The conserved kinetochore protein shugoshin protects centromeric cohesion during meiosis. Nature 427: 510–7.

Kitajima, T. S., Y. Miyazaki, M. Yamamoto and Y. Watanabe 2003a. Rec8 cleavage by separase is required for meiotic nuclear divisions in fission yeast. Embo J 22: 5643–53.

Kitajima, T. S., T. Sakuno, K. Ishiguro, S. Iemura, T. Natsume, S. A. Kawashima and Y. Watanabe 2006. Shugoshin collaborates with protein phosphatase 2A to protect cohesin. Nature 441: 46–52.

Kitajima, T. S., S. Yokobayashi, M. Yamamoto and Y. Watanabe 2003b. Distinct cohesin complexes organize meiotic chromosome domains. Science 300: 1152–5.

Klein, F., F. Mahr, M. Galova, S. B. Buonomo, C. Michaelis, K. Nairz and K. Nasmyth 1999. A central role for cohesins in sister chromatid cohesion, formation of axial elements, and recombination during yeast meiosis. Cell 98: 91–103.

Knippschild, U., F. Gocht, S. Wolff, N. Huber, J. Lohler and M. Stoter 2005. The casein kinase 1 family: participation in multiple cellular processes in eukaryotes. Cell Signal 17: 675–89.

Koch, B., F. Kueng, C. Ruckenbauer, K. S. Wendt and J. M. Peters 2008. The Suv39h-HP1 histone methylation pathway is dispensable for enrichment and protection of cohesin at centromeres in mammalian cells. Chromosoma 117: 199–210.

Koehler, K. E., C. L. Boulton, H. E. Collins, R. L. French, K. C. Herman, S. M. Lacefield, L. D. Madden, C. D. Schuetz and R. S. Hawley 1996. Spontaneous X chromosome MI and MII nondisjunction events in Drosophila melanogaster oocytes have different recombinational histories. Nat Genet 14: 406–14.

Kouznetsova, A., F. Lister, M. Nordenskjold, M. Herbert and C. Hoog 2007. Bi-orientation of achiasmatic chromosomes in meiosis I oocytes contributes to aneuploidy in mice. Nat Genet.

Kudo, N. R., K. Wassmann, M. Anger, M. Schuh, K. G. Wirth, H. Xu, W. Helmhart, H. Kudo, M. McKay, B. Maro, J. Ellenberg, P. de Boer and K. Nasmyth 2006. Resolution of chiasmata in oocytes requires separase-mediated proteolysis. Cell 126: 135–46.

Kueng, S., F. Hegemann, B. H. Peters, J. J. Lipp, A. Schleiffer, K. Mechtler and J. M. Peters 2006. Wapl controls the dynamic association of cohesin with chromatin. Cell 127: 955–67.

Laloraya, S., F. Guacci and D. Koshland 2000. Chromosomal addresses of the cohesin component Mcd1p. J Cell Biol 151: 1047–56.

Lamb, N. E., E. Feingold and S. L. Sherman 1997. Estimating meiotic exchange patterns from recombination data: an application to humans. Genetics 146: 1011–7.

Lamb, N. E., S. B. Freeman, A. Savage-Austin, D. Pettay, L. Taft, J. Hersey, Y. Gu, J. Shen, D. Saker, K. M. May, D. Avramopoulos, M. B. Petersen, A. Hallberg, M. Mikkelsen, T. J. Hassold and S. L. Sherman 1996. Susceptible chiasmate configurations of chromosome 21 predispose to non-disjunction in both maternal meiosis I and meiosis II. Nat Genet 14: 400–5.

Lamb, N. E., S. L. Sherman and T. J. Hassold 2005. Effect of meiotic recombination on the production of aneuploid gametes in humans. Cytogenet Genome Res 111: 250–5.

Lambie, E. J. and G. S. Roeder 1986. Repression of meiotic crossing over by a centromere (CEN3) in Saccharomyces cerevisiae. Genetics 114: 769–89.

Lambie, E. J. and G. S. Roeder 1988. A yeast centromere acts in cis to inhibit meiotic gene conversion of adjacent sequences. Cell 52: 863–73.

Lee, B. H. and A. Amon 2003. Role of Polo-like kinase CDC5 in programming meiosis I chromosome segregation. Science 300: 482–6.

Lee, B. H., A. Amon and S. Prinz 2002. Spo13 regulates cohesin cleavage. Genes Dev 16: 1672–81.

Lee, B. H., B. M. Kiburz and A. Amon 2004a. Spo13 maintains centromeric cohesion and kinetochore coorientation during meiosis I. Curr Biol 14: 2168–82.

Lee, J., F. Iwai, T. Yokota and M. Yamashita 2003. Temporally and spatially selective loss of Rec8 protein from meiotic chromosomes during mammalian meiosis. J Cell Sci 116: 2781–90.

Lee, J., F. S. Kitajima, Y. Tanno, K. Yoshida, T. Morita, T. Miyano, M. Miyake and Y. Watanabe 2008. Unified mode of centromeric protection by shugoshin in mammalian oocytes and somatic cells. Nat Cell Biol 10: 42–52.

Lee, J., F. Okada, S. Ogushi, T. Miyano, M. Miyake and M. Yamashita 2006. Loss of Rec8 from chromosome arm and centromere region is required for homologous chromosome separation and sister chromatid separation, respectively, in mammalian meiosis. Cell Cycle 5: 1448–55.

Lee, J. Y., K. J. Dej, J. M. Lopez and T. L. Orr-Weaver 2004b. Control of centromere localization of the MEI-S332 cohesion protection protein. Curr Biol 14: 1277–83.

Lengronne, A., F. Katou, S. Mori, S. Yokobayashi, G. P. Kelly, T. Itoh, Y. Watanabe, K. Shirahige and F. Uhlmann 2004. Cohesin relocation from sites of chromosomal loading to places of convergent transcription. Nature 430: 573–8.

Li, X. and R. B. Nicklas 1995. Mitotic forces control a cell-cycle checkpoint. Nature 373: 630–2.

Losada, A., F. Hirano and T. Hirano 2002. Cohesin release is required for sister chromatid resolution, but not for condensin-mediated compaction, at the onset of mitosis. Genes Dev 16: 3004–16.

Losada, A. and T. Hirano 2005. Dynamic molecular linkers of the genome: the first decade of SMC proteins. Genes Dev 19: 1269–87.

MacDonald, M., F. Hassold, J. Harvey, L. H. Wang, N. E. Morton and P. Jacobs 1994. The origin of 47,XXY and 47,XXX aneuploidy: heterogeneous mechanisms and role of aberrant recombination. Hum Mol Genet 3: 1365–71.

Marston, A. L., W. H. Tham, H. Shah and A. Amon 2004. A genome-wide screen identifies genes required for centromeric cohesion. Science 303: 1367–70.

Martinez-Perez, E., F. Shaw and G. Moore 2001. The Ph1 locus is needed to ensure specific somatic and meiotic centromere association. Nature 411: 204–7.

McGuinness, B. E., T. Hirota, N. R. Kudo, J. M. Peters and K. Nasmyth 2005. Shugoshin prevents dissociation of cohesin from centromeres during mitosis in vertebrate cells. PLoS Biol 3: e86.

Meraldi, P. and P. K. Sorger 2005. A dual role for Bub1 in the spindle checkpoint and chromosome congression. Embo J 24: 1621–33.

Milutinovich, M. and D. E. Koshland 2003. Molecular biology. SMC complexes–wrapped up in controversy. Science 300: 1101–2.

Monje-Casas, F., F. R. Prabhu, B. H. Lee, M. Boselli and A. Amon 2007. Kinetochore orientation during meiosis is controlled by Aurora B and the monopolin complex. Cell 128: 477–90.

Moore, D. P. and T. L. Orr-Weaver 1998. Chromosome segregation during meiosis: building an unambivalent bivalent. Curr Top Dev Biol 37: 263–99.

Moore, D. P., A. W. Page, T. T. Tang, A. W. Kerrebrock and T. L. Orr-Weaver 1998. The cohesion protein MEI-S332 localizes to condensed meiotic and mitotic centromeres until sister chromatids separate. J Cell Biol 140: 1003–12.

Morton, N. E., P. A. Jacobs, T. Hassold and D. Wu 1988. Maternal age in trisomy. Ann Hum Genet 52: 227–35.

Musacchio, A. and E. D. Salmon 2007. The spindle-assembly checkpoint in space and time. Nat Rev Mol Cell Biol 8: 379–93.

Nasmyth, K. 2001. Disseminating the genome: joining, resolving, and separating sister chromatids during mitosis and meiosis. Annu Rev Genet 35: 673–745.

Nasmyth, K. 2005. How might cohesin hold sister chromatids together? Philos Trans R Soc Lond B Biol Sci 360: 483–96.

Nasmyth, K. and C. H. Haering 2005. The structure and function of SMC and kleisin complexes. Annu Rev Biochem 74: 595–648.

Nicklas, R. B. 1967. Chromosome micromanipulation. II. Induced reorientation and the experimental control of segregation in meiosis. Chromosoma 21: 17–50.

Nicklas, R. B. and C. A. Koch 1969. Chromosome micromanipulation. 3. Spindle fiber tension and the reorientation of mal-oriented chromosomes. J Cell Biol 43: 40–50.

Nicklas, R. B., J. C. Waters, E. D. Salmon and S. C. Ward 2001. Checkpoint signals in grasshopper meiosis are sensitive to microtubule attachment, but tension is still essential. J Cell Sci 114: 4173–83.

Nonaka, N., F. Kitajima, S. Yokobayashi, G. Xiao, M. Yamamoto, S. I. Grewal and Y. Watanabe 2002. Recruitment of cohesin to heterochromatic regions by Swi6/HP1 in fission yeast. Nat Cell Biol 4: 89–93.

Oelschlaegel, T., F. Schwickart, J. Matos, A. Bogdanova, A. Camasses, J. Havlis, A. Shevchenko and W. Zachariae 2005. The yeast APC/C subunit Mnd2 prevents premature sister chromatid separation triggered by the meiosis-specific APC/C-Ama1. Cell 120: 773–88.

Paliulis, L. V. and R. B. Nicklas 2000. The reduction of chromosome number in meiosis is determined by properties built into the chromosomes. J Cell Biol 150: 1223–32.

Parra, M. T., A. Viera, R. Gomez, J. Page, R. Benavente, J. L. Santos, J. S. Rufas and J. A. Suja 2004. Involvement of the cohesin Rad21 and SCP3 in monopolar attachment of sister kinetochores during mouse meiosis I. J Cell Sci 117: 1221–34.

Parra, M. T., A. Viera, R. Gomez, J. Page, M. Carmena, W. C. Earnshaw, J. S. Rufas and J. A. Suja 2003. Dynamic relocalization of the chromosomal passenger complex proteins inner centromere protein (INCENP) and aurora-B kinase during male mouse meiosis. J Cell Sci 116: 961–74.

Pasierbek, P., F. Jantsch, M. Melcher, A. Schleiffer, D. Schweizer and J. Loidl 2001. A Caenorhabditis elegans cohesion protein with functions in meiotic chromosome pairing and disjunction. Genes Dev 15: 1349–60.

Penkner, A. M., S. Prinz, S. Ferscha and F. Klein 2005. Mnd2, an essential antagonist of the anaphase-promoting complex during meiotic prophase. Cell 120: 789–801.

Peter, M., F. Castro, T. Lorca, C. Le Peuch, L. Magnaghi-Jaulin, M. Doree and J. C. Labbe 2001. The APC is dispensable for first meiotic anaphase in Xenopus oocytes. Nat Cell Biol 3: 83–7.

Petronczki, M., F. Matos, S. Mori, J. Gregan, A. Bogdanova, M. Schwickart, K. Mechtler, K. Shirahige, W. Zachariae and K. Nasmyth 2006. Monopolar attachment of sister kinetochores at meiosis I requires casein kinase 1. Cell 126: 1049–64.

Pinsky, B. A., C. Kung, K. M. Shokat and S. Biggins 2006. The Ipl1-Aurora protein kinase activates the spindle checkpoint by creating unattached kinetochores. Nat Cell Biol 8: 78–83.

Pouwels, J., F. M. Kukkonen, W. Lan, J. R. Daum, G. J. Gorbsky, T. Stukenberg and M. J. Kallio 2007. Shugoshin 1 plays a central role in kinetochore assembly and is required for kinetochore targeting of Plk1. Cell Cycle 6: 1579–85.

Prieto, I., F. Pezzi, J. M. Buesa, L. Kremer, I. Barthelemy, C. Carreiro, F. Roncal, A. Martinez, L. Gomez, R. Fernandez, A. C. Martinez and J. L. Barbero 2002. STAG2 and Rad21 mammalian mitotic cohesins are implicated in meiosis. EMBO Rep 3: 543–50.

Prieto, I., F. A. Suja, N. Pezzi, L. Kremer, A. C. Martinez, J. S. Rufas and J. L. Barbero 2001. Mammalian STAG3 is a cohesin specific to sister chromatid arms in meiosis I. Nat Cell Biol 3: 761–6.

Rabitsch, K. P., J. Gregan, A. Schleiffer, J. P. Javerzat, F. Eisenhaber and K. Nasmyth 2004. Two fission yeast homologs of Drosophila Mei-S332 are required for chromosome segregation during meiosis I and II. Curr Biol 14: 287–301.

Rabitsch, K. P., M. Petronczki, J. P. Javerzat, S. Genier, B. Chwalla, A. Schleiffer, T. U. Tanaka and K. Nasmyth 2003. Kinetochore recruitment of two nucleolar proteins is required for homolog segregation in meiosis I. Dev Cell 4: 535–48.

Resnick, T. D., D. L. Satinover, F. MacIsaac, P. T. Stukenberg, W. C. Earnshaw, T. L. Orr-Weaver and M. Carmena 2006. INCENP and Aurora B promote meiotic sister chromatid cohesion through localization of the Shugoshin MEI-S332 in Drosophila. Dev Cell 11: 57–68.

Riedel, C. G., V. L. Katis, Y. Katou, S. Mori, T. Itoh, W. Helmhart, M. Galova, M. Petronczki, J. Gregan, B. Cetin, I. Mudrak, E. Ogris, K. Mechtler, L. Pelletier, F. Buchholz, K. Shirahige and K. Nasmyth 2006. Protein phosphatase 2A protects centromeric sister chromatid cohesion during meiosis I. Nature 441: 53–61.

Risch, N., F. Stein, J. Kline and D. Warburton 1986. The relationship between maternal age and chromosome size in autosomal trisomy. Am J Hum Genet 39: 68–78.

Rockmill, B., F. Voelkel-Meiman and G. S. Roeder 2006. Centromere-proximal crossovers are associated with precocious separation of sister chromatids during meiosis in Saccharomyces cerevisiae. Genetics 174: 1745–54.

Rogers, E., F. D. Bishop, J. A. Waddle, J. M. Schumacher and R. Lin 2002. The aurora kinase AIR-2 functions in the release of chromosome cohesion in Caenorhabditis elegans meiosis. J Cell Biol 157: 219–29.

Ross, L. O., S. Rankin, M. F. Shuster and D. S. Dawson 1996. Effects of homology, size and exchange of the meiotic segregation of model chromosomes in Saccharomyces cerevisiae. Genetics 142: 79–89.

Ross-Macdonald, P. and G. S. Roeder 1994. Mutation of a meiosis-specific MutS homolog decreases crossing over but not mismatch correction. Cell 79: 1069–80.

Ruchaud, S., F. Carmena and W. C. Earnshaw 2007. Chromosomal passengers: conducting cell division. Nat Rev Mol Cell Biol 8: 798–812.

Salah, S. M. and K. Nasmyth 2000. Destruction of the securin Pds1p occurs at the onset of anaphase during both meiotic divisions in yeast. Chromosoma 109: 27–34.

Salic, A., F. C. Waters and T. J. Mitchison 2004. Vertebrate shugoshin links sister centromere cohesion and kinetochore microtubule stability in mitosis. Cell 118: 567–78.

Scherthan, H. 2006. Factors directing telomere dynamics in synaptic meiosis. Biochem Soc Trans 34: 550–3.

Sears, D. D., J. H. Hegemann, J. H. Shero and P. Hieter 1995. Cis-acting determinants affecting centromere function, sister-chromatid cohesion and reciprocal recombination during meiosis in Saccharomyces cerevisiae. Genetics 139: 1159–73.

Shonn, M. A., R. McCarroll and A. W. Murray 2000. Requirement of the spindle checkpoint for proper chromosome segregation in budding yeast meiosis. Science 289: 300–3.

Shonn, M. A., R. McCarroll and A. W. Murray 2002. Spo13 protects meiotic cohesin at centromeres in meiosis I. Genes Dev 16: 1659–71.

Shonn, M. A., A. L. Murray and A. W. Murray 2003. Spindle checkpoint component Mad2 contributes to biorientation of homologous chromosomes. Curr Biol 13: 1979–84.

Siomos, M. F., A. Badrinath, P. Pasierbek, D. Livingstone, J. White, M. Glotzer and K. Nasmyth 2001. Separase is required for chromosome segregation during meiosis I in Caenorhabditis elegans. Curr Biol 11: 1825–35.

Smith, J. S., E. Caputo and J. D. Boeke 1999. A genetic screen for ribosomal DNA silencing defects identifies multiple DNA replication and chromatin-modulating factors. Mol Cell Biol 19: 3184–97.

Stein, K. K., E. S. Davis, T. Hays and A. Golden 2007. Components of the spindle assembly checkpoint regulate the anaphase-promoting complex during meiosis in Caenorhabditis elegans. Genetics 175: 107–23.

Stemmann, O., F. Zou, S. A. Gerber, S. P. Gygi and M. W. Kirschner 2001. Dual inhibition of sister chromatid separation at metaphase. Cell 107: 715–26.

Sumara, I., F. Vorlaufer, P. T. Stukenberg, O. Kelm, N. Redemann, E. A. Nigg and J. M. Peters 2002. The dissociation of cohesin from chromosomes in prophase is regulated by Polo-like kinase. Mol Cell 9: 515–25.

Sym, M. and G. S. Roeder 1994. Crossover interference is abolished in the absence of a synaptonemal complex protein. Cell 79: 283–92.

Taieb, F. E., S. D. Gross, A. L. Lewellyn and J. L. Maller 2001. Activation of the anaphase-promoting complex and degradation of cyclin B is not required for progression from Meiosis I to II in Xenopus oocytes. Curr Biol 11: 508–13.

Tanaka, T. U., N. Rachidi, C. Janke, G. Pereira, M. Galova, E. Schiebel, M. J. Stark and K. Nasmyth 2002. Evidence that the Ipl1-Sli15 (Aurora kinase-INCENP) complex promotes chromosome bi-orientation by altering kinetochore-spindle pole connections. Cell 108: 317–29.

Tang, Z., F. Shu, W. Qi, N. A. Mahmood, M. C. Mumby and H. Yu 2006. PP2A is required for centromeric localization of Sgo1 and proper chromosome segregation. Dev Cell 10: 575–85.

Tang, Z., F. Sun, S. E. Harley, H. Zou and H. Yu 2004. Human Bub1 protects centromeric sister-chromatid cohesion through Shugoshin during mitosis. Proc Natl Acad Sci U S A 101: 18012–7.

Terret, M. E., K. Wassmann, I. Waizenegger, B. Maro, J. M. Peters and M. H. Verlhac 2003. The meiosis I-to-meiosis II transition in mouse oocytes requires separase activity. Curr Biol 13: 1797–802.

Topp, C. N. and R. K. Dawe 2006. Reinterpreting pericentromeric heterochromatin. Curr Opin Plant Biol 9: 647–53.

Toth, A., F. P. Rabitsch, M. Galova, A. Schleiffer, S. B. Buonomo and K. Nasmyth 2000. Functional genomics identifies monopolin: a kinetochore protein required for segregation of homologs during meiosis I. Cell 103: 1155–68.

Tsubouchi, T. and G. S. Roeder 2005. A synaptonemal complex protein promotes homology-independent centromere coupling. Science 308: 870–3.

Vanoosthuyse, V., F. Prykhozhij and K. G. Hardwick 2007. Shugoshin 2 regulates localization of the chromosomal passenger proteins in fission yeast mitosis. Mol Biol Cell 18: 1657–69.

Vaur, S., F. Cubizolles, G. Plane, S. Genier, P. K. Rabitsch, J. Gregan, K. Nasmyth, V. Vanoosthuyse, K. G. Hardwick and J. P. Javerzat 2005. Control of Shugoshin function during fission-yeast meiosis. Curr Biol 15: 2263–70.

Waizenegger, I. C., S. Hauf, A. Meinke and J. M. Peters 2000. Two distinct pathways remove mammalian cohesin from chromosome arms in prophase and from centromeres in anaphase. Cell 103: 399–410.

Wang, X., F. Yang, Q. Duan, N. Jiang, Y. Huang, Z. Darzynkiewicz and W. Dai 2008. sSgo1, a major splice variant of Sgo1, functions in centriole cohesion where it is regulated by Plk1. Dev Cell 14: 331–41.

Warburton, P. E. 2004. Chromosomal dynamics of human neocentromere formation. Chromosome Res 12: 617–26.

Watanabe, Y. and P. Nurse 1999. Cohesin Rec8 is required for reductional chromosome segregation at meiosis. Nature 400: 461–4.

Watanabe, Y., F. Yokobayashi, M. Yamamoto and P. Nurse 2001. Pre-meiotic S phase is linked to reductional chromosome segregation and recombination. Nature 409: 359–63.

Weber, S. A., J. L. Gerton, J. E. Polancic, J. L. DeRisi, D. Koshland and P. C. Megee 2004. The kinetochore is an enhancer of pericentric cohesin binding. PLoS Biol 2: E260.

Winey, M., F. L. Mamay, E. T. O'Toole, D. N. Mastronarde, T. H. Giddings, Jr., K. L. McDonald and J. R. McIntosh 1995. Three-dimensional ultrastructural analysis of the Saccharomyces cerevisiae mitotic spindle. J Cell Biol 129: 1601–15.

Winey, M., F. P. Morgan, P. D. Straight, T. H. Giddings, Jr. and D. N. Mastronarde 2005. Three-dimensional ultrastructure of Saccharomyces cerevisiae meiotic spindles. Mol Biol Cell 16: 1178–88.

Xu, H., F. Beasley, S. Verschoor, A. Inselman, M. A. Handel and M. J. McKay 2004. A new role for the mitotic RAD21/SCC1 cohesin in meiotic chromosome cohesion and segregation in the mouse. EMBO Rep 5: 378–84.

Xu, H., F. D. Beasley, W. D. Warren, G. T. van der Horst and M. J. McKay 2005. Absence of mouse REC8 cohesin promotes synapsis of sister chromatids in meiosis. Dev Cell 8: 949–61.

Yeh, E., J. Haase, L. V. Paliulis, A. Joglekar, L. Bond, D. Bouck, E. D. Salmon and K. S. Bloom 2008. Pericentric chromatin is organized into an intramolecular loop in mitosis. Curr Biol 18: 81–90.

Yokobayashi, S. and Y. Watanabe 2005. The kinetochore protein Moa1 enables cohesion-mediated monopolar attachment at meiosis I. Cell 123: 803–17.

Yokobayashi, S., F. Yamamoto and Y. Watanabe 2003. Cohesins determine the attachment manner of kinetochores to spindle microtubules at meiosis I in fission yeast. Mol Cell Biol 23: 3965–73.

Yu, H. 2007. Cdc20: a WD40 activator for a cell cycle degradation machine. Mol Cell 27: 3–16.

Yu, H. G. and R. K. Dawe 2000. Functional redundancy in the maize meiotic kinetochore. J Cell Biol 151: 131–42.

Yu, H. G. and D. Koshland 2005. Chromosome morphogenesis: condensin-dependent cohesin removal during meiosis. Cell 123: 397–407.

Yu, H. G. and D. Koshland 2007. The Aurora kinase Ipl1 maintains the centromeric localization of PP2A to protect cohesin during meiosis. J Cell Biol 176: 911–8.

# Chapter 14
# The Kinetochore-Cancer Connection

**Takeshi Tomonaga**

**Abstract** An abnormal chromosome number and gross structural aberrations of chromosomes are hallmarks of human cancers. These chromosomal changes are thought to occur due to the accelerated rate of gains or losses of whole or large portions of chromosomes, termed chromosomal instability (CIN), as the result of continuous chromosome missegregation during mitosis. Recently, the mechanism of proper mitotic processes has been unraveled and the aberrant function of factors involved in equal chromosome segregation has been reported in various cancers. Among them, the centromere and kinetochore have a pivotal role in ensuring accurate chromosome segregation; thus, defects in kinetochore function are candidate sources of CIN and the generation of aneuploidy. In this chapter, recent progress in our understanding of how kinetochore dysfunction underlies CIN is introduced and described. Furthermore, I will discuss how it leads to the development of cancer.

## 14.1 Introduction

The morphologic criteria of malignant tumors include large nuclei containing abundant DNA and the high frequency of abnormal mitotic figures such as multipolar spindles, chromosome bridges and lagging chromosomes. Although these criteria are widely used for the diagnosis of cancer, these atypical mitotic figures were described more than a century ago by David Hansemann. He observed that a fraction of the chromosomes, which fail to segregate properly in tissue sections of various cancers and poorly differentiated tumors show asymmetric divisions most clearly (Hansemann, 1890). A quarter of a century after Hansemann's observations, Theodor Boveri considered that aneuploidy is a direct cause of cancer (Boveri, 1914). He proposed a model for tumorigenesis

T. Tomonaga (✉)
Department of Molecular Diagnosis, Graduate School of Medicine, Chiba University, 1-8-1 Inohana, Chuo-ku, Chiba 260-8670, Japan
e-mail: tomonaga@faculty.chiba-u.jp

P. De Wulf, W.C. Earnshaw (eds.), *The Kinetochore*,
DOI 10.1007/978-0-387-69076-6_14, 

in which a tumor originates from a single cell that has inherited a particular set of incorrectly distributed chromosomes.

Although there was general agreement that a high frequency of chromosome missegregation is a common phenomenon in tumors, its connection with carcinogenesis was often rejected because abnormal nuclear features were observed in benign tumors as well as in regenerating tissues (Stroebe, 1893). Furthermore, Boveri's hypothesis was unappreciated because the current molecular view focused on somatic gene mutation hypothesis, which argued that tumors arise by activation of oncogenes and/or inactivation of tumor suppressor genes (Bishop, 1987).

Recently, however, several observations breathed new life into Hansemann and Boveri's hypotheses that an imbalance in the number of chromosomes (aneuploidy) is the direct cause of cancer. First, Lengauer and colleagues found that 85% of colorectal cancers, and an even larger proportion of solid tumors, show aneuploidy (Lengauer et al., 1997, 1998; Rajagopalan et al., 2003; Rajagopalan and Lengauer, 2004). Since aneuploidy results from an accelerated rate of gains or losses of whole or large portions of chromosomes, this type of tumor was termed chromosomal instability (CIN). In contrast, tumors that are defective in mismatch repair but do not show aneuploidy were termed microsatellite instability (MIN), which represents less than 15% of solid tumors. Second, aneuploidy arises early in tumor development as a result of the mutation of genes involved in maintaining chromosome stability (Shih et al., 2001; Nowak et al., 2002). This indicates that aneuploidy is an important step in the initiation and/or progression of cancer. Third, experimentally transformed Chinese hamster cells using nongenotoxic carcinogens showed 100% aneuploidy (Li et al., 1997). Fourth, individuals suffering from mosaic variegated aneuploidy (MVA), a genetic disorder in which more than 25% of the cells in the body are aneuploid, frequently develop cancers at a young age. MVA results from mutations in both alleles of the *hBUBR1* gene (Hanks et al., 2004).

One means of how CIN may contribute to cancer promotion is that it could accelerate the rate of loss of heterozygosity (LOH) of a tumor suppressor gene and/or effectively amplify an oncogene by duplicating the chromosome. Inactivation of both alleles of a tumor suppressor gene must occur for a cell to acquire a growth advantage; thus, accelerated LOH is an apparent mechanism by which CIN can contribute significantly to the inactivation of a tumor suppressor gene. In contrast, it was proposed that cancer development does not necessarily require mutations in oncogenes and tumor suppressor genes, but rather that dosage imbalance of thousands of normal genes caused by CIN is important (Duesberg et al., 1998; Duesberg and Li, 2003). In either case, CIN can alter the expression of thousands of genes, which may or may not be essential for tumor development. In agreement with this idea, a recent report showed that analysis of genes in two common human malignancies, 11 breast and 11 colorectal cancers, revealed 189 mutant genes, most of which are not cancer-related genes but are involved in a wide range of cellular functions including transcription, cell adhesion, and invasion (Sjoblom et al., 2006). More recently, yeast strains that bear an extra copy of one or more of almost all of the yeast chromosomes revealed numerous intracellular functions to become aberrant (Torres et al., 2007). Although how CIN contributes to the initiation and/or progression of cancer is still a matter of debate, a clear correlation between the degree of aneuploidy and patient survival time is observed, which is useful for cancer diagnosis and treatment (Auer et al., 1980; Zetterberg and Esposti, 1980).

## 14.2 Mitotic Targets Involved in Chromosomal Instability

Both Hansemann (1890) and Boveri (1914) believed that the origin of malignant tumors is a defined but incorrectly combined set of chromosomes; however, their ideas for the mechanisms that create aneuploidy are different. Boveri emphasized that multipolar spindle formation is the most important mechanism for abnormal chromosome segregation. In contrast, Hansemann postulated that unequal chromatin distribution results from asymmetric forms of bipolar segregation such as chromosome bridges and lagging chromosomes. In recent years, the molecular basis for the segregation of chromosomes has been unraveled through analysis of mitosis in yeast and other eukaryotes, which has lead to the identification of processes that may contribute to CIN. Potential mitotic targets that could be involved in CIN include those such as chromosome condensation, sister-chromatid cohesion, kinetochore structure and function, centrosome /microtubule formation and dynamics, as well as 'checkpoint' genes that monitor the proper progression of the cell cycle (Table 14.1; Lengauer et al., 1998).

Altered behavior of any of the mitotic genes can cause CIN and quite a few genetic alterations in the activity or levels of those genes have been implicated in CIN (Kops et al., 2005; Perez de Castro et al., 2007; Table 14.1). One of the first-reported candidate genes responsible for CIN is the centrosome-associated kinase, STK15/BTAK/Aurora A. It was found to be amplified or overexpressed in multiple human tumor cell lines or in primary cancers including breast, colorectal, pancreatic, ovarian, esophageal, gastric, and bladder cancers (Bischoff et al., 1998; Katayama et al., 1999; Miyoshi et al., 2001; Sakakura et al., 2001; Sen et al., 2002; Gritsko et al., 2003; Li et al., 2003). Exogenous expression of the kinase in rodent and human cells induced unequal partitioning of chromosomes during mitosis and tumorigenic transformation of cells (Bischoff et al., 1998; Zhou et al., 1998). Overexpression of Aurora A in the mammary gland results in breast hyperplasia (Zhang et al., 2004; Wang et al., 2006). Other centrosome kinases (PLK1 and NEK2) or centrosome regulators (LATS2, PCM1, PIN1) are aberrantly expressed in various human tumors (Holtrich et al., 1994; Yuan et al., 1997; Armes et al., 2004; Bao et al., 2004; Hayward et al., 2004; Eckerdt et al., 2005; Jimenez-Velasco et al., 2005; Pils et al., 2005; Takahashi et al., 2005; Jiang et al., 2006; Strebhardt and Ullrich, 2006). Overexpression of PIN1 induced centrosome amplification,

**Table 14.1** Mitotic proteins that are mutated or aberrantly expressed in human cancers

| Molecular function | Name | Cancer-associated mutation | Cancer-associated altered expression | References |
|---|---|---|---|---|
| Centrosome / microtubule formation and dynamics | Aurora A | Amplification | Overexpression | Zhou, 1998; Bischoff, 1998; Katayama, 1999; Miyoshi, 2001; Sakakura, 2001; Sen, 2002; Gritsko, 2003; Li, 2003; Zhang, 2004; Wang, 2006 |
| | PLK1 | Mutation | Overexpression | Holtrich, 1994; Yuan, 1997; Eckerdt, 2005; Strebhardt and Ullrich 2006 |
| | NEK2 | none | Overexpression | Hayward, 2004; |
| | LATS2 | Hyper-methylation | Suppression | Takahashi, 2005; Jiang, 2006; Jimenez-Velasco, 2005 |
| | PCM1 | Deletion | Suppression | Armes, 2004; Pils, 2005 |
| | PIN1 | none | Overexpression | Bao, 2004; Suizu, 2006 |
| | ChTOG | none | Overexpression | Charrasse, 1995 |
| | EB1 | none | Overexpression | Wang, 2005 |
| Sister chromatid cohesion | PTTG1 | none | Overexpression | Grutzmann, 2004; Heaney, 2000; Puri, 2001; Shibata, 2002; Ramaswamy, 2003; Genkai, 2006; Fujii, 2006; Zhu, 2006; Su, 2006; Zou, 1999 |
| | RAD21 | Amplification | Overexpression | Porkka, 2004; |
| | STAG1 | Amplification | Overexpression | Giannini, 2003; Rae, 2001; |
| Kinetochore structure and function | CENP-A | none | Overexpression | Tomonaga, 2003 |
| | CENP-H | none | Overexpression | Tomonaga, 2003; Shigeishi, 2006; Liao, 2007 |
| | CENP-F | none | Overexpression | Clark, 1997; Liu, 1998; Erlanson, 1999; Pimkhaokham, 2000; de la Guardia, 2001; Esguerra, 2004; Shigeishi , 2005 |
| | Hec 1/KNTC2 | none | Overexpression | Hayama, 2006, Kirschner-Schwabe, 2006; Leupin, 2006 |
| | hSPC105/ Blinkin/ AF15q14 | Chromosome translocation | none | Hayette, 2000 |
| | APC | Mutation | none | Kinzler and Vogelstein 1996; Polakis 2000 |

**Table 14.1** (continued)

| Molecular function | Name | Cancer-associated mutation | Cancer-associated altered expression | References |
|---|---|---|---|---|
| Spindle checkpoint | BUB1 | Mutation or Hypermethylation | Overexpression or Supression | Cahill, 1998; Gemma, 2000; Ohshima, 2000; Ru, 2002; Shichiri, 2002; Hempen, 2003; Lin, 2002; Yuan, 2006; Moreno-Bueno, 2003; Grabsch, 2003; Matsuura, 2006 |
| | BUBR1 | Mutation or Hypermethylation | Overexpression | Ohshima, 2000; Shichiri, 2002; Grabsch, 2003, Hanks, 2004 |
| | BUB3 | none | Overexpression | Grabsch, 2003 |
| | MAD1 | Mutation | Suppression | Nomoto, 1999; Tsukasaki, 2001; Nishigaki,. 2005; Han, 2000 |
| | MAD2 | Mutation | Overexpression or Supression | Percy, 2000; Hernando, 2001, 2004; Kim, 2005; Li and Benezra, 1996; Wang, 2000; Wang, 2002; Yuan, 2006; Sotillo, 2007 |
| | Mps1 | none | Overexpression | Yuan, 2006 |
| | hRod/KNTC1 | Mutation | none | Wang, 2004 |
| | hZW10 | Mutation | none | Wang, 2004 |
| | HZwilch | Mutation | none | Wang, 2004 |
| Chromosome passenger proteins | Aurora B | none | Overexpression | Bischoff,. 1998; Tatsuka, 1998; Chieffi, 2004, 2006; Araki, 2004; Smith, 2005 |
| | Survivin | none | Overexpression | Ambrosini et al. 1997; Li, 2005 |
| | Borealin | none | Overexpression | Chang, 2006 |
| | INCENP | none | Overexpression | Adams, 2001 |
| Cell cycle control | CDC28 | Amplification | Overexpression | Shaughnessy, 2005; Chang, 2006; Kawakami, 2006; Stanbrough, 2006; Wong, 2006; Li, 2004 |
| | CDK1 | none | Overexpression | Kallakury, 1997; Soria, 2000; Takeno, 2002 |
| | CDC20 | none | Overexpression | Yuan, 2006; Ouellet, 2006; Singhal, 2003; Kim, 2005 |
| | Cyclin B1 | none | Overexpression | Kettunen, 2004; Koon, 2004; Wikman, 2002 |
| | Cyclin B2 | none | Overexpression | Moreno-Bueno, 2003 |

chromosomal instability, and tumorigenesis (Charrasse et al., 1995; Wang et al., 2005; Suizu et al., 2006). Factors involved in sister-chromatid cohesion such as RAD21, STAG1 and human securin (PTTG1) have been reported to be overexpressed in a wide range of tumors (Heaney et al., 2000; Puri et al., 2001; Rae et al., 2001; Shibata et al., 2002; Giannini et al., 2003; Ramaswamy et al., 2003; Grutzmann et al., 2004; Porkka et al., 2004; Fujii et al., 2006; Genkai et al., 2006; Su et al., 2006; Zhu et al., 2006). PTTG1 also exhibits transforming activity in NIH3T3 cells (Zou et al., 1999).

Altered expression of several mitotic cell-cycle control proteins such as CDC28, CDC20, CDK1 and A- or B-type cyclins have been associated with CIN (Table 14.1; Kallakury et al., 1997; Soria et al., 2000; Takeno et al., 2002; Wikman et al., 2002; Moreno-Bueno et al., 2003; Singhal et al., 2003; Kettunen et al., 2004; Koon et al., 2004; Li et al., 2004; Kim et al., 2005b; Shaughnessy, 2005; Chang et al., 2006a; Kawakami et al., 2006; Ouellet et al., 2006; Stanbrough et al., 2006; Wong et al., 2006; Yuan et al., 2006). For example, amplifications of cyclin E and mutations in *hCDC4*, both previously implicated in G1-S phase transitions, were identified in aneuploid cancers. Inactivation of hCDC4 in karyotypically stable colorectal cancer cells caused CIN (Rajagopalan et al., 2004).

Cytokinesis failure is also a key component for the generation of aneuploidy (Table 14.1). It has been hypothesized that aneuploidy develops from a tetraploid/polyploid intermediate during tumorigenesis (Shackney et al., 1989; Levine et al., 1991; Galipeau et al., 1996; Andreassen et al., 2001). Recent reports demonstrated that the failure of cytokinesis yields tetraploid cells, and that it increases with the frequency of chromosome missegregation (Storchova and Pellman, 2004; Fujiwara et al., 2005; Shi and King, 2005). Tetraploid cells were transformed in vitro after exposure to a carcinogen and generated tumors in vivo. The precise mechanism of aberrant cytokinesis is not clear; however, a protein complex called the chromosome passenger complex (composed of the Aurora B kinase, Survivin, INCENP, and Borealin) may be involved in the process. This complex shows dynamic localization change during mitosis. They associate along chromosomes during prophase, concentrate to inner centromeres until anaphase and transfer to a central spindle where a cleavage furrow will form; thus, they were considered to coordinate chromosome segregation and cytokinesis. Moreover, the perturbation of normal Aurora B function inhibits cleavage furrow formation and cytokinesis, resulting in multinucleation, which strongly suggests the important role of chromosome passenger proteins for cytokinesis (Cooke et al., 1987; Mackay et al., 1998; Schumacher et al., 1998; Terada et al., 1998; Adams et al., 2000). Conversely, overexpression of chromosome passenger proteins has been observed in various cancers (Bischoff et al., 1998; Tatsuka et al., 1998; Adams et al., 2001; Araki et al., 2004; Chieffi et al., 2004; Li, 2005; Smith et al., 2005; Chang et al., 2006b; Chieffi et al., 2006). However, if anomalous levels of chromosome passenger complex subunits are associated with the defect of cytokinesis and polyploidy remains unclear.

## 14.3 Kinetochore Dysfunction and Cancer

Centromeric DNA and kinetochores are essential for microtubule spindle attachments, spindle checkpoint activity and, sister chromatid separation, and segregation. The spindle checkpoint ensures that all chromosomes are correctly aligned at the metaphase plate and are properly attached to the metaphase spindle prior to chromosome separation. The spindle checkpoint is activated and induces the recruitment of checkpoint proteins such as MAD and BUB when one or more kinetochores are unattached or misattached to the spindle. Several types of cancer contain mutations in or aberrantly express either of the *hBUB1, hBUBR1, hBUB3* checkpoint genes (Cahill et al., 1998; Gemma et al., 2000; Ohshima et al., 2000; Lin et al., 2002; Ru et al., 2002; Shichiri et al., 2002; Grabsch et al., 2003; Hempen et al., 2003; Moreno-Bueno et al., 2003; Matsuura et al., 2006; Yuan et al., 2006). Inhibition of the *BUB1* gene induced genomic instability and cellular transformation in human fibroblasts (Musio et al., 2003). Germline biallelic mutations in the *hBUBR1* gene are associated with inherited predispositions to cancer (Hanks et al., 2004). Carcinogen-induced tumor development was enhanced in mice with *hBUBR1* or *BUB3* haploinsufficiency (Babu et al., 2003; Dai et al., 2004). Another spindle checkpoint protein MAD1, which recruits MAD2 to unattached kinetochores, is also mutated or downregulated in various cancers (Nomoto et al., 1999; Han et al., 2000; Tsukasaki et al., 2001; Nishigaki et al., 2005). *hMAD2* is also mutated in several human cancers (Percy et al., 2000; Hernando et al., 2001; Kim et al., 2005a). Furthermore, analysis of human homologs of CIN genes identified in yeast and flies found mutations in kinetochore proteins, hRod/KNTC1, hZw10 and hZwilch, which account for ~2% of colorectal cancers (Wang et al., 2004b). Recently, KNTC2, members of the evolutionarily conserved centromere protein complex, was activated in non-small cell lung carcinomas (Hayama et al., 2006). Hec1, a kinetochore protein involved in spindle checkpoint signaling, was also overexpressed in human leukemias (Martin-Lluesma et al., 2002; Kirschner-Schwabe et al., 2006; Leupin et al., 2006). Upregulation of Hec1 gene was observed in patients with very early relapse of acute lymphoblastic leukemia (Kirschner-Schwabe et al., 2006). On the other hand, suppression of Hec1 reduced tumor growth (Gurzov and Izquierdo, 2006). These results indicate that Hec1 might be a good prognostic marker and a therapeutic target of cancer. Another kinetochore-based checkpoint protein hSPC105/blinkin/AF15q14 was shown to be a partner gene fused to the MLL gene in an acute myeloid leukemia with a t(11;15)(q23;q14), however the precise role of the fusion protein for carcinogenesis is not known (Hayette et al., 2000; Kiyomitsu et al., 2007).

Mounting evidence suggests that mutations in mitotic checkpoint proteins may induce CIN, although such mutations are only rarely found in human cancers (Cahill et al., 1998; Imai et al., 1999; Gemma et al., 2000; Percy et al., 2000; Hernando et al., 2001). Interestingly, MAD2 is either up- or down-regulated depending on the tumor type and, in both cases, these alterations result in chromosomal instability and tumor development. (Li and Benezra, 1996; Wang et al., 2000; Wang et al., 2002; Hernando et al., 2004; Yuan et al., 2006; Sotillo et al., 2007). Heterozygous *MAD2* mice develop lung tumors at high rates, suggesting that biallelic expression of MAD2 is important for its function (Michel et al., 2001). On the other hand, overexpression of MAD2 in transgenic mice leads to a wide variety of tumors, as well as the acceleration of myc-induced lymphomagenesis (Sotillo et al., 2007). These data suggest that subtle changes in mitotic regulators might have oncogenic effects, whereas complete elimination of the proteins may not be compatible with cell survival.

Centromere-associated protein E (CENP-E), an essential mitotic kinesin that is required for stable microtubule capture by kinetochores, is also implicated as an amplifier of the mitotic spindle checkpoint by stimulating BUBR1 kinase activity (Cleveland et al., 2003). Inhibition of CENP-E function leads to a failure of metaphase alignment of chromosomes, resulting in aneuploidy either in primary mouse fibroblasts in vitro or in regenerating hepatocytes in vivo (Putkey et al., 2002; Weaver et al., 2003). Moreover, Weaver and colleagues showed that mice with reduced levels of CENP-E develop aneuploidy and drive spontaneous lymphomas and lung tumors in aged animals (Weaver et al., 2007). Controversially, the same group also found that CENP-E heterozygosity inhibits tumorigenesis in p19/ARF null or carcinogen-treated mice, which indicates that a higher level of unstable chromosomes might act as a tumor suppressor. This is another example showing that the aberrant expression of kinetochore proteins can induce CIN (e.g., by hampering the structural and functional integrity of kinetochores), resulting in either tumorigenesis, the stimulation of tumor progression, or in cell death if CIN is too deleterious for cell survival.

Some chromosomes in transformed rat cells and somatic cell hybrids are known to lack kinetochore proteins as revealed by imaging with antikinetochore antibodies. These cells contain micronuclei and are aneuploid (Kirchner et al., 1993). Thus, abnormal kinetochore structure is a possible source of CIN. Mutations in structural kinetochore proteins have not yet been identified in cancer, whereas aberrant expression of kinetochore proteins has recently been reported in various cancers. CENP-A, a centromeric histone H3 variant, which recruits all other kinetochore components to centromeres, is overexpressed and mistargeted in colorectal cancer tissues (Tomonaga et al., 2003). Overexpressed CENP-A localizes along the entire chromosome and recruits a subset of kinetochore proteins to noncentromeric chromatin. The formed ectopic prekinetochore complex may represent a potential link between CENP-A mislocalization and CIN in cancer (Van Hooser et al., 2001; Tomonaga et al., 2003; Heun et al., 2006). Overexpressed *Drosophila* CENP-A also mislocalizes to noncentromeric chromatin and promotes the formation of ectopic centromeres and multicentric chromosomes, which causes chromosome missegregation, aneuploidy, and growth defects (Heun et al., 2006). Thus, CENP-A mislocalization is one possible mechanism for CIN during cancer progression, as well as centromere plasticity during evolution.

Another inner kinetochore protein, CENP-H, which is important for kinetochore organization, is also upregulated in colorectal cancer tissue (Tomonaga et al., 2005). Transfection of a CENP-H expression plasmid into either diploid colorectal cancer cells or normal mouse fibroblasts induces an aberrant mitosis, suggesting that the upregulation of CENP-H can lead to a CIN phenotype (Tomonaga et al., 2005). Contrary to cells overexpressing CENP-A, in cells that overexpress CENP-H, it completely disappeared from the centromere of mitotic chromosomes, possibly by disturbing the stoichiometry of kinetochore components. CENP-H expression was also elevated in oral squamous cell carcinomas and nasopharyngeal carcinoma (Shigeishi et al., 2006; Liao et al., 2007). The mRNA level of CENP-H correlated positively with the clinical stage and/or poor survival time of patients with these cancers. Overexpression of CENP-A or CENP-H is not a general phenomenon of kinetochore proteins because the expression of other kinetochore protein such as hMis12 did not increase in colorectal cancer tissues (Tomonaga et al., 2005).

Overexpression of another kinetochore protein, CENP-F, has been reported to correlate with tumor proliferation and metastasis, although the precise function of CENP-F at kinetochores is not yet understood (Clark et al., 1997; Liu et al., 1998; Erlanson et al., 1999; Pimkhaokham et al., 2000; de la Guardia et al., 2001; Esguerra et al., 2004; Shigeishi et al., 2005). The evidence above suggests that overexpression of kinetochore proteins may induce CIN by disrupting the stoichiometry of the multiprotein kinetochore complex, interfere with normal mitosis, and hence contribute to tumor development. Moreover, the presence of dicentric and multicentric chromosomes is a common feature in certain cancers. Centromeres in some of the multicentric cells are located in the immediate vicinity of each other, forming a so called "compound centromere," where a large number of kinetochore proteins accumulate (Paweletz et al., 1989). The presence of compound centromeres may interfere with chromosome segregation and cause chromosome breakage, possibly contributing to the cancer formation. As the expression of many kinetochore proteins is cell cycle regulated and most of them have higher expression at G2/M phase, it is really important to differentiate "real" overexpression of kinetochore proteins from the enrichment of G2/M cells in cancer samples due to their fast division. Further investigations are needed to clarify whether the aberrant expression of kinetochore proteins is directly involved in carcinogenesis.

As mentioned above, chromosome passenger proteins localize to the inner kinetochore during prometaphase and several studies have shown that they are involved in accurate chromosome segregation through control of microtubule–kinetochore attachment. INCENP, conserved among eukaryotes, cause defects in the mitotic congression of chromosomes to the metaphase plate as well as chromosome segregation, if the C-terminal domain is deleted (Cooke et al., 1987; Mackay et al., 1998). Overexpression of chromosome passenger proteins has also been observed in various cancers (Ambrosini et al., 1997;

Bischoff et al., 1998; Tatsuka et al., 1998; Adams et al., 2001; Araki et al., 2004; Chieffi et al., 2004; Smith et al., 2005; Chang et al., 2006b; Chieffi et al., 2006). Furthermore, Aurora B-overexpressing cells exhibit CIN and injection of these cells into nude mice induces tumor formation (Ota et al., 2002; Nguyen et al., 2005). Conversely, inhibition of Aurora B expression reduces the proliferation rate of thyroid cancer cells, suggesting the association between Aurora-B expression and cancer initiation and/or progression.

The precise mechanism of how overexpression of Aurora-B causes aneuploidy and tumorigenesis is not clear. Aurora B, for example, phosphorylates histone H3 and its variant CENP-A and regulates proper chromosome segregation (Hsu et al., 2000; Kaitna et al., 2002; Kallio et al., 2002; Murata-Hori and Wang, 2002; Tanaka et al., 2002). There is a strong correlation between Aurora B overexpression, hyperphosphorylation of histone H3 at Ser 10, and increased mitotic Ser 10 phosphorylation induced lagging chromosomes during mitosis (Ota et al., 2002). On the other hands, prevention of CENP-A Ser 7 phosphorylation led to chromosome misalignment during mitosis as a result of a defect in kinetochore attachment to microtubules (Kunitoku et al., 2003). In addition, dominant negative phosphorylation mutants of CENP-A result in cytokinesis defect (Zeitlin et al., 2001). These results suggest that either hyperphosphorylation of histone H3 by Aurora B or hypophosphorylation of CENP-A might be involved in CIN.

Recent studies showed that the gene product of *adenomatous polyposis coli (APC),* the most frequently mutated gene in colorectal tumors, is involved in CIN. APC has been recognized as a suppressor of the Wnt/$\beta$-catenin signaling pathway, which is important for cell proliferation. Aberrant activation of the signaling pathway due to mutation of the *APC* gene is believed to be the initial event in colorectal carcinogenesis (Kinzler and Vogelstein, 1996; Polakis, 2000). Recently, several reports showed that the aberrant Wnt/$\beta$-catenin signaling pathway induces CIN, either by compromising the mitotic spindle checkpoint and/or inhibiting mitotic arrest and apoptosis (Hadjihannas et al., 2006; Aoki et al., 2007; Dikovskaya et al., 2007), although the precise mechanism remains to be elucidated. Meanwhile, interactions between APC and microtubules, either directly or indirectly through EB1, have been reported (Zumbrunn et al., 2001; Su et al., 2006). Moreover, APC was observed to accumulate at the kinetochore during mitosis. Colorectal tumor cell lines with *APC* mutations, similar to the mutations found in tumor cells, exhibit inefficient microtubule plus end attachments to kinetochore during mitosis, accompanied by impairment of chromosome alignment in metaphase (Fodde et al., 2001; Kaplan et al., 2001; Green and Kaplan, 2003). APC can also associate with the microtubule-destabilizing protein KinI, *Xenopus* mitotic centromere-associated kinesin (XMCAK; Banks and Heald, 2004). Taken together, these observations imply that loss of APC function induces CIN via inefficient microtubule–kinetochore attachment and failure of the mitotic spindle checkpoint.

## 14.4 Clinical Applications

Diagnosis of malignancy is made by a pathologist based on morphologic changes such as pleomorphism, abnormal nuclear morphology, and mitotic index. The distinction between benign and malignant tumors is not difficult. However, in between both lies a gray zone where even pathologists are unable to decide if a tumor is one or the other. In addition, a morphology-based diagnosis cannot predict the biological behavior or clinical course of a neoplasm, such as metastasis or prognosis, thus more objective evaluation criteria of malignancy that are based on molecular signatures is needed. In this regard, the detection of aneuploidy or CIN might permit an early prognosis of susceptibility to cancer (Hartwell et al., 1994). Moreover, aneuploid and/or chromosomally unstable cancers are likely to have a poorer prognosis than diploid cancers and the degree of aneuploidy correlates with the severity of the disease (Watanabe et al., 2001; Zhou et al., 2002; Kronenwett et al., 2004; Sinicrope et al., 2006). A recent study showed that a signature of CIN inferred from gene expression profiles predicted the clinical outcome and metastatic potential of the tumor (Carter et al., 2006). The accelerated rate of mutability of chromosome alterations, which defines CIN, dramatically alters the expression of thousands of genes and increases the possibility of cellular dedifferentiation and malignancy. This could explain why CIN tumors have a poorer prognosis than MIN tumors; therefore, examining the ploidy of tumors may represent novel and powerful prognostic indicators for cancers.

Overexpression of the CIN signature may reflect a compensatory mechanism for impaired functioning of the machinery responsible for chromosome stability. This possibility would provide potential target strategies for cancer treatment. Many antitumor drugs that are currently used, such as taxanes and vinca alkaloids, are drugs that alter microtubule dynamics and cause a mitotic arrest (Jordan and Wilson, 2004). Thus, inhibitors of protein involved in mitosis would be useful for the treatment of cancers. Indeed, a complete inhibition of mitotic checkpoint proteins such as MAD2 and BUBR1 caused an inactivation of the checkpoint, resulting in chromosome missegregation and cell death (Kops et al., 2004; Michel et al., 2004). Further, inhibition of Aurora B kinase also diminishes checkpoint signaling and cause mitosis with unequal chromosome separation (Keen and Taylor, 2004). A small molecule inhibitor of Aurora B kinases suppresses tumor growth when injected in nude mice (Harrington et al., 2004). These observations indicate that mitotic checkpoint genes involved in CIN could be potential targets for cancer therapy. Likewise, inhibitors of kinetochore proteins that are overexpressed in various cancers can also become effective antitumor drugs. This kind of inhibitor, however, must be used with caution because inhibiting the mechanism responsible for chromosome stability also increases the risk of CIN and tumorigenesis in healthy cells. They also promote the resistance of cancer cells to chemotherapy, as described below, if inhibition is incomplete. (Kops et al., 2005; Weaver and Cleveland, 2005).

CIN could also contribute to a cancer's ability to acquire chemoresistance (Sawyers, 2001; Wang et al., 2004a). CIN might increase the rate of Darwinian adaptation to changing intracellular and extracellular environments. In this way, CIN is thought to contribute to cellular resistance to chemotherapeutic drugs such as imatinib or 5-FU. Finally, mutated or upregulated CIN genes, such as kinetochore/spindle checkpoint genes, might allow the selective killing of tumor cells by applying genetic approaches to the discovery of new anticancer drugs (Hartwell et al., 1997). Yeast genetics has been used to identify pairs of non-allelic gene mutants that are each individually viable, but lethal in combination. If synthetic lethal interactions are conserved in humans, then synthetic lethal interactors that are common to CIN mutants may be potential targets for anticancer drugs (Yuen et al., 2005).

## References


Adams, R.R., Eckley, D.M., Vagnarelli, P., Wheatley, S.P., Gerloff, D.L., Mackay, A.M., Svingen, P.A., Kaufmann, S.H., and Earnshaw, W.C. (2001). Human INCENP colocalizes with the Aurora-B/AIRK2 kinase on chromosomes and is overexpressed in tumour cells. *Chromosoma 110*, 65–74.

Adams, R.R., Wheatley, S.P., Gouldsworthy, A.M., Kandels-Lewis, S.E., Carmena, M., Smythe, C., Gerloff, D.L., and Earnshaw, W.C. (2000). INCENP binds the Aurora-related kinase AIRK2 and is required to target it to chromosomes, the central spindle and cleavage furrow. *Curr Biol 10*, 1075–1078.

Ambrosini, G., Adida, C., and Altieri, D.C. (1997). A novel anti-apoptosis gene, survivin, expressed in cancer and lymphoma. *Nat Med 3*, 917–921.

Andreassen, P.R., Lohez, O.D., Lacroix, F.B., and Margolis, R.L. (2001). Tetraploid state induces p53-dependent arrest of nontransformed mammalian cells in G1. *Mol Biol Cell 12*, 1315–1328.

Aoki, K., Aoki, M., Sugai, M., Harada, N., Miyoshi, H., Tsukamoto, T., Mizoshita, T., Tatematsu, M., Seno, H., Chiba, T., Oshima, M., Hsieh, C.L., and Taketo, M.M. (2007). Chromosomal instability by beta-catenin/TCF transcription in APC or beta-catenin mutant cells. *Oncogene 26*, 3511–3520.

Araki, K., Nozaki, K., Ueba, T., Tatsuka, M., and Hashimoto, N. (2004). High expression of Aurora-B/Aurora and Ipll-like midbody-associated protein (AIM-1) in astrocytomas. *J Neurooncol 67*, 53–64.

Armes, J.E., Hammet, F., de Silva, M., Ciciulla, J., Ramus, S.J., Soo, W.K., Mahoney, A., Yarovaya, N., Henderson, M.A., Gish, K., Hutchins, A.M., Price, G.R., and Venter, D.J. (2004). Candidate tumor-suppressor genes on chromosome arm 8p in early-onset and high-grade breast cancers. *Oncogene 23*, 5697–5702.

Auer, G.U., Caspersson, T.O., and Wallgren, A.S. (1980). DNA content and survival in mammary carcinoma. *Anal Quant Cytol 2*, 161–165.

Babu, J.R., Jeganathan, K.B., Baker, D.J., Wu, X., Kang-Decker, N., and van Deursen, J.M. (2003). Rael is an essential mitotic checkpoint regulator that cooperates with Bub3 to prevent chromosome missegregation. *J Cell Biol 160*, 341–353.

Banks, J.D., and Heald, R. (2004). Adenomatous polyposis coli associates with the microtubule-destabilizing protein XMCAK. *Curr Biol 14*, 2033–2038.

Bao, L., Kimzey, A., Sauter, G., Sowadski, J.M., Lu, K.P., and Wang, D.G. (2004). Prevalent overexpression of prolyl isomerase Pin1 in human cancers. *Am J Pathol 164*, 1727–1737.

Bischoff, J.R., Anderson, L., Zhu, Y., Mossie, K., Ng, L., Souza, B., Schryver, B., Flanagan, P., Clairvoyant, F., Ginther, C., Chan, C.S., Novotny, M., Slamon, D.J., and Plowman, G.D. (1998). A homologue of Drosophila aurora kinase is oncogenic and amplified in human colorectal cancers. *Embo J 17*, 3052–3065.

Bishop, J.M. (1987). The molecular genetics of cancer. *Science 235*, 305–311.

Cahill, D.P., Lengauer, C., Yu, J., Riggins, G.J., Willson, J.K., Markowitz, S.D., Kinzler, K.W., and Vogelstein, B. (1998). Mutations of mitotic checkpoint genes in human cancers. *Nature 392*, 300–303.

Carter, S.L., Eklund, A.C., Kohane, I.S., Harris, L.N., and Szallasi, Z. (2006). A signature of chromosomal instability inferred from gene expression profiles predicts clinical outcome in multiple human cancers. *Nat Genet 38*, 1043–1048.

Chang, H., Yeung, J., Xu, W., Ning, Y., and Patterson, B. (2006a). Significant increase of CKS1B amplification from monoclonal gammopathy of undetermined significance to multiple myeloma and plasma cell leukaemia as demonstrated by interphase fluorescence in situ hybridisation. *Br J Haematol 134*, 613–615.

Chang, J.L., Chen, T.H., Wang, C.F., Chiang, Y.H., Huang, Y.L., Wong, F.H., Chou, C.K., and Chen, C.M. (2006b). Borealin/Dasra B is a cell cycle-regulated chromosomal passenger protein and its nuclear accumulation is linked to poor prognosis for human gastric cancer. *Exp Cell Res 312*, 962–973.

Charrasse, S., Mazel, M., Taviaux, S., Berta, P., Chow, T., and Larroque, C. (1995). Characterization of the cDNA and pattern of expression of a new gene over-expressed in human hepatomas and colonic tumors. *Eur J Biochem 234*, 406–413.

Chieffi, P., Cozzolino, L., Kisslinger, A., Libertini, S., Staibano, S., Mansueto, G., De Rosa, G., Villacci, A., Vitale, M., Linardopoulos, S., Portella, G., and Tramontano, D. (2006). Aurora B expression directly correlates with prostate cancer malignancy and influence prostate cell proliferation. *Prostate 66*, 326–333.

Chieffi, P., Troncone, G., Caleo, A., Libertini, S., Linardopoulos, S., Tramontano, D., and Portella, G. (2004). Aurora B expression in normal testis and seminomas. *J Endocrinol 181*, 263–270.

Clark, G.M., Allred, D.C., Hilsenbeck, S.G., Chamness, G.C., Osborne, C.K., Jones, D., and Lee, W.H. (1997). Mitosin (a new proliferation marker) correlates with clinical outcome in node-negative breast cancer. *Cancer Res 57*, 5505–5508.

Cleveland, D.W., Mao, Y., and Sullivan, K.F. (2003). Centromeres and kinetochores: from epigenetics to mitotic checkpoint signaling. *Cell 112*, 407–421.

Cooke, C.A., Heck, M.M., and Earnshaw, W.C. (1987). The inner centromere protein (INCENP) antigens: movement from inner centromere to midbody during mitosis. *J Cell Biol 105*, 2053–2067.

Dai, W., Wang, Q., Liu, T., Swamy, M., Fang, Y., Xie, S., Mahmood, R., Yang, Y.M., Xu, M., and Rao, C.V. (2004). Slippage of mitotic arrest and enhanced tumor development in mice with BubR1 haploinsufficiency. *Cancer Res 64*, 440–445.

de la Guardia, C., Casiano, C.A., Trinidad-Pinedo, J., and Baez, A. (2001). CENP-F gene amplification and overexpression in head and neck squamous cell carcinomas. *Head Neck 23*, 104–112.

Dikovskaya, D., Schiffmann, D., Newton, I.P., Oakley, A., Kroboth, K., Sansom, O., Jamieson, T.J., Meniel, V., Clarke, A., and Nathke, I.S. (2007). Loss of APC induces polyploidy as a result of a combination of defects in mitosis and apoptosis. *J Cell Biol 176*, 183–195.

Duesberg, P., and Li, R. (2003). Multistep carcinogenesis: a chain reaction of aneuploidizations. *Cell Cycle 2*, 202–210.

Duesberg, P., Rausch, C., Rasnick, D., and Hehlmann, R. (1998). Genetic instability of cancer cells is proportional to their degree of aneuploidy. *Proc Natl Acad Sci USA 95*, 13692–13697.

Eckerdt, F., Yuan, J., and Strebhardt, K. (2005). Polo-like kinases and oncogenesis. *Oncogene 24*, 267–276.

Erlanson, M., Casiano, C.A., Tan, E.M., Lindh, J., Roos, G., and Landberg, G. (1999). Immunohistochemical analysis of the proliferation associated nuclear antigen CENP-F in non-Hodgkin's lymphoma. *Mod Pathol 12*, 69–74.

Esguerra, R.L., Jia, L., Kaneko, T., Sakamoto, K., Okada, N., and Takagi, M. (2004). Immunohistochemical analysis of centromere protein F expression in buccal and gingival squamous cell carcinoma. *Pathol Int 54*, 82–89.

Fodde, R., Kuipers, J., Rosenberg, C., Smits, R., Kielman, M., Gaspar, C., van Es, J.H., Breukel, C., Wiegant, J., Giles, R.H., and Clevers, H. (2001). Mutations in the APC tumour suppressor gene cause chromosomal instability. *Nat Cell Biol 3*, 433–438.

Fujii, T., Nomoto, S., Koshikawa, K., Yatabe, Y., Teshigawara, O., Mori, T., Inoue, S., Takeda, S., and Nakao, A. (2006). Overexpression of pituitary tumor transforming gene 1 in HCC is associated with angiogenesis and poor prognosis. *Hepatology 43*, 1267–1275.

Fujiwara, T., Bandi, M., Nitta, M., Ivanova, E.V., Bronson, R.T., and Pellman, D. (2005). Cytokinesis failure generating tetraploids promotes tumorigenesis in p53-null cells. *Nature 437*, 1043–1047.

Galipeau, P.C., Cowan, D.S., Sanchez, C.A., Barrett, M.T., Emond, M.J., Levine, D.S., Rabinovitch, P.S., and Reid, B.J. (1996). 17p (p53) allelic losses, 4 N (G2/tetraploid) populations, and progression to aneuploidy in Barrett's esophagus. *Proc Natl Acad Sci USA 93*, 7081–7084.

Gemma, A., Seike, M., Seike, Y., Uematsu, K., Hibino, S., Kurimoto, F., Yoshimura, A., Shibuya, M., Harris, C.C., and Kudoh, S. (2000). Somatic mutation of the hBUB1 mitotic checkpoint gene in primary lung cancer. *Genes Chromosomes Cancer 29*, 213–218.

Genkai, N., Homma, J., Sano, M., Tanaka, R., and Yamanaka, R. (2006). Increased expression of pituitary tumor-transforming gene (PTTG)-1 is correlated with poor prognosis in glioma patients. *Oncol Rep 15*, 1569–1574.

Giannini, G., Ambrosini, M.I., Di Marcotullio, L., Cerignoli, F., Zani, M., MacKay, A.R., Screpanti, I., Frati, L., and Gulino, A. (2003). EGF- and cell-cycle-regulated STAG1/PMEPA1/ERG1.2 belongs to a conserved gene family and is overexpressed and amplified in breast and ovarian cancer. *Mol Carcinog 38*, 188–200.

Grabsch, H., Takeno, S., Parsons, W.J., Pomjanski, N., Boecking, A., Gabbert, H.E., and Mueller, W. (2003). Overexpression of the mitotic checkpoint genes BUB1, BUBR1, and BUB3 in gastric cancer–association with tumour cell proliferation. *J Pathol 200*, 16–22.

Green, R.A., and Kaplan, K.B. (2003). Chromosome instability in colorectal tumor cells is associated with defects in microtubule plus end attachments caused by a dominant mutation in APC. *J Cell Biol 163*, 949–961.

Gritsko, T.M., Coppola, D., Paciga, J.E., Yang, L., Sun, M., Shelley, S.A., Fiorica, J.V., Nicosia, S.V., and Cheng, J.Q. (2003). Activation and overexpression of centrosome kinase BTAK/Aurora-A in human ovarian cancer. *Clin Cancer Res 9*, 1420–1426.

Grutzmann, R., Pilarsky, C., Ammerpohl, O., Luttges, J., Bohme, A., Sipos, B., Foerder, M., Alldinger, I., Jahnke, B., Schackert, H.K., Kalthoff, H., Kremer, B., Kloppel, G., and Saeger, H.D. (2004). Gene expression profiling of microdissected pancreatic ductal carcinomas using high-density DNA microarrays. *Neoplasia 6*, 611–622.

Gurzov, E.N., and Izquierdo, M. (2006). RNA interference against Hec1 inhibits tumor growth in vivo. *Gene Ther 13*, 1–7.

Hadjihannas, M.V., Bruckner, M., Jerchow, B., Birchmeier, W., Dietmaier, W., and Behrens, J. (2006). Aberrant Wnt/beta-catenin signaling can induce chromosomal instability in colon cancer. *Proc Natl Acad Sci USA 103*, 10747–10752.

Han, S., Park, K., Kim, H.Y., Lee, M.S., Kim, H.J., Kim, Y.D., Yuh, Y.J., Kim, S.R., and Suh, H.S. (2000). Clinical implication of altered expression of Mad1 protein in human breast carcinoma. *Cancer 88*, 1623–1632.

Hanks, S., Coleman, K., Reid, S., Plaja, A., Firth, H., Fitzpatrick, D., Kidd, A., Mehes, K., Nash, R., Robin, N., Shannon, N., Tolmie, J., Swansbury, J., Irrthum, A., Douglas, J., and Rahman, N. (2004). Constitutional aneuploidy and cancer predisposition caused by biallelic mutations in BUB1B. *Nat Genet 36*, 1159–1161.

Hansemann, D. (1890). Ueber asymmetrische Zelltheilung in Epithelkrebsen und deren biologische Bedeutung. *Virschown Arch Pathol Anat* 119, 299–326.

Harrington, E.A., Bebbington, D., Moore, J., Rasmussen, R.K., Ajose-Adeogun, A.O., Nakayama, T., Graham, J.A., Demur, C., Hercend, T., Diu-Hercend, A., Su, M., Golec, J.M., and Miller, K.M. (2004). VX-680, a potent and selective small-molecule inhibitor of the Aurora kinases, suppresses tumor growth in vivo. *Nat Med 10*, 262–267.

Hartwell, L., Weinert, T., Kadyk, L., and Garvik, B. (1994). Cell cycle checkpoints, genomic integrity, and cancer. *Cold Spring Harb Symp Quant Biol 59*, 259–263.

Hartwell, L.H., Szankasi, P., Roberts, C.J., Murray, A.W., and Friend, S.H. (1997). Integrating genetic approaches into the discovery of anticancer drugs. *Science 278*, 1064–1068.

Hayama, S., Daigo, Y., Kato, T., Ishikawa, N., Yamabuki, T., Miyamoto, M., Ito, T., Tsuchiya, E., Kondo, S., and Nakamura, Y. (2006). Activation of CDCA1-KNTC2, members of centromere protein complex, involved in pulmonary carcinogenesis. *Cancer Res 66*, 10339–10348.

Hayette, S., Tigaud, I., Vanier, A., Martel, S., Corbo, L., Charrin, C., Beillard, E., Deleage, G., Magaud, J.P., and Rimokh, R. (2000). AF15q14, a novel partner gene fused to the MLL gene in an acute myeloid leukaemia with a t(11;15)(q23;q14). *Oncogene 19*, 4446–4450.

Hayward, D.G., Clarke, R.B., Faragher, A.J., Pillai, M.R., Hagan, I.M., and Fry, A.M. (2004). The centrosomal kinase Nek2 displays elevated levels of protein expression in human breast cancer. *Cancer Res 64*, 7370–7376.

Heaney, A.P., Singson, R., McCabe, C.J., Nelson, V., Nakashima, M., and Melmed, S. (2000). Expression of pituitary-tumour transforming gene in colorectal tumours. *Lancet 355*, 716–719.

Hempen, P.M., Kurpad, H., Calhoun, E.S., Abraham, S., and Kern, S.E. (2003). A double missense variation of the BUB1 gene and a defective mitotic spindle checkpoint in the pancreatic cancer cell line Hs766T. *Hum Mutat 21*, 445.

Hernando, E., Nahle, Z., Juan, G., Diaz-Rodriguez, E., Alaminos, M., Hemann, M., Michel, L., Mittal, V., Gerald, W., Benezra, R., Lowe, S.W., and Cordon-Cardo, C. (2004). Rb inactivation promotes genomic instability by uncoupling cell cycle progression from mitotic control. *Nature 430*, 797–802.

Hernando, E., Orlow, I., Liberal, V., Nohales, G., Benezra, R., and Cordon-Cardo, C. (2001). Molecular analyses of the mitotic checkpoint components hsMAD2, hBUB1 and hBUB3 in human cancer. *Int J Cancer 95*, 223–227.

Heun, P., Erhardt, S., Blower, M.D., Weiss, S., Skora, A.D., and Karpen, G.H. (2006). Mislocalization of the Drosophila centromere-specific histone CID promotes formation of functional ectopic kinetochores. *Dev Cell 10*, 303–315.

Holtrich, U., Wolf, G., Brauninger, A., Karn, T., Bohme, B., Rubsamen-Waigmann, H., and Strebhardt, K. (1994). Induction and down-regulation of PLK, a human serine/threonine kinase expressed in proliferating cells and tumors. *Proc Natl Acad Sci USA 91*, 1736–1740.

Hsu, J.Y., Sun, Z.W., Li, X., Reuben, M., Tatchell, K., Bishop, D.K., Grushcow, J.M., Brame, C.J., Caldwell, J.A., Hunt, D.F., Lin, R., Smith, M.M., and Allis, C.D. (2000). Mitotic phosphorylation of histone H3 is governed by Ipl1/aurora kinase and Glc7/PP1 phosphatase in budding yeast and nematodes. *Cell 102*, 279–291.

Imai, Y., Shiratori, Y., Kato, N., Inoue, T., and Omata, M. (1999). Mutational inactivation of mitotic checkpoint genes, hsMAD2 and hBUB1, is rare in sporadic digestive tract cancers. *Jpn J Cancer Res 90*, 837–840.

Jiang, Z., Li, X., Hu, J., Zhou, W., Jiang, Y., Li, G., and Lu, D. (2006). Promoter hypermethylation-mediated down-regulation of LATS1 and LATS2 in human astrocytoma. *Neurosci Res 56*, 450–458.

Jimenez-Velasco, A., Roman-Gomez, J., Agirre, X., Barrios, M., Navarro, G., Vazquez, I., Prosper, F., Torres, A., and Heiniger, A. (2005). Downregulation of the large tumor suppressor 2 (LATS2/KPM) gene is associated with poor prognosis in acute lymphoblastic leukemia. *Leukemia 19*, 2347–2350.

Jordan, M.A., and Wilson, L. (2004). Microtubules as a target for anticancer drugs. *Nat Rev Cancer 4*, 253–265.

Kaitna, S., Pasierbek, P., Jantsch, M., Loidl, J., and Glotzer, M. (2002). The aurora B kinase AIR-2 regulates kinetochores during mitosis and is required for separation of homologous Chromosomes during meiosis. *Curr Biol 12*, 798–812.

Kallakury, B.V., Sheehan, C.E., Ambros, R.A., Fisher, H.A., Kaufman, R.P., Jr., and Ross, J.S. (1997). The prognostic significance of p34cdc2 and cyclin D1 protein expression in prostate adenocarcinoma. *Cancer 80*, 753–763.

Kallio, M.J., McCleland, M.L., Stukenberg, P.T., and Gorbsky, G.J. (2002). Inhibition of aurora B kinase blocks chromosome segregation, overrides the spindle checkpoint, and perturbs microtubule dynamics in mitosis. *Curr Biol 12*, 900–905.

Kaplan, K.B., Burds, A.A., Swedlow, J.R., Bekir, S.S., Sorger, P.K., and Nathke, I.S. (2001). A role for the Adenomatous Polyposis Coli protein in chromosome segregation. *Nat Cell Biol 3*, 429–432.

Katayama, H., Ota, T., Jisaki, F., Ueda, Y., Tanaka, T., Odashima, S., Suzuki, F., Terada, Y., and Tatsuka, M. (1999). Mitotic kinase expression and colorectal cancer progression. *J Natl Cancer Inst 91*, 1160–1162.

Kawakami, K., Enokida, H., Tachiwada, T., Gotanda, T., Tsuneyoshi, K., Kubo, H., Nishiyama, K., Takiguchi, M., Nakagawa, M., and Seki, N. (2006). Identification of differentially expressed genes in human bladder cancer through genome-wide gene expression profiling. *Oncol Rep 16*, 521–531.

Keen, N., and Taylor, S. (2004). Aurora-kinase inhibitors as anticancer agents. *Nat Rev Cancer 4*, 927–936.

Kettunen, E., Anttila, S., Seppanen, J.K., Karjalainen, A., Edgren, H., Lindstrom, I., Salovaara, R., Nissen, A.M., Salo, J., Mattson, K., Hollmen, J., Knuutila, S., and Wikman, H. (2004). Differentially expressed genes in nonsmall cell lung cancer: expression profiling of cancer-related genes in squamous cell lung cancer. *Cancer Genet Cytogenet 149*, 98–106.

Kim, H.S., Park, K.H., Kim, S.A., Wen, J., Park, S.W., Park, B., Gham, C.W., Hyung, W.J., Noh, S.H., Kim, H.K., and Song, S.Y. (2005a). Frequent mutations of human Mad2, but not Bub1, in gastric cancers cause defective mitotic spindle checkpoint. *Mutat Res 578*, 187–201.

Kim, J.M., Sohn, H.Y., Yoon, S.Y., Oh, J.H., Yang, J.O., Kim, J.H., Song, K.S., Rho, S.M., Yoo, H.S., Kim, Y.S., Kim, J.G., and Kim, N.S. (2005b). Identification of gastric cancer-related genes using a cDNA microarray containing novel expressed sequence tags expressed in gastric cancer cells. *Clin Cancer Res 11*, 473–482.

Kinzler, K.W., and Vogelstein, B. (1996). Lessons from hereditary colorectal cancer. *Cell 87*, 159–170.

Kirchner, S., Stopper, H., Papp, T., Eckert, I., Yoo, H.J., Vig, B.K., and Schiffmann, D. (1993). Cytogenetic changes in primary, immortald and malignant mammalian cells. *Toxicol Lett 67*, 283–295.

Kirschner-Schwabe, R., Lottaz, C., Todling, J., Rhein, P., Karawajew, L., Eckert, C., von Stackelberg, A., Ungethum, U., Kostka, D., Kulozik, A.E., Ludwig, W.D., Henze, G., Spang, R., Hagemeier, C., and Seeger, K. (2006). Expression of late cell cycle genes and an increased proliferative capacity characterize very early relapse of childhood acute lymphoblastic leukemia. *Clin Cancer Res 12*, 4553–4561.

Kiyomitsu, T., Obuse, C., and Yanagida, M. (2007). Human Blinkin/AF15q14 is required for chromosome alignment and the mitotic checkpoint through direct interaction with Bub1 and BubR1. *Dev Cell 13*, 663–676.

Koon, N., Schneider-Stock, R., Sarlomo-Rikala, M., Lasota, J., Smolkin, M., Petroni, G., Zaika, A., Boltze, C., Meyer, F., Andersson, L., Knuutila, S., Miettinen, M., and El-Rifai, W.

(2004). Molecular targets for tumour progression in gastrointestinal stromal tumours. *Gut 53*, 235–240.

Kops, G.J., Foltz, D.R., and Cleveland, D.W. (2004). Lethality to human cancer cells through massive chromosome loss by inhibition of the mitotic checkpoint. *Proc Natl Acad Sci USA 101*, 8699–8704.

Kops, G.J., Weaver, B.A., and Cleveland, D.W. (2005). On the road to cancer: aneuploidy and the mitotic checkpoint. *Nat Rev Cancer 5*, 773–785.

Kronenwett, U., Huwendiek, S., Ostring, C., Portwood, N., Roblick, U.J., Pawitan, Y., Alaiya, A., Sennerstam, R., Zetterberg, A., and Auer, G. (2004). Improved grading of breast adenocarcinomas based on genomic instability. *Cancer Res 64*, 904–909.

Kunitoku, N., Sasayama, T., Marumoto, T., Zhang, D., Honda, S., Kobayashi, O., Hatakeyama, K., Ushio, Y., Saya, H., and Hirota, T. (2003). CENP-A phosphorylation by Aurora-A in prophase is required for enrichment of Aurora-B at inner centromeres and for kinetochore function. *Dev Cell 5*, 853–864.

Lengauer, C., Kinzler, K.W., and Vogelstein, B. (1997). Genetic instability in colorectal cancers. *Nature 386*, 623–627.

Lengauer, C., Kinzler, K.W., and Vogelstein, B. (1998). Genetic instabilities in human cancers. *Nature 396*, 643–649.

Leupin, N., Kuhn, A., Hugli, B., Grob, T.J., Jaggi, R., Tobler, A., Delorenzi, M., and Fey, M.F. (2006). Gene expression profiling reveals consistent differences between clinical samples of human leukaemias and their model cell lines. *Br J Haematol 135*, 520–523.

Levine, D.S., Sanchez, C.A., Rabinovitch, P.S., and Reid, B.J. (1991). Formation of the tetraploid intermediate is associated with the development of cells with more than four centrioles in the elastase-simian virus 40 tumor antigen transgenic mouse model of pancreatic cancer. *Proc Natl Acad Sci USA 88*, 6427–6431.

Li, D., Zhu, J., Firozi, P.F., Abbruzzese, J.L., Evans, D.B., Cleary, K., Friess, H., and Sen, S. (2003). Overexpression of oncogenic STK15/BTAK/Aurora A kinase in human pancreatic cancer. *Clin Cancer Res 9*, 991–997.

Li, F. (2005). Role of survivin and its splice variants in tumorigenesis. *Br J Cancer 92*, 212–216.

Li, M., Lin, Y.M., Hasegawa, S., Shimokawa, T., Murata, K., Kameyama, M., Ishikawa, O., Katagiri, T., Tsunoda, T., Nakamura, Y., and Furukawa, Y. (2004). Genes associated with liver metastasis of colon cancer, identified by genome-wide cDNA microarray. *Int J Oncol 24*, 305–312.

Li, R., Yerganian, G., Duesberg, P., Kraemer, A., Willer, A., Rausch, C., and Hehlmann, R. (1997). Aneuploidy correlated 100% with chemical transformation of Chinese hamster cells. *Proc Natl Acad Sci USA 94*, 14506–14511.

Li, Y., and Benezra, R. (1996). Identification of a human mitotic checkpoint gene: hsMAD2. *Science 274*, 246–248.

Liao, W.T., Song, L.B., Zhang, H.Z., Zhang, X., Zhang, L., Liu, W.L., Feng, Y., Guo, B.H., Mai, H.Q., Cao, S.M., Li, M.Z., Qin, H.D., Zeng, Y.X., and Zeng, M.S. (2007). Centromere protein H is a novel prognostic marker for nasopharyngeal carcinoma progression and overall patient survival. *Clin Cancer Res 13*, 508–514.

Lin, S.F., Lin, P.M., Yang, M.C., Liu, T.C., Chang, J.G., Sue, Y.C., and Chen, T.P. (2002). Expression of hBUB1 in acute myeloid leukemia. *Leuk Lymphoma 43*, 385–391.

Liu, S.C., Sauter, E.R., Clapper, M.L., Feldman, R.S., Levin, L., Chen, S.Y., Yen, T.J., Ross, E., Engstrom, P.F., and Klein-Szanto, A.J. (1998). Markers of cell proliferation in normal epithelia and dysplastic leukoplakias of the oral cavity. *Cancer Epidemiol Biomarkers Prev 7*, 597–603.

Mackay, A.M., Ainsztein, A.M., Eckley, D.M., and Earnshaw, W.C. (1998). A dominant mutant of inner centromere protein (INCENP), a chromosomal protein, disrupts prometaphase congression and cytokinesis. *J Cell Biol 140*, 991–1002.

Martin-Lluesma, S., Stucke, V.M., and Nigg, E.A. (2002). Role of Hec1 in spindle checkpoint signaling and kinetochore recruitment of Mad1/Mad2. *Science 297*, 2267–2270.
Matsuura, S., Matsumoto, Y., Morishima, K., Izumi, H., Matsumoto, H., Ito, E., Tsutsui, K., Kobayashi, J., Tauchi, H., Kajiwara, Y., Hama, S., Kurisu, K., Tahara, H., Oshimura, M., Komatsu, K., Ikeuchi, T., and Kajii, T. (2006). Monoallelic BUB1B mutations and defective mitotic-spindle checkpoint in seven families with premature chromatid separation (PCS) syndrome. *Am J Med Genet A 140*, 358–367.
Michel, L., Diaz-Rodriguez, E., Narayan, G., Hernando, E., Murty, V.V., and Benezra, R. (2004). Complete loss of the tumor suppressor MAD2 causes premature cyclin B degradation and mitotic failure in human somatic cells. *Proc Natl Acad Sci USA 101*, 4459–4464.
Michel, L.S., Liberal, V., Chatterjee, A., Kirchwegger, R., Pasche, B., Gerald, W., Dobles, M., Sorger, P.K., Murty, V.V., and Benezra, R. (2001). MAD2 haplo-insufficiency causes premature anaphase and chromosome instability in mammalian cells. *Nature 409*, 355–359.
Miyoshi, Y., Iwao, K., Egawa, C., and Noguchi, S. (2001). Association of centrosomal kinase STK15/BTAK mRNA expression with chromosomal instability in human breast cancers. *Int J Cancer 92*, 370–373.
Moreno-Bueno, G., Sanchez-Estevez, C., Cassia, R., Rodriguez-Perales, S., Diaz-Uriarte, R., Dominguez, O., Hardisson, D., Andujar, M., Prat, J., Matias-Guiu, X., Cigudosa, J.C., and Palacios, J. (2003). Differential gene expression profile in endometrioid and nonendometrioid endometrial carcinoma: STK15 is frequently overexpressed and amplified in nonendometrioid carcinomas. *Cancer Res 63*, 5697–5702.
Murata-Hori, M., and Wang, Y.L. (2002). The kinase activity of aurora B is required for kinetochore-microtubule interactions during mitosis. *Curr Biol 12*, 894–899.
Musio, A., Montagna, C., Zambroni, D., Indino, E., Barbieri, O., Citti, L., Villa, A., Ried, T., and Vezzoni, P. (2003). Inhibition of BUB1 results in genomic instability and anchorage-independent growth of normal human fibroblasts. *Cancer Res 63*, 2855–2863.
Nguyen, H.G., Chinnappan, D., Urano, T., and Ravid, K. (2005). Mechanism of Aurora-B degradation and its dependency on intact KEN and A-boxes: identification of an aneuploidy-promoting property. *Mol Cell Biol 25*, 4977–4992.
Nishigaki, R., Osaki, M., Hiratsuka, M., Toda, T., Murakami, K., Jeang, K.T., Ito, H., Inoue, T., and Oshimura, M. (2005). Proteomic identification of differentially-expressed genes in human gastric carcinomas. *Proteomics 5*, 3205–3213.
Nomoto, S., Haruki, N., Takahashi, T., Masuda, A., Koshikawa, T., Takahashi, T., Fujii, Y., Osada, H., and Takahashi, T. (1999). Search for in vivo somatic mutations in the mitotic checkpoint gene, hMAD1, in human lung cancers. *Oncogene 18*, 7180–7183.
Nowak, M.A., Komarova, N.L., Sengupta, A., Jallepalli, P.V., Shih Ie, M., Vogelstein, B., and Lengauer, C. (2002). The role of chromosomal instability in tumor initiation. *Proc Natl Acad Sci U S A 99*, 16226–16231.
Ohshima, K., Haraoka, S., Yoshioka, S., Hamasaki, M., Fujiki, T., Suzumiya, J., Kawasaki, C., Kanda, M., and Kikuchi, M. (2000). Mutation analysis of mitotic checkpoint genes (hBUB1 and hBUBR1) and microsatellite instability in adult T-cell leukemia/lymphoma. *Cancer Lett 158*, 141–150.
Ota, T., Suto, S., Katayama, H., Han, Z.B., Suzuki, F., Maeda, M., Tanino, M., Terada, Y., and Tatsuka, M. (2002). Increased mitotic phosphorylation of histone H3 attributable to AIM-1/Aurora-B overexpression contributes to chromosome number instability. *Cancer Res 62*, 5168–5177.
Ouellet, V., Guyot, M.C., Le Page, C., Filali-Mouhim, A., Lussier, C., Tonin, P.N., Provencher, D.M., and Mes-Masson, A.M. (2006). Tissue array analysis of expression microarray candidates identifies markers associated with tumor grade and outcome in serous epithelial ovarian cancer. *Int J Cancer 119*, 599–607.

Paweletz, N., Vig, B.K., and Finze, E.M. (1989). Evolution of compound centromeres. A new phenomenon. *Cancer Genet Cytogenet 42*, 75–86.
Percy, M.J., Myrie, K.A., Neeley, C.K., Azim, J.N., Ethier, S.P., and Petty, E.M. (2000). Expression and mutational analyses of the human MAD2L1 gene in breast cancer cells. *Genes Chromosomes Cancer 29*, 356–362.
Perez de Castro, I., de Carcer, G., and Malumbres, M. (2007). A census of mitotic cancer genes: new insights into tumor cell biology and cancer therapy. *Carcinogenesis 28*, 899–912.
Pils, D., Horak, P., Gleiss, A., Sax, C., Fabjani, G., Moebus, V.J., Zielinski, C., Reinthaller, A., Zeillinger, R., and Krainer, M. (2005). Five genes from chromosomal band 8p22 are significantly down-regulated in ovarian carcinoma: N33 and EFA6R have a potential impact on overall survival. *Cancer 104*, 2417–2429.
Pimkhaokham, A., Shimada, Y., Fukuda, Y., Kurihara, N., Imoto, I., Yang, Z.Q., Imamura, M., Nakamura, Y., Amagasa, T., and Inazawa, J. (2000). Nonrandom chromosomal imbalances in esophageal squamous cell carcinoma cell lines: possible involvement of the ATF3 and CENPF genes in the 1q32 amplicon. *Jpn J Cancer Res 91*, 1126–1133.
Polakis, P. (2000). Wnt signaling and cancer. *Genes Dev 14*, 1837–1851.
Porkka, K.P., Tammela, T.L., Vessella, R.L., and Visakorpi, T. (2004). RAD21 and KIAA0196 at 8q24 are amplified and overexpressed in prostate cancer. *Genes Chromosomes Cancer 39*, 1–10.
Puri, R., Tousson, A., Chen, L., and Kakar, S.S. (2001). Molecular cloning of pituitary tumor transforming gene 1 from ovarian tumors and its expression in tumors. *Cancer Lett 163*, 131–139.
Putkey, F.R., Cramer, T., Morphew, M.K., Silk, A.D., Johnson, R.S., McIntosh, J.R., and Cleveland, D.W. (2002). Unstable kinetochore-microtubule capture and chromosomal instability following deletion of CENP-E. *Dev Cell 3*, 351–365.
Rae, F.K., Hooper, J.D., Nicol, D.L., and Clements, J.A. (2001). Characterization of a novel gene, STAG1/PMEPA1, upregulated in renal cell carcinoma and other solid tumors. *Mol Carcinog 32*, 44–53.
Rajagopalan, H., Jallepalli, P.V., Rago, C., Velculescu, V.E., Kinzler, K.W., Vogelstein, B., and Lengauer, C. (2004). Inactivation of hCDC4 can cause chromosomal instability. *Nature 428*, 77–81.
Rajagopalan, H., and Lengauer, C. (2004). Aneuploidy and cancer. *Nature 432*, 338–341.
Rajagopalan, H., Nowak, M.A., Vogelstein, B., and Lengauer, C. (2003). The significance of unstable chromosomes in colorectal cancer. *Nat Rev Cancer 3*, 695–701.
Ramaswamy, S., Ross, K.N., Lander, E.S., and Golub, T.R. (2003). A molecular signature of metastasis in primary solid tumors. *Nat Genet 33*, 49–54.
Ru, H.Y., Chen, R.L., Lu, W.C., and Chen, J.H. (2002). hBUB1 defects in leukemia and lymphoma cells. *Oncogene 21*, 4673–4679.
Sakakura, C., Hagiwara, A., Yasuoka, R., Fujita, Y., Nakanishi, M., Masuda, K., Shimomura, K., Nakamura, Y., Inazawa, J., Abe, T., and Yamagishi, H. (2001). Tumour-amplified kinase BTAK is amplified and overexpressed in gastric cancers with possible involvement in aneuploid formation. *Br J Cancer 84*, 824–831.
Sawyers, C.L. (2001). Research on resistance to cancer drug Gleevec. *Science 294*, 1834.
Schumacher, J.M., Golden, A., and Donovan, P.J. (1998). AIR-2: An Aurora/Ipl1-related protein kinase associated with chromosomes and midbody microtubules is required for polar body extrusion and cytokinesis in Caenorhabditis elegans embryos. *J Cell Biol 143*, 1635–1646.
Sen, S., Zhou, H., Zhang, R.D., Yoon, D.S., Vakar-Lopez, F., Ito, S., Jiang, F., Johnston, D., Grossman, H.B., Ruifrok, A.C., Katz, R.L., Brinkley, W., and Czerniak, B. (2002). Amplification/overexpression of a mitotic kinase gene in human bladder cancer. *J Natl Cancer Inst 94*, 1320–1329.
Shackney, S.E., Smith, C.A., Miller, B.W., Burholt, D.R., Murtha, K., Giles, H.R., Ketterer, D.M., and Pollice, A.A. (1989). Model for the genetic evolution of human solid tumors. *Cancer Res 49*, 3344–3354.

Shaughnessy, J. (2005). Amplification and overexpression of CKS1B at chromosome band 1q21 is associated with reduced levels of p27Kip1 and an aggressive clinical course in multiple myeloma. *Hematology 10 Suppl 1*, 117–126.

Shi, Q., and King, R.W. (2005). Chromosome nondisjunction yields tetraploid rather than aneuploid cells in human cell lines. *Nature 437*, 1038–1042.

Shibata, Y., Haruki, N., Kuwabara, Y., Nishiwaki, T., Kato, J., Shinoda, N., Sato, A., Kimura, M., Koyama, H., Toyama, T., Ishiguro, H., Kudo, J., Terashita, Y., Konishi, S., and Fujii, Y. (2002). Expression of PTTG (pituitary tumor transforming gene) in esophageal cancer. *Jpn J Clin Oncol 32*, 233–237.

Shichiri, M., Yoshinaga, K., Hisatomi, H., Sugihara, K., and Hirata, Y. (2002). Genetic and epigenetic inactivation of mitotic checkpoint genes hBUB1 and hBUBR1 and their relationship to survival. *Cancer Res 62*, 13–17.

Shigeishi, H., Higashikawa, K., Ono, S., Mizuta, K., Ninomiya, Y., Yoneda, S., Taki, M., and Kamata, N. (2006). Increased expression of CENP-H gene in human oral squamous cell carcinomas harboring high-proliferative activity. *Oncol Rep 16*, 1071–1075.

Shigeishi, H., Mizuta, K., Higashikawa, K., Yoneda, S., Ono, S., and Kamata, N. (2005). Correlation of CENP-F gene expression with tumor-proliferating activity in human salivary gland tumors. *Oral Oncol 41*, 716–722.

Shih, I.M., Zhou, W., Goodman, S.N., Lengauer, C., Kinzler, K.W., and Vogelstein, B. (2001). Evidence that genetic instability occurs at an early stage of colorectal tumorigenesis. *Cancer Res 61*, 818–822.

Singhal, S., Amin, K.M., Kruklitis, R., DeLong, P., Friscia, M.E., Litzky, L.A., Putt, M.E., Kaiser, L.R., and Albelda, S.M. (2003). Alterations in cell cycle genes in early stage lung adenocarcinoma identified by expression profiling. *Cancer Biol Ther 2*, 291–298.

Sinicrope, F.A., Rego, R.L., Halling, K.C., Foster, N., Sargent, D.J., La Plant, B., French, A.J., Laurie, J.A., Goldberg, R.M., Thibodeau, S.N., and Witzig, T.E. (2006). Prognostic impact of microsatellite instability and DNA ploidy in human colon carcinoma patients. *Gastroenterology 131*, 729–737.

Sjoblom, T., Jones, S., Wood, L.D., Parsons, D.W., Lin, J., Barber, T.D., Mandelker, D., Leary, R.J., Ptak, J., Silliman, N., Szabo, S., Buckhaults, P., Farrell, C., Meeh, P., Markowitz, S.D., Willis, J., Dawson, D., Willson, J.K., Gazdar, A.F., Hartigan, J., Wu, L., Liu, C., Parmigiani, G., Park, B.H., Bachman, K.E., Papadopoulos, N., Vogelstein, B., Kinzler, K.W., and Velculescu, V.E. (2006). The consensus coding sequences of human breast and colorectal cancers. *Science 314*, 268–274.

Smith, S.L., Bowers, N.L., Betticher, D.C., Gautschi, O., Ratschiller, D., Hoban, P.R., Booton, R., Santibanez-Koref, M.F., and Heighway, J. (2005). Overexpression of aurora B kinase (AURKB) in primary non-small cell lung carcinoma is frequent, generally driven from one allele, and correlates with the level of genetic instability. *Br J Cancer 93*, 719–729.

Soria, J.C., Jang, S.J., Khuri, F.R., Hassan, K., Liu, D., Hong, W.K., and Mao, L. (2000). Overexpression of cyclin B1 in early-stage non-small cell lung cancer and its clinical implication. *Cancer Res 60*, 4000–4004.

Sotillo, R., Hernando, E., Diaz-Rodriguez, E., Teruya-Feldstein, J., Cordon-Cardo, C., Lowe, S.W., and Benezra, R. (2007). Mad2 overexpression promotes aneuploidy and tumorigenesis in mice. *Cancer Cell 11*, 9–23.

Stanbrough, M., Bubley, G.J., Ross, K., Golub, T.R., Rubin, M.A., Penning, T.M., Febbo, P.G., and Balk, S.P. (2006). Increased expression of genes converting adrenal androgens to testosterone in androgen-independent prostate cancer. *Cancer Res 66*, 2815–2825.

Storchova, Z., and Pellman, D. (2004). From polyploidy to aneuploidy, genome instability and cancer. *Nat Rev Mol Cell Biol 5*, 45–54.

Strebhardt, K., and Ullrich, A. (2006). Targeting polo-like kinase 1 for cancer therapy. *Nat Rev Cancer 6*, 321–330.

Stroebe, H. (1893). Veber Vorkommen und Bedeutung der asymmetrischen Karyokinese, nebst Bemerkungen uber die 'Schlummerzellen' in der verletzten Cornea. *Beitr Pathol Anat Allg Pathol* 14, 154–173.

Su, M.C., Hsu, H.C., Liu, Y.J., and Jeng, Y.M. (2006). Overexpression of pituitary tumor-transforming gene-1 in hepatocellular carcinoma. *Hepatogastroenterology 53*, 262–265.

Suizu, F., Ryo, A., Wulf, G., Lim, J., and Lu, K.P. (2006). Pin1 regulates centrosome duplication, and its overexpression induces centrosome amplification, chromosome instability, and oncogenesis. *Mol Cell Biol 26*, 1463–1479.

Takahashi, Y., Miyoshi, Y., Takahata, C., Irahara, N., Taguchi, T., Tamaki, Y., and Noguchi, S. (2005). Down-regulation of LATS1 and LATS2 mRNA expression by promoter hypermethylation and its association with biologically aggressive phenotype in human breast cancers. *Clin Cancer Res 11*, 1380–1385.

Takeno, S., Noguchi, T., Kikuchi, R., Uchida, Y., Yokoyama, S., and Muller, W. (2002). Prognostic value of cyclin B1 in patients with esophageal squamous cell carcinoma. *Cancer 94*, 2874–2881.

Tanaka, T.U., Rachidi, N., Janke, C., Pereira, G., Galova, M., Schiebel, E., Stark, M.J., and Nasmyth, K. (2002). Evidence that the Ipl1-Sli15 (Aurora kinase-INCENP) complex promotes chromosome bi-orientation by altering kinetochore-spindle pole connections. *Cell 108*, 317–329.

Tatsuka, M., Katayama, H., Ota, T., Tanaka, T., Odashima, S., Suzuki, F., and Terada, Y. (1998). Multinuclearity and increased ploidy caused by overexpression of the aurora- and Ipl1-like midbody-associated protein mitotic kinase in human cancer cells. *Cancer Res 58*, 4811–4816.

Terada, Y., Tatsuka, M., Suzuki, F., Yasuda, Y., Fujita, S., and Otsu, M. (1998). AIM-1: a mammalian midbody-associated protein required for cytokinesis. *Embo J 17*, 667–676.

Tomonaga, T., Matsushita, K., Ishibashi, M., Nezu, M., Shimada, H., Ochiai, T., Yoda, K., and Nomura, F. (2005). Centromere protein H is up-regulated in primary human colorectal cancer and its overexpression induces aneuploidy. *Cancer Res 65*, 4683–4689.

Tomonaga, T., Matsushita, K., Yamaguchi, S., Oohashi, T., Shimada, H., Ochiai, T., Yoda, K., and Nomura, F. (2003). Overexpression and mistargeting of centromere protein-A in human primary colorectal cancer. *Cancer Res 63*, 3511–3516.

Torres, E.M., Sokolsky, T., Tucker, C.M., Chan, L.Y., Boselli, M., Dunham, M.J., and Amon, A. (2007). Effects of aneuploidy on cellular physiology and cell division in haploid yeast. *Science 317*, 916–924.

Tsukasaki, K., Miller, C.W., Greenspun, E., Eshaghian, S., Kawabata, H., Fujimoto, T., Tomonaga, M., Sawyers, C., Said, J.W., and Koeffler, H.P. (2001). Mutations in the mitotic check point gene, MAD1L1, in human cancers. *Oncogene 20*, 3301–3305.

Van Hooser, A.A., Ouspenski, II, Gregson, H.C., Starr, D.A., Yen, T.J., Goldberg, M.L., Yokomori, K., Earnshaw, W.C., Sullivan, K.F., and Brinkley, B.R. (2001). Specification of kinetochore-forming chromatin by the histone H3 variant CENP-A. *J Cell Sci 114*, 3529–3542.

Wang, T.L., Diaz, L.A., Jr., Romans, K., Bardelli, A., Saha, S., Galizia, G., Choti, M., Donehower, R., Parmigiani, G., Shih Ie, M., Iacobuzio-Donahue, C., Kinzler, K.W., Vogelstein, B., Lengauer, C., and Velculescu, V.E. (2004a). Digital karyotyping identifies thymidylate synthase amplification as a mechanism of resistance to 5-fluorouracil in metastatic colorectal cancer patients. *Proc Natl Acad Sci USA 101*, 3089–3094.

Wang, X., Jin, D.Y., Ng, R.W., Feng, H., Wong, Y.C., Cheung, A.L., and Tsao, S.W. (2002). Significance of MAD2 expression to mitotic checkpoint control in ovarian cancer cells. *Cancer Res 62*, 1662–1668.

Wang, X., Jin, D.Y., Wong, Y.C., Cheung, A.L., Chun, A.C., Lo, A.K., Liu, Y., and Tsao, S.W. (2000). Correlation of defective mitotic checkpoint with aberrantly reduced expression of MAD2 protein in nasopharyngeal carcinoma cells. *Carcinogenesis 21*, 2293–2297.

Wang, X., Zhou, Y.X., Qiao, W., Tominaga, Y., Ouchi, M., Ouchi, T., and Deng, C.X. (2006). Overexpression of aurora kinase A in mouse mammary epithelium induces genetic instability preceding mammary tumor formation. *Oncogene 25*, 7148–7158.

Wang, Y., Zhou, X., Zhu, H., Liu, S., Zhou, C., Zhang, G., Xue, L., Lu, N., Quan, L., Bai, J., Zhan, Q., and Xu, N. (2005). Overexpression of EB1 in human esophageal squamous cell carcinoma (ESCC) may promote cellular growth by activating beta-catenin/TCF pathway. *Oncogene 24*, 6637–6645.

Wang, Z., Cummins, J.M., Shen, D., Cahill, D.P., Jallepalli, P.V., Wang, T.L., Parsons, D.W., Traverso, G., Awad, M., Silliman, N., Ptak, J., Szabo, S., Willson, J.K., Markowitz, S.D., Goldberg, M.L., Karess, R., Kinzler, K.W., Vogelstein, B., Velculescu, V.E., and Lengauer, C. (2004b). Three classes of genes mutated in colorectal cancers with chromosomal instability. *Cancer Res 64*, 2998–3001.

Watanabe, T., Wu, T.T., Catalano, P.J., Ueki, T., Satriano, R., Haller, D.G., Benson, A.B., 3rd, and Hamilton, S.R. (2001). Molecular predictors of survival after adjuvant chemotherapy for colon cancer. *N Engl J Med 344*, 1196–1206.

Weaver, B.A., Bonday, Z.Q., Putkey, F.R., Kops, G.J., Silk, A.D., and Cleveland, D.W. (2003). Centromere-associated protein-E is essential for the mammalian mitotic checkpoint to prevent aneuploidy due to single chromosome loss. *J Cell Biol 162*, 551–563.

Weaver, B.A., and Cleveland, D.W. (2005). Decoding the links between mitosis, cancer, and chemotherapy: The mitotic checkpoint, adaptation, and cell death. *Cancer Cell 8*, 7–12.

Weaver, B.A., Silk, A.D., Montagna, C., Verdier-Pinard, P., and Cleveland, D.W. (2007). Aneuploidy acts both oncogenically and as a tumor suppressor. *Cancer Cell 11*, 25–36.

Wikman, H., Kettunen, E., Seppanen, J.K., Karjalainen, A., Hollmen, J., Anttila, S., and Knuutila, S. (2002). Identification of differentially expressed genes in pulmonary adenocarcinoma by using cDNA array. *Oncogene 21*, 5804–5813.

Wong, Y.F., Cheung, T.H., Tsao, G.S., Lo, K.W., Yim, S.F., Wang, V.W., Heung, M.M., Chan, S.C., Chan, L.K., Ho, T.W., Wong, K.W., Li, C., Guo, Y., Chung, T.K., and Smith, D.I. (2006). Genome-wide gene expression profiling of cervical cancer in Hong Kong women by oligonucleotide microarray. *Int J Cancer 118*, 2461–2469.

Yuan, B., Xu, Y., Woo, J.H., Wang, Y., Bae, Y.K., Yoon, D.S., Wersto, R.P., Tully, E., Wilsbach, K., and Gabrielson, E. (2006). Increased expression of mitotic checkpoint genes in breast cancer cells with chromosomal instability. *Clin Cancer Res 12*, 405–410.

Yuan, J., Horlin, A., Hock, B., Stutte, H.J., Rubsamen-Waigmann, H., and Strebhardt, K. (1997). Polo-like kinase, a novel marker for cellular proliferation. *Am J Pathol 150*, 1165–1172.

Yuen, K.W., Montpetit, B., and Hieter, P. (2005). The kinetochore and cancer: what's the connection? *Curr Opin Cell Biol 17*, 576–582.

Zeitlin, S.G., Shelby, R.D., and Sullivan, K.F. (2001). CENP-A is phosphorylated by Aurora B kinase and plays an unexpected role in completion of cytokinesis. *J Cell Biol 155*, 1147–1157.

Zetterberg, A., and Esposti, P.L. (1980). Prognostic significance of nuclear DNA levels in prostatic carcinoma. *Scand J Urol Nephrol Suppl 55*, 53–58.

Zhang, D., Hirota, T., Marumoto, T., Shimizu, M., Kunitoku, N., Sasayama, T., Arima, Y., Feng, L., Suzuki, M., Takeya, M., and Saya, H. (2004). Cre-loxP-controlled periodic Aurora-A overexpression induces mitotic abnormalities and hyperplasia in mammary glands of mouse models. *Oncogene 23*, 8720–8730.

Zhou, H., Kuang, J., Zhong, L., Kuo, W.L., Gray, J.W., Sahin, A., Brinkley, B.R., and Sen, S. (1998). Tumour amplified kinase STK15/BTAK induces centrosome amplification, aneuploidy and transformation. *Nat Genet 20*, 189–193.

Zhou, W., Goodman, S.N., Galizia, G., Lieto, E., Ferraraccio, F., Pignatelli, C., Purdie, C.A., Piris, J., Morris, R., Harrison, D.J., Paty, P.B., Culliford, A., Romans, K.E., Montgomery, E.A., Choti, M.A., Kinzler, K.W., and Vogelstein, B. (2002). Counting alleles to predict recurrence of early-stage colorectal cancers. *Lancet 359*, 219–225.

Zhu, X., Mao, Z., Na, Y., Guo, Y., Wang, X., and Xin, D. (2006). Significance of pituitary tumor transforming gene 1 (PTTG1) in prostate cancer. *Anticancer Res 26*, 1253–1259.

Zou, H., McGarry, T.J., Bernal, T., and Kirschner, M.W. (1999). Identification of a vertebrate sister-chromatid separation inhibitor involved in transformation and tumorigenesis. *Science 285*, 418–422.

Zumbrunn, J., Kinoshita, K., Hyman, A.A., and Nathke, I.S. (2001). Binding of the adenomatous polyposis coli protein to microtubules increases microtubule stability and is regulated by GSK3 beta phosphorylation. *Curr Biol 11*, 44–49.

# Chapter 15
# The Kinetochore as Target for Cancer Drug Development

**Song-Tao Liu and Tim J. Yen**

## 15.1 Introduction

Due to diverse and complex genetic abnormalities in different cancers, there are no silver bullets in cancer treatment. According to the NCI (www.cancer.gov/), chemotherapy, together with radiotherapy and surgery, are the three major modalities of cancer treatment.

As uncontrolled proliferation is the most distinctive characteristic of cancer cells, many anticancer drugs directly inhibit growth. These include drugs that target DNA or nucleotide metabolism and cell division. In a cell division cycle, mitosis is the phase during which duplicated sister chromatids physically separate from each other to produce two genetically identical daughter cells. Chemical compounds that interfere with this process should in principle be efficient inhibitors of cell proliferation. However, currently available anticancer drugs of this class are limited to two types of plant alkaloids: taxanes (docetaxel, paclitaxel) and vinca alkaloids (vinblastine, vincristine, vindesine, vinorelbine). The molecular target of these drugs is microtubules which form the spindle that is essential for accurate chromosome segregation. By interfering with proper formation of the spindle, these drugs interfere with the mitotic process that ultimately kills the cell (Wilson and Jordan, 1995).

Although microtubule poisons are very efficient and are used extensively to treat cancer patients (especially those suffering from solid tumors), drug resistance is a serious problem (Orr et al., 2003; Jablonski et al., 2003). In an effort to overcome this problem, several structurally unrelated microtubule-targeted drugs are currently being tested in clinical trials (Rowinsky and Calvo, 2006). As these drugs target tubulin, a cytoskeleton protein involved not only in mitosis, but also in other important cellular functions outside of mitosis, the inhibition of all the microtubule functions in a patient produces dose-limiting toxicities such as peripheral neuropathy. This has stimulated the search for drugs that specifically target proteins that work exclusively in mitosis (Garber, 2005). As the

T.J. Yen (✉)
Fox Chase Cancer Center, Philadelphia, PA19111, U.S.A
e-mail: TJ_Yen@fccc.edu

P. De Wulf, W.C. Earnshaw (eds.), *The Kinetochore*,
DOI 10.1007/978-0-387-69076-6_15, © Springer Science+Business Media, LLC 2009 

centromere/kinetochore complex plays essential roles in chromosome segregation and mitotic checkpoint signaling (Chan et al., 2005; Cleveland et al., 2003; Rieder and Salmon, 1998), it has naturally become an attractive target for developing mitosis-specific anticancer drugs.

In this review, we will briefly present current knowledge about "the centromere/kinetochore complex," and then we will discuss the connections between kinetochore proteins and cancer development. We then describe in more details the strategies and current available data on targeting kinetochore proteins for cancer treatment. Emphasis will be placed on how kinetochore proteins are targeted by drug candidates and how the interaction affects the fate of cancer and normal cells. In the last section we present our views on how to improve the therapeutic index of anticancer drugs targeting kinetochore proteins.

## 15.2 The Centromere/Kinetochore Complex

As discussed in other chapters of this book, the centromere/kinetochore proteins play essential roles in maintaining faithful chromosome segregation during mitosis in somatic cells and meiosis in germ line cells. The centromere may be considered to be the chromatin component of this complex while the kinetochore is a multiprotein complex built upon the centromere and is involved in attaching the chromosome to the spindle (Cleveland et al., 2003). So far there are more than 100 proteins reported to associate with the centromere/kinetochore structure in human cells (Chan et al., 2005; Fig. 15.1, also see labs.fccc.edu/yen/reference.html). Under electron microscope (EM), kinetochores appear as trilamilar plates that are situated on opposite sides of the centromere (Cleveland et al., 2003; Rieder, 1982). The centromeric heterochromatin of each sister chromatid is held together by cohesion proteins.

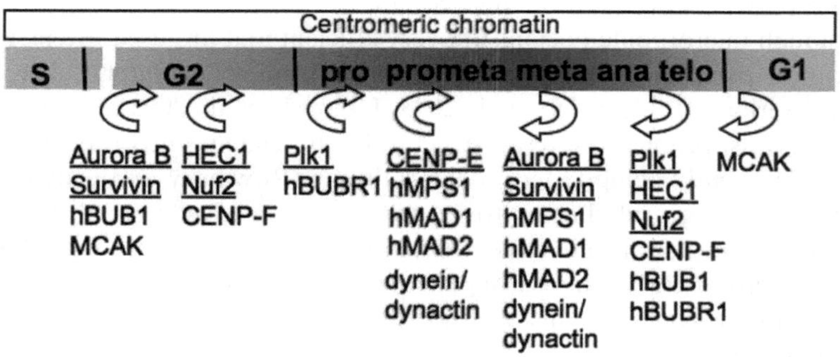


**Fig. 15.1** Temporal assembly and disassembly (indicated by arrows) of representative kinetochore proteins. Underlined proteins have been targeted for cancer drug development (also see Table 15.1) by modulating their expression level, or enzymatic activities, or their binding to partner proteins

The molecular information accumulated in recent years has blurred the cytological distinction between the "centromere" and the "kinetochore" as discrete subcellular structures. For example, Mitotic Centromere-Associated Kinesin (MCAK) (Andrews et al., 2004), Sgo1 and Sgo2 (Watanabe and Kitajima, 2005; Huang et al., 2007), and the chromosomal passenger proteins Aurora B, inner centromere protein (INCENP), survivin and borealin (Carmena and Earnshaw, 2003; Vagnarelli and Earnshaw, 2004; Vader et al., 2006), which localize to the inner centromere, provide essential kinetochore functions such as microtubule attachments and checkpoint signaling. On the other hand, the centromere specific histone H3 variant, CENP-A, is concentrated at the inner plate of the trilaminar kinetochore rather than being associated with the bulk of the centromeric heterochromatin that spans between the sister kinetochores. For simplicity, we will generally refer to all proteins of the centromere/kinetochore complex as kinetochore proteins.

## 15.3 Kinetochore Proteins and Cancer Development

As mentioned in the introduction, cancer is a disease of uncontrolled cell proliferation. As the functions, and in some cases the expression patterns, of kinetochore proteins are linked with cell proliferation, it is not a surprise that abnormalities in some kinetochore proteins have been linked to cancer development (Liu et al., 2003; Giet et al., 2005; Yuen et al., 2005). These abnormalities can be found at both genetic and epigenetic levels.

Inherited (Germ line) or somatic mutations in kinetochore proteins may contribute to cancer development. Mutations residing in the mitotic checkpoint kinase budding uninhibited by benzimidazole protein (BUB1) were first identified in two colon cancer cell lines displaying chromosomal instability (CIN; Cahill et al., 1998). Reintroduction of these mutant alleles of BUB1 into diploid cells with a normal mitotic checkpoint response led to CIN phenotype. Germ line mutations in the mitotic checkpoint kinase BUBR1 were reported in patients with a rare recessive condition called mosaic variegated aneuploidy (MVA; Hanks et al., 2004). The patients display truncating and missense mutations of BUBR1, and exhibit increased cancer risks because of constitutional mosaicism for chromosomal gains and losses. A recent systematic search discovered somatic mutations in eight candidate CIN genes in colon cancers: APC (adenopolyposis coli), CDC4, hBUB1, hRod, hZW10, hZwilch, MRE11, and DING (Wang et al., 2004). Amongst these, hBUB1, hRod, hZW10, hZwilch are localized at kinetochores independent of spindle microtubules, while APC is localized at the microtubule plus ends and can be seen at kinetochores that are attached to microtubules (Kaplan et al., 2001).

It is perhaps a surprise that after intensive efforts only a handful of genetic alterations have been discovered in kinetochore proteins in human cancers. Furthermore, for most of the mutations, it remains to be clarified whether they actually affect the fidelity of chromosome segregation or are silent polymorphisms. As has been discussed by several earlier reviews, mutations that completely abolish the functions of a kinetochore protein are likely to be lethal and cells harboring such mutations are probably eliminated early on (Kops et al., 2005). A corollary to this conclusion is that the mutations that were found in cancer cells may be hypomorphic alleles that may increase the rate of chromosome missegregation but not to the extent that affects cell survival. These alleles may end up providing some evolutionary advantages for the host cells to gain a transformed phenotype through accelerated gain or loss of chromosomes (Kops et al., 2005; Weaver and Cleveland, 2005, 2006). That "weakened" alleles (or "weakened" proteins if epigenetic changes or chemical inhibitors are involved, see below) may lead to tumorigenesis or accelerate cancer development is a serious issue when selecting kinetochore proteins as targets for drug design. Epigenetic changes that alter kinetochore protein expression are more commonly observed in both primary tumors and cell lines (Yuen et al., 2005). CENP-A and CENP-H have been shown to be upregulated in colorectal cancer tissues (Tomonaga et al., 2003, 2005). Overexpression of Polo-like kinase (Plk1) and Aurora B kinases and survivin has been observed in many types of cancers (Giet et al., 2005; Ahmad, 2004; Dai and Cogswell, 2003; Carvajal et al., 2006; Li and Brattain, 2006). Different labs has reported either overexpression or reduced expression of mitotic checkpoint proteins in different cancers (Yuen et al., 2005). As the expression of many kinetochore proteins is cell cycle regulated and usually peak during G2/M phase, it is important to document that changes in expression levels are not merely due to alterations in the cell cycle distribution of a population of cancer cells.

## 15.4 The Strategies: How can Kinetochore Proteins be Used as Drug Targets

At the cellular level, cancer can only be cured if all tumor cells are removed by surgery, or killed by radiotherapy or chemotherapy, or induced to senesce or differentiate so as to lose their transformed state. Most of the chemotherapy agents aim to kill (e.g., alkylating agents) or starve (e.g., angiogenesis inhibitors) cancer cells.

Two non-mutually exclusive strategies can be envisioned to target kinetochore proteins for the development of anticancer drugs. First, the protein can be targeted as a means to disrupt mitosis in cancer cells. Although the underlying mechanisms are not fully appreciated yet, disruption of mitosis usually leads to cell death either during mitosis or in the ensuing G1 (Giet et al., 2005; Keen and Taylor, 2004). In rare cases, senescence or cell death in polyploid cells may be involved (Giet et al., 2005; Keen and Taylor, 2004; Roninson et al., 2001). Side effects seem unavoidable as the drugs may also affect actively dividing normal cells. Nevertheless, these drugs may be used as alternatives for patients who have developed resistance against antimicrotubule drugs. The benefit of mitosis-specific drugs is that they should not interfere with non-mitotic processes whose disruption would result in undesirable side effects. The second strategy is to exploit potential genetic or epigenetic differences in kinetochore and spindle associated proteins between cancer and normal cells. This strategy has been suggested for kinetochore proteins (e.g., BUB1; Jallepalli and Lengauer, 2001) in an approach similar to what was used to target tumors that express the *K-Ras* oncogene (Torrance et al., 2001). To date, the most common alteration of mitotic proteins in cancers is overexpression of proteins such as Plk1 and Aurora B kinases (Giet et al., 2005; Dai and Cogswell, 2003). Small molecules targeting both of these kinases have been tested in preclinical and clinical trials and hold great promise to become novel anticancer drugs (see below). Some BUB1 alleles may only exist in cancer cells but not in normal cells. By identifying biochemical differences between the mutant and normal BUB1 kinase, it may be possible to selectively inhibit the mutant form (Jallepalli and Lengauer, 2001). Despite the increase in specificity for cancer cells, the effectiveness of such drugs in patients may be limited by pharmacogenetic differences amongst individuals. Thus, there is a great need to improve in vitro assays so that they could reliably reflect patient responses. This is a serious issue when planning early phase clinical trials because a drug may not be pursued if a positive response is not achieved even when only a very limited number of patients are tested.

We have summarized in Fig. 15.1 and Table 15.1 the most up-to-date information about novel mitotic proteins that have been selected as drug targets for cancer therapy. Aurora B and Plk1 have been at the center of attention not only because they are overexpressed in many cancers (Ota et al., 2002; Tatsuka et al., 2005), but also because they are enzymes that can be easily assayed in vitro (Dai and Cogswell, 2003; Carvajal et al., 2006; Keen and Taylor, 2004; Jackson et al., 2007; Warner et al., 2006). Mitotic checkpoint kinases, BUB1, BUBR1, and hMPS1 are also attractive targets (Warner et al., 2006). Chk1 kinase, which is a critical component of the G2/M DNA damage checkpoint, has been recently shown to be also involved in mitosis (Zachos et al., 2007) and may be a new antimitotic target in addition to being a sensitizer of DNA damaging drugs. It will be of interest to further examine the effects of several known Chk1 inhibitors on various aspects of mitotic functions (Prudhomme, 2006). Mitotic kinesins, a group of molecular motors that hydrolyze ATP to generate force have also been proposed as novel drug targets (Jackson et al., 2007; Warner et al., 2006). One example in this group is Eg5, to which several specific inhibitors have been developed. At least one of these is in clinical trials. Although Eg5 is not specifically localized at kinetochores, it acts specifically in mitosis by contributing to bipolar spindle formation and thus fulfills the criteria of selectively acting during mitosis. Inhibitors targeting a *bona fide* kinetochore associated kinesin, CENP-E, are also being tested (Jackson et al., 2007). Additional kinetochore proteins that are enzymes include Nek2A protein kinase (Lou et al., 2004) and the SUMO1 E3 ligase, RanBP2 (Pichler et al., 2002;

**Table 15.1** Examples of clinical and preclinical agents targeting centromere/kinetochore proteins or a spindle protein KSP (updated from (Jackson et al., 2007))

| Target | Compound | Company | Comments |
|---|---|---|---|
| Aurora kinase | MK-0457 (VX680) | Vertex/Merck | Phase I/II; Pan Aurora inhbitors; Typical anti-proliferation IC50s of 10-100 nM. |
| | AZD1152 | Astra Zeneca | Phase I/II; Aurora B selective |
| | ZM447439 | Astra Zeneca | |
| | PF-03814735 | Pfizer | Phase I |
| | MLN8054 | Millennium | Phase I; Aurora A selective |
| | PHA-680632 | Nerviano Medical Sciences | Phase I |
| | PHA-739358 | Nerviano Medical Sciences | Phase II |
| | AT9283 | Astex | Phase I |
| | R763 | Rigel Pharmaceuticals | preclinical |
| | CYC116 | Cyclacel | preclinical |
| | SNS314 | Sunesis | preclinical |
| | Hesperadin | Boehringer Ingelheim | (Hauf et al., 2003) |
| Plk1 kinase | BI2536 | Boehringer Ingelheim | Phase I, >10,000-fold selective for Plk1 versus 43 other kinases |
| | ON01910 | Onconova | Phase I, inhibits several kinases in addition to PLKs. Not ATP competitive, peptide substrate competitive. |
| | GSK461364 | GlaxoSmithKline | Phase I |
| | CYC800 | Cyclacel | Preclinical |
| CENP-E | GSK-923295A | GlaxoSmithKline/ Cytokinetics | Phase I |
| survivin | YM155 | Astellas Pharma | Phase II, |
| Chk1 | UCN-01(KW-2401) | Kyowa Hakko Kogyo | Phase II |
| Eg5 (KSP, kinesin 5); not a kinetochore protein | Ispinesib (SB-715992) | Cytokinetics | Phase I/II |
| | Monastrol | | Mayer et al., 1999] |
| | *S*-trityl-L-cysteine | | DeBonis et al., 2004 |
| | SB-743921 | Cytokinetics | Phase I |
| | MK-0731 | Merck | Phase I |
| | ARRY-520 | Array Biopharma | Phase I |
| | ARRY-649 | Array Biopharma | Preclinical |
| | CRx-026 | CombinatoRx | Phase I, KSP/PRL phosphatase inhibitor |

Joseph et al., 2004). Their contribution to mitosis is less clear and their potentials as novel drug targets should be further explored.

Disruption of protein–protein interactions within the kinetochore is also a viable strategy as identification of small molecules that interfere with specific protein complexes is now technically possible. For example, the nutlins, which disrupt the MDM2-p53 interaction are being actively studied (Vassilev, 2007). Given that a functional kinetochore may consist of over 100 proteins, these interactions provide a rich source for drug development. Efforts to reveal the structural basis for various protein–protein interactions, as for the HEC1–Nuf2 complex (Ciferri et al., 2005; Hayama et al., 2006; Wei et al., 2007), should provide new strategies to develop inhibitors of kinetochore functions.

Although this review focuses primarily on small molecule inhibitors as drug candidates, we believe that RNA interference (RNAi) therapy targeting kinetochore proteins also holds potential for the future.

## 15.5 Preclinical and Clinical Research on Kinetochore Proteins as Anticancer Drug Targets

Kinetochore proteins were identified as novel drug targets by two different routes. In the first situation, detailed mechanistic studies of several drugs under clinical trials have uncovered unsuspected effects on the assembly or functions of kinetochore proteins. These drugs include heat shock protein 90 (Hsp90) inhibitors and farnesyl transferase inhibitors (FTI). These drugs have additional targets and modulate multiple biological processes. In the second situation, kinetochore proteins were selected as the targets of rational drug design based on extensive information about their biological and biochemical activities in mitosis. Inhibitors of key kinetochore proteins that include several protein kinases and kinesins have been discovered. These inhibitors were usually discovered through high throughput screening of chemical libraries for compounds that inhibit the enzymatic activity of the target. In the most advanced case, Aurora B kinase inhibitors are currently being studied in multiple clinical trials.

### *15.5.1 From Drugs to Kinetochore Proteins*

#### 15.5.1.1 HSP90 Inhibitors

The ansamycin antibiotics, herbimycin A (HA) and geldanamycin (GM), bind to a conserved ATP binding pocket in Hsp90 and alter the function of this chaperone protein. Hsp90 is a ubiquitous molecular chaperone critical for the folding, assembly, and activity of multiple mutated and overexpressed signaling proteins that promote the growth and/or survival of tumor cells. Hsp 90 target proteins include mutated p53, Raf-1, Akt, ErbB2, and hypoxia-inducible factor

1a (HIF-1a) and also kinetochore proteins Plk1 and survivin. Binding of both HA or GM to Hsp 90 usually causes destabilization and degradation of its client proteins (Cullinan and Whitesell, 2006; Powers and Workman, 2006; Goetz et al., 2003).

Despite its potent antitumor potential, GM has several major disadvantages that led to the development of GM analogs (www.geldanamycin.com). These analogs contain a substitution on the 17 position: 17-AAG, 17-DMAG, and others. There are more than 20 phase I or phase II clinical trials testing the antineoplastic activities of 17-AAG and 17-DMAG in patients with different cancers (Cullinan and Whitesell, 2006).

The connection between Hsp90 and centromere/kinetochore proteins came from studies of the evolutionarily conserved protein, SGT1. In budding yeast, SGT1 was isolated as a dosage suppressor of a specific *skp1* mutant allele. SKP1 is an essential component of the Centromere binding factor 3 (CBF3) of centromeric protein complex (Kitagawa et al., 1999). Hsp90 was found to bind to SGT1 in both yeast and human cells (Steensgaard et al., 2004; Niikura et al., 2006; Bansal et al., 2004). HeLa cells depleted of SGT1 by RNAi displayed dramatic reduction in the amount of CENP-I, CENP-E, HEC1, CENP-F, and checkpoint proteins MAD1, MAD2 (mitotic arrest deficient proteins) and BUBR1 at kinetochores. These cells experience a transient mitotic delay then enter G1 phase followed by cell death (Steensgaard et al., 2004). These results have been confirmed recently by directly testing the effects of 17-AAG in cancer cell lines. After 17-AAG treatment, several kinetochore proteins including CENP-I and CENP-H but not CENP-B and CENP-C were delocalized (Niikura et al., 2006).

These discoveries are consistent with earlier reports that Hsp90 inhibitors induced a G1 arrest in Rb positive cell lines, while Rb negative cells would instead progress through S phase then die after a prolonged mitotic arrest (Srethapakdi et al., 2000, Munster et al., 2001). When cells were treated with both 17-AAG and taxol, Rb positive cells were blocked in G1 and did not undergo apoptosis within the duration of their experiment. In contrast, cells with mutated Rb were blocked in mitosis and then died. Thus, the combination of taxol (or another antimitotic drug) and 17-AAG may selectively target cancer cells with defective Rb (Munster et al., 2001).

#### 15.5.1.2 Farnesyltransferase Inhibitors (FTIs)

FTIs were developed to specifically inhibit the activity of oncogenic Ras proteins in tumor cells by preventing their association with membranes (Brunner et al., 2003; Cesario et al., 2005). But interests in these molecules were further stimulated by the unexpected efficacy of FTIs even in tumor cells with normal *ras*. FTIs inhibit the enzymatic activity of farnesyltransferase (FTase) that normally catalyzes the addition of C15-farnesyl isoprenoid group to proteins. Another form of prenylation is the addition of C20-geranylgeranyl group catalyzed by geranylgeranyltransferases (GGTase-1 and GGTase-2). Under

FTI treatment, alternative prenylation of certain proteins, including RhoB, K-Ras, and N-Ras, can be added by GGTase-1. The recognition site for prenylation by both FTase and GGTase-1 is called the *CAAX* box, a COOH-terminal tetrapeptide sequence composed of cysteine, two aliphatic amino acids, and a terminal amino acid that serves as the primary determinant of farnesylation or geranylgeranylation. The prenyl group is covalently linked to the cysteine residue. Multiple FTIs are presently in clinical trials, have excellent preclinical anticancer activity and impressive lack of toxicity. However their precise mechanism of action with respect to cancer treatment is not well understood (Brunner et al., 2003; Cesario et al., 2005).

Interestingly, several groups have reported that FTIs induce a G2/M delay in a number of different cancer cell lines (Ashar et al., 2000; Crespo et al., 2001, 2002; Hussein and Taylor, 2002; Schafer-Hales et al., 2007). All reports described defects in spindle formation and chromosome alignment that result in a prometaphase delay. As kinetochore proteins CENP-E and CENP-F have a *CAAX* motif and are found to be farnesylated in cancer cells (Ashar et al., 2000), the roles of their farnesylation in the mitotic defects were studied. Early results suggested that the localization of CENP-E and CENP-F were not affected in the presence of FTIs, so the authors concluded neither proteins were responsible for the mitotic defects caused by FTIs (Ashar et al., 2000; Crespo et al., 2001). However, in one report, delayed progression from G2 to M was observed by overexpression of the C-terminal kinetochore-targeting domain of CENP-F. This delay was found to require the *CAAX* box and FTase activity (Hussein and Taylor, 2002). Localization of CENP-F to nuclear envelope and kinetochores at G2-M, as well as its degradation after mitosis, also appears to require farnesylation as evidenced by the changes after treatment with FTI SCH66336 (Hussein and Taylor, 2002). A recent study found that FTI lonafarnib (aka. SCH66336) caused chromosome alignment defects that blocked cells in a pseudometaphase state (Schafer-Hales et al., 2007). The authors found CENP-E and CENP-F were indeed still localized at kinetochores in prometaphase cells but they were abolished from metaphase kinetochores and may affect chromosome alignment in cancer cell lines. These contradictory results suggest either FTIs have additional mitotic targets or the genetic background of cancer cells may affect their responses to the treatment of FTIs (Crespo et al., 2002).

## *15.5.2 From Kinetochore Proteins to Drugs*

### 15.5.2.1 Aurora B Kinase

Aurora B belongs to the group of chromosomal passenger proteins (Aurora B, INCENP, survivin, borealin, and TD60) that are localized to the inner centromere from late G2 until metaphase, and are then redistributed to the midzone and midbody during later stages of mitosis. These proteins play critical regulatory roles during mitosis ranging from chromosome modification (phosphorylation of Ser10 in histone H3 and Ser7 in CENP-A), chromosome alignment, kinetochore assembly, correcting aberrant kinetochore attachments, modulating mitotic checkpoint responses, to cytokinesis (Carmena and Earnshaw, 2003; Vagnarelli and Earnshaw, 2004; Vader et al., 2006; Giet et al., 2005; Keen and Taylor, 2004; Andrews et al., 2003). It has been shown that mice injected with p53 (Gly245Ser) fibroblast cells overexpressing Aurora B form aggressive tumors that metastasize, in contrast to those injected with the same cells that do not ectopically express Aurora B (Ota et al., 2002). As mentioned earlier, overexpression of Aurora B has been demonstrated in some cancer cell lines (Ota et al., 2002; Tatsuka et al., 2005). These findings along with those cited above provided the rationale for the development of Aurora kinase inhibitors.

Aurora B is but one of the three Aurora kinases that are expressed in human cells: Aurora A, Aurora B, and Aurora C, which share a conservative catalytic core but differ in their N-terminal domains (Carmena and Earnshaw, 2003; Andrews et al., 2003). Aurora A kinase localizes on duplicated centrosomes from the end of S phase, through M and to the beginning of the following G1, and has been shown to be critical in bipolar spindle assembly and centrosome separation and maturation. Most people think Aurora A is a bona fide oncogene as ectopic overexpression of Aurora A alone was sufficient to transform NIH3T3 and Rat1 cells that formed tumors when implanted in nude mice (Zhou et al., 1998; Bischoff et al., 1998). Transformation required kinase activity which further validates the development of inhibitors to this family of kinases. Overexpression of Aurora A in primary mouse embryonic fibroblast (MEF) cells does not induce colony formation indicating that Aurora A overexpression alone is not enough to induce carcinogenesis (Anand et al., 2003). Overexpression of Aurora A has been observed in many cancer cells, and may induce mitotic abnormalities through failure of cytokinesis and centrosome duplication especially in p53-deficient cells (Meraldi et al., 2002). Aurora C kinase is predominantly expressed in testis, but is also highly expressed in several cancer cell lines. Recent data indicate that Aurora C is also a chromosomal passenger protein and resembles Aurora B in many aspects (Li et al., 2004; Sasai et al., 2004). It is important to note that Aurora C does not act redundantly with Aurora B given that Aurora B provides essential functions in mitosis.

Due to their important roles in mitosis progression and cancer connections, Aurora kinases have become very attractive targets for novel anticancer drug development. Many small molecule inhibitors of Aurora kinases are currently being tested in clinical trials (Table 15.1). Most were originally developed against Aurora A, but when they were applied to cells, the phenotype cells displayed was more similar to Aurora B inhibition. A molecule capable of inhibiting the function of Aurora A was expected to induce a mitotic arrest due to the formation of monopolar mitotic spindles and this is followed by the rapid induction of apoptosis (Glover et al., 1995; Rojanala et al., 2004). Instead, data on four Aurora kinase inhibitors, ZM447439, Hesperadin, AZD1152-HQPA, and VX-680 all showed that drug-treated tumor cells establish seemingly normal bipolar spindles, and enter and exit mitosis with normal kinetics. Despite failing to complete cytokinesis, these cells go on to re-replicate their genomes, resulting in tetraploidy (Warner et al., 2006; Hauf et al., 2003; Ditchfield et al., 2003; Harrington et al., 2004; AACR 2007 Annual Meeting Abstract #4359). This distinguishes their behavior from that of classic "antimitotic" agents. One explanation for the dominant Aurora B phenotype in drug-treated cells is that inhibition of Aurora B caused a defective checkpoint and thus the spindle defects induced by loss of Aurora A activity could not cause mitotic delay any more (Giet et al., 2005; Warner et al., 2006; Yang et al., 2005).

The long-term effects of these drugs on cancer cells seem to depend on the genetic background. Some cells will continue DNA endoreplication despite failed cytokinesis producing polyploid cells; some tetraploid cells will arrest in a pseudo G1 state in a p53 dependent manner; other cells may go into apoptosis. The status of p53 in the cells may be important in determining cell fates after treatment of Aurora inhibitors, but the relationship has not been firmly established (Giet et al., 2005; Keen and Taylor, 2004). However, both ZM447439 and VX-680 eventually kill tumor cells in vitro and treatment with VX-680 induces tumor regression in vivo (Ditchfield et al., 2003; Harrington et al., 2004), proving the validity of Aurora kinases as drug targets. Moreover, although the selectivity for proliferating normal cells versus proliferating tumor cells has not yet been reported, VX-680 seems well tolerated in animal models (Harrington et al., 2004).

#### 15.5.2.2 Plk1 Kinase

Plk1, as with Aurora B, has many roles in mitosis. It is required for entry into mitosis, centrosome separation, spindle assembly, chromosome alignment and cytokinesis (Dai and Cogswell, 2003; Strebhardt and Ullrich, 2006; Burkard et al., 2007). Consistent with these functions, the expression and activity of Plk1 is low throughout G0, G1 and S phase, begins to rise in G2 and peaks during mitosis (Golsteyn et al., 1995). Its connection with cell division is also reflected by its subcellular localization at centrosomes, kinetochores, spindle midzone and midbodies as cells progress through mitosis (Golsteyn, 2001; Arnaud et al., 1998). Similar to Aurora kinases, there are four related Plk kinases in human cells, Plk1-4 (Barr et al., 2004). In addition to the kinase domain, all Plks have so called PBD (polo box) domains at the C-terminal of the proteins. These are phosphospecific binding motifs that anchor Plks to specific subcellular localizations (Barr et al., 2004). Plk1 is the most extensively studied and is overexpressed in a broad spectrum of cancers that include breast, colorectal, non-small cell lung, oesophageal, and ovarian cancers. Furthermore, its expression level often correlates with poor prognosis (Eckerdt et al., 2005).

Plk1 emerged as an attractive anticancer drug target because experiments to interfere with its expression or activity, including antibody injection, dominant negative mutants, siRNA and antisense oligonucleotides, all showed inhibition of cancer cell proliferation but minimal effects on normal cell growth (Strebhardt and Ullrich, 2006; Eckerdt et al., 2005). Typical effects include a delay in mitotic entry and prolonged prometaphase arrest with unseparated centrosomes and abnormal spindles that result in chromosomes arranged in a rosette pattern. The arrested cells then died without exiting from mitosis (Peters et al., 2006; Lenart et al., 2007; Steegmaier et al., 2007). This is different from Aurora B inhibition which usually overrides the spindle checkpoint and produces polyploid cells. Since polyploid cells may survive (Ganem et al., 2007), it has been proposed that inhibition of Plk1 may provide some advantages over inhibition of Aurora B kinase (Plyte and Musacchio, 2007).

In recent years, several small molecules have been reported to inhibit Plk1 kinase activity (Strebhardt and Ullrich, 2006). The first reported was scytonemin (Stevenson et al., 2002), a natural product isolated from cyanobacteria, which inhibits the ability of Plk1 to phosphorylate CDC25C in a concentration-dependent manner with an in vitro IC50 of $2 \pm 0.1$ μM. Later, it was found that wortmannin (Liu et al., 2007), a PI-3 kinase inhibitor that has also been used extensively in studies involving DNA-PK, ATM or ATR, surprisingly also inhibits Plk1 at nanomolar concentrations. The more recent development is BI2536, a very potent and selective ATP competitive inhibitor that has entered phase I clinical trials (Lenart et al., 2007; Steegmaier et al., 2007). The IC50 of BI2536 for Plk1 is about 0.8 nM, and it showed at least 10,000-fold selectivity for Plk1 against a panel of tyrosine-and serine/threonine-kinases. Functional studies showed that addition of BI2536 to cancer cells inhibits both activities and localization of Plk1, recapitulating most if not all phenotypes demonstrated before using other methods to inhibit Plk1 (Lenart et al., 2007). Importantly, the compound potently causes a mitotic arrest and induces apoptosis in human cancer cell lines of diverse tissue origins and oncogenomic signature. BI 2536 inhibits growth of human tumor xenografts in nude mice and induces regression of large tumors with well-tolerated intravenous dose regimens (Steegmaier et al., 2007). Results on another potent and selective Plk1 inhibitor, GSK461364, have been reported in 2007 AACR annual meeting (abstracts #4171, #5388 and #5389). The compound is currently in phase I clinical trial. It displays an IC50 at about 3 nM with greater than 100-fold selectivity across a panel of 47 other kinases. When tested for antiproliferative activity against >120 cancer cell lines, it inhibited the proliferation of greater than 83% of them with IC50s lower than 50 nM and 91% with IC50 lower than 100 nM. GSK461364 had no measurable effect on human non-proliferating normal cells. Depending on its concentration, GSK461364 may cause G2/M arrest, prometaphase-like mitotic arrest or cytokinesis inhibition, with the later two leading to cell death. GSK461364 also showed positive antitumor activities on xenograft animal models. Another Plk1 inhibitor, ON01910, also currently in clinical trials, was proposed to compete at the substrate peptide binding site instead of the ATP binding site of Plk1 with an IC50 at 9–10 nM (Gumireddy et al., 2005). But this compound also inhibits several other tyrosine kinases as well as CDK1. Even though it was shown to inhibit tumor cell proliferation, others reported that it did not show any Plk1 inhibitory activity either in vitro or in cells (Steegmaier et al., 2007). In addition to these two inhibitors, there are several other Plk1 inhibitors (Peters et al., 2006; McInnes et al., 2006) that have been described recently (Table 15.1) which we will not discuss further here.

#### 15.5.2.3 Spindle Checkpoint Proteins

As discussed earlier, the mitotic checkpoint is a signaling cascade that detects defective kinetochore–microtubule interactions and delays the metaphase–anaphase transition by inhibiting the anaphase promoting complex/cyclosome (APC/C; Liu et al., 2003; Chan and Yen, 2003). Of great interest are BUB1, BUBR1 and hMPS1, the only known kinases to be critically important for the spindle checkpoint. However, as the spindle checkpoint could be considered as a tumor suppression mechanism, concerns have mitigated efforts to target these kinases as anticancer therapy. Inhibition of these kinases might compromise the mitotic checkpoint thus promoting aneuploidy and increasing the risk for tumorigenesis. As discussed below, this concern is not baseless but the underlying reasons are more complicated.

Recent literature has suggested that proper activity of the mitotic checkpoint depends on threshold levels of specific proteins and by inference, their biochemical activities. In line with the classical definition of tumor suppressors, the haploinsufficiency of BUBR1, MAD2, BUB3 and Rae1 all resulted in increased rates of chromosome missegregation in mice. In the cases of MAD2, BUB3, and Rae1, but not BUBR1, heterozygous mutant mice showed a slight increase in cancer after a long latency or after chemical induction (Babu et al., 2003; Michel et al., 2001; Baker et al., 2004; Wang et al., 2004). Consistent with the animal studies, loss-of-function mutations of these proteins, although not as common as originally thought, have been identified in various types of tumors (Yuen et al., 2005; Cahill et al., 1998).

However, mitotic checkpoint genes differ from classical tumor suppressors (e.g., Rb) in at least two ways. First, unlike classic tumor suppressors, mitotic checkpoint genes are essential for cell survival. Homozygous deletion of various mitotic checkpoint genes result in early embryonic lethality (Babu et al., 2003; Michel et al., 2001; Baker et al., 2004; Wang et al., 2004). Human cell lines experimentally depleted of BUBR1 or MAD2 also cannot maintain clonogenic survival (Kops et al., 2004; Michel et al., 2004). Secondly, overexpression of mitotic checkpoint proteins including the three checkpoint kinases in cancer tissues is well documented in the literature (Liu et al., 2003; Sotillo et al., 2007). Along this line, transgenic mice transiently overexpressing MAD2 have produced a wide variety of neoplasias, appearance of broken chromosomes, anaphase bridges, and whole-chromosome gains and losses, as well as acceleration of myc-induced lymphomagenesis (Sotillo et al., 2007; Hernando et al., 2004). Unlike the majority of oncogenes studied to date, chronic overexpression of Mad2 is not required for tumor maintenance. Actually, prolonged overexpression of Mad2 delays mitotic exit and eventually generates polyploid/aneuploid cells with very low viability. These results demonstrate that transient Mad2 overexpression and chromosome instability can be an important stimulus in the initiation and progression of different cancer subtypes. Thus, Mad2 overexpression may stabilize Securin and Cyclin B, delay exit from mitosis, and increase nondisjunction events and aneuploidy. It is not yet known whether overexpression of BUB1, BUBR1 or MPS1 can induce tumorigenesis similar to MAD2, but it is very intriguing that mitotic checkpoint proteins can promote cancer development when either under- or over-expressed, although through different mechanisms. Thus, in addition to the "weakened checkpoint" discussed earlier (Kops et al., 2005; Weaver and Cleveland, 2005, 2006; Weaver et al., 2007), the "transient over-active mitotic checkpoint" may also result in chromosomal instability that provides evolutionary advantages for pre-neoplastic cells to maintain malignant cell division and growth.

The collective data reinforces the notion that accurate chromosome segregation depends critically on delicately balanced levels of mitotic checkpoint proteins. This leaves a narrow window to target these proteins for cancer treatment. In addition, since only transient overexpression (probably also inactivation) of mitotic checkpoint proteins is sufficient to induce cancer-promoting chromosomal instability, it may be of no treatment value to simply reduce the level of overexpressed mitotic checkpoint proteins that are observed in many cancer types. Currently, the importance of mitotic checkpoint proteins as anticancer drug targets remains an open question. So far there are only limited reports about chemical inhibitors of the mitotic checkpoint kinases. One example is cincreasin, a small molecule inhibitor of budding yeast MPS1 that was discovered by phenotype based chemical genetic screening (Dorer et al., 2005). Unfortunately, cincreasin does not inhibit human MPS1. However, SP600125, a JNK kinase inhibitor, was found to inhibit hMPS1 at micromolar concentration, providing a pharmacophore for further development of more potent inhibitors (Schmidt et al., 2005).

Studies of these proteins especially the three mitotic checkpoint kinases may provide insights related to tumorigenesis and cancer prevention as we can imagine transient inhibition (or activation) of these kinases by environmental or dietary factors, even in the absence of genetic or epigenetic alterations of their genes, may contribute to the loss of fidelity in chromosome segregation and tumorigenesis.

#### 15.5.2.4 Other Enzymes: Chk1 and CENP-E

Chk1 is a major component in the kinase cascade in the DNA damage signal transduction pathways. Among other substrates, it phosphorylates p53 and CDC25C, and blocks CDK activity that is important for intra-S or G1/S or G2/M transitions (Bartek and Lukas, 2003). Several small molecule inhibitors have been found to inhibit Chk1 kinase activity, the best characterized being UCN01(7-Hydroxystaurosporine; Prudhomme, 2006; Zhou and Sausville, 2003; Zhou et al., 2003; Tao and Lin, 2006). UCN01 also inhibits the kinase activities of CDKs, protein kinase C, Chk2, in vitro. Recently, Chk1, at least the gfp:chk1 fusion protein, was detected at kinetochores of human cells during mitosis, where it is thought to protect cells from chromosome missegregation. It was shown to delay mitosis when spindle function is disrupted by taxol, but not when microtubules are depolymerized by nocodazole (Zachos et al., 2007). It may exert its effects through regulating the activity and localization of Aurora B and BUBR1. Considering Chk1 inhibitors like UCN01 have already been extensively tested in clinical trials (Prudhomme, 2006), it will be interesting to test whether this inhibitor enhances the tumor killing potential of paclitaxel.

Kinesins are a large family of microtubule dependent ATPases. Only a limited number of kinesins (approximately 10) are essential for cell division (www.proweb.org/kinesin/). Among them KSP/hEg5 has recently stood out as a promising target for anticancer therapy (Table 15.1; Garber, 2005; Jackson et al., 2007; Warner et al., 2006). Although KSP is not a kinetochore protein, for reasons of comparison with another kinetochore-localized kinesin, CENP-E, we will briefly review several features of this protein. Firstly, KSP is essential for establishing a bipolar spindle, and no role outside mitosis has been found. Consistent with this, the expression profiles of KSP mRNA in normal tissues are restricted primarily to proliferating cells and its overexpression was observed in tumor tissue compared with normal adjacent tissue. Secondly, the KSP inhibitors reported so far all display a high selectivity for this kinesin subtype. Specificity is achieved as the inhibitor binds to an allosteric site that is not present in other kinesins. Thirdly, preclinical experiments showed that inhibition of KSP activity results in mitotic delay and death of tumor cells. Even though the exact mechanisms how the cells died may differ from cell line to cell line, many of them clearly showed antiproliferative activity in tumor cell lines and significant efficacy in several murine tumor models.

CENP-E is a kinetochore associated kinesin-like protein that has been shown in many ways to be critical for proper kinetochore attachments to the spindle (Putkey et al., 2002; McEwen et al., 2001; Tanudji et al., 2004). A CENP-E inhibitor (GSK-923295A) is in phase I clinical trial (Jackson et al., 2007). A recent report (2007 AACR annual meeting abstract #3179) has suggested that GSK923295A is uncompetitive for both ATP and MT, indicating that this compound is an allosteric inhibitor of the motor domain of human CENP-E. It appears to "lock" CENP-E onto MTs and inhibit the movement of this motor by stabilizing the ADP-Pi state of the MT–CENP-E complex. This is distinct from allosteric inhibitors of KSP (like Monastrol and *ispinesib*). *Ispinesib* altered the ability of KSP to bind microtubules and inhibited its movement by slowing the release of ADP without trapping the microtubule-KSP-ADP intermediate. The effects of inhibiting CENP-E on the mitotic progression are probably different from those resulting from KSP inhibition. Inhibition of CENP-E does not immediately lead to cell death, rather cells with a few mono-oriented chromosomes exit from mitosis after a brief mitotic delay. As with the mitotic checkpoint proteins, CENP-E depletion will cause aneuploidy on one hand that can be lethal because of genetic imbalance. Yet, aneuploidy may increase the risk of tumorigenesis (Weaver et al., 2007). It remains to be seen how CENP-E inhibitors tip the balance between cell death and survival.

#### 15.5.2.5 Other Kinetochore Proteins: Survivin and the HEC1–Nuf2 Complex

Several other kinetochore proteins with no known enzymatic activities are also being investigated as potential drug targets. Survivin and the HEC1–Nuf2 complex are examples of essential mitotic proteins that are targeted for drug development.

Survivin is unique among kinetochore proteins as it also belongs to the Inhibitor of Apoptosis (IAP) gene family. The expression of survivin is cell cycle regulated and peak levels are seen from G2 through M phases. Survivin is undetectable in most normal adult tissues, and becomes abundantly expressed in a variety of human cancers in vivo (Li and Brattain, 2006; Lens et al., 2006; Altieri, 2006). Interestingly, the overexpression of survivin in cancer tissues seems independent of cell cycle stages. Two recent reviews presented their respective views of survivin as either a mitotic regulator or as an apoptosis inhibitor (Lens et al., 2006; Altieri, 2006). For its roles in mitosis, survivin forms the chromosomal passenger complex with Aurora B, INCENP, and borealin. Depletion of survivin results in defects in chromosome alignment, failure of cytokinesis, and eventually cell death. On the other hand there is evidence that loss and/or interference with survivin can result in spontaneous cell death by enhancing activities of proapoptotic proteins. Although mechanisms of action of survivin are not completely resolved, its ascribed roles in mitosis and apoptosis provides a novel target for cancer therapy.

Distinct from all the other inhibitors we have discussed so far, the small molecule targeting survivin that is currently in clinical trials, YM155, does not directly bind to survivin protein, instead it induces the downregulation of mRNA and protein level of survivin by undisclosed mechanisms (*Journal of Clinical Oncology*, 2006 ASCO Annual Meeting Proceedings Part I. Vol 24, No. 18S (June 20 Supplement), 2006: 3014). It has demonstrated potent antitumor activities in experimental human hormone refractory prostate cancer (HRPC) models (*Journal of Clinical Oncology*, 2007 ASCO Annual Meeting Proceedings Part I. Vol 25, No. 18S (June 20 Supplement), 2007: 3536). There are several other survivin inhibitors reported in the literature. One example is M4N (tetra-O-methyl-nordihydroguaiaretic acid) that binds to the consensus binding site of transcription factor Sp1 thus preventing Sp1 binding to the survivin promoter and activating its transcription (Huang et al., 2006). The other example is shepherdin, a cell-permeable peptidomimetic that disrupts the association between survivin and molecular chaperone, Hsp90, leading to apoptosis of tumor cells (Plescia et al., 2005).

The HEC1–Nuf2 heterodimer is associated with two other subunits, SPC24 and SPC25, and has recently been shown to display microtubule binding activities (Cheeseman et al., 2006; DeLuca et al., 2006; Wei et al., 2006). Elevated expression of HEC1 and Nuf2 have been demonstrated in several types of cancers (Hayama et al., 2006; Chen et al., 1997). Because the expression of both proteins peaks in G2/M phase in normal dividing cells, it is still not clear whether the increased expression in cancer cells is the result of having a higher proportion of cells in G2/M. Many labs have reported that disruption of the complex by either antibody injection or siRNA caused extensive cell death after a mitotic arrest (Hayama et al., 2006; DeLuca et al., 2002; Martin-Lluesma et al., 2002). Several groups have tested the idea that this complex is a valuable anticancer target by RNAi technology (Hayama et al., 2006; Li et al., 2007; Gurzov and Izquierdo, 2006). Extensive depletion of HEC1 or Nuf2 leads to inactivation of the mitotic checkpoint and cells prematurely exiting from mitosis (Meraldi et al., 2004), but the fates of these prematurely divided cells were not tracked. Thus, it is unclear whether premature exit is an explanation for the high incidence of cell death. Some animal xenograft experiments to target the complex have given exciting results. One group found adenocarcinomas induced in the flanks of nude mice show significant reduction in size compared with control when treated with either Hec1-shRNA retroviruses or adenoviruses (Gurzov and Izquierdo, 2006). The second report used the invertebrate *Spodoptera frugiperda* (Sf9) cell line to produce large amount of high-titer recombinant adeno-associated viruses (rAAV) that express short hairpin RNA (shRNA) that could efficiently deplete Hec1 in human cell lines (Li et al., 2007). On the basis of the results of screening 56 human tumor cell lines with three different serotype vectors, the authors also tested the effects in a tumor xenograft model. The effects of the shHec1 vector were evident in sectioned and stained tumors. Another report did not present any in vivo data, but was of interest because of the strategy developed to inhibit Hec1 and Nuf2. They mapped the interaction domain between HEC1 (aka. KNTC2) and Nuf2 (aka. CDCA1) and found that a cell-permeable peptide carrying the Nuf2-derived 19-amino-acid peptide (11R-CDCA1398-416) that corresponds to the binding domain to HEC1 effectively suppressed growth of NSCLC cells (Hayama et al., 2006). This provides proof of concept that disruption of protein–protein interactions can be achieved not only in vitro but also in cells.

## 15.6 Challenges and Future Directions

The primary goal of all anticancer drugs is to kill cancer cells while doing no or minimal harm to normal non-dividing and dividing cells. The kinetochore proteins that have been selected as drug targets function only in mitosis, and their inhibition should not adversely affect non-dividing cells. Thus, reduced side effects may provide an advantage over taxanes and vinca alkaloids. The outstanding question is how to use these novel drugs in ways that increase their selectivity for cancer cells over normal dividing cells. Clinical investigators can optimize drug action by manipulating dosage and delivery schedules. From the basic sciences standpoint, this can be achieved by additional efforts to study mechanisms of mitotic progression.

The first and foremost problem is that we lack mechanistic insights about how disruption of mitosis kills cells. The term "mitotic catastrophe" is often used to describe the fates of cells that undergo an aberrant mitosis. However, this is purely a cytological description where the presence of multinucleated cells is often used as indicator of a catastrophic mitosis. Death that ensues is more likely the result of gross chromosome imbalance rather than through an apoptotic program that is specifically activated by aneuploidy. The heterogeneous nature of aneuploidy may in fact be a mechanism to select for survivors. This could lead to a recurrence of a tumor that may be resistant to chemotherapy. Some recent evidence suggests that there is a mechanism that induces mitotic cells with misaligned chromosomes to undergo apoptosis. Interestingly, the apoptotic program is independent of caspases and has been named CIMA (caspase independent mitotic apoptosis; Niikura et al., 2007). CIMA is activated in cells whose BUB1 levels are reduced but not eliminated. The residual amount of BUB1 provides cells with a limited capacity to maintain a checkpoint arrest in response to kinetochore attachment defects. During the transient delay, CIMA is activated and cells die before they have the chance to exit mitosis. While the connections between BUB1 and CIMA remain obscure, the apoptotic program also appears to be dependent on the p73 status of the cell. These findings confirm the existence of pathways that directly lead a cell that is blocked in mitosis to cell death. Understanding the details of this mechanism should provide pharmacological opportunities to enhance killing of cells that have been treated with antimitotic drugs.

Understanding how antimitotic drugs kill cells may also help to prevent the side effects on inhibiting the proliferation of normal cells. Even many of the newer antimitotic agents discussed in this review have shown dose-limiting toxicities involving myelosuppression (Jackson et al., 2007). This is undoubtedly related to the effects of these drugs on the proliferating progenitor cells in the bone marrow. Because the hematopoietic turnover is so sensitive to transient inhibition of proliferation, it may significantly limit the treatment period that can be used for these drugs to a point where the antitumor activity is insufficient. Unfortunately there are currently no good ways to completely understand this preclinically, so we are forced to take the agents into clinical trials to find out if there is a therapeutic window. Preclinical solutions to address this would represent a major breakthrough, and the combination of drugs with different mechanisms of action may help improving the therapeutic index against cancer cells (as shown for Hsp90 inhibitor 17-AAG and paclitaxel (Munster et al., 2001), also see the next two paragraphs).

Related to the dearth of knowledge of how antimitotic drugs kill cells is the issue as to whether the potency of a highly specific drug is superior to existing drugs that are less specific. For example, low concentrations of taxanes have been shown to induce apoptosis not exclusively by interfering with mitosis, but other cellular activities that induce cell death (Torres and Horwitz, 1998). By limiting their actions to mitosis, antikinetochore drugs may lack the added punch that affects clinical outcome. As cited in recent reviews, high selectivity of some novel drugs clearly reduced neurotoxicity posed by conventional antimicrotubule drugs while maintaining strong antitumor activity (Jackson et al., 2007; Plyte and Musacchio, 2007). While these observations might ameliorate such concerns, one strategy to overcome the drawbacks of drugs with high selectivity is to take a combinatorial route. Optimally, the use of two or more chemotherapeutic agents, with non-overlapping mechanisms of action may produce a synergistic outcome.

Finally, significant improvement in deciding the choice of chemotherapy will come once we are able to factor in genetic variability of patients and their tumors. We need to develop a standard set of pharmacodynamic biomarkers to evaluate mitotic index, cell proliferation, and apoptosis. In this regard, developing advanced noninvasive functional imaging methods will be very beneficial. For example, fluorothymidine-positron emission tomography can be used to assess cell proliferation or cell cycle arrest; annexin IV coupled Cy5.5 can be used to detect apoptosis for noninvasive infrared optical imaging (Steegmaier et al., 2007; Kenny et al., 2005). Genetic profiling of primary tumors by microarray combined with functional genomics studies of established cancer cell lines can identify collections of genes ("signatures") that dictate drug sensitivity (Glinsky et al., 2005). This information can then be used to correlate drug response and gene expression in patient tumors. Patterns that correlate positively or negatively may be used to improve clinical trials and to predict patient response to specific drugs.

In summary, efforts to understand fundamental mechanisms of mitosis have uncovered many proteins that have been targeted for drug development. While the rationale for using these targets to inhibit mitosis is clear, there is a gap in our understanding about how these drugs kill cells. There is also a gap in understanding why some cells do not respond even though they share the same target as cells that do respond. Efforts to address these questions will be critical to realizing the full potential of mitosis-specific drugs in cancer treatment.

## References

Ahmad, N., *Polo-like kinase (Plk) 1: a novel target for the treatment of prostate cancer*. Faseb J, 2004. **18**(1): 5–7.

Altieri, D.C., *The case for survivin as a regulator of microtubule dynamics and cell-death decisions*. Curr Opin Cell Biol, 2006. **18**(6): 609–15.

Anand, S., S. Penrhyn-Lowe, and A.R. Venkitaraman, *AURORA-A amplification overrides the mitotic spindle assembly checkpoint, inducing resistance to Taxol*. Cancer Cell, 2003. **3**(1): 51–62.

Andrews, P.D., et al., *Aurora B regulates MCAK at the mitotic centromere*. Dev Cell, 2004. **6**(2): 253–68.

Andrews, P.D., et al., *Mitotic mechanics: the auroras come into view*. Curr Opin Cell Biol, 2003. **15**(6): 672–83.

Arnaud, L., J. Pines, and E.A. Nigg, *GFP tagging reveals human Polo-like kinase 1 at the kinetochore/centromere region of mitotic chromosomes*. Chromosoma, 1998. **107**(6–7): 424–9.

Ashar, H.R., et al., *Farnesyl transferase inhibitors block the farnesylation of CENP-E and CENP-F and alter the association of CENP-E with the microtubules*. J Biol Chem, 2000. **275**(39): 30451–7.

Babu, J.R., et al., *Rae1 is an essential mitotic checkpoint regulator that cooperates with Bub3 to prevent chromosome missegregation*. J Cell Biol, 2003. **160**(3): 341–53.

Baker, D.J., et al., *BubR1 insufficiency causes early onset of aging-associated phenotypes and infertility in mice*. Nat Genet, 2004. **36**(7): 744–9.

Bansal, P.K., R. Abdulle, and K. Kitagawa, *Sgt1 associates with Hsp90: an initial step of assembly of the core kinetochore complex*. Mol Cell Biol, 2004. **24**(18): 8069–79.

Barr, F.A., H.H. Sillje, and E.A. Nigg, *Polo-like kinases and the orchestration of cell division*. Nat Rev Mol Cell Biol, 2004. **5**(6): 429–40.

Bartek, J. and J. Lukas, *Chk1 and Chk2 kinases in checkpoint control and cancer*. Cancer Cell, 2003. **3**(5): 421–9.

Bischoff, J.R., et al., *A homologue of Drosophila aurora kinase is oncogenic and amplified in human colorectal cancers*. Embo J, 1998. **17**(11): 3052–65.

Brunner, T.B., et al., *Farnesyltransferase inhibitors: an overview of the results of preclinical and clinical investigations*. Cancer Res, 2003. **63**(18): 5656–68.

Burkard, M.E., et al., *Chemical genetics reveals the requirement for Polo-like kinase 1 activity in positioning RhoA and triggering cytokinesis in human cells*. Proc Natl Acad Sci USA, 2007. **104**(11): 4383–8.

Cahill, D.P., et al., *Mutations of mitotic checkpoint genes in human cancers*. Nature, 1998. **392**(6673): 300-3.

Carmena, M. and W.C. Earnshaw, *The cellular geography of aurora kinases*. Nat Rev Mol Cell Biol, 2003. **4**(11): 842–54.

Carvajal, R.D., A. Tse, and G.K. Schwartz, *Aurora kinases: new targets for cancer therapy*. Clin Cancer Res, 2006. **12**(23): 6869–75.

Cesario, A., et al., *Farnesyltransferase inhibitors and human malignant pleural mesothelioma: a first-step comparative translational study*. Clin Cancer Res, 2005. **11**(5): 2026–37.

Chan, G.K., S.T. Liu, and T.J. Yen, *Kinetochore structure and function*. Trends Cell Biol, 2005. **15**(11): 589–98.

Chan, G.K., Yen, T. J., *The mitotic checkpoint: a signaling pathway that allows a single unattached kinetochore to inhibit mitotic exit*, in *Progress in Cell Cycle Research*, L. Meijer, Jezequel, A., and Roberge, M., Editor. 2003, Editions "Life in Progress": Roscoff, France. 431–439.

Cheeseman, I.M., et al., *The conserved KMN network constitutes the core microtubule-binding site of the kinetochore*. Cell, 2006. **127**(5): 983–97.

Chen, Y., et al., *HEC, a novel nuclear protein rich in leucine heptad repeats specifically involved in mitosis*. Mol Cell Biol, 1997. **17**(10): 6049–56.

Ciferri, C., et al., *Architecture of the human ndc80-hec1 complex, a critical constituent of the outer kinetochore*. J Biol Chem, 2005. **280**(32): 29088–95.

Cleveland, D.W., Y. Mao, and K.F. Sullivan, *Centromeres and kinetochores: from epigenetics to mitotic checkpoint signaling*. Cell, 2003. **112**(4): 407–21.

Crespo, N.C., et al., *The farnesyltransferase inhibitor, FTI-2153, blocks bipolar spindle formation and chromosome alignment and causes prometaphase accumulation during mitosis of human lung cancer cells*. J Biol Chem, 2001. **276**(19): 16161–7.

Crespo, N.C., et al., *The farnesyltransferase inhibitor, FTI-2153, inhibits bipolar spindle formation during mitosis independently of transformation and Ras and p53 mutation status*. Cell Death Differ, 2002. **9**(7): 702–9.

Cullinan, S.B. and L. Whitesell, *Heat shock protein 90: a unique chemotherapeutic target*. Semin Oncol, 2006. **33**(4): 457–65.

Dai, W. and J.P. Cogswell, *Polo-like kinases and the microtubule organization center: targets for cancer therapies*. Prog Cell Cycle Res, 2003. **5**: 327–34.
DeBonis, S., et al., *In vitro screening for inhibitors of the human mitotic kinesin Eg5 with antimitotic and antitumor activities*. Mol Cancer Ther, 2004. **3**(9): 1079–90.
DeLuca, J.G., et al., *hNuf2 inhibition blocks stable kinetochore-microtubule attachment and induces mitotic cell death in HeLa cells*. J Cell Biol, 2002. **159**(4): 549–55.
DeLuca, J.G., et al., *Kinetochore microtubule dynamics and attachment stability are regulated by Hec1*. Cell, 2006. **127**(5): 969–82.
Ditchfield, C., et al., *Aurora B couples chromosome alignment with anaphase by targeting BubR1, Mad2, and Cenp-E to kinetochores*. J Cell Biol, 2003. **161**(2): 267–80.
Dorer, R.K., et al., *A small-molecule inhibitor of Mps1 blocks the spindle-checkpoint response to a lack of tension on mitotic chromosomes*. Curr Biol, 2005. **15**(11): 1070–6.
Eckerdt, F., J. Yuan, and K. Strebhardt, *Polo-like kinases and oncogenesis*. Oncogene, 2005. **24**(2): 267–76.
Ganem, N.J., Z. Storchova, and D. Pellman, *Tetraploidy, aneuploidy and cancer*. Curr Opin Genet Dev, 2007. **17**(2): 157–62.
Garber, K., *Divide and conquer: new generation of drugs targets mitosis*. J Natl Cancer Inst, 2005. **97**(12): 874–6.
Giet, R., C. Petretti, and C. Prigent, *Aurora kinases, aneuploidy and cancer, a coincidence or a real link*? Trends Cell Biol, 2005. **15**(5): 241–50.
Glinsky, G.V., O. Berezovska, and A.B. Glinskii, *Microarray analysis identifies a death-from-cancer signature predicting therapy failure in patients with multiple types of cancer*. J Clin Invest, 2005. **115**(6): 1503–21.
Glover, D.M., et al., *Mutations in aurora prevent centrosome separation leading to the formation of monopolar spindles*. Cell, 1995. **81**(1): 95–105.
Goetz, M.P., et al., *The Hsp90 chaperone complex as a novel target for cancer therapy*. Ann Oncol, 2003. **14**(8): 1169–76.
Golsteyn, R.M., et al., *Cell cycle regulation of the activity and subcellular localization of Plk1, a human protein kinase implicated in mitotic spindle function*. J Cell Biol, 1995. **129**(6): 1617–28.
Golsteyn, R.M., *The position of Chk1 in cancer research*. Trends Cell Biol, 2001. **11**(5): 191.
Gumireddy, K., et al., *ON01910, a non-ATP-competitive small molecule inhibitor of Plk1, is a potent anticancer agent*. Cancer Cell, 2005. **7**(3): 275–86.
Gurzov, E.N. and M. Izquierdo, *RNA interference against Hec1 inhibits tumor growth in vivo*. Gene Ther, 2006. **13**(1): 1–7.
Hanks, S., et al., *Constitutional aneuploidy and cancer predisposition caused by biallelic mutations in BUB1B*. Nat Genet, 2004. **36**(11): 1159–61.
Harrington, E.A., et al., *VX-680, a potent and selective small-molecule inhibitor of the Aurora kinases, suppresses tumor growth in vivo*. Nat Med, 2004. **10**(3): 262–7.
Hauf, S., et al., *The small molecule Hesperadin reveals a role for Aurora B in correcting kinetochore-microtubule attachment and in maintaining the spindle assembly checkpoint*. J Cell Biol, 2003. **161**(2): 281–94.
Hayama, S., et al., *Activation of CDCA1-KNTC2, members of centromere protein complex, involved in pulmonary carcinogenesis*. Cancer Res, 2006. **66**(21): 10339–48.
Hernando, E., et al., *Rb inactivation promotes genomic instability by uncoupling cell cycle progression from mitotic control*. Nature, 2004. **430**(7001): 797–802.
Huang, H., et al., *Tripin/hSgo2 recruits MCAK to the inner centromere to correct defective kinetochore attachments*. J Cell Biol, 2007. **177**(3): 413–24.
Huang, R.C., C.C. Chang, and D. Mold, *Survivin-dependent and -independent pathways and the induction of cancer cell death by tetra-O-methyl nordihydroguaiaretic acid*. Semin Oncol, 2006. **33**(4): 479–85.
Hussein, D. and S.S. Taylor, *Farnesylation of Cenp-F is required for G2/M progression and degradation after mitosis*. J Cell Sci, 2002. **115**(Pt 17): 3403–14.

Jablonski, S.A., S.T. Liu, and T.J. Yen, *Targeting the kinetochore for mitosis-specific inhibitors*. Cancer Biol Ther, 2003. **2**(3): 236–41.
Jackson, J.R., et al., *Targeted anti-mitotic therapies: can we improve on tubulin agents*? Nat Rev Cancer, 2007. **7**(2): 107–17.
Jallepalli, P.V. and C. Lengauer, *Chromosome segregation and cancer: cutting through the mystery*. Nat Rev Cancer, 2001. **1**(2): 109–17.
Joseph, J., et al., *The RanGAP1-RanBP2 complex is essential for microtubule-kinetochore interactions in vivo*. Curr Biol, 2004. **14**(7): 611–7.
Kaplan, K.B., et al., *A role for the Adenomatous Polyposis Coli protein in chromosome segregation*. Nat Cell Biol, 2001. **3**(4): 429–32.
Keen, N. and S. Taylor, *Aurora-kinase inhibitors as anticancer agents*. Nat Rev Cancer, 2004. **4**(12): 927–36.
Kenny, L.M., et al., *Quantification of cellular proliferation in tumor and normal tissues of patients with breast cancer by [18F]fluorothymidine-positron emission tomography imaging: evaluation of analytical methods*. Cancer Res, 2005. **65**(21): 10104–12.
Kitagawa, K., et al., *SGT1 encodes an essential component of the yeast kinetochore assembly pathway and a novel subunit of the SCF ubiquitin ligase complex*. Mol Cell, 1999. **4**(1): 21–33.
Kops, G.J., B.A. Weaver, and D.W. Cleveland, *On the road to cancer: aneuploidy and the mitotic checkpoint*. Nat Rev Cancer, 2005. **5**(10): 773–85.
Kops, G.J., D.R. Foltz, and D.W. Cleveland, *Lethality to human cancer cells through massive chromosome loss by inhibition of the mitotic checkpoint*. Proc Natl Acad Sci U S A, 2004. **101**(23): 8699–704.
Lenart, P., et al., *The small-molecule inhibitor BI 2536 reveals novel insights into mitotic roles of polo-like kinase 1*. Curr Biol, 2007. **17**(4): 304–15.
Lens, S.M., G. Vader, and R.H. Medema, *The case for Survivin as mitotic regulator*. Curr Opin Cell Biol, 2006. **18**(6): 616–22.
Li, F. and M.G. Brattain, *Role of the Survivin gene in pathophysiology*. Am J Pathol, 2006. **169**(1): 1–11.
Li, L., et al., *Development of recombinant adeno-associated virus vectors carrying small interfering RNA (shHec1)-mediated depletion of kinetochore Hec1 protein in tumor cells*. Gene Ther, 2007. **14**(10): 814–27.
Li, X., et al., *Direct association with inner centromere protein (INCENP) activates the novel chromosomal passenger protein, Aurora-C*. J Biol Chem, 2004. **279**(45): 47201–11.
Liu, S.T., J.M. van Deursen, and T.J. Yen, *The role of mitotic checkpoint in maintaining genomic stability*. Curr Top Dev Biol, 2003. **58**: 27–51.
Liu, Y., et al., *Polo-like kinases inhibited by wortmannin. Labeling site and downstream effects*. J Biol Chem, 2007. **282**(4): 2505–11.
Lou, Y., et al., *NEK2A interacts with MAD1 and possibly functions as a novel integrator of the spindle checkpoint signaling*. J Biol Chem, 2004. **279**(19): 20049–57.
Martin-Lluesma, S., V.M. Stucke, and E.A. Nigg, *Role of hec1 in spindle checkpoint signaling and kinetochore recruitment of mad1/mad2*. Science, 2002. **297**(5590): 2267–70.
Mayer, T.U., et al., *Small molecule inhibitor of mitotic spindle bipolarity identified in a phenotype-based screen*. Science, 1999. **286**(5441): 971–4.
McEwen, B.F., et al., *CENP-E is essential for reliable bioriented spindle attachment, but chromosome alignment can be achieved via redundant mechanisms in mammalian cells*. Mol Biol Cell, 2001. **12**(9): 2776–89.
McInnes, C., et al., *Inhibitors of Polo-like kinase reveal roles in spindle-pole maintenance*. Nat Chem Biol, 2006. **2**(11): 608–17.
Meraldi, P., R. Honda, and E.A. Nigg, *Aurora-A overexpression reveals tetraploidization as a major route to centrosome amplification in p53-/- cells*. Embo J, 2002. **21**(4): 483–92.
Meraldi, P., V.M. Draviam, and P.K. Sorger, *Timing and checkpoints in the regulation of mitotic progression*. Dev Cell, 2004. **7**(1): 45–60.

Michel, L., et al., *Complete loss of the tumor suppressor MAD2 causes premature cyclin B degradation and mitotic failure in human somatic cells.* Proc Natl Acad Sci U S A, 2004. **101**(13): 4459–64.

Michel, L.S., et al., *MAD2 haplo-insufficiency causes premature anaphase and chromosome instability in mammalian cells.* Nature, 2001. **409**(6818): 355–9.

Munster, P.N., et al., *Inhibition of heat shock protein 90 function by ansamycins causes the morphological and functional differentiation of breast cancer cells.* Cancer Res, 2001. **61**(7): 2945–52.

Munster, P.N., et al., *Modulation of Hsp90 function by ansamycins sensitizes breast cancer cells to chemotherapy-induced apoptosis in an RB- and schedule-dependent manner. See: E. A. Sausville, Combining cytotoxics and 17-allylamino, 17-demethoxygeldanamycin: sequence and tumor biology matters.* Clin Cancer Res, 2001. **7**(8): 2228–36.

Niikura, Y., et al., *17-AAG, an Hsp90 inhibitor, causes kinetochore defects: a novel mechanism by which 17-AAG inhibits cell proliferation.* Oncogene, 2006. **25**(30): 4133–46.

Niikura, Y., et al., *BUB1 mediation of caspase-independent mitotic death determines cell fate.* J Cell Biol, 2007. **178**(2): 283–96.

Orr, G.A., et al., *Mechanisms of Taxol resistance related to microtubules.* Oncogene, 2003. **22**(47): 7280–95.

Ota, T., et al., *Increased mitotic phosphorylation of histone H3 attributable to AIM-1/Aurora-B overexpression contributes to chromosome number instability.* Cancer Res, 2002. **62**(18): 5168–77.

Peters, U., et al., *Probing cell-division phenotype space and Polo-like kinase function using small molecules.* Nat Chem Biol, 2006. **2**(11): 618–26.

Pichler, A., et al., *The nucleoporin RanBP2 has SUMO1 E3 ligase activity.* Cell, 2002. **108**(1): 109–20.

Plescia, J., et al., *Rational design of shepherdin, a novel anticancer agent.* Cancer Cell, 2005. **7**(5): 457–68.

Plyte, S. and A. Musacchio, *PLK1 inhibitors: setting the mitotic death trap.* Curr Biol, 2007. **17**(8): R280–R3.

Powers, M.V. and P. Workman, *Targeting of multiple signalling pathways by heat shock protein 90 molecular chaperone inhibitors.* Endocr Relat Cancer, 2006. **13**( Suppl 1): S125–S35.

Prudhomme, M., *Novel Checkpoint 1 Inhibitors.* Recent Patents on Anti-Cancer Drug Discovery, 2006. **1**(1): 55–68.

Putkey, F.R., et al., *Unstable kinetochore-microtubule capture and chromosomal instability following deletion of CENP-E.* Dev Cell, 2002. **3**(3): 351–65.

Rieder, C.L. and E.D. Salmon, *The vertebrate cell kinetochore and its roles during mitosis.* Trends Cell Biol, 1998. **8**(8): 310–8.

Rieder, C.L., *The formation, structure, and composition of the mammalian kinetochore and kinetochore fiber.* Int Rev Cytol, 1982. **79**: 1–58.

Rojanala, S., et al., *The mitotic serine threonine kinase, Aurora-2, is a potential target for drug development in human pancreatic cancer.* Mol Cancer Ther, 2004. **3**(4): 451–7.

Roninson, I.B., E.V. Broude, and B.D. Chang, *If not apoptosis, then what? Treatment-induced senescence and mitotic catastrophe in tumor cells.* Drug Resist Updat, 2001. **4**(5): 303–13.

Rowinsky, E.K. and E. Calvo, *Novel agents that target tublin and related elements.* Semin Oncol, 2006. **33**(4): 421–35.

Sasai, K., et al., *Aurora-C kinase is a novel chromosomal passenger protein that can complement Aurora-B kinase function in mitotic cells.* Cell Motil Cytoskeleton, 2004. **59**(4): 249–63.

Schafer-Hales, K., et al., *Farnesyl transferase inhibitors impair chromosomal maintenance in cell lines and human tumors by compromising CENP-E and CENP-F function.* Mol Cancer Ther, 2007. **6**(4): 1317–28.

Schmidt, M., et al., *Ablation of the spindle assembly checkpoint by a compound targeting Mps1.* EMBO Rep, 2005. **6**(9): 866–72.

Sotillo, R., et al., *Mad2 overexpression promotes aneuploidy and tumorigenesis in mice*. Cancer Cell, 2007. **11**(1): 9–23.
Srethapakdi, M., et al., *Inhibition of Hsp90 function by ansamycins causes retinoblastoma gene product-dependent G1 arrest*. Cancer Res, 2000. **60**(14): 3940–6.
Steegmaier, M., et al., *BI 2536, a potent and selective inhibitor of polo-like kinase 1, inhibits tumor growth in vivo*. Curr Biol, 2007. **17**(4): 316–22.
Steensgaard, P., et al., *Sgt1 is required for human kinetochore assembly*. EMBO Rep, 2004. **5**(6): 626–31.
Stevenson, C.S., et al., *The identification and characterization of the marine natural product scytonemin as a novel antiproliferative pharmacophore*. J Pharmacol Exp Ther, 2002. **303**(2): 858–66.
Strebhardt, K. and A. Ullrich, *Targeting polo-like kinase 1 for cancer therapy*. Nat Rev Cancer, 2006. **6**(4): 321–30.
Tanudji, M., et al., *Gene silencing of CENP-E by small interfering RNA in HeLa cells leads to missegregation of chromosomes after a mitotic delay*. Mol Biol Cell, 2004. **15**(8): 3771–81.
Tao, Z.F. and N.H. Lin, *Chk1 inhibitors for novel cancer treatment*. Anticancer Agents Med Chem, 2006. **6**(4): 377–88.
Tatsuka, M., et al., *Overexpression of Aurora-A potentiates HRAS-mediated oncogenic transformation and is implicated in oral carcinogenesis*. Oncogene, 2005. **24**(6): 1122–7.
Tomonaga, T., et al., *Centromere protein H is up-regulated in primary human colorectal cancer and its overexpression induces aneuploidy*. Cancer Res, 2005. **65**(11): 4683–9.
Tomonaga, T., et al., *Overexpression and mistargeting of centromere protein-A in human primary colorectal cancer*. Cancer Res, 2003. **63**(13): 3511–6.
Torrance, C.J., et al., *Use of isogenic human cancer cells for high-throughput screening and drug discovery*. Nat Biotechnol, 2001. **19**(10): 940–5.
Torres, K. and S.B. Horwitz, *Mechanisms of Taxol-induced cell death are concentration dependent*. Cancer Res, 1998. **58**(16): 3620–6.
Vader, G., R.H. Medema, and S.M. Lens, *The chromosomal passenger complex: guiding Aurora-B through mitosis*. J Cell Biol, 2006. **173**(6): 833–7.
Vagnarelli, P. and W.C. Earnshaw, *Chromosomal passengers: the four-dimensional regulation of mitotic events*. Chromosoma, 2004. **113**(5): 211–22.
Vassilev, L.T., *MDM2 inhibitors for cancer therapy*. Trends Mol Med, 2007. **13**(1): 23–31.
Wang, Q., et al., *BUBR1 deficiency results in abnormal megakaryopoiesis*. Blood, 2004. **103**(4): 1278–85.
Wang, Z., et al., *Three classes of genes mutated in colorectal cancers with chromosomal instability*. Cancer Res, 2004. **64**(9): 2998–3001.
Warner, S.L., P.J. Gray, and D.D. Von Hoff, *Tubulin-associated drug targets: Aurora kinases, Polo-like kinases, and others*. Semin Oncol, 2006. **33**(4): 436–48.
Watanabe, Y. and T.S. Kitajima, *Shugoshin protects cohesin complexes at centromeres*. Philos Trans R Soc Lond B Biol Sci, 2005. **360**(1455): 515–21, discussion 521.
Weaver, B.A. and D.W. Cleveland, *Decoding the links between mitosis, cancer, and chemotherapy: The mitotic checkpoint, adaptation, and cell death*. Cancer Cell, 2005. **8**(1): 7–12.
Weaver, B.A. and D.W. Cleveland, *Does aneuploidy cause cancer*? Curr Opin Cell Biol, 2006. **18**(6): 658–67.
Weaver, B.A., et al., *Aneuploidy acts both oncogenically and as a tumor suppressor*. Cancer Cell, 2007. **11**(1): 25–36.
Wei, R.R., et al., *Structure of a central component of the yeast kinetochore: the Spc24p/Spc25p globular domain*. Structure, 2006. **14**(6): 1003–9.
Wei, R.R., J. Al-Bassam, and S.C. Harrison, *The Ndc80/HEC1 complex is a contact point for kinetochore-microtubule attachment*. Nat Struct Mol Biol, 2007. **14**(1): 54–9.
Wilson, L. and M.A. Jordan, *Microtubule dynamics: taking aim at a moving target*. Chem Biol, 1995. **2**(9): 569–73.
Yang, H., et al., *Mitotic requirement for aurora A kinase is bypassed in the absence of aurora B kinase*. FEBS Lett, 2005. **579**(16): 3385–91.

Yuen, K.W., B. Montpetit, and P. Hieter, *The kinetochore and cancer: what's the connection*? Curr Opin Cell Biol, 2005. **17**(6): 576–82.

Zachos, G., et al., *Chk1 is required for spindle checkpoint function.* Dev Cell, 2007. **12**(2): 247–60.

Zhou, B.B. and E.A. Sausville, *Drug discovery targeting Chk1 and Chk2 kinases.* Prog Cell Cycle Res, 2003. **5**: 413–21.

Zhou, B.B., H.J. Anderson, and M. Roberge, *Targeting DNA checkpoint kinases in cancer therapy*. Cancer Biol Ther, 2003. **2**(4 Suppl 1): S16–S22.

Zhou, H., et al., *Tumour amplified kinase STK15/BTAK induces centrosome amplification, aneuploidy and transformation.* Nat Genet, 1998. **20**(2): 189–93.

# Index

## E



## F

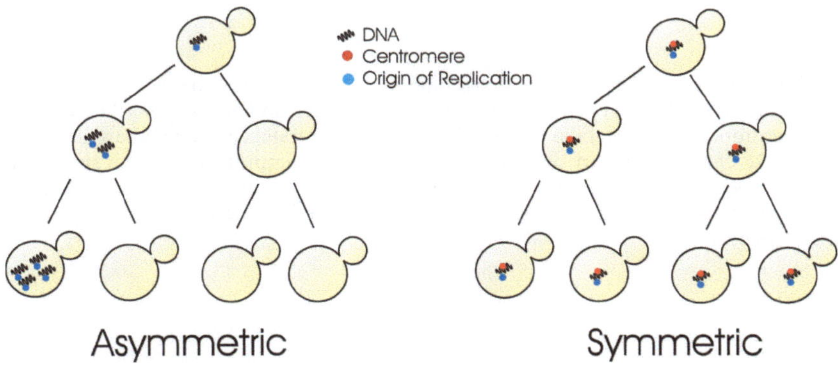


**Fig. 1.1** Pedigree analysis of chromosome segregation in budding yeast. The DNA, original or replication (*red*), and centromere (*blue*) are indicated.

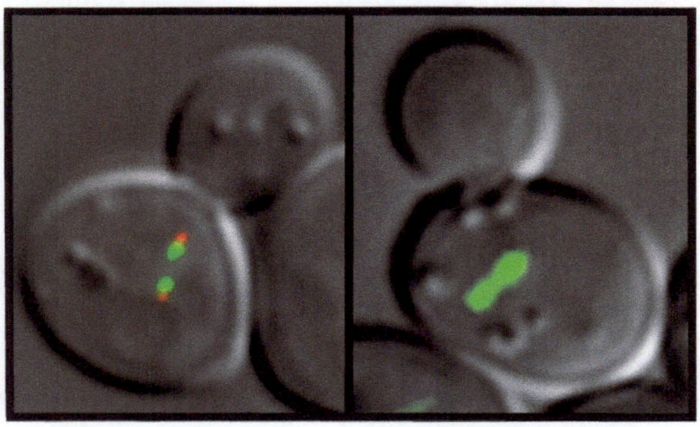

**Fig. 1.2** Centromere DNA, spindle poles and microtubules in mitosis. (*left*) Centromere (CEN) DNA visualized with lac-repressor-GFP bound to lac operator integrated 1 kb from CEN3 on chromosome III (*green*) and spindle poles (*red*). Sister centromeres are separated in mitosis (*right*) Microtubules in the mitotic spindle visualized with tubulin-GFP. The spindle is thicker at the ends versus the middle. Sixteen kinetochore microtubules originate from each spindle pole and extend approximately a quarter the length of the spindle.

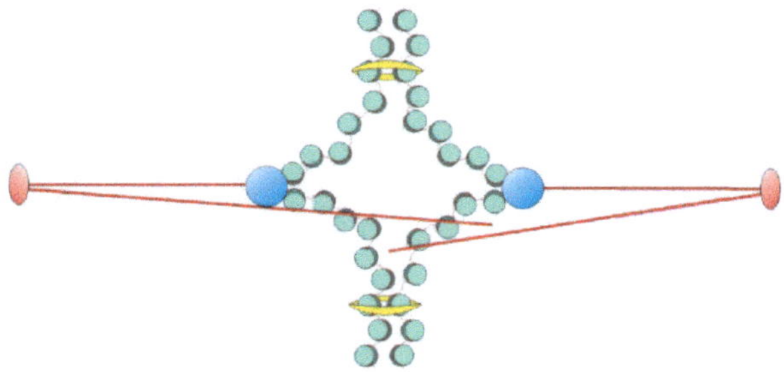

**Fig. 1.3** Organization of a mitotic chromosome. Chromosome arms are closely apposed and held together via cohesin (*yellow rings*). Sister kinetochores (in *blue*) are attached to kinetochore microtubules (*red*) and the pericentric chromatin is stretched toward the spindle poles. There are 16 chromosomes in yeast, and 16 kinetochore microtubules in each spindle half. Cohesin between sister chromatids provides a mechanism to resist microtubule pulling forces and generate tension at centromeres. The function of cohesin in pericentric chromatin is not well understood.

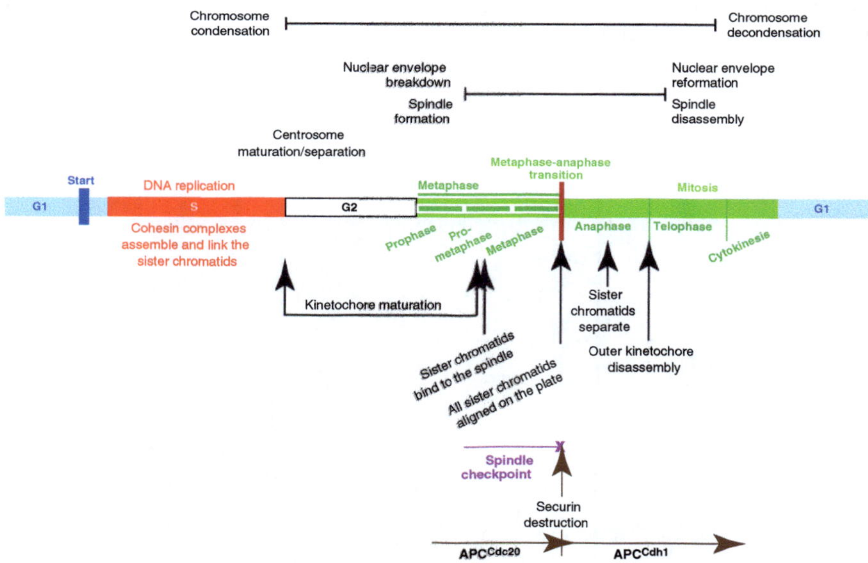


**Fig. 2.1** The mitotic cell division, kinetochore and chromosome cycles. Schematic outline of the cell cycle stages (G1: *blue*, S-phase: *red*, G2: white, mitosis: *green*), and of the state/activity of chromosomes, the spindle, kinetochores and sister chromatids (shown in *black*). The activities of the mitotic spindle checkpoint and the anaphase-promoting complex (APC) are shown in pink and brown, respectively. In metazoan, kinetochores undergo a cycle of maturation and partial disassembly. In the yeasts, kinetochores are tethered to microtubules throughout the cell cycle and shortly detach from the microtubule tips during replication of the centromeric regions in S phase.

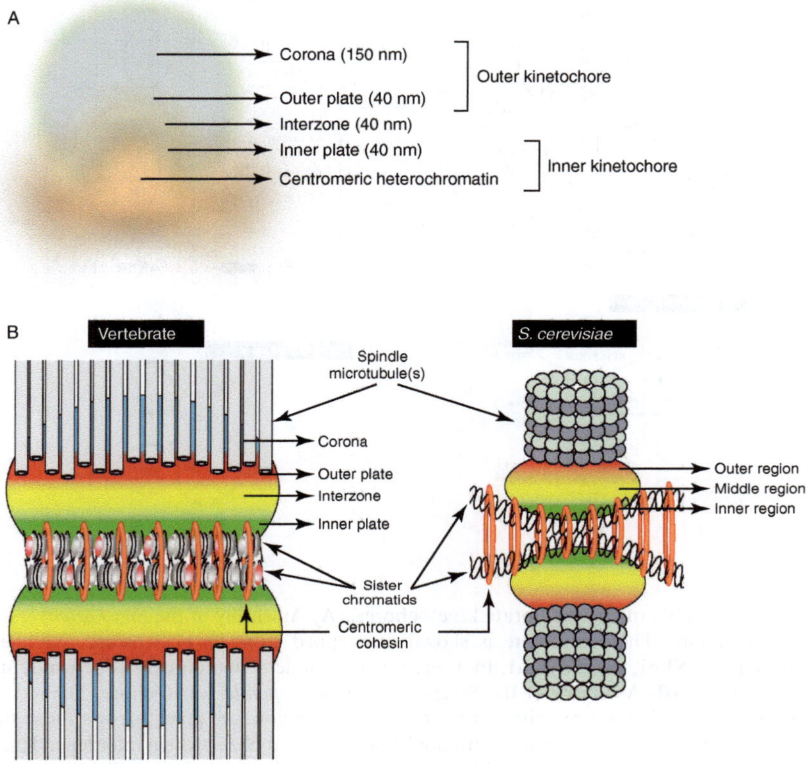


**Fig. 6.1** Kinetochores (**A**) Model of the vertebrate kinetochore (not bound to spindle microtubules) based on electron micrographs (Brinkley and Stubblefield 1966, Jokelainen 1967, Comings and Okada 1971, McEwen et al. 1998). (**B**) Models of metaphase vertebrate and budding yeast kinetochores bipolarly attached to spindle microtubules. Intra-kinetochore regions are shown in color.

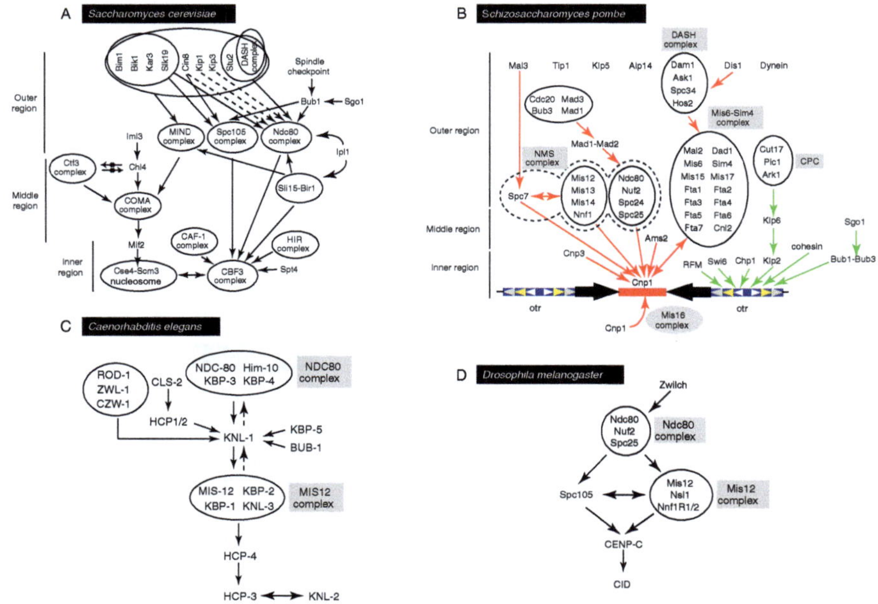

**Fig. 6.4** Assembly of nonvertebrate kinetochores. (**A**) Assembly of the *Saccharomyces cerevisiae* kinetochore. The centromere is shown unwrapped from its nucleosome. Middle layer proteins (e.g., Nkp1, Nkp2, Cnn1, Plc1, etc.) whose mode of recruitment is presently unclear are not shown. (**B**) Assembly of the *Schizosaccharomyces pombe* kinetochore. The red arrows indicate proteins that are recruited to the central core region, the green arrows those that are recruited to the outer repeats. Lines with double arrows symbolize protein interdependencies for localization to the centromere. RFM: proteins required for methylation (Rdp1, Dcr1, Ctr4, Chp1, Tas3, Ago1, Clr3, Clr6, Sir2, Abp1, Cbh1, Cbh2, Rik1). RFM proteins Rdp1, chp1, Tas3, and Abp1 were shown to physically associate with a certain domain via ChIP. The other RFM proteins are required for the proper function of the domain but for them no ChIP data have been published. Imr: inner most repeat, otr: outer repeat. (**C**) Assembly of the *Caenorhabditis elegans* kinetochore. The dashed lines represent partial recruitment dependencies. (**D**) Assembly of the *Drosophila melanogaster* kinetochore. Conserved Ndc80 complex member Spc24 has not yet been identified in *Drosophila*. Nnf1R1 and Nnf1R2 are two Nnf1 homologues. It is currently unclear which one of the two is part of the Mis12 complex.

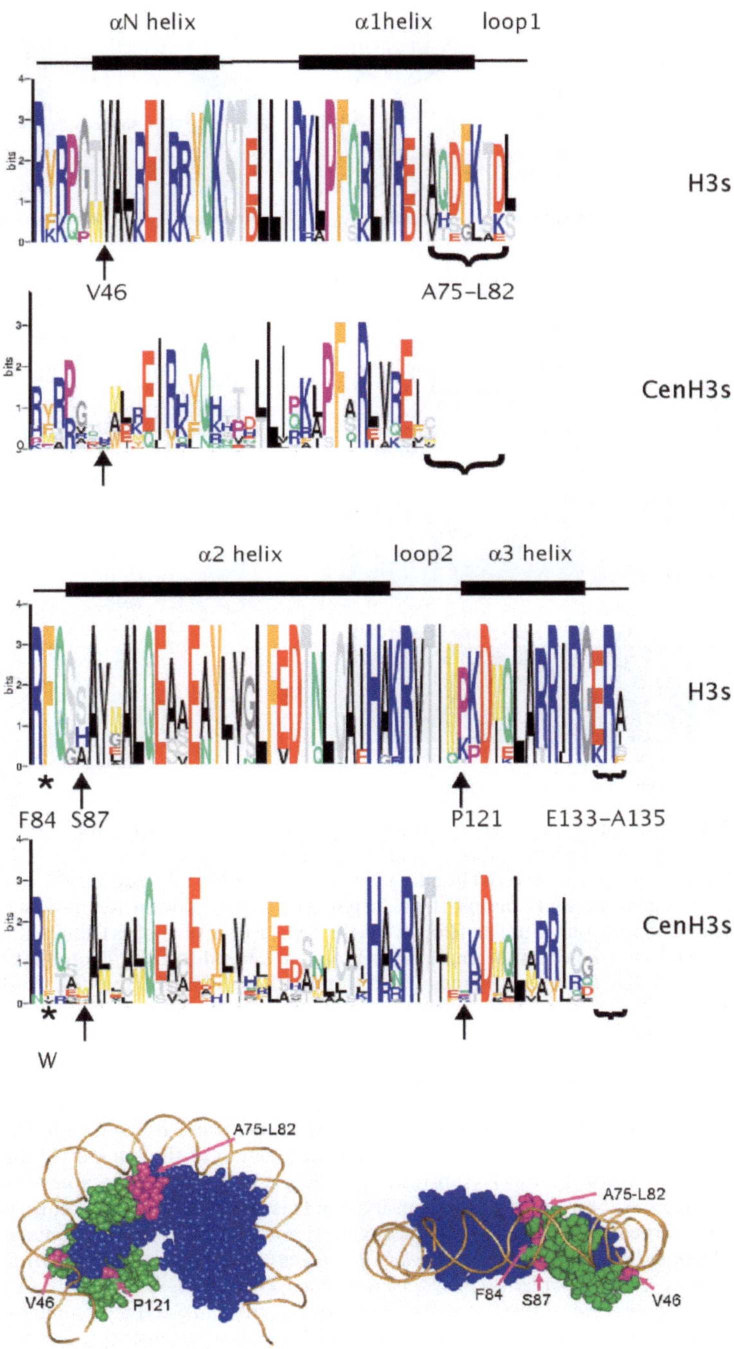


**Fig. 7.2** CenH3 conservation. (*Top*) Alignment of histone fold domains of CenH3s and packaging H3s in Logos format, in which the height of each letter reflects the conservation

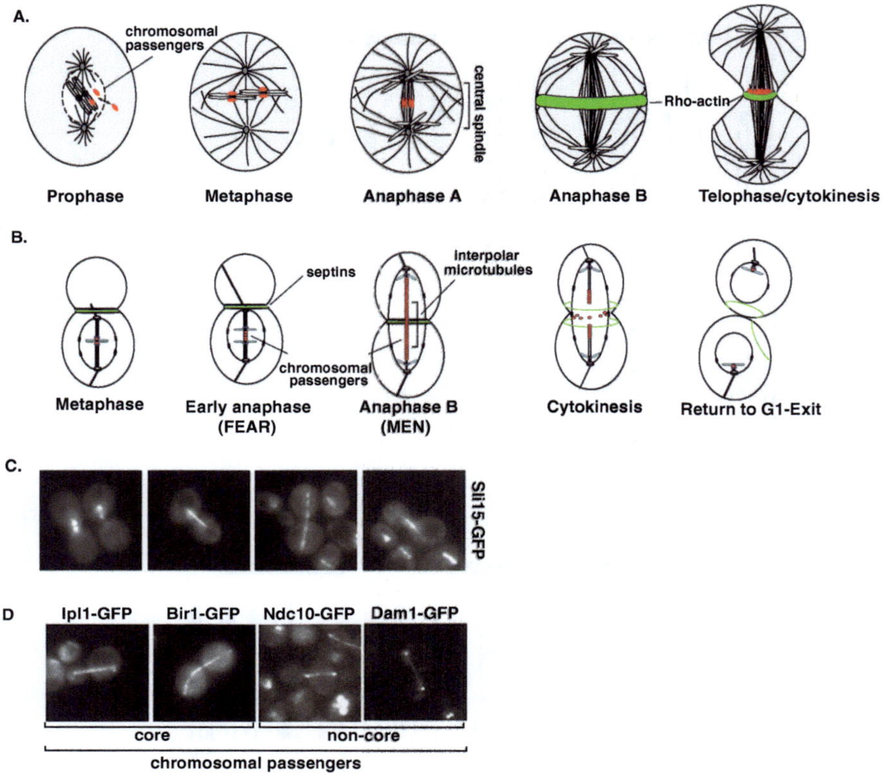


**Fig. 12.3** Chromosomal passengers during anaphase in animal cells and budding yeast. Chromosomal passenger localization is shown by the red ovals and related to each stage of mitosis for (**A**) animal cells and (**B**) budding yeast. (**C**) Examples of budding yeast expressing Sli15-GFP; images represent examples from different cells that show the temporal sequence of localization from metaphase to late telophase. (**D**) Representative images from budding yeast expressing the indicated chromosomal passenger protein fused to GFP. Note that the grey-scale values for the Dam1-GFP image were adjusted to highlight the anaphase spindle localization.

**Fig. 7.2** (continued) of the corresponding amino acid. Upper panel: $\alpha$N and $\alpha$1 helices. Lower panel: $\alpha$2 and $\alpha$3 helices. Residue numbers indicate position in the human H3 amino acid sequence. (*) indicates the F84W substitution in CenH3s. Individual residues (arrows) or unalignable regions (brackets) that are conserved in H3s but show little or no conservation in CenH3s are marked. Sequences of both packaging H3s and CenH3s are from animals (*Homo sapiens, Drosophila melanogaster*, *Caenorhabditis elegan*); fungi (*Schizosaccharomyces pombe*, *Saccharomyces cerevisiae*); plants (*Arabidopsis thaliana*, *Oryza sativa*; *Luzula nivea*); red algae (*Cyanidioschyzon merolae*); alveolates (*Tetrahymena thermophila*); and diplomonads (*Giardia lamblia*). (Bottom) Spatial arrangement in a H3/H4/H2B/H2A tetramer (hemisome) of select H3 amino acids that are altered in CenH3s. Histone tails are not shown. Amber: DNA double helix. Green: H3; Blue: H2A, H2B, and H4. Magenta: selected H3 residues as labeled. Upper panel: 'front' view. Lower panel: 'side' view.

MIX
Papier aus verantwortungsvollen Quellen
Paper from responsible sources
FSC® C105338

If you have any concerns about our products,
you can contact us on
**ProductSafety@springernature.com**

In case Publisher is established outside the EU,
the EU authorized representative is:
**Springer Nature Customer Service Center GmbH**
**Europaplatz 3, 69115 Heidelberg, Germany**

Printed by Libri Plureos GmbH
in Hamburg, Germany